3rd International Conference on Mechanical Engineering Research (ICMER 2015)

IOP Conference Series: Materials Science and Engineering
Volume 100

Kuantan, Malaysia
18 - 19 August 2015

ISBN: 978-1-5108-1809-5
ISSN: 1757-8981

Printed from e-media with permission by:

Curran Associates, Inc.
57 Morehouse Lane
Red Hook, NY 12571

Some format issues inherent in the e-media version may also appear in this print version.

Printed by Curran Associates, Inc. (2016)

For permission requests, please contact the Institute of Physics
at the address below.

Institute of Physics
Dirac House, Temple Back
Bristol BS1 6BE UK

Phone: 44 1 17 929 7481
Fax: 44 1 17 920 0979

techtracking@iop.org

Additional copies of this publication are available from:

Curran Associates, Inc.
57 Morehouse Lane
Red Hook, NY 12571 USA
Phone: 845-758-0400
Fax: 845-758-2634
Email: curran@proceedings.com
Web: www.proceedings.com

Table of contents

Volume 100

3rd International Conference of Mechanical Engineering Research (ICMER 2015)
18?19 August 2015, Kuantan, Pahang, Malaysia

Accepted papers received: 27 October 2015
Published online: 22 December 2015

Preface ICMER2015

The 3rd ICMER2015 is the continuity of the NCMER2010. The year 2010 represents a significant milestone in the history for Faculty of Mechanical Engineering, Universiti Malaysia Pahang (UMP) Malaysia with the organization of the first and second national level conferences (1st and 2nd NCMER) at UMP on May 26-27 and Dec 3-4 2010.

The Faculty then changed the name from National Conference on Mechanical Engineering Research (NCMER) to International Conference on Mechanical Engineering Research (ICMER) in 2011 and this year, 2015 is our 3rd ICMER. These proceedings contain the selected scientific manuscripts submitted to the conference.

It is with great pleasure to welcome you to the "International Conference on Mechanical Engineering Research (ICMER2015)" that is held at Zenith Hotel, Kuantan, Malaysia. The call for papers attracted submissions of over two hundred abstracts from twelve different countries including Japan, Iran, China, Kuwait, Indonesia, Norway, Philippines, Morocco, Germany, UAE and more.

The scientific papers published in these proceedings have been revised and approved by the technical committee of the 3rd ICMER2015. All of the papers exhibit clear, concise, and precise expositions that appeal to a broad international readership interested in mechanical engineering, combustion, metallurgy, materials science as well as in manufacturing and biomechanics. The reports present original ideas or results of general significance supported by clear reasoning and compelling evidence, and employ methods, theories and practices relevant to the research. The authors clearly state the questions and the significance of their research to theory and practice, describe how the research contributes to new knowledge, and provide tables and figures that meaningfully add to the narrative.

In this edition of ICMER representatives attending are from academia, industry, governmental and private sectors. The plenary and invited speakers will present, discuss, promote and disseminate research in all fields of mechanical engineering. Topics cover synthesis, applications, and fundamental studies of the topics related to mechanical engineering. In addition, booths for industries to showcase their state-of-the-art products are also provided.

The organizing committee of the conference thanks all the participants for their fruitful work and personal contribution to the development of these conference proceedings.

Associate Professor Dr. Rizalman Mamat

The Chairman of ICMER2015

Peer review statement

All papers published in this volume of *IOP Conference Series: Materials Science and Engineering* have been peer reviewed through processes administered by the proceedings Editors. Reviews were conducted by expert referees to the professional and scientific standards expected of a proceedings journal published by IOP Publishing.

IOP Conf. Series: Materials Science and Engineering **100** (2015) 012001 doi:10.1088/1757-899X/100/1/012001

Rheological study of feed stock for NiTi alloy molded parts

I Subuki[1], Z Abdullah[1], R Razali[1] and M H Ismail[2]
[1]Faculty of Chemical Engineering, UniversitiTeknologi MARA (UiTM), 40450 Shah Alam Selangor, Malaysia
[2]Centre for Advanced Materials Research (CAMAR), Faculty of Mechanical Engineering, UniversitiTeknologi MARA (UiTM), 40450 Shah Alam Selangor, Malaysia

Email: istikamah@salam.uitm.edu.my

Abstract. A rheological behaviour of the powder-binder mixture is one of essential analysis upon to success of Metal Injection Moulding (MIM) process. The purpose of this experimental work is to investigate the rheological behavior of feedstock containing mixtures of elemental Ni and Ti powders mixed with composite binder of palm stearin (PS) and polyethylene (PE) binder system. An equiatomic Ni-Ti (50-50) ratio was used in the present work for all formulations owing to excellent shape memory behaviour. The experimental rheological result indicated that all the feedstocks exhibited pseudo plastic flow behaviour; viscosity decreasing with temperature and shear rate. Increasing the powder loading resulted in higher viscosity, particularly at the low-range of shear rate. Owing to pseudo-plastic flow, it was found that the feedstock prepared exhibit promising rheological properties, thus resulting successfully injection moulding at an optimum temperature of 130°C.

1. Introduction

NiTi is categorized as a Shape Memory Alloy (SMA) and has been commercially studied and used in biomedical industry due to two main unique properties, pseudo-elastic (PE) and shape memory effect (SME). Several techniques for producing NiTi alloys by powder metallurgy (PM) have been investigated, emphasizing towards reproducibility and microstructural homogeneity [1]. One of the remarkable PM techniques is metal injection moulding (MIM) which is capable in producing near-net-shape component with a relatively low cost for mass production. However, processing of NiTi by MIM is quite challenging due to restriction in controlling the impurity contents to the lowest level, particularly oxygen and carbon which is mainly originated from inhomegeneity of the powders mixture and residual binder left on particles during debinding stage prior sintering process[1, 2]. Besides that, the powder-binder mixture or generally termed as feedstock should also exhibit good thermal stability and rheological properties to ensure its flowability to fill the mould completely and the shape of moulded part can be retained until the sintering begins[2-4].

The feedstock rheological properties are the key features that affect the steady flow and the uniform filling into the mould. The evaluation of the feedstock rheological properties is based on the viscosity shear and temperature sensitivities, which were used to define a mouldability index [3, 4]. Rheological properties of NiTi alloy feedstock has never been addressed by researchers, since most of them emphasize only the final properties of the SME and PE. It should be borne in mind that the viscosity of MIM feedstock plays a very important role in MIM as it is essential for allowing the particles to flow into the die cavity.

Therefore, specific rheological characteristic is required to establish some insights on the quality of injection moulded samples[5]. The MIM feedstock has often been characterized rheologically by a capillary rheometer especially at low shear rates. Good rheological properties of binder and feedstock are one of the keys to getting green parts with uniform density and defect-free, besides achieving successful debinding and sintering and obtaining high quality products [6].

During injection moulding practice, the feedstock must exhibit pseudo-plastic flow with decreasing viscosity at increased shear rate. This helps to reduce the temperature and pressure required for successful moulding. Generally, during moulding, the shear rate usually ranges between 10^2 and 10^4 s^{-1}, and the maximum useful viscosity for the mixture is 1000 Pa.s or lower at the moulding temperature[7].

The main aim of the present study is to investigate the rheological properties of the feedstock containing elemental Ni-Ti powder blend mixed with palm stearin (PS) based binder. Besides, some experimental on injection moulding was also carried out to demonstrate the feedstock prepared is suitable to be injection moulded.

2. Experimental setup

2.1. Feedstock preparation
Initially, elemental Nickel and Titanium powder were mixed at composition Ti-50.0at% Ni using a stainless steel ball mill for 4 hours at 100 rpm with a ball-to-powder ratio of 2:1 to ensure uniform distribution of the as-mixed powders. Three different powder loadings were studied; 65.5, 67.5 and 69.5 vol% and the binder system consisted of 60 wt% palm stearin (PS) and 40 wt% polyethylene (PE) [8]. The mixing process was carried out using an internal mixer, HAAKE Rheomix at 160°C with a rotational speed of 50 rpm for 2 hrs until homogenous mixtures was observed. The feedstock was then examined using a Hitachi Scanning Electron Microscope (SEM) to observe bonding morphology between the powder and binder.

2.2. Capillary Rheometer
The flow behavior of the feedstock was characterized using a Rosand RH2000 capillary rheometer manufactured by Malvern Instruments TM. The pelletized feedstock was loaded into a preheated chamber at a temperature range of 110-150°C and increased shear rate from 5–8000 s^{-1}. The die orifice used was made of tungsten carbide and having a dimension of 1 mm in diameter and 16 mm in length.

2.3. Injection Mouding
Injection moulding was carried out using a manually operated vertical injection moulding with a tensile shape cavity according to the ASTM 638 type V as shown in figure 1(a) and (b)[6]. The moulding temperature range used was based on the rheological result obtained which in range of 110 to 150°C.

3. Results and discussion
Rheological test is used to determine the viscosity level of feedstock when it passes through a die orifice. Figure 2 shows the rheological results for all the feedstock with different powder loading at different temperatures, clearly shear thinning behaviour or pseudo-plastic curve is observed. The selected temperature was referred to the previous successful study of MIM for NiTi [7].

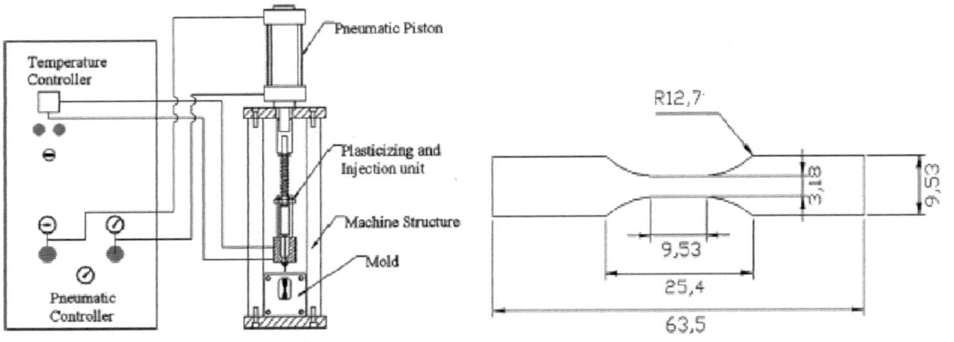

(a) (b)

Figure 1. (a) Schematic diagram of manually operated vertical injection moulding and (b) ASTM D638 Type V specimen. [6]

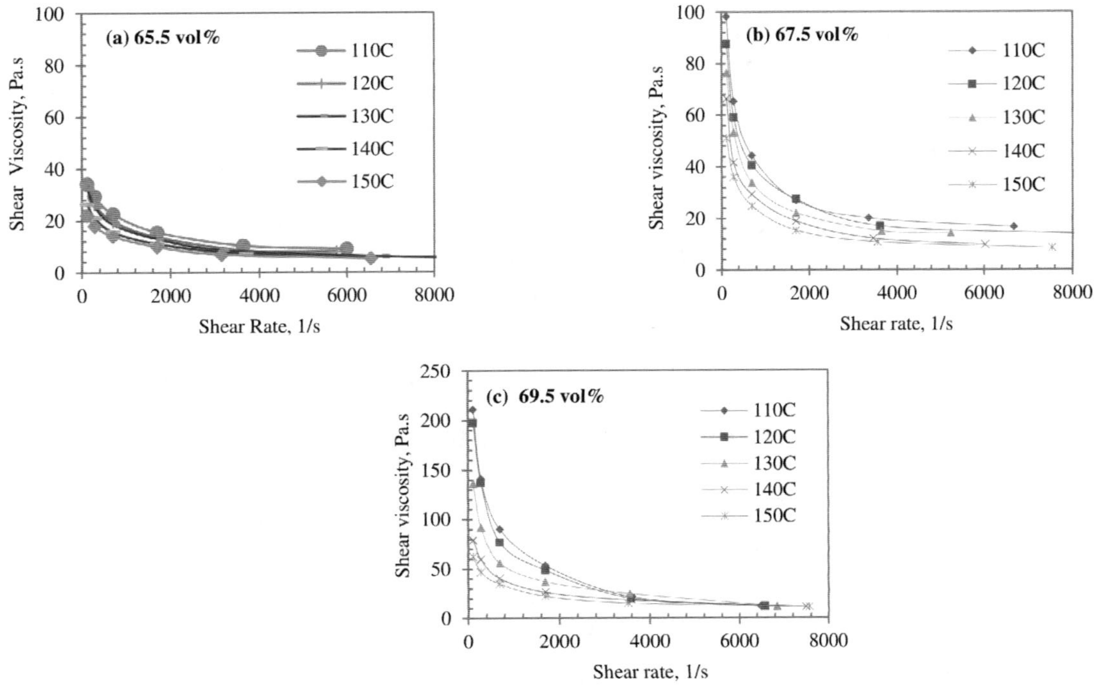

Figure 2. Viscosity versus shear rate at different temperature for different powderloading; (a) 65.5 vol%, (b) 67.5 vol% and (c) 69.5 vol%.

From figure 2, it shows that increasing powder loading results in increasing viscosity, particularly at lower shear rates with maximum viscosity of 250 Pa.s at the maximum powder loading of 69.5%. The trend of shear thinning is clearly seen with increased shear rate and temperature, thus promoting better fluidity during feedstock flow through mould cavity. The maximum viscosity of 250 Pa.s was achieved for the highest powder loading, which were considerably lower than 1000 Pa.s,[7]. A higher viscosity at low shear rates, particularly lower than $100s^{-1}$ was attributed from less particle orientation and ordering with the flow. As the shear rate increased, the viscosity also decreased, corresponding to improved fluidity and homogeneity.

The relationship between viscosity (η)and shear rate (γ) for the MIM feedstock can be expressed as equation (1) [7, 8].

$$\eta = K\gamma^{n-1} \quad (1)$$

where η is a viscosity
K is a low yielding stress
γ is a shear rate
n is a consistent flow index

From figure 2, the shear sensitivity (n) was determined according to equation (1) and the values are summarized in Table 1. It clearly shows that of the n values calculated are within the range of 0.50 to 0.83, which are slightly higher than the previous research [8-10], owing to different powder characteristics, binder system used and method of mixing From the MIM point of view, the lowest the n value, the most sensitive viscosity to shear rate changes. However, too high n values will result in instability of the process and difficulties in controlling the quality of the green parts due to greater mixture viscosity [7, 8]. In the present work, the feedstock prepared shows the values of n lower than 1 for all temperature applied, thus suggesting appropriate range of temperature for injection moulding.

Table 1. Shear sensitivity value (n).

Feedstock	n				
	110	120	130	140	150
65.5 vol%	0.83	0.72	0.73	0.77	0.79
67.5 vol%	0.50	0.56	0.58	0.52	0.60
69.5 vol%	0.7	0.72	0.68	0.63	0.68

Figure 3 shows the effect of temperature on viscosities at different powder loading. According to the Arrhenius equation, the flow activation energy, E can be expressed by equation (2).

$$\eta = \eta_o \exp\left(\frac{E}{RT}\right) \quad (1)$$

where η_o is a relative viscosity and R constant (8.314 kJ/mol).

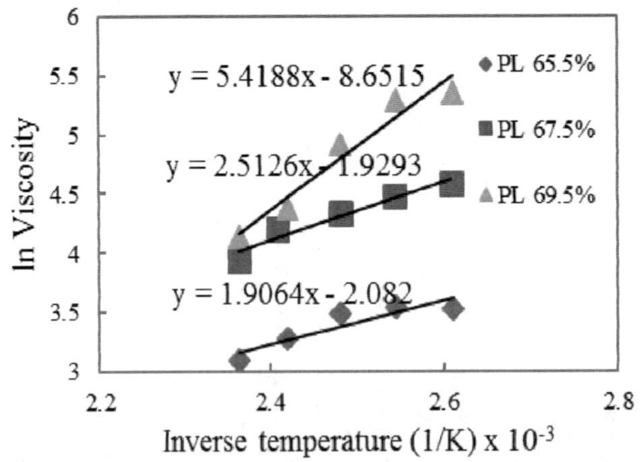

Figure 3. Arrhenius plots of different powder loading feedstock.

The activation energy E was obtained from the slope of the curve as shown in Figure 3; $E_{65.5}$ = 15.8 kJ/mol, $E_{67.5}$ = 20.9 kJ/mol and $E_{69.5}$ = 45.2kJ/mol, clearly the E increases as the powder loading increases From the standpoint of MIM flow during injection, the lower E value is required to ensure viscosity stability as a result of changes in temperature and shear rate[10]. This is particularly important when the part geometries consist of sharp edges and significant changes in cross sectional areas, thus, defects-free moulded parts can be minimized[7]. It means the viscosity is not so sensitive to temperature variation and will not result sudden viscosity which may cause cracking and distortion.The activation energy of feedstock $E_{65.5}$ is the lowest which is implies the best from the standpoint of temperature sensitivity.Compare to the previous research, the current study achieved considerably lower E for NiTi feedstock with the same powder loading employed. This is probably due to the binder system used, in which the one used by Ismail [11] was based on PEG-PMMA system.

Figure 4 shows the morphologies of the feedstock observed under SEM with different powder loadings. It shows that increasing powder loading resulted in greater particles packing with less fraction of binder network observed resulting greater viscosity of the feedstock. The use of palm stearin and polyethylene as a binder in the present work is believed could reduce the feedstock viscosity significantly; when comparing the same powder laoding of 69% conducted by Ismail[11], with maximum viscosity observed at $10s^{-1}$ was approximately 500 Pas. Therefore, less friction between feedstock and die wall could be expected during injection moulding and parts' surface defects could be minimized considerably. It also attributed a deviation viscosity with increasing temperatures. Similar results were reported by several researchers,claiming the same trend of viscosity[7, 8].

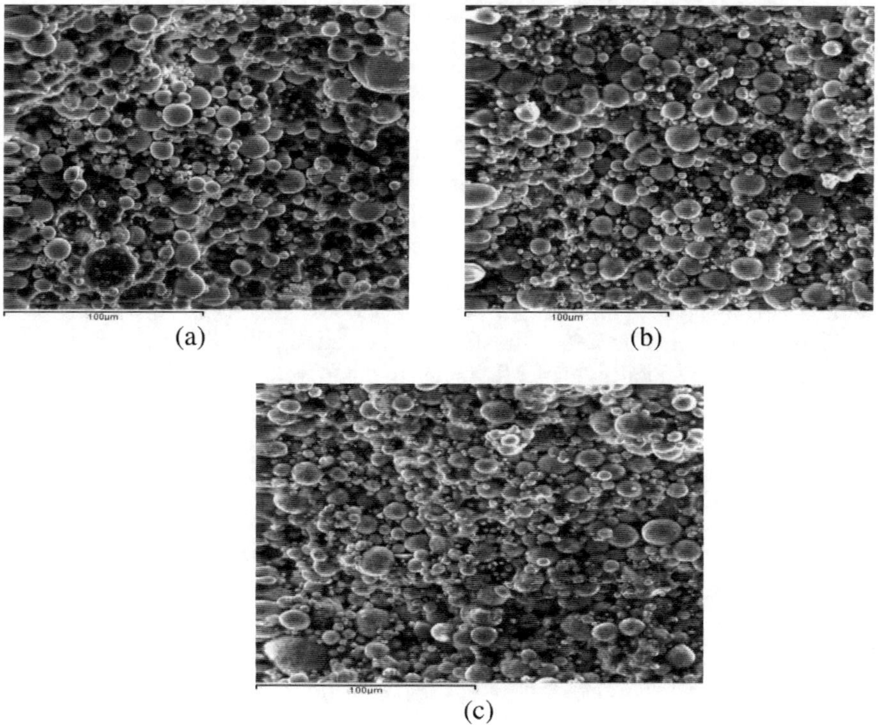

Figure 4. SEM micrograph of different powder loading; (a) 65.5 vol%, (b) 67.5 vol% and (c) 69.5 vol%.

The rheological behaviour explicitly shows how important the development of optimized feedstock composition is. Even though rheological test has been carried out, it is difficult when translating the data during injection moulding practice. This is because the arbitrary value must depend on machine pressure and sprue, gate and mould design, to make it possible for higher viscosity to be tolerated. However, the data would give some insights on the scope of the parameter condition to be applied during injection moulding practice. Initial experiments of rheological behaviour showed that the choice of a suitable binder system is a major challenge.

During injection moulding process, it was observed that the samples were successfully injection moulded at the barrel temperature of 120°C to 150°C. Figure 5 shows the image of the defect free moulded samples. The total cycle time for each sample was approximately 5 minutes owing to manually operated process, taken into account the time to place the mould underneath the nozzle, injection moulding (~ 1-2 seconds), cooling stage, mould removal and removal of samples. It was observed that the green parts exhibited adequate strength to be handled after they were removed from the mould cavity.

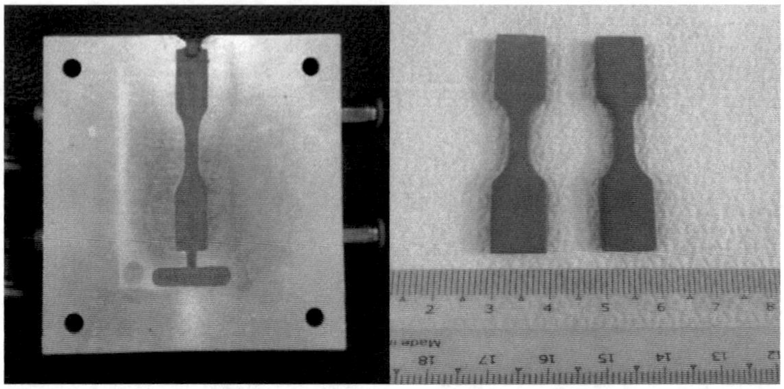

(a) (b)

Figure 4. a) Quality of moulded and b) Green parts.

4. Conclusions

All the prepared NiTifeedstocks containing PS/PE binder indicated pseudo-plastic behaviour that facilitated the moulding process. It was found that the viscosity increased as the powder loading increased owing to greater amount of powder in the feedstock. Increasing the powder loading resulted in increased flow activation energy instead of flow behaviour gives fluctuate value.

Acknowledgement

The authors would like to be obliged to Ministry of Education (MOE) and UniversitiTeknologi MARA for providing laboratory facilities and financial assistance under project no 600-RMI/RACE 16/6/2 (5/2012) and RMI/FRGS 5/3 (38/2012).

References

[1] Krone L, Schüller E, Bram M, Buchkremer H-P, Stöver D 2004 *Material Science Engineering A* **378** 185-90

[2] Bram M, Ahmad-Khanlou A, Heckmann A, Fuchs B, Buchkremer H-P 2002 *Material Science Engineering A* **337** 254–63

[3] Li Y, Huang B, Qu X 1999 *Powder Metallurgy* **41** 86–90

[4] Zauner R, Binet C, Heaney D F, Piemme J 2004 *Powder Metallurgy* **47**

[5] Nor N H M, Ismail M H, Kasim N A A, Muhamad N 2014 *Applied Mechanics Materials* **466** 709–14

[6] Azuddin M, Choudhury I A, Taha Z 2012 *Journal of the Brazillian Society of Mechanical Sciences and Engineering* **37** 79-86

[7] Li Y, Li L, Khalil K A 2007 *Journal of Materials Processing Technology* **183** 432–9

[8] Omar M A, Subuki I, Abdullah S, Ismail M F 2010 *Journal of Science and Technology* **2** 1-14

[9] Huang B, Fan, J. , Liang S, Qu X 2003 *Journal of Materials Processing Technology* **137** 177-82

[10] Liu L, Loh N H, Tay B Y, Tor S B, Murakoshi Y, Maeda R 2005 *Materials Characterisation* **54** 230-8

[11] Ismail M H. Porous NiTi Alloy by Metal Injection Moulding (MIM) Using Partly Water Soluble Binder System [PhD Thesis]: University of Sheffield; 2012.

Feature combination analysis in smart grid based using SOM for Sudan national grid

Z H Bohari[1], M A M Yusof[1], M H Jali[1], M F Sulaima[1] and M N M Nasir[1]

[1]Faculty of Electrical Engineering, Universiti Teknikal Malaysia Melaka (UTeM), HTJ, 76100 Durian Tunggal, Melaka, Malaysia

Email: azulhasrizal@utem.edu.my

Abstract. In the investigation of power grid security, the cascading failure in multi-contingency situations has been a test because of its topological unpredictability and computational expense. Both system investigations and burden positioning routines have their limits. In this project, in view of sorting toward Self Organizing Maps (SOM), incorporated methodology consolidating spatial feature (distance)-based grouping with electrical attributes (load) to evaluate the vulnerability and cascading impact of various part sets in the force lattice. Utilizing the grouping result from SOM, sets of overwhelming stacked beginning victimized people to perform assault conspires and asses the consequent falling impact of their failures, and this SOM-based approach viably distinguishes the more powerless sets of substations than those from the conventional burden positioning and other bunching strategies. The robustness of power grids is a central topic in the design of the so called "smart grid". In this paper, to analyze the measures of importance of the nodes in a power grid under cascading failure. With these efforts, we can distinguish the most vulnerable nodes and protect them, improving the safety of the power grid. Also we can measure if a structure is proper for power grids.

1. Introduction

Power systems are worked with the goal that over-burdens don't happen either progressively or under any measurably likely contingency. This is regularly called keeping up framework “security". Test system is outfitted with devices for dissecting possibilities in a programmed manner. Contingency can comprise of a few activities or components that are straightforward case for blackout of a solitary transmission line and intricate for blackout of single of a few lines, various generators, and the conclusion of typically open transmission line. The Power grid security is one of the huge perspectives, where the correct move needs to be made by the operational specialists for the unseen contingency. In this way the contingency investigation is key for the power grid security. The contingency positioning utilizing the execution list is a strategy for the line blackouts in a power grid, which positions the most noteworthy execution record line first and returns in a plummeting way focused around the computed PI for all the line blackouts [1].

A self organizing map (SOM) is a sort of artificial neural network (ANN) that is prepared utilizing unsupervised figuring out how to create a low-dimensional (ordinarily two-dimensional), discretized representation of the information space of the preparation examples, called a map. "Self Organizing" is on the grounds that no supervision is needed [2]. SOMs learn all alone through unsupervised aggressive learning. "Maps" is because they attempt to map their weights to conform to the given input data [3, 4]. Smart grid is conveys electrical power to the shoppers utilizing two way computerized engineering. Monitors are the supply to the customers and estimations. Numerous nations and power markets are taking a gander at Smart Grid as progressive arrangements in conveying blend of upgraded qualities going from higher security, dependability and power quality, lower expense of conveyance, interest streamlining and vitality productivity. Smart grid arrangements empower utilities to build vitality profit and power dependability while permitting the clients to deal with the use and expenses through on going data trade [5]. It affects all the parts of the power grid like generation, transmission and distribution.

As the result, the main of this research are understand the various types of transients and its analysis in power system and illustrate the concept and design of self organizing maps in smart grid.

2. Methodology

The flowchart as shown in figure 1 describes all the activities or tasks to be done at each stage of the project. It is important to make sure the project complete at the specific time.

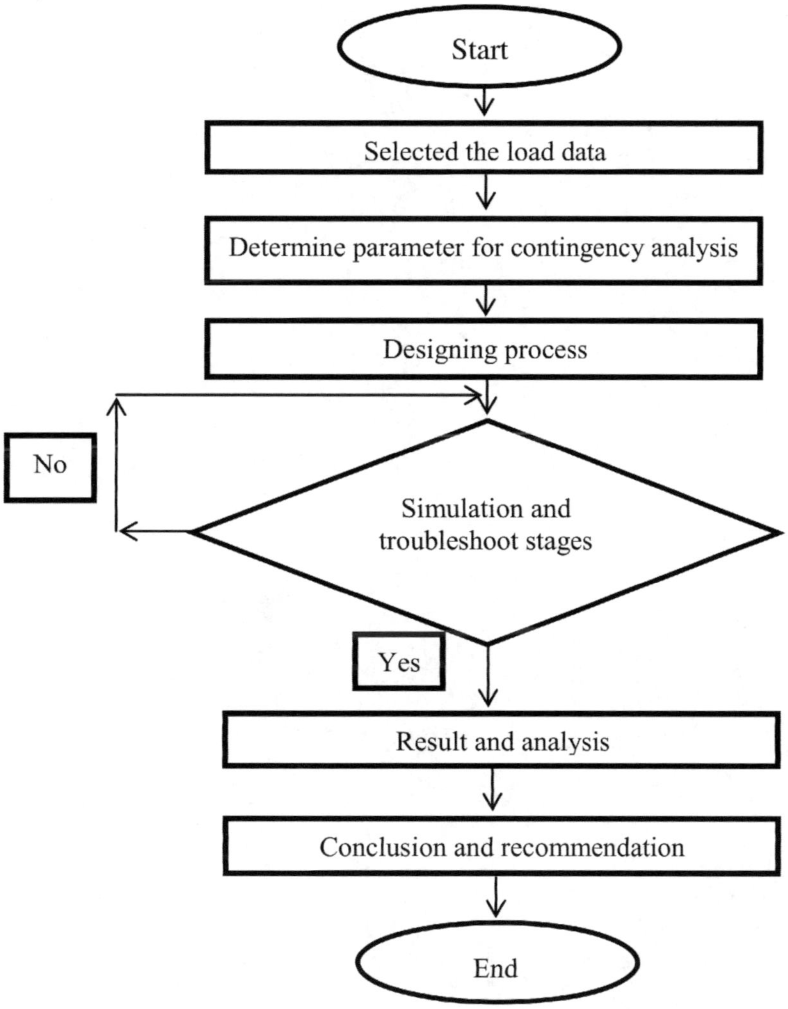

Figure 1. Flow Chart of Methodology.

3. Result

Analysis and simulation there were using MATLAB software to determine the analysis. From the analysis, based on the Self Organizing Map, U-matrix is shown and easy to see that the top three rows of the SOM form a very clear cluster[6]. There are four types of normalization method that are Log, Logistic, Range and Var. For the data, previous work paper from Sudan National Grid will be selected to determine the analysis [7]. Optimized number of neurons value at 200 neurons is selected for analysis and simulate to determine the result.

From this load data table, the comparison U-matrix result between parameters'P and Q', 'P,Q and S', and 'P,Q,S and θ' with four types of normalization method that is Log, Logistic, Range and Varwill be shown in figure 2, figure 3 and figure 4 below :

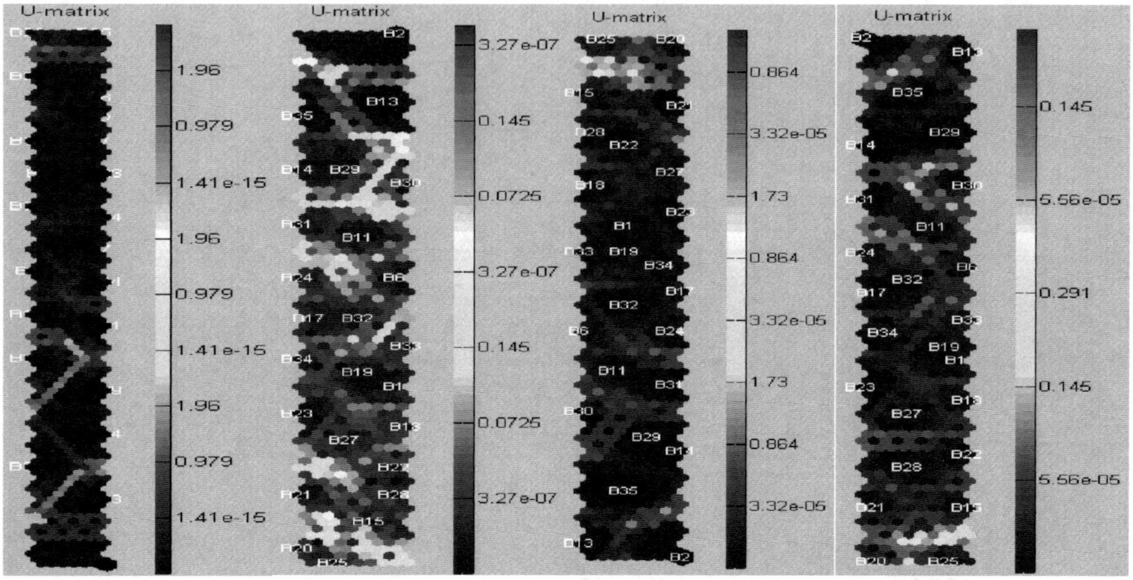

Figure 2.U-matrix result using parameters P and Q.

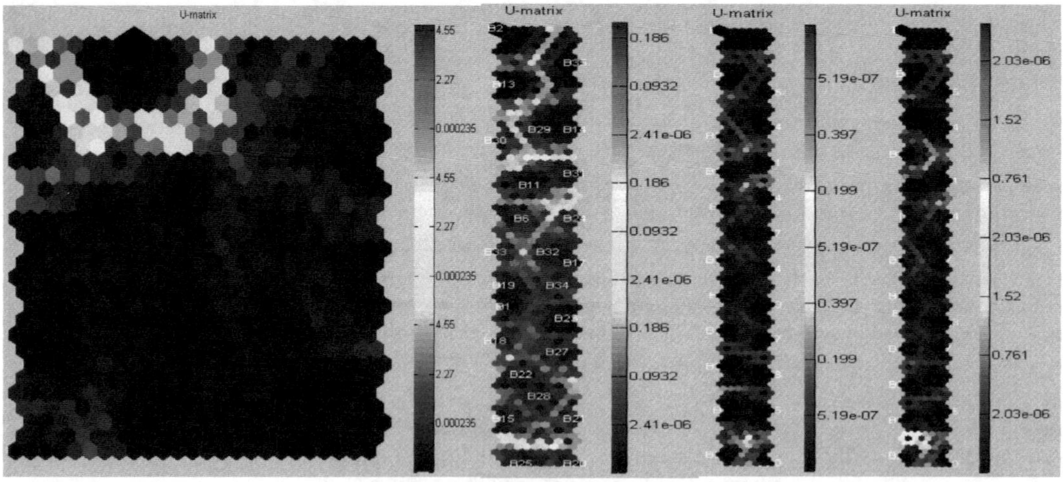

Figure 3. U-matrix result using parameters P,Q and S.

Table 1. Load Data.

Bus Number	Active Power (MW)	Reactive Power (MVAR)	Apparent Power (MVA)	Angle (θ)
1	35.5	21	41.3	30.6
2	0	0	0	0
3	0	0	0	0
4	0	0	0	0
5	0	0	0	0
6	30	15	33.5	26.6
7	0	0	0	0
8	0	0	0	0
9	0	0	0	0
10	0	0	0	0
11	24	12	28.8	26.6
12	0	0	0	0
13	6	3	6.7	26.6
14	15	6	16.2	21.8
15	49	33	59.1	33.96
16	0	0	0	0
17	40	15	42.7	20.6
18	35	25	43	35.54
19	35.1	20.2	40.5	30
20	93	52	106.6	29.2
21	52	32	61.1	31.6
22	43	28	51.3	33.1
23	40	21	45.2	27.7
24	39	12	40.8	17.1
25	83	45	94.4	28.5
26	40	21	45.2	27.7
27	40	24	46.7	31
28	44	30	53.3	34.3
29	14	7	15.7	26.6
30	10	14	17.2	54.5
31	24	10	26	22.6
32	37	15	40	22.1
33	30	20	36	33.7
34	38	18	42	25.4
35	12	5	13	22.6

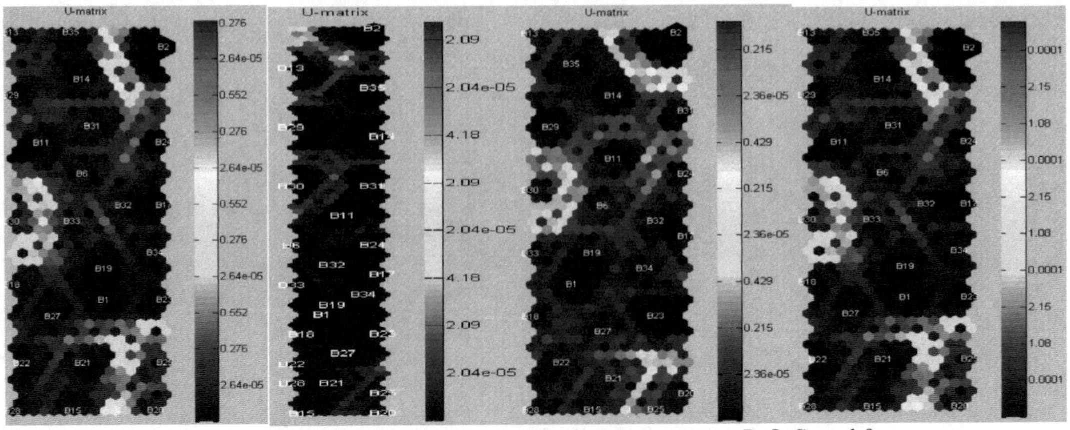

Figure 4.U-matrix result using parameters P,Q,S and θ.

Table 2.SOM Result Using Hexagonal Topology using parameters P and Q.

Normalization Method	Classification Result (P,Q)			
	Map Size	Quantization Error	Topographic Error	Training Time (sec)
Log	[50, 4]	0.005	0.029	0
Logistic	[40, 5]	0.002	0.086	0
Range	[40, 5]	0.002	0.000	0
Var	[40, 5]	0.007	0.029	0

Table 3.SOM Result Using Hexagonal Topology using parameters P,Q and S.

Normalization Method	Classification Result (P,Q,S)			
	Map Size	Quantization Error	Topographic Error	Training Time (sec)
Log	[15, 13]	0.011	0.086	0
Logistic	[40, 5]	0.002	0.057	0
Range	[50, 4]	0.001	0.000	0
Var	[50, 4]	0.005	0.000	0

Table 4.SOM Result Using Hexagonal Topology using parameters P,Q,S and θ.

Normalization Method	Classification Result (P,Q,S, θ)			
	Map Size	Quantization Error	Topographic Error	Training Time (sec)
Log	[40, 5]	0.009	0.000	0
Logistic	[25, 8]	0.002	0.057	0
Range	[25, 8]	0.002	0.029	0
Var	[25, 8]	0.009	0.029	0

4. Discussion

Based on theoretical, the best value of Quantization error must nearly '0', Topographic error also must nearly '0' and Training time must be '5' and below The recreation of SOM programming comprises of the mix between the different normalization methods ('var','range','log' or 'logistic') and the optimum number of neurons. The 'var' data input will normalize the variance variable to unity and the means to zero. For the 'range' input data will scale the variable values between zero and one. The 'log' is a logarithmic change and the 'logistic' softmax change scales all expected values somewhere around zero and one[8]. The challenges of multi-contingency analysis for smart grid attacks can be investigation in the following perspectives that the restricted scalability of parameter P,Q,S and θ many contingency analyses which are validated mostly on relatively small power system benchmarks. The constrained learning of attackers on the complex dynamics in real-time power systems, in contrast to the power system managers, that restricts their strength in modeling the power system and the estimation of the impact of their attacks. The difficulties in solving both linear and nonlinear equations in bulk power systems with inadequate data and intensive computational burden.

5. Conclusion

This paper proposes a topological technique to study the vulnerability of subsets of substations in power grids focused around SOM clustering. While the physical attributes is considered as the premise in the assessment of power grid security, partner falling investigation with spatial feature based clustering demonstrates that the joined methodology has the capacity spot the more basic segments in a substantial scale power lattice than conventional routines, giving a proficient device to the contingency analysis. In our methodology, the potential victimized people are prepared by the powerful SOM clustering so the applicants of pursuit are refined to a restricted extent, which essentially decreases the computational cost while keeping the capacity to recognize the absolute most powerless sets or assault conspires in the lattice. SOM is a compelling stage for visualization of high-dimensional information. However, to have the capacity to completely comprehend substance of an information set, it is basic to see whether the information has cluster structure. In the event that this is the situation, the clusters need to be concentrated to have the capacity to completely abuse the properties of the information set by delivering synopsis data.

References

[1] Glover J D, Sarma M, Overbye T 2011 *Power System Analysis and Design, Fifth Edition* Cengage Learning)
[2] Jali M H, Bohari Z H, Sulaima M F, Nasir M, Na'im M, Jaafar H I 2014 *Research Journal of Applied Sciences, Engineering and Technology* **8** 235-42
[3] Kohonen T 1990 *Proceedings of the IEEE* **78** 1464-80
[4] Jali M H, Bohari Z H, Sulaima M F, Nasir M, Na'im M 2014 *Journal of Theoretical and Applied Information Technology* **68** 125-37
[5] Pin-Yu C, Shin-Ming C, Kwang-Cheng C 2012 *Communications Magazine, IEEE* **50** 24-9
[6] Bohari Z H, Ab Ghani S, Baharom M F, Nasir M, Jali M, Thayoob Y M 2014 *Jurnal Teknologi* **67**
[7] Hassan M. Transmission Line overload Alleviation by Generation Rescheduling and Load Shedding in the National Grid of Sudan. Khartoum, Sudan: Karary Academy of Technology; 2005.
[8] Ambroise C, Govaert G 1996 *Journal of Classification* **13** 299-313

Statistical analysis of imperfection effect on cylindrical buckling response

M S Ismail[1, 2], J Purbolaksono[1,2], N Muhammad[1], A Andriyana[2] and H L Liew[2]
[1]Department of Mechanical Engineering, Politeknik Sultan Salahuddin Abdul Aziz Shah, Shah Alam 40150, Selangor, Malaysia
[2]Centre of Advanced Manufacturing and Materials Processing, Department of Mechanical Engineering, Faculty of Engineering, University of Malaya, Kuala Lumpur 50603, Malaysia

E-mail: judha@um.edu.my

Abstract. It is widely reported that no efficient guidelines for modelling imperfections in composite structures are available. In response, this work evaluates the imperfection factors of axially compressed Carbon Fibre Reinforced Polymer (CFRP) cylinder with different ply angles through finite element (FE) analysis. The sensitivity of imperfection factors were analysed using design of experiment: factorial design approach. From the analysis it identified three critical factors that sensitively reacted towards buckling load. Furthermore empirical equation is proposed according to each type of cylinder. Eventually, critical buckling loads estimated by empirical equation showed good agreements with FE analysis. The design of experiment methodology is useful in identifying parameters that lead to structures imperfection tolerance.

1. Introduction

Thin-walled cylindrical shells are susceptible to buckling failures caused by the axial compressive loading. During the design process or the buckling failure evaluation of axially compressed cylindrical shells, initial geometric and loading imperfections are of important parameters for the analyses. Therefore, the engineers/designers are expected to well understand the physical behaviours of shell buckling to prevent unexpected serious failure in structures. At the same time it is believe that most of cylinder imperfection is caused by defect and uncertainties that can be traced from the manufacturing process. As this process involved with a lot of variables and parameters to control consequently lead to manufacturing defects. These defects involved with various factors such as micro dented on the shell surface, fibre degree disorientations and fibre volume misalignment. As a result it can probably say that these aforementioned factors give large contribution towards structural imperfection.

To date, the issue corresponding to the discrepancies between the manufactured and computationally design finite element (FE) cylinder has become a vital point on establishing better design guideline. For example, in reality during curing process, the composite cylinder can be highly exposed to the distributed of residual stresses while this process is hardly replicate through FE model. As the FE model can be perfectly designed through finite element environment without considering the related issues of uncontrollable process variables [1]. As reported by Lee [2], initial geometry is one of a parameter that gives a significant different towards imperfection. This can be connected to the curing process that led the cylinder shell into variation of nominal radii and curvature and eventually changes the cylinder to be into slightly oval-shaped geometry. Indeed the deviated geometry somehow would affect the cylinder buckling response and its characteristic in final post buckling collapse.

The design of experiment (DOE) is a statistical based technique that used on conducting and analysing the sensitivity of parameters to the response output. Initially developed by Fisher [3], it has been promoted by various researchers for the use of experimental design strategy and sampling planning. Simultaneously, the interaction of single or combined parameter to the output response is then evaluated by analysis of variance (ANOVA). As this method is ideally used to reduce and statistically signify the number of necessary experiments and parameters. Up until now there is still unanswered question about to the precision knowledge of structure and material properties for composite cylinder. In response, the DOE technique is a suitable method to randomly vary the suspected variables to depict the output response thus link with its sensitivity. For instance, deviations in material properties, geometric correction factors and loading and boundary conditions are one of a few candidates that attribute towards the issue. While according to the manufacture, a typical composite fibre angles can be scatter deviated around ±2°.

In previous work, random imperfection based stochastic eigen-affine modes buckling is conducted by [4, 5]. It followed with the study conducted by [6-8] to generate eigen mode imperfection randomly using DOE method. While a study by [9-12] demonstrated a probabilistic imperfection analysis towards cylinders shells buckling through FE analysis. From their results it directs a possibility to improve the cylinder knockdown factor. However some limitation was found from their investigation, where the geometry and composite material is identical with a little different on fibres angle.

In response, new investigation comprised of different cylinder geometry and material characteristic is needed. As it is essential to re-evaluated the factors that affected the cylinder buckling response is more or less to be identical as previous reported in [9]. So far, it found that there is still shortage of study has been done on cylinder imperfection by means of design of experiment (DOE) method. The proposed area of investigation may confirmed of what been found from the previous studies with some extended work for different cylinders. Thus a better design guideline on predicting the buckling load that describes the imperfection from manufacture defect can be withdrawn.

2. Material and Methods

All the analyses were carried by using a finite element software package of ABAQUS v6.10. The schematic of the axially compressed cylindrical shell is shown in figure 1. As shown in figure 1, the displacement and rotation constraints are applied on one end of the cylindrical shell, and the load-controlled displacement ΔU with slow quasi-static compression is applied on the other end of the cylinder.

The cylinder geometries used in this study referring to those experimentally conducted by [13, 14] for the CFRP laminated shell, and the details are presented in table 1. The material properties of the monolithic laminates of carbon fibre composites reported by [14] were also summarised in table 1. Nevertheless two cases of cylinders with different layers are performed in this work which is four-ply with stacking sequences of [0/45/-45/0] and [45/-45] CFRP laminated shell.

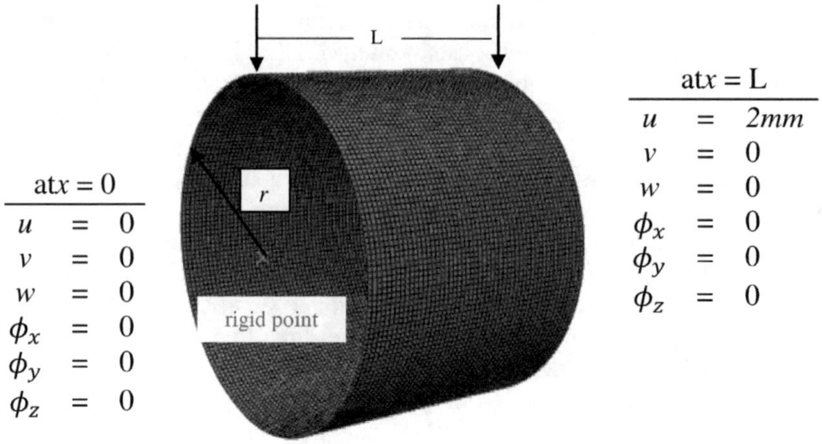

Figure 1. Boundary condition of FE modelling.

Table 1. Nominal cylinder geometries and material properties of CFRP.

Description	Unit	Bisagni [13]
Cylinder designation		CFRP
Radius, r	[mm]	350
Free length, L	[mm]	520
Nominal thickness, t	[mm]	1.32
r/t		265
L/r		1.45
Controlled displacement, ΔU	[mm]	2
Stiffness		
Elastic modulus E_{11}	[N/mm^2]	52000
Elastic modulus E_{22}	[N/mm^2]	52000
Shear modulus G_{12}	[N/mm^2]	2350
Poission's ratio υ		0.302
Density, ρ	[kg/mm^3]	1.32 x 10^{-8}
Ply thickness t_{ply}	[mm]	0.33

3. Design of Experiment: Theoretical Background

The response surface method is statistical based mathematical method that gives an effective and practical way on optimizing design [15]. The theory is based on the output response function, y which react in the function of designs variables, x_i. The function of designs variables, x_i is mostly designed as feasible domain. On the other hand, the polynomial approximation expressed by, $y = f(x_i)$ is usually described to the function on the basis of observation data. However in case of quadratic response function with two variables the polynomial reads as equation (1).

$$y = \beta_0 + \beta_1 x_1 + \beta_2 x_2 + \beta_3 x_1^2 + \beta_4 x_2^2 + \beta_5 x_1 x_2 + \varepsilon \quad (1)$$

These terms is defined as β_i are the regression coefficient and ε is error approximate function. Consequently when equation (1) is substitute with such variables as, $x_1^2 = x_3$, $x_2^2 = x_4$ and $x_1x_2 = x_5$ it yields to be multivariable equation as shows in equation (2). As this substitution shows the regression model that contained higher terms can often reduce to a linear regression model. Then by taking into account of k regression coefficient, the following equation obtained:

$$y = \beta_0 + \beta_1 x_1 + \beta_2 x_2 + \beta_i x_i + \ + \beta_k x_k + \ \varepsilon \tag{2}$$

Let says if n sets of observation data and the n corresponding to combination of design variables are known. Then the response can be expressed by the representation of matrix as shows in equation (3).

$$\begin{Bmatrix} y_1 \\ y_2 \\ \vdots \\ y_n \end{Bmatrix} = \begin{bmatrix} 1 & x_{11} & x_{12}\cdots & x_{1k} \\ 1 & x_{21} & x_{22}\cdots & x_{2k} \\ \vdots & \vdots & \vdots \ \ddots & \vdots \\ 1 & x_{n1} & x_{n2}\cdots & x_{nk} \end{bmatrix} \begin{Bmatrix} \beta_1 \\ \beta_2 \\ \vdots \\ \beta_k \end{Bmatrix} + \begin{Bmatrix} \varepsilon_1 \\ \varepsilon_2 \\ \vdots \\ \varepsilon_n \end{Bmatrix} \tag{3}$$

Then equation (3) can be rewrite in matrix form (4):

$$y = X\beta + \varepsilon \tag{4}$$

4. Design of Experiment: Factorial Design

The number of samples can be derived by summed the 2 levels and 7 factors (2^7) of the designed numerical experiments that yields to a set of 128 samples. However to minimised the numbers of runs, only a fraction of one-eight (1/8) from the whole samples is taken thus yields to 16 samples. Then for accuracy purposed, the 16 factorial designs is augmented by 5 centre points that lead to in total of 21 cases was used to study the effect of factors for each cylinder. Yet from the minimised work the samples is still adequate for design of experiment (DOE) prospect. These factors namely Young's modulus (longitudinal), E_{11}, cylinder thickness, t, load imperfection, θ_{load} and each of fibres degree angle orientation (θ_1, θ_2, $\theta_3$andθ_4) are consider as a primary candidate that would give a significant effect on cylinder buckling load response. The selected investigated factors are based on the report by Degenhardt et al. [9] which identified these four parameters having high sensitivity towards buckling response compared to the others. Therefore the other parameters were set to be constant throughout the study. Moreover the replicates run at centre point are design to take into account of quadratic effect. Thus lead to an independent estimate of error to be obtained.

Table 2 summarised the factors and range applied for the 21 factorial designs. In this study, the possibility range of the factors was determined by varying in the range of ±5% from the nominal value. Also the statistical significant of each parameters and their combinations were evaluated by 5% of their level of significant. On the other hand a multiple regression analysis was performed based on the regression model to obtain the second-order response function [16]. From equation (1) y is the output response while $\beta_0, \beta_i, \beta_{ii}$ and β_{ij} terms are the coefficients to intercept, linear, quadratic and interaction variables respectively. In additional x_i and x_j are the independent variables in coded unit and ε is the error terms that account the effect of excluded parameters. During analysis, the least squared method is been used to generate the prediction model equation (1) for a best fit of collected response data.

Table 2. Factors and range applied of 2^7 factorial designs.

Parameters	Symbols	Units	Range	
Load imperfection, θ	A	[°]	$0 \leq 0.33$	
Shell thickness, t	B	[mm]	$1.254 \leq 1.38$	
Elastic modulus, E_{11}	C	[N/mm^2]	$49.4 \leq 54.6$	
Ply orientation			[0/45/-45/0]	[45/-45]$_S$
Fibre, θ_1	D	[°]	$-1 \leq 1$	$44 \leq 46$
Fibre, θ_2	E	[°]	$44 \leq 46$	$-46 \leq -44$
Fibre, θ_3	F	[°]	$-46 \leq -44$	$-46 \leq -44$
Fibre, θ_4	G	[°]	$-1 \leq 1$	$44 \leq 46$

5. Result and discussion

In this section the response of load-end shortening from FE analyses is presented and focus is highlighted to [0/45/-45/0] ply stack-up. The CFRP of [0/45/-45/0] ply with its load shortening response, modeshape and final collapse deformation is presented in figure 2 (a)-(c). The load-displacement curve is given in figure 2 (a) indicated as a guide of shell compressive response. The axial load, P and end-shortening is normalized to the linear eigenvalue critical buckling load, $Pcrit$ and nominal shell wall thickness, t of 240 kN and 1.32 mm respectively. General instability occurs at normalized axial load of 0.75, labelled by letter A. The instability response at point A is followed by a sudden reduction of axial load at point B - D that associated with shell transient collapse. Then it reached to steady state value at point D – E. The reduction of shell axial stiffener is associated with the slope at postbuckling region. The shell reached to it stable-equilibrium postbuckling condition after approximately 0.007 seconds. Furthermore the buckles wave pattern illustrates in figure 2 (c) and (c) shows different characteristic of its buckles bay occurred around the cylinder circumferential by comparing both eigenvalue linear and non-linear analyses. It should point out that the difference of buckling characteristic is due to the dissimilar of solver method. The eigenvalue solve the problem based on bifurcation manner while the non-linear analysis solves the problem based on incremental time until the solution reaches its equilibrium path.

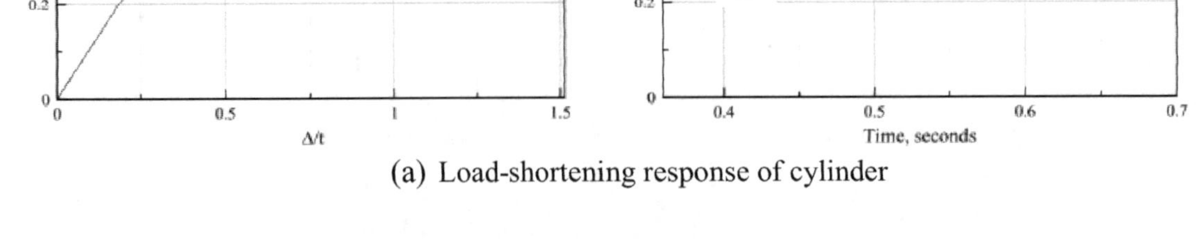

(a) Load-shortening response of cylinder

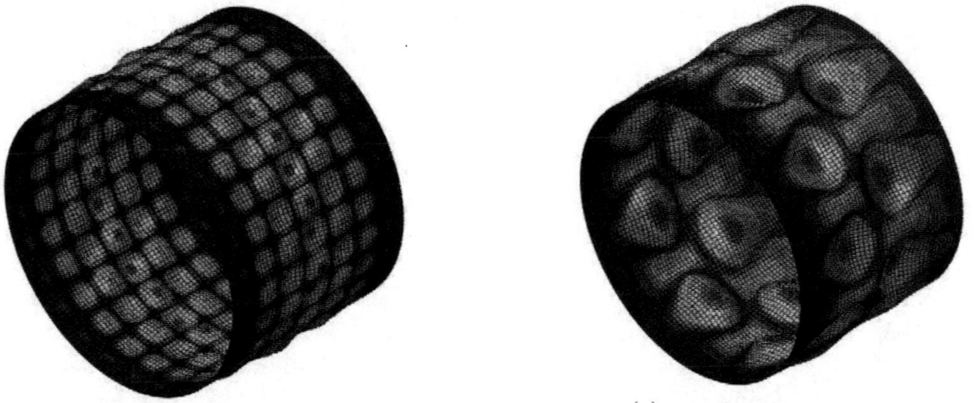

(b) 1[st]Eigenmode from linear analysis.

(c) Final collapse from non-linear analysis.

Figure 2. FE results of CFRP [0/45/-45/0] cylinder.

The significant effect of each parameter was evaluated by standardized effect of 5% significance level. In particular the interaction analysis of variance (ANOVA) is conductedand summarised in table 3 for the case of cylinder [0/45/-45/0].In which the approximated regressed equation throughout the factorial design analysis is derived from the second order polynomial. Herein the unknown parameters, β has been identified from the numerical experiment and solved by least square method. Therefore from 21 sampling points of each case, the cylinder buckling load prediction can be estimated by the following empirical expression;

For case of cylinder [0/45/-45/0];

$$Y = -307.875 - 28.4205A + 345.069B + 1.59591C + 0.353125D - 0.061875E - 0.234375F + 0.301875G$$

For case of cylinder $[45/-45]_S$;

$$Y = -135.494 - 12.2917A + 178.383B + 0.194471C - 0.175625D - 0.353125E + 0.351875F + 0.306525G$$

Table 3. ANOVA for cylinder [0/45/-45/0].

Source	DF	Seq SS	F	P	Significant
Load imperfection,	7	351.84	378.61	0.000	Significant
Shell thickness,	1	7561.61	8136.94	0.000	Significant
Elastic modulus E_{11}	1	275.48	296.44	0.000	Significant
Fibre, θ_1	1	2	2.15	0.169	
Fibre, θ_2	1	0.06	0.07	0.802	
Fibre, θ_3	1	0.88	0.95	0.35	
Fibre, θ_4	1	1.46	1.57	0.234	

The design matrix and results of 21 factorial designs are depicted in figure 3 (a) and (b) for each case of cylinder. Overall the cylinder buckling load are obtained in the range of 200 kN to 260 kN for cylinder [0/45/-45/0] and 100 kN to 128 kN for cylinder $[45/-45]_S$. The results also provide the ratio of maximum to minimum equal to 1.3 and 1.28 respectively. The smaller ratio indicated no further requirement to apply any kind of data transformation. The ANOVAof both cylinder cases are then evaluated to determine the accuracy of the empirical models. Consequently the coefficient of determination, $R^2$are found for the case of [0/45/-45/0] and $[45/-45]_S$to be 0.99 and 0.96. The R^2 value indeed showed the level of model accuracy. This is due to the fact that the closer value to 1 indicates that the regression model fits well with the data.

At the same time the Pareto's chart shows in figure 4 (a) and (b) indicated the sensitivity of candidate factors that possibly affect the cylinder buckling response for each case. From the results, it observed that the main factors that affect the cylinder buckling response were identified as shell thickness, t (A), load imperfection (B) and Young modulus, E_{11} (C) for the case of [0/45/-45/0] plies. However for cylinder with ply angle of $[45/-45]_S$ shows a different pattern, whereas the Young modulus, E_{11} (C) is not taking effect on cylinder response as previous case. This outcome has been expected due to the axial stiffness of $[45/-45]_S$ cylinder is considerably lower than [0/45/-45/0] cylinders. In fact they shared similar value of Young modulus and the rest of material properties. This behaviour is believed to be connected with direct consequences of the missing 0° of fibre angle that perpendicular to the loading axis. Eventually lead to the relatively low cylinder stiffness. Yet with this indication, it confirmed the finding of Degenhardt et al. [9], that these three factors posses high sensitivity towards cylinder buckling response as it with different material properties and fibre direction. In return, it can be assumed that these factors can be more consistent with the following isotropic material.

Figure 5 (a) and (b) illustrated the effect of selected factors toward buckling load for each cylinders. The relationship shows in the results indicate the level of lower and upper bound on the buckling response. While the straight horizontal line marked as the mean data distribution. A part from that it distinguished the relationships of shell thickness towards buckling load are linear for both cylinders accordingly. Interestingly the effect of loading imperfection and material stiffness are remarks to be less important. As this can be seen where a steady decreases was found by increasing the compression rotation angle. This behaviour is assumed that single factor like loading imperfection could not affect the cylinder buckling alone. Whereas a combination between either shell thickness or elastic stiffness should provide more reasonable effect on cylinder buckling load sensitivity.

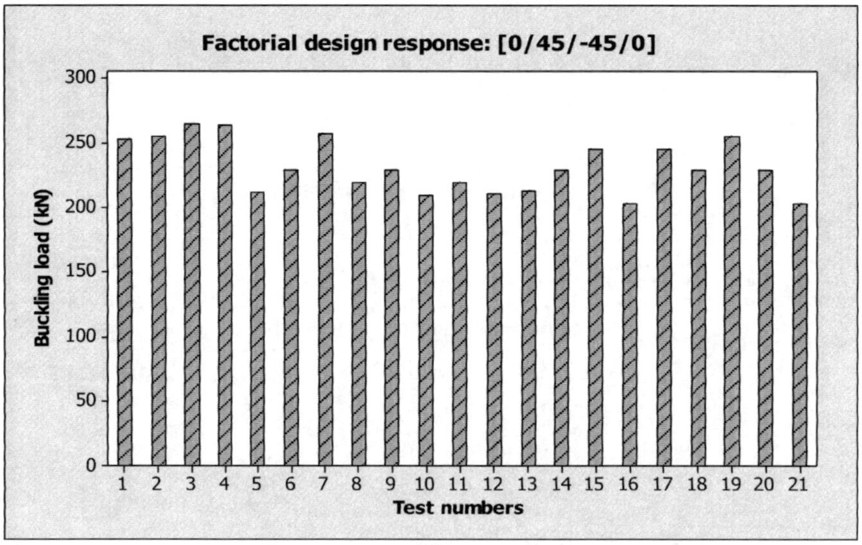

(a)

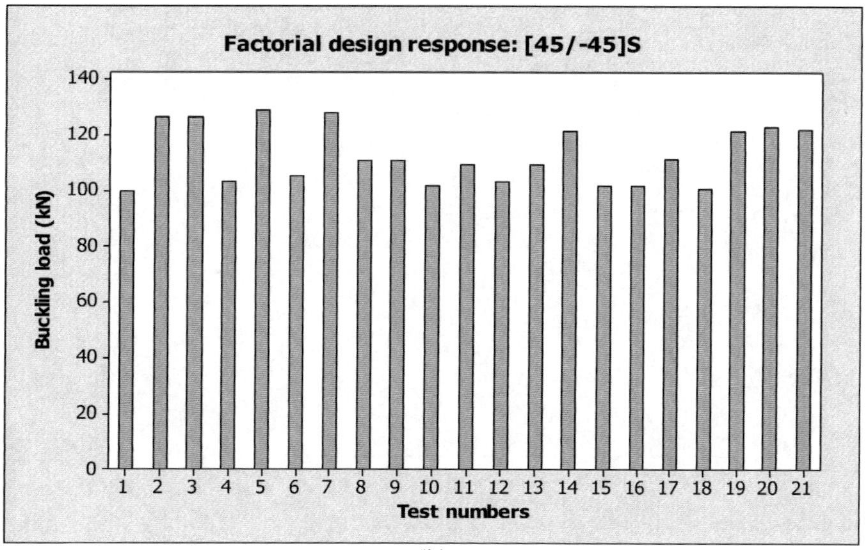

(b)

Figure 3. Factorial design response of cylinders buckling load distribution.

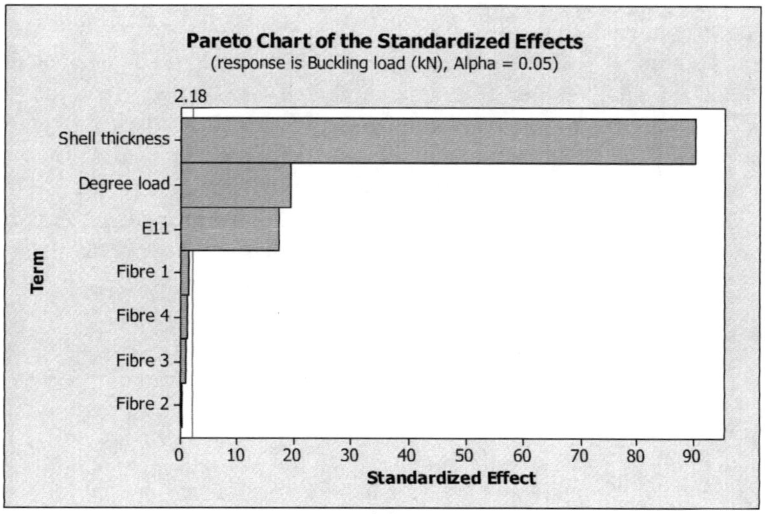

(a)

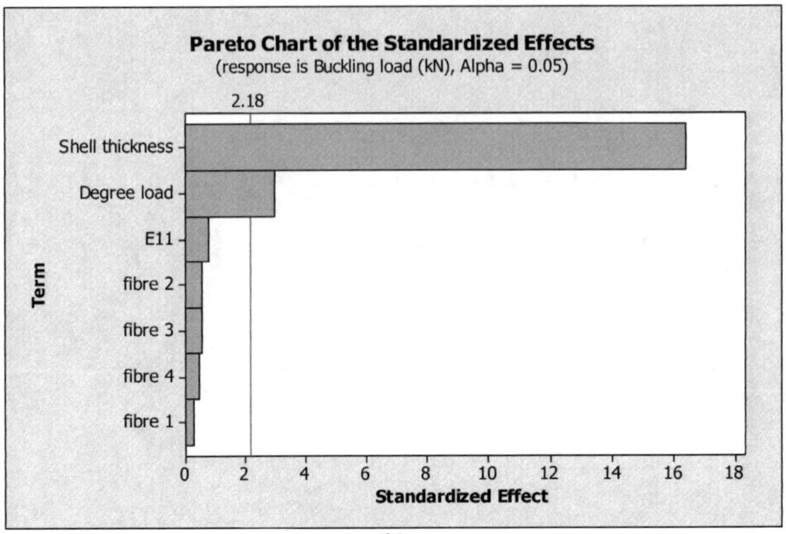

(b)

Figure 4. Pareto chart of CFRP composite cylinder standardised effect for (a) [0/45/-45/0] and (b) $[45/-45]_S$ attribute to each factors.

Figure 6 shows how well the empirical model prediction by comparing it with FE analyses for both cylinders. The data distribution is based on the ratio of empirical model over FE model prediction against number of test performs. According to the distribution it tells the accuracy of the results by closer it is to the line. In addition the marks below the benchmark line indicated that the empirical model is much lower than FE analysis. At the same time the marks over the line region show otherwise. However few observations for test number (11 and 13) gives a little high values of ratio to be nearly 5% for the case of cylinder $[45/-45]_S$. Finally most of the input values for both cylinders point out to have less than 5% of discrepancy to the unified value (benchmark line). Finally simple empirical formulation has been derived to predict the cylinder buckling load. Yet the proposed formulation is still bound with the range of considered variables to ensure it accuracy.

IOP Conf. Series: Materials Science and Engineering **100** (2015) 012003 doi:10.1088/1757-899X/100/1/012003

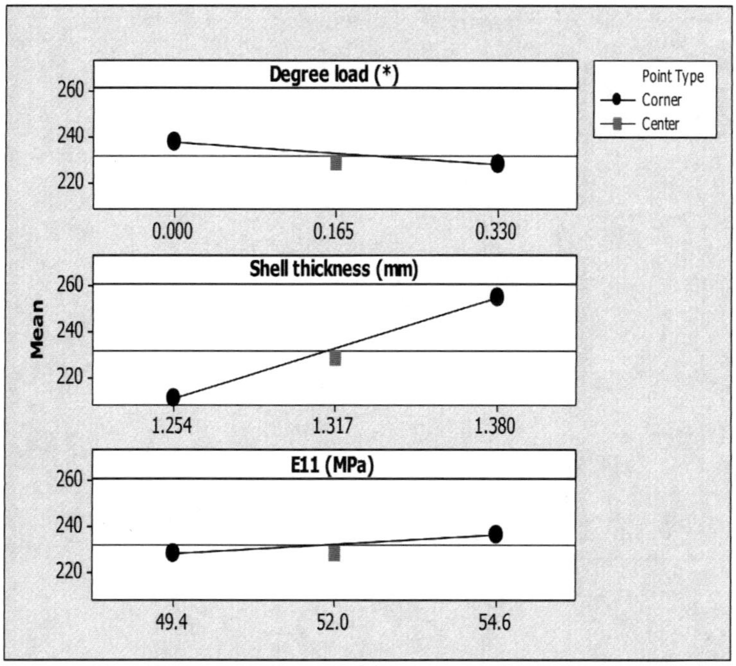

(a)

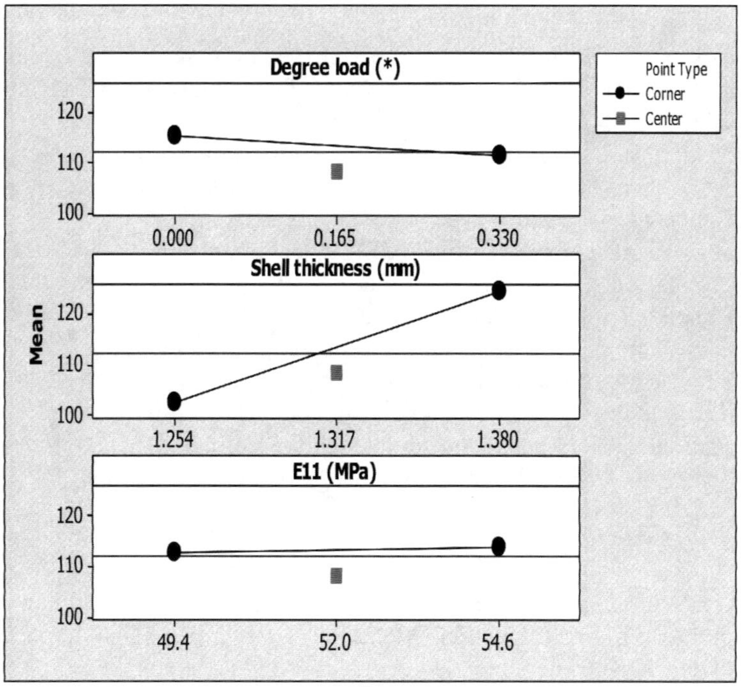

(b)

Figure 5. Main effect plot for cylinder (a) [0/45/-45/0] and (b) $[45/-45]_S$.

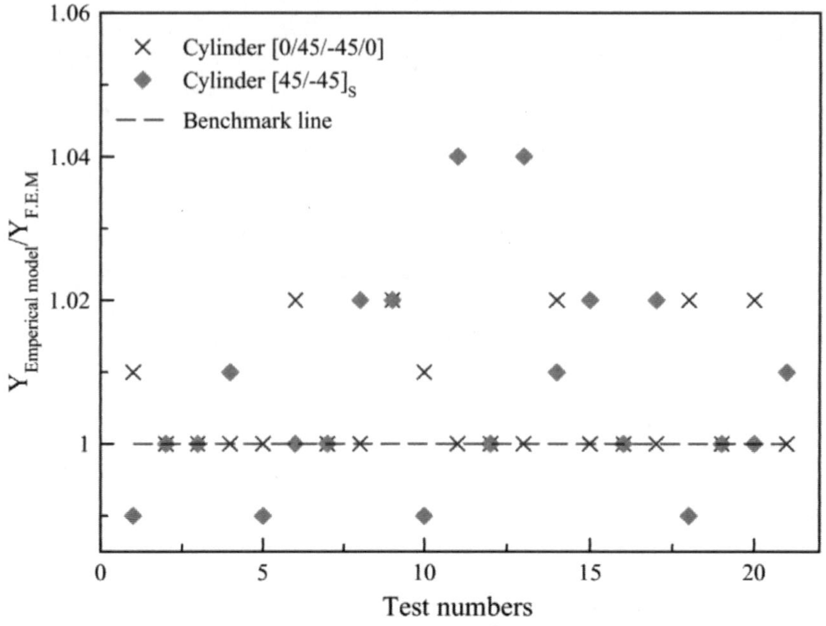

Figure 6. Comparison result for both cylinders using empirical model with FE analysis.

6. Conclusion

Statistical based design of experiment analyses on axially compressed CFRP cylindrical shells with different ply angle were demonstrated through the finite element simulations. The finding of this work may be summarized as follows:

- The characteristic of the composite cylinder buckling event is briefly describe and identified and suggest that non-linear analysis is better approach on capturing the real buckling mechanism compared to the linear one.
- The design of experiment approach made a significant contribution on identifying the factors that sensitively reacted towards buckling load. As the sensitivity factors is drawn to be load imperfection, shell thickness and elastic modulus.
- From the statistical studies it may suggest that loading imperfection and material stiffness are less sensitive towards structural imperfection compared to shell thickness during the buckling event. However it delivers greater effect of structural imperfectionby coupled of each other.
- Empirical equation is proposed according to each type of cylinder and critical buckling loads estimated by empirical equation showed good agreements with FE analysis.

Acknowledgement

The authors would like to express sincere gratitude to Politeknik Sultan Salahuddin Abdul Aziz Shah and thank the Ministry of Higher Education, Malaysia, through the High Impact Research Grant (UM.C/625/1/HIR/MOHE/ENG/33).

References

[1] Lee M C, Kelly D W, Degenhardt R, Thomson R S 2010 *Composite Structures* **92** 223-32
[2] Lee M C. Stochastic analysis and robust design of stiffened composite structures: The University of New South Wales; 2009.
[3] Fisher R A 1935 *The Design of Experiment* (Edinburgh: Oliver and Boyd)
[4] Chryssanthopoulos M, Poggi C 1995 *Thin-walled structures* **23** 179-200

[5] Prabu B, Raviprakash A, Venkatraman A 2010 *Thin-Walled Structures* **48** 639-49
[6] Prabu B 2010
[7] Warren Jr J. Nonlinear stability analysis of frame-type structures with random geometric imperfections using a total-lagrangian finite element formulation: Virginia Polytechnic Institute and State University; 1997.
[8] Tran K, Douthe C, Sab K, Dallot J, Davaine L 2014 *Thin-Walled Structures* **79** 129-37
[9] Degenhardt R, Kling A, Bethge A, Orf J, Kärger L, Zimmermann R, et al. 2010 *Composite Structures* **92** 1939-46
[10] Degenhardt R, Bethge A, Kärger L. Probabilistic aspects of buckling knockdown factors—test and analysis. nal report. Technical report, ESA Contract 19709/06/NL/IA, 2007.
[11] J. Orf L K, R. Degenhardt, and A. Bethge. The Influence of imperfection on the buckling behaviour of unstiffened CFRD-cylinders. In proceeding of 2nd International Conference on Buckling and Postbuckling Behaviour of Composite Laminated Shell Structures2008. p. 2-5.
[12] J Kepple G P, G Pearce, D KElly, R Thomson and R Degenhardt, Influence of Imperfection on Axial Buckling Load of Composite Cylindrical Shells. 19th International Conference on Composite Materials; 2013.
[13] Bisagni C, Cordisco P 2003 *Composite Structures* **60** 391-402
[14] Bisagni C 2000 *Composites Part B: Engineering* **31** 655-67
[15] Bahloul R, Mkaddem A, Dal Santo P, Potiron A 2006 *International journal of mechanical sciences* **48** 991-1003
[16] Myers R H, Montgomery D C, Anderson-Cook C M 2009 *Response surface methodology: process and product optimization using designed experiments* **705** John Wiley & Sons)

Effect of phase advance on the brushless dc motor torque speed respond

M S Mohd[1], M N Karsiti[1] and M S Mohd[2,3]
[1]Electrical and Electronics Engineering Department, Universiti Teknologi PETRONAS, 32610 Bandar Seri Iskandar, Perak, Malaysia
[2]Mechanical Engineering Department, Universiti Teknologi PETRONAS, 32610 Bandar Seri Iskandar, Perak, Malaysia
[3]MOR Energy, Universiti Teknologi PETRONAS, 32610 Bandar Seri Iskandar, Perak, Malaysia

E-mail: m_syafiq_mohd18@yahoo.com

Abstract. Brushless direct current (BLDC) motor is widely used in small and medium sized electric vehicles as it exhibit highest specific power and thermal efficiency as compared to the induction motor. Permanent magnets BLDC rotor create a constant magnetic flux, which limit the motor top speed. As the back electromotive force (EMF) voltage increases proportionally with motor rotational speed and it approaches the amplitude of the input voltage, the phase current amplitude will reach zero. By advancing the phase current, it is possible to extend the maximum speed of the BLDC motor beyond the rated top speed. This will allow smaller BLDC motor to be used in small electric vehicles (EV) and in larger applications will allow the use of BLDC motor without the use of multispeed transmission unit for high speed operation. However, increasing the speed of BLDC will affect the torque speed response. The torque output will decrease as speed increases. Adjusting the phase angle will affect the speed of the motor as each coil is energized earlier than the corresponding rise in the back emf of the coil. This paper discusses the phase advance strategy of Brushless DC motor by phase angle manipulation approaches using external hall sensors. Tests have been performed at different phase advance angles in advance and retard positions for different voltage levels applied. The objective is to create the external hall sensor system to commutate the BLDC motor, to establish the phase advance of the BLDC by varying the phase angle through external hall sensor manipulation, observe the respond of the motor while applying the phase advance by hall sensor adjustment.

1. Introduction

In the recent years increase in energy prices as well as global climate changes have increase public awareness and government effort to seek alternative technology for clean energy transportation [1]. There is also a rapid increase in the adoption of a single seat or two-seat lightweight commuter electric vehicle [2]. Figure 1 shows small commuter electric vehicles, EV, in the form of an electric scooter and a prototype single seat electric vehicle. In the segment of small to medium sized hybrid electric and electric vehicles, majority of electric motor utilized is typically of the BLDC motor type controlled by a three phase motor controller [3].When the amplitude of back EMF reaches the amplitude of input voltage of the coils [4]. In this study, single seat electric vehicle will be used in the field test to verify the phase advance control algorithm.

Figure 1. Small commuter electric vehicles – Vectrix E Scooter and a UTP prototype single seat EV.

Study shows that it is possible to increase the operating speed of the motor by advancing the phase current with reference to back EMF [5-9]. According to H. Kong et al., transformers EMF in the voltage equation can restrain the increase of the peak back EMF when phase current lead the phase voltage back EMF [7]. There are several research related the phase advance. H C. Kim and J M. Kim was used in-phase concept which phase current compensates lagging angle based on back EMF because of increasing reactance at high speed. Conventional phase advance control method was optimized by compensating the difference of lagging angle between phase current and back EMF. S.M. Sue et al. in their paper stated that inverter gating time has been advanced so that phase angle of the stator current can lead the corresponding back emf and by that the reluctance torque can be developed properly [10]. Cambier et al. claim improvements in phase timing advancement to maintain constant power output during high speed use.

Phase advance angle manipulation is one of the methods in applying the phase advance. From literature, there are several approaches that been used to utilize phase angle manipulation. Manipulation of the external hall sensor systems has been used as the approach in applying the phase advance through varying the phase angle. Some preliminary works on motor characterization and modeling has been done. Experiment was established by testing the motor in varying phase advance angle. Different input voltage level also been applied in the test. Output speed and the output torque of the motor have been observed.

Most of the effects, benefits and drawbacks of the phase advanced control algorithm are not widely discussed in most literatures during partial and transitional loads expected such as in the operation of an electric vehicle. In actual vehicle operation, the drive cycle characteristics of the vehicle will determine the actual required operating points of the motor and majority of these points are located below the peak torque curve and are transitional in nature [11].

1.1. Problem formulation

Installing multispeed transmission unit in order to allow the electric drivetrain to be able to deliver high torque at low speed and at the same time are able to operate at a high cruising speed [10] will has additional driveline loss of around 10 percent [12]. The BLDC motor is capable of energy conversion at a peak rate of over 90 percent [13]. The top operational speed of this type of motor is largely limited by the constant magnetic field presence due to the existence of permanent magnets in the rotor. The torque speed respond will effect as the phase advance is applied. As the speed increase, the torque value will be decreased.

1.2. Effect of applying the phase advance to a bldc

Hypothesis:

There are several effects of applying the phase advance to a BLDC:

1. The top speed of BLDC can be increased by advancing the phase current.
2. The peak torque will be reduced below the base speed when advancing the phase current.

3. The optimum angle for phase advance operation has to be optimized as a function of motor speed and motor load.

Figure 2 shows the proposed hypothesis of the correlation between motor peak torque, motor speed and phase advance angle. The motor peak torque at zero phase advance angle is a slice of the curve intersecting the xz-plane (torque-phase advanced angle).

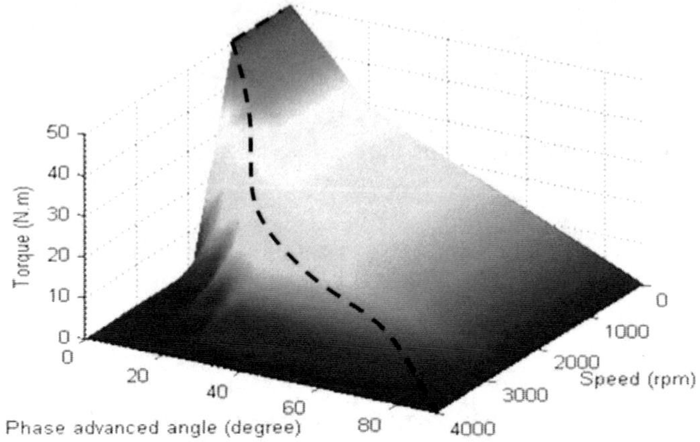

Figure 2. Torque vs Speed for different phase advance angles.

2. Preliminary work

2.1. Axial Flux Motor Characterization And Modelling

A custom made 200 W axial flux BLDC motor was used for the test. A MATLAB model from the existing model in the MATLAB has been used and modified as shown in figure 4. Equations (1) and (2) describe the relevant equations for back emf per unit speed and electrical torque for axial flux BLDC motor. All the measured and assumed values are entered into a MATLAB script, which is used to calculate the back emf rms amplitude and to calculate the motor electrical torque constant.
Axial flux BLDC motor equations (1) [14].
Back *emf* per unit speed:

$$E_f = \pi\sqrt{2} f N_1 k_{w1} \phi_f = \pi\sqrt{2} p\, k_{w1} \phi_f n_s = k_E n \quad (1)$$

where:

f = frequency
N_1 = no of turns
ϕ_f = magnetic flux density
k_{w1} = winding factor
p = no of pole pair
k_E = EMF constant
n_s = motor speed
Torque_{rms}, T_d:

$$T_d = \frac{m1}{\sqrt{2}} p N_1\, k_{w1} \phi_f I_a = k_T \quad (2)$$

where:

m_1 = no of phase
k_T = torque constant

$I_a = I_{phaserms}$

Using the equations (1) and (2), a MATLAB script is generated to calculate the motor properties as shown in figure 3. The user will have to select motor topologies of either radial flux BLDC or axial flux BLDC and also the physical dimensions and other motor design properties. The resultant torque constant, coil resistance, coil induction and back emf constant are then used by a dynamic model in the motor drive simulation.

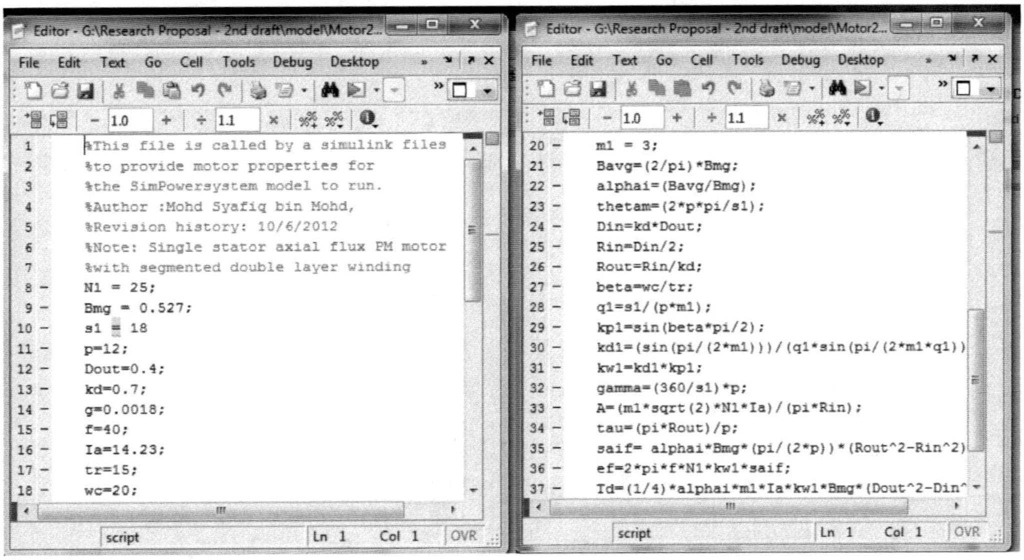

Figure 3. MATLAB script to calculate back *emf* and torque constant for the BLDC motor.

2.2. *Simulation*

A MATLAB Simulink model is then created using Power System block sets to simulate the permanent magnet BLDC motor as in figure 5. The simulation model incorporate phase current control, a constant voltage power supply and the analytical model of the motor as determined by equations (1) and (2) for radial flux BLDC motor (or equations (3) and (4) for axial flux BLDC motor). The main goal of the preliminary simulation work is to create a model for the permanent BLDC motor, which then can be validated using the actual test data. The validated model can then be used to study the effect of phase advanced control algorithms.

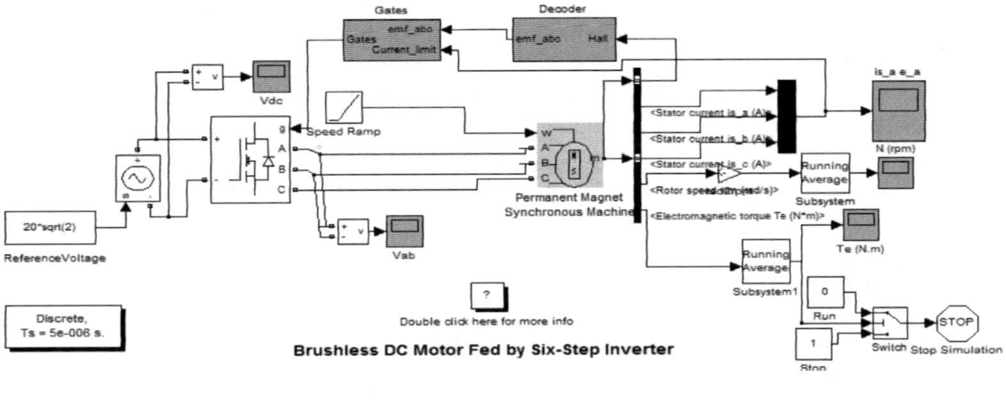

Figure 4. MATLAB Simulink model of the BLDC motor.

2.3. Simulation results

The preliminary result obtained before from the simulation is shown in figures 5 and figure 6. The torque constant and back emf constant is similar to actual torque constant, kt provided by the manufacturer. The model is not entirely accurate, since there is a lack of information about the magnetic strength of the magnets and the actual number of turns of the coils, similar output characteristics between the model and the actual motor specifications allows for useful comparison between the model operation and actual motor operation.

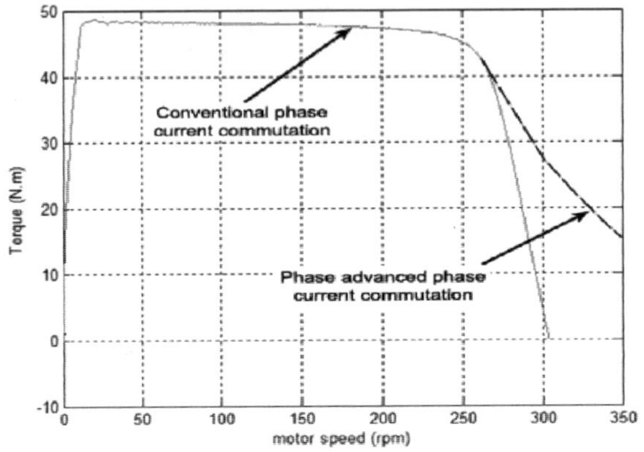

Figure 5. MATLAB Simulink model of the BLDC motor.

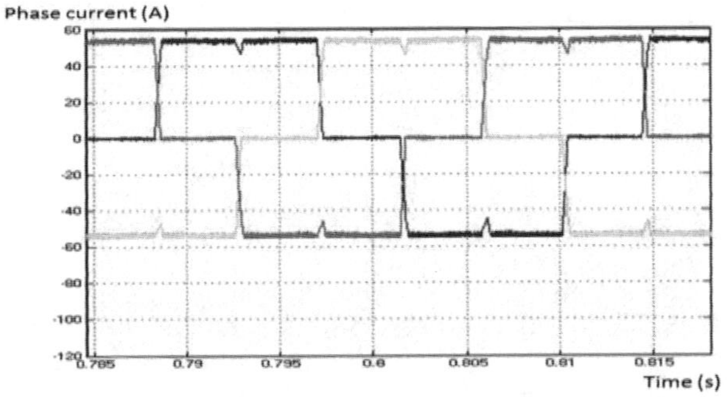

Figure 6. MATLAB Simulink model of the BLDC motor.

Based on the simulation results, the model can be used to study the effects of phase advanced control on the motor torque output. However, an accurate comparison between the simulated model and the actual motor requires the installation of a rotary torque meter and an adjustable load to control the operating points (torque and speed) of the BLDC motor.

Experiments have been conducted using Kelly BLDC and Electrocraft BLDC controllers. A programmable Sorensen Power supply provides a constant voltage supply to the motor controller. The test was done in a no- load test condition.

Figure 8 show the details of the external hall sensor set-up. The main reason of using the external adjustable hall sensors is to physically vary the phase advanced angle. At present, this is the only viable solution as the programming code of the motor controller is not accessible. In the future, it is possible to control a motor drive unit using custom made programmable logic board to replace the motor controller altogether for full control of the phase advance angle. Moreover, the implementation of the phase advanced control might require additional set of hall sensors located perhaps at certain angle from the default set of hall sensors.

2.4. External Hall Sensor

The test has been done using a custom made axial flux BLDC motor that fabricated by other researcher. A set of hall sensor used to commutate that motor. The holder of that hall sensor is designed so that the position can be changed to different phase angle. Figure 7 shows the hall sensors holder used for the motor.

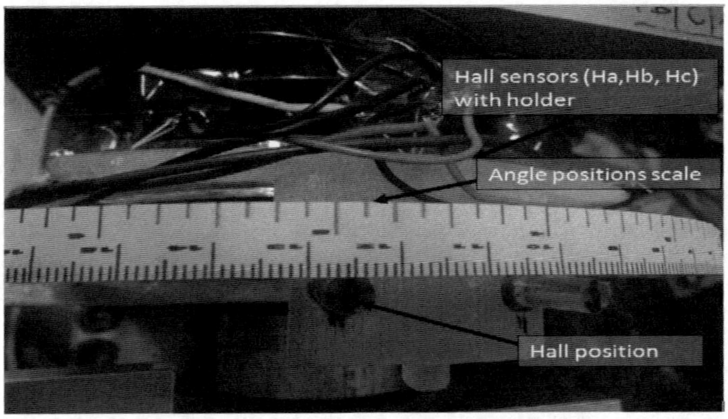

Figure 7. Hall sensor positioning.

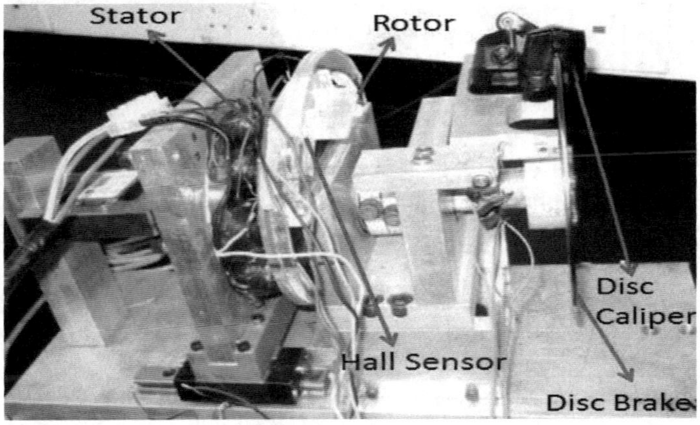

Figure 8. Custom made axial flux motor test bench [15].

The test has been done by using test bench that run by custom made axial flux motor as shown in figure 8.The test has been established with no load and load applied.

3. Results and discussion

In An axial flux BLDC motor has been tested using Kelly BLDC controller. The test has been done in no load condition for different phase advance angle and different voltage applied. The objective is to observe the effect of varying phase advance angle at different voltage applied. The angle test is at advance angle and also at retard angle. Table 1 shows the axial flux motor parameters.

Table 1. Axial Flux Motor Parameters.

Parameter	Values
Pole pairs	3
Number of coil	9 coils
Number of turns	30 turns
Magnet size	OD = 114 mm, ID = 46 mm
Magnet type	NdFE N38 (1.125 T)

The results and operating characteristics were also achieved by another researcher [17] working with a custom made axial flux permanent magnet motor as shown in figure 11. However, it can be seen that the torque constant is reduced in inverse proportion manner to the increase in speed.

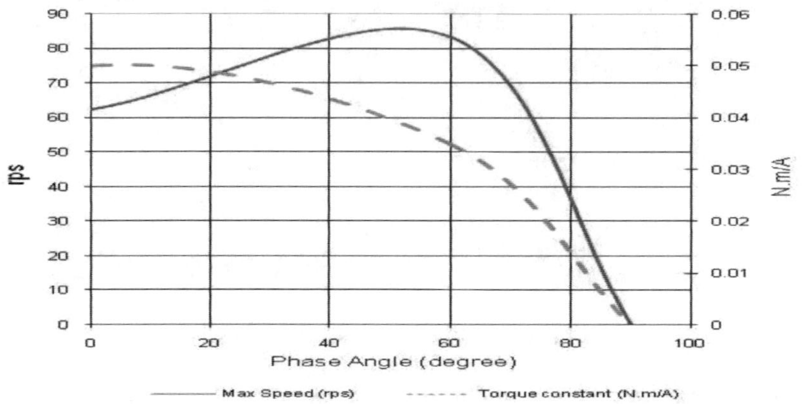

Figure 9. Output speed and torque constant versus phase angle for a 200 W axial flux PM motor [15].

As shown in figure 9, the top speed of the PM motor can be increased to around 50 percent over the baseline top speed. It is expected that the torque constant will also be reduced in the similar manner shown in the figure 9. The phase current is offset from the optimum 0 angle to generate maximum torque. Too much increase in the phase advance will results in lower amount of torque generation, thus it is necessary to investigate the appropriate control method of phase advanced to generate useful torque output while increasing the operating speed.

Figure 10 shows effect of phase advance on motor top speed. Maximum speed test results of the motor operating at different phase angles between -30 and 45 degrees and at maximum supply voltage range between 24 and 48 V. That range been has been used because of minimum voltage required to turn on the Kelly BLDC controller and at higher voltage more than 48, motor become unstable. The maximum speed of the motor is observed to increase with an increase in the phase advanced angle until it reaches maximum speed at around 30 degrees.

When the angle is retarded, the maximum motor speed is observed to reduce in a linear manner with almost a constant slope. If all the lines of the graph are extrapolated, the maximum speed of the motor will reach zero rpm as the phase angle reaches -90 degrees during phase retardation of 90 degrees during phase advancement.

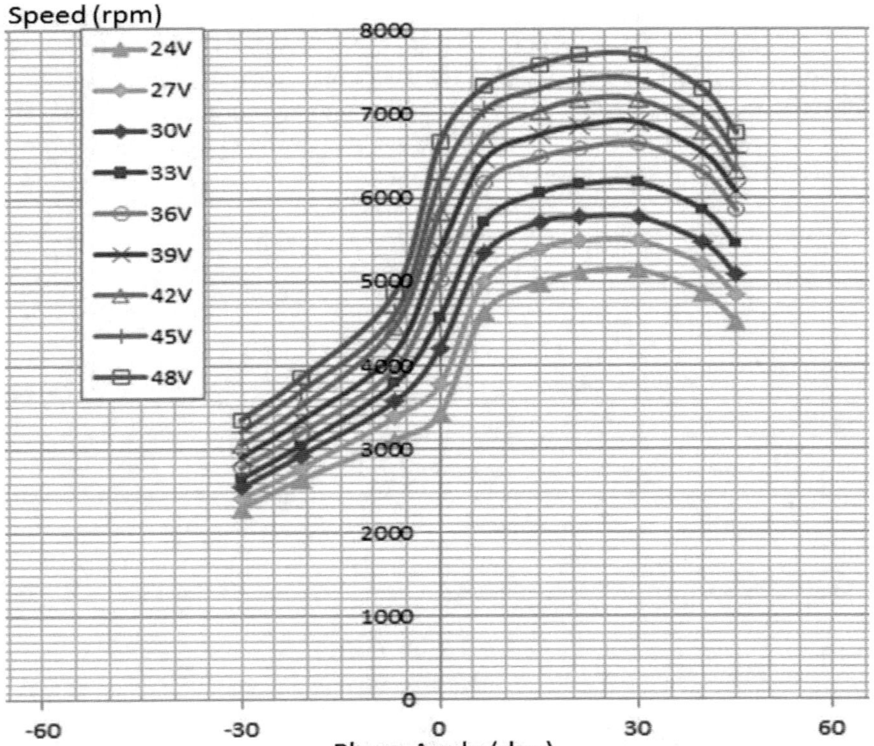

Figure 10. Effect of Phase Advance on Motor Top Speed.

These characteristics can be explained by six step phase current waveform advanced angle. As the phase current is advanced from zero position, it will obtain maximum speed at zero loads at 30 degrees advanced angle. As the advanced angle is above 30 degrees, the phase current will respond to negative back emf, which will results in a gradual increase of opposing torque. Thus, at 90 degrees advanced angle, there will be equivalent amount of clockwise and counter clockwise torque hence the motor will produce a zero amount of net torque.

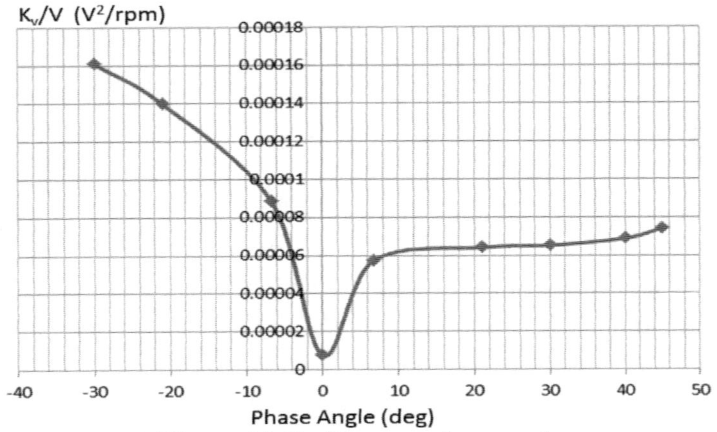

Figure 11. Kv/V versus phase angle.

The change of the speed depend on the back emf constant value, Kv. Different input voltage applied will resulted different Kv value. Figure 11 shows a graph of Kv/V applied as a function of phase angle. It can be observed that Kv has an almost zero slope at zero phase angles.

Torque is the one of the parameter that be effected when advancing the phase angle. The torque value depend on the torque constant of the motor and it change when phase angle varying. Figure 12 shows the torque constant, Kt versus phase advance angle.

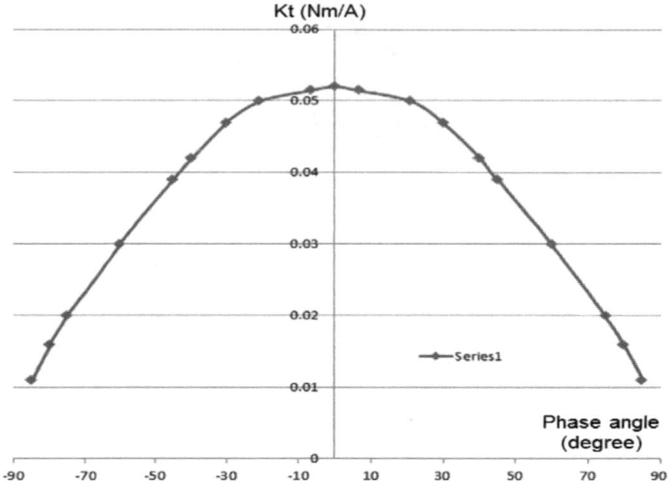

Figure 12. Torque constant, Kt vs phase angle.

From the shape of graph, it shown that, the value of Kt is high at lower phase advance angle and this Kt value become decrease as phase advance angle increasing and become zero at 90 degree phase advance angle. Similar shape also been shown for retarding angle. This Kt related to the torque value as torque value affected by the Kt value.

From the speed constant data, Kv that indicate the change of speed and also the torque constant that related to the torque value, a model of torque speed respond has been developed for different phase advance angle as shown in figure 13. Equation (3) and equation (4) that are the electrical torque equation been used to establish this model. These equations describe the relation of electrical torque with terminal or input voltage, Vt and the speed of the motor.

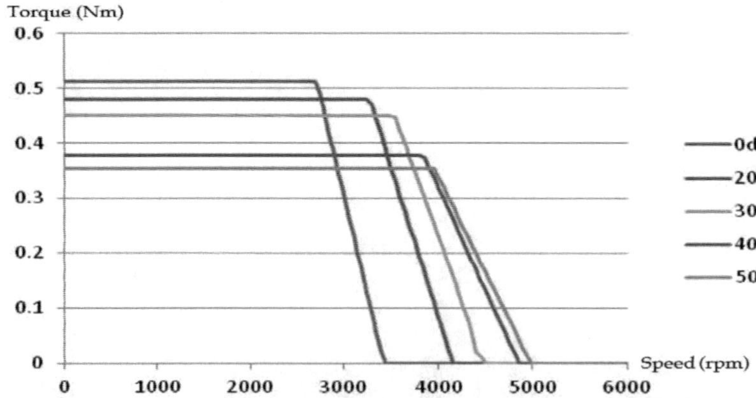

Figure 13. Torque Speed respond base on Kv/V and Kt data.

Electrical torque equation:

$$Te = \frac{(Vt - K_{E\omega_T}) K_T}{R_S} \quad (3)$$

$$T = K_T * I \quad (4)$$

Te = Electrical torque
Vt = Supply voltage
KE = Back emf constant
ωt = Motor speed
KT = Torque constant
Rs = Phase resistance
I= phase current

Based on the results the speed of the motor keeps increasing as the phase angle increases until certain point that is about 30 degree advance angle. Beyond this angle, the speed decreases gradually. The optimum angle is determined by considering significant increase of speed at that angle and also considering the current that indicated the lost. The results show as in figure 13 slope of Kv/V increases from zero degree until 10 degree. After phase angle 30 degree, the Kv/V value becomes almost constant. That means the increasing of the speed becomes small and insignificant. For different voltages applied, it shows the back *emf* constant value, Kv is almost constant at 0.01 rpm/volt. Increasing the speed will affect the motor torque speed respond that caused the output torque of the motor reduced. The actual load level, operating torque, operating speed and motor efficiency will be the determining factor in finding the optimum phase advanced control method.

4. Conclusion

The increasing in energy prices and environment issue caused several research in new technology for electric vehicles be established. The heavy and huge in gearbox system cause the high in power over weight ratio for EV. The use of BLDC motor becomes popular in EV application because of small size and high power given. The constant magnetic flux of permanent magnet rotor in BLDC cause the limitation of speed range. Phase advanced control allows the permanent magnet BLDC motor to extend the operating range without the usage of a transmission unit. The absent of a mechanical gearing system will result in higher system efficiency and allows for the usage of smaller motor hence improving the power to weight ratio of the electric drive unit.

The phase advance could be established by varying the phase advance angle. The external hall sensor system can be utilized to apply phase advance by manipulating the hall sensors positions. Increasing the phase advance angle caused the speed of the motor increase. The optimum angle determined based on the application used need for high speed or high torque. The external hall sensors system is reliable to utilize in EV application by creating suitable design to deploy it. A suitable control system also needs to be developed to control the hall positions.

Acknowledgement

The research project is supported under UTP GA scheme, with the small single seat electric vehicle developed under a STIRF research grant for Shell EcoMarathon 2012 and several equipment used are from MOR Energy of UTP.

References

[1] Bedi G. Plug-in Electric Vehicle: A Practical Plan for Progress, The Report of an Expert Panel. Feb; 2011.
[2] Kurani K S, Sperling D, Lipman T, Stanger D, Turrentine T, Stein A 2001 *University of California Transportation Center*
[3] Herman S L. Industrial Motor Control. Delmar: Cengage Learning; 2010.
[4] DiRenzo M 2000 *Texas Instrument, SPRA420A*

[5] Kong H, Liu J, Cui G, Study on field-weakening theory of brushless DC motor based on phase advance method. Measuring Technology and Mechatronics Automation (ICMTMA), 2010 International Conference on; 2010: IEEE.

[6] Luo X, Qin D, Lipo T, A novel two phase doubly salient permanent magnet motor. Industry Applications Conference, 1996 Thirty-First IAS Annual Meeting, IAS'96, Conference Record of the 1996 IEEE; 1996: IEEE.

[7] Chiu C-L, Chen Y-T, Shen Y-H, Liang R-H 2009 *Magnetics, IEEE Transactions on* **45** 120-6

[8] Sue S-M, Wu K-L, Syu J-S, Lee K-C, A phase advanced commutation scheme for IPM-BLDC motor drives. Industrial Electronics and Applications, 2009 ICIEA 2009 4th IEEE Conference on; 2009: IEEE.

[9] Ehsani M, Gao Y, Emadi A. Modern Electric, Modern Hybrid, and Fuel Cell Vehicles. Book; 2010.

[10] Rahman Z, Ehsani M, Butler K. An investigation of electric motor drive characteristics for EV and HEV propulsion systems. SAE Technical Paper, 2000 0148-7191.

[11] Jordan H E 2013 *Energy-efficient electric motors and their applications* Springer Science & Business Media)

[12] German J. Vehicle Technology: How Far and How Fast', Transportation Efficiency in 21st Century. 2010.

[13] Abourida S, Belanger J, Real-time platform for the control prototyping and simulation of power electronics and motor drives. Proceedings of 3rd International Conference on Modeling, Simulation, and Applied Optimization; 2009.

[14] Hanselman D C 2003 *Brushless permanent magnet motor design* The Writers' Collective)

[15] Mohd M S. Mechanical Field Weakening of Axial Flux PM Motor by Manipulation of Rotor-Stator Attractive Force. PG Symposium Progress Report (Mechanical Engineering Department, Universiti Teknologi PETRONAS), 2012.

IOP Conf. Series: Materials Science and Engineering **100** (2015) 012005 doi:10.1088/1757-899X/100/1/012005

A review on simple assembly line balancing type-e problem

M Jusop[1] and M F F Ab Rashid[1]
[1]Manufacturing Focus Group, Faculty of Mechanical Engineering, Universiti Malaysia Pahang, 26600 Pekan, Pahang, Malaysia

E-mail: masitahjusop@yahoo.com

Abstract. Simple assembly line balancing (SALB) is an attempt to assign the tasks to the various workstations along the line so that the precedence relations are satisfied and some performance measure are optimised. Advanced approach of algorithm is necessary to solve large-scale problems as SALB is a class of NP-hard. Only a few studies are focusing on simple assembly line balancing of Type-E problem (SALB-E) since it is a general and complex problem. SALB-E problem is one of SALB problem which consider the number of workstation and the cycle time simultaneously for the purpose of maximising the line efficiency. This paper review previous works that has been done in order to optimise SALB-E problem. Besides that, this paper also reviewed the Genetic Algorithm approach that has been used to optimise SALB-E. From the reviewed that has been done, it was found that none of the existing works are concern on the resource constraint in the SALB-E problem especially on machine and tool constraints. The research on SALB-E will contribute to the improvement of productivity in real industrial application.

1. Introduction

An assembly line is a manufacturing process comprises of a sequence of workstations in which a set of necessary task to assemble a product are performed. The development of assembly is system usually used in the production of goods in the industry. The idle time and the number of workstations on the production line have to be minimised whereas the line efficiency has to be maximised so as to achieve a balance line.

The decision problem of optimally partitioning the assembly task among the workstations with respect to some objective is known as Simple Assembly Line Balancing (SALB) [1]. This problem intends at grouping assembly operations which have to be performed to produce final products, and assigning the groups of operations to workstations, so as to make sure the total assembly time required at each station is nearly the same and the precedence constraints between operations are respected [2]. SALB is a type of NP-hard optimisation problems which means that when the number of assembly task is increased, the feasible solution will rise staggeringly [3-5]. Advanced approach of algorithm is necessary to solve large-scale problems.

SALB can be classified into two categories (i) Simple Assembly Line Balancing Problems (SALBP) (ii) General Assembly Line Balancing Problems (GALBP) [6, 7]. The most well-known assembly line is called simple assembly line balancing problem. Simple assembly line balancing is considered when the same product is running on the line. This type of problem is classified into four groups with respect to the objectives function [6, 8].

- Simple assembly line balancing Type-1 (SALB-1) aims to minimise the number of workstations on the line for a fixed cycle time.
- Simple assembly line balancing Type-2 (SALB-2) aims to minimise the cycle time for fixed number of workstations on the line.
- Simple assembly line balancing Type-E (SALB-E) aims to maximise the efficiency of the line simultaneously minimising the number of workstations and the cycle time.
- Simple assembly line balancing Type-F (SALB-F) aims to determine a feasible line for a combination of the number of workstations and cycle time.

Other problems which are not included in simple assembly line are considered as generalised assembly line balancing problems. Mixed-model assembly line balancing (MALBP) or mixed-model sequencing problem (MSP) and also U-line balancing problem (UALBP) are categorised as GALBP [7]. The classification of assembly line balancing problems is illustrated as in figure 1.

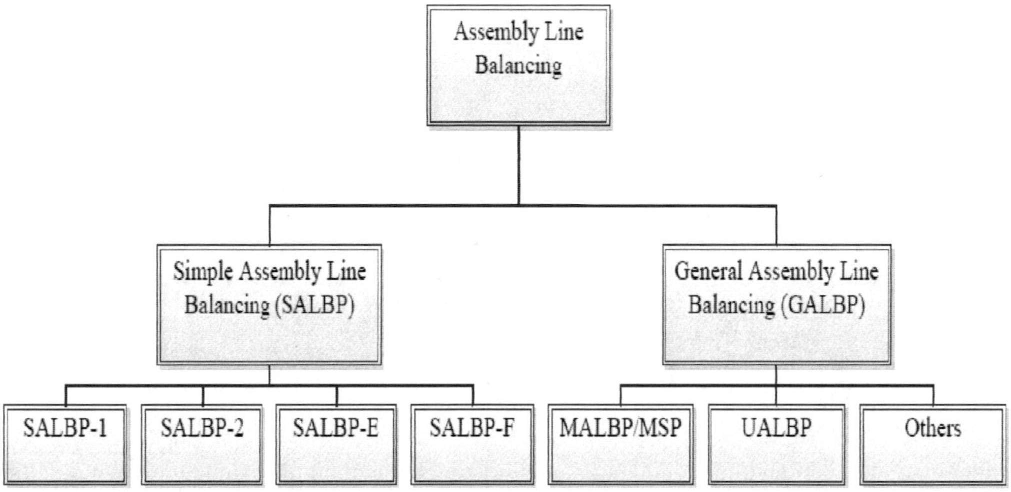

Figure 1. Classification of assembly line balancing problems.

Most of previous researches are focusing on SALB-1 [5, 6, 9, 10] and SALB-2 [2, 11-13]. Only a small number of previous research study on SALB-E as it is more complicated compare with SALB-1 and SALB-2. Study on SALB-E need to consider multi-objective functions instead of single objective in both SALB-1 and SALB-2. In real manufacturing scenario, it is better if we consider both parameters; minimised the number of workstations and minimised the cycle time for the purpose to maximise the assembly efficiency.

This paper reviews the previous study on simple assembly line balancing Type-E. The rest of the paper consists of problem modelling and objective function, SALB-E optimisation algorithm, and genetic algorithm for SALB-E. Finally, conclusion and suggestion for future research are addressed.

2. Problem modelling and objective function

Simple Assembly Line Balancing of Type-E Problem (SALBP-E) has been reviewed by Gurevsky et al. under dissimilarities of task processing times [14]. The research on stability of feasible and optimal solutions for SALBP-E is presented in this paper. Two heuristic procedures are proposed and evaluated on certain targets in order to find a concession between the two goal functions. Polynomial time algorithm has been proposed so as to compute the stability radius of feasible balances.

The paper presented by Suwannarongsri & Puangdownreong proposed a combination of partial random permutation (PRP) method and an adaptive tabu search (ATS) in an attempt to specify the optimum solutions for the assembly line balancing problem [15]. The researcher has considered the simple assembly line balancing in the work with four objective functions (i) minimise the number of workstations, (ii) minimise the idle time, (iii) minimise the workload variance and (iv)maximise the line efficiency. The equation (1) is used to represent the line efficiency.

$$E = \sum_{i=1}^{m} T_i /(mc) \quad (1)$$

where E : Line efficiency
m: Number of workstations
c : Cycle time
T_i: processing time of the ith workstation

A test against three benchmark single-model SALB problems such as Buxey, Sawyer, and Warnecke on actual SALB problem has been conducted by the researcher to assure the efficiency of the proposed multiple-objective method. The results shows that the proposed method is efficient for multiple-objective compare to the single-objective.

Previous study by Scholl & Becker stated that there is no direct method to solve the SALBP-E [6]. That type of model can be solved by a search method; the combination of the number of stations m and the cycle time c which is feasible for the efficient line is chosen among the others or, the value of required line capacity as in equation (2) should be minimal.

$$T = m.c \quad (2)$$

where T is line capacity

The review published by Wei & Chao are focused on SALBP-E in order to optimise the line balancing efficiency as well as minimising the idle time [16]. This objective can be achieved by minimising the number of stations and the cycle time. SALBP-1 and SALBP-2 models are combined by the researcher in order to develop the SALBP-E model. In SALBP-1, the number of stations is minimised with fixed cycle time. This model is re-defined to SALBP-1-i with the intention of determining the minimum number of stations. The goal of modified model SALBP-2 is to ensure the minimisation of cycle time ct with a fixed number of workstations m. The efficiency of the line is formulated as equation (3):

$$E = \frac{t_{sum}}{m.ct} \quad (3)$$

where t_{sum} is the total time of all tasks

In order to maximise the line efficiency, the optimal number of workstation must be obtained by a given ct_{max}. The value of ct_{max} must be less than or equal to the total task times and at the same time it also should be greater than or equivalent to the largest task time in data. Only one workstation will be required whenever the value of ct_{max} is exceed or the same as total task times. No solution will obtained as the value for ct_{max} is less than or equivalent to the largest task time in data. The respecting conditions are used for ct_{max}.

$$\max t_i \leq ct_{max} \leq \sum t_i$$

If $ct_{max} \geq \sum t_i$ then $m = 1, E = \frac{T_{total}}{1.T_{total}} = 1$ thus, Balance loss = 0

If $ct_{max} \geq \max t_i$, no solution

After the value of ct_{max} has been set, the optimal number of workstations m can be attained by using the spreadsheet. The value of m lies between m_{min} and m_{max} and it has been calculated as equation (4) and equation (5):

$$m_{min} = \left\lceil \sum_{i=1}^{n} \frac{t_i}{ct_{max}} \right\rceil \tag{4}$$

$$m_{max} = \left\lfloor \frac{\sum t_i}{\max t_i} \right\rfloor \tag{5}$$

where $m_{min} \leq m \leq m_{max}$

In another work, Zacharia & Nearchou minimised the number of workstations m and cycle time c using fuzzy task processing times so-called as f-SALBP-E [17]. The objective functions of the problem are to maximise the efficiency of the line, simultaneously minimising the number of workstations m and the cycle time c. The fuzzy efficiency $\check{e}$ of the line is linearly dependent with summation of fuzzy processing times of all the task $\dot{t}_{sum}$. It is also can be attained by minimising the product of number of workstations and fuzzy cycle time of the line. The line efficiency function is represented by equation (6):

$$\check{e} = \frac{\dot{t}_{sum}}{m.\check{c}} \tag{6}$$

where $\dot{t}_{sum}$: total sum of the fuzzy processing time of all the tasks
$\check{c}$: fuzzy cycle time of the line

The uncertainty and variability of task processing time and cycle time are presented by triangular fuzzy numbers (TFNs). A heuristic method based on Genetic Algorithm (GA) has been developed to solve the *f*-SALBP-E as it is a type of NP-hard optimisation problems. A two-phase GA is used for the purpose to solve the problem. In this approach, the optimal solution found from the first run is used to generate the early population of the binary run. There is no resource constraint being stated in the study. By considering the fuzzy processing time for the single assembly line balancing problem, a formulated mathematical model is performed and thus minimised the number of workstations and the fuzzy cycle time on the line.

A new genetic algorithm has been presented by Al-Hawari et al. to solve multi-objective simple assembly line balancing problem [18]. Minimisation of number of workstations, minimisation of workload variation, and maximisation of line efficiency are considered as the objective functions in the study. A Multi-Assignment Genetic Algorithm (MA-GA) has been proposed by the researcher with the combination of forward, backward, and bidirectional methods. The researcher concluded that the proposed algorithm has shown a better performance in solving multi-objective simple assembly line balancing for a larger size of problem. Equation (7) represents the line efficiency, E which is supposed to be maximised.

$$\max E = \frac{\sum_{i=1}^{n} t_i}{m.C_a} \tag{7}$$

The efficiency of the line can be maximised by minimising both variables; the actual number of workstations m and the actual cycle time of the assembly line $= \max_{1 \leq k \leq m}\{t(S_k)\}$ whereas the sum of handling time of task i is fixed. The minimum number of actual workstations m can be obtained using the mathematical formulation as stated in equation (8):

$$\min m = \sum_{k=1}^{M} \max_{1 \le i \le n} \{x_{ik}\} \tag{8}$$

$$x_{ik} = \begin{cases} 1, & \text{if task i is assigned to station k} \\ 0 & \text{otherwise} \end{cases}$$

$$t(S_k) = \text{the total time assigned to workstation k}$$

Suwannarongsri et al. has proposed a combination of tabu search (TS) and genetic algorithm (GA) to identify the solution for simple assembly line balancing problem [13]. The goals of the problem are to (i) minimise the number of workstations, (ii) minimise the workload variance, (iii) minimise the idle time and (iv) maximise the efficiency of the line. The maximum line efficiency can be calculated by using equation (9):

$$max\ L_{eff} = max \frac{\sum_{i=1}^{n} T_i}{(n \times ct\_r)} \times 100 \tag{9}$$

where n : number of workstations
T_i : processing time of ith workstation
ct_r : actual cycle time
L_{eff} : line efficiency

3. SALB-E optimisation algorithm

A two-part genetic algorithm (GA) is established to solve *f*-SALBP-E [17]. The first part of GA started with generating initial population, followed by performing the best solutions until it reached termination conditions. The optimal solution achieved from the first attempt is used as the source for the early population in the binary part for the purpose to find a better performance. The algorithm rises in a good feasible solution which is approximately to the exact solution in an acceptable time period.

The algorithm proposed by Al-Hawari et al. uses the combination of forward, backward, and bidirectional methods of task assignment [18]. These methods are used to assign each of tasks in a chromosome to workstations. Priority-based encoding, crossover, mutation, sequence encoding, decoding (assignment), evaluation, and selection are the primary procedures in MA-GA. As mentioned previously, the researcher simplified that the proposed MA-GA can solve problem for a larger size. It provides many feasible solutions of task assignments by combining the three methods simultaneously instead of combine using the only forward method. MA-GA will also increase the probability of identifying the optimal solution.

Suwannarongsri et al. used TSGA-based method which is the combination of TS and GA method to find the solutions for simple assembly line balancing problem. The researchers have performed a test of all type of SALBP problems from a literature against the proposed method. The result showed that the proposed TSGA-based method is capable in producing better solutions compared with conventional method [13].

Most of previous researcher used genetic algorithms (GAs) as an optimisation technique especially in SALB problem [2, 11, 12, 19, 20]. However, only a small number of researchers are focusing on simple assembly line balancing of Type-E problem [13, 14, 17]. As a consequence, the implementation of GA method has not been widely publicised in SALB-E itself.

4. Genetic algorithm for SALB-E

Genetic Algorithms (GAs) are mainly used by researcher for optimising large and complex problem specifically in SALB problem [2, 21-24]. GAs used a direct random search as an optimisation method for complex problem with the aim of finding optimum solutions [21]. The application of genetic algorithm is quite popular compare with the simulated annealing and ant colony optimisation [25].

In [17], the design of GA comprises of chromosome's encoding, a decoding mechanism, an evaluation mechanism, generation of early population, and generation of offspring. The solution for *f*-SALBP-E is characterised by chromosome's encoding, which is consists of tasks priorities (first part of the chromosome) and number of workstations on the line (binary part of chromosome). The tasks are then assigned to workstations by using a suitable decoding scheme. In evaluation mechanism, an individual chromosome with higher fitness value tends to have higher probability to be selected. The feasible tasks provide a better solution for the problem as it has low values of total fuzzy idle time.

The early random population undergoes selection, crossover, and mutation process to produce new generation. The optimum solution obtained from the first part is used as the source for the early population in second part for the aim of finding a better solution. A roulette wheel method is used in selection process. Chromosomes with higher fitness value will be selected to produce new population. Crossover operator is developed to produce new chromosomes from two parents' chromosomes by changing the tasks order. In GA, mutation mechanism worked by flipping or swapping an only chromosome to produce a single new chromosome.

Previous paper presented by Al-Hawari et al. used three assignment methods (i) forward (ii) backward and (iii) bidirectional in Multi-Assignment Genetic Algorithm (MA-GA) [18]. A forward assignment method is the mainly used for solving SALBP. By using this technique, the works are allocated sequentially to workstations by taking into consideration the cycle time constraint. In backward assignment, a flipping method is used. The task sequence chromosome is flipped to be assigned using forward assignment method whereas, the bidirectional assignment method used both forward and backward directions. From the acquired result, bidirectional assignment attained the best solution.

Three genetic operators that have been used in GA are (i) crossover (ii) mutation and (iii) selection. The researcher used weight mapping crossover operator (WMX), swap mutation operator, and roulette wheel selection (RWS). The crossover operates two chromosomes (parent) to produce a new chromosome. One-point WMX is used in the proposed MA-GA and one crossover cut has been pointed at anyplace along the length of the parent, producing two offspring that have their genes. In the research, the swap mutation operator is used in order to keep the genetic diversity. In selection step, the roulette wheel selection method has been applied to produce a new population. The chromosomes with higher fitness value get more chances to be selected. To avoid the loss of the best chromosome(s), an elitism approach is adopted while using the RWS.

Suwannarongsri et al. used TS method to determine the number of tasks assign in each workstation whereas GA is employed to assign the sequence of tasks for each workstation by considering the precedence constraints [13]. The searching process of the GA is comparable to the nature development of biological beings. The flowchart of GA is summarised as in figure 2.

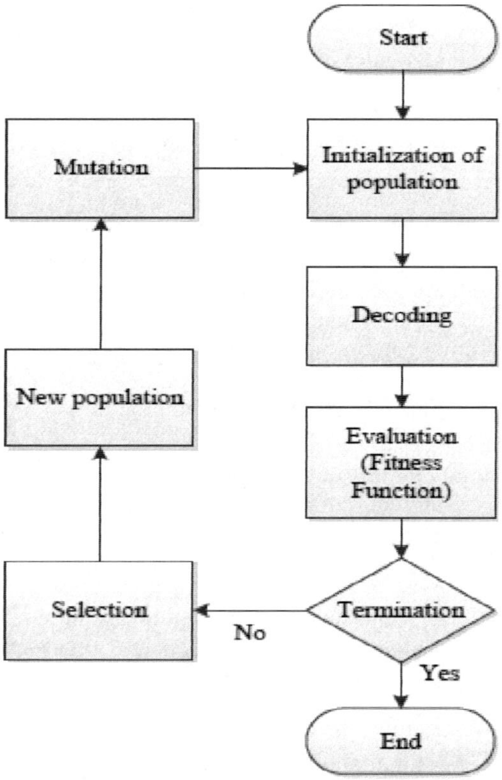

Figure 2. Flowchart of genetic algorithm.

5. Conclusion

This paper reviewed the optimisation algorithm and techniques used by the previous researcher on SALB-E. From literature review that have been conducted, it can be concluded that the application of genetic algorithm (GA) as an optimisation technique are on the rise due to its ability to solve a large-scale optimisation problem as well as searching near optimal solution.

Only a few studies are focusing on SALB-E as it is a general and complex problem. Up till now, none of them are concern on the resource constraint in the problem especially machine and tool constraint. Future research direction could be to consider recourse constraint in the optimisation of SALB-E itself.

Acknowledgement

This study was supported by a research grant from the Ministry of Higher Education under RDU140103. The authors would like to thank Universiti Malaysia Pahang for the facilities provided.

References

[1] Rashid MFF, Hutabarat W, Tiwari A. A review on assembly sequence planning and assembly line balancing optimisation using soft computing approaches. The International Journal of Advanced Manufacturing Technology. 2012;59:335-349.

[2] Gu L, Hennequin S, Sava A, Xie X. Assembly line balancing problems solved by estimation of distribution. Automation Science and Engineering, 2007 CASE 2007 IEEE International Conference on: IEEE; 2007. p. 123-127.

[3] Kriengkorakot N, Pianthong N. The Assembly Line Balancing Problem. KKU Enginieering Journal. 2007;34:133-140.

[4] Nearchou AC. Multi-objective balancing of assembly lines by population heuristics. International Journal of Production Research. 2008;46:2275-2297.

[5] Emeke Great O, Offiong A. Productivity Improvement In Breweries Through Line Balancing Using Heuristic Method. International Journal of Engineering Science & Technology. 2013;5.

[6] Scholl A, Becker C. State-of-the-art exact and heuristic solution procedures for simple assembly line balancing. European Journal of Operational Research. 2006;168:666-693.

[7] Boysen N, Fliedner M, Scholl A. A classification of assembly line balancing problems. European Journal of Operational Research. 2007;183:674-693.

[8] Becker C, Scholl A. A survey on problems and methods in generalized assembly line balancing. European Journal of Operational Research. 2006;168:694-715.

[9] Chong KE, Omar MK, Bakar NA. Solving assembly line balancing problem using genetic algorithm with heuristics-treated initial population. 2008.

[10] Ponnambalam S, Aravindan P, Naidu GM. A multi-objective genetic algorithm for solving assembly line balancing problem. The International Journal of Advanced Manufacturing Technology. 2000;16:341-352.

[11] Zhang W, Gen M, Lin L. A multiobjective genetic algorithm for assembly line balancing problem with worker allocation. Systems, Man and Cybernetics, 2008 SMC 2008 IEEE International Conference on: IEEE; 2008. p. 3026-3033.

[12] RuiJun Z, DingFang C, Yong W, ZhongHua Y, Xinxin W. Study on line balancing problem based on improved genetic algorithms. Wireless Communications, Networking and Mobile Computing, 2007 WiCom 2007 International Conference on: IEEE; 2007. p. 2033-2036.

[13] Suwannarongsri S, Limnararat S, Puangdownreong D. A new hybrid intelligent method for assembly line balancing. Industrial Engineering and Engineering Management, 2007 IEEE International Conference on: IEEE; 2007. p. 1115-1119.

[14] Gurevsky E, Battaïa O, Dolgui A. Balancing of simple assembly lines under variations of task processing times. Annals of Operations Research. 2012;201:265-286.

[15] Suwannarongsri S, Puangdownreong D. Multi-objective assembly line balancing via adaptive tabu search method with partial random permutation technique. Industrial Engineering and Engineering Management, 2008 IEEM 2008 IEEE International Conference on: IEEE; 2008. p. 312-316.

[16] Wei N-C, Chao I-M. A solution procedure for type E simple assembly line balancing problem. Computers & Industrial Engineering. 2011;61:824-830.

[17] Zacharia PT, Nearchou AC. A meta-heuristic algorithm for the fuzzy assembly line balancing type-E problem. Computers & Operations Research. 2013;40:3033-3044.

[18] Al-Hawari T, Ali M, Al-Araidah O, Mumani A. Development of a genetic algorithm for multi-objective assembly line balancing using multiple assignment approach. The International Journal of Advanced Manufacturing Technology. 2014:1-14.

[19] Gonçalves JF, De Almeida JR. A hybrid genetic algorithm for assembly line balancing. Journal of Heuristics. 2002;8:629-642.

[20] Sabuncuoglu I, Erel E, Tanyer M. Assembly line balancing using genetic algorithms. Journal of intelligent manufacturing. 2000;11:295-310.

[21] Ranjan R, Pawar P. Assembly Line Balancing Using Real Coded Genetic Algorithm. International Journal of Scientific Research in Computer Science and Engineering. 2014;2:1-5.
[22] Matondang MZ, Jambak MI. Soft computing in optimizing assembly lines balancing. Journal of Computer Science. 2010;6:141.
[23] Tasan SO, Tunali S. A review of the current applications of genetic algorithms in assembly line balancing. Journal of intelligent manufacturing. 2007;19:49-69.
[24] Mohd Razali N, Geraghty J. Biologically inspired genetic algorithm to minimize idle time of the assembly line balancing. Nature and Biologically Inspired Computing (NaBIC), 2011 Third World Congress on: IEEE; 2011. p. 105-110.
[25] Battaïa O, Dolgui A. A taxonomy of line balancing problems and their solutionapproaches. International Journal of Production Economics. 2013;142:259-277.

Experimental study on the effect of varying syngas composition on the emissions of dual fuel CI engine operating at various engine speeds

B K M Mahgoub[1], S A Sulaiman[1], Z A A Karim[1] and F Y Hagos[2]
[1]Department of Mechanical Engineering, Universiti Teknologi PETRONAS, 31750 Tronoh, Perak, Malaysia
for Automotive Research and Energy Management, Universiti Teknologi PETRONAS, 32610 Bandar Seri Iskandar, Perak, Malaysia.
[2]Automotive Engineering Centre, Faculty of Mechanical Engineering, Universiti Malaysia Pahang, 26300 Kuantan, Pahang, Malaysia

E-mail: m.bahaa02@uofk.edu

Abstract. Using syngas as a supplement fuel of diesel in dual fuel mode is a proposed solution in the effort to protect the environment and control the serious threats posed by greenhouse gas emissions from compression ignition engines. The objective of this study was to experimentally examine the effect of syngas composition on the exhaust emission of dual fuel compression ignition (CI) engine at various engine speeds, and to compare the operating ranges of imitated syngas versus pure diesel. The study was conducted using a naturally aspirated, two strokes, single cylinder 3.7 kW diesel engine operated at speeds of 1200, 2000 and 3000 rpm. The engine was tested with three different syngas compositions. Diesel fuel was partially substituted by syngas through the air inlet. The test results disclose the impact of using syngas in CI engines on emission of CO_2, NO_x, unburned hydrocarbons and carbon monoxide. The experimental measurements confirmed that all syngas compositions are capable of reducing the emissions of CO_2 and NO_X compared with diesel fuel. Wide range of diesel replacement ratios (up to 72%) was attained without any penalty. Syngas with composition of 49% N_2, 12% CO_2, 25% CO, 10% H_2, and 4% CH_4 reduced the emissions of CO_2 and NO_x at engine speed of 1200 rpm up to 1% and 108 ppm, respectively. The lowest emission of UHC and NO_x was emitted when the engine was operating at speed of 2000 rpm and 3000 rpm, respectively with composition of 38% N_2, 8% CO_2, 29% CO, 19% H_2, and 6% CH_4. Therefore, syngas could be a promising technique for controlling NO_x emissions in CI engines. However, hydrogen content in syngas is important parameter that needs to be further investigation for its effect.

1. Introduction

Many compression ignition (CI) engines have been converted to run on gaseous fuel through dual fueling in order to improve the combustion process and achieve low exhaust emission. This was done by maintaining the engines' existing design without any modification. Dual fuel CI engines could be easily exchanged from dual fuel operation to operate on pure diesel at any time. Selection of the type of gaseous fuel to be used in dual fuel for CI engine should be on the basis of environmental considerations and economy [1]. For environmental, usage of syngas in dual fuel CI engine has the capability to control the emission of nitrogen oxides (NO_X) and carbon dioxide (CO_2) due to the lower heating value of syngas and the effective presence of nitrogen and carbon dioxide in its composition. For the economic purpose, using of syngas in CI engines allows substituting the diesel fuel and reduce the cost effectively as this fuel is produced by converting readily available natural materials [2].

Great interests have emerged in using of syngas produced from biomass gasification as alternative fuel for CI engines. However, the conditions of syngas produced from gasification of solid fuels are not consistent [3-5]. This instability affects the engine combustion and emission, leading to several researchers to focus on operation of CI engines with syngas in dual fuel mode. Sadykov et al. [6] operated a water-cooled CI engine fueled with diesel and syngas generated from the common gas. A steady speed of 1300 rpm was used while testing the engine. The operation was tested at standard conditions of an unadulterated diesel and at partial substitution of diesel by syngas on dual fuel mode. Measurements of engine exhaust emission showed that, reduction of NO_X emission has occurred with the addition of syngas to the pilot fuel in each of the ultra-lean and moderately rich mixtures; while an increment of NO_X emission was acquired in the moderate mixture. Hence, the decrease of NO_X substance with the expansion of syngas could be allocated to the weakening impact of nitrogen and carbon oxides together with the left over water held in syngas. While hydrocarbons emitted slightly with syngas expansion staying at a level (lower than 40 ppm). CO emission built extensively, this inferred that observed impact was because of an extensive CO content in syngas. Besides, because of the weakening impact of inert segments of syngas, the diminishing of the combustion temperature and relating weakening of combustion efficiency, particularly at level identical proportions, could be dependable for the increment of CO substance in exhaust emission as well.

Sridhar et al. [7] concentrated on the exhaust emission from engines working on producer gas as fuel. The emissions of NO_X and CO were measured for dual fuel operation mode and pure diesel mode. Firstly, when a CI dual fuel engine was operated utilizing high speed diesel and producer gas fuel, it was discovered that the NO_X levels were lower compared with pure diesel fuel, while the CO levels were higher due to ignition inefficiencies. At that point a 100% ultra-clean producer gas was utilized as a flash ignition engine. It was discovered to be ecologically harmful in terms of emissions; NO_X and CO levels were discovered to be much lower than a large portion of the existing emissions standards of different countries including the United States and European Union.

Ahrenfeldt et al. [8] operated a gas engine with producer gas from biomass as fuel for more than 2000 hours. Two diverse control ways were followed and investigated. Firstly, the stream rate of the producer gas was set to be steady and the engine worked with distinctive measures of air because of gas composition variation and hence stoichiometry. Secondly, the amount of air in the exhaust gas was set to be constant and distinctive amount of producer gas was utilized. Measurement was reported in term of standard emissions, load and efficiency at diverse loads ranging from 50% to 90%. Generally, the engine operation with producer gas produced high flow of unburned CO whatsoever conditions compared with current regulation for CO emission. The outflows of aldehydes were considerably lower for producer gas operation as compared to the natural gas.

Mahgoub et al. [9] studied the emission from dual fuel CI engine running on three different controllable compositions and conditions of imitated syngas at engine speed of 2000 rpm. A reduction on CO_2 and NO_X emission level was recorded for syngas/diesel dual fuel operation when compared with the operation of the engine on pure diesel fuel. A disadvantages concerning UHC and CO were observed for dual fuel operation throughout all syngas compositions used in their study due to poor combustion efficiency of syngas-diesel dual fuel operation. It can be concluded from previous studies that, dual fuel combustion using syngas as a supplement for diesel fuel affects positively (reduction) NO_x and CO_2 emissions, while it causes a considerable increase of CO and UHC emissions compared to the ones observed under normal diesel operation. This study is an extension of the previous work by Mahgoub et al. [6]. The emissions of the three imitated syngas compositions was determined at higher and lower than engine speed of 2000 rpm in order to study the effect of engine speed on the emission from dual fuel CI engine running on different syngas compositions and to find the operating ranges of imitated syngas versus pure diesel.

2. Experimental setup

The engine used in this study was a naturally aspirated, two stroke, single cylinder, 3.7 kW diesel engine with a displacement volume of 230 cc and a compression ratio of 17.6:1. This engine was used to investigate its emission by fueling different compositions of imitated syngas at engine speed of 1200, 2000, and 3000 rpm for various diesel replacement ratios. Experimental rig were incorporated into the engine setup to deliver the syngas through the air intake manifold. The rig and properties of used syngas compositions were described previously by Mahgoub et al. [10]. The schematic of experimental rig are shown in Figure 1. The properties of used syngas compositions are described in table 1.

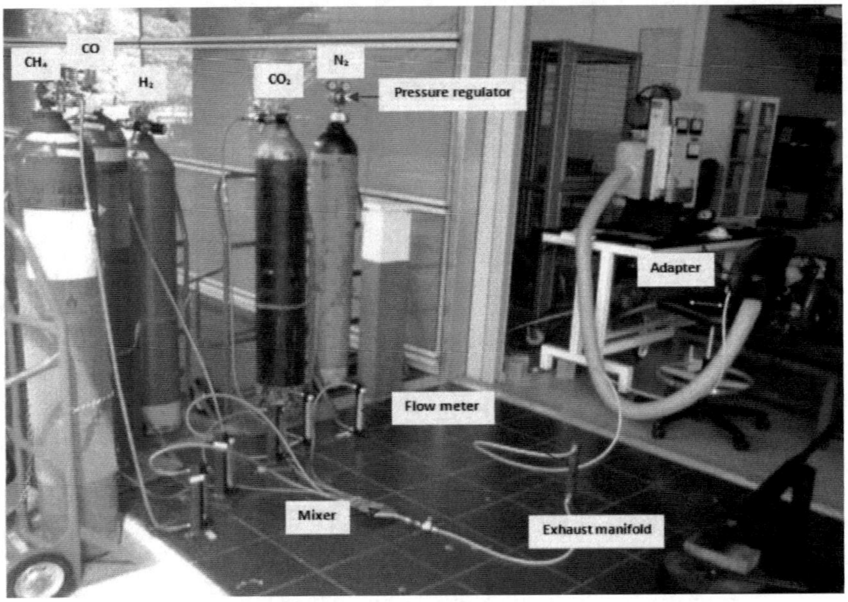

Figure 1. Schematic of experimental rig.

Table 1. Properties of selected syngas compositions.

Composition	N_2	CO_2	CO	H_2	CH_4	LHV (kJ/ kg)	ρ (kg/m^3)
A	49	12	25	10	4	4726.19	1.1
B	51	6	22	18	3	5418.4	0.94
C	38	8	29	19	6	7444.13	0.93

A multi-component FTIR gas analyzer (GASMET Cr-4000) was used to analyze the concentration level of CO_2, CO, UHC and NO_X emission at each diesel replacement ratio for all compositions of syngas at engine speed of 1200, 2000 and 3000 rpm.

3. Results and discussions

The engine exhaust emission evaluated for the concentration of CO_2, CO, UHC and NO_X. CO_2 concentration versus diesel replacement ratios for each syngas composition at different engine speed is given in figure 2.

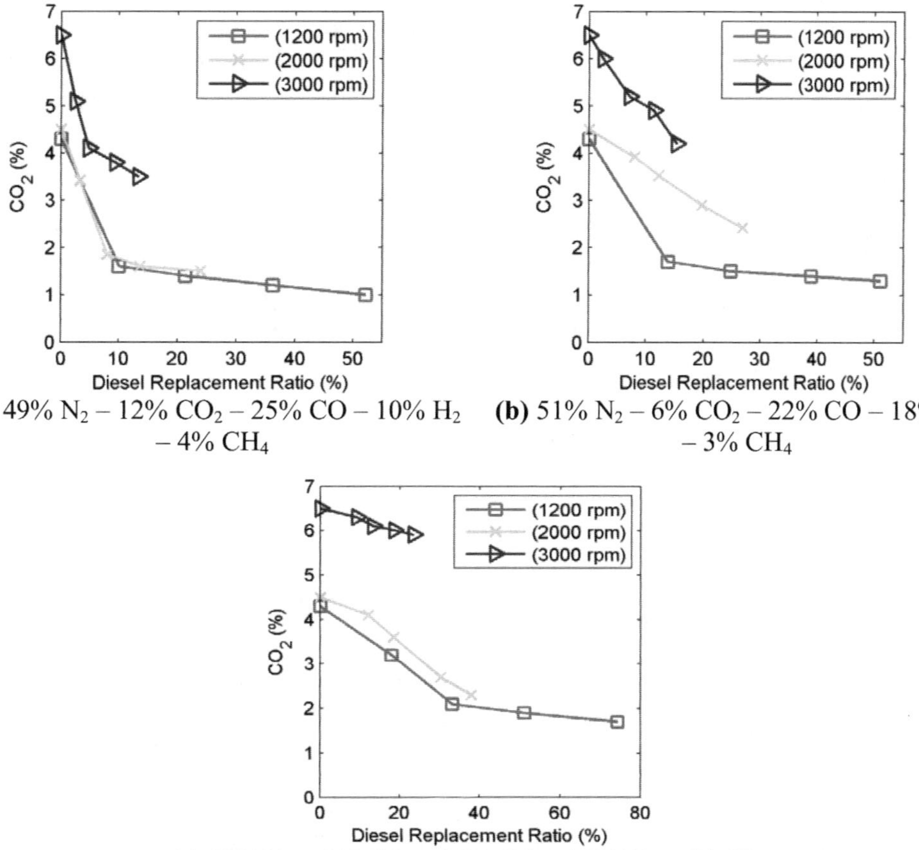

(a) 49% N_2 – 12% CO_2 – 25% CO – 10% H_2 – 4% CH_4

(b) 51% N_2 – 6% CO_2 – 22% CO – 18% H_2 – 3% CH_4

(c) 38% N_2 – 8% CO_2 – 29% CO – 19% H_2 – 6% CH_4

Figure 2. CO_2 concentration for various ratios of diesel replacement at various engine speeds.

CO_2 emission increased with the increase of engine speed for all syngas compositions due to increase of in-cylinder temperature and the amount of air at higher engine speeds. This is interpreted from the measured exhaust gas temperature. This has increased oxidation rate of carbon to CO_2. The highest amount of CO_2 was at 3000 rpm for all syngas compositions which is considered as an indicator to good combustion and high power output. The trends in Figure 2 show that, the emission of carbon dioxide reduces as the amount of syngas increases. This is largely due to the displacement of the incoming air by syngas in syngas dual fuel mode which then leading to decrease on the oxidation rate and reduce the carbon dioxide formation. Although composition A contains higher amount of inert carbon dioxide compared to the other compositions, lower amount of carbon dioxide is emitted at all engine speeds. This is might be because there was no enough incoming air leading to a reduction on the oxidation rate of carbon to carbon dioxide. Figure 3 provides the variation of CO concentration as a function of diesel replacement ratios for each syngas composition at different engine speeds. CO emission increased with the increase of engine speed for all syngas compositions as an increase in engine speed shortens the oxidization time of CO to CO_2. An increase in carbon monoxide content is also explained by incomplete combustion in the cylinder because of the average cylinder temperature decline due to the lower heating value of syngas. Similar results were obtained when different producer gas to air flow ratios has been inducted into diesel engine by Ramadhas et al. [11].

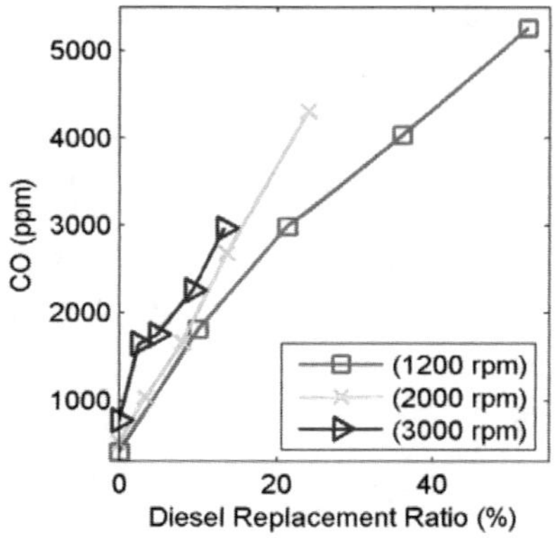

(a) 49% N_2 – 12% CO_2 – 25% CO – 10% H_2 – 4% CH_4

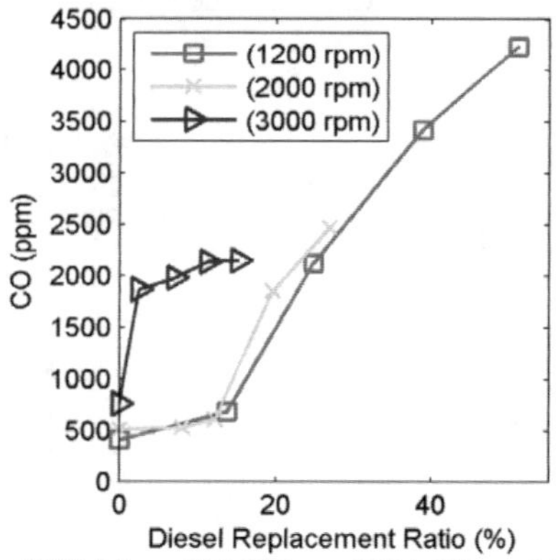

(b) 51% N_2 – 6% CO_2 – 22% CO – 18% H_2 – 3% CH_4

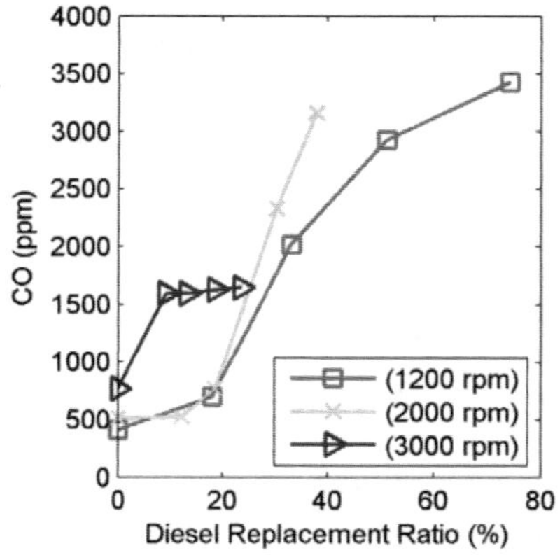

(c) 38% N_2 – 8% CO_2 – 29% CO – 19% H_2 – 6% CH_4

Figure 3. CO concentration for various ratios of diesel replacement at various engine speeds.

Composition A emitted the highest amount of carbon monoxide at all engine speeds due to its lower heating value which leads to an average cylinder temperature decline and incomplete combustion. While compositions B and C showing lower carbon monoxide level due to the rising in the heating value, the high level of carbon monoxide emission from composition A is due an incomplete combustion and due to the presence of higher carbon monoxide in composition A compared to other compositions.

Figure 4 shows the concentration of UHC versus diesel replacement ratio for each syngas composition at different engine speed. UHC emissions are the results of an incomplete combustion fuels. Similar to CO, UHC emissions are the result from flame quenching in crevice regions and at cylinder walls. Other causes of UHC emissions formation are due to the running of the engine on too rich fuel air ratio with insufficient oxygen and the incomplete combustion of lube oil. UHC emissions also increase due to decline of the cylinder peak temperature causing incomplete combustion. Similar results were obtained when different loads were applied by Saha et al. [12] to study the effect of syngas composition on exhaust emission of diesel engine. The poor combustion of syngas with air resulted in the HC emissions at a higher level.

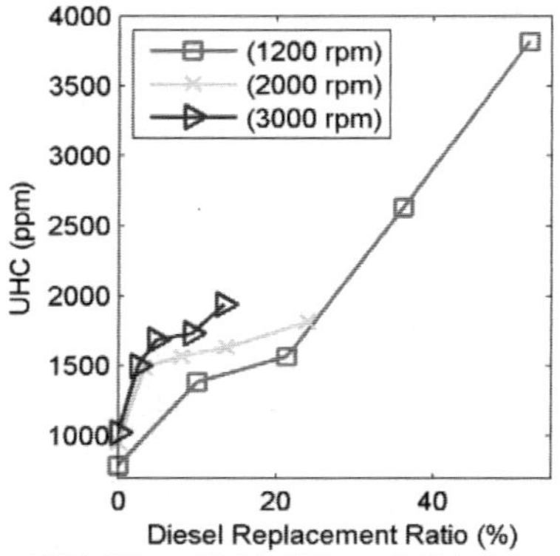

(a) 49% N_2 – 12% CO_2 – 25% CO – 10% H_2 – 4% CH_4

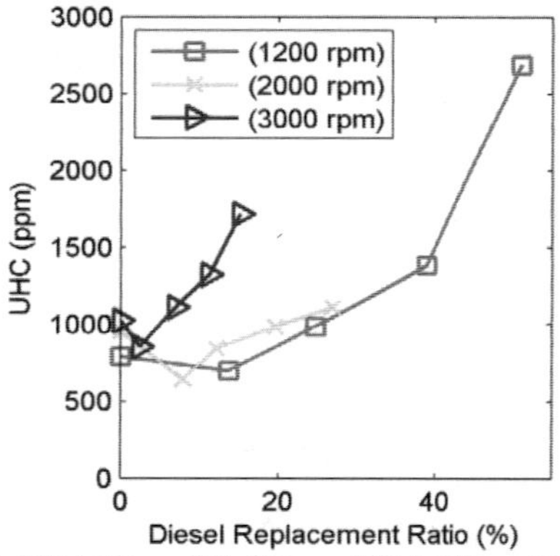

(a) 51% N_2 – 6% CO_2 – 22% CO – 18% H_2 – 3% CH_4

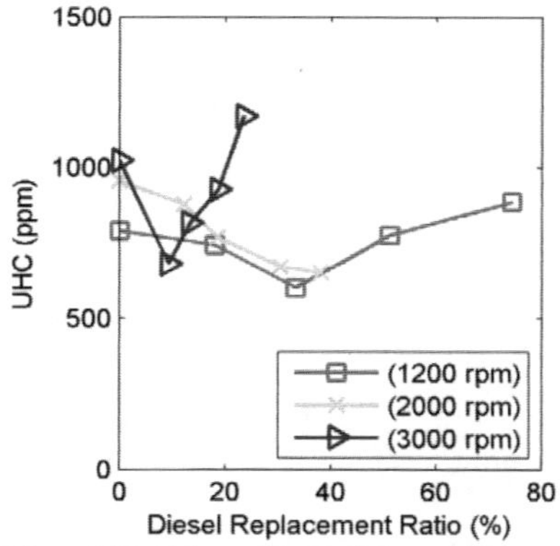

(c) 38% N_2 – 8% CO_2 – 29% CO – 19% H_2 – 6% CH_4

Figure 4. UHC concentration for various ratios of diesel replacement at various engine speeds.

Although the engine emitted the highest amount of CO_2 at 3000 rpm as shown in Figure 2 for all syngas compositions, which considered as an indicator to the good combustion, the highest amount of UHC was also emitted at 3000 rpm for all syngas compositions. This is because there was not enough oxygen for combustion inside the combustion chamber, or most of the oxygen consumed on the oxidation process of carbon to carbon monoxide. This can be inferred from the higher amount of CO_2 emitted from the engine as shown in Figure 3.Composition A emitted the highest UHC due to the excess oxygen concentration drop in the intake mixture inside the combustion chamber which leads to improperly combustion and higher UHC emission. The trends show that, compositions B and C emitted less amount of UHC and started to increase gradually this can be interpreted that rigorous combustion occurred in early stage then the combustion rate decreased at the later stage.

NO_X emissions are shown in Figure 5 for each syngas composition at different engine speeds. It can be seen that the NO_X emission increased with the increase of engine speed. This is due to increase of in-cylinder combustion temperature and the availability of more amount of oxygen inside the cylinder which facilitated the thermal formation of NO_X content. The engine emitted the lowest amount of NO_X at 3000 rpm for all syngas compositions. This is because insufficient oxygen remaining for thermal oxidation of nitrogen due to the consumption of the oxygen inside the combustion chamber into the process of oxidation of carbon to carbon monoxide as shown in Figure 3.

IOP Conf. Series: Materials Science and Engineering **100** (2015) 012006 doi:10.1088/1757-899X/100/1/012006

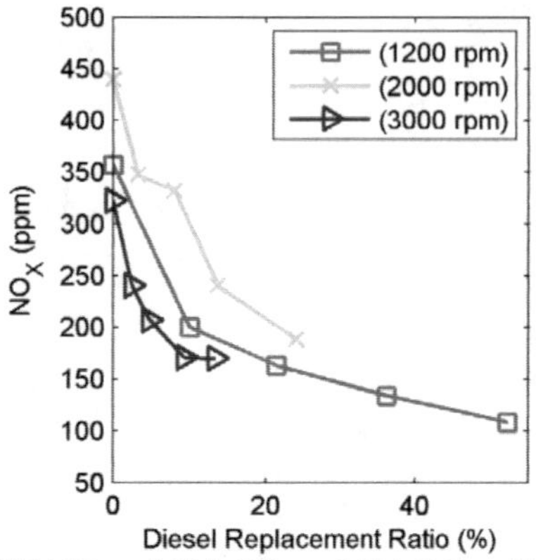

(a) 49% N_2 – 12% CO_2 – 25% CO – 10% H_2 – 4% CH_4

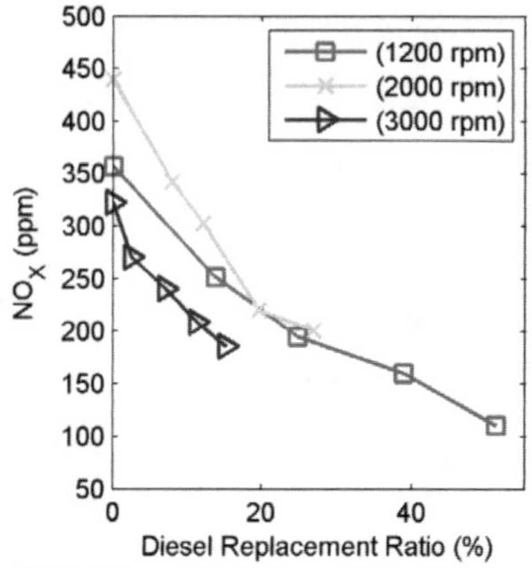

(b) 51% N_2 – 6% CO_2 – 22% CO – 18% H_2 – 3% CH_4

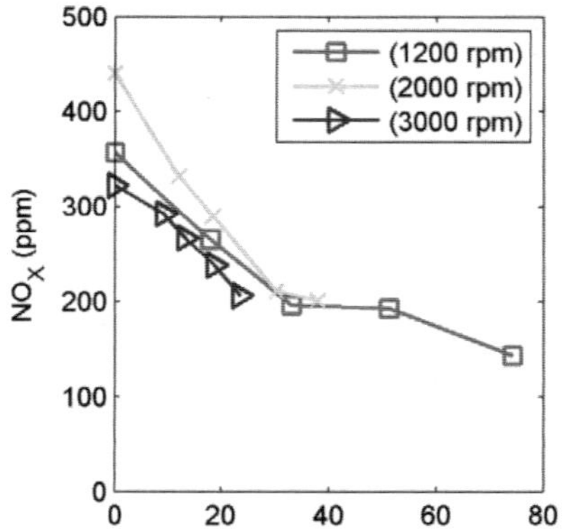

(c) 38% N_2 – 8% CO_2 – 29% CO – 19% H_2 – 6% CH_4

Figure 5. NO_X concentration for various ratios of diesel replacement at various engine speeds.

Figure 5 presented that the dual fuel effect could decrease the nitrogen oxides from combustion process in cylinders. The diminishing of nitrogen oxides in the case of syngas dual fuel mode occurred from the dilution of nitrogen in the air by syngas mixing, that it reduces the thermal nitrogen oxides effect in cylinders. The decrease in NO_X levels also may be attributed to diluting effect of nitrogen and carbon oxide components of the syngas and the lower hydrogen content which leads to lower cylinder peak temperature. More pronounced decline of NO_X emission with addition of syngas apparently is caused by the oxygen concentration drop in the intake mixture. Hence, in this study, composition A emitted the lowest NO_X content, which attributed to the temperature-lowering effect of the syngas, owing to the composition lower heating value compared to other compositions or due to oxygen concentration drop. The higher hydrogen presence of composition C have resulted the highest combustion temperature as indicated by the measured exhaust temperatures. Hence, the NO_X emission for this composition was higher as compared to other compositions at all conditions.

4. Conclusions

In the present work, an experimental study was conducted to determine the emissions of a CI engine dual fuelled with diesel and imitated syngas at various engine speeds. From the analysis of experimental data and the comparison between normal diesel and syngas dual fuel operation, the main conclusions from the study are as follows:

i. Dual fuel operation considered as a promising technique for controlling NO_X emissions on existing CI diesel engines and it affects positively carbon dioxide emissions, while it causes a considerable increase carbon monoxide and UHC emissions compared to the observed under normal diesel operation.
ii. The relative content of hydrogen in syngas composition is very important to overcome the negative effects of the presence of CO and UHC in the exhaust emission. So the engine operating on Composition C showed the lowest level of CO and UHC.
iii. Composition C can be considered the most appropriate composition compared to the other compositions because it showed the lowest level of CO and UHC.

Acknowledgements
The authors are grateful for the contributions of members at Power Generation Laboratory in Universiti Teknologi PETRONAS.

References
[1] Kalogirou S A 2004 *Progress in energy and combustion science* **30** 231-95
[2] Gera D, Mathur M P, Freeman M C, Robinson A 2002 *Energy & Fuels* **16** 1523-32
[3] Hagos F Y, Aziz A R A, Sulaiman S A 2014 *Energy Procedia* **61** 2567-71
[4] Hagos F Y, Aziz A R A, Sulaiman S A 2014 *Advances in Mechanical Engineering* **2014**
[5] Hagos F Y, Rashid A A A, Sulaiman S A 2013 *Asian Journal of Scientific Research* **6** 187-96
[6] Sobyanin V, Sadykov V, Kirillov V, Kuzmin V, Kuzin N, Vostrikov Z, et al., Syngas as a fuel for IC and diesel engines: efficiency and harmful emissions cut-off. Proceedings of International Hydrogen Energy Congress and Exhibition; 2005: Istanbul Turkey.
[7] Sridhar G, Dasappa S, Sridhar H V, Paul P J, Rajan N K S 2005 *SAE Technical Paper* **10.4271/2005-01-1732.**
[8] Ahrenfeldt J, Jensen T K, Henriksen U, Schramm J 2005 *SAE transactions* **114** 1464-75
[9] Mahgoub B K M, Sulaiman S A, Karim Z A B A, Emission of a compression ignition engine fuelled by diesel and imitated syngas. AIP Conference Proceedings.
[10] Mahgoub B K M, Sulaiman S A, Karim Z A A 2013 *Asian Journal of Scientific Research* **6** 456-66
[11] Ramadhas A, Jayaraj S, Muraleedharan C 2008 *Renewable Energy* **33** 2077-83
[12] Sahoo B B, Sahoo N, Saha U K *Applied thermal engineering* **49** 139-46

IOP Conf. Series: Materials Science and Engineering **100** (2015) 012007 doi:10.1088/1757-899X/100/1/012007

Leak detection in medium density polyethylene (MDPE) pipe using pressure transient method

M M Amin[1], M F Ghazali[1], M A PiRemli[1], A M A Hamat[1] and N F Adnan[1]
[1]Faculty of Mechanical Engineering, Universiti Malaysia Pahang, 26600 Pekan, Pahang, Malaysia

E-mail: makeenamin@gmail.com

Abstract. Water is an essential part of commodity for a daily life usage for an average person, from personal uses such as residential or commercial consumers to industries utilization. This study emphasizes on detection of leaking in medium density polyethylene (MDPE) pipe using pressure transient method. This type of pipe is used to analyze the position of the leakage in the pipeline by using Ensemble Empirical Mode Decomposition Method (EEMD) with signal masking. Water hammer would induce an impulse throughout the pipeline that caused the system turns into a surge of water wave. Thus, solenoid valve is used to create a water hammer through the pipelines. The data from the pressure sensor is collected using DASYLab software. The data analysis of the pressure signal will be decomposed into a series of wave composition using EEMD signal masking method in matrix laboratory (MATLAB) software. The series of decomposition of signals is then carefully selected which reflected intrinsic mode function (IMF). These IMFs will be displayed by using a mathematical algorithm, known as Hilbert transform (HT) spectrum. The IMF signal was analysed to capture the differences. The analyzed data is compared with the actual measurement of the leakage in term of percentage error. The error recorded is below than 1% and it is proved that this method highly reliable and accurate for leak detection.

1. Introduction

Water is an important part of an economic growth of a country that would provide. Even through precaution steps have been taken, technical faulty is still exist during their transmission upon reaching the users as a part of the supplied water are loss due to the various problems for example leakage [1]. Leakage is due to high pressure of water flowing through the underground pipe. In addition, such catastrophes can have significant social and environmental influences [2]. Non-revenue water (NRW) program has been implemented in Malaysia for the precaution from water lost. NRW is the difference between the water supply volume in water distribution system and the volume that is billed to consumer [3].

To serve the purpose of supplying the demand to consumer, water is supplied from production facilities i.e. rivers, reservoir, damns, tanks and it is important to ensure that the supplied water would reach to the consumer accordingly [4, 5].

NRW could be a result from theft, pipe leakage, meter inaccuracy, and other uncountable losses once its leaves the water treatment. The problem that caused from the pipeline leakage would contribute a big loss to the company as it would increment maintenance and operating costs. In fact, it will give negative impacts on the water supply services and to the nation's total economic growth [6].

Since there are various methods that can be used to detect pipe leaks, this paper is focused on the method that based on internal inspection method which required a vibration approach upon conducting the leakage in the pipeline [3]. It is proved that EEMD method can obtain significant results on the pipe leaks experiments by using Piezoceramic Transducer [7].

2. Experimental setup

The experiment used various types of the pipeline network which can be constructed in various methods in collecting the acquired data. It is almost impossible to experimentally fabricate and simulate the actual situation as it requires a very big and longer pipeline test rig that consumed massive space. Therefore, most of the researchers are conducting experiment on pipeline model based on the actual size of the pipeline that is easy to fabricate and easy to be manipulated throughout the test setup.

The pipeline was connected to a series of pipe features such as 90°-angle, T-junction, and different types of pipe diameters. The outlet of the pipe is kept connected to a free surface tank such that the water from the pipe ends discharged underwater as shown in figure 1. This is to prevent the sudden expansion of the pressure waves and the negative pressure of the waves could be minimized as it will affect the collected data from the transducer.

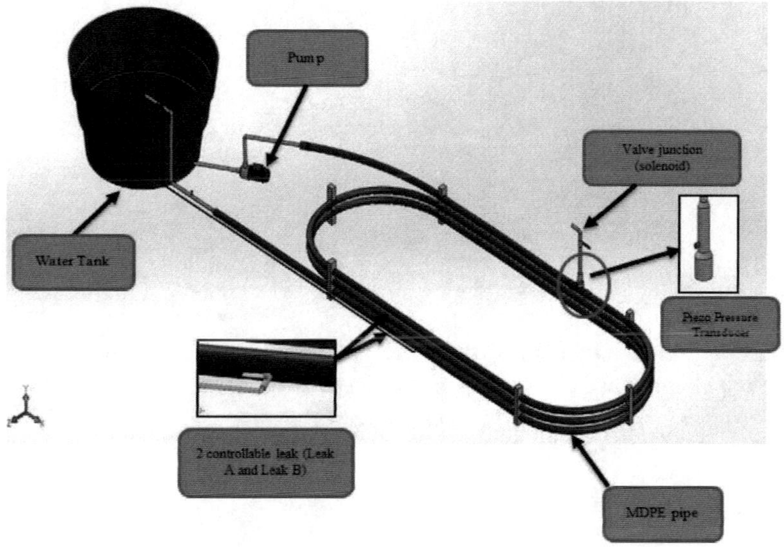

Figure 1. Design of test rig.

In some cases the unit will shut down in crisis, and the "water hammer" existence will inevitable occur in the pressure pipeline. This phenomenon may take place in all of pressure pipe system, frequently causing about strong vibration and destruction on the pipeline system [8]. Solenoid valve used as to create a water hammer through the pipe. This is order to create a same phenomenon happen in real life when there is pressure disorder occur through the underground pipe. Solenoid valve was controlled by a switch so that people in charge can control whenever to create the water hammer.

PresSystems (NI-DAQ) to collect data when the pressure changes occurred in the pipelines. The water is commuted in the system as long as the pump operates continuously. The total length of the MDPE pipe is 57.90m; where it started from the inlet of the pipeline i.e. from the pump until the outlet of the pipe that is located at back of the water tank.

The pipeline is attached with a few pipe features that will affect the pipeline behaviour. The signal response from this behaviour will be retrieved from the transducer. Even though this signal response will influence the analysis of the experiment data, this study is mainly focused on the leakage that is exist in the pipeline. The distance of the leakage "A" is measured to be 27.08m. The leak was created by drilling a hole at some distance in the pipeline where this leakage can be control either to be opened or closed.

2.1. Ensemble empirical mode decomposition (EEMD)

Ensemble Empirical Mode Decomposition (EEMD) has more suitability in extracting a broad range of signals for a rich data of non-linear signal and non-stationary systems. The key idea on EEMD relies on averaging the modes obtained by EMD applied to several realizations of Gaussian white noise added to the original signal. The resulting decomposition solves the EMD mode mixing problem, however it introduces new ones. In the proposed method, a particular noise is added at each stage of the decomposition and a unique residue is computed to obtain each mode. It can be concluded that EEMD method is basically based on Empirical Mode Decomposition (EMD) but the problems of the mode mixing have been alleviate by introducing noise-assisted method [4].

The single Intrinsic Mode Function (IMF) that acquired is either having signals of widely disparate scales or a signal with a same scale reading in different IMF components. The mode mixing could result in a serious data aliasing with the time frequency distribution, and caused IMF data to be suppressed and unclear [9]. In other word, that can be summarized in the form of equations are stated as below:-

$$x_m(t) = x(t) + w_m(t), \quad m = 1,2,\dots,n \tag{1}$$

$$x_m(t) = \sum_{i=1}^{L} c_{m,i}(t) + r_{m,L}(t), \quad m = 1,2,3,\dots,n. \tag{2}$$

$$x(t) = \frac{1}{N}\sum_{i=1}^{L}\sum_{m=1}^{n} c_{m,i}(t) + \frac{1}{N}\sum_{m=1}^{n} r_{m,L}(t) \tag{3}$$

where $x(t)$ is the original signal, $wm(t)$ is the mth term added white noise, $xm(t)$ is the noisy signal of the mth trial, L is the number of IMFs from the EMD method, and N is the ensemble number of the EEMD method.

3. Results and discussion

3.1. Single leak detection signal using water hammering method

This signal response is recorded using the solenoid valve to induce a water hammering method. Compared to the manual method of opening and closing the valve, this method shows a better way to obtain the signal as it regulates the water flow and reduce the oscillations in the pipeline during water hammering process. The condition of the solenoid valve is in a normally closed condition valve and thus caused water hammer by closing the valve with three consecutive times for each of the data were recorded by the DASYLab Software.

3.2. Signal response for closed valve on the leakage of the pipeline

The experiment data that is obtained from the transducer response shows the behaviour response of the pipeline. The signal that is recorded does also retrieve the characteristic of the whole pipeline network itself. This pressure pulse travels in both directions away from the burst origin at the speed of sound in water (wave speed of the pipe) [10].

Figures 2(a), 2(b), and 2(c) indicate on the original responses of the pressure transducer that were obtained from the experimental process with a variation of pressures (P) with data retrieved from the transducer. The original response was recorded at three different level of pressures: 1mmHg, 2mmHg, and 3mmHg. The original responses are contrast due to the differences in pressure levels. The higher the pressure creates higher frequency of signal and this causes the higher amplitude of signal.

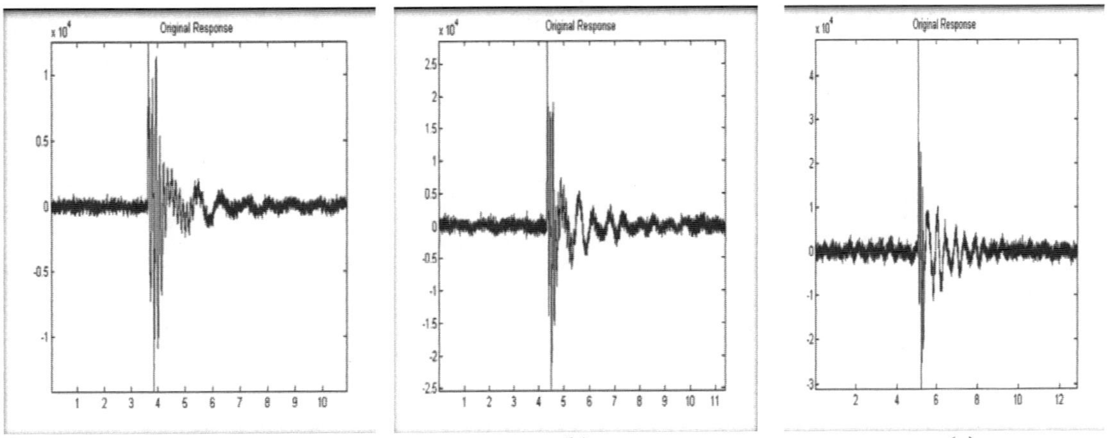

(a) (b) (c)

Figure 2. Signal responses for (a) 1mmHg (b) 2mmHg and (c) 3 mmHg with no leakage in the pipeline.

Results based on the original responses obtained from analysis on EEMD method with signal masking showed inconclusive value which indicates that the pipeline is associated with any kind of leakage. Thus, signal response from 2mmHg data is then analyzed through the EEMD with signal masking method using the MATLAB software where this signal is being decomposed into a series of IMFs that would be later converted into instantaneous frequency (IF) for further analytical calculation.

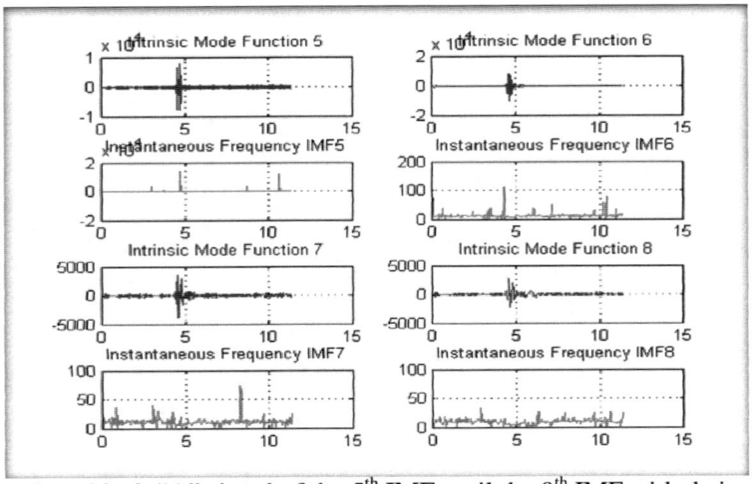

Figure 3. The analysed leak "A" signal of the 5th IMF until the 8th IMF with their respective IF (2mmHg).

Figure 3 shows the analysis of signal based on IMF with their respective IF for 2mmHg pressure level. This analysis was done using the MATLAB software in order to decompose the signal from the original response. The results obtained from 5th IMF until 8th IMF with respective IF were observed in order to select the best IF for further analysis. Compared to all IF, the 5th IF shows the most appropriate response to be used as reference for the further analytical calculation.

Figure 4 shows the response from the 5th IF with the detected leakage on leak "A" at pressure level of 2mmHg. All peaks in 5th IF response was observed because when there is a leakage through the pipe, the pressure disorder will occur and create fluctuated response in detected signal. So, the time difference is calculated for every peaks in 5th IF response.

Based on the collected data and analyses, each signal can have a few different readings compared to the actual measurement of the leakage distance.

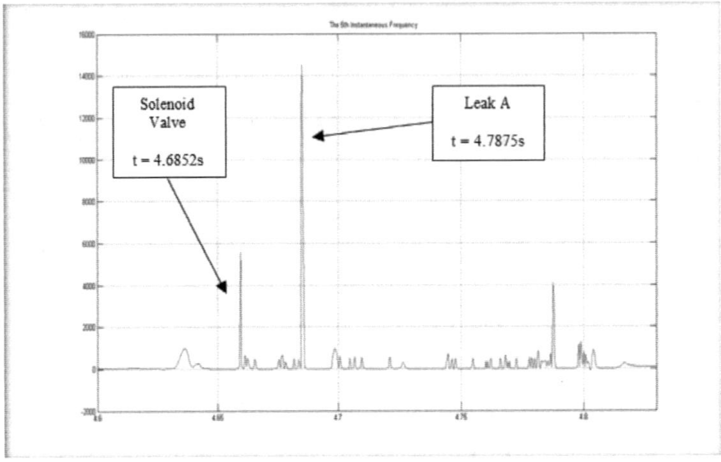

Figure 4. The 5th instantaneous frequency with the detected leakage on leak "A" (2mmHg).

With the calculated speed of sound of the fluid in the pipe are 524.3005 ms^{-1}. The time difference for 2mmHg read is 0.1044s and is identified at 26.82m from the located pressure sensor. The percentage error is 0.96%. Thus, table 1 summarized the data gathered for 2mmHg pressure during experiment and the calculated data from signal response obtained from the transducer.

Table 1. Theoretical and actual leak distance.

Pressure (mmHg)	Leak distance (m)		Percentage error (%)
	Theoretical	Actual	
2	27.08	26.82	0.96

4. Conclusion

In this paper, it was endeavoured to detect the leaking through medium density polyethylene (MDPE) pipe by using a transient method. This method was verified due to the percentage error between the analytical and theoretical just below than 1%. In this study, the ensemble empirical mode decomposition (EEMD) method was used in order to decompose the response from the original response. The leakage was successfully detected through the pipe with only small tolerance of error.

Acknowledgements

The authors would like to be obliged to Universiti Malaysia Pahang for providing laboratory facilities and financial assistance under project no. RDU121410

References

[1] Mpesha W, Gassman S L, Chaudhry M H 2001 *Journal of Hydraulic Engineering* **127** 134-47
[2] Berardi L, Kapelan Z, Giustolisi O, Savic D 2008 *Journal of Hydroinformatics* **10** 113-26
[3] Thakur G, Wu H-T 2011 *SIAM Journal on Mathematical Analysis* **43** 2078-95
[4] Thakur G, Brevdo E, Fučkar N S, Wu H-T 2013 *Signal Processing* **93** 1079-94
[5] Brunone B 1999 *Journal of water resources planning and management*
[6] Brunone B, Ferrante M 2001 *Journal of hydraulic research* **39** 539-47
[7] Beck S, Curren M, Sims N, Stanway R 2005 *Journal of hydraulic engineering* **131** 715-23
[8] Li J, Wu P, Jiandong Y, CFD Numerical Simulation of Water Hammer in Pipeline based on the Navier-Stokes Equation. Fifth European Conference on Computational Fluid Dynamics; 2010.
[9] Ferrante M, Brunone B 2003 *Advances in Water Resources* **26** 95-105
[10] Misiūnas D 2008 *Failure Monitoring and Asset Condition Asssessment in Water Supply Systems* Vilniaus Gedimino technikos universitetas)

Combustion parameters of spark ignition engine using waste potato bioethanol and gasoline blended fuels

B Ghobadian[1], G Najafi[1], M Abasian[1] and R Mamat[2]
[1]Faculty of Biosystem Engineering, Tarbiat Modares University, Tehran, Iran
[2]Faculty of Mechanical Engineering, Universiti Malaysia Pahang, 26600 Pekan, Pahang, Malaysia

Email: rizalman@ump.edu.my

Abstract. The purpose of this study is to investigate the combustion parameters of a SI engine operating on bioethanol-gasoline blends (E0-E20: 20% bioethanol and 80% gasoline by volume). A reactor was designed, fabricated and evaluated for bioethanol production from potato wastes. The results showed that increasing the bioethanol content in the blend fuel will decrease the heating value of the blended fuel and increase the octane number. Combustion parameters were evaluated and analyzed at different engine speeds and loads (1000-5000 rpm). The results revealed that using bioethanol-gasoline blended fuels will increase the cylinder pressure and its 1st and 2nd derivatives ($P(\theta)$, $P^{\bullet}(\theta)$ and $P^{\bullet\bullet}(\theta)$). Moreover, using bioethanol-gasoline blends will increase the heat release ($Q\bullet(\theta)$) and worked of the cycle. This improvement was due to the high oxygen percentage in the ethanol.

1. Introduction

The SI engines are widely utilized due to its reliable operation and economy. As the petroleum reserves are depleting at a faster rate, an urgent need for a renewable alternative fuel arise. Also the threat of global warming and the stringent government regulation made the engine manufacturers and the consumers to follow the emission norms to save the environment from pollution. Among the many alternative fuels, bioethanol is considered as a most desirable fuel extender and fuel additive due to its high oxygen content and renewable in nature [1]. Using gasoline-ethanol blend fuel in SI engines caused higher engine performance and lower emissions than gasoline fuel [2]. According to Yücesu, Sozen [3], using ethanol‑gasoline blend fuel in a spark ignition (SI) engine caused a higher engine torque than that of gasoline fuel. The maximum torque was obtained at 0.9 relative air‑fuel ratios. The effects of ethanol‑gasoline blends (E0, E10, E20, E40 and E60) on engine exhaust emissions and performance has been investigated by Yücesu, Topgül [4]. According to the results of the experiment, engine torque increased. It was also reported that blends with ethanol allowed the compression ratio to increase without any knock [5]. Celik [6] reported that the most suitable ethanol‑gasoline fuel blend in terms of performance and emissions was E50 in a small gasoline engine with low efficiency. Engine power increased by about 29% running with E50 fuel at high compression ratio compared to running with E0 fuel. The specific fuel consumption was reduced by approximately 3% [6, 7]. With increasing the ethanol content in gasoline fuel, the heating value of the blended fuels is decreased, while the octane number of the blended fuels increases. Al Hasan [8] reported that blending unleaded gasoline with ethanol increases the brake power, torque, volumetric and brake thermal efficiencies and fuel consumption, while it decreases the brake specific fuel consumption (bsfc; the same conclusions can be reached from [9, 10]. The 20 vol.% ethanol in fuel blend gave the best results for all measured parameters at all engine speeds [8]. Using E40 and E60 blends led to a significant reduction of CO and HC emissions [4]. For E50, the CO, CO_2, and HC emissions were reduced by 53%, 10% and 12%, respectively [6, 7]. NOx emission depends on the engine operating condition rather than the ethanol content [11]. NOx concentrations are increased due to rising of the cylinder temperature with increasing ethanol percentage in the blends [9]. The effect of ethanol blended gasoline fuels on emissions was investigated in a spark-ignition engine with an electronic fuel injection (EFI) system [10]. The addition of ethanol to gasoline fuel enhances the octane number of the blended fuels and changes distillation temperature. Ethanol is reported to be an important contributor to decreased engine-out regulated emissions and decreased brake specific energy consumption [10]. Wu et al. [12] reported that using E10 blended fuel at a relative air-fuel ratio (k) slightly greater than one can generally reduce pollutant emission components. The effects of using ethanol-unleaded gasoline blends on cyclic variability and emissions in a spark-ignited engine have been investigated by Cevis [13]. It was reported that using ethanol-unleaded gasoline blends as a fuel decreased the coefficient of variation in indicated mean effective pressure. The 10 vol.% ethanol in fuel blend gave the best results [13].

2. Experimental Set Up

2.1. Design of Experiment

The combustion parameters from the engine running on bioethanol (derived from potato waste) and blended with gasoline (E5, E10, E15 and E20) were evaluated and compared with gasoline fuel. The properties of ethanol fuel investigated and it has been proved that all specifications are in the standard range based on the ASTM standard. Above 20% ethanol, engine could not run smoothly, therefore, only experimental results obtained up to this percentage of bioethanol will be presented. The fuel blends were prepared just before starting the experiment to ensure that the fuel mixture was homogenous and to avoid the reaction of ethanol with water. A series of experiments were carried out using gasoline, and the various bioethanol blends. All the blends were tested under varying engine speed conditions. The engine was started using gasoline fuel and it was operated until it reached the steady state condition. The engine speed, fuel consumption, and load were measured. After the engine reached the stabilized working condition, combustion parameters were measured. All experiments have been carried out at full throttle setting. To adjust ignition timing, electronic ignition system was used. Before obtaining data from the engine operated with a new blended fuel, the engine was operated using the new fuel for sufficient time to clean out the remaining fuel from the previous blend. Fuel properties were determined at the laboratories of Research Institute of Petroleum Industry (RIPI) in Iran. In this paper, the quantity EX represents a blend consisting of X% bioethanol by volume, e.g., E5 indicates a blend consisting of 5% ethanol in 95% gasoline. Five test fuels were used in this study: 0% ethanol (E0); 5% ethanol (E5); 10% ethanol (E10); 15% ethanol (E15); and 20% ethanol (E20). In this study, the experiments were performed on four cylinder, four-stroke, and spark-ignition (SI) gasoline engine. A 190 kW SCHENCK-WT190 eddy-current dynamometer was used in the experiments. Fuel consumption rate was measured in the range of 0.4-45 kg/h by using laminar type flow meter, Pierburg model. Air consumption was measured using an air flow meter. Five separate fuel tanks were fitted to the gasoline engine and these contained gasoline and the bioethanol-gasoline blends. The engine control unit (ECU) that was used in this engine was a Johnson Controls JCAE S2000. ECU function is to control the quantity of fuel, injection timing, ignition timing and engine speed by receiving signals from seven sensors. These sensors are oxygen sensor, knock sensor, manifold air pressure sensor, intake air temperature sensor, throttle position sensor, water temperature sensor and engine speed sensor. A multipoint fuel injection (MPFI) system with top-feed injectors is used to inject the fuel into the combustion chamber. The ignition system was semi-static distributor less ignition (DLI). A schematic diagram and real of the experimental setup is shown in figure 1. A reactor was designed, fabricated and evaluated for bioethanol production from potato wastes (figure 2). To save the cost and energy for bioethanol production, a continuous solid fermentation system has been developed and by using of this system bioethanol was produced from potato wastes. The basic step for production of bioethanol is fermentation of sugars (figure 2). The purity of bioethanol from this machine was measured around 99% that is suitable for blend with gasoline fuel.

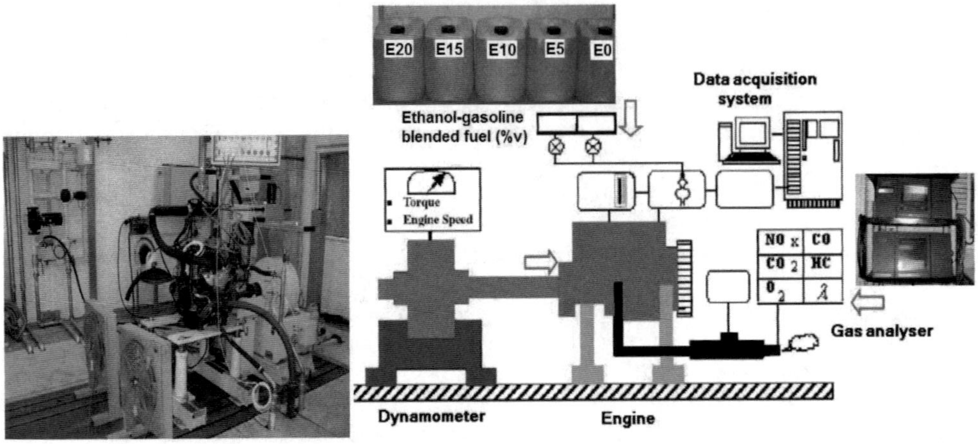

(a) (b)

Figure 1. Engine test set-up and test instruments (a) real and (b) schematic.

3. Results and Discussion

Figure 3 shows the effect of various fuels on engine cylinder pressure. When the bioethanol content in the blended fuel is increased, the engine cylinder pressure slightly increased for all engine speeds. The gain of the cylinder pressure can be attributed to the increase of the indicated mean effective pressure for higher ethanol content blends [9]. The heat of evaporation of ethanol is higher than that gasoline, this provides fuel-air charge cooling and increases the density of the charge, and thus higher power output is obtained [6]. With the increase in bioethanol percentage, the density of the mixture and the engine volumetric efficiency increases and this causes the increase of power [8].

Under time domain analysis various operations are performed on digital time history records of the signals in the time domain itself to yield information on rates, commencement, duration and end of events, and identification of different phases of any process. This kind of analysis often required some intermediate stages of data processing. In this paper, analysis of combustion parameters in a SI engine which has been performed in the time domain yielded the followings: estimates of cylinder pressure derivatives, estimates for start of injection and ignition delay, estimates of the duration of fuel injection along with the start of dynamic fuel injection, estimates of the heat release descriptors, identification of the phases of combustion and estimates of their duration, identification of the phases of liner vibrations vis-à-vis their origin. The derived parameters identified combustion parameters are peak cylinder pressure [Pc]max, maximum rate of cylinder pressure rise [P•c]max, maximum acceleration of cylinder pressure [P••c]max, peak rate of heat release [Q•]max which have been investigated in this paper. Figure 4 and 5 show the influence of different bioethanol-gasoline blended fuels on rate of cylinder pressure rise [P•c]max and maximum acceleration of cylinder pressure [P••c]max. The increase of bioethanol content increases the rate of cylinder pressure rise and maximum acceleration of cylinder pressure. Added bioethanol produces lean mixtures that increase the relative air-fuel ratio (k) to a higher value and makes the burning more efficient [11]. The improved antiknock behavior (due to the addition of ethanol, which raises the octane number) allowed a more advanced timing those results in higher combustion pressure and thus higher pressure [7].

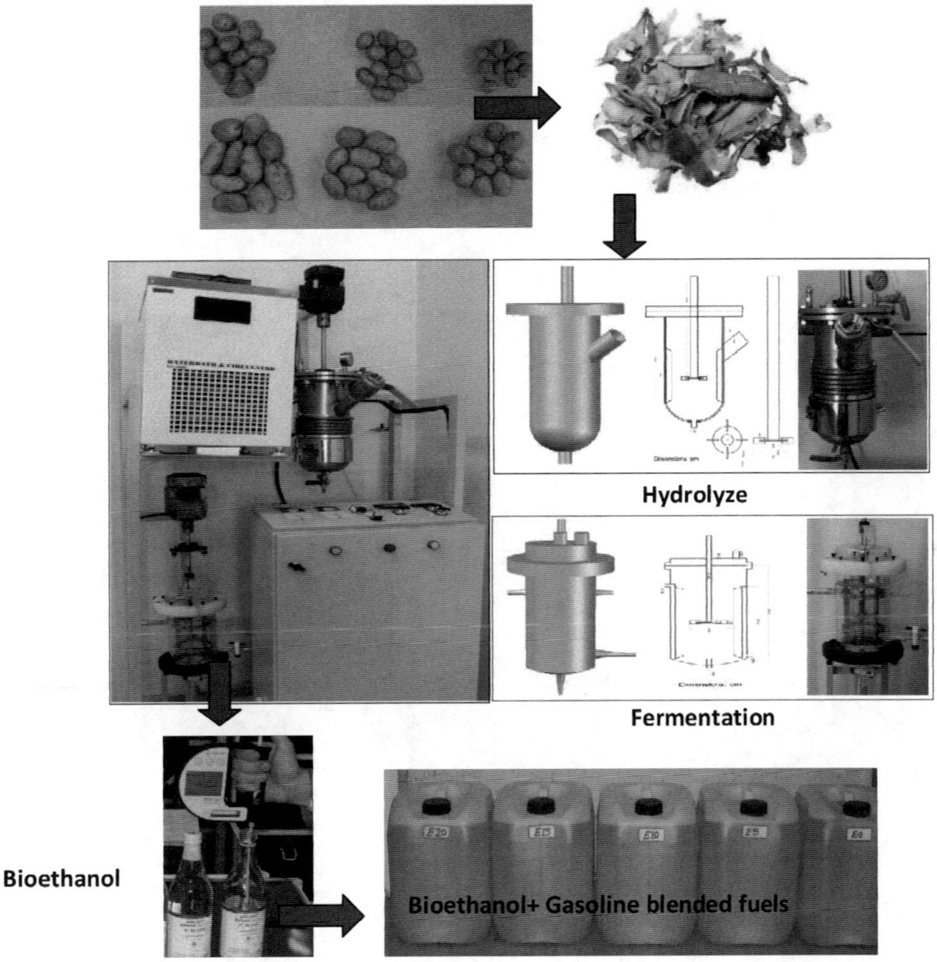

Figure 2. Bioethanol, gasoline blended fuels preparation process.

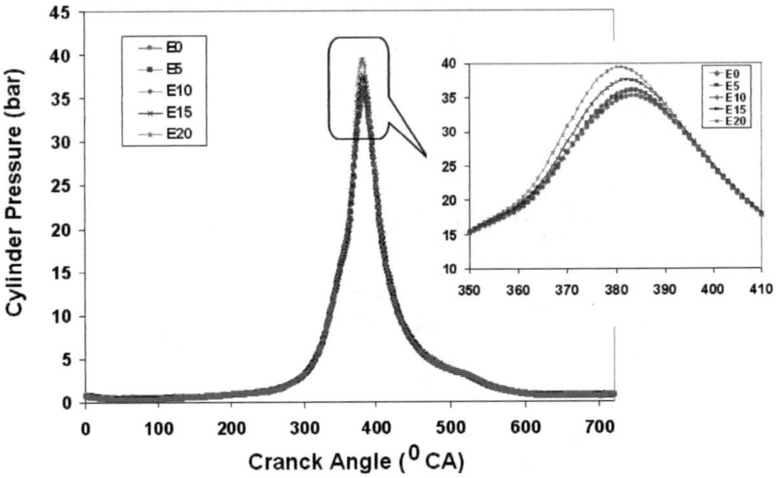

Figure 3. Cylinder pressure for Bioethanol, gasoline blended fuels.

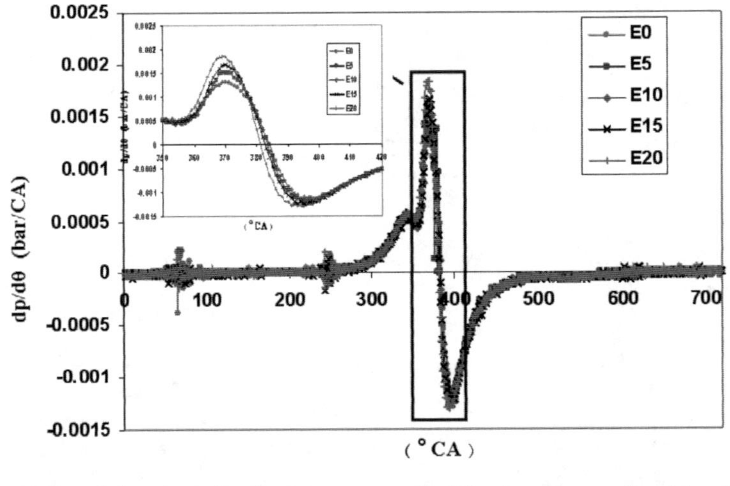

Figure 4. First derivation of cylinder pressure for Bioethanol, gasoline blended fuels.

For identification of the phases of combustion viz. premixed, mixing controlled and tail end of combustion, their duration and the proportion of the heat release, it was necessary to estimate the rate of heat release and the total heat release. Though the start of combustion could be identified from other criteria like $\dot{p}_c(t)$ and $\ddot{p}_c(t)$, and the end of the mixing controlled phase of combustion (ECC) by $\dot{p}_c(t)$ but the end of premixed phase (ERC) and the end of combustion (EC) could not be identified by any criterion other than the rate of heat-release history. For this purpose an approximate estimate of the rate of heat-release was obtained from the knowledge of measured cylinder pressure history and the cylinder volume change due to piston motion. Disregarding the minor effects of wall heat transfer, blow by and crevice region gas motion, the rate of heat release is given by [2]:

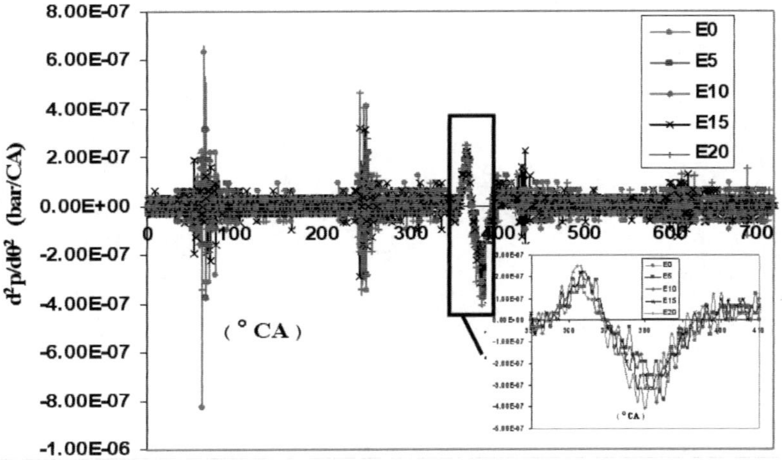

Figure 5. Second Der. Of Cylinder pressure for Bioethanol, gasoline blended fuels.

$$\frac{dQ}{dt} = \frac{\gamma}{\gamma - 1} p_c \frac{dv}{dt} + \frac{1}{\gamma - 1} v \frac{dp_c}{dt} \quad (1)$$

where dQ/dt is the rate of heat release, kJ/s. p_c is the instantaneous cylinder pressure, kPa, v is the instantaneous cylinder volume, m^3, γ is the ratio of specific heat for the air charge ($\cong$ 1.35). The instantaneous cylinder volume v is given by:

$$v = v_c \left[1 + \frac{1}{2}(r_c - 1)\left\{ 1 + R - \cos\sqrt{R^2 - \sin^2\theta} \right\} \right] \quad (2)$$

where,

r_c is the compression ratio,

R is the ratio of the length of the connecting rod to the crank radius,

v_c is the clearance volume $[v_s/(r_c - 1)], m^3$,

v_s is the swept volume, m^3, and

θ is the crank angle referred to compression bdc.

Accordingly, a calculation scheme was developed and typical plots of the rate of heat release are shown in figure 6. Figure 6 show the influence of different bioethanol-gasoline blended fuels on rate of heat release. The increase of bioethanol content increases the rate of heat release rise. Effect of bioethanol blending with gasoline fuel on P-V diagram and output work has been indicated in figures 7 and 8. As shown in this figures, the maximum value of these diagrams increases as the ethanol percentage increases. The maximum amount was detected when 20% ethanol was in the fuel blend. As the E% increases in the fuel blend, the indicated work increases (i.e., the indicated efficiency increases).

Considering the temperatures for burned and unburned gases (figures 9 and 10) show that the temperature is higher when ethanol percentage increases. It shows that as the percentage of ethanol in the blends increased, gas temperatures were increased. When the combustion process is closer to stoichiometric, flame temperature increases, therefore, the gas temperature is increased, due to oxygenated blended fuels and completed combustion [2].

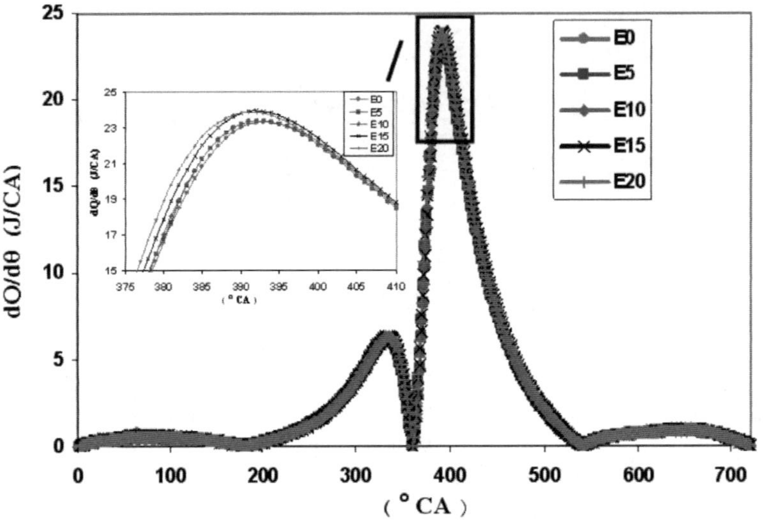

Figure 6. Heat Release for Bioethanol, gasoline blended fuels.

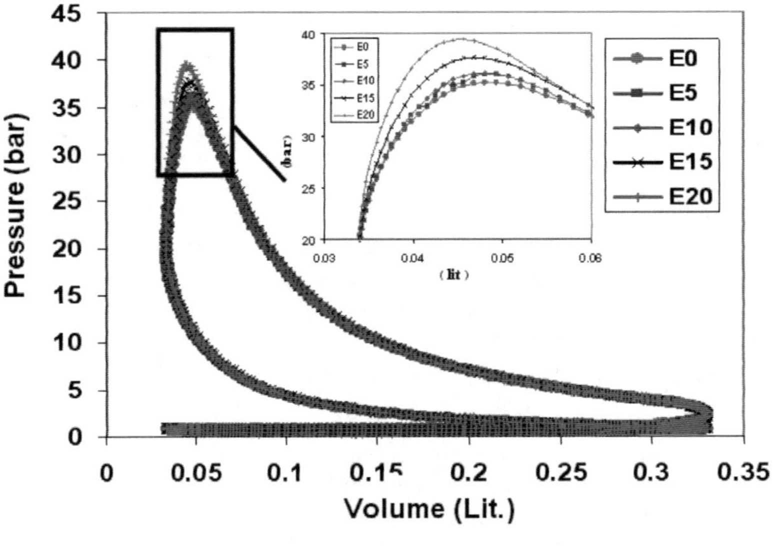

Figure 7. P-V diagram for Bioethanol, gasoline blended fuels.

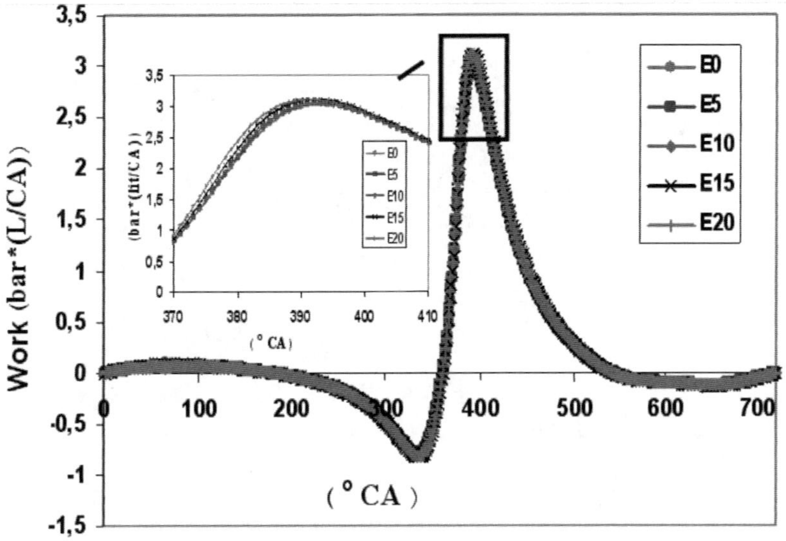

Figure 8. Work for Bioethanol, gasoline blended fuels.

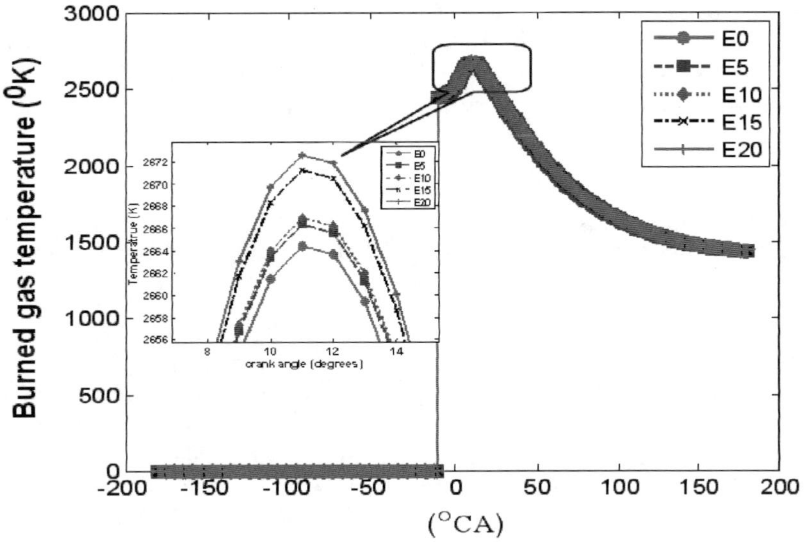

Figure 9. Burned gas temperature for Bioethanol, gasoline blended fuels.

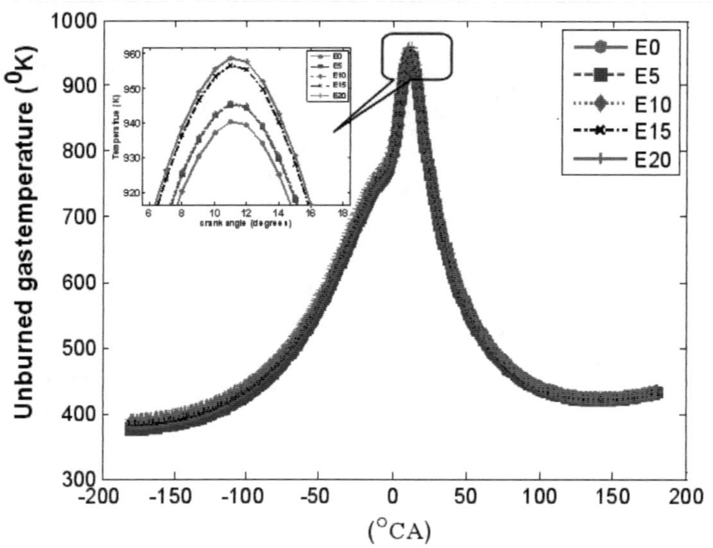

Figure 10. Unburned gases temperature for Bioethanol, gasoline blended fuels.

4. Conclusions

The results revealed that using bioethanol-gasoline blended fuels will increase the cylinder pressure and its increase of bioethanol content increases the rate of cylinder pressure rise and maximum acceleration of cylinder pressure (P(θ), P•(θ) and P••(θ)). Moreover, using bioethanol-gasoline blends will increase the heat release (Q•(θ)) and worked of the cycle. This improvement was due to the high oxygen percentage in the ethanol.

Acknowledgements

The authors wish to thank the Iranian Fuel Conservation Organization (IFCO) of NIOC for the research grant provided to complete this project and Mega Motor Company for providing of laboratory facilities.

References

[1] Selvan V A M, Anand R, Udayakumar M 2009 *J Eng Appl Sci* **4** 1819-6608
[2] Najafi G, Ghobadian B, Tavakoli T, Buttsworth D, Yusaf T, Faizollahnejad M 2009 *Applied Energy* **86** 630-9
[3] Yücesu H S, Sozen A, Topgül T, Arcaklioğlu E 2007 *Applied Thermal Engineering* **27** 358-68
[4] Yücesu H S, Topgül T, Cinar C, Okur M 2006 *Applied Thermal Engineering* **26** 2272-8
[5] Topgül T, Yücesu H S, Cinar C, Koca A 2006 *Renewable energy* **31** 2534-42
[6] Celik M B 2008 *Applied Thermal Engineering* **28** 396-404
[7] Agarwal A K 2007 *Progress in energy and combustion science* **33** 233-71
[8] Al-Hasan M 2003 *Energy Conversion and Management* **44** 1547-61
[9] Bayraktar H 2005 *Renewable Energy* **30** 1733-47
[10] He B-Q, Wang J-X, Hao J-M, Yan X-G, Xiao J-H 2003 *Atmospheric Environment* **37** 949-57
[11] Hsieh W-D, Chen R-H, Wu T-L, Lin T-H 2002 *Atmospheric Environment* **36** 403-10
[12] Wu C-W, Chen R-H, Pu J-Y, Lin T-H 2004 *Atmospheric Environment* **38** 7093-100
[13] Ceviz M, Yüksel F 2005 *Applied Thermal Engineering* **25** 917-25

IOP Conf. Series: Materials Science and Engineering **100** (2015) 012009 doi:10.1088/1757-899X/100/1/012009

Design of single phase inverter using microcontroller assisted by data processing applications software

K Ismail[1*], A Muharam[1], Amin[1] and S Widodo Budi[1]
[1]FResearch Centre for Electrical Power and Mechtronics, Indonesian Institute of Sciences Kampus LIPI Gd.10 Basement, Jl. Cisitu No.21/154D, Bandung, Indonesia.

[*]E-mail: ksansan@yahoo.com

Abstract. Inverter is widely used for industrial, office, and residential purposes. Inverter supports the development of alternative energy such as solar cells, wind turbines and fuel cells by converting dc voltage to ac voltage. Inverter has been made with a variety of hardware and software combinations, such as the use of pure analog circuit and various types of microcontroller as controller. When using pure analog circuit, modification would be difficult because it will change the entire hardware components. In inverter with microcontroller based design (with software), calculations to generate AC modulation is done in the microcontroller. This increases programming complexity and amount of coding downloaded to the microcontroller chip (capacity flash memory in the microcontroller is limited). This paper discusses the design of a single phase inverter using unipolar modulation of sine wave and triangular wave, which is done outside the microcontroller using data processing software application (Microsoft Excel), result shows that complexity programming was reduce and resolution sampling data is very influence to THD. Resolution sampling must taking ½ degree to get best THD (15.8%).

1. Introduction

Inverter converts DC voltage into AC voltage [1]. There are two types of inverters commonly used in electric power system which are inverter with variable voltage variable frequency (VVVF) and inverter with constant voltage constant frequency (CVCF) [2]. VVVF inverters are generally used for specific usage such as 3-phase AC motor control for pumps. Inverters with CVCF are widely used for general purpose that does not require changes in frequency. There will be no discussion on CVCF inverter in this paper.

There are several methods for inverter switching including square wave signal method, multiple pulse width modulation (PWM), unipolar sine pulse width modulation and bipolar sine pulse width modulation [3]. This paper discusses the design of a single phase inverter using unipolar sine pulse width modulation. The advantages of unipolar switching is lower output harmonics [4].

Many variations of hardware, software and algorithms have been widely used for the inverter, such as the use of multi vibrator circuit using IC CD4047 as a purely electronic control without software [5]. There could be generation of AC wave modulation done in the microcontroller [6, 7], but it may increase the complexity of programming and number of coding. Basically, generating unipolar sine PWM can be performed by comparing two signals/waves namely sine wave as fundamental signal and triangular wave as carrier signal [8]. The result is a noncontinuous square wave PWM, so that the frequency of the sine wave is equal to the frequency used by the load (usually about 50 Hz and 60 Hz [9].

2. Method

2.1. Unipolar Modulation with Data Processing Software

Software used to generate unipolar modulation is Microsoft Excel, which meets all requirements to generate the computation of triangular and sine waves. Sine wave was generated using the equation (1):

$$\text{Sine wave} = \sin(A_1 * PI()/180) \tag{1}$$

where:

sin : sine
A1 : number of columns and rows (0 - 360)
PI()/180 : convert radians to degrees [10]

The wavelength of the triangular wave must be determined first to generate a triangular wave. Determining pulse width of triangular wave is the same as determining the carrier frequency. Modulation ratio is the ratio of the fundamental frequency (sine wave) with the carrier frequency (triangular wave) [11]. Optimum modulation ratio is 22.5 [12]. Triangular wavelength (λ) calculation was obtained using equation (2):

$$\lambda\,\text{Triangular} = \frac{\lambda\,\text{sine}}{modulation\ ratio} \tag{2}$$

$$= \frac{360}{22.5} = 16°$$

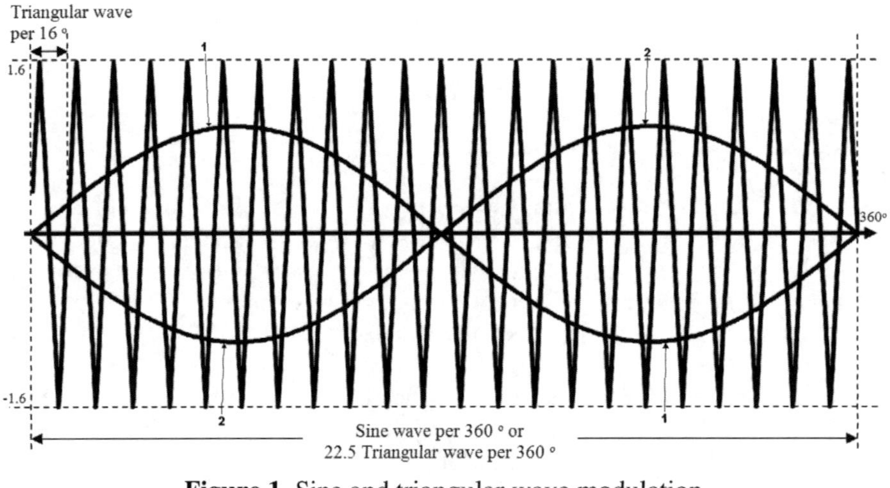

Figure 1. Sine and triangular wave modulation.

From equation (2), it was known that λ of triangular wave is 16 degrees (figure 1) which means in range of 16 degrees there are positive half wave and negative half wave. The wavelength of each half wave is 8 degrees, hence for each 8 degrees there would be half to rise and fall. Wavelength to peak is 4 degrees, as shown in figure 2.

Modulation index is ratio between the amplitudes of triangular and sine waves with optimum modulation index of 0.6 [11, 13]. Resolution per degree for the triangular using equation (3):

$$\text{resolution per degree} = \frac{1}{\textit{modulation index} \times \lambda \textit{ to peak}} \quad (3)$$

$$= \frac{1}{0.6 \times 4}$$

$$= 0.416 \approx 0.4$$

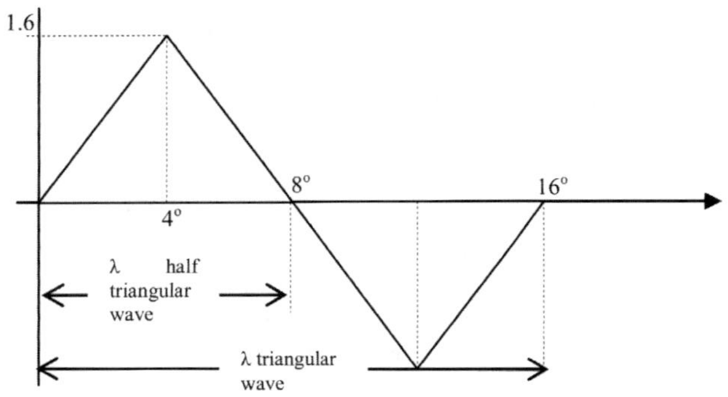

Figure 2. Creating triangular wave on Microsoft Excel.

Resolution per degree of triangular wave was 0.4, meaning each degree of triangular wave is changed by 0.4 units as shown in figure 6. Both sine and triangular waves were compared (with the subtraction) and transformed into zeros and ones depending on the result of the comparison. To transform the result of modulated sine wave with a triangular wave into a digital form can use flowchart as shown in figure 3.

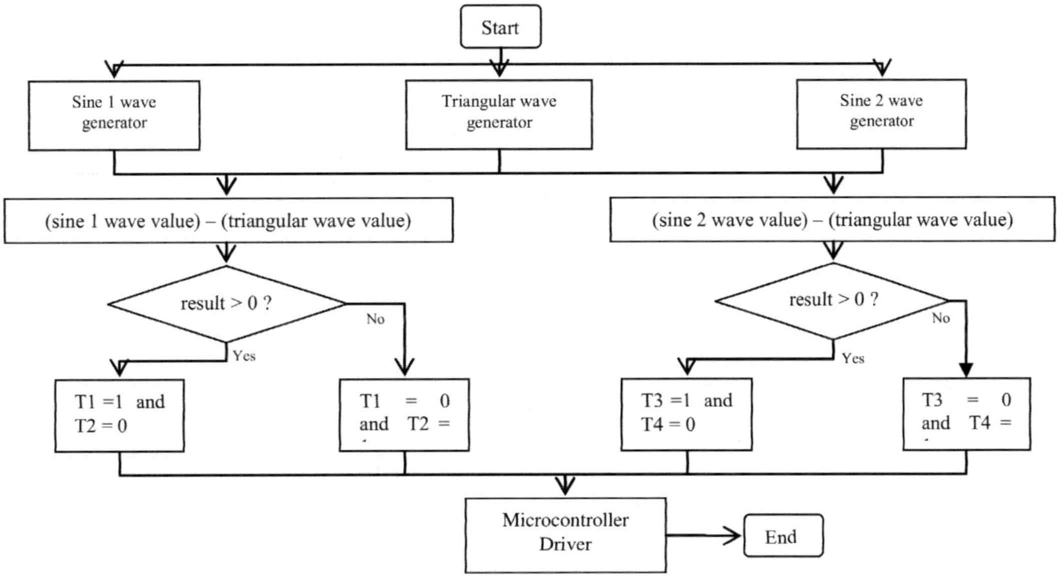

Figure 3. Flowchart modulating sine wave with a triangular wave into a digital.

In this paper we used single phase full bridge inverter (figure 4) which can be viewed as two half bridge inverters. Single phase full bridge inverter uses 4 units power electronic components (T1, T2, T3, and T4) so that binary have 16 possible different combinations of active to cut off conditions [14]. Only two combinations produce an AC voltage at load (T1, T4 = active T2, T3 = cut off, and T1, T4 = cut off T2, T3 = active). Other combinations will result in short circuit, freewheeling or does not produce AC voltage (see table 1). The column is numbered based on binary, "1" indicated the power electronic components will be in a state of active, while "0" indicated power electronic components in a "cut off". Thus power electronic components T1 with T2 and T3 with T4 were always in opposite number. To get opposite binary column can used the equation "= IF (D2> 0,01)" written in Microsoft Excel. Equation functioned to give number "1" when column condition was colorless or less than zero and to give number "0" when column condition is colored or greater than zero.

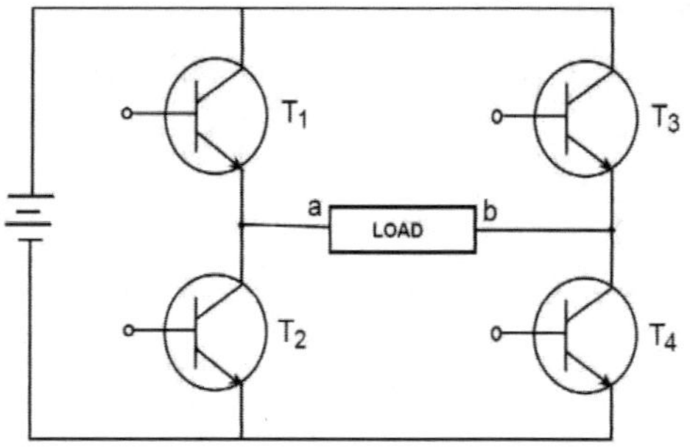

Figure 4. Full bridge inverter circuit.

2.2. Programming Look Up Table at Microcontroller

In this paper, trigger for power electronic components was gained from microcontrollers MCS 51 (base 8051 Intel processor) from ATMEL corp. This microcontroller has features for indexed data store (such as tables). Indexed addressing mode is used to access data elements (look-up table entries) located at the ROM in 8051 [15]. Instructions used for was "MOVC A, @ A + DPTR. DPTR" was used to access the ROM. Look-up table was used to access elements in the table for a fast operation. Microcontroller works as infinitive loops. During power supply applied microcontroller will continue reading the data on the table from the beginning until the end repeatedly. Microcontroller reads/ processes the program with binary code, however writes data/programs that are downloaded with hexadecimal code [15]. Therefore, the combination of triggers for the power electronic components needed to be converted into hexadecimal. According to figure 3, the conversion can be seen in table 1.

Table 1. Binary sequence to trigger power electronic components

T4	T3	T2	T1	Hex	Annotation
0	0	0	0	0	Not used
0	0	0	1	1	Not used
0	0	1	0	2	Not used
0	0	1	1	3	Restricted
0	1	0	0	4	Not used
0	1	0	1	5	Freewheeling high side
0	1	1	0	6	Negative cycle
0	1	1	1	7	Restricted
1	0	0	0	8	Not used
1	0	0	1	9	Positive cycle
1	0	1	0	A	Freewheeling low side
1	0	1	1	B	Restricted
1	1	0	0	C	Restricted
1	1	0	1	D	Restricted
1	1	1	0	E	Restricted
1	1	1	1	F	Extremely restricted

From table 1 it can be concluded that besides generating AC voltage at the load a-b (figure 4) in value hex 6 and hex 9. While some values of hex did not result short circuit this hex would not

IOP Conf. Series: Materials Science and Engineering **100** (2015) 012009 doi:10.1088/1757-899X/100/1/012009

produce AC voltage at the load a-b (value 5 hex A hex as freewheeling). Whereas values hex 3, hex 7, hex B, hex C, hex D and hex E had to be restricted to prevent short circuit on one side. Value hex F was extremely restricted because it enables short circuit on both sides. Only three data were entered into look-up table which are hex 3, hex 9 and hex 0, the number of data equivalent to the number of data in Microsoft Excel tables. The same data in series indicated pulse width, the more data to be found same in the series, the wider the pulse is.

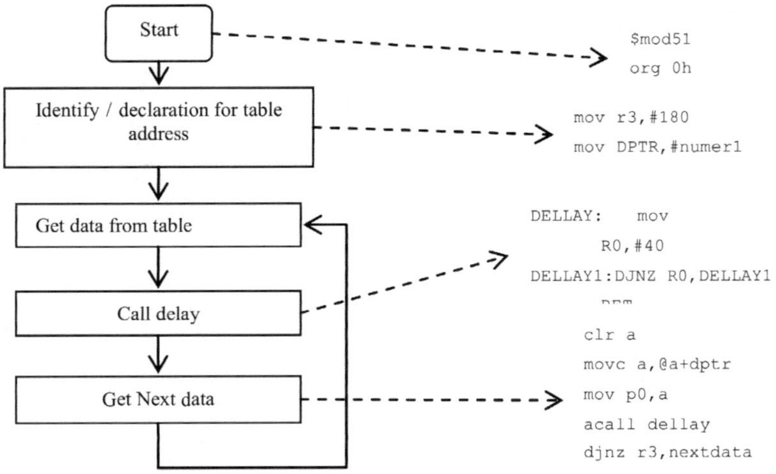

Figure 5. Assembler programming flow chart.

In figure 5, "$ mod51" was used to initialize MCS51 family of microcontrollers. "org 0h" meant the program would begin at address hex 0. "Mov r3, # 180" was used to enter the number 180 into register r3 in an attempt to determine the number of data in a single table. MCS51 was only able to accommodate 255 data, whereas the design requires 360 data. Therefore it took two tables (each consisted of 180 data with different names). "DELLAY: mov R0, # 40" was used to create a delay time between data collections. This delay time could determine the final outcome frequency in which the smaller the gap the larger the frequency, vice versa. "DJNZ R0, DELLAY1" was for reducing data in registers until the value of the data was 0. "clr a" aimed to clear data in the accumulator since the accumulator would be filled with new data. The process of moving data from the table into the accumulator was done using the command "MOVC A, @ A + DPTR". Finally, to read the next data in the table the given command was: "inc DPTR" which reduced the addressing DPTR.

3. Results and discussion

Data processing stated in assembler programs is shown in figure 6. It implied that the data in columns was arranged in rows and indexed in the look-up table. Program that contained the data was converted into hex using MCS51 compiler then downloaded to memory in microcontrollers (flash Memory). This paper compare three resolution sample of degrees. The final waveform output of the modulation was presented in figure 7. Figure 7(a) shows the waveform output across the high side and the low side, figure 7(b) displays THD (70.0 %) at 2 degree resolution sampling, figure 7(c) exhibits THD (47.0 %) at 1 degree resolution sampling, and figure 7(d) shows THD (15.8 %) at ½ degree resolution sampling.

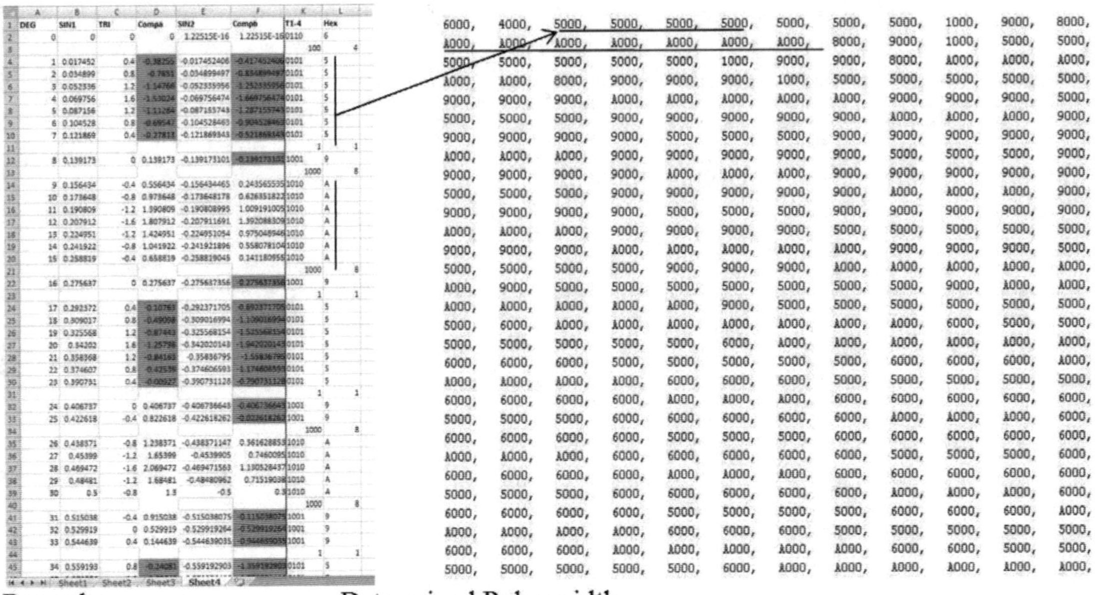

Figure 6. Processing of data as outlined in the assembler program.

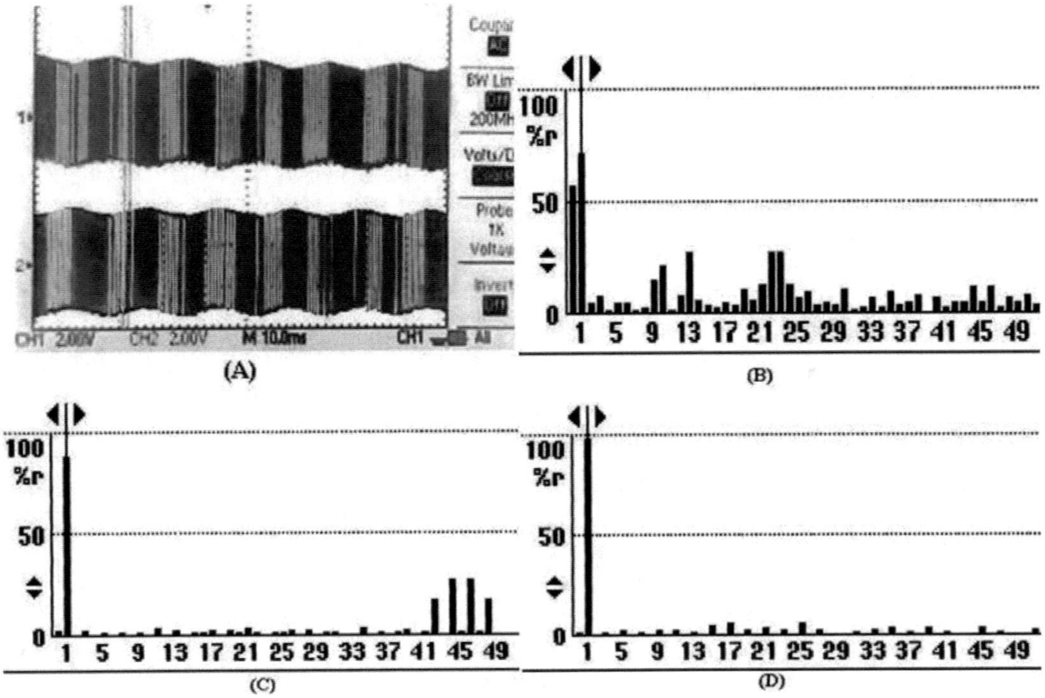

Figure 7. Modulating waveform and THD at the output hardware.

Calculation for the frequency as explained by Hennessy and Patterson [16] in paper written by Bhandarkar and Clark [17] is:

$$\frac{Time}{Program} = \frac{Time}{Cycle}\ x\ \frac{Cycle}{Instruction}\ x\ \frac{Instrution}{Program} \quad (4)$$

where:

$\frac{Time}{Program}$: time for one wave (second)

$\frac{Time}{Cycle}$: depend on frequency of crystal (12 MHz) [15].

$\frac{Cycle}{Instruction}$: depend on Architecture (CISC = 12) [4].

$\frac{Instrution}{Program}$: number label of code (3), number data lookup table (405), and delay (16)

$$\frac{Time}{Program} = \frac{1}{12.000.000}\ x\ 12\ x\ \frac{405\ x\ 3\ x\ 16}{1} = 0{,}01944\ s$$

so the frequency is: 1/0.01944 = 51.44 Hz.
The measurement with oscilloscope (figure 7) and manual calculation show the same result.

4. Conclusion

In this paper the single phase inverter design with unipolar method is proposed. Inverter single phase unipolar method could be obtained by modulating the ratio of sine and triangular waves, where the fundamental sine wave was 50Hz while carrier triangular waves were as much as 22.5 times. Both sine wave and triangular wave generations were done using Microsoft Excel, because the software has features that can accommodate the required waveform processing. The data generated in Microsoft excel was transferred to look up tables using assembler programs. AC voltage frequency could be adjusted using delay subroutine in assembler programs. THD at 2 degree resolution sampling are high but its decrease significantly at ½ degree resolution sampling. Future research for this paper is to secure high side and low side of electrical component with logical calculation.

Acknowledgement

We would like to thank to the publishers for the opportunity to publish the results of this study as well as colleagues who have provided assistance in this study.

References

[1] Saputra A. Rancang Bangun Inverter Satu Phase Sebagai Penggerak Motor Induksi Satu Phase. Surabaya: Institut Teknologi Sepuluh Nopember; 2009.

[2] Tim 2003 *Teknik Dasar Rectifier dan Inverter* (Yogyakarta: Fakultas Teknik Universitas Negeri Yogyakarta)

[3] Setiawan B. Pembangkitan Optimum Pulse Width Modulation (PWM) untuk Inverter Satu Fasa. Surabaya: ITS; 2003.

[4] Bose B K. Academic Press is an imprint of Elsevier, USA. 2006.

[5] Nugraha H. Rangkaian Inverter Satu Fasa dengan Frekuensi 50 Hertz. Depok: Universitas Gunadharma; 2010.

[6] Rusdiyanto A, Susanto B 2008 *INKOM, Jurnal Informatika Sistem Kendali dan Komputer* **2** 75-81

[7] Muhida R, Zaidi A F A, Tamsir A, Irawan R 2012 *Mechatronics, Electrical Power, and Vehicular Technology (MEV)* **3** 95-102

[8] Arisandy M. Analisa Penggunaan PWM Inverter Satu Fasa Unipolar pada Motor Induksi Satu Fasa Menggunakan Software Matlab di Laboratorium Konversi Energy Elektrik. Malang: ITN; 2008.

[9] NECA. Reliability Panel. Frequency Operating Standards Determination: National Electricity Code Administrator Limited; 2001.

[10] Team. Bekerja dengan Rumus dalam Excel Yogyakarta: Lab Komputer Jurusan Pendidikan Teknik Mesin; [cited 2013 February 12th, 2013]. Available from: http://www.staff.uny.ac.id/sites/default/files/Excel%201%20-%20TIK.pdf.

[11] Ismail B, Taib S, Isa M, Daut I, Saad A R M, Fauzy F, Microcontroller Implementation of Single Phase Inverter Switching Strategies. International Conference on Control Instrumentation and Mechatronics Engineering; 2007; Johor Bahru, Johor-Malaysia.

[12] Nagy I a, Bartal P, Quasi-Subharmonics in PWM Inverters. 14th International Power Electronics and Motion Control Conference, EPE-PEMC; 2010; Ohrid-Republic of Macedonia: IEEE.

[13] Zope P H, G.Bhangale P, Sonare P, R.Suralkar S 2012 *International Journal of Advanced Research in Electrical, Electronics and Instrumentation Engineering* **1** 230-6

[14] Sugiyono A. Teknik Penghilangan Harmonisa pada Inverter Satu Fasa dengan Mikroprosesor Sebagai Pembangkit Sinyal Kendali. Bandung: Institut Teknologi Bandung; 1988.

[15] Intel. 8x931AA, 8x931HA Universal Serial Bus Peripheral Controller User's Manual. In: Intel, editor. 1997.

[16] Hennessy J L, Patterson D A 1998 *Computer organization and design (2nd ed.): the hardware/software interface* Morgan Kaufmann Publishers Inc.)

[17] Bhandarkar D, Clark D W 1991 *SIGPLAN Not* **26** 310-9

IOP Conf. Series: Materials Science and Engineering **100** (2015) 012010 doi:10.1088/1757-899X/100/1/012010

Performance and energy saving analysis of a refrigerator using hydrocarbon mixture (HC-R134a) as working fluid

M N Mohtar[1], H Nasution[2] and A A Aziz[2]
[1]Deparment of Mechanical Engineering Faculty, Faculty of Mechanical Engineering Universiti Teknologi Malaysia, Skudai, 81310 Johor Bahru, Johor, Malaysia
[2] Automotive Development Centre, Universiti Teknologi Malaysia, Skudai, 81310 Johor Bahru, Johor, Malaysia

E-mail: henry@fkm.utm.my

Abstract. The use of hydrocarbon mixture as a working fluid in a refrigerator system is rarely explored. Almost all domestic refrigerators use hydroflourocarbon R134a (HFC-R134a) as refrigerants. In this study, hydrocarbon gas (HC-R134a) is used as the alternative refrigerant to replace HFC-R134a. It has a composition of R290 (56%), R600a (54.39%) and additive (0.1%wt) blended for the trials. The experiments were conducted with 105 g and 52.5 g refrigerant mass charge, subjected to internal heat load of 0, 1, 2, 3 and 4 kg respectively. The study investigates the coefficient of performance of the refrigerator (COPR) and energy consumption. The results show that the use of HC-R134a as the replaceable refrigerant can save energy ranging from 2.04% to 7.09%, as compared to the conventional HFC-R134a refrigerant. Naturally, the COPR improvement and temperature distribution using HC-R134a are much better than HFC-R134a

1. Introduction

The refrigerant releasing to the environment will eventually have an impact on the ozone depletion and indirectly enhances global warming. Montreal Protocol [1] was formed in 1987 with the purpose to phase-out the use of CFCs and the formulation of new environmentally-friendly. The importance of the stratospheric ozone layer is to absorb ultraviolet radiation produced by the sun and to protect human and other living creatures from this exposure [2]. Kyoto Protocol [3] comes into existence in 1997 to gather efforts to curb greenhouse warming. One of the objectives of this protocol is the requirement to phase out all substances that promote greenhouse warming such as HFCs. Over several decades there were global consensus and effort expedited to promote awareness on this global warming issue and to strategize ways to formulate solutions.

Hydrocarbon mixture is the combination of two or more hydrocarbon refrigerants by blending. Hydrocarbon mixture can be categorized in two i.e. azeotrope and zeotrope. Azeotrope can be defined as a stable mixture of refrigerants and its vapour and liquid phase retain identical compositions over a wide-range temperature [4]. Basically, azeotropic refrigerant mixtures blended from two or more refrigerants with similar boiling points respectively. Zeotrope or non azeotrope phase change process is non-isothermal, and the composition does not remain constant due to temperature lead and composition shift [5]. Non azeotrope refrigerant mixtures (NARMs) consist of two or more refrigerants with different boiling points for each refrigerant.

A sample of the hydrocarbon mixture was provided for to the researchers by a local chemical company for trial. The hydrocarbon mixture is an alternative drop in for HFC-R134a and R12 with a code name OS-12a and is also referred as HC-R134a [6]. The mixture is a non-zoetrope refrigerant mixture (NARMs), since R290 present in the mixture is of lower boiling point, the other constituent, iso-butane (R600a), is of higher boiling point. In addition, R290 is having higher pressure and smaller suction vapour volume, whereas R600a is of lower pressure and high suction vapour volume [7]. The percentage of hydrocarbon mixture is 56% of Propane, 54.39% of isobutane and 0.1%wt of additive. The primary objective of this study is to investigate the performance of a refrigerator on an instrumented test rig as well as energy consumption using the prescribed hydrocarbon as working fluid. A refrigerator for trial has a volume capacity of 153 litres.

2. Experimental setup

2.1. Test rig design

Figure 1 shows the positions of the power meter, flow meter and pressure gauge setup. The refrigerant will flow from the condenser at high pressure and temperature. The flow meter is installed between the condenser and a drier to measure the mass flow rate of refrigerant which also circulate in whole of the refrigeration system. The refrigerant will pass through the evaporator to the compressor and will be at low pressure and temperature. The pressure line in the evaporator pressure is located between the evaporator and compressor tubing and is installed the low-pressure gauge at that region. As obvious, the refrigerant will be at high pressure and temperature between compressor and condenser. The components installed are fitted with pressure gauges at the respective locations. Power meter is installed just before the refrigerator for input power measurement. Figure 2 shows the eight points in which the temperatures are monitored in the refrigeration circuit.

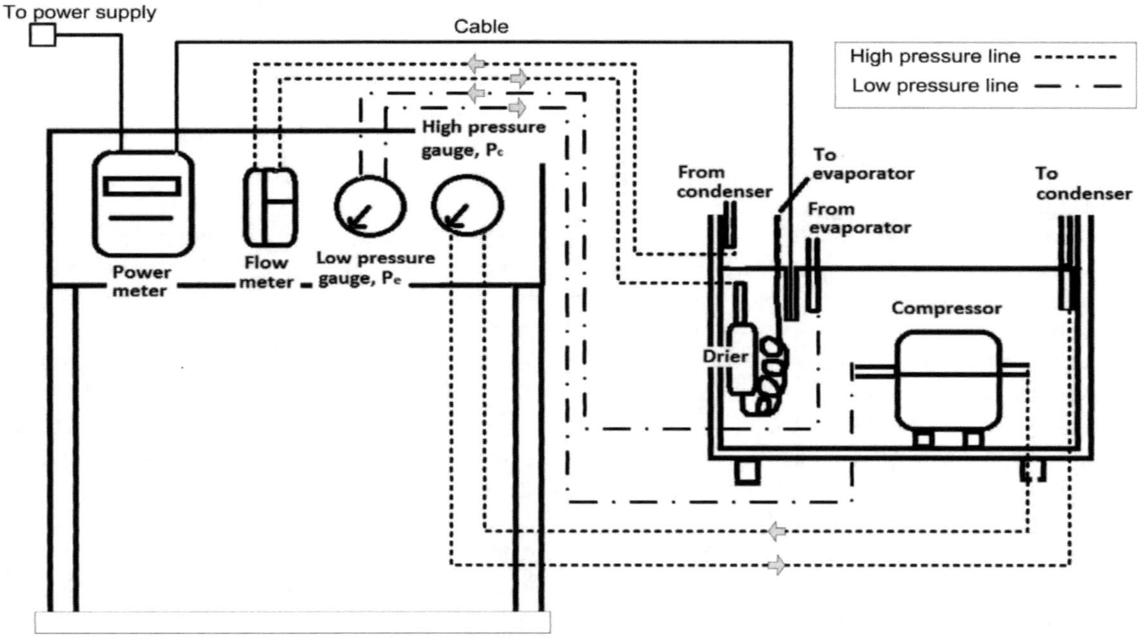

Figure 1. Monitoring panels fitted onto the experimental set-up.

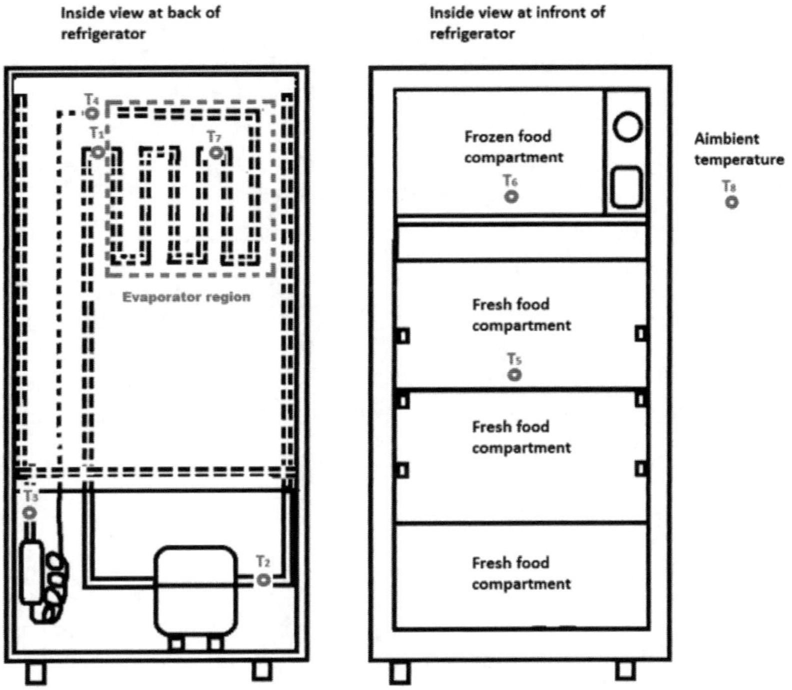

Figure 2. Temperature probes fitted onto the refrigerator during trials.

2.2. Mass charging of the alternative refrigerant

The information gathered from the supplier, indicated that the mass charge of HC-R134a (OS-12) is only half of that of HFC-R134a. Although, the mass charge is half, its pressure is almost similar to that produced by HFC-R134a when subjected to similar working conditions. Thus for the experiment undertaken the quantity of the mass charge for HFC-R124a is 105g meanwhile mass charging for HC-R134a is 52.5g. Figure 3 illustrates the technique of charging the refrigerants into the refrigeration circuit.

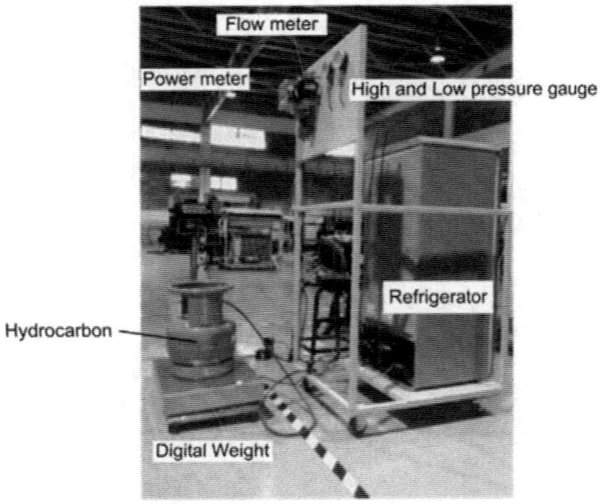

Figure 3. Refrigerant being charged into the refrigerator.

2.3. Experiment procedures

Table 1 is the manipulated variables for the COPR and energy consumption meanwhile the constant variables for these experiments are thermostat setting, mass of refrigerant and duration of experiment. The load being used in this study is water. For the thermostat, it should be set at 3 for both experiments

Table 1. Experimental parameters.

Parameter	Range of variation
Type of refrigerant	HFC-R134a with mass charging 105.g HC-R134a with mass charging 52.5 g
Mass of water as load for COP_R experiment	0 kg, 1 kg, 2 kg
Mass of water for energy consumption experiment	0 kg, 1 kg 2 kg, 3 kg, 4 kg

3. Performance and energy analysis

3.1. Coefficient of performance for refrigerator

The efficiency of a refrigerator is expressed in terms of its coefficient of performance, denoted by COP_R [8].Equation (1) is COP_R equation for a typical refrigerator.

$$COP_R = \frac{h_1 - h_4}{h_2 - h_1} = \frac{Q_L}{W_{net,in}} \tag{1}$$

where ;

Q_L is latent heat or cooling capacity (kJ/kg),
$W_{net,in}$ is compressor work net (kJ/kg), h_1 is enthalpy at after expansion valve (kJ/kg)
h_2 is enthalpy at after compressor (kJ/kg) and
h_4 is enthalpy at after condenser (kJ/kg). The higher the value to more efficient the refrigerator becomes.

3.2. Energy Consumption

Basically, the energy consumed by refrigerator is measured by the power meter. Eq. 2 is the energy consumption equation used when the refrigerator is tested for 24 hours.

$$Power\ per\ day\ (kWh/day) = Final\ power\ reading - Initial\ power\ reading \tag{2}$$

3.3. Energy Saving

The compressor work basically the current supply consumed and is to operate the refrigerator for cooling the freezer and food compartment. The type of refrigerant will also has an effect on the energy consumed by the compressor as well as the energy saving which can be gained. Eq. 3 is the energy saving formula in calculating the energy saving.

$$\text{Energy saving} = \frac{\text{Energy by HFC} - \text{Energy by HC}}{\text{Energy by HFC}} \times 100\% \tag{3}$$

4. Result and discussion

4.1. Coefficient of performance for refrigerator (COP_R)
Figure 4 shows the trend of COPR for each load which is generally dissimilar, and they do not look linear. At load 0 kg (refer Figure 4a), the COPR of HFC-R134a dramatically dropped from 3.06 to 2.20 and then gradually touches 1.83. Meanwhile the COPR for HC-R134a gradually drop from 2.7 to 1.80. In the freezer food compartment, there is only air present and it has a small heat capacity as there is no load. HFC-R134a at -2oC seems quite high to absorb heat from the freezer and food compartment which is the latent heat, QL. However the compressor work input, Wnet, in is the lowest at beginning of the cooling process. At load 1 kg, the distances between two lines are afar, when evaporator temperature is from -2oC to -14oC. The gap will get closer when the evaporator temperature is from -16oC to -20oC. When the evaporator temperature is at -14oC, the COPR of HFC-R134a dropped dramatically from 2.11 to 1.98. At load 2 kg, the evaporator temperature is from -2oC to -6oC, while the distance between two lines (of these refrigerants) is quite afar where the evaporator temperature is from -8oC to -12oC, i.e. the distance is too close to each other. When evaporator temperature at -14oC, these two lines will start to drift for a steady length until -20oC. However, for all loads tested, it shows that the COPR of HC-R134a is higher than the COPR of HFC-R134a.

4.2. Temperature distribution in the freezer
Figure 5 shows the temperature distribution against time. It shows that when a freezer is in the HC-R134a mode, the temperature distribution is quite similar to that of HFC-R134a for all loads. At 0 kg, HC-R134a has the tendency to generate the lower of the two temperatures which is at -8.88oC compared with HFC-R134a is -8.22oC. At 1 kg, the lowest temperature is at -9.72oC when in HC-R134a mode as compared to HFC-R134a mode, which generate -9.58oC. At 2 kg, the lowest temperature produced by HC-R134 is -5.88oC but the lower of the two is by HFC-R134a, which is -8.12oC. When the compressor influence in the temperature drop of the freezer, obtained that the compressor stop early at minute 47, 86 and 76 when using HC- R134a at 0 kg, 1 kg and 2 kg respectively compared using HFC-R134a. Besides that, HC-R134a recorded higher number of compressor fluctuation (on/off) with 4, 2 and 3 cycles at 0 kg, 1 kg and 2 kg respectively, as compared to HFC-R134a. At load 0 kg, the compressor stopped early i.e. at the 56th minute and yielded two fluctuation of compressor (on/off) which is at 1 kg and 2 kg load settings. The compressor will continue to influence in the freezer temperature until 120th minute, when using HFC-R134a without any cyclic variation of the compressor (on/off).

4.3. Power Consumption by Load Quantity Effect
The energy consumption increased proportionally to the mass flow rate of water, as much of the energy needed by the compressor is rejected as latent heat, QL from the mass flow rate of the refrigerant. Figure 6 shows that on the basis of the load quantity effect, HFC-R134a demonstrate higher energy consumption than HC-R134a for all of the water mass flow rate. Table 2 on the other hand shows the result of energy saving for HC-R134a with varying mass of water in the freezer on a daily basis.

Table 2. Energy saving for HC-R134a refrigerant.

Mass of water (kg)	0	1	2	3	4
Energy saving (%) per day	7.09	3.05	4.44	5.52	2.04

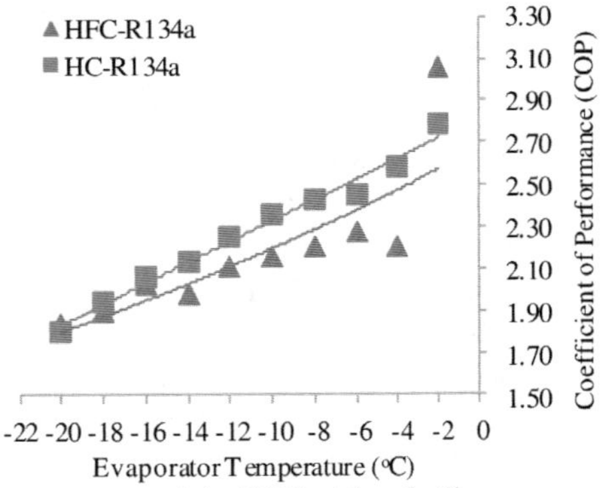

(a) COP at load=0kg

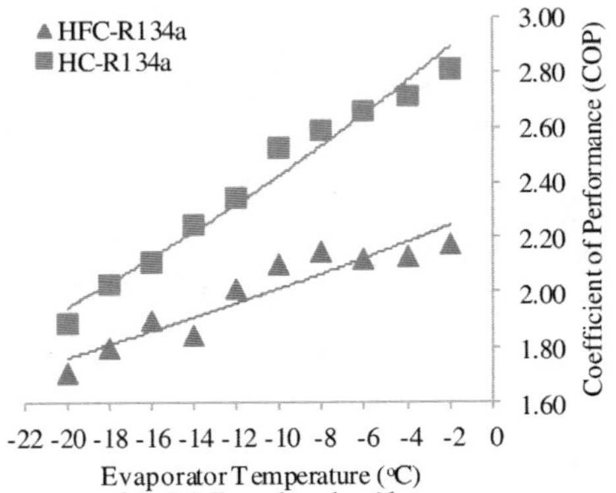

(b) COP at load= 1kg

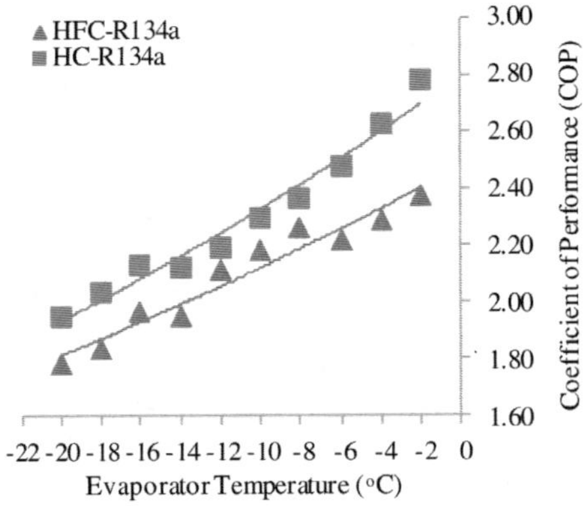

(c) COP at load = 2 kg

Figure 4. COP against evaporator temperature with varying load.

From table 2, the highest energy saving that can be achieved is 5.52 percent per day when the mass of water in the freezer is 3 kg. However, the energy saving can be considered to fluctuate from 2.04 percent to 7.09 percent per day to measure for one-unit refrigerator. In general, using HC-R134a as refrigerant can save energy of a domestic refrigerator ranging from 2.04 percent (at 4 kg of water as load) up to 7.09 percent (zero load) per day while the timing of the compressor is set to on/off mode

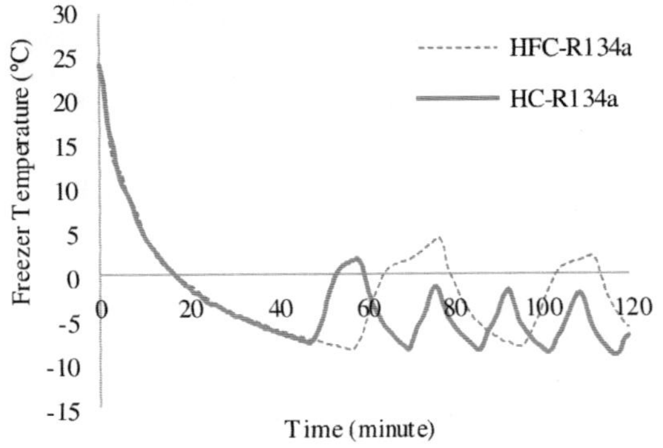

(a) Load = 0 kg

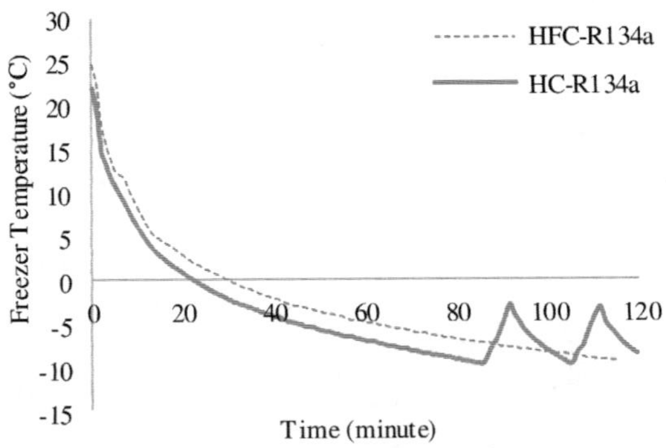

(b) Load = 1 kg

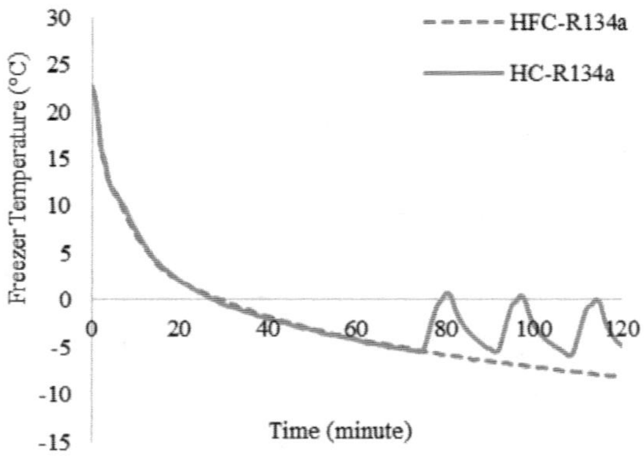

(c) Load = 2 kg

Figure 5. Freezer temperature distribution with varying load.

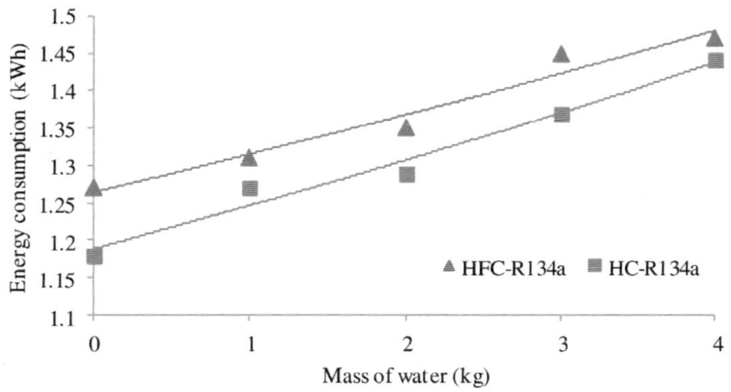

Figure 6. Energy consumption upon load quantity effect.

5. Conclusion

Based on the experimental investigation, the hydrocarbon mixture of HC-R134a has shown to improve COPR when used in a typical domestic refrigerator at ambient condition. The improved COPR for the refrigerator can be interpreted as utilizing minimum energy by the compressor yet yielding maximum latent heat, QL I which the heat is rejected thus lowering the freezer and food compartment temperatures. The hydrocarbon mixture, HC-R134a can be considered as one of the many solutions in overcoming this problem, as it has shown to save energy consumption by as much as 2 to 7 percent, from high to low load. Hydrocarbon mixture of HC-R134a can be an alternative refrigerant to replace the existing HFC-R134a in domestic refrigerator units.

Acknowledgements

The research work is supported partially by the Knowledge Transfer Program (KTP), Vote No. R.J130000.7809.4L509 and the Automotive Development Centre (ADC), Universiti Teknologi Malaysia (UTM). The assistance of the technicians involved, their guidance and assistance are gratefully acknowledged.

IOP Conf. Series: Materials Science and Engineering **100** (2015) 012010 doi:10.1088/1757-899X/100/1/012010

References

[1] Montreal Protocol. Montreal protocol on substances that deplete the ozone layer. 1987.
[2] Wongwises S, Chimres N 2005 *Energy conversion and management* **46** 85-100
[3] Breidenich C, Magraw D, Rowley A, Rubin J W 1998 *American Journal of International Law* 315-31
[4] Khurmi R S, Gupta J K 2005 *A textbook of Refrigeration and Air Conditioning* Eurasia Publishing House)
[5] Rajapaksha L 2007 *Energy conversion and Management* **48** 539-45
[6] Dahlan A A 2012 *Falkulti Kejuruteraan Mekanikal: UTM - Malaysia*
[7] Arora C P 2003 *Refrigeration and Air Conditioning* (New Delhi: Tata McGraw-Hill Publishing Company)
[8] Cengel Y A, Boles M A. Thermodynamics, An Engineering Approach Sixth Edition (SI Units). Mc Graw Hill, Singapore; 2007.

IOP Conf. Series: Materials Science and Engineering **100** (2015) 012011 doi:10.1088/1757-899X/100/1/012011

Mechanical behavior of Kenaf/Epoxy corrugated sandwich structures

S Bakhori[1], M Z Hassan[1], Y Daud[1], S Sarip[1], N Rahman[1], Z Ismail[2], and S A Aziz[1]
[1]UTM Razak School of Engineering and Advanced Technology, Universiti Teknologi Malaysia Kuala Lumpur, Jalan Sultan Yahya Petra,54100 Kuala Lumpur, Malaysia
[2]FPSK, Tower B, Persiaran MPAJ, Jalan Pandan Utama, 55100, Kuala Lumpur

Email:mzaki.kl@utm.my
Phone: +60326154629

Abstract. This study presents the response of kenaf/epoxy corrugated sandwich structure during quasi-static test. Force-displacements curves have been deducted to determine the deformation pattern and collapse behavior of the structure. Kenaf/epoxy sandwich structures skins fabricated by using hand layup technique and the corrugated core were moulded by using steel mould. Different thicknesses of corrugated core web with two sizes of kenaf fibers were used. The corrugated core is then bonded with the skins by using poly-epoxy resin and has been cut into different number of cells. The specimens then tested under tensile and compression at different constant speeds until the specimens fully crushed. Tensile tests data showed the structure can be considered brittle when it breaking point strain, ε less than 0.025. In compression test, the specimens fail due to dominated by stress concentration that initiated by prior cracks. Also, the specimens with more number of cells and thicker core web have higher strength and the ability to absorb higher energy.

1. Introduction

Application of composite sandwich structure in industries is getting expanded. The characteristics of this structure can be considered good performance in specific modulus and specific strength also known as stiffness-to-weight ratio [1, 2] and strength-to-weight ratio respectively [3-5]. Sandwich structure has also shown best properties in shear strength and good ability to absorb energy [4-6]. In various types of applied load condition, bonding between the core and the skins exhibits different mechanical response. This is due to the optimisation of the energy absorption of the sandwich structure [7]. Also, geometrical composition of the core members has important features that influenced the deformation pattern and the strength of the structure [8]. As the structure reached the peak stress in a compression test, the micro-buckling appeared and the failure load of the sandwich structure can be expressed as,

$$\frac{1}{P_{cr}} = \frac{1}{P_E} + \frac{1}{P_S} \tag{1}$$

where, P_{cr} is the critical buckling load, P_E is the Euler buckling load and P_S is the core shear buckling load. A previous study [9] suggested that the corrugated core failure was caused by short strut, subsequently decreased the stability of the core that caused micro-buckling during compression test.

The study also suggested that by increasing the thickness of the core, the stiffness of the structure also increasedthen the deflection had decrease. Another study by Zhang [[10] reported that the sandwich structure capable to absorb high energy also demonstrated good mechanical properties. Another investigation of composite rod reinforced PVC foam panels using easy rules of mixtures also showed a good capability to absorb energy during compression test [7].

Kenaf is a natural material that has potential to replace conventional fibre in composite structure. It offer at competitive cost and easily grow in tropical country. This study aims to investigate quasi-static response, energy absorbing and failure modes of kenaf/epoxy sandwich structures. The overall deformation pattern and collapse behaviour will be characterised by force-displacement data.

2. Experiment Set Up

2.1. Manufacturing process

Kenaf/epoxy sandwich structure consists of skin layer and corrugated core have been prepared using kenaf natural fibre and poly-epoxy resin. The epoxy resin used as binder to bind the skin and core. Both the skin and core were layup by hand. The skins were initially prepared using 300mm × 300 mm flat metal mould. The corrugated core structures were prepared using corrugated mould.The mould consists of uniform corrugation with trapezoidal profile. The mould was made by steel with the dimension of 510 mm × 300 mm using computer-controlled machine (CNC). The top and bottom cavity plate of the mould illustrated as in figure 1.

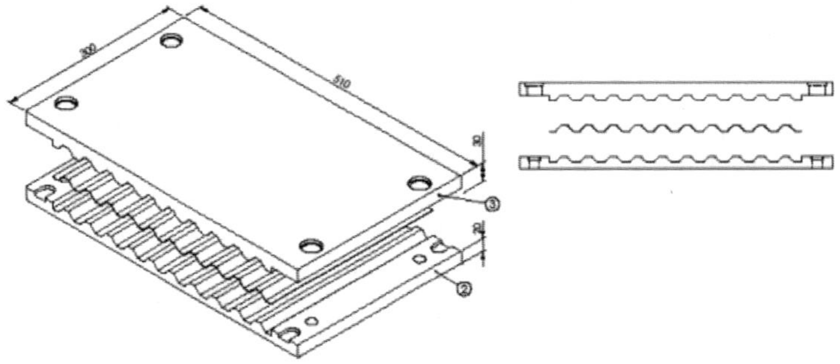

Figure 1. Schematic diagram of top and bottom cavity plate of corrugated steel mould.

Table 1. Specimen properties of kenaf/epoxy sandwich panels.

Material	Specimen ID	No. of unit cell	Kenaf thickness (mm)	Corrugated core thickness (mm)	Skin thickness (mm)	Mass (g)
Kenaf/epoxy	2A, 2B, 2C	2	4	3.5	5	30
	3A, 3B, 3C	3	4	3.5	5	40
	4A, 4B,4C	4	4	3.5	5	51
	4C, 4D, 4E	4	8	6.1	5	59

Both the flat metal sheet and corrugated mould surface were layered by wax before poly-epoxy resins were applied. Then, the kenaf fibre was laid on the surface with poly-epoxy resin. Two different thicknesses of corrugated core, 4 mm and 8 mm kenaf fibre were used. To reduce voids, roller or brush was used to remove air bubbles that trapped within the layers. Then, the top plate and upper metal sheet were placed on the laid kenaf fibre with poly-epoxy resin and being clamped together with the lower plates, and left for 24 hours to cure. Finally, the sandwich structures cut into various sizes as shown in Figure 1 for subsequent tests.

2.2. Experimental procedure

Tensile and compression tests were carried out using Shimadzu AG-X Universal Testing Machine. For tensile test, each material was tested in tension according to ASTM D638 Standards. Three specimens were tested for each different tensile speed (1 mm/min, 150 mm/min and 300 mm/min). For the compression test, the sandwich panels were tested in different number of cells (2 cells; 3 cells; 4 cells) and different thickness of the core web (4mm and 8mm). Cell is the void produced by repetitive trapezoidal corrugated core. In this test, the sandwich panel placed between the compression plates of the testing machine and uniform lateral load applied by giving a constant speed of the machine. The experiment stops when the sandwich panels are fully crushed. Data was recorded in the force-displacement traces through the machine recording system.

3. Results and Discussion

3.1. Tensile strength of kenaf/epoxy structure

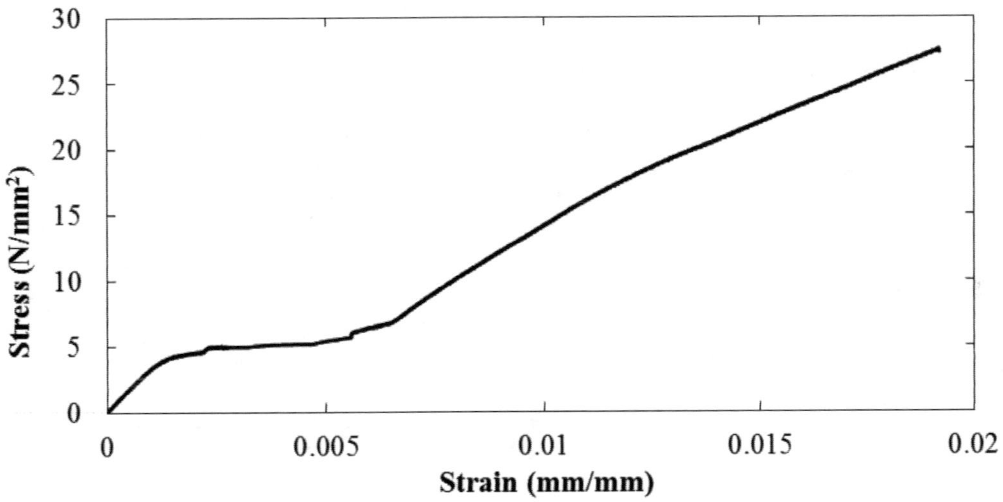

Figure 2. Stress-strain curve for kenaf/epoxy tensile test.

Figure 2 show the stress-strain curve for kenaf/epoxy specimen in tensile test at the speed of 1 mm/min. Based on the graph, the specimen initially shows elastic behaviour and followed by plastic behaviour. In the elastic region, the proportional limit stress, $\sigma_{pl}$is recorded at 2.05 MPa and the proportional limit strain, ϵ_{pl} is 0.0006 mm/mm. The upper yield point, $(\sigma_Y)_u$of 2.88MPa and the lower yield point $(\sigma_Y)_l$of 2.85 MPa also recorded. The strain value where the yield ends is at 0.006 that is 10 times greater than the strain of the proportional limit of 0.0006 MPa. The specimen seem to be hardened until it reaches the ultimate stress, σ_u at 27.58 MPa and the stress total failure, $\sigma_f$at 27.33 MPa. The strain failure value is 0.019 mm/mm which is 31 times greater compared to strain of the proportional limit. During the tensile test, the matrix already brooks at the end of second region but the data keep to continue as the strain increases. This could be explained by strain hardening of the fibre due to pulling out condition, detachment and debonding. Therefore, large amount of overall force needed to break the kenaf fibres.

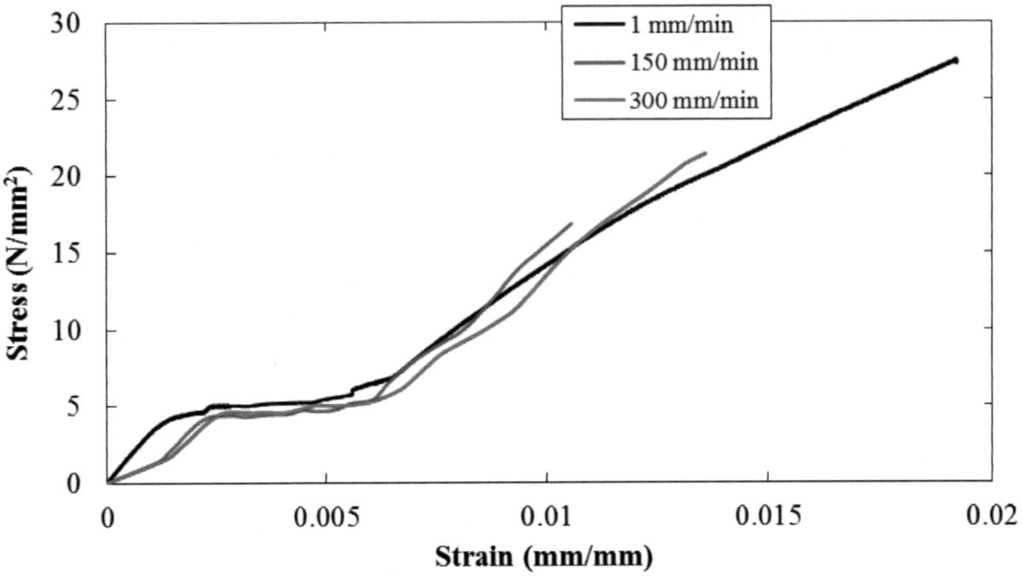

Figure 3. Stress-strain curve for kenaf/epoxy tensile test with different loading rate.

Figure 3 shows the failure modes of the specimens when three different loading rates applied under tensile test. It is observed that specimens undergo uniform deformation and localise necking. A uniform deformation starts when loading rate applied the stress reached a maximum value. Then, the stress is reduced as the cross-sectional area of the specimen reduced. At this point, kenaf natural fibre achieved the maximum force. Elastic modulus for these specimens is 3.50 GPa (for the loading rate of 1 mm/min), 1.85 GPa (for the loading rate of 150 mm/min) and 1.75 GPa (for the loading rate of 300 mm/min). The figure, shows that an increase of loading rate is resulted an increase in elastic modulus of elastic for the kenaf/epoxy laminates. As we can be seen from the figure, all the specimen break at strain that are less than 0.025mm/mm and the stress required to break all specimens are less that 30 N/mm^2. From the observation it could be suggested that the kenaf/epoxy specimens are considerably brittle.

3.2. Quasi-static response ofkenaf/epoxy sandwich structure

The force-displacement relationship of quasi-static compression test of kenaf/epoxy is shown in figure 4. The compression test data have determined three comparable phases; they are elastic deformation, compression shear-coupled progressive buckling and densification [11]. Firstly, the curve shows non-linear trend. This happened because the load is not applied yet and the compression plate does not touch the upper skin of the sandwich structures. Therefore, a large displacement is recorded because the thickness of the sandwich panel is not lateral to each other [4]. Non-linear trend of the stress is kept to increase during the compression until the peak stress achieved which indicates that the specimens are having elastic deformation as shown in Figure 5(a) and stops at the peak load. At this point, the structure is having compressive shear-coupled progressive buckling upon the failure of the corrugated-cell structure as in figure 5(b). This could decrease its stability and the core wall starts to buckle up as in figure 5(c). Subsequently the applied force at this phase has been reduced. At the densification stages, the core is break off and the shape of structure is changed as the load increases as in figure 5(d).

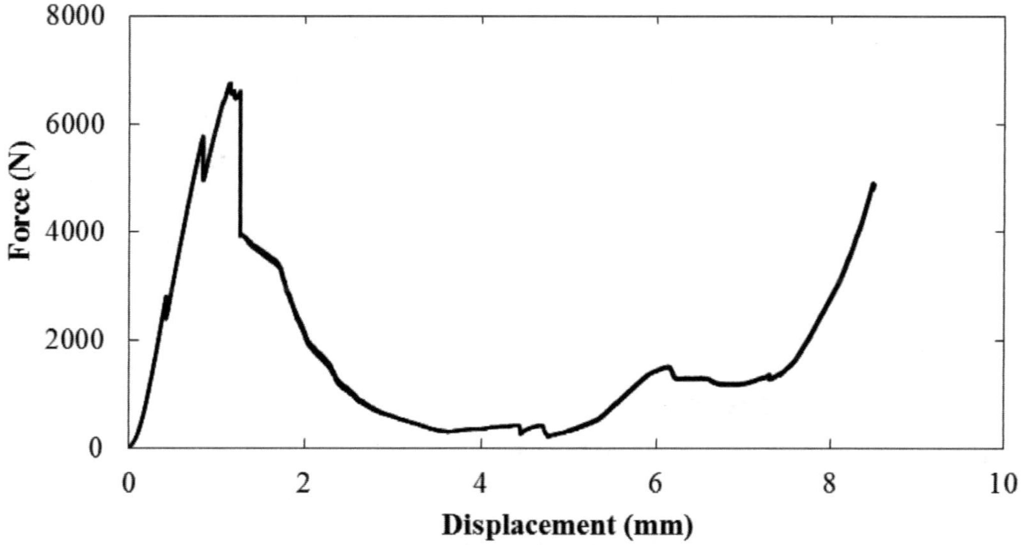

Figure 4. Force-displacement curve for 4mm kenaf/epoxy sandwich structure.

Figure 5. Photograph of kenaf/epoxy sandwich panel structure failure during compression.

The compression tests results of kenaf/epoxy sandwich structure have also been carried out with different number of cells. Force-displacement data have been recorded and plotted in one graph as shown figure 6. This experimental data showed that the specimen that has more number of cells indicate higher peak load compared to other specimen that has less number of cells. The highest peak load of 6.75 kN for four cells specimen, followed by 4.65 kN and 2.9 kN for three cells and two cells specimen. During the compression test, the load slightly decreases before reaching the peak stress. It can be explained by edge debonding between the core and the skins that reduce the contact area as showed in figure 5(b). The major load reduction after the peak stress is due to instability of the core where the structure begins to fracture. At the end, the normal force regain until the compression stopped. At this point, the delamination of the sandwich structure occurred. The failure modes of the specimens were dominated by stress concentration and cracks were initiated.

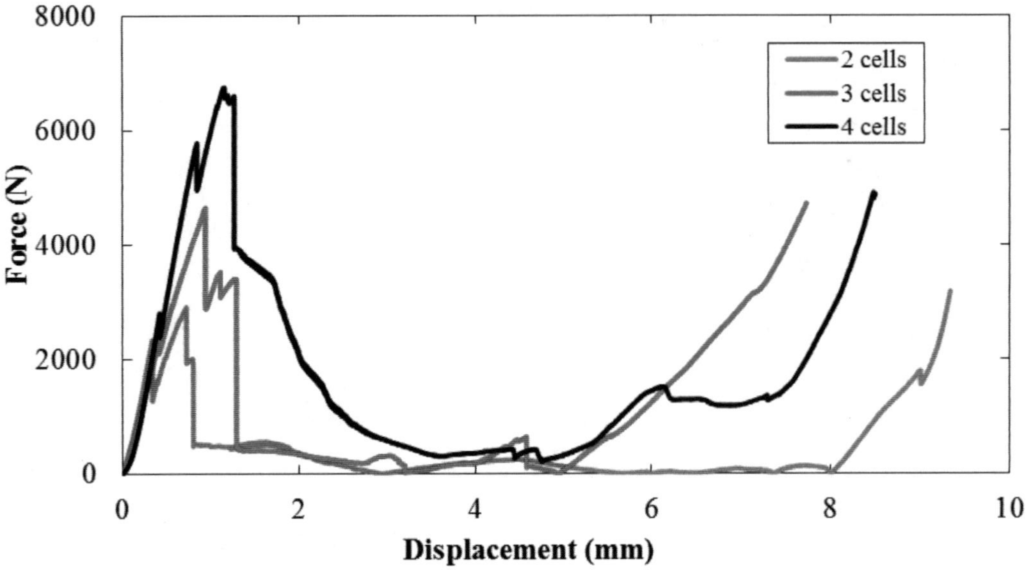

Figure 6. Force-displacement curve of kenaf/epoxy sandwich panel with different number of cells.

Figure 7 shows the comparison of kenaf/epoxy force-displacement curves with different core-web thickness. The peak load for 8 mm and 4 mm core-web sandwich panel are 17.55 kN and 6.76 kN respectively. This means 8mm corrugated core panel needs higher stress to make it lose core stability and to initially buckle [4]. This result is clearly shown that by increasing the thickness of the corrugated core is giving significantly increased the peak strength of the sandwich panel during the quasi-static compression test [11].

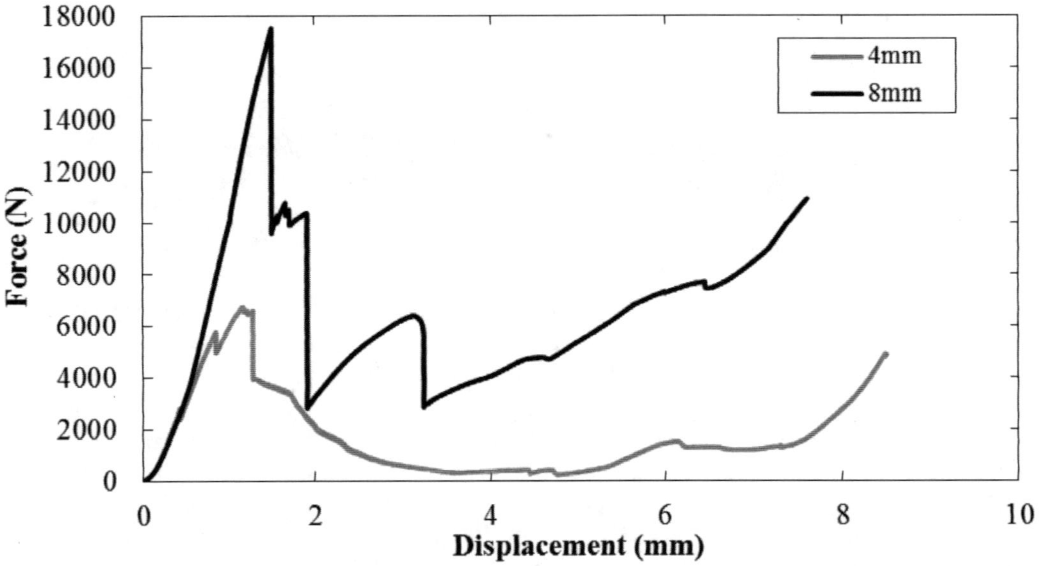

Figure 7. Force-displacement curve of kenaf/epoxy sandwich structure with different core web thickness during compression test.

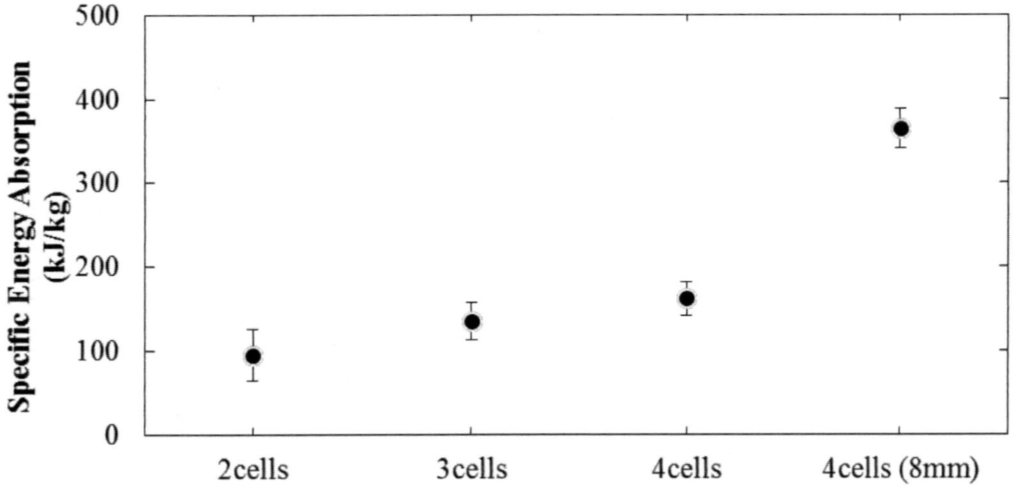

Figure 8. Total energy absorption for corrugated kenaf/epoxy sandwich structures.

Total energy absorption was calculated by measuring area under force-displacement curve. The total energy absorption for this study is shown as in figure 8. Sandwich panel that have 2 cells have total energy absorbed of 94.51 kJ/kg whereas for panel with 3 cells is 134.59 kJ/kg and panel with 4 cells have the total energy absorbed of 161.63 kJ/kg. Kenaf/epoxy sandwich panels with more number of cells have absorbed more energy. The panel that has highest energy absorption, 365.07 kJ/kg is the panel that consists of 4 cells with doubled core thickness of 8 mm. By comparing the panel that have 4 cells with 4mm core thickness, the total energy absorption of this panel has 2.25 times higher. This data suggest that by increasing the core thickness could increase the stiffness and strength of the panel structure.

4. Conclusions

Series of tensile and compression tests have been carried out to investigate the response of kenaf/epoxy corrugated sandwich structure. Follow are the conclusions that can be deducted from the current study.

(a) The tensile test showed that the kenaf/epoxy specimen can be considered as brittle material.
(b) Failure mode of the sandwich panels have been nominated by stress concentration that begins with initial cracks.
(c) Sandwich structure with more number of cells and have thicker corrugated core have the higher strength.
(d) Panels that have more number of cells and have thicker core web showed good ability to absorb more energy.

Acknowledgements

The authors would like to acknowledge Universiti Teknologi Malaysia Kuala Lumpur for providing laboratory facilities and aids under project no. PY/2014/03023.

References

[1] Castanié B, Bouvet C, Aminanda Y, Barrau J-J, Thévenet P 2008 *International Journal of Impact Engineering* **35** 620-34
[2] Bartolozzi G, Baldanzini N, Pierini M 2014 *Composite Structures* **108** 736-46
[3] Yahaya M, Ruan D, Lu G, Dargusch M 2015 *International Journal of Impact Engineering* **75** 100-9

[4] Rejab M, Cantwell W 2013 *Composites Part B: Engineering* **47** 267-77
[5] Kazemahvazi S, Zenkert D 2009 *Composites Science and Technology* **69** 913-9
[6] Jin F, Chen H, Zhao L, Fan H, Cai C, Kuang N 2013 *Composite Structures* **98** 53-8
[7] Zhou J, Hassan M Z, Guan Z, Cantwell W J 2012 *Composites science and Technology* **72** 1781-90
[8] Kazemahvazi S, Tanner D, Zenkert D 2009 *Composites Science and Technology* **69** 920-5
[9] Russell B, Malcom A, Wadley H, Deshpande V 2010 *Journal of Mechanics of Materials and Structures* **5** 477-93
[10] Zhang P, Liu J, Cheng Y, Hou H, Wang C, Li Y 2015 *Materials & Design* **65** 221-30
[11] Hou S, Zhao S, Ren L, Han X, Li Q 2013 *Materials & Design* **51** 1071-84

IOP Conf. Series: Materials Science and Engineering **100** (2015) 012012 doi:10.1088/1757-899X/100/1/012012

Performance and emission parameters of single cylinder diesel engine using castor oil bio-diesel blended fuels

A Rahimi[1], B Ghobadian[1], G Najafi[1], F Jaliliantabar[1] and R Mamat[2]
[1]Faculty of Biosystem Engineering, Tarbiat Modares University, Tehran, Iran
[2]Faculty of Mechanical Engineering, Universiti Malaysia Pahang,Malaysia

E-mail: ghobadib@modares.ac.ir

Abstract. The purpose of this study is to investigate the performance and emission parameters of a CI single cylinder diesel engine operating on biodiesel-diesel blends (B0, B5, B10, B15 and E20: 20% biodiesel and 80% diesel by volume). A reactor was designed, fabricated and evaluated for biodiesel production. The results showed that increasing the biodiesel content in the blend fuel will increase the performance parameters and decrease the emission parameters. Maximum power was detected for B0 at 2650 rpm and maximum torque was belonged to B20 at 1600 rpm. The experimental results revealed that using biodiesel–diesel blended fuels increased the power and torque output of the engine. For biodiesel blends it was found that the specific fuel consumption (sfc) was decreased. B10 had the minimum amount for sfc. The concentration of CO_2 and HC emissions in the exhaust pipe were measured and found to be decreased when biodiesel blends were introduced. This was due to the high oxygen percentage in the biodiesel compared to the net diesel fuel. In contrast, the concentration of CO and NOx was found to be increased when biodiesel is introduced.

1. Introduction

The CI engines are widely utilized due to its reliable operation and economy. As the petroleum reserves are depleting at a faster rate, an urgent need for a renewable alternative fuel arise [1, 2]. Also the threat of global warming and the stringent government regulation made the engine manufacturers and the consumers to follow the emission norms to save the environment from pollution [3]. Among the many alternative fuels, biodiesel is considered as a most desirable fuel extender and fuel additive due to its high oxygen content and renewable in nature [4]. Alternative fuels for diesel engines are becoming increasingly important due to diminishing petroleum reserves and the environmental consequences of exhaust gases from petroleum fuelled engines [5, 6].

A number of studies have shown that triglycerides hold promise as alternative diesel engine fuels. So, many countries are interested in that, for example, evaluation of the production of biodiesel in Europe since 1992 shows an increasing trend [7]. Waste vegetable oil methyl ester is a biodiesel. Biodiesel is defined as the mono alkyl esters of long chain fatty acids derived from renewable lipid sources [8]. Biodiesel, as defined, is widely recognized in the alternative fuel industry. Biodiesel is typically produced through the reaction of a vegetable oil or animal fat with methanol in the presence of a catalyst to yield glycerin and methyl esters [9]. The blend of 75:25 ester/diesel (B25) gave the best performance [10].

Among the attractive features of biodiesel fuel are: (i) it is plant, not petroleum-derived, and as such its combustion does not increase current net atmospheric levels of CO_2 a ''greenhouse'' gas; (ii) it can be domestically produced, offering the possibility of reducing petroleum imports; (iii) it is biodegradable; and (iv) relative to conventional diesel fuel, its combustion products have reduced levels of particulates, carbon monoxide, and, under some conditions, nitrogen oxides.

It is well established that biodiesel affords a substantial reduction in SOx emissions and considerable reductions in CO, hydrocarbons, soot, and particulate matter (PM) [11]. There is a slight increase in NOx emissions, which can be positively influenced by delaying the injection timing in engines [12].

Castor (*Ricinus communis L.*) is one of the most promising non-edible oil crops, due to its high annual seed production and yield, and since it can be grown on marginal land and in semi-arid climate. Currently, edible oils account to more than 95% of the world biodiesel production. This is due to the fact that they are easily available from large scale agricultural production activities; however, this is not a desirable situation due to the potential impacts on food supply leading to the so called "food versus fuel dispute".

In addition to the social aspects, it is known that the cost of the raw-materials can account for 70–95% of total biodiesel production cost; therefore, alternative low cost raw materials are also essential to overcome economic constraints to its production. Amongst the most studied crops are: jatropha, karanja, castor rice bran, linseed, coriander, tobacco seed, cotton, rubber and wild mustard.

The comparison of plantation costs and oil yields of the mentioned crops, place castor clearly on the top of the most interesting non-edible crops. The castor oil seeds usually contain 40–55% of oil, a very high potential as compared to other most commonly used oil crops (soybean: 15–20% (*w*/*w*), sunflower: 25–35% (*w*/*w*), rapeseed: 38–46% (*w*/*w*), and palm: 30–60% (*w*/*w*)). In addition, the cost of plantation can be 50% of the cost of planting rapeseed and 25% of the cost of planting jatropha [13].

2. Experimental set up

2.1. Design of experiment

The performance and emission parameters from the engine running on biodiesel (derived from castor oil) and blended with diesel fuel (B5, B10, B15 and B20) were evaluated and compared with diesel fuel (B0) (figure 1). The properties of biodiesel fuel investigated and it has been proved that all specifications are in the standard range based on the ASTM standard (table 1). Gas Chromatography (GC) machine was utilized to biodiesel product analysis (figure 2). Above 20% biodiesel, engine could not run smoothly, therefore, only experimental results obtained up to this percentage of biodiesel will be presented. The fuel blends were prepared just before starting the experiment to ensure that the fuel mixture was homogenous.

A series of experiments were carried out using diesel, and the various biodiesel blends. All the blends were tested under varying engine speed conditions (1600, 1950, 2300, 2650 and 3000 rpm). The engine was started using diesel fuel and it was operated until it reached the steady state condition. The engine speed, fuel consumption, and load were measured. After the engine reached the stabilized working condition, performance and emission parameters were measured. All experiments have been carried out at full throttle setting. To adjust ignition timing, electronic ignition system was used.

Before obtaining data from the engine operated with a new blended fuel, the engine was operated using the new fuel for sufficient time to clean out the remaining fuel from the previous blend. In this paper, the quantity BX represents a blend consisting of X% biodiesel by volume, e.g., B5 indicates a blend consisting of 5% biodiesel in 95%diesel. Five test fuels were used in this study: 0% biodiesel (B0); 5% biodiesel (B5); 10% biodiesel (B10); 15% biodiesel (B15); and 20% biodiesel (B20). The experiments were performed on one cylinder, four-stroke, and compression-ignition (CI) Diesel engine (specification of engine has been illustrated at table 2).

A SCHENCK eddy–current dynamometer was used in the experiments. Fuel consumption rate was measured by using laminar type flow meter, Pierburg model. Air consumption was measured using an air flow meter. Five separate fuel tanks were fitted to the gasoline engine and this contained diesel and the biodiesel–diesel blends. A schematic diagram and real of the experimental setup is shown in figure 3. An ultrasonic reactor was used for biodiesel production from castor oil (figure 4).

Figure 1. Diesel and biodiesel blended fuels.

Table 1. Specifications of produced castor oil biodiesel and standard levels.

Spec.	Standard method	Levels	Unit
		Biodiesel	
Flash point	EN 14214	178	°C
Kinematic viscosity (40⁰ C)	EN 14214	13.30	mm^2/s
Density *(40⁰ C)*	EN 14214	0.908	g/cm^3

Table 2. Specifications of diesel engine.

Spec.	
Model	3LD 510, Italy
No. of cylinder	1
Bore×Stroke (mm)	85×90
Cyl. Vol. (cm^3)	510
Comp. Ratio	17.5:1
Max. Power (kW)	9
Max. Torque (Nm) @1800 rpm	33

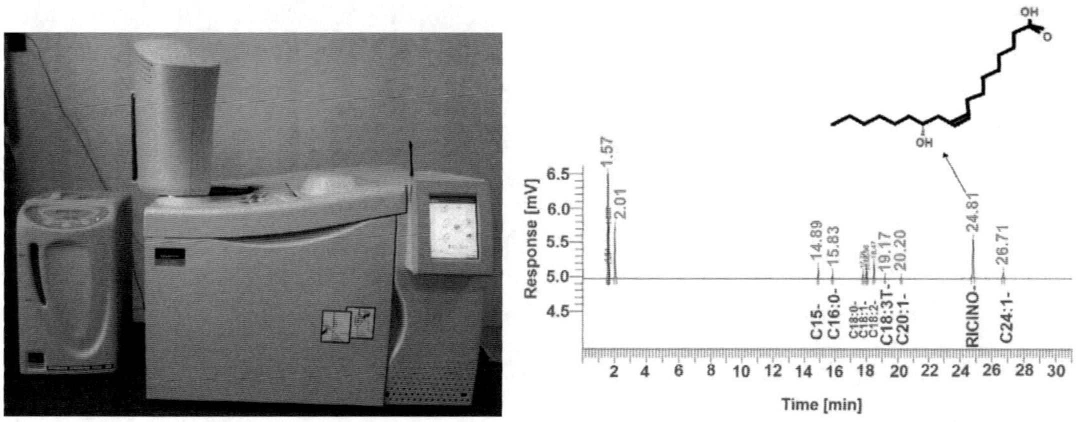

Figure 2. (a) GC Perkin Elmer Clarus 580 and (b) GC diagram for castor biodiesel.

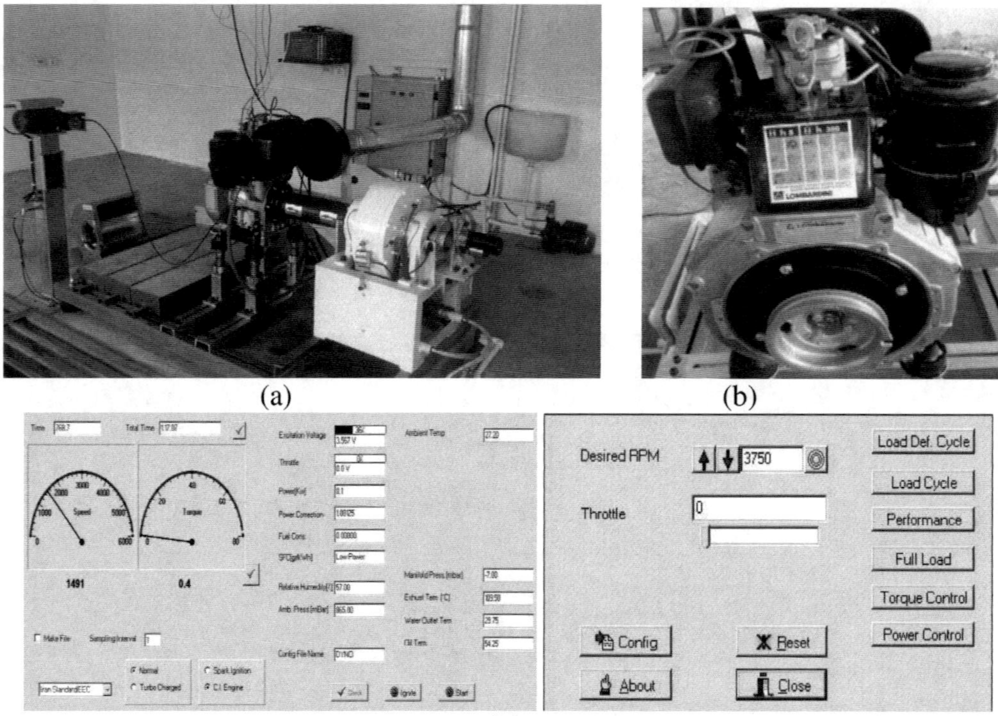

(c)

Figure 3. (a) Engine test set-up, (b) Single cylinder diesel engine and (c) controlling setup

(a) (b) (c) (d) (e)

Figure 4. (a) Castor oil, (b) Ultrasonic machine for transestrification reaction, (c) biodiesel+ glycerol, (d) washing process and (e) biodiesel.

3. Results and discussion

Figure 5 shows the effect of various fuels on engine performance parameters. When the biodiesel content in the blended fuel is increased, the engine power and torque slightly increased for all engine speeds. The gain of the phenomena can be attributed to the increase of the indicated mean effective pressure for higher biodiesel content blends. With the increase in biodiesel percentage, the density of the mixture and the engine volumetric efficiency increases and this causes the increase of power.

The trend of performance curves (power and torque) are very common like those mentioned in valid concerned existing literature. Range of speed was selected between 1600 rpm and 3000 rpm. Engine test results with net diesel fuel showed that maximum torque was 27.2 Nm which occurred at 1600 rpm. The maximum power was 8 kW at 2650 rpm.

Power and torque for fuel blends at full load are shown in Figure 5. Considering power and torque performance with fuel blends, one can say that the trend of these parameters vs. speed is perfectly similar to net diesel fuel. Figure 5 (a) shows engine speed and engine power relationship at full load condition using net diesel fuel and fuel blends. The net diesel fuel was used as a base for comparison. The fuel blend behavior was similar to that of net diesel fuel in developing power. Figure 5 (c) and (d) shows fuel consumption and specific fuel consumption with various fuels blend percentage. The curves show that brake specific fuel consumption of fuel blends trends is very similar to net diesel fuel. Specific fuel consumption of fuel blends is higher than net diesel fuel. In other words, increasing fuel blend percentage, a mild increase in brake specific fuel consumption is observed. The gain of the engine power can be attributed to the increase of the indicated mean effective pressure for higher biodiesel content blends [14] .

The heat of evaporation of biodiesel is higher than that diesel, this provides fuel–air charge cooling and increases the density of the charge, and thus higher power and torque output is obtained [15]. With the increase in biodiesel percentage, the density of the mixture and the engine volumetric efficiency increases and this causes the increase of power. Added biodiesel produces lean mixtures that increase the relative air–fuel ratio to a higher value and makes the burning more efficient. The improved anti-detonation behavior) allowed a more advanced timing that results in higher combustion pressure and thus higher torque [16] .

For biodiesel blends it was found that the specific fuel consumption (sfc) was decreased. B10 had the minimum amount for sfc. The sfc decreases as the biodiesel percentage increases. This is due to the increase in brake thermal efficiency due to increase of indicated work [17]. The concentration of CO_2 and HC emissions in the exhaust pipe were measured and found to be decreased when biodiesel blends were introduced. This was due to the high oxygen percentage in the biodiesel compared to the net diesel fuel. As a result of the lean burning associated with increasing biodiesel percentages, the CO2 emission decreased because of the improved combustion [11].

In contrast, the concentration of CO and NOx was found to be increased when biodiesel is introduced. The CO concentration increases which means the combustion is tuned to be completed [18]. The reduction in CO concentration using blended fuels is due to the fact that biodiesel has higher carbon than diesel. Another significant reason of this increase is that the oxygen content in the blended fuels increases. The oxygen-to-fuel ratio in the fuel-rich regions. Relative air–fuel ratio approaches 1 as the biodiesel content of the blended fuel increases, and consequently combustion be comes complete.

The NOx concentration is higher when biodiesel percentage increases. When the combustion process is closer to stoichiometric, flame temperature increases, therefore, the NOx emission is increased, particularly by the increase of thermal NO. Due to higher performance and lower emissions, it has been recommended to use 10 Volume% biodiesel in diesel fuel in diesel engines [18, 19].

Biodiesel contains oxygen in its structure. When biodiesel is added to diesel fuel, the oxygen content of fuel blend is increased and thus less oxygen is needed for combustion. However, oxygen content of fuel is the main reason for more complete combustion and hence HC emission reduction. Figure 6 show the relationship between the fuel blends and CO and HC concentrations respectively. The concentration of CO_2 and HC emissions in the exhaust pipe were measured and found to be decreased when biodiesel blends were introduced. This was due to the high oxygen percentage in the biodiesel compared to the net diesel fuel. In contrast, the concentration of CO and NOx was found to be increased when biodiesel is introduced.

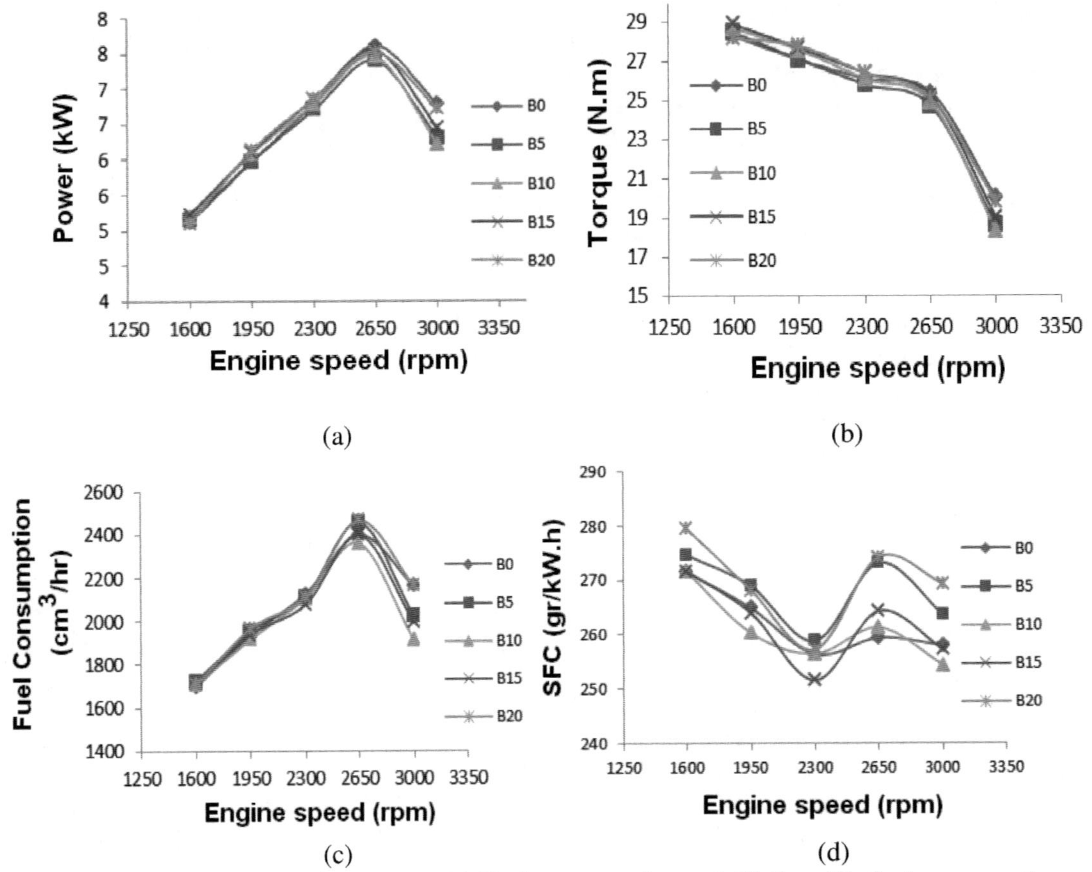

Figure 5. (a) Power, (b) Torque, (c) Fuel consumption and (d) Specific fuel consumption.

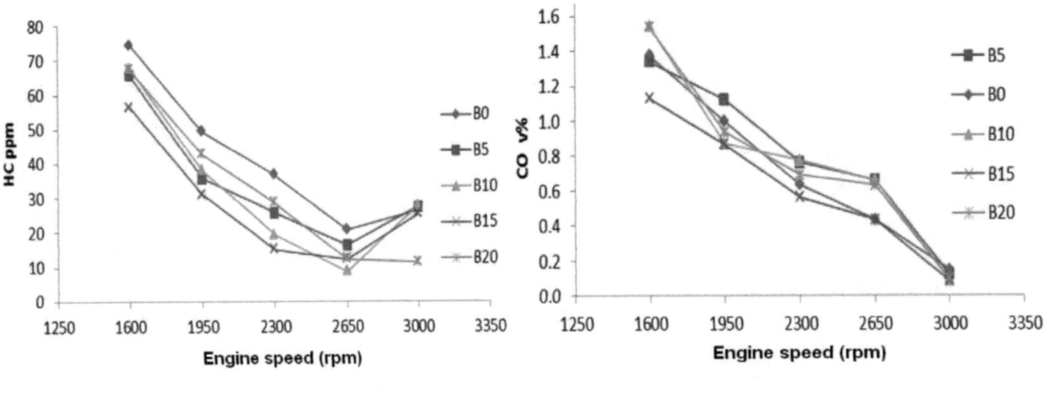

(a) (b)

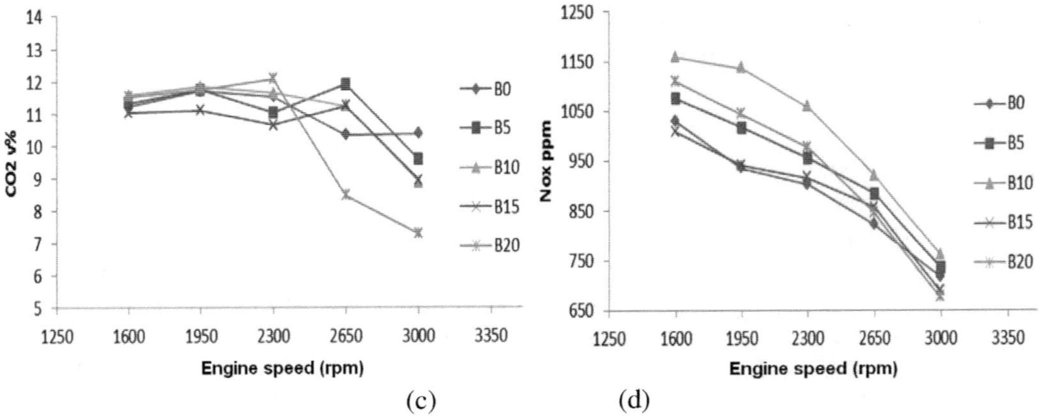

Figure 6. (a) HC, (b) CO, (c) CO_2 and (d) NOx emissions.

4. Conclusions

The results revealed that increasing the biodiesel content in the blend fuel will increase the performance parameters and decrease the emission parameters. Maximum power was detected for B0 at 2650 rpm and maximum torque was belonged to B20 at 1600 rpm. The experimental results revealed that using biodiesel–diesel blended fuels increased the power and torque output of the engine. The gain of the engine power can be attributed to the increase of the indicated mean effective pressure for higher biodiesel content blends. The heat of evaporation of biodiesel is higher than that diesel, this provides fuel–air charge cooling and increases the density of the charge, and thus higher power output is obtained.

For biodiesel blends it was found that the specific fuel consumption (sfc) was decreased. B10 had the minimum amount for sfc. The sfc decreases as the biodiesel percentage increases. This is due to the increase in brake thermal efficiency due to increase of indicated work. The concentration of CO_2 and HC emissions in the exhaust pipe were measured and found to be decreased when biodiesel blends were introduced. This was due to the high oxygen percentage in the biodiesel compared to the net diesel fuel. As a result of the lean burning associated with increasing biodiesel percentages, the CO_2 emission decreased because of the improved combustion.

In contrast, the concentration of CO and NOx was found to be increased when biodiesel is introduced. The CO concentration increases which means the combustion is tuned to be completed. The NOx concentration is higher when biodiesel percentage increases. When the combustion process is closer to stoichiometric, flame temperature increases, therefore, the NOx emission is increased. Due to higher performance and lower emissions, it has been recommended to use 10 Volume% biodiesel in diesel fuel in diesel engines.

Acknowledgements

The authors wish to thank the Iranian Fuel Conservation Organization (IFCO) of NIOC for the research grant provided (project number IFCO-1393) to complete this project and Mega Motor Company for providing of laboratory facilities.

References

[1] Demirbas A 2007 *Progress in Energy and Combustion Science* **33** 1-18
[2] Gumus M 2010 *Fuel* **89** 2802-14
[3] Aydin H, Bayindir H 2010 *Renewable Energy* **35** 588-92
[4] Prakash R, Singh R K, Murugan S 2013 *Energy* **55** 610-8
[5] Abu-Elyazeed O S M 2015 *Fuel* **144** 157-63

[6] Sánchez N, Sánchez R, Encinar J M, González J F, Martínez G 2015 *Fuel* **147** 95-9
[7] Shahir V K, Jawahar C P, Suresh P R 2015 *Renewable and Sustainable Energy Reviews* **45** 686-97
[8] Tsai S-Y, Tseng W-S, Wu C-T, Lin C-P 2014 *Procedia Engineering* **84** 940-7
[9] Meller E, Green U, Aizenshtat Z, Sasson Y 2014 *Fuel* **133** 89-95
[10] Mejía J D, Salgado N, Orrego C E 2013 *Industrial Crops and Products* **43** 791-7
[11] Panwar N L, Shrirame H Y, Rathore N S, Jindal S, Kurchania A K 2010 *Applied Thermal Engineering* **30** 245-9
[12] Wander P R, Altafini C R, Colombo A L, Perera S C 2011 *Energy* **36** 3917-23
[13] Dias J M, Araújo J M, Costa J F, Alvim-Ferraz M C M, Almeida M F 2013 *Energy* **53** 58-66
[14] Lin Y-S, Lin H-P 2011 *Renewable Energy* **36** 3507-16
[15] Thomas T P, Birney D M, Auld D L 2012 *Industrial Crops and Products* **36** 267-70
[16] Valente O S, Pasa V M D, Belchior C R P, Sodré J R 2011 *Fuel* **90** 1700-2
[17] López J M, Cota T d N J G, Monterrosas E E G, Martínez R N, de la Cruz González V M, Flores J L A, et al. 2011 *Chemical Engineering Journal* **178** 391-7
[18] Yahyaee R, Ghobadian B, Najafi G 2013 *Renewable and Sustainable Energy Reviews* **17** 312-9
[19] Atabani A E, Silitonga A S, Ong H C, Mahlia T M I, Masjuki H H, Badruddin I A, et al. 2013 *Renewable and Sustainable Energy Reviews* **18** 211-45

Leak detection in gas pipeline by acoustic and signal processing – A review

N F Adnan[1], M F Ghazali[1], M M Amin[1] and A M A Hamat[1]
[1]Faculty of Mechanical Engineering, Universiti Malaysia Pahang, 26600 Pekan, Pahang, Malaysia

Email: fairusham@ump.edu.my

Abstract. The pipeline system is the most important part in media transport in order to deliver fluid to another station. The weak maintenance and poor safety will contribute to financial losses in term of fluid waste and environmental impacts. There are many classifications of techniques to make it easier to show their specific method and application. This paper's discussion about gas leak detection in pipeline system using acoustic method will be presented in this paper. The wave propagation in the pipeline is a key parameter in acoustic method when the leak occurs and the pressure balance of the pipe will generated by the friction between wall in the pipe. The signal processing is used to decompose the raw signal and show in time-frequency. Findings based on the acoustic method can be used for comparative study in the future. Acoustic signal and HHT is the best method to detect leak in gas pipelines. More experiments and simulation need to be carried out to get the fast result of leaking and estimation of their location.

1. Introduction

The pipelines operated for a number of years and tend to corrode. The slow process will lead to the big explosion if the gas is the main fluid filled in the pipeline system. The leak detection in gas pipeline is more difficult compared to water because of the poor signal noise to ratio (SNR). The large noise comes from gas flow itself. In addition, the geometrical features also one of the noise and make the leak detection in gas pipeline more complicated [1].

The leak detection techniques introduced in order to minimize the losses of medium transported by the pipeline. Three major categories can be divided to classify the technique. The first technique is automated detection, which can detect by monitoring system of pipeline network without human operator after the installation while, semi-automated detection need a certain input to perform some task in order to detect the leakage. The third category is manual detection, which is using the system and the device. It always needs to operate by human [2].

Classification of technique in order to detect leak in the pipeline also can be described as optical and non-optical methods as by Batzias [3]. Optical method more specific and sensitive compared to non-optical method, but their faces problem with the background noise. So, they need low reflection background material such as green grass, coarse snow, frost and sandy soil. But, direct and indirect leak detection also gives some idea in the classification of techniques. Direct technique normally uses visual inspection or device for detects the gas that escapes from pipeline network [4, 5].

Normally they classify the method based on their technical nature[6]. Technical nature divided in two categories which are hardware based method and software based method. Non-technical method also can be categorized in technical nature [2]. These categories sometimes mentioned as internal and external method in leak detection system [7]. In this paper, the study of gas leak detection using acoustic, sound propagation, signal analysis and some simulation synthetic signal will be discussed in detail.

2. Sound wave propagation and acoustic method

Basically, the leak can described as gas or liquid flow through the wall with the imperfection such as hole, crack or ruptures. The pressure difference in pipeline always flows from higher pressure to lower pressure. The imbalance pressure detected and analysed to detect the leak. The wave propagation or speed of sound, a will be dependent on the material and pipe diameter based on Thorley [8]. The formula of speed of wave propagation in the thin walled pipes as follows:

$$a = \sqrt{\frac{1}{\rho\left[\frac{1}{K} + \frac{D}{Ee}\emptyset\right]}} \tag{1}$$

Where a = wave propagation speed in m/s, ρ = fluid density (kg/m^3), E = Young's modulus of elasticity of pipe wall material (N/m^2), K = bulk modulus of the liquid (N/m^2), and ø = a restraining factor dependent on the Poisson's ratio of the wall material and how well the pipe is supported. Equation (1) is only true for single phase fluids in rigid pipes, whereas for the flexible pipe or the fluid is multiphase, the speed of the wave will be decreased [9].

When a sound wave comes into contact with a surface, some part of the energy is reflected from it, some are transmitted through it and some is absorbed. If absorption and transmission are low, and thus most of the sound energy incident on the surface is reflected, it is supposed to be acoustically hard and can be regarded as mirror reflect the illumination [10].

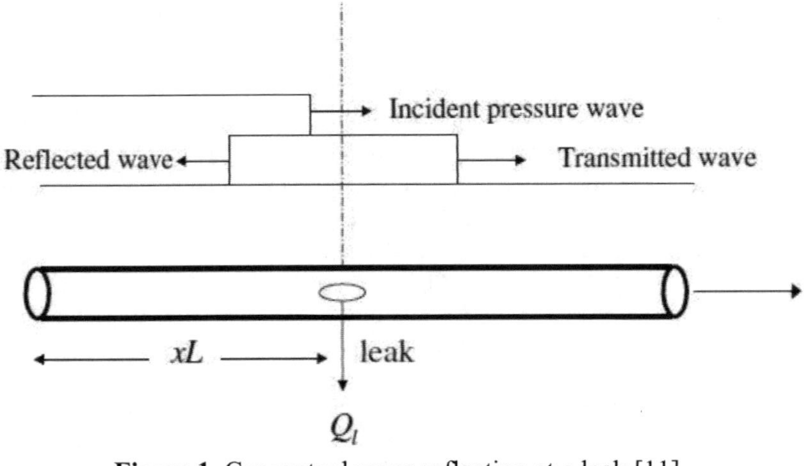

Figure 1. Conceptual wave reflection at a leak [11].

Dozens of subjects have been performed in a leaking gas pipeline using acoustic methods. The basic rule behind this acoustic method is when leak or ruptures occur, the friction between the rampart in the pipeline happen because of the pressure balance was broken. The acoustic signal will propagate to upstream and downstream [12]. The stress wave that transmitted through the pipeline can be recorded by acoustic sensor or accelerometer [13].Sound wave exhibits four different phenomena which are reflection, interference, refraction and diffraction. Three effects that occur to the incident wave mentioned by [14, 15] are reflection, transmission and absorption as shown in figure 1 [11].

Acoustic sensor normally installed outside the pipeline network. The leak generates noise and pick up by these acoustic sensors and the most important thing is to minimize the background noise [6, 16]. The location and size can be determined by using acoustic methods. This method can be used on new as well as on existing pipe network. The fast and high sensitivity give advantages in this method compared to others. In addition, this method also offer accurate leak location and low false alarm rate [12]. But, high background and noise condition will affect the actual leak and also produce false alarm when the leak is too small [2, 6, 17, 18].

Early studies in acoustic wave was done by Watanabe [19] by measured the pipeline and its shows that sharp positive and negative pulse at a certain time using mathematical model of the pipeline acoustic. From there, they can detect the location of leak. Hunaidi and Chu [20] investigated about the characteristic of frequency content as function of leak type, flow rate, pressure of the pipeline system and also season in sound and vibration signals. The details will help in order to choose appropriate instrumentation, design of procedures, and accurate locating leaks using cross-correlation method. The technique using pulse reflectometry involving injection of sound pulse into pipeline system and the resultant of reflection will be recorded was done by Sharp and Campbell [21]. The value of the reflection coefficient for a small leak depends on leak area except for the crack. Experimental and theoretical analysis of acoustic properties of a finite elastic pipe filled with water or air was investigated by Feng [22].

A patented acoustic technique known as Acoustic to detect features such as blockage and leakage by injecting pressure pulse in the pipeline was developed [18]. The development and test of gas leakage present by Sousa [23] based on acoustic method. They used microphone to catch the leak noise by operating and non-operating of gas flow in 60 m long pipeline. Besides, the high-temperature microphones developed in nuclear power plant and successfully operating to detect the leaks with sensitivity 0.23 m3/h and also their location [24].

An acoustic measurement method was proposed to non-invasively detect L-bends and T-branches. This pipeline is buried under floors or in the walls of buildings [25]. Acoustic sensor used to detect the leaking in the pipeline and handle by operators or intelligent pigs hold the device along the pipeline to inspect the leak. Placing sensor to far will increase the risk of undetected leak while placing too close will increased the cost.

The range and dynamic pressure transducer discussed by [26]. Rocha [27] used pressure sensor to record the appearance of acoustic pressure waves. Time-frequency technique [28, 29] or low frequency impulse [26] method used in detection the leaking involves the measurement of two acoustic signals on each of a pipe segment. A recent study of experimental focus on distinguishing between signals make by leak and background noise using time-frequency analysis and added by location formula to increase accuracy [12] while Brodetsky and Savic present a new continuous acoustic leak detection system by using k nearest neighbour classifier to distinguish leaks in underground high pressure gas pipelines [30].

3. Signal processing

Dealing with dynamic signal is challenging because not-deterministic type of signal normally occurs in real application. So, signal processing was used in order to reveal the hidden information. This problem can be solved and understand by statistical behaviour or any pattern recognition. Commonly, they are applied for time domain signals. But, in the leak detection issue, the signal needs to be transformed to solve the hidden result behind the raw data. Hence, the transformation methods were used such as Fast Fourier transforms (FFT), Wavelet transforms (WT), and Hilbert Huang transforms (HHT).

Years ago, a study using cepstrum analysis was developed by using closure and open solenoid valve. The time domain obtained by identifying the delaying time between initial wave with and their reflections [31, 32]. A wavelet-based filtering was introduced for detection in gas pipeline. This main idea is based on the echoes reflected from turbulences induced by a leak [1]. The leakage detection algorithm was combined with the filtering method which is to transform, filtering to remove noise the and also cross-correlation technique to locate the leak by using time difference between two signals [33, 34]. An algorithm for gas pipeline leak detection based on Hilbert-Huang Transform studied and they use acoustic wave measured at two sensors at different positions. The other study was done using HHT and obtain Hilbert marginal spectrum. The result comparing between the normal condition and leak condition [35]. The leakage can be detected by feature of acoustic signal extraction [36]. Mostafapour and Davoudi [13] using Donnell's non-linear theory for cylindrical shell to derive motion equation and then solve by Galerkin method.

Some simulation of synthetic data used to show the process of signal processing with different method. The signal with different frequency and amplitude is defined by the equation (2) and as shown in figure 2.

$$\left.\begin{array}{ll} x(t)=5\sin(2\pi f1t) & 0 < t \leq 1.0s \\ x(t)=10\sin(2\pi f2t) & 1 < t \leq 2.0s \end{array}\right\} \quad (2)$$

where f1 is equal to 20Hz and f2 is 80Hz respectively.

Both signals are combined with impulse at two places which are at 0.4 s and 1.6 s. The impulse represents the irregularity in the signal which means reflection of the leaks, or features in the pipeline system.

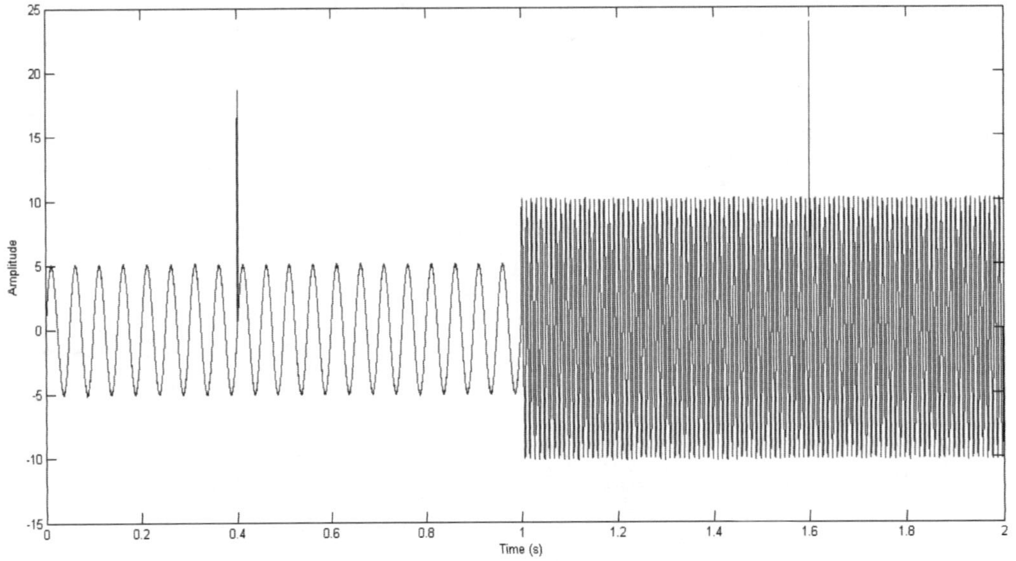

Figure 2. Time domain of the synthetic signal with 20Hz and 100Hz.

The simplest method of analysing is FFT, but there are too many drawbacks. From figure 3, both frequencies clearly seen in frequency domain. They can show the frequency, but not the time happens. The time is important since detection of leak location is based on time and speed of sound.

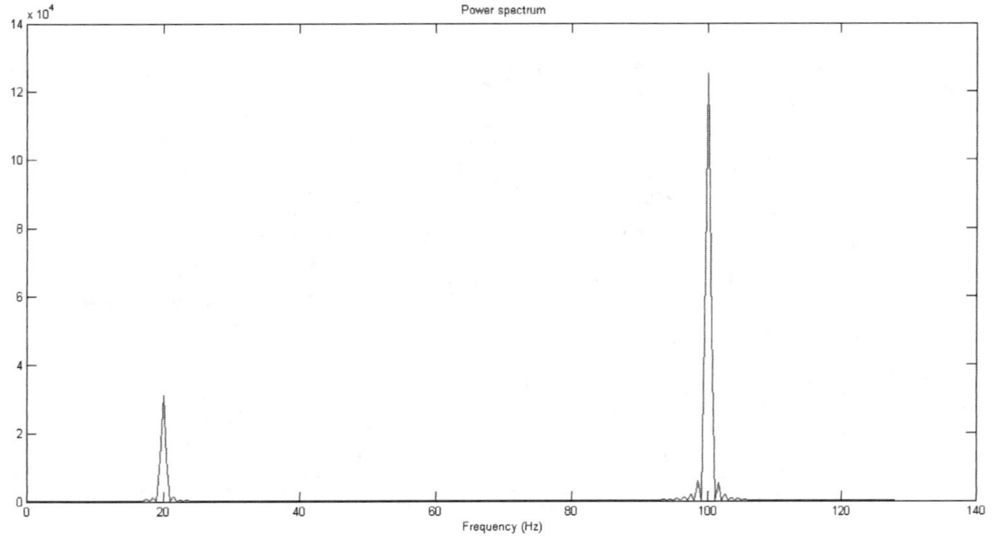

Figure 3. Frequency domain of the synthetic signal with 20Hz and 100Hz.

Compared to HHT, the result is more accurate and reliable. Figure 4 illustrates the both frequencies which are at 20Hz and 100Hz. The irregularity clearly shows at 0.4 s and 1.6 s. This plot provides a high resolution time-frequency representation and gives details of the time and frequency. It is good to provide an excellent result for the leakage in the pipeline and the location can be calculated based on the speed of sound and time delay from this plot.

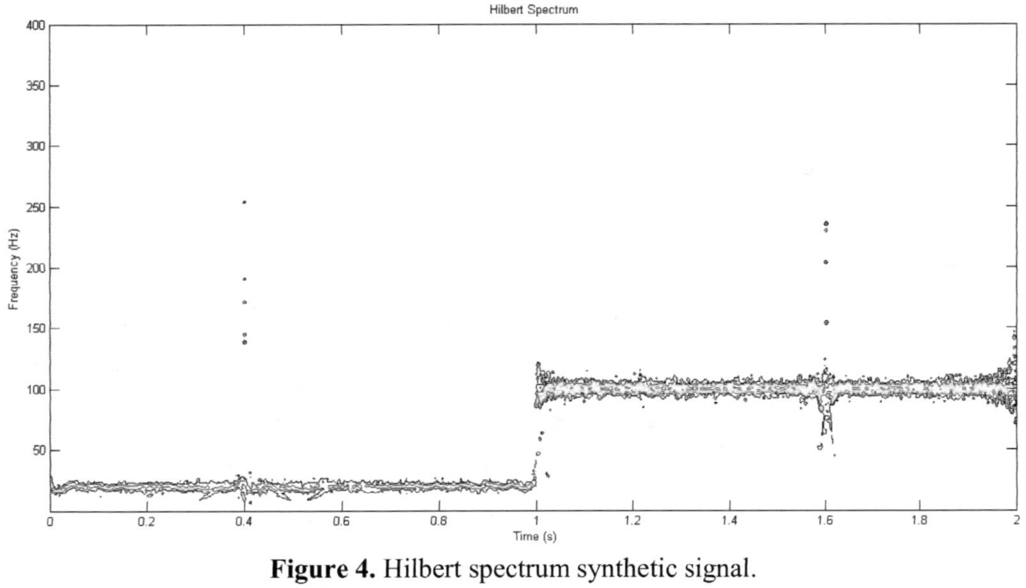

Figure 4. Hilbert spectrum synthetic signal.

The synthetic signal also compared between their advantages and disadvantages. Figure 5 shows the scalogram with Mexican hat. The wavelet analysis is also one of the powerful methods to show the frequency at the certain time. Based on the plotting in scalogram, the irregularity cannot be seen clearly. Only the spike at 1.6 s can be seen but not really good to show the result.

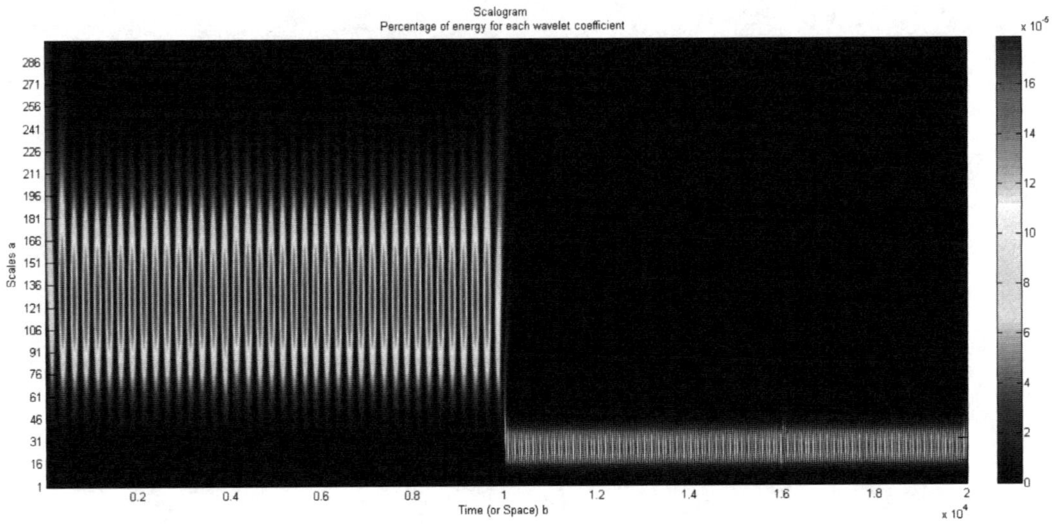

Figure 5. Scalogram of synthetic signal in Mexican hat as mother wavelet.

The spectrogram corresponding to the STFT for the synthetic signal was shown in Figure 6 below. The spike can be detected at 0.4 s and 1.6 s but unfortunately, the frequency show all over the time. There is no indicator when the frequency was happening at that time. Obviously, it fails to show the spike.

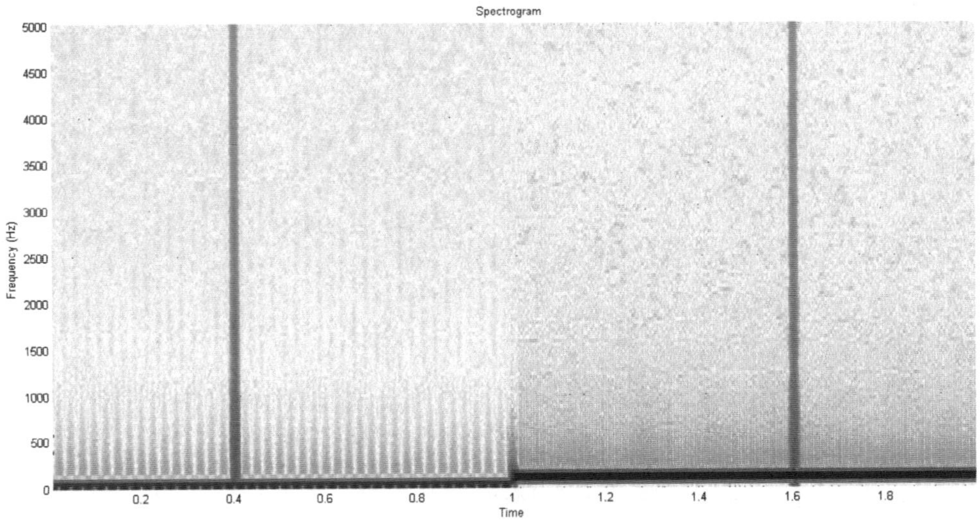

Figure 6. Spectrogram of synthetic signal with N/4.

4. Result and discussion

The finding of the present study suggests that acoustic signal and HHT analysis is the best way to detect leak in gas pipelines. Table 1 shows the common method with their advantages and disadvantages while table 2 shows the advantages and disadvantages between FFT, STFT, wavelet and HHT.

Table 1. Comparison of different method for leak detection.

Method	Speed of detection	Leak localization	Leak size estimation	Cost
Visual inspection	Slow	Yes	Yes	Cheap
Soap bubble	Slow	Yes	Yes	Cheap
Cable sensor	Fast	Yes	No	Expensive
Acoustic	Fast	Yes	Yes	Expensive

Table 2. Comparison of different method for signal processing.

Method	FFT	STFT	Wavelet	HHT
Advantages	Easy to implement		Features extraction	
		Can use for non-stationary signal	Non-stationary data analysis Multiresolution Basis function obtained by sifting and scaling a particular function Analytic form for the result Allow filters to be constructed for stationary and non-stationary signals	Can use for all types of signal High time-frequency resolution Both local and adaptive method in frequency analysis Suitable for non-linear and non-stationary data analysis Sharper spectrum Adaptive – data driven basis
Disadvatages	Not appropriate for nonlinear and non-stationary signals Feature extraction impossible	Non-adaptive Time-frequency resolution limited by the Heisenberg principle	Leakage generated by the limited length of the basic wavelet function Non-adaptive nature Unable to resolve intrawave frequency modulation	Lack of theoretical analysis No mathematical formulation No physical meaning of some IMFs

5. Conclusions

This review paper acts as a quick reference to readers, who are interested in leak detection by acoustic and signal processing. In this paper, the basic of wave propagation in gas pipeline introduced and some method were discussed to reveal the hidden information of the signal with simulation of synthetic signal. In the end, the acoustic method is the comparative method for gas leak detection study. The propagation of waves in the pipeline system will generate the friction and give important output to detect the leak. The finding provides evidence that acoustic signal as the wave propagation and HHT is the best combination to detect the leak in gas pipelines. More experiments and simulation need to be done to get the fast result of leakage and estimation of their location accurately and control the decrease false alarm.

Acknowledgements

The authors would like to thank the Ministry of Higher Education Malaysia and Universiti Malaysia Pahang for providing laboratory facilities under Advance Structural Integrity and Vibration Research (ASIVR) group and financial assistance via FRGS under project grant RDU130111.

References

[1] Urbanek J, Barszcz T, Uhl T, Staszewski W, Beck S, Schmidt B 2012 *Structural Health Monitoring* **11** 405-12
[2] Murvay P-S, Silea I 2012 *Journal of Loss Prevention in the Process Industries* **25** 966-73
[3] Batzias F A, Siontorou C G, Spanidis P M 2011 *Journal of hazardous materials* **186** 35-58
[4] Folga S M. Natural Gas Pipeline Technology Overview. Tech. rep Argonne National Laboratory: 2007.
[5] Liu A E. Overview: Pipeline Accounting And Leak Detection By Mass Balance, Theory And Hardware Implementation. 2008.
[6] Scott S L, Barrufet M A. Worldwide Assessment of Industry Leak Detection Capabilities for Single & Multiphase Pipelines. Department of Petroleum Engineering Texas A&M University, 2003.
[7] Geiger G, Werner T, Matko D. Leak Detection and Locating – A Survey. 2003.
[8] Thorley A R D 2004 *Fluid Transients in Pipeline Systems* John Wiley & Sons)
[9] Wang H, Priestman G H, Beck S B M, Boucher R F 2000 *International Journal of Heat and Fluid Flow* **21** 104-11
[10] Hassall J R, Zaveri K 1988 *Acoustic noise measurements* Brüel & Kjaer)
[11] Colombo A, Karney B W 2003 *Water Science and Technology: Water Supply* **3** 163-9
[12] Meng L, Yuxing L, Wuchang W, Juntao F 2012 *Journal of Loss Prevention in the Process Industries* **25** 90-102
[13] Mostafapour A, Davoudi S 2013 *Applied Acoustics* **74** 335-42
[14] Burn S, Desilva D, Eiswirth M, Speers A, Thornton J 1999 *Proceedings of the Pipes Wagga Wagga Conference, New South Wales, Australia* 1–18
[15] Beck S B M, Staszewski W J, Cepstrum analysis for identifying reflection points in pipeline networks. Proceedings Of The International Conference On Pressure Surges; 2004.
[16] Chis T, Saguna A 2007 *Annals Computer Science Series, 5th Tome 1st Fasc* 25-34
[17] Sivathanu Y. Natural Gas Leak Detection in Pipelines. U.S. Department of Energy, National Energy Technology Laboratory, 2003.
[18] Wang X, Short G, Dawson K, Lennox B 2009
[19] Watanabe K, Himmelblau D M 1986 *AIChE Journal* **32** 1690-701
[20] Hunaidi O, Chu W T 1999 *Applied Acoustics* **58** 235-54
[21] Sharp D B, Campbell D M 1997 *Acta Acustica united with Acustica* **83** 560-6
[22] Feng L 1996 *Journal of Sound and Vibration* **189** 511-24

[23] Sousa E O, Cruz S L, Pereira J A F R 2007 *4o PDPETRO, Campinas, SP*
[24] Shimanskii S B, Strelkov B P, Anan'ev A N, Lyubishkin A M, Iijima T, Mochizuki H, et al. 2005 *Atomic Energy* **98** 89-96
[25] Yunus U, Tsunasaki M, He Y, Kominami M, Yamashita K. Detection of Leak Location in a Pipeline by Acoustic Signal. IEICE TRANS FUNDAMENTALS2008. p. 2053-61.
[26] Loth J, Morris, G., & Palmer, G. Technology assessment of on-line acoustic monitoring for leaks/infringements in underground natural gas transmission lines. USA: West Virginia University: 2003.
[27] Rocha M 1989 *ISA Calgary 1989 Symposium-Paper* 283-90
[28] Kim M-S, Lee, S.-K. 2008 *Journal of Loss Prevention in the Process Industries* 1-5
[29] Jing-pin J, Ren-yuan F, Cun-fu H, Bin W 2006 *Frontiers of Mechanical Engineering in China* **1** 146-50
[30] Brodetsky I, Savic M. Leak Monitoring System For Gas Pipelines. IEEE Int Conf Acoustics, Speech, and Signal Processing; Minneapolis, USA1993. p. 17-20.
[31] Taghvaei M, Beck S B M, Staszewski W J 2006 *Measurement Science and Technology* **17** 367-72
[32] Taghvaei M. Leakage detection in pipeline networks: Sheffield University; 2007.
[33] Mostafapour A, Davoodi S 2012 *Journal of Nondestructive Evaluation* **32** 113-23
[34] Davoodi S, Mostafapour A 2012 *Journal of Nondestructive Evaluation* **32** 67-80
[35] Yang H, Zhang L, Liang W, Ye Y, Ren Y, Study of Gas Pipeline Leak Detection Based on Hilbert Marginal Spectrum. Computational and Information Sciences (ICCIS), 2012 Fourth International Conference on; 2012 17-19 Aug. 2012.
[36] Jia Y, Gao B, Jiang C, Chen S. Leak Diagnosis of Gas Transport Pipelines Based on Hilbert-Huang Transform. International Conference on Measurement, Information and Control (MIC)2012. p. 614-7.

Numerical studies on the separation performance of liquid-liquid Hydrocyclone for higher water-cut wells

H Osei[1], H H Al-Kayiem[1] and F M Hashim[1]
[1]Mechanical Engineering Department, Universiti Teknologi PETRONAS, 32610 Bandar Seri Iskandar, Perak, Malaysia

E-mail: harrisdecany@yahoo.com

Abstract. Liquid-liquid hydrocyclones have nowadays become very useful in the oil industry because of their numerous applications. They can be installed downhole in the case of a well that produces higher water-oil ratios. The design of a liquid-liquid hydrocyclone for such a task is critical and every geometric part of the hydrocyclone has a part to play as far as separation is concerned. This work, through validated numerical technique, investigated the liquid-liquid hydrocyclone performance for the cases of single-inlet and dual-inlets, with different upper cylindrical lengths, specifically, 30mm and 60mm.It was observed that the hydrocyclones with the 30mm upper cylindrical section perform better than the ones with 60 mm upper cylindrical section. It was again noted that, even though higher number of tangential inlets increases the swirl intensity, they have the tendency to break up the oil droplets within the hydrocyclone because of increasing shear and jet flow interaction.

1. Introduction

The influx of water into the well has been known for more than 100 years and is regarded as one of the biggest problem during oil production [1]. The fact is that, there is a time in the production life of a reservoir where production reaches its peak and the produced oil is often associated with increased water production. This is normally the case when the field is matured and has produced for many years [2]. A contributing factor to this increased water influx is the upward dynamic force resulting from wellbore drawdown which causes the water underlying the oil zone to rise to a height after which it becomes unstable and finally breaks into the well [3]. As a result of that, higher water-cut wells therefore have the tendency of depleting fast the pressure of an oil reservoir and indirectly causing low oil productivity. A suitable way is therefore needed to limit the production of water and at the same time increase oil production.

Hydrocyclone is one of the separation devices that have found application in industry for many years and its use and installation downhole can help overcome this problem by separating the oil and water, after which the oil is pumped to the surface and the water injected back into the formation to maintain the reservoir pressure and boost the oil inflow into the well. Hydrocyclone has proved to be an important device when it comes to the separation of liquid-liquid, gas-liquid and liquid-solid streams or products. It is simple in design, low in cost, easy to operate, low in maintenance and high in separation efficiency. It is very compact, has no moving parts and does not need any chemical additives. No wonder it provides a competitive method for oily water treatment. In spite of the wide application of hydrocyclones, their accurate design is often difficult because there are many mathematical relationships for predicting the separation efficiency and the pressure drop based on semi-empirical models which are restricted to hydrocyclone geometry and operating conditions from which they are deduced [4].

In the oil production industry, the use of liquid-liquid hydrocyclone (LLHC) for the treatment of oily water cannot be underrated especially when production is from higher water-cut wells. Some matured fields can produce higher water volumes in the range of more than 70 % when compared with the oil. This can lead to early well abandonment and inability to recover hydrocarbon. In such cases, wellbore separation of produced fluids (oil/water) by the use of liquid-liquid hydrocyclone and same well disposal of the separated water is an environmentally friendly tool that provides a unique opportunity to reduce operating costs and enhance the economic viability of such higher water-cut wells while at the same time reducing pollution [5]. The selection and design of hydrocyclones for downhole oil-water separation are therefore important and needed.

This paper seeks to show how numerical simulation can be an effective tool in the design of liquid-liquid hydrocyclone to accomplish oil/water separation in the downhole. It presents the effects of the length of the upper cylindrical section and the number of tangential inlets of the hydrocyclone on the separation performance. The results would lead to improvement in the design of the LLHC, and in turn, reduce the water production at the surface and the oil production cost in higher water-cut wells.

2. LLHC geometry

There are many possibilities in the design of LLHC geometry for optimizing its structure and improving the oil–water separation efficiency. LLHC for higher water-cut wells consists of a set of cylindrical and conical sections. Figure 1 shows a conventional liquid-liquid hydrocyclone. It is very important to ensure that the inlet chamber and the reducing sections are designed to achieve higher fluid tangential acceleration, reduce both the pressure drop and the shear stress to an acceptable level. The shear stress has to be minimized to avoid droplet breakup which can lead to reduction in separation efficiency. The tapered section is where most of the separation is achieved. The low angle at the tapering section maintains the swirl intensity with high residence time. The smallest fluid droplets migrate to join the reverse flow at the axis of the LLHC so as to be separated through the overflow exit [6, 7].

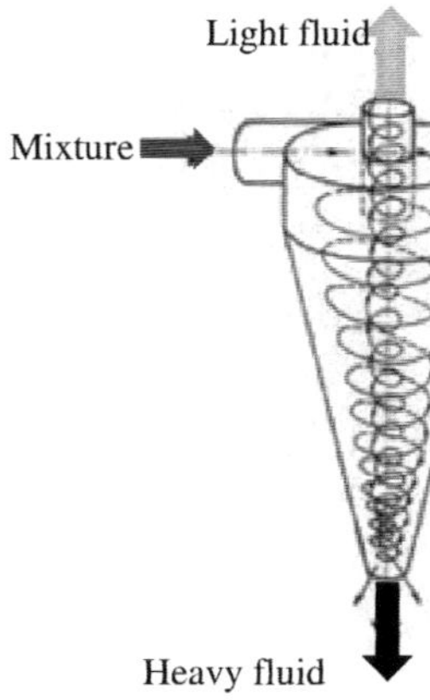

Figure 1. Conventional LLHC [8].

3. Flow dynamic of LLHC

A typical LLHC is illustrated in figure 2. The feed enters into the upper cylindrical section through the tangential inlet(s) which cause(s) the feed to swirl inside the hydrocyclone. As the fluid moves through the LLHC to the apex, the centrifugal force and the fluid angular velocity increase because of the nature of the narrowing cross-sectional area of the cyclone. The centrifugal force developed accelerates the settling rate of the fluid particles thereby separating them according to size, shape, and differential density. The fluid segregation is witnessed in the creation of two spiral movements within the hydrocyclone – primary spiral towards the walls of the hydrocyclone and secondary spiral at the core of the hydrocyclone. The development of the secondary spiral is also as a result of the high swirling intensity at the inlet. This causes an increase in pressure near at the walls of the cyclone with a decrease in pressure at the core and therefore by ensuring a higher pressure at the underflow outlet than that at the vortex finder, the fluid at the core is forced to flow countercurrent to the main [7, 9]

The secondary spiral constitutes the reverse flow and carries the lighter fraction of the fluid (oil) to the vortex finder which is the opening through the cover for the top cylindrical part of the hydrocyclone. Fluid through the vortex finder are termed overflow. The heavier fraction of the fluid (water) is carried by the primary spiral and drags down against the walls of the LLHC to be discharged at the spigot as underflow [8]. There is the tendency of the creation of some recirculation zones at the inlet region and this is associated with the high swirling. These zones will have long residence time, very low axial velocity and normally diminish as the flow enters the lower conical section of the LLHC [7].

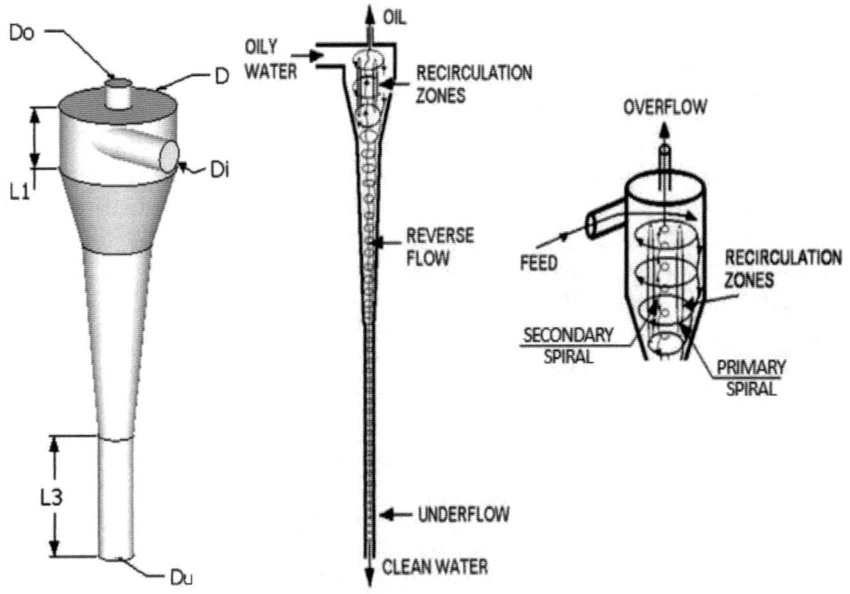

Figure 2. Typical LLHC (a) showing parts and (b) showing flow dynamics [6, 10].

4. Numerical simulation

Numerical studies were carried out using ANSYS–CFX software to test the efficiency of the hydrocyclone designs. Two hydrocyclone cases with the upper cylindrical section, L1 = 60 mm and 30 mm, with all other geometrical lengths maintained were considered. Study was also made into the situation where there are one and two tangential inlets. 3–D computational model was selected for the study and figure 3 shows the meshing of the hydrocyclone types. The optimized mesh density for a reasonable solution for the different cyclone designs was around 160,000 computational cells. Different oil-water mixture compositions were considered. The total inlet flow rate was assumed to be 2 m/s. SSG Reynolds Stress Model was used because it is more suited for complex flows and is also more accurate for swirling and rotational flow. The simulation was run and convergence was achieved at a residual target of 1x10–4.

IOP Conf. Series: Materials Science and Engineering **100** (2015) 012014 doi:10.1088/1757-899X/100/1/012014

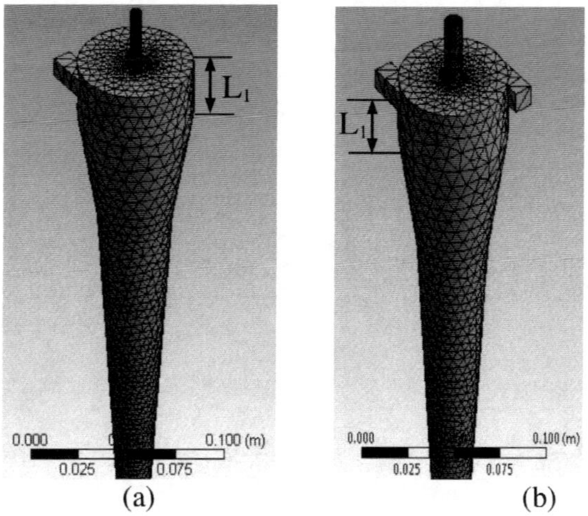

Figure 3. Meshing of the LLHC types (a) one inlet LLHC and (b) twin inlet LLHC.

5. Result and discussion

This section shows figures having plots of oil volume separated versus radial distance at the oil outlet. The fluid mixture compositions considered were 25% oil-75% water and 30% oil-70% water. This section will show results of these different fluid mixtures as they are introduced into the one inlet type and twin inlet type of hydrocyclones having different upper cylindrical lengths. The oil separation volume for the hydrocyclone with L1=60 mm is read from the primary y-axis whereas the oil separation volume for the hydrocyclone with L1=30 mm is read from the secondary y-axis.The use of sections to divide the text of the paper is optional and left as a decision for the author. Where the author wishes to divide the paper into sections the formatting shown in table 2 should be used.

The plot in figure 4 shows the case where the LLHC has only one tangential inlet with 25% oil-75% water as the inlet fluid composition. From the figure 4, it can be seen that separation is better in the case of the LLHC with L1=30 mm. While it records efficiencies in the range of 90.5–92.3 %, the cyclone with L1=60 mm records efficiencies between 70.00–75.16 %.

The results from the same one tangential inlet LLHCs but in this case having 30% oil-70% water as the inlet fluid composition is also presented in figure 5. The separation efficiency patterns here follow the same trend like that in figure 4 since the same sets of cyclones are been used. However, there are some slight differences in the separation volumes. The LLHC with L1 = 30 mm still outperforms the L1=60 mm type. The efficiencies are in the range of 99.14–99.27 % for L1=30 mm type and 71.00–76.02 % for the L1=60 mm type.

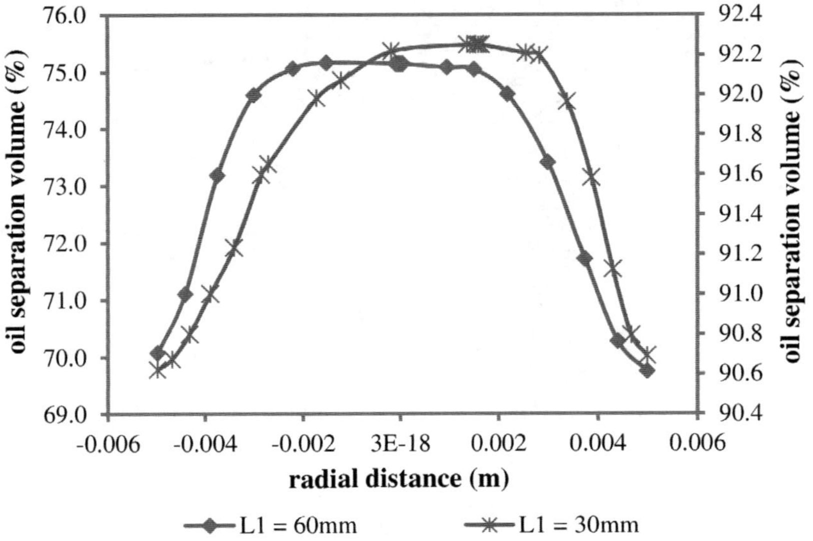

Figure 4. Oil separation efficiency in LLHC with one tangential inlet using 25% oil 75% water mixture.

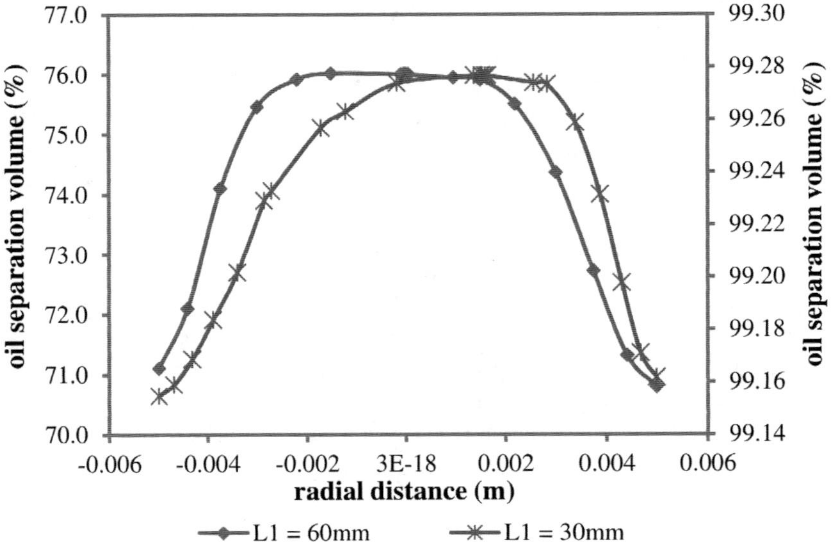

Figure 5. Oil separation efficiency in LLHC with one tangential inlet using 30% oil 75% water mixture.

Investigation was also made for twin tangential inlets carrying the fluids. The plots in figure 6 are for the case of using 25% oil-75% water as the inlet fluid composition. From the figure, the performance of the hydrocyclone with longer upper cylindrical column lags behind that with the shorter upper cylindrical column. The L1=30 mm type produces separation volume values that range between 89.2–92.1 % against 76.79–79.43 % separation values for the L1=60 mm type.

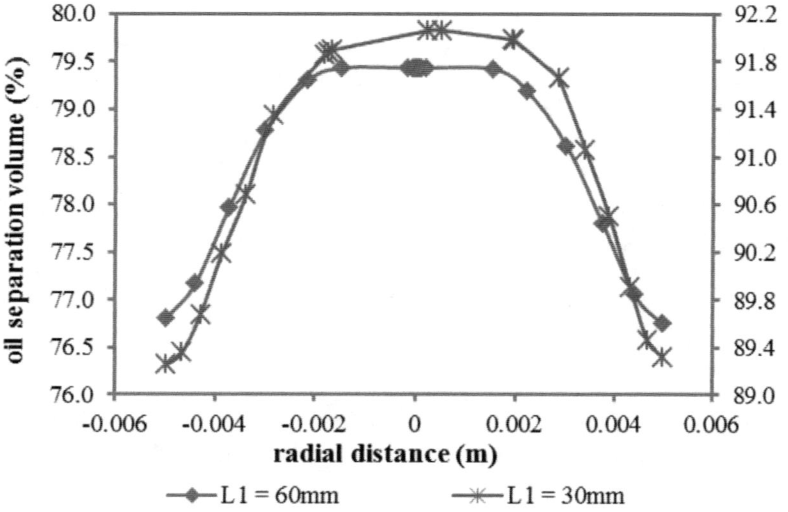

Figure 6. Oil separation efficiency in LLHC with two tangential inlets using 25% oil 75% water mixture.

Figure 7 shows plots made after having used the 30% oil-70% water as the inlet fluid composition in the doubled inlet LLHC. The LLHC with L1=30 mm in this case again performs better than the L1=60 mm type producing efficiencies in the range of 84.59–88.84 % and 76.75– 9.42 % respectively. The drop in the efficiency values recorded here is due to some fluids in the recirculation zones escaping without been classified and also some oil droplets breaking down because of the turbulence.

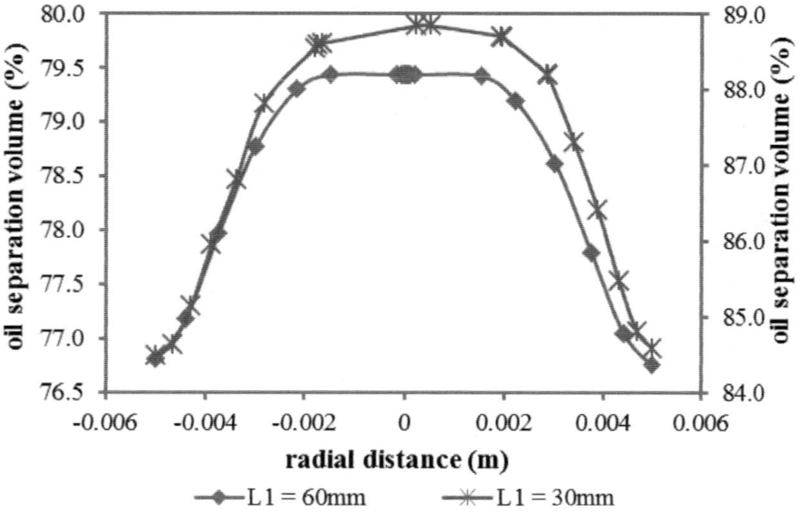

Figure 7. Oil separation efficiency in LLHC with two tangential inlets using 30% oil 70% water mixture.

The results obtained from this work provide a useful guide in the design of liquid-liquid hydrocyclone to control excessive water production from oil wells. Due to the limited wellbore space, LLHCs used in downhole oil-water separation (DOWS) are narrow and tall in design requiring a minimum of 0.1397 m casing size.

6. Conclusions

A liquid-liquid hydrocyclone has successfully been modelled and simulated computationally using ANSYS-CFX commercial software. The simulation results demonstrate that the LLHCs with the 30 mm upper cylindrical column provide higher separation efficiency compared to the ones with 60 mm upper cylindrical height. The separated oil volumes from the hydrocyclones with L1 = 30 mm are all above 90 %, giving way to approximately 10 % of water to be produced with the oil. 100 % separation efficiency is not possible as some water will escape with the oil and vice versa. Again, the separation performance was not appreciable in the case of using twin tangential inlets and even dropped in the case of the hydrocyclones with L1 = 30 mm which is attributable to shear and break down of the oil droplets. In conclusion, liquid-liquid hydrocyclones with one tangential inlet and short upper cylinder would lead to higher separation efficiency of oil/water mixture.

Acknowledgments

The authors acknowledge Universiti Teknologi PETRONAS for the financial and technical support to produce this paper under the research grant YUTP – FRG 015-3AA-A73.

References

[1] Ju B, Qiu X, Dai S, Fan T, Wu H, Wang X 2008 *Journal of Energy Resources Technology* **130** 033102

[2] Ogunsina O O, Wiggins M L. A Review of Downhole Separation Technology. 2005/1/1/. SPE: Society of Petroleum Engineers.

[3] Boyun G, Lee R-H 1993 *SPE reservoir engineering* **8** 249-55

[4] Zhang F L, Deng S S, Zhang P F, Numerical Study of Oil-Water Two Phase Separation in Hydrocyclone. Advanced Materials Research; 2011: Trans Tech Publ.

[5] Shaw C, Downhole separation as a strategic water and environmental management tool. SPE International Conference on Health, Safety and Environment in Oil and Gas Exploration and Production; 2000: Society of Petroleum Engineers.

[6] Gomez C, Caldentey J, Wang S, Gomez L, Mohan R, Shoham O 2002 *Spe Journal* **7** 353-72

[7] Gómez C H. OIL-WATER SEPARATION IN LIQUID-LIQUID HYDROCYCLONES (LLHC) œ EXPERIMENT AND MODELING: The University of Tulsa; 2001.

[8] Bradley D 2013 *The Hydrocyclone: International Series of Monographs in Chemical Engineering* **4** Elsevier)

[9] Bowers B, Brownlee R, Schrenkel P, Development of a Downhole Oilwater Separation and Reinjection System for Offshore Application. Offshore Technology Conference; 1998: Offshore Technology Conference.

[10] Amini S, Mowla D, Golkar M 2012 *Desalination* **285** 131-7

A review on model updating in structural dynamics

N A Z Abdullah[1], M S M Sani[1,2], M M Rahman [1,2] and I Zaman[3]
[1]Faculty of Mechanical Engineering, Universiti Malaysia Pahang, 26600 Pekan, Pahang, Malaysia,
[2]Automotive Engineering Centre, Universiti Malaysia Pahang, 26600 Pekan, Pahang, Malaysia
[3]Department of Engineering Mechanics, Faculty of Mechanical and Manufacturing Engineering,Universiti Tun Hussein Onn Malaysia, 86400 Parit Raja, Batu Pahat, Johor, Malaysia

Email:znooramzura@hotmail.com

Abstract. This paper is a review of information and related studies concerning topic of model updating in structural dynamics. It is purposed to introduce and explain the important concept of the discussed subject in model updating area as well as to summarize the development and available guidance and method of conducting the study in this area. The review is structured by presenting an overview of general concept of finite element model updating in structural dynamics and the capabilities of finite element method as a tool in model updating. After the concept introduction, the reliable methodology to perform model updating, the limitation and the critical issues in model updating is discussed. The limitation and problems arise concerning correcting inaccurate finite element model is also discussed. This lead to issue of parameterization and regularization which the limitation and uncertainty in choosing the updating parameter is shown able to be overcome. Further studies on the elimination of systematic errors from both the measurements and the finite element model so that it can provide more reliable model updating is recommended.

1. Introduction

Model updating techniques are about updating a finite element model of a structure so that it can assume more accurate dynamics of a structure. Finite element method is widely used to model the dynamics of a certain structures as it is considered as reliable of providing accurate results. It is considered by Zienkiewics and Taylor [1], as the most appropriate tool for numerical modeling in structural engineering as it is capable of handling complex structural geometry, large complex assemblies of structural components and can perform many different types of analysis.

However, problems of inaccuracies in the finite element model will arise and sometimes viewed as poor reflection of actual structure. It may be caused by poorly known boundary conditions of the structure, the unknown material properties of the structure or because of the simplification in the modeling of very complex structural systems [2, 3]. For example, modelling joints is particularly not easy as it is difficult to predict its finite stiffness. These uncertainties in the modelling process cause the predicted dynamics of a structure to be different from the measured dynamics of the real structure. If there is accurate measured data, then general improvement on the numerical model and uncertain parameters of the model can be made based on this data.

Many model updating techniques have been proposed, tested and published. Mottershead [2] stated that the methods can be divided according to the type of measured data they used and the model parameters that are updated. The measured data may be in form of frequency response function (FRF) data or natural frequencies and mode shapes. The updating process may estimate physical parameters, complete mass, damping and stiffness matrices or groups of individual matrix elements. So far, research has concentrated on updating physical parameters using either FRF or modal data.

Parameter uncertainty has become an important research topic when discussing model updating according to Mottershead and Friswell [2, 4]. This is because the key to success in model updating is always the choice of parameters. The critical issues that have been stressed by Friswell [5] in modal updating also is about deciding the way finite element model of a structure is to be parameterized and how to estimate the unknown parameters from ill-conditioned equations. Therefore consideration on these issues is really crucial.

There are a few issues that also important and being studied in model updating apart from parameterization such as parameter uniqueness, efficient computation, ill-conditioning and use of incomplete data. Regularization method is needed in model updating to modify the poor or ill conditioned system of equations toward a well-conditioned one so that parameters can be identified.

In this paper, an overview of general concept of finite element model updating in structural dynamics and the capabilities of finite element method as a tool in model updating is presented. The first part will discuss the methods of model updating. The issue of parameterization and regularization which the limitation and uncertainty in choosing the updating parameter can be overcome are also included.

2. Finite element model updating

2.1. Methods of model updating

There have been several extensive reviews [2, 4, 6] of different kinds of model updating methods that have been developed.

Based on the comparative study of damped finite element model updating methods by Arora et al. [7], model updating technique can be classified into non-iterative or also called direct methods and iterative methods which also called sensitivity method.

The direct method shows precise results. Thus, the model assumptions match the experimental modal data. The way the measured modal data is reproduced is more computationally cheaper. Nevertheless, the updated finite element model result may have poor physical meaning as the methods violate the structural connectivity and updated structural matrices are difficult to interpret. This is the reason why the methods have not been generally used in practice.

Iterative methods or the sensitivity methods, which concern of reducing an objective function that is generally a non-linear function of selected updating parameters, are carried out by either using Eigen data or frequency response function (FRF) data. Therefore it provides wider choice of parameters for updating. These methods considered as capable of overcoming the limitations of the direct methods [3]. It also has been applied successfully to large-scale industrial problems.

A brief explanation and tutorial on this sensitivity method in finite element model updating is provided by Mottershead, Link and Friswell [8]. Example of model updating of a helicopter airframe is also showed in the paper. The sensitivity method is based upon linearization of the generally non-linear relationship between the measurable outputs such as natural frequencies, mode shapes or displacement responses, of the model's parameters in need of emendation. The most important quality is to define an error function of modal data obtained from computer simulation and experimental. The estimated parameters are attained by minimizing the error function with respect to the updating parameters.

The simplified flow diagram for the model updating method is shown in figure 1 below.

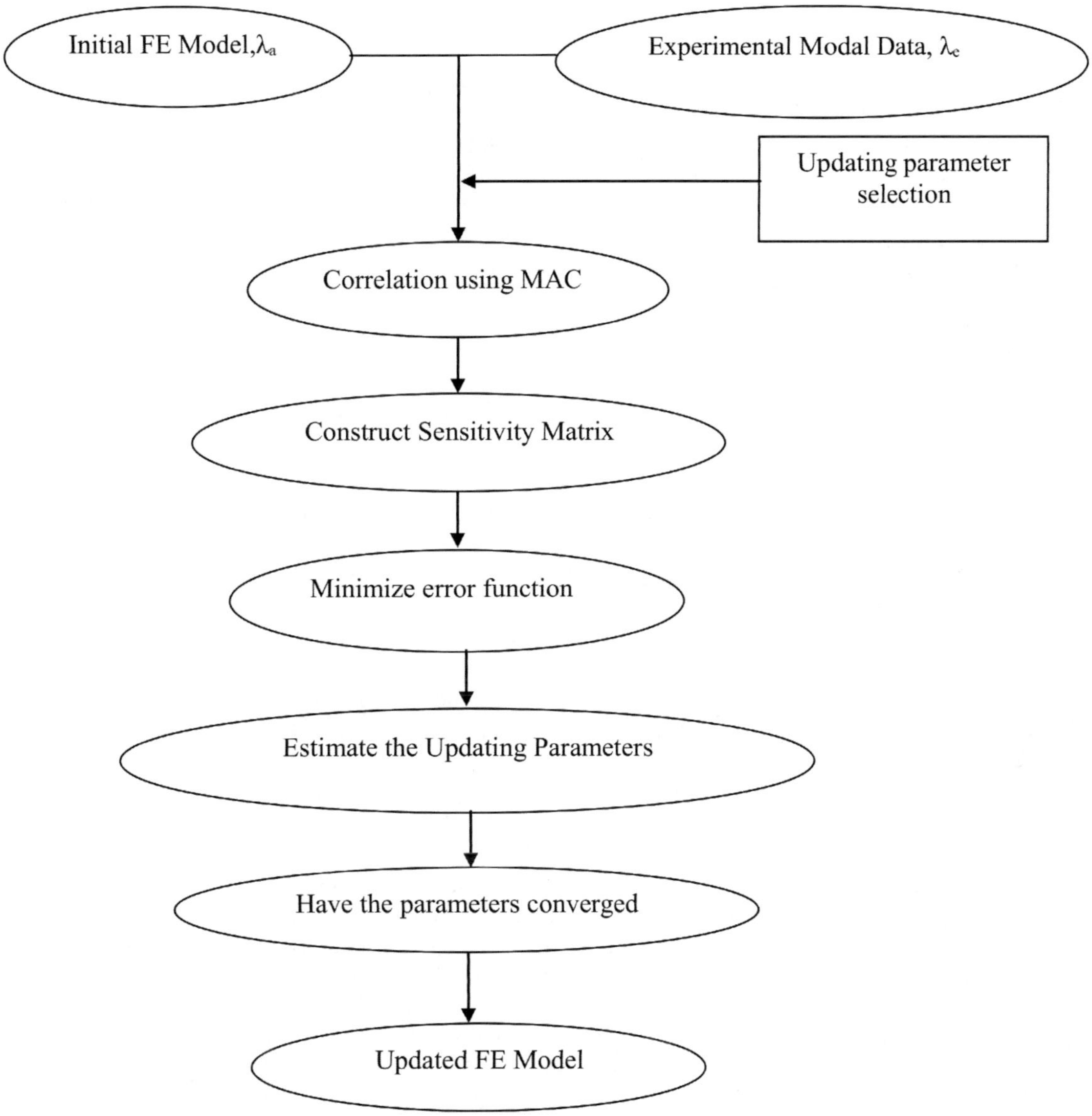

Figure 1. Simplified flow diagram for the model updating method [3].

Another study on other model updating method called 'coupled inverse Eigen-sensitivity method' is carried out [9]. This study had shown concern on updating a finite element model that take into account the acoustic loading on the structure. The study was applied to a 3D rectangular cavity backed by a flexible plate in order to investigate the effectiveness of the approach to obtain an accurate structural finite element model. It was also to study the influence of ignoring the effect of acoustic loading in updating process. Based on the results obtained, the approach shown to be effective in estimating the updating parameters and in predicting the natural frequencies, the eigenvectors and the frequency response function in case of complete and incomplete data accurately.

Model reduction technique can usually be used in model updating process. An investigation of an iterative method associated the model updating method with the model reduction technique is carried out by Li [10]. Using existing iterative method, the errors out coming from replacing the reduction matrix of the experimental model with the finite element model are not fully considered, which needs more iterations and computing time. A new iterative method with correction term related to errors is added to reduce to error produced in the replacement.

There is also a study to identify limitation of the existing iterative methods of model updating [11]. The paper raises a solution to overcome the limitation in the form of new method of finite element model updating that accepts both correlated and uncorrelated modes of updating. This is different to the existing iterative modal data based on methods of modal updating that are based on the prediction of availability of correlated mode pairs and hence cannot use uncorrelated mode shapes and corresponding natural frequencies in the updating process. Formulation of this new method was explained and a couple of numerical examples based on a beam structure were presented to validate the method.

2.2. Parameterization of Finite Element Model

In model updating, there is a process of estimating certain parameters of the models on the basis of dynamic tests carried out on the corresponding actual structures. Physical variables are measured at several points of the structures and recorded in the time domain during tests. These data will be transformed to the frequency and modal domains. The parameters are then obtained when the discrepancies between the experimental data and estimation of simulated model are minimized. In the method of minimizing the error function, the process has been formed in simple algorithm that is capable of solving both global and local phases of updating.

The aim or parameterization in mode updating is fitting the parameters of a given initial analytical model so that the model behaviour corresponds as closely as possible to the measured behavior. In finite element model updating, the selected parameters should be able to explain the uncertainties of the model. It is a requirement for the model output to be sensitive to the parameters. Joint in model is always facing difficulties in doing adequate parameterization using physical design variables such as stiffness and dimensions. Thus, geometric or generic parameterizations are views as significant application area.

As stated previously, direct methods of model updating are not well-approved because elements in the mass and stiffness matrices perform very poorly as candidate parameters [4]. This poor performance is because the stiffness matrix element values are dominated by the high-frequency modes, but instead the low-frequency modes are measured [5]. Mottershead et al. in their paper [12] had used geometric parameters like beam offsets for the updating of mechanical joints and boundary conditions. Ahmadian et al. [13] in separate paper had showed the effectiveness of parameterizing the modes at the element level and used both geometric parameters and element-modal parameters to update mechanical joints.

There is also a study by Terrell et al. [14] that shows parameters obtained from generic element and substructure transformations are able to increase the choices in parameters and therefore capable to correct structural errors. The method proposed assumes that substructure eigenvector matrix is to be made as effective to enforce the connectivity constrains. This method has been successfully demonstrated on a simple L shape structure with substructure is the corner. Approach with respect of this method also conducted by Weng et al. [15] in the study of substructure-based finite element model updating. Applications of the proposed substructure-based model updating to a frame structure and a bridge show that the method is computationally efficient.

A review on finite element model updating of spot welds in structural dynamics [16] also discuss the parameterization issue. Analysis on structures with spot welds, for example car body, will contain too much degree of freedom to be used in practice. Study of geometric non-linear characteristics of spot welded joints [17-19] is carry out by using finite element models. In order to validate the model, a six spot weld model is created as benchmark model. The results show that most models reach same level of accuracy after updating as they are all able to approximate the local stiffness due to the spot weld.

A method for finite element model updating in field of structural dynamics is proposed by Valle et al. [20] which focus on minimization of error function in the time domain. An error function is representing the discrepancies between the experimental data and the simulation test on the model. It is defined and minimized with respect to the parameters.

The paper that discuss physical realization of generic-element parameters in model updating [21], generic element models for updating are developed by forcing the model to have appropriate null space, positivity properties, total mass and moments of inertia and geometric symmetry. The parameters are obliged to follow the requirements of internal-force equilibrium at each node. Then the generic-element models obtained by this approach can be used in updating the model. In this study, the methodology for parameterization is carried out by updating a model using experimental data obtained from a cracked beam.

Esfandiari [22] had published a paper on finite element model updating of structures using frequency response function (FRF) data. Response sensitivities related to mass changes and stiffness parameters are indirectly determined using the decomposed form of frequency response function data. Findings of this study show the reliability of this method to identify the location and severity of parameter change at the elemental level in a structure. Furthermore, the finite element parameter estimation results are improved using higher excitation frequencies.

2.3. Regularization

There is a study on regularization method for finite element model updating by Ahmadian et al. [23] that highlights the problem of choosing side constraint and determining the regularization parameter in model updating. Author stated that noise contamination in test data is a problem in finite element model updating. Therefore, he demonstrated the regularization methods that can be used for modifying the ill-condition and noisy equations systems that arise during correcting the finite element models by using vibration measurements. Selection of good side constraint was shown to be important and lead to updated parameters with physical understanding. The methods considered in this study were based on singular value decomposition, cross-validation, and L-curves. The outcome received by applying these methods to a numerical example has provided the basis for a comparative study.

An overview of two approaches, a non-probabilistic fuzzy approach and a probabilistic Bayesian approach of dealing with uncertainty in model updating is presented by Simoen et al. in their review paper [24]. This work shows interest to the treatment of uncertainties in model updating problems with more focus on vibration-based finite element model updating. Both approaches stated in this paper are fundamentally different naturally because of their contrasting interpretations of uncertainty. This mean that the results are only can be compared qualitatively. Gathering results from probabilistic method can be quite challenging and computationally demanding. Generally, the most suitable method is dependent on the nature size of model updating problem, the available information and the needed end purpose of uncertainty quantification procedure.

Hansen [25] in his study proposed regularization methods for obtaining a solution of the inverse problem. He applied the theory that the conventional output error which is the vector if differences between the computed and measured responses, can be made subjectively small if the process of damage identification is allowed to behave poorly.

Tikhonov regularization is then applied [26] and from the experiences gained in model updating with simulated structures, it was found that Tikhonov regularization can give optimal solution when there is noise in measurement. An adaptive regularization approached for solving the nonlinear model updating inverse problem was presented by Li and Law [27]. The results obtained from proposed method in this study are well improved over the one obtained from Tikhonov regularization even though there are larger noise contamination in the measurements.

A paper of study of an approach for directly updating finite element model from measured incomplete vibration modal data with regularized algorithm is presented by Chen [28]. The suggested method is based on the relationship between the perturbation of structural parameters such as stiffness change and the modal data measurements of the tested structure which in this case is the measured mode shape readings. Structural updating parameters were selected at a critical point level to represent the modelling errors at the joints of structural elements. These parameters were then evaluated by iterative and direct solution procedure, which in the end gives optimize solutions. The Tikhonov regularization method incorporating the L-curve criterion is then applied to produce more effective solutions for the chose updating parameters in order to reduce the influence of modal measurements uncertainty. The findings of this study had demonstrated a reliable estimation method for finite element model updating using measurements of incomplete modal data.

References

[1] Zienkiewicz O C, Taylor R L, Zienkiewicz O C, Taylor R L 1977 *The finite element method* **3** McGraw-hill London)
[2] Mottershead J, Friswell M 1993 *Journal of sound and vibration* **167** 347-75
[3] Sinha J K, Friswell M I 2003 *Nuclear Engineering and Design* **223** 11-23
[4] Friswell M, Mottershead J E 1995 *Finite element model updating in structural dynamics* **38** Springer Science & Business Media)
[5] Friswell M I, Mottershead J E, Ahmadian H 2001 *Philosophical Transactions of the Royal Society of London A: Mathematical, Physical and Engineering Sciences* **359** 169-86
[6] Imregun M, Visser W 1991 *The Shock and vibration digest* **23** 9-20
[7] Arora V, Singh S, Kundra T 2009 *Mechanical Systems and Signal Processing* **23** 2113-29
[8] Mottershead J E, Link M, Friswell M I 2011 *Mechanical Systems and Signal Processing* **25** 2275-96
[9] Nehete D, Modak S, Gupta K 2015 *Mechanical Systems and Signal Processing* **50** 362-79
[10] Li W-M, Hong J-Z 2011 *Mechanical Systems and Signal Processing* **25** 180-92
[11] Modak S 2014 *Journal of Sound and Vibration* **333** 2297-322
[12] Mottershead J, Friswell M, Ng G, Brandon J 1996 *Mechanical Systems and Signal Processing* **10** 171-82
[13] Ahmadian H, Gladwell G, Ismail F 1997 *Journal of Vibration and Acoustics* **119** 37-45
[14] Terrell M, Friswell M, Lieven N 2007 *Journal of sound and vibration* **300** 265-79
[15] Weng S, Xia Y, Xu Y-L, Zhu H-P 2011 *Computers & Structures* **89** 772-82
[16] Palmonella M, Friswell M I, Mottershead J E, Lees A W 2005 *Computers & structures* **83** 648-61
[17] Sonsino C, Fricke W, De Bruyne F, Hoppe A, Ahmadi A, Zhang G 2012 *International Journal of Fatigue* **34** 2-16
[18] Sonsino C, Radaj D, Brandt U, Lehrke H 1999 *International Journal of Fatigue* **21** 985-99
[19] Chang B, Shi Y, Dong S 1999 *Journal of Materials Processing Technology* **87** 230-6
[20] Zapico-Valle J L, Alonso-Camblor R, González-Martínez M P, García-Diéguez M 2010 *Mechanical Systems and Signal Processing* **24** 2137-59
[21] Ahmadian H, Mottershead J, Friswell M 2002 *Journal of Vibration and Acoustics* **124** 628-33
[22] Esfandiari A, Bakhtiari-Nejad F, Sanayei M, Rahai A 2010 *Computers & structures* **88** 54-64
[23] Ahmadian H, Mottershead J, Friswell M 1998 *Mechanical Systems and Signal Processing* **12** 47-64
[24] Simoen E, De Roeck G, Lombaert G 2015 *Mechanical Systems and Signal Processing* **56** 123-49
[25] Hansen P C 1998 *Rank-deficient and discrete ill-posed problems: numerical aspects of linear inversion* **4** Siam)
[26] Weber B, Paultre P, Proulx J 2009 *Mechanical Systems and Signal Processing* **23** 1965-85
[27] Li X, Law S 2010 *Mechanical Systems and Signal Processing* **24** 1646-64
[28] Chen H-P, Maung T S 2014 *Journal of Sound and Vibration* **333** 5566-82

Mechanical and microstructural characterization of single and double pass Aluminum AA6061 friction stir weld joints

N H Othman[1], L H Shah[1], and M Ishak[1]
[1]Manufacturing Focus Group, Faculty of Mechanical Engineering, Universiti Malaysia Pahang, 26600, Pekan, Pahang, Malaysia

E-mail: nurul_hidayah213@yahoo.com

Abstract. This study focuses on the effect of single pass (SP), double sided pass (DSP) and normal double pass (NDP) method on friction stir welding of aluminum AA6061. Two pieces of AA6061 alloy with thickness of 6 mm were friction stir welded by using conventional milling machine. The rotational speeds that were used in this study were 800 rpm, 1000 rpm and 1200 rpm, respectively. The welding speed is fixed to 100 mm/min. Microstructure observation of welded area was studied by using optical microscope. Tensile test and Vickers hardness test were used to evaluate the mechanical properties of this specimen. Mechanical property analysis results indicate that at low rotational speeds, defects such as surface lack of fill and tunneling in the welded area can be observed. Vickers hardness of specimens however did not vary much when rotational speed is varied. Welded specimens using single pass method shows higher tensile strength and hardness value compared to both double pass methods up to 180.61 MPa. Moreover, DSP showed better tensile test and hardness test compared to NDP method. The optimum parameters were found to be single pass method with 1200 rpm of rotational speed. Therefore economically sound to only perform SP method to obtain maximum tensile strength for AA6061 FSW with thickness of 6 mm.

1. Introduction

Friction stir welding (FSW) is a welding process that involves solid state joining technique. FSW has expended rapidly since its development in 1991 by The Welding Institute (TWI), UK. This application can be found in a wide variety of industries such as automotive, aerospace, maritime and railway. This joining process has been regarded as one of the most significant joining process in recent years due to its many advantages like energy efficiency, environmental friendliness and versatility. As compared to conventional welding process, FSW requires less energy and uses no gas and flux that makes the process environmentally friendly. FSW joining process does not involve any use of filler metals and unlike fusion welding; any aluminum alloy can be joined without concern for the suitability of composition[1-8].

The FSW method that is commonly used in most research is single pass method (SP) where pin tool is inserted into the abutting edge of plate to be joined and transverse along the line of joint, passing only once. Several papers have suggested variations on this approach, namely the number of welding passes for a single joint. An alternative approach to this research is by using double sided pass (DSP) or multiple pass name normal double pass (NDP). Double sided pass (DSP) method is the process when a single pass process is made on both sides of the plate. On the other hand, multiple pass or normal double pass method (NDP) is a process where the pin passes along the same welding path twice or more. The illustrations of passes method are shown as figure 1.

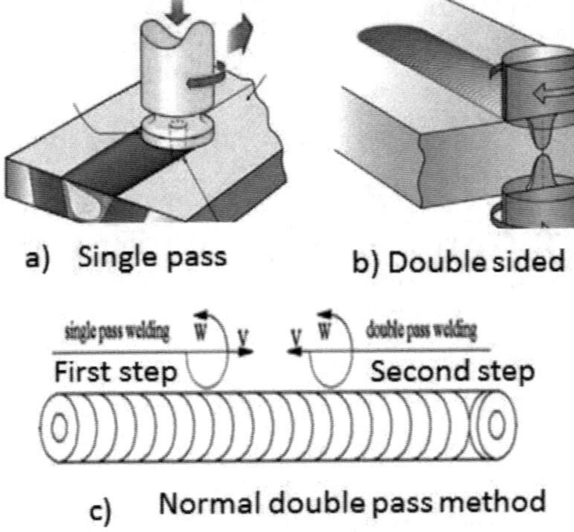

Figure 1. Welding passes method. a) Single pass, b) Double sided pass method and c) Normal double pass method [9, 10].

Several papers have reported the effect of double or multiple passes on aluminium alloy's mechanical property. Mehra (2012) conducted single and double sided FSW on AL19000 aluminium alloy and stated that higher ultimate tensile strength can be obtained from joints fabricated using double pass method [11]. This is also supported by a recent study by Sathari (2013) using AA1100 aluminium alloy, where the tensile strength have even doubled at a certain rotational speed [12]. In addition to that, Johannes and Mishra (2006) succeeded in creating a larger superplastic region of 7075 aluminium alloy after undergoing multiple FSW process, despite the single pass materials having the largest elongations [13]. Some researchers have even opt for utilizing multiple pass FSW method to improve the mechanical properties of die cast aluminium alloys such as ADC12 and A356 [14, 15]. However, Rebecca Brown (2009) contradicts the above claims and concluded that single pass have higher tensile strength compared to double and triple pass method when investigating multi-pass FSW on 7050-T7451 aluminium alloy [9].

2. Experimental method

The base material used in this study is AA6061 aluminium alloy plate with thickness of 6 mm. A pair of plate with dimensions of 100 mm × 50 mm were abutted and clamped rigidly at the backing plate for welding. Table 1 shows the chemical compositions of AA6061 alloy that was check using Foundry Mass Spectrometers machine.

Table 1. Chemical compositions of AA6061 (wt.%).

Element	Al	Si + Fe	Mg	Cu	Mn	Cr	Other
AA6061	97.8	0.80	0.84	0.22	0.096	0.1	0.144

The FSW was carried out by using Milltronics Partner milling machine with square butt joint configuration of samples. The welding tool material is H13 steel with a normal cylindrical pin shape. The tool was processed by using a conventional lathe machine.

Table 2 presents the details of the welding parameters and tool dimensions used for the FSW. For both double passes, the specimen was cooled down after the first pass to ambient temperature before initiating the second pass to eliminate the effect of accumulative heating. For the DSP method, the plates were flipped over about an axis along the weld after the first pass was made, so that the two welding passes start at the same position along the joint interface, and the advancing side of the second pass was over the retreating side of the first pass weld. The tilt angle is kept constant at a 3° angle.

Mechanical characterizations were conducted using tensile test and Vickers hardness test. The tensile test was prepared according to the ASTM-E8 standard. Figure 2 shows the dimension for tensile specimen. Micro hardness test was done by using Vickers hardness test with 200 gf load for 10 seconds. The results were recorded and analyzed accordingly. The microstructure of the weld specimens was observed using an optical microscope

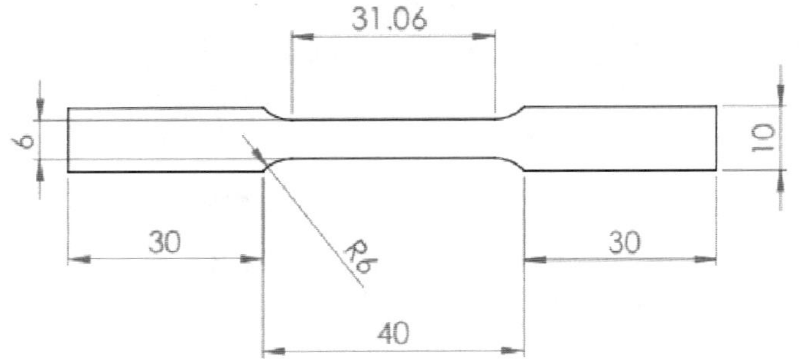

Figure 2. Tensile specimen dimensions.

Table 2. Welding parameters and tool dimensions for FSW.

Process parameters	Values
Rotational speed(rpm)	800, 1000, 1200
Welding speed (mm/min)	100
D/d ratio of tool	3.0
Pin length, l (mm)	5.5(single pass (SP), normal double pass (NDP)), 3.0(double sided pass (DSP))
Tool shoulder diameter, D(mm)	18
Pin diameter, d(mm)	6
Holder diameter, a(mm)	10
Holder length, b(mm)	20
Shoulder length, c(mm)	24

3. Result

3.1. Microstructure Analysis

Figure 3 shows the cross section of the FSW joint of SP, NDP and DSP specimen, respectively. The microstructural image was taken by using an optical microscope with $\times 10$ magnifications. The major FSW regions can be observed, namely the base metal (BM), heat affected zone (HAZ), thermo mechanically affected zone (TMAZ) and weld zone (WZ). The DSP specimen cannot be labeled with advancing or retreating side, since both sides experienced the advancing and retreating stir flow. The WZ and HAZ for this specimen show a larger region due to the double sided pass. Moreover, the TMAZ region can be seen in both upper and lower part of the figure 3.

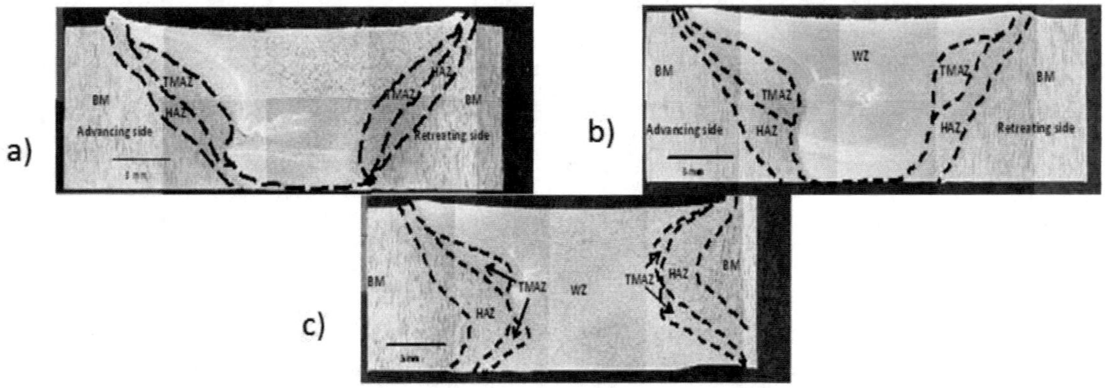

Figure 3. Cross section area for FSW joint. a) SP joint, b) NDP joint and c) DSP joint.

Figure 4, 5 and 6 show the grain structures of each FSW region for NDP and DSP, respectively. Figure 4(a), figure 5(a) and figure 6(a) shows similar elongated grain structure for BM that involved negligible heat input. Grain structures at the WZ for both groups (figure 4(b), 5(b) and figure 6(b)) show fully recrystallization, where the heat generated from the interaction of the pin tool deformed the grain boundaries [16]. In other words, the recrystallization nuclei can be created more easily at the high speed welding and it will get more refinement grain structure that lead to the higher tensile strength [10].

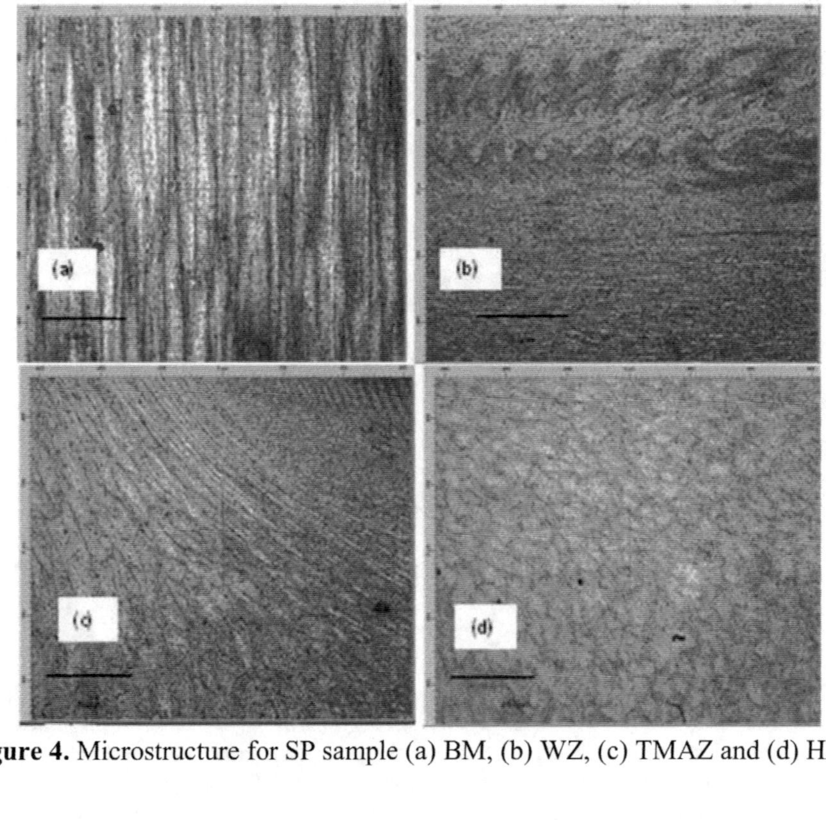

Figure 4. Microstructure for SP sample (a) BM, (b) WZ, (c) TMAZ and (d) HAZ.

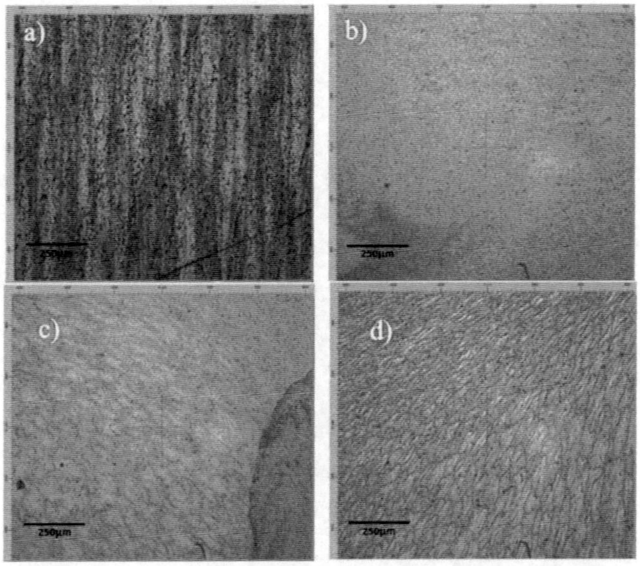

Figure 5. Microstructure for NDP sample (a) BM, (b) WZ, (c) TMAZ and (d) HAZ.

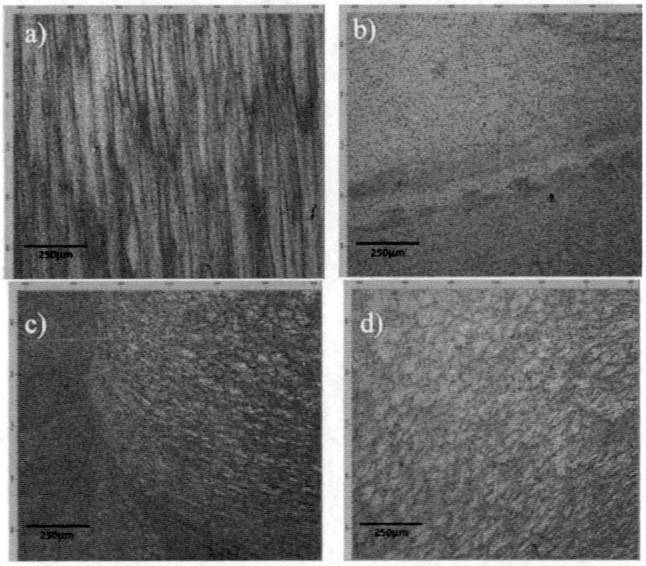

Figure 6. Microstructure for DSP sample (a) BM, (b) WZ, (c) TMAZ and (d) HAZ.

In the TMAZ, it can be observed that the grain size was slightly bigger than the weld nugget. Grain structure at TMAZ as shown in figure 5(c) and figure 6(c) have slightly changed in grain structure due to the heat input that is sufficiently heated to undergo deformation from its original grain boundaries. While TMAZ for figure 4(c) shows smaller grain size compare to others. TMAZ region undergo plastic deformation but recrystallization did not take place in this region due to inadequate deformation strain. Grain structure in figure 5(c) shows coarser grain structure compared to grain structure in figure 6(c).

In HAZ, the aluminum alloy experienced thermal cycle but only little plastic deformation occurred. It can be seen that the average grain size of SP result as shown in figure 4(d) give small grain size compare to other passes and HAZ for NDP (Figure 5(d)) slightly increased compared to DSP grain structure (Figure 6(d)). Grain size for NDP shows coarsening grain that will degrade the mechanical property. The increase in grain size consequently decreased the tensile strength [17].

3.2. Microhardness Evaluation

Figure 7 shows the Vickers hardness profile of the cross section welded area for the SP, NDP and DSP method with distinctive welding regions. All groups showed relatively similar results, indicating that the welding methods have little affect the specimen's Vickers hardness. The based metal hardness for AA6061 is 70.5 HV. It can be seen from the figure that the hardness decreases from base metal to the HAZ region, followed by a slight increase at the WZ. The highest hardness at the WZ is 50.6 HZ from the SP group. The lowest value for SP, NDP and DSP is 39.3 HV, 40.8 HV and 35.5 HV, respectively. All the lowest values are typically located in the HAZ due to recrystallization. The HAZ regions have coarsening of the grain size due to the plastic deformation and fracture occur at HAZ region [18].

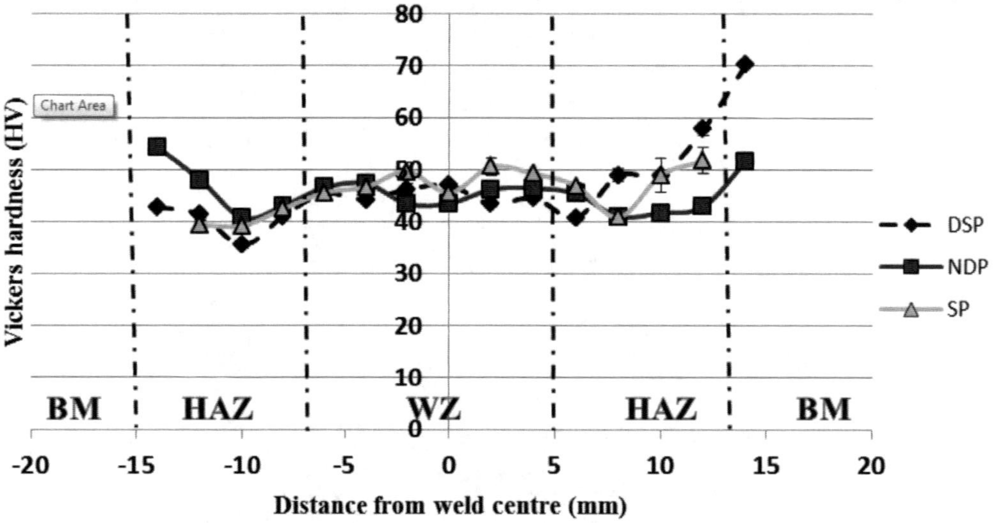

Figure 7. Vickers hardness profile for single pass (SP), normal double pass (NDP) and double sided pass (DSP) method.

3.3. Tensile Test

Figure 8 shows the tensile test results for all the groups with varying rotational speed. All specimens fractured at the HAZ. The highest tensile strength is 180.61 MPa which occurred at rotational speed of 1200 rpm using the single pass method. The highest tensile strength for NDP and DSP is 164.37 MPa at 800 rpm and 166.13 MPa at 1000 rpm, respectively.

From the graph observation, it can be seen that SP collectively has higher tensile value compared to other groups due to the smaller grain size compare to other passes method. This finding is in line with a previous report that states that the tensile value will decrease slightly at the second pass, increasing again at the fourth pass [9].Furthermore, no conclusive pattern can be seen linking the rotational speed to tensile value, since each group yielded different results.

The DSP has a slightly higher tensile value compared to NDP. It is probably due to the DSP having both advancing and retreating side at either specimen sides, while NDP only have advancing or retreating at either side of the plate. Since it is proven from the results that SP method shows highest tensile strength compared to both double passes, it is therefore economically sound to only perform SP method to obtain maximum tensile strength for AA6061 FSW with thickness of 6 mm.

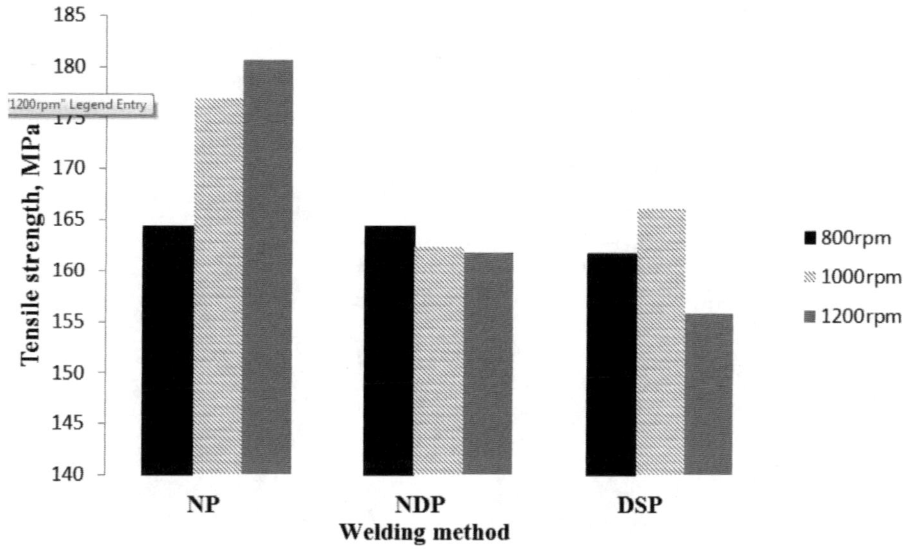

Figure 8. Tensile test results for group 1, group 2 and group 3.

4. Conclusion

6 mm thick aluminum 6061 plates were successfully welded by using friction stir welding process with varying welding method and rotational speed. The conclusions are as follows:
All specimens resulted in sound quality joints with minimal defects. The best weld appearance from single pass, normal double pass and double sided passes, is by using rotational speeds of 1000 rpm, 800 rpm and 800 rpm, respectively. The best weld appearance can be observed in lower rotational speeds. The specimen microstructure shows similar result for all groups. Fine grain size can be seen in WZ and TMAZ and coarser grain size can be seen in the HAZ. The WZ and HAZ for the double sided pass show a larger region due to the stirring of both sides. Moreover, a distinguished TMAZ region can be seen in both upper and lower part of the sample.

No distinct change can be seen from the Vickers hardness of all groups. The highest value of the WZ is 50.6 HV from the SP group. All groups show lowest value at the HAZ. The SP method yielded the highest tensile value of 180.61 MPa compared to NDP and DSP method with ultimate tensile values of 164.37 MPa and 166.13 MPa, respectively. No conclusive pattern can be observed from the tensile test results connecting the rotational speed and strength value.

Acknowledgements

The authors would like to be obliged to Universiti Malaysia Pahang for providing laboratory facilities and financial assistance.

References

[1] Gibson B T, Lammlein D H, Prater T J, Longhurst W R, Cox C D, Ballun M C, et al. 2014 *Journal of Manufacturing Processes* **16** 56-73
[2] Mishra R S, Ma Z Y 2005 *Mater Sci Eng* **50** 1-78
[3] Burford D 2006 *6th AIAA Aviation Technology, Integration and Operations Conference (ATIO)* (Wichita, Kansas)
[4] Xu X, Yang X, Zhou G, Tong J 2012 *Materials and Design* **35** 175–83
[5] Shanmuga Sundaram N, Murugan N 2010 *Materials & Design* **31** 4184-93
[6] Pourahmad P, Abbasi M 2013 *Transactions of Nonferrous Metals Society of China* **23** 1253-61
[7] Coelho R S, Kostka A, dos Santos J F, Kaysser-Pyzalla A 2012 *Materials Science and Engineering: A* **556** 175-83

[8] Assidi M, Fourment L, Guerdoux S, Nelson T 2010 *International Journal of Machine Tools and Manufacture* **50** 143-55
[9] Brown R, Tang W, Reynolds A P 2009 *Materials Science and Engineering: A* **513–514** 115-21
[10] Song K H, Fujii H, Nakata K 2009 *Materials Transactions* **50** 832-6
[11] Mehra S, Dhanda P, Khanna R, Goyat N S, Verma S 2012 *International Journal of Scientific & Engineering Research* **3** 2229-5518
[12] Sathari N A A, Shah L H, Razali A R 2014 *Journal of Mechanical Engineering and Sciences (JMES)* **7** 1053-61
[13] Johannes L B, Mishra R S 2007 *Materials Science and Engineering: A* **464** 255-60
[14] Nakata K, Kim Y G, Fujii H, Tsumura T, Komazaki T 2006 *Materials Science and Engineering: A* **437** 274-80
[15] Ma Z Y, Sharma S R, Mishra R S 2006 *Scripta Materialia* **54** 1623-6
[16] Dawood H I, Mohammed K S, Rajab M Y 2014 *Advances in Materials Science and Engineering* **2014** 10
[17] Rajakumar S, Muralidharan C, Balasubramanian V 2010 *Transactions of Nonferrous Metals Society of China* **20** 1863-72
[18] Mathers G 2002 *The welding of aluminium and its alloys* **32** (Cambridge, England: Woodhead publishing)

IOP Conf. Series: Materials Science and Engineering **100** (2015) 012017 doi:10.1088/1757-899X/100/1/012017

Dynamic performances analysis of a real vehicle driving

M A Abdullah[1,2], J F Jamil[1] and M A Salim[1,2]
[1]Faculty of Mechanical Engineering, Universiti Teknikal Malaysia Melaka, Hang Tuah Jaya, 76100, Durian Tunggal, Melaka, Malaysia.
[2]Centre for Advanced Research on Energy, Universiti Teknikal Malaysia Melaka, Hang Tuah Jaya, 76100 Durian Tunggal, Melaka, Malaysia.

Email: mohdazman@utem.edu.my

Abstract. Vehicle dynamic is the effects of movement of a vehicle generated from the acceleration, braking, ride and handling activities. The dynamic behaviours are determined by the forces from tire, gravity and aerodynamic which acting on the vehicle. This paper emphasizes the analysis of vehicle dynamic performance of a real vehicle. Real driving experiment on the vehicle is conducted to determine the effect of vehicle based on roll, pitch, and yaw, longitudinal, lateral and vertical acceleration. The experiment is done using the accelerometer to record the reading of the vehicle dynamic performance when the vehicle is driven on the road. The experiment starts with weighing a car model to get the center of gravity (COG) to place the accelerometer sensor for data acquisition (DAQ). The COG of the vehicle is determined by using the weight of the vehicle. A rural route is set to launch the experiment and the road conditions are determined for the test. The dynamic performance of the vehicle are depends on the road conditions and driving maneuver. The stability of a vehicle can be controlled by the dynamic performance analysis.

1. Introduction

Vehicle riding comfort is important on automobile industries which have impact on the drivers of the vehicle, the safety on road and the efficient of the transportation. A running vehicle causes vibration and affecting the riding performance of a vehicle. The vibration from the vehicle causes the noise and reduces the riding comfort of the vehicle. Suspension system is needed to control the vibration influence by the road irregularity and the speeds of the vehicle [1, 2]. The vehicle vibration also generated by the running engine, the forces of actuators and the resonant frequency of the body [3]. Modeling of vehicle dynamics performance is important to absorb the unneeded vibration from the vehicle to increase the dynamic performance as well as the drivers comfort. The critical part of the dynamics is the suspension system including the tires of the vehicle because of the road irregularity so the vibration is absorbs by the system [4].

The suspension system is designed by considering the human effect because human weight can change the riding comfort of a vehicle. The acceleration on the vehicle dynamics performance is different when the vehicle has passengers. This causes the suspension plays the important role to control the riding performance of the vehicle [5].There are several researchers has done improvement to the vehicle dynamic performance of a vehicle such as K. Dongshin that introduces the electronic stability program (ESP) system for vehicle when in emergency situation and A. Sohel introduces the predictive control algorithm for yaw stability management of a vehicle [6].

The analysis of the vehicle dynamic performances is also important for the vibration study for the energy regenerative suspension system (EReSS). The EReSS is applicable on the suspension system that uses the vertical vibration of the vehicle [7]. Vehicle parameters are important for variety of purposes such as the center of gravity (COG) of the vehicle. The COG of a vehicle is rarely provided by the automobile manufacturers. The estimation process needs some knowledge of the COG to have a correct bias for gyroscope measurement [8]. The COG location is important considering the passenger and load distribution. The location is difficult to be estimated because the COG positions changes when the weight on the vehicle changes. Vehicles that have higher COG are more tends to rollover accidents [9]. The COG of vehicle is important to be calculated for producing vehicle in automobile industries for the safety of the user and the public traffic [10].

2. Methodology

2.1. Experimental set up

The experiment done is for determining the longitudinal, lateral, and vertical accelerations, roll, pitch and yaw effect on the vehicle while driving on irregular road surface. Firstly, the route for the experiment is determined to ease the experiment process. Figure 1 illustrates the route that is used for the experiment. It is a combination of urban and highway roads with several corners, roundabouts bumpers and turns. The length is 9.6 km.

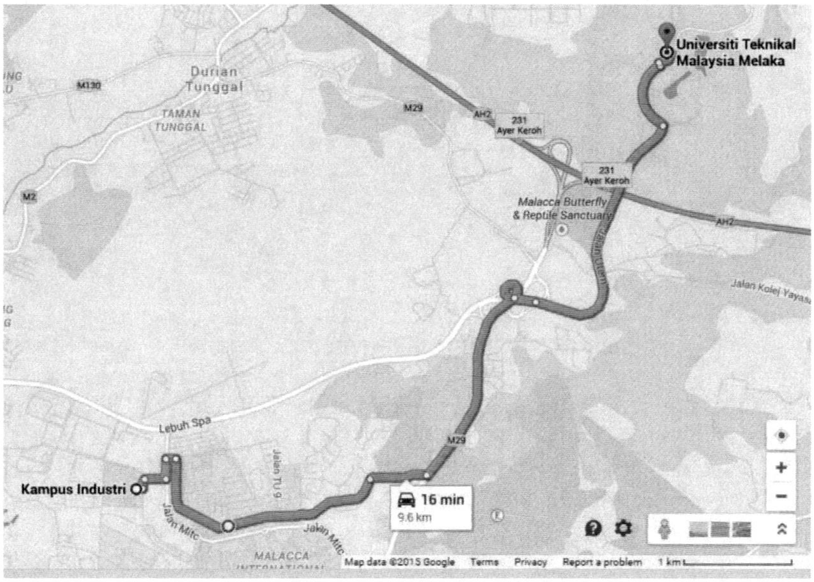

Figure 1. The route use for the experiment.

Next, the car model for the experiment is selected. The specifications of the vehicle are shown in table 1.

Table 1. Vehicle specifications.

Criteria	Specifications
Model Code	4G13
Transmission	F5M21
Fuel system	MPI
Max.power	55kW(74hp)@6000rpm
Max.torque	108Nm@3000rpm
Bore & Stroke	71x82
Displacement	1299cc
Max.speed	164 km/h
Overall length	4360mm
Overall width	1680mm
Overall height	1385mm
Wheelbase	2500mm
Kerb weight	980 kg

After that, the vehicle will be weight to find the center of gravity (COG). The DAQ used for the experiment is Lego Mindstorms EV3 as shown in figure 2 (a). Figure 2 (b) shows the car is being weighted using load sensor. The weighting process starts with adding extra load on the load sensor so that the car is in inclined position which is 5 cm and the data is recorded. Then, the car is weighted on normal condition.

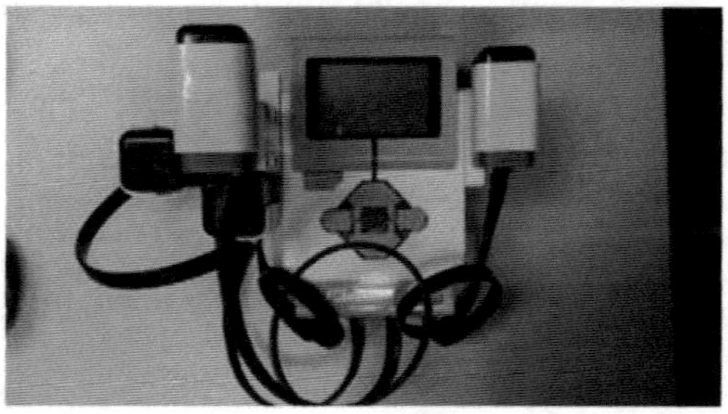

(a)

(b)

Figure 2. Experimental set up (a) Lego Mindstorms EV3; (b) car being weighted.

2.1. Experimental procedure

The experiment process is done according to the road set for the experiment. The data is recorded by the DAQ Lego Mindstorm EV3 data logger. Figure 3 shows the Lego Mindstorms EV3 is put on the COG of the car. The equipment is ensured to be fixed on the COG only to avoid any error on recording the data acquisition. While driving, the road condition and the velocity of the car every 30 s is recorded. The road

IOP Conf. Series: Materials Science and Engineering **100** (2015) 012017 doi:10.1088/1757-899X/100/1/012017

used is according to the route that has been set up. All the data required for the experiment is recorded by the Lego Mindstorms EV3 equipment as all sensors needed for the reading is attached on it. The road condition for the experiment is identified to know the effect of the reading of the dynamic performance of the car. The time of the experiment is recorded.

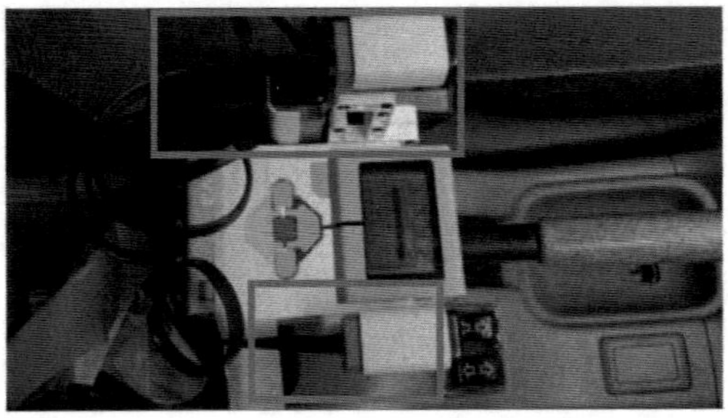

Figure 3. Lego Mindstorms EV3 put on the center of gravity (COG). Top box shows the gyro meters and bottom box shows the accelerometer.

3. Result and discussions

The center of gravity (COG) is obtained from calculation and used as the place to put the Lego Mindstorms EV3. The calculation starts from the car in normal condition as shown in figure 4. The car is at horizontal condition with no slope with the force distribution acting on the vehicle. The force is solved using equation (1) which is moment law of equilibrium.

$$\sum M = Fd \tag{1}$$

where

F is the force in N,
d is the distance in m.

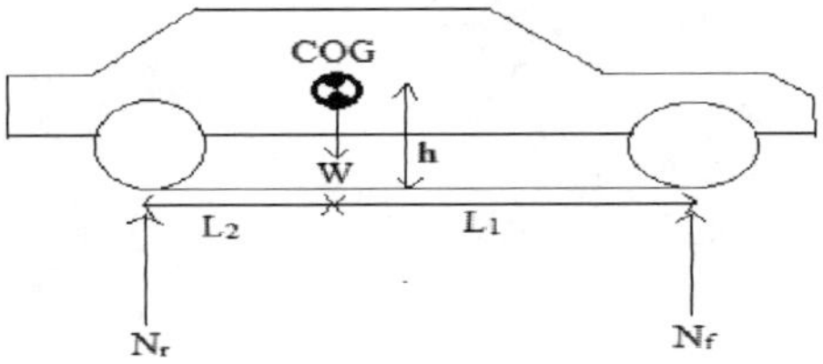

Figure 4. Force acting on car in normal condition.

The calculation to find L_1:

Wheelbase, $L_1 + L_2 = 2.5m$

Mass of vehicle, $m = 923.49kg$, $W = mg = (923.49kg)(9.81m/s^2) = 9059.44N$

Rear wheel mass, $N_r = 339.27kg = (339.27kg)(9.81m/s^2) = 3328.24N$

$$N_r.(L_1 + L_2) + W.L_1 = 0$$
$$L_1 = 0.918m$$

The calculation to find L_2 :

$$L_2 = 2.5m - 0.918m = 1.582m$$

The calculation of the COG position continues by using the car weight on inclined position which have slope when weighing the car. Figure 5 shows the car on inclined position.

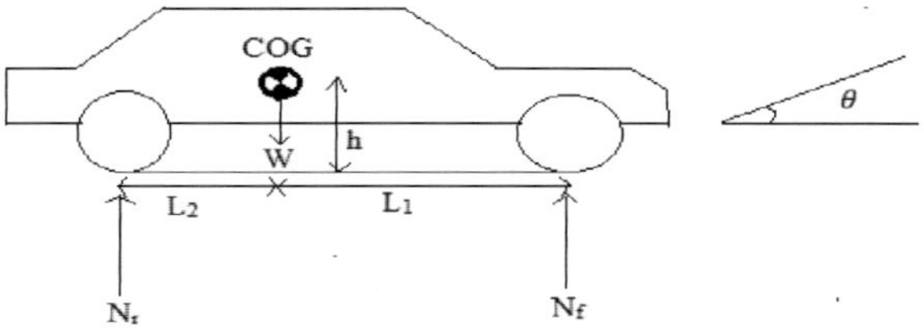

Figure 5. Force acting on car in inclined position.

The calculation to find h :

$$N_r \cos\theta + N_f \cos\theta - W \cos\theta = 0$$

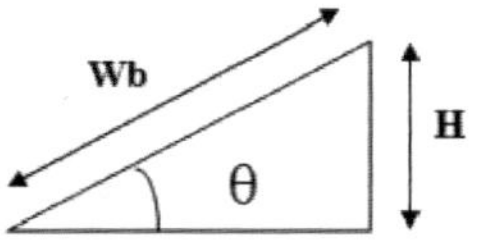

$$\sin\theta = H / Wb$$
$$\theta = \sin^{-1} \frac{0.05}{2.5}$$
$$\theta = 1.15°$$
$$N_f \cos\theta(L_1 + L_2) + W \sin\theta(h_a) - W \cos\theta(L_2) = 0$$
$$h = 0.34m$$

From the calculation, the COG is located at L_1 = 1.582 m from the front wheel and h_a = 0.34 m from the horizontal. The Lego Mindstorm EV3 is then put on the COG for the test. Figure 6 shows the data for roll of the vehicle. The roll is the movement or rotational motion on the x-axis. At 39.34 s, the value of the roll moment is 3 °/s where the vehicle is cornering at the roundabout. When the vehicle is stop in a traffic light at 134.4 s, the value of the moment is -2 °/s. The roll moment is high when the vehicle is moving through a bumper with reading 6 °/s. It is get that the roll moment will affect when the vehicle is at cornering phase.

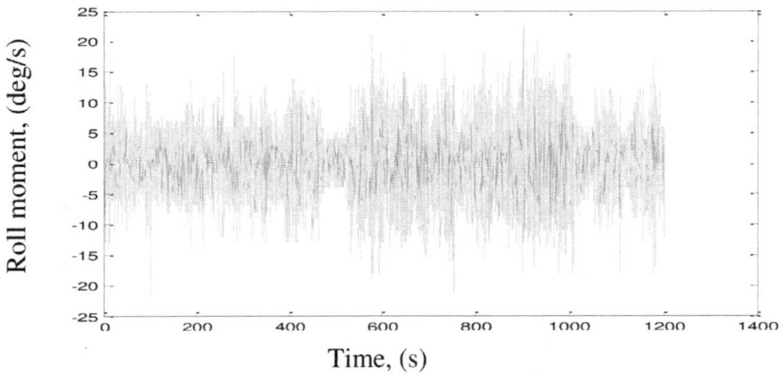

Figure 6. Roll moment of the vehicle.

Figure 7 shows the pitch moment of the vehicle that acting at y-axis. The value of pitch is high when the vehicle is passing through the bumper at 48.14 s with 32 °/s. The pitch is 28 °/s at 124.3 s as the vehicle stop at the traffic light and passing through a bumper again with the value of 38 °/s. The pitch of the vehicle is affected when the car is passing through the bumper.

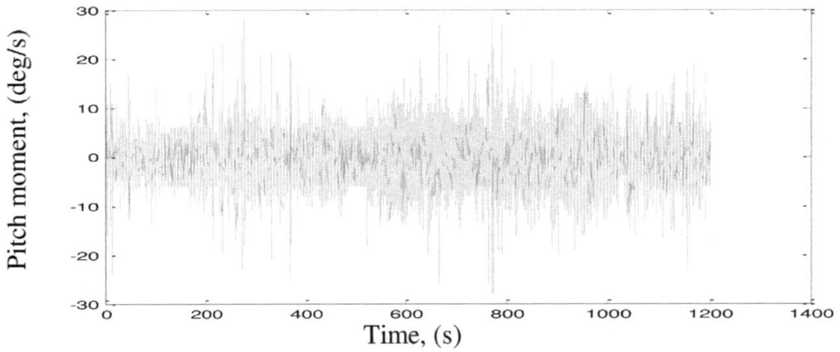

Figure 7. Pitch moment of the vehicle.

Figure 8 illustrates the data for yaw of the vehicle. The yaw moment is the moment acting on the z-axis. The yaw of the vehicle is 32 °/s at 40.34 s when the vehicle is turn in a roundabout and at 68.92 s the value of the yaw is 27 °/s where the vehicle was turn to a right junction. At 122.1 s the vehicle stopped at the traffic light and give out the yaw value of 8 °/s. The higher the yaw value, the vehicle will overseer.

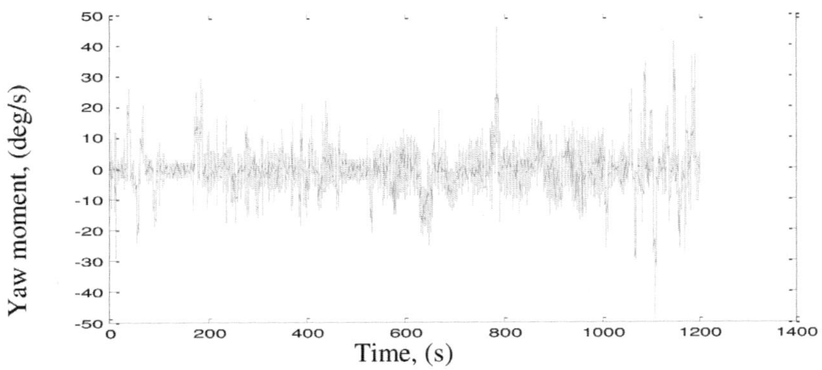

Figure 8. Yaw moment of the vehicle.

Figure 9 shows the longitudinal acceleration of the vehicle which is acting on the x-axis (A_x). The longitudinal acceleration is affected by various conditions of the road surface and including the acceleration, braking and aerodynamics of the vehicle. The acceleration is 1.029 m/s^2 when the vehicle is taken a corner at 172 s and 1.372 m/s^2 when the vehicle is passing through a bumper at 283.7 s.

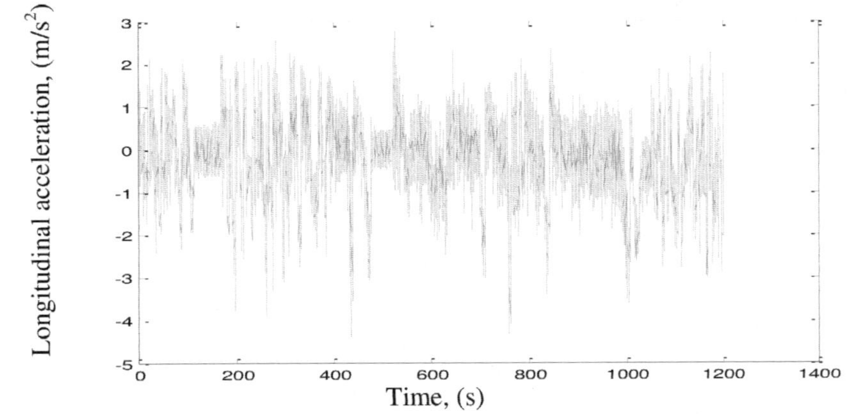

Figure 9. Longitudinal acceleration of the vehicle (A_x).

Figure 10 shows the lateral acceleration on the vehicle acting on the y-axis (A_y). At 57.16 s, the reading is 2.646 m/s^2 which the vehicle is taking a junction to the right. When the vehicle taking a junction to the left at 93.9 s, the acceleration is 2.156 m/s^2 and the acceleration is smaller when the vehicle stops at a traffic light at 125.4 s with reading of 0.098 m/s^2. The vehicle is passing through a bumper and the reading of the lateral motion is higher with 4.165 m/s^2 at 257.3 s.

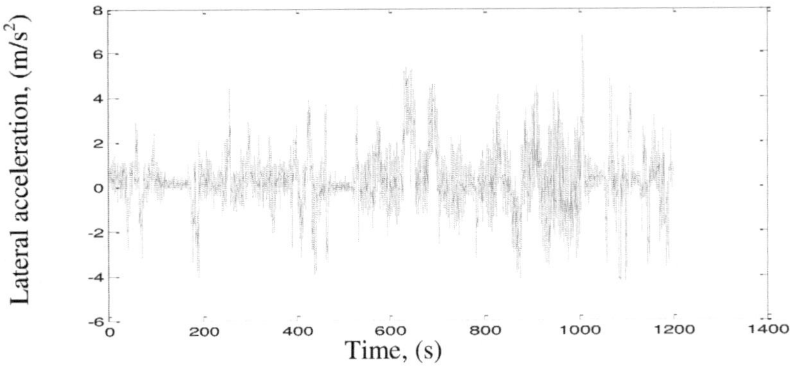

Figure 10. Lateral acceleration of the vehicle (A_y).

Figure 11 shows the vertical acceleration of the vehicle that acts on the z-axis (A_z). The acceleration of the vehicle is 12.1 m/s^2 at 23.86 s when the vehicle is moving on a straight road with irregular road surfaces that causes bumping and the vertical vibration is higher. The vehicle is moving and stops at a traffic light at 114.3 s with the acceleration of 10.93 m/s^2. The value is also high because when braking, the vertical motion is affected. At 334.2 s, the vehicle is on a straight road which accelerates after taking a corner that made the vertical acceleration higher with 14.26 m/s^2.

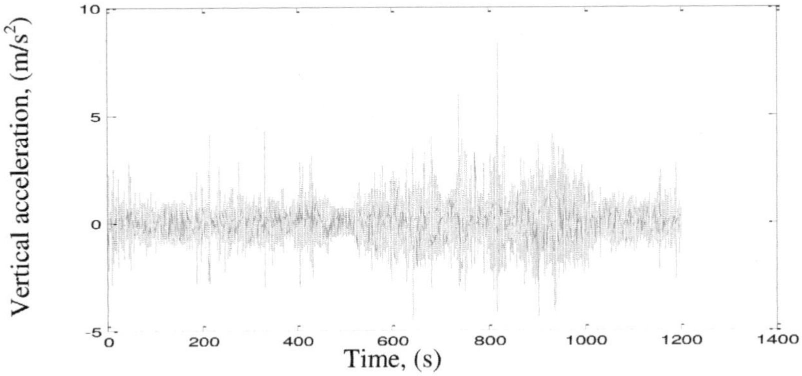

Figure 11. Vertical acceleration of the vehicle (A_z).

Figure 12 shows the vehicle speed during the experiment. The speed of the vehicle is inconsistent with disturbance such as road bumps and corners. The vehicle is at highest speed when on a straight road.

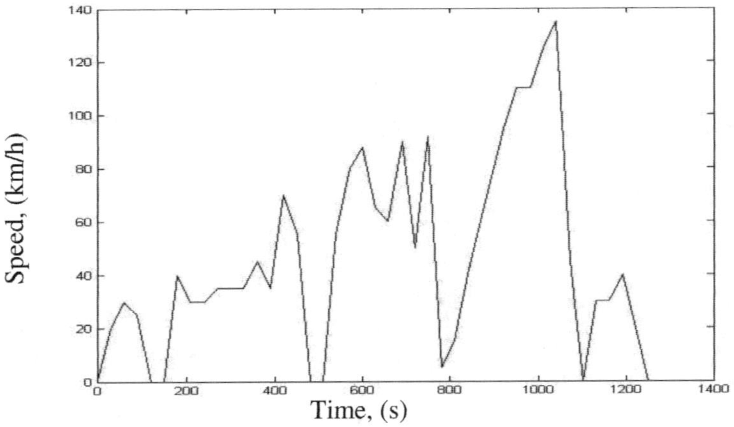

Figure 12. Vehicle speed during experiment.

4. Conclusion

In this research, the dynamic performance of the vehicle is analyzed with the effect of the various road conditions. The center of gravity (COG) of the car is important to determine the dynamic performance as all the forces and moments are acting at the COG. The stability of the vehicle depends on the road conditions and driving maneuvers. Other than that, this performance will help the manufacturer to investigate the safety and comfort of the vehicle itself. Besides, it helps in controlling the stability of the vehicle while driving on different road inputs. The results are also potential to be benefited for energy harvesting from vibration.

Acknowledgements

The authors gratefully acknowledged the Advanced Vehicle Technology (AcTiVe) research group of Centre for Advanced Research on Energy (CARe), the financial support from Universiti Teknikal Malaysia Melaka and The ministry of Education, Malaysia under Short Term Research Grant, Grant no. PJP/2014/FKM(10A)/S01330 and Fundamental Research Grant Scheme (FRGS), grant no.: FRGS/2013/FKM/TK06/02/2/F00165.

References

[1] He L, Qin G, Zhang Y, Chen L. Non-stationary Random Vibration Analysis of Vehicle with Fractional Damping. International Conference on Intelligent Computation and Automation; 20-22 October; Hunan: IEEE; 2008. p. 150-7.
[2] Abdullah M A, Ramli F R, Lim C S 2014 *Applied Mechanics and Materials* **645-466** 13-7
[3] Nagai N, Hashimoto S, Hiroki Y, Fujikura Y, Takahashi J, Kumagai S, et al., Construction of Multi-Axis Vibration Control System for Vibration Reproduction in Vehicles. International Conference on Advanced Mechatronic Systems; 2012 18-21 September; Tokyo: IEEE.
[4] Abdullah M A, Salim M A, Nasir M Z M, Dynamic performances of Malaysian Passenger Vehicle. International conference on Automotive, Innovation and Green Energy Vehicle; 2014 26-27 August; Pahang: Universiti Malaysia Pahang.
[5] He P, Wang Y, Zhang Y, Liu Y. Integrated Control of Semi-active Auspension and Vehicle Dynamics Control System. International Conference on Computer Application and System Modeling; 22-24 October; Taiyuan: IEEE; 2010. p. 63-8.

[6] Sivaramakrishnan S. Simultaneous Identification of Tire Cornering Stiffnesses and Vehicle Center of Gravity. American Control Conference; 11-13 June; Seattle: IEEE; 2008. p. 2846-51.

[7] Jin-qiu Z, Zhi-zhao P, Lei Z, Yu Z, A Review on Energy-Regenerative Suspension System for Vehicles. World Congress on Engineering; 2013 3-5 July; London: IEEE.

[8] Solmaz S, Akar M, Shorten R. Online Center of Gravity Estimation in Automotive Vehicles using Multiple Models and Switching. 9th International Conference on Control, Automation, Robotics and Vision; 5-8 December; Singapore: IEEE; 2006. p. 1-7.

[9] Shraim H, Ananou B, Fridman L, Noura H, Ouladsine M. Sliding Mode Observers for the Estimation of Vehicle Parameters, Forces and States of the Center of Gravity. 45th IEEE Conference on Decision and Control; 13-15 December; San Diego, CA: IEEE; 2006. p. 1635-40.

[10] Peng Y, Lang M. Calculation of Height Limit of Gravity Center of Loaded Wagon Based on Vertical Dynamic Load Coefficient. International Conference on Transportation, Mechanical and Electrical Engineering; 16-18 December; Changchun: IEEE; 2011. p. 1321-4.

IOP Conf. Series: Materials Science and Engineering **100** (2015) 012018 doi:10.1088/1757-899X/100/1/012018

Energy regenerative suspension test for EEV and hybrid vehicle

M A Abdullah[1,2], J F Jamil[1] and N S Muhammad[1]
[1]Faculty of Mechanical Engineering, UniversitiTeknikal Malaysia Melaka, Hang Tuah Jaya, 76100, Durian Tunggal, Melaka, Malaysia.
[2]Centre for Advanced Research on Energy, UniversitiTeknikal Malaysia Melaka, Hang Tuah Jaya, 76100 Durian Tunggal, Melaka, Malaysia.
Phone: +6062346845; Fax:+6062346845

Email: mohdazman@utem.edu.my

Abstract. The world is demanding on the alternative fuel and reducing the fuel consumption of land transportation especially in the automotive industries. This paper emphasizes the development of the energy regenerative suspension system (EReSS) for energy efficient vehicle (EEV) or hybrid. The EReSS product is fabricated and tested on the laboratory and real vehicle. The test is conducted to test the function of the EReSS system on real vehicle. The test is done using the multimeter to record the reading of voltage produces by the EReSS system that is attached to the vehicle suspension system. The experiment starts by setting the parameters in the EReSS system which is the number of windings with a standard magnet. Road irregularity is one of the important parts of the experiment which is set to be various types of road condition and driving style. A domestic car model is selected for the EReSS test that the system can be installed. The test of the EReSS gives out the maximum output voltage of 5.6 V with 530 windings. Improvement on the material can increase the output voltage. The EReSS is function on the real vehicle by producing voltage by harvesting the kinetic energy from the suspension vibration.

1. Introduction

There are bumpy roads, the acceleration or reduction of the driving speed and the operations of the steering wheel when driving a land vehicle. This can cause the relative shock between the sprung mass and the unsprung mass. The shock is a part of mechanical power that can be recycled and reduced the energy consumption that can achieves the energy saving vehicle demands [1]. The purpose of suspension system is to isolate the road irregularity and absorbs vibration from the road to the vehicle body by using damper and spring element. The element must be designed properly and develop because of the limitation on the classic type of suspension system [2].

Regeneration of waste energy is one of the important elements in mechanical engineering fields. The mechanical element is always plays as an important role on designing a new system in mechanical. A suspension system on land vehicle is one of them which dissipate energy to facilitate passenger isolation by using viscous dampers. The analysis of energy regenerative suspension must be done properly by selecting a suitable suspension model because the model will predict the power stored by the damper [3]. The energy regenerative suspension system is already been researched by several automotive industries since the 1990s and several papers have been presented and proposed

[4]. The regenerative suspension system functions by reciprocating part of the system. When the vehicle travels on rough road, it will operate the regenerative system and harvest the kinetic energy from the suspension vibration and convert to electricity [5]. The displacement between the coil windings and magnet causes magnetic induction cutting and produces current[6].

There are many studies and researched on the regenerative suspension system and improvement done to the proposed model but until now the development is still far away from the demands on the commercial applications. There are only some researchers that discovered and made a concrete achievement in specific test [7]. The energy regenerative suspension system is introduced to reduce the energy consumption without losing damping efficiency and electric energy is generated during a high speed motion of the suspension system. It is function with the help of the vibrating system using single-degree-of-freedom (SDOF) system. The suspension system has high dissipation energy that is suitable for harvested for the regeneration system [8].

Electromagnetic suspension system are one of the system that attracts many researchers to replace the current passive, semi-active and hydraulic suspension system because the improvement in efficiencies and reducing the costs of production. The regeneration system ability to harvest energy is limited compared to the hydraulic system [9]. The regeneration system limitation can be prevented by a suitable selection of design, materials and geometrical optimization that will result in high force density. The electromagnetic suspension system is easier to apply because hybrid or electric car have their own energy storage system that can control the peak power and regenerative power of the system [10]. The voltage produces by energy regenerative suspension system is about 8 V for a single system and power available is about 100 W to 400 W [11, 12].

2. Methodology

The design of the energy regenerative suspension (EReSS) is selected. The design is selected by analysis of several concepts that have been proposed. The selected design is the best concept that can give out the highest voltage reading theoretically. Other than that, the selected EReSS concept is easier to be fabricated. Figure 1 shows the selected design concept of the EReSS. The concept is using a single barrel housing that contains all the important part for the electromagnetic regenerative suspension system. The system is retrofit that will be attached on the original suspension system of a vehicle.

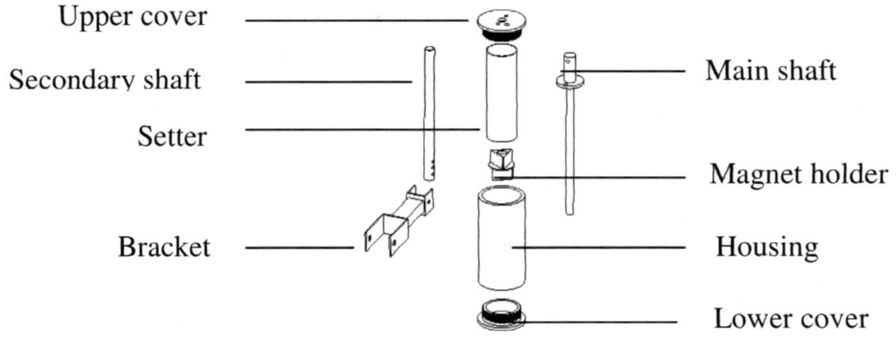

Figure 1. Selected design concept of the EReSS.

After the selection of the concept, the EReSS is fabricated. The fabricated EReSS system is shown in figure 2 and figure 3. The fabrication process is done by following the design concept that has been drawn on computer-aided-drafting (CAD) software. The CAD drawing is done to ensure the product component can fit to each other and ease the process of fabrication.

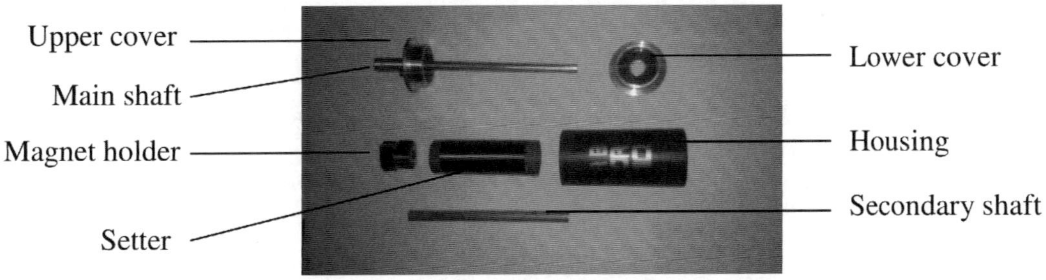

Figure 2. Fabricated EReSS exploded view.

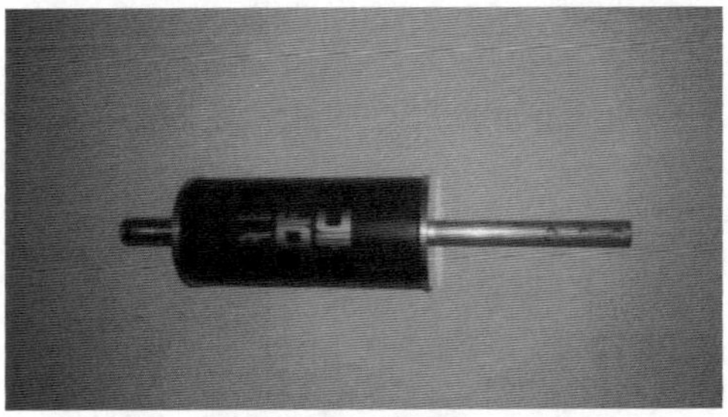

Figure 3. Fabricated EReSS assembly view.

Next, the EReSS product is tested on the laboratory by setting up different parameter such as the number of winding and diameter of the coil. The experiment on the laboratory is done on a test rig that moves vertically direction same as the suspension system on a vehicle. In the laboratory experiment, the frequency of the system can be set and read using DasyLab software. Figure 4 shows the EReSS is tested on the laboratory experiment.

Figure 4.EReSS system on test rig for laboratory experiment.

After the test on the laboratory, the EReSS will be tested on the real vehicle. Firstly, a suitable car model is selected. The selected car must have a proper suspension system that the EReSS system can be attached. The car selected is PeroduaMyvi 1.3 SE AT. The EReSS system is then attached to the car rear suspension system as the space is suitable for the EReSS to be attached. Then, the parameter of the EReSS is set for the testing on the car. Table 1 shows the parameters of EReSS used for the experiment.

Table 1. Parameters of EReSS for on car experiment.

Criteria	Description
Number of windings	530,350,250
Magnetic flux density, [T]	0.2
Magnet type	Isotropic Ferrite
Coil diameter, [mm]	0.29
Coil material	Copper

After the parameters have been set up for the EReSS system, the system is then attached to the car. The EReSS is attached to the car rear suspension system with proper set up to get the reading of voltage. Figure 5 shows the EReSS is attached to the car suspension system. From the system, a multimeter is equipped to record the voltage output of the system. The equipped multimeter is put in the car to ease for recording the reading of voltage for the system.

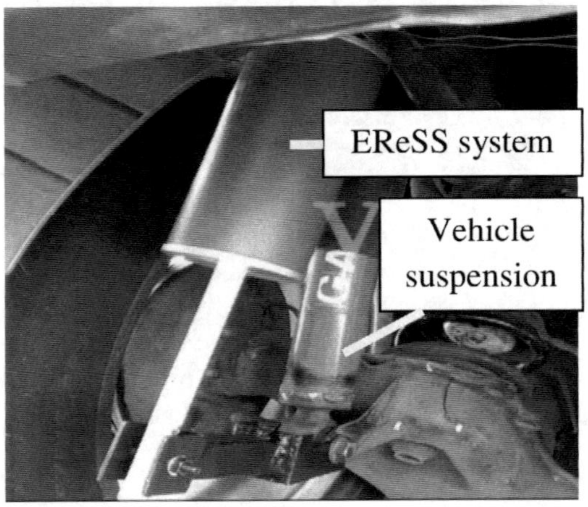

Figure 5.EReSS is attached to the car suspension system.

3. Results and discussion

The selected car is installed with the EReSS system and run on the road. There are many road irregularity surfaces that affect the reading of voltage for the system. The reading of voltage is observed during the experiment process when the car is starts to be driven.

Figure 6 shows the result for EReSS testing for 530 windings, figure 7 for 350 windings and figure 8 for 250 windings. The maximum reading of the EReSS is get when the car is passing through a road bumper. This is because the vertical movement of the car is higher when passing through the road bumper. Frequency is one of the other reasons that cause the voltage reading is high. A higher frequency will affect the reading of the EReSS system. As on the experiment that already done on the laboratory, the higher the frequency the higher the voltage produces by the EReSS system. All the reading cannot be recorded as the frequency cannot be set to a constant. The frequency also cannot be recorded as there is no equipment for the frequency reading. This experiment is done to test the function of EReSS when attached to a real vehicle and recording the voltage produces by the system.

Other than that, the reading of the voltage is high when taking a corner with a high speed. A slalom type of driving will also give out higher voltage reading compared to normal driving style. When the car is on braking process, the voltage is also given out a higher voltage as the suspension system vertical movement is greater that straight driving. During a constant driving on a straight road, the voltage reading is slightly lower but in a constant charge. This charge of voltage can be stored for alternative power for the car electronics system such as the lighting and electronics. The use of this voltage charge can reduces the demands on the alternator and in the same time will reduce the fuel consumption of the car as the work load for the engine is also reduced.

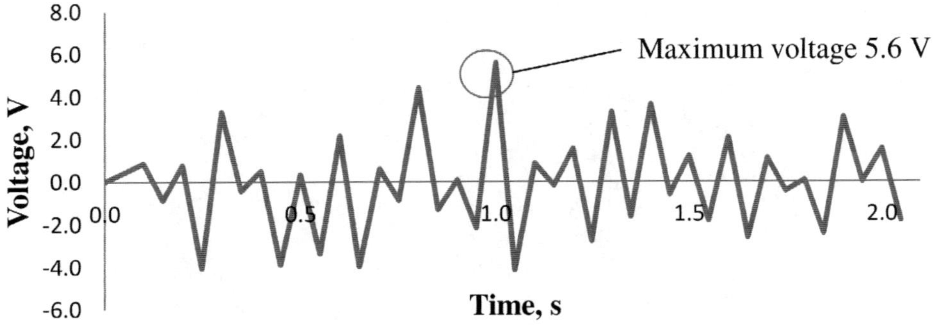

Figure 6. EReSS testing with 530 windings.

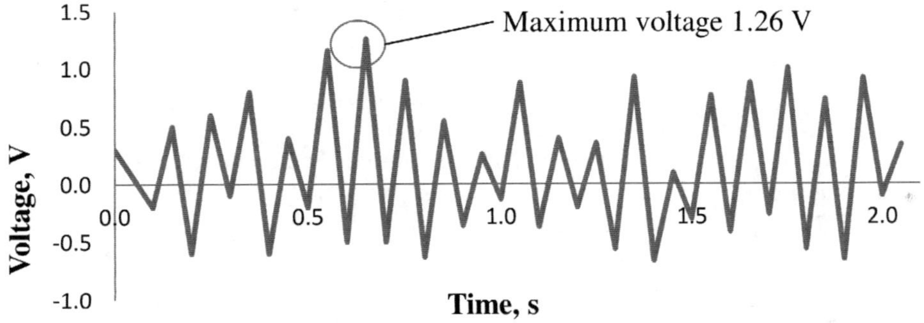

Figure 7. EReSS testing with 350 windings.

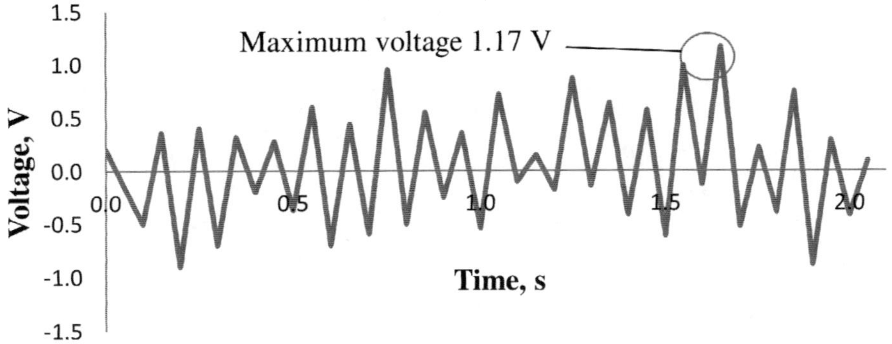

Figure 8. EReSS testing with 250 windings.

The highest voltage reading is recorded during the experiment. The maximum reading of voltage during experiment is |-5.6| V with 530 windings. Figure 9 shows the recorded maximum voltage reading.

Figure 9. Maximum voltage reading of EReSS during experiment on car.

The number of windings in the EReSS system is also plays a role on the test. The higher the number of windings, the higher the voltage produces by the EReSS system as shown in table 2.The EReSS system also can be improved by using different type of magnet and coil material. The higher the magnetic flux density of the magnet, the voltage charge for the system will be higher. This is proved by the theoretical calculation for the system. This current system is only using standard type of magnet and copper coil so the voltage produced is limited. The diameter of the coil is also important for the system as the smaller the diameter of the coil the greater the voltage will be produced. Test is done on the laboratory experimentation and the result is shown in table 3. Moreover, the smaller the coil diameter, the number of windings can be added.

Table 2. Maximum voltage reading of different number of windings on car testing.

Number of windings	Voltage, [V]
530	5.60
350	1.26
250	1.17

Table 3. Maximum voltage reading of different coil diameter on laboratory testing.

Coil diameter, [mm]	Voltage, [V]
0.29	6.30
0.40	0.90
0.8	0.78

4. Conclusion

In this research, the EReSS system that has been designed and fabricated is tested on a real vehicle and proved that the system is function on the real vehicle. The driving style of the vehicle correlates to the voltage produced by the EReSS system installed on the car. From the test, it is found that the maximum voltage reading for standard parameters set for the experiment is 5.6 V. Other than that, the voltage output can be improved by improving the material of the electromagneticsystem of the EReSS. Equipment should be installed for the reading of frequency of the car so that all the different data of the voltage can be recorded with different value of frequency. The voltage output of the EReSS is affected by different number of windings and diameter of the coil.The road conditions are also affecting the voltage output of the EReSS system.As the voltage is produced during on car experimentation, the EReSS is proved to be function when installed on a real vehicle by harvesting kinetic energy from the vehicle suspension system.

Acknowledgement

The authors gratefully acknowledged the Advanced Vehicle Technology (AcTiVe) research group of Centre for Advanced Research on Energy (CARe), the financial support from UniversitiTeknikal Malaysia Melaka and The ministry of Education, Malaysia under Short Term Research Grant, Grant no. PJP/2014/FKM(10A)/S01330 and Fundamental Research Grant Scheme (FRGS), grant no.: FRGS/2013/FKM/TK06/02/2/F00165.

References

[1] Lin X, Xuexun G. Hydraulic Transmission Electromagnetic Energyregenerative Active Suspension and Its Working Principle. 2nd International Workshop on Intelligent Systems and Applications; Wuhan: IEEE; 2010. p. 1-5.
[2] Zuscikova M, Loksik M, Huertas V V, Belavy C. Adaptive Control Design for Vehicle Suspension. International Conference on Process Control; 18-21 June; Slovakia: IEEE; 2013. p. 235-40.
[3] Fador M G, Redfield R C, The Variable Linear Transmission for Regenerative Damping in Vehicle Suspensio Control. American Control Conference; 1992; Chicago: IEEE.
[4] Zhang Y, Huang K, Yu F, Gu Y, Li D. Experimental Verification of Energy-regenerative Feasibility for an Automotive Electrical Suspension System. International Conference on Vehicular Electronics and Safety; 13-15 December; Beijing: IEEE; 2007.
[5] Patil R U, Gawade D S S 2012 *International Journal of Advenaced Engineering Technology* **3** 54-9
[6] Zhen L, Wei X 2010 *Journal of Networks* **5**
[7] Lin X, Bo Y, Xuexun G, Jun Y. Simulation and Performance Evaluation of Hydraulic Transmission Electromagnetic Energy-Regenerative Active Suspension. Second WRI Global Congress on Intelligent System2010. p. 58-61.
[8] Okada Y, Harada H, Regenerative Control of Active Vibration Damper and Suspension Systems. Decision and Control 1996 11-13 December; Kobe: IEEE.
[9] Gysen B L J, Van der Sande T P J, Paulides J J H, Lomonova E A. Efficiency of a Regenerative Direct-drive Electromagnetic Active Suspension. Vehicle Power and Propulsion Conference; 1-3 September; Lille: IEEE; 2010. p. 1384-93.
[10] Gysen B L J, Van der Sande T P J, Paulides J J H, Lomonova E A. Efficiency of a Regenerative Direct-drive Electromagnetic Active Suspension. Transactions on Vehicular Technology; 22 March 2011: IEEE; 2011. p. 1384-93.
[11] Zuo L, Zhang P-S 2013 *Journal of Vibration and Acoustics* **135** 011002
[12] Tonoli A, Amati N, Detoni J G, Galluzzi R, Gasparin E 2013 *Vehicle System Dynamics* **51** 1186-99

Critical evaluation on structural stiffness of porous cellular structure of cobalt chromium alloy

N M S Abd Malek[1], S R Mohamed[1], S A Che Ghani[1] and W S Wan Harun[1]
[1]Manufacturing Focus Group, Faculty of Mechanical Engineering, Universiti Malaysia Pahang, 26600, Pekan Pahang Malaysia

Email: syamsimalek83@yahoo.com

Abstract. In order to improve the stiffness characteristics of orthopedic devices implants that mimic the mechanical behavior of bone need to be considered. With the capability of Additive layer manufacturing processes to produce orthopedic implants with tailored mechanical properties are needed. This paper discusses finite element (FE) analysis and mechanical characterization of porous medical grade cobalt chromium (CoCr) alloy in cubical structures with volume based porosity ranging between 60% to 80% produced using direct metal laser sintering (DMLS) process. ANSYS 14.0 FE modelling software was used to predict the effective elastic modulus of the samples and comparisons were made with the experimental data. The effective mechanical properties of porous samples that were determined by uniaxial compression testing show exponential decreasing trend with the increase in porosity. Finite element model shows good agreement with experimentally obtained stress-strain curve in the elastic regions. The models prove that numerical analysis of actual prosthesis implant can be computed particularly in load bearing condition

1. Introduction

The major challenge in developing medical implant devices is to produce biocompatible alloys which possess good tensile and fatigue properties. The technique used to manufacture these alloy can markedly affect the mechanical and metallurgical properties of the resulting components [1]. Additive manufacturing (AM) was proposed as a novel candidate for the fabrication of customized biomedical implants with the cobalt alloys. Among them, is direct metal laser sintering (DMLS) which has been a high demand since it offers a lower time-to-market, a near-net-shape production, a higher material utilization rate, along with its ability to produce functional metallic parts with mechanical properties comparable to those in bulk materials [2]. From the previous studies, SLM has the potential of controlling porosities according to the capacity of providing different energy inputs by its processing parameters while in DMLS process all the parameters were constant. A study on cobalt chromium molybdenum (CoCrMo) powder using DMLS shown that minor porosity produced by the highest layer thickness, which provided a better particle packing. Thus thermal conductivity among the particles was enhanced and subsequently permitted successful densification, and finally removed or shrunk pores.

Cobalt (Co) in CoCrMo alloys exhibits high corrosion resistance and has excellent wear resistance which makes Co, often employed as artificial joints and body implants. Co also shows stability of allotropic transformation at room and elevated temperatures. In biomedical applications, CoCrMo alloys (ASTM F75) were much preferred to be used in human since they are free of Nickel (Ni). Ni element is well known as highly toxic element [3].

Finite element analysis (FEA) can be used to identify a suitable replacement for conventional materials and designs through practical applications in self-programmed and commercial simulation software, using e.g. MATLAB, COMSOL, ABAQUS or ANSYS. FEA method is common in the biomechanical study to predict implant performance. Prediction of macroscopic mechanical properties of porous structures using (FEA) can reduce amount of experimental works needed in tailoring special properties of porous implants hence allows researchers to predict stress distribution in the contact area of the implants with the bones by solving structural mechanics problems using numerical approximation [4]. Recent study shows that the effective elastic modulus of porous structure of CoCr biomedical grade alloy produced by selective laser melting (SLM) can be predicted using FEA [5] to some extent. However, the discrepancies in result for a wide range of porosities are still huge as compared to experiment due to heterogeneity structure resulted from rapid melting and cooling of SLM process [6].

In this study, focuses on applying the method of direct metal laser sintering (DMLS) process to CoCrMo which expected have a capability as an orthopaedic implant with tailored mechanical properties. This research combines the simulation n and experimental method in result to find out the stiffness of the samples. Therefore, it can meet up effectively to predict the mechanical properties by computing analysis.

2. Methodology

2.1. Component design

CAD models individual open cellular structures were generated using Solidworks 2013 software. The porosity of cellular structure determined using equation (1). Wall thickness of open cellular structures samples were fixed at 0.80 mm with pore sizes in square shape ranging between 0.60 and 1.80 mm within 15 mm cubic. Detailed of individual CAD models are shown in figure 1(a). In this design, only porosity volumes were designed at 60%, 70% and 80%.

$$\text{Porosity}, \varphi = \frac{V_p}{V_b} \qquad (1)$$

Where

Vp = Volume of pores
Vb = Volume of bulk

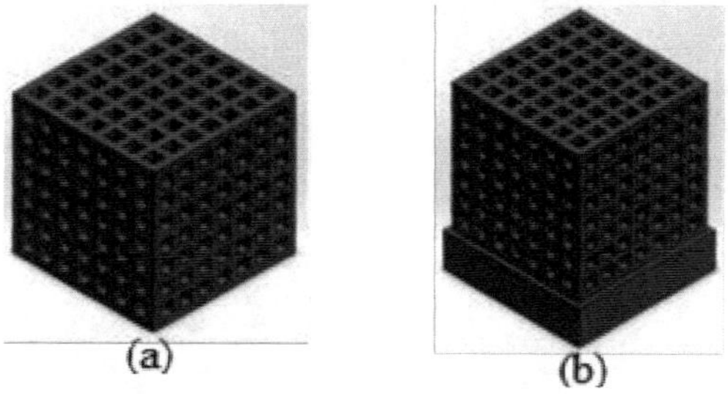

Figure 1. CAD model (a) individual and (b) extension basement.

IOP Conf. Series: Materials Science and Engineering **100** (2015) 012019 doi:10.1088/1757-899X/100/1/012019

The solid models were exported from the CAD software in STL file format into Materialise Magics 17.0 software. The components were then positioned on the build platform and slice files were generated to enable manufacture. A typical CAD model of the component shows extension basement to ease detachment on completing fabrication as show in figure 1(b).

2.2. Component design

In this study, the actual structures were produced by EOSINT M280 Direct Metal Laser Sintering machine using metal powder from EOS Cobalt Chrome MP1 which is a Cobalt Chromium Molybdenum, CoCrMo powder super alloy biomaterial. Parts build from EOS Cobalt Chrome MP1 complied with the chemical composition of UNS R31538 of high carbon CoCrMo alloy as shown in table 1. The alloy is nickel-free (< 0.1 % nickel content) and was characterized for fine, uniform crystal grain structure. As built EOS Cobalt Chrome MP1 meets the chemical and mechanical specifications of ISO 5832-4 and ASTM F75 for cast CoCrMo implant alloys, and the specifications of ISO 5832-12 and ASTM F1537 for wrought CoCrMo implants alloys except remaining elongation. Ideally, elastic modulus, Poisson's ratio and full density of a stress relieved EOS Cobalt Chrome MP1 were given as 200 GPa, 0.3 mm and 8.3 g/ cm3, respectively.

Table 1. Physical and composition of EOS Cobalt Chrome MP1 (CoCrMo) [7].

Material composition (wt-%)	Co(60-65) Cr(26-30) Mo(5-7) Si(≤1.0) Mn(≤1.0) Fe(≤0.75) C(≤0.16) Ni((≤0.10)
Relative density (%)	Approximately 100
Density (g/cm^3)	Approximately 8.3

2.3. Sample manufacturing

The structures made from EOS EOS Cobalt Chrome MP1 powder was filled up and sieved to a particle size of no greater than 25µm into dispenser platform and the components were manufactured using the DLMS EOSINT M280 Direct Metal Laser Melting machine. A schematic diagram of SLM process is illustrated in figure 2(a). The machine uses laser system with spot diameter of 100µm to selectively melt the powder. The components were built using 20µm layers. The parameters were; laser power of 195W, laser step over distance of 0.08 mm and a laser scan speed of 800 mm/s. The fabrication chamber was filled up with argon gas with less than 1.3 % oxygen present. All completed samples were then detached from the substrate using electro discharge machining (EDM) wire-cut prior to geometrical measurement.

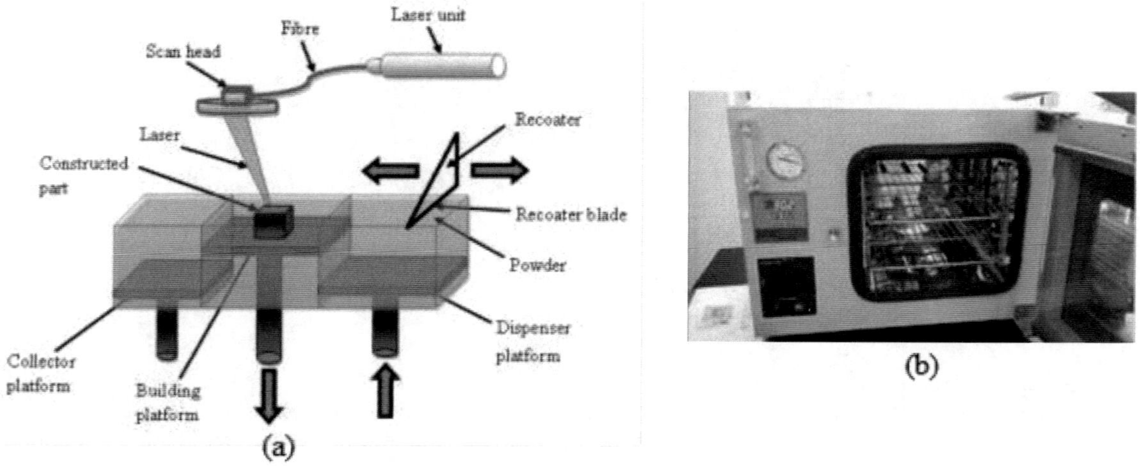

Figure 2. (a) Schematic diagram of SLM and (b) Stress relieve process.

2.4. Correlation of published empirical equation

In order to obtain an additional comparison with the physical test results and numerical analysis, a model proposed in [4, 6] were utilized. According to Gibson and Ashby, equation can be rewritten to calculate the effective elastic modulus as follows:

$$E_{eff} = E_s(1-\varphi)^2 \quad (2)$$

for Hazlehurst, Wang and Stanford, as in equation (3) where:

$$E_{eff} = E_s(1-\varphi)e^{(-2.376\varphi)} \quad (3)$$

Where;
E_s is the elastic modulus of the solid CoCrMo alloy taken as 200 GPa and
φ is the porosity.

2.5. Finite element analysis

ANSYS Workbench 14.0 finite element modelling software was used to predict the effective elastic modulus of the individual structures and comparisons were made with the physical test data and analytical predictions. The software was also used to perform a simulation that considered the effect of structural variation and heterogeneities within a cellular structure, in order to investigate what effect this had on the effective stiffness. Three dimensional models created in Solidworks® 2013 that had been used for specimen manufacturing by SLM were transferred into ANSYS Workbench 14.0 environment for structural simulation. Linear static structural analysis was chosen to solve the static problem in order to obtain equivalent stress, equivalent strain and compressive deformation of each model within the set boundary conditions.

The full simulations with different load conditions were carried out using ANSYS DesignXplorer for obtaining mean stress-strain relationship of every sample with different porosity. Triangular meshes with minimum mesh size of 0.2mm are set for the FEA. Figure 3 shows that all meshes are together connected and all the struts and pores are defined as one geometry in a three dimensional model with the assumptions that all struts have identical size and the model is homogenous.

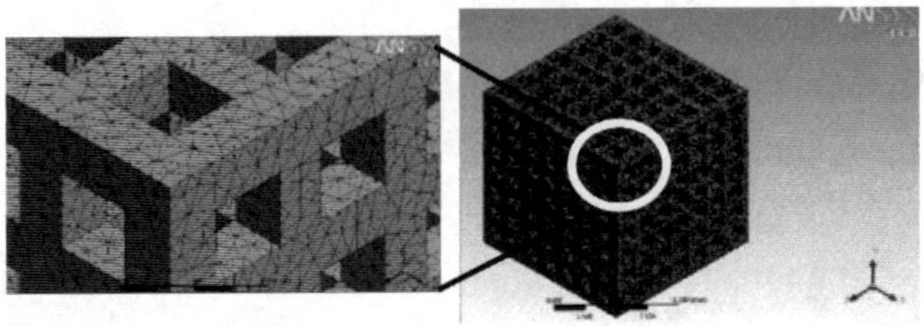

Figure 3. Mesh elements are together connected within the struts and pores with minimum mesh size of 0.2 mm of tetrahedral meshes.

In Static Structural Analysis function, the model was fixed at the bottom surface and a downward force was applied on the top surface as depicted in figure 4(a). Compression test on cellular structure were modelled for different porosity of unit cells in simulation prior to fabrication on the DMLS machine. Upon the convergence, three output variables were recorded, i.e. equivalent stress, equivalent strain and total deformation in Z-direction. The applied force increased for every load cycle until 1 mm structure deformation was achieved.

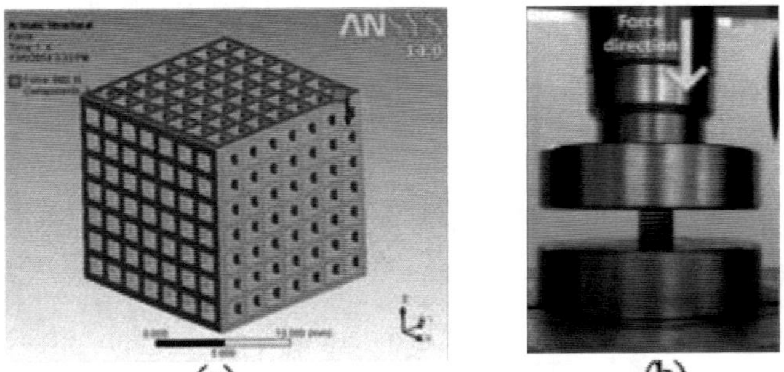

Figure 4. Load applied (a) FEA and (b) Experiment.

2.6. Compression test

Uniaxial compression tests were performed using a Shimadzu materials testing machine with a maximum load capacity of 100kN. Tests were performed to physically determine the compressive properties of the cellular structures in the orientation which they were built. The open cell specimens were elastic tested at stroke rate of 0.5 mm/min and the acquired real time mechanical testing data were recorded in a PC for further analysis (in accordance with ISO 13314:2011) with efforts being made to ensure uniaxial loading by using a machined fixture as shown in figure 4(b). The components were loaded to failure or until the maximum load capacity of the machine was reached. The stress strain relationship for each individual component was calculated from the real time force versus displacement data obtained from the test machine.

3. Results and Discussion

The specimens chosen are in cubic shape with controlled variations of porosity and strut size. The detail samples specification shown in table 2.

Table 2. Samples specification of 60%-80% porosity.

Sample	Cube size (mm x mm x mm)	Porosity (%)	Strut size (mm)	Pore size (mm x mm)
Sample 1	15x15x15	60	0.80	1.22x1.22
Sample 2	15x15x15	70	0.77	1.59x1.59
Sample 3	15x15x15	80	0.52	1.80x1.80

3.1. Samples fabrication

Open cellular structure CoCrMo alloy samples with various designed volume porosity have been successfully fabricated by DMLS process in this work. Sintered density of each sample attained higher than 7.96 g/cm3 that was 96 % of the theoretical density. Samples of different designed porosity are demonstrated in figure 5(a). An observation under optical microscope showed that there were structural variations found on external surfaces of each sample especially on the side (horizontal) faces. As shown in figure 5(b), the strut structures with irregular surfaces forfeited geometry integrity of the sample. This is different to ideal CAD design with smooth and uniform regular surface [7].

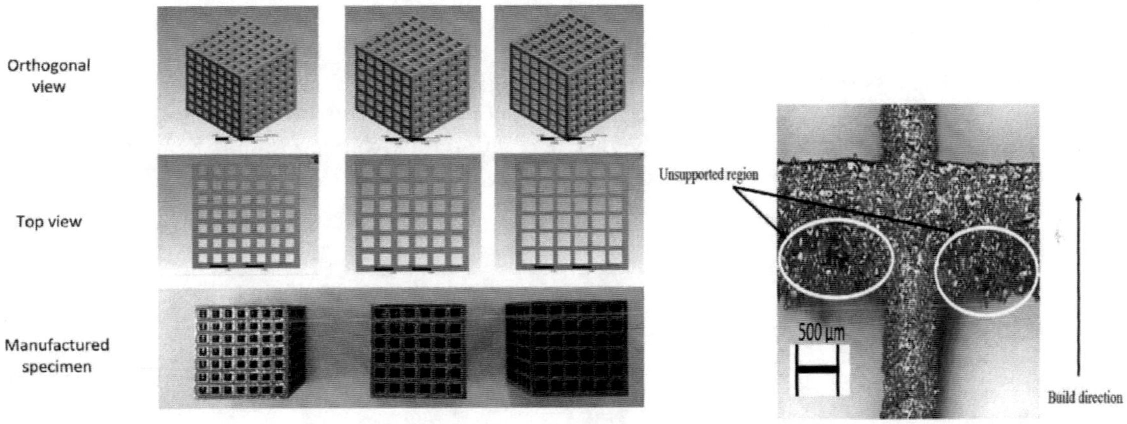

Figure 5. (a) Samples of different designed porosity and (b) Irregular surfaces.

3.2. Finite element analysis

In the FE Analysis, the representative volume element (RVE) method is selected to represent the porous structure of the implant and the mechanical properties, such as Total Deformation as on figure 6 (a). Equivalent Stress and Equivalent Static Strain are then calculated. Once the effective mechanical properties are determined for a range of structures and porosities, the value can then be the material input into the FEA analysis for assessing the performance of the final part. Periodic boundary conditions were assumed at all other faces. The RVE was meshed using tetrahedral elements. From the Equivalent Stress in finite element analysis, it show that stress concentration at the vertical strut and nearly the edge of pore area. Figure 6 (b) show the maximum stress concentration in RVE indicated by red colour area.

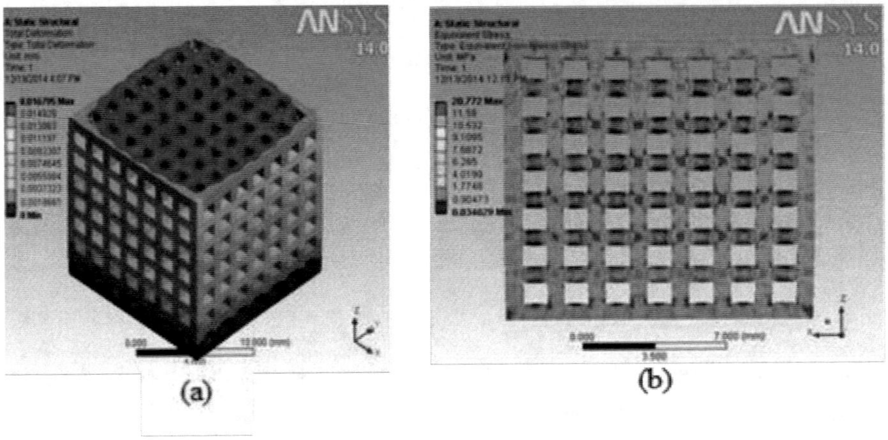

Figure 6. FEA contour plot for (a) total deformation and (b) maximum stress concentration.

The FEA was then used to predict the effective stiffness of the porous structure. The result obtained by FEA were used to draw relationships between stresses and strains of all cellular structures of different porosities using the applied forces and strokes obtained from the compressive test.

3.3. Experiment compression test

The experimental results reveal that the elastic modulus of open cell structure reduces tremendously as compared to solid material when produced by DMLS. Figure 7 shows the graph of compression test for the samples experienced 20 hours stress relieve process.

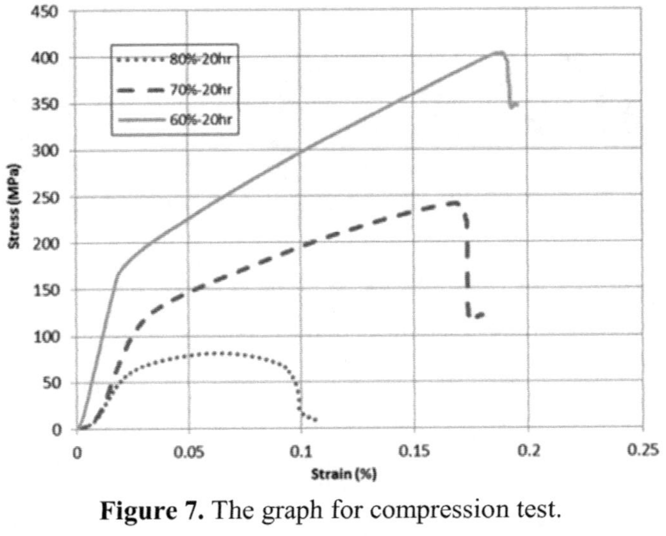

Figure 7. The graph for compression test.

3.4. Comparison FEA, experimental and published equation

The consideration to use the finite element method to predict the effective mechanical properties of cellular structures manufactured using DMLS. This would help to reduce the need for expensive and time consuming physical testing. The effective mechanical properties of cobalt chromium alloy cellular structures manufactured using DMLS with a rhombic dodecahedron unit cell in both uniaxial compression and flexure have been investigated and comparison have been carried out with the numerical results and the physical test data. Further tests considering the effective stiffness of Cobalt chromium porous cellular structures, manufactured using DMLS with porosity ranging from 60% and 80%, were investigated when subject to finite element simulation and again the published data of empirical equations results significantly over estimated the stiffness. In this study, the focus is more to the linear region on the graph to obtain its effective elastic modulus as shown in figure 8. From the predicting effective stiffness value from simulation result graph, it shown that the close agreement of linear gradient in this work was achieved with the experimental results, showing the minimum different as show in table 3.

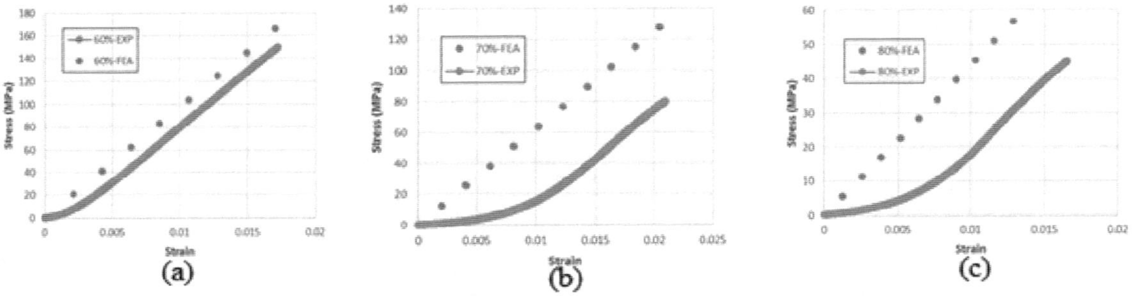

Figure 8. The linear region on the graph to obtain effective elastic modulus between FEA and experiment.

Table 3. Comparison result by simulation and experiment method.

Porosity (%)	Method	Elastic modulus (GPa)	Error
60	FEA	9.760	0.0085
	EXP	9.844	
70	FEA	6.260	0.0340
	EXP	6.481	
80	FEA	4.410	0.0640
	EXP	4.143	

The effective elastic modulus obtained in the experiment and simulation for different porosities was compared with the published correlations from [6] and modified Gibson [4]. It can be seen in Table 4 that the effective elastic modulus computed from the experiment are much lower than the published models. These might be due to the struts and pores sizes used in those models.

Table 4. Numerical and experiment effective elastic moduli comparison.

Porosity	Elastic modulus (GPa)			
%	FEA	EXP	Gibson Ashby	Hazlehurst et al.
0	200	200	200	200
60	9.76	9.844	32	19.23
70	6.26	6.481	18	11.27
80	4.41	4.143	8	5.98

IOP Conf. Series: Materials Science and Engineering **100** (2015) 012019 doi:10.1088/1757-899X/100/1/012019

The coefficient of determination R2 is another measure of how well the least squares equation performs as a predictor of stiffness. In this study, the relationship between the effective elastic modulus and porosity has been analysed using nonlinear regression due to the numerical and analytical results also showing non linearity. The derived expression has different coefficient of determination value (R2) based on the various method and the relationship can be observed from figure 9.

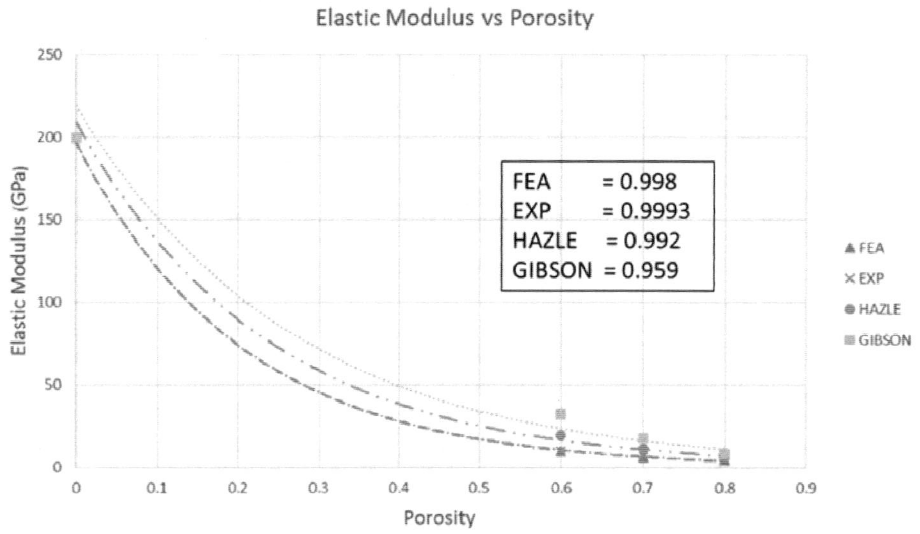

Figure 9. Different coefficient of determination values.

4. Conclusion

The development of Additive Layer Manufacturing (ALM) processes such as Selective Laser Melting (SLM) and Direct Metal Laser Sintering (DMLS) have introduced the capability to produce orthopaedic implants from common biomedical alloys, with tailored mechanical properties which can potentially emulate human bone and can be considered as viable alternative to manufacturing using traditional methods. This study evaluates the stiffness of the three different porosities (60%, 70% and 80%) of square pore CoCrMo cellular structures. These specimens are powder based and manufactured through DMLS processes particularly for orthopaedic applications. Influences of structural variation and heterogeneities on cellular structures stiffness are justified by using a finite element analysis. The analyses conducted in FEA are tally with the result produced in experiments. This include boundary conditions, mesh shapes and sizes and material properties in the function of sample porosities. The study contributes effective boundary conditions for advance simulation in predicting the stress distribution in real bone shape. The effective mechanical properties of porous samples that were determined by uniaxial compression testing show exponential decreasing trend with the increase in porosity. The models prove that numerical analysis of actual prosthesis implant can be computed particularly in load bearing condition.

Acknowledgements

The authors would like to be obliged to Universiti Malaysia Pahang for providing laboratory facilities and Gramslab members for support in completing the study. The study is funded by AEC internal grant: RDU140377.

References

[1] Mat Taib Z A W H W S, Che Ghani S A, Ab Rashid M F F, Omar M A and Ramli H. Dimensional accuracy study of open cellular structure CoCrMo alloy fabricated by selective laser melting process. Langkawi2014.
[2] Monroy K, Delgado J, Ciurana J 2013 *Procedia Engineering* **63** 361-9
[3] Niinomi M, Nakai M, Hieda J 2012 *Acta Biomaterialia* **8** 3888-903
[4] Kayabaşı O, Yüzbasıoğlu E, Erzincanlı F 2006 *Advances in Engineering Software* **37** 649-58
[5] Hazlehurst K, Wang C J, Stanford M 2013 *Materials & Design* **51** 949-55
[6] Che Ghani S A W H W S, Mat Taib Z A, Ab. Rashid M F F , Ramli M H and Omar M A 2015 *Applied Mechanics and Materials* 1-5
[7] EOS. Electro Optical System (Ed)2011.

Analysis of semi-active vehicle suspension system using airspring and MR damper

A H Tesfay[1,2] **and V K Goel**[3]
[1]Department of Mechanical Engineering, Ethiopian Institute of Technology-Mekelle, Mekelle University, P.O.box 231 Mekelle, Ethiopia
[2]Deparment of Energy and Process Engineering, Norwegian University of Science and Technology, 7491 Trondheim, Norway
[3]Deparment of Mechanical and Industrial Engineering, Indian institute of Technology Roorkee, Roorkee, India

E-mail: asfafaw.hailselassie@ntnu.no

Abstract. With the new advancements in vibration control strategies and controllable actuator manufacturing, semi-active actuators and dampers are finding their way as an essential part of vibration isolators, particularly in vehicle suspension systems. This is attributed to the fact that in a semi-active system, the damping coefficients can be adjusted to improve ride comfort and road handling performances. The currently available semi-active damper technology uses MR fluid to control the damping characteristics of the suspension system. In addition to MR dampers, combining air springs in a semi-active suspension system leads to better handling and ride performance in vehicles. Furthermore, the use of air spring in semi-active suspension system helps to ease design of variable spring stiffness. This easy design opportunity leads to independent control of stiffness and ride height of the vehicle. This paper deals with the design and modelling of variable stiffness air spring for semi-active suspension system, modelling of semi-active suspension systems with variable stiffness and MR damper, and study their

1. Introduction

The evolution of suspension is merely close to the evolution of transportation technology. For centuries, carts were not equipped with any sort of suspension at all. In the eighth century, a primitive suspension based on an iron chain system was developed [1]. Today's vehicle suspension systems can be categorized as passive, active, and semi-active system. Passive suspension has design and cost related advantages, however, it has fixed parameters and limited performance on variable operating conditions. In contrast, active suspension utilizing hydraulic or pneumatic actuators to provide high control performance. Nevertheless, it requires high power and sophisticated control implementation. On the other hand, semi-active suspension system use variable dampers or variable spring stiffness with small power requirement and simpler control implementation. Today's rapid development of semi-active suspension systems have improved ride, handling and safety of vehicles and occupants [2]. Fuzzy logic controllers incorporated in semi-active suspension systems gave better road handling and rid comfort [3]. Semi-active suspension system with modulated damping coefficient value has shown minimum percentage of overshoot and fast stabilizing time with fuzzy logic controller[4]. One of the effective semi-active suspension system that coupled air-springs with variable axillary volume are liable to give lower amplitude transmissibility and lower system natural frequency [5].

The objective of this paper is to develop a conceptual design of variable stiffness air spring and MR model with a fuzzy logic controller. The developed model is used to analyze effects of variable stiffness in various operating frequencies, simulate semi-active suspension system using variable damper and variable air spring stiffness using Math lab Simulink and study the advantage of Fuzzy logic controller in semi-active suspension system to smooth and ease operation of MR damper.

2. Magnetorheological Damper

The semi-active suspension system considered in this paper contains variable stiffness airspring and MR damper as its suspension elements. The magnetorheological (MR) damper consists of a hydraulic cylinder containing a solution that can reversibly change from a free flowing, linear viscous fluid to a semi-solid with controllable yield strength. This solution is termed MR fluid and is composed of micron-sized magnetically polarizable particles dispersed in a carrier medium such as water, mineral or synthetic oil. Out of the many developed MR damper models, Bouc-Wen's [6] modified model gives best results portraying the hysteretic behavior of MR dampers as shown in figure1.

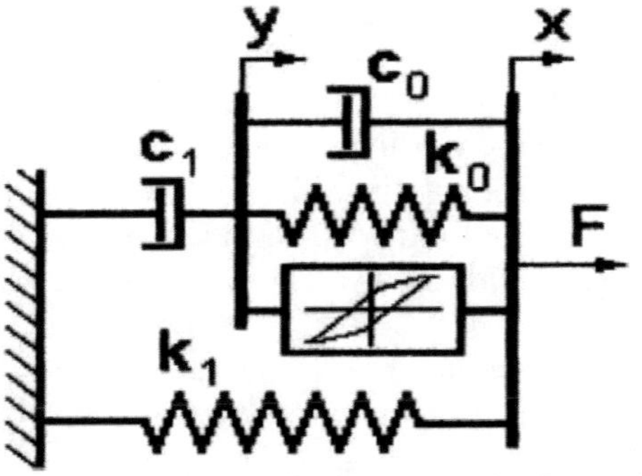

Figure 1. Bouc-Wen modified model of MR damper.

In this model, the damping force generated by the device is:

$$F = c_1\dot{y} + k_1(x - x_0) \tag{1}$$

Where x is the total relative displacement and $x_0$the initial displacement of spring k_1 associated with the force due to the accumulator; k_1 represents the accumulator stiffness; *y* is an internal displacement of the MR fluid damper and *z* is the evolutionary variable that accounts for history dependence of the response and are governed by the coupled differential equations:

$$\dot{y} = \frac{1}{c_0 + c_1}[\, c_0\dot{x} + k_0(x - y) + kz] \tag{2}$$

$$\dot{z} = -\gamma|\dot{x} - \dot{y}|z|z|^{n-1} - v(\dot{x} - \dot{y})|z|^n + A(\dot{x} - \dot{y}) \tag{3}$$

Where k is a stiffness coefficient associated with the displacement *z*, c_0 represents the viscous damping observed at large velocities; and k_0 is to control the stiffness at large velocities. The parameters n, γ, ν, A, and k_1 are 'loop parameters' which control the shape and the magnitude of the hysteretic loops [6]. The parameters c_0 , c_1 , k and k_0 are assumed to be functions of the delayed current *i*= *i*(*t*) applied to the MR damper: $c_0 = c_0$ (i) , $c_1 = c_1$(i) , k = k(i) , $k_0 = k_0$ (i) . For numerical simulations the following values of the fixed coefficients were chosen: n = 2, γ = 500 N/m, υ= 613000 m^{-2}, A = 30.56, k_1 = 540 N/m and k_0=1050 N/m.

The functional approximations (equation (4)) obtained for the remaining parameters c_0, c_1, k, are:

$$c_0(i)=26.134i+5.164 \tag{4}$$

$$c_1(i)=1150\tanh 1.95i \tag{5}$$

$$k(i)=1297.2\tanh 1.3i \tag{6}$$

3. Conceptual design of variable stiffness air spring

The motivation to design an automotive suspension system with independent control of stiffness, damping and ride height comes from the trade-offs involved in conflicting requirements of ride and handling. Hrishikesh Deo and Nam P Suh [7] have proposed an electromechanical suspension system capable of independent control of stiffness, damping and ride height and have discussed its application to improved vehicle dynamics[8]. The application to improved vehicle dynamics requires the stiffness change to be instantaneous and no change in ride-height during stiffness change. In the electromechanical suspension system [7], stiffness change requires power-input and is not instantaneous. This paper focuses on design of pneumatic automotive suspension system capable of instantaneous stiffness change with no power input, and no ride-height change due to stiffness change. In this section we discuss how to format the title, authors and affiliations. Please follow these instructions as carefully as possible so all articles within a conference have the same style to the title page. This paragraph follows a section title so it should not be indented.

3.1. Design for Variable Stiffness Suspension

Assuming adiabatic compression and expansion of air, the nominal stiffness K of an air-spring is:

$$K = \gamma A^2 \frac{P_0}{V_1} \tag{7}$$

Where P_0, T_0, and $V_1$are the equilibrium values of pressure, temperature and volume of air in the air-spring, A is the effective area and γ = Cp/Cv = heat capacity ratio of air. In some existing adaptive suspensions [8], air-springs are employed to achieve variable stiffness and variable ride-height by pumping air into (or out of) the air bags. But this leads to coupling as the design parameter (DP): amount of air affects both, functional requirements (FR), FR1: controlled stiffness and FR2: controlled ride-height.

$$\begin{Bmatrix} FR_1\text{:Controlled Ride Height} \\ FR_2\text{:Controlled Stiffness} \end{Bmatrix} = \begin{bmatrix} X \\ X \end{bmatrix} \{DP\text{:Amount of air in} V_1\} \tag{8}$$

Coupling due to insufficient number of DPs is a common mistake made by designers [8]. Suh proposed decoupling of such a coupled system by addition of new DPs to make the number of FRs and DPs equal. Here additional DPs have been introduced through the following proposed design modification.

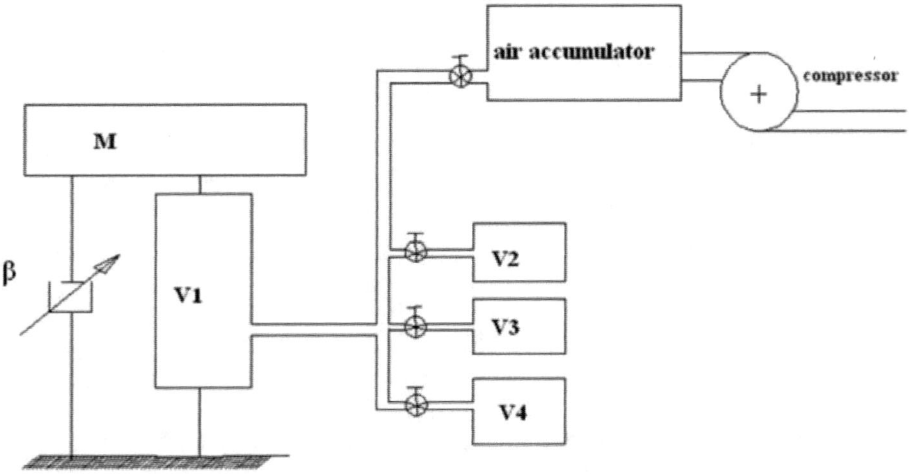

Figure 2. Proposed design of airspring.

The proposed modification lets us to connect the airspring volume V1 to auxiliary volumes, represented by V2 to V4 in figure 2, through on-off valves. When all of the three valves are closed, the effective volume is V1 and the stiffness is:

$$K_{max} = \gamma A^2 \frac{P_0}{V_1} \tag{9}$$

When all of the three valves are open and communicate with the airspring volume, the effective volume is (V1+V2+V3+V4) and the stiffness as given by:

$$K_{min} = \gamma A^2 \frac{P_0}{(V_1 + V_2 + V_3 + V_4)} \tag{10}$$

By efficient choices of non-equal volumes V2 ≠ V3 ≠ V4, we can get 2^3=8 stiffness band settings between the Kmin and Kmax, through different combinations of valves open and closed. In general, by adequately choosing N different volumes, V2 ≠ V3 ≠... ≠VN+1, we can get 2N different stiffness bands. Note that all the volumes are at the same pressure. As a result, there will not be any change in pressure or ride-height. When the valves are opened or closed, to adjust the stiffness, they do not use any effort and they are instantaneous. Damping control is achieved using magneto-rheological damper [8] that is mounted in parallel with the air spring arrangement. The coupled design matrix in equation (11).

$$\begin{Bmatrix} FR_1 : \text{Controled Rideheight} \\ FR_2 : \text{Controled Stiffness} \\ FR_3 : \text{Controled Damping} \end{Bmatrix} = \begin{bmatrix} X & 0 & 0 \\ 0 & X & 0 \\ 0 & 0 & X \end{bmatrix} \begin{Bmatrix} DP_1 : \text{Compressor Control} (V_1) \\ DP_2 : \text{Valve Control} \left(\sum V_i\right) \\ DP_3 : \text{MR damping} \end{Bmatrix} \tag{11}$$

The set-up in figure 2 with only one auxiliary volume connected to the main volume through an on-off valve can be schematically represented as shown in figure3, where β2 = ∞ (for valve closed) results in stiffness $K = K_{max}$; and β2 = 0 (for valve open) results in stiffness $K = K_{min.}$ β denotes the variable damper in parallel with the air spring.

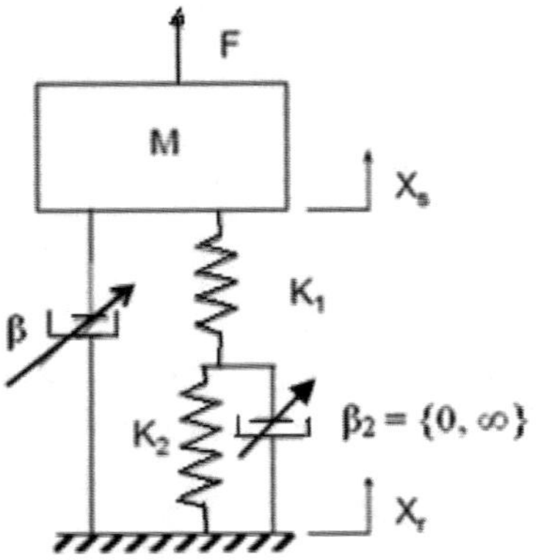

Figure 3. Schematic representation of on-off valve control to vary stiffness.

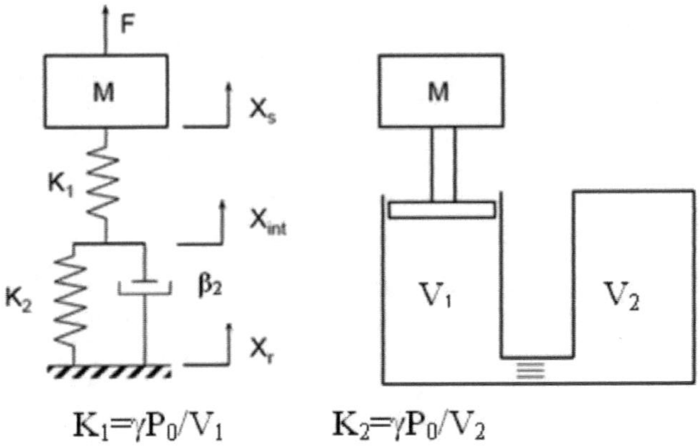

Figure 4. Airspring with one auxiliary volume and flow resistance across the valve.

3.2. Thermodynamic Modelling to Estimate Damping due to the Valve

With the valve closed, β2 = ∞ is an accurate assumption. But for the valve open, β2 = 0 is not accurate. There is always restriction to the flow across the valve. Pressure drop in most valves is proportional to the square of the volume flow through the valve. This introduces damping in the system and in this section it has been tried to determine the damping through thermodynamic modelling of the system. The system is characterized by the P1, V1, m1, T1 for CV1 and P2, V2, m2, T2 for CV2.We have made an assumption that all the gas in CV1 is in the same state and all the gas in CV2 is in the same state. It is reasonable to assume that the gas properties are not a function of position within the chamber for modelling the air spring, as the corner frequency for the suspension system is in order of magnitude lower than the frequency at which pressure wave propagates with in the chamber to equalize the pressure at different points in the chamber. The damping in the system is due to the pressure drop "effort" across the valve which is proportional to the square of the volume flow through the valve. There is an additional source of irreversibility in the system, namely the mixing of air at the outlet of the valve as it flows from one chamber to another. There are eight parameters and we need a relation of the form $\Delta P = f(\Delta V, \omega)$, i.e., pressure variation in response to a sinusoidal forcing of magnitude ΔV and frequency ω.

$$V_1 = V_0 + \Delta V \sin(\omega t); \quad \Delta \dot{V} = A_e \cdot x_s \tag{12}$$

$$V_2 = Constant \tag{13}$$

3.3. Conservation of mass

The total mass of a control volume is conserved and net rate of change of mass of the system is zero since there is no mass flowing out of the system boundary. The relation is given by:

$$\begin{aligned} m_1 + m_2 &= m \\ \dot{m}_1 + \dot{m}_2 &= 0 \\ \text{i.e. } \dot{m}_1 &= -\dot{m}_2 \end{aligned} \tag{14}$$

3.4. Constitutive relation for gas in control volume

For isentropic processes where **K** is a constant, When this relationship is combined with the equation of state for a perfect gas given by, where ***R*** is the gas constant, we can write the following relationships linking the variables at two different states (or stations) of an isentropic flow

$$\frac{P_1}{\rho_1 T_1}=\frac{P_2}{\rho_2 T_2}, \quad and \quad \frac{P_1}{\rho_1^{\gamma}}=\frac{P_2}{\rho_2^{\gamma}} \tag{15}$$

From these relations it follows that:

$$\frac{T_2}{T_1}=\left(\frac{\rho_2}{\rho_1}\right)^{\gamma-1}=\left(\frac{P_2}{P_1}\right)^{(\gamma-1)/\gamma} \tag{16}$$

From the equations of state for quasi-one dimensional adiabatic flow, constitutive relation for gas in CV1 in algebraic and differential form becomes:

$$\frac{d(P/\rho T)}{dx}=0 \tag{17}$$

Expanding equation (15) and rearranging gives $\frac{dP}{P}-\frac{d\rho}{\rho}-\frac{dT}{T}=0$, further simplification and rearrangement gives the following relation:

$$\begin{gathered} P_1=\frac{m_1 R T_1}{V_1} \quad and \quad \frac{d\rho}{\rho}=\frac{d(mV^{-1})}{mV^{-1}} \\ \frac{\dot{P}_1}{P_1}+\frac{\dot{V}_1}{V_1}=\frac{\dot{m}_1}{m_1}+\frac{\dot{T}_1}{T_1} \Rightarrow \dot{P}_1=P_1\left(\frac{\dot{m}_1}{m_1}+\frac{\dot{T}_1}{T_1}-\frac{\dot{V}_1}{V_1}\right) \end{gathered} \tag{18}$$

In a similar fashion the derivation for constitutive relation for gas in CV_2 in algebraic and differential form is given in by:

$$P_2=\frac{m_2 R T_2}{V_2}, \quad \frac{\dot{P}_2}{P_2}=\frac{\dot{m}_2}{m_2}+\frac{\dot{T}_2}{T_2}, \quad \dot{P}_2=P_2\left(\frac{\dot{m}_2}{m_2}+\frac{\dot{T}_2}{T_2}\right) \tag{19}$$

The mass/volume flow rate is related to the pressure drop across the valve by the valve flow equation. Mass flow rate can be calculated from the density of the substance, the cross sectional a rea through which the substance is flowing, and its velocity relative to the area of interest.

$$\dot{M}=\rho\cdot v\cdot A \tag{20}$$

This is equivalent to multiplying the volume flow rate by the density.

$$\dot{M}=\rho\cdot q \tag{21}$$

The symbol for mass flow rate is Newton's notation for a derivative: $\dot{M}=\frac{dm}{dt}$

$$\begin{gathered} \dot{M}_1=-\dot{M}_2=f(P_1-P_2) \\ q=c_v\frac{1}{11.7}\sqrt{\frac{P_1-P_2}{SG}} \\ \dot{M}_1=\rho q \end{gathered} \tag{22}$$

Equations (10 to 20) are the complete set of equations that describe the behaviour of this spring. These are a set of non-linear differential equations and it is not possible to obtain an analytical solution for the purpose of determining the effect of the DPs on FRs.

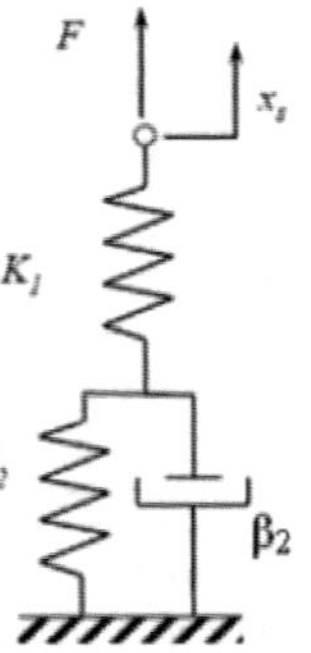

Figure 5. Model of air spring connected to auxiliary volume through a valve.

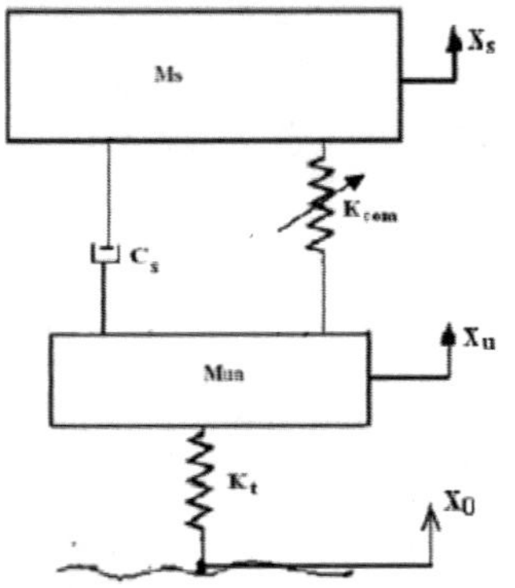

Figure 6. Model of variable air spring stiffness in semi active suspension system.

This qualitative behaviour of the airspring with auxiliary volumes can be represented as lead compensator element, employing a spring in series with a parallel arrangement of spring and damper as shown in figure 5. The complex stiffness of the element is given as:

$$K_{comp} = \frac{K_1(K_2 + \beta_2 s)}{K_1 + K_2 + \beta_2 s} \tag{23}$$

Numerical simulation based on the following vehicle parameters has been performed for a quarter car. The sprung mass stiffness is a variable stiffness of air spring; the parameters used for this analysis are taken from [9] and they are presented in table 1.

Table 1. Airspring model parameter.

Parameters	Amount	Unit
Sprung mass (Ms)	8650	kg
Unsprang mass(Mun)	1350	kg
Operating pressure (Pg)	7 *105	N/m2
Absolute pressure (P)	8 *105	N/m2
Tire stiffness (Kt)	6.5*106	N/m
Suspension damping coefficient (Cs)	43100	Ns/m
Maximum height (H)	0.3	m
Maximum diameter (D)	0.548	m
V1 (main airspring volume)	0.070758	m3
V2 airspring auxiliary volume	0.112287	m3
V3 airspring auxiliary volume	0.074858	m3
V4 airspring auxiliary volume	0.024953	m3
High stiffness	8.8*105	N/m
Medium stiffness	3.6483*105	N/m
Low stiffness	2.2*105	N/m

The volume ratio of V2, V3, and V4 is assumed as 1: 1.5: 4.5. Hence their value from the above analysis is given in table 1. In this part, for analysis purpose the damper element has been considered as a fixed value damper and the airspring stiffness varies with volume. The effect of the valve restriction needs a laboratory investigational justification and is not included in this paper. Based on the numerical manipulation this valve restriction pressure drop is found to be small and it is disregarded as it has a negligible effect. Hence the compound spring stiffness is treated as an equivalent stiffness of springs connected in series.

4. Modelling of semi-active suspension system

The complexity to develop a mathematical model for a semi-active suspension system of a vehicle depends on the suspension system components model sophistication i.e. the model developed for MR damper and airspring. Hence, the mathematical model for semi-active suspension system involves analysis of a non-linear system. To simplify this model, it is assumed that the model is combination of linear and non-linear parts.

4.1. Mathematical modelling of Semi-Active suspension with Variable Spring Stiffness

The physical model of semi-active suspension system with variable spring stiffness is shown in figure 6. The variable stiffness of the spring is due to the arrangement of variable air bag volumes together with the main volume. The curve between stiffness and airbag volume is piece wise linear.

The damping coefficient (Cs) remains fixed. Thus, the equations of motion for the TDF quarter car model shown in figure 8 are:

$$M_s\ddot{x}_s + K_{com}(x_s - x_u) + C_s(\dot{x}_s - \dot{x}_u) = 0 \tag{24}$$

$$M_{un}\ddot{x}_u + K_{com}(x_u - x_s) + C_s(\dot{x}_u - \dot{x}_s) + K_t x_u = K_t x_0 \tag{25}$$

Assume the following state variables: $x_1 = x_s$, $x_2 = \dot{x}_s$, $x_3 = x_u$, and $x_4 = \dot{x}_u$ its state space equations of motion are:

$$\begin{bmatrix}\dot{x}_1\\ \dot{x}_2\\ \dot{x}_3\\ \dot{x}_4\end{bmatrix} = \left\{\begin{bmatrix}0 & 1 & 0 & 0\\ 0 & -\frac{C_s}{M_s} & 0 & \frac{C_s}{M_s}\\ 0 & 0 & 0 & 1\\ 0 & \frac{C_s}{M_u} & -\frac{K_t}{M_u} & -\frac{C_s}{M_u}\end{bmatrix} + \begin{bmatrix}0 & 0 & 0 & 0\\ -\frac{K_{com}}{M_s} & 0 & \frac{K_{com}}{M_s} & 0\\ 0 & 0 & 0 & 0\\ \frac{K_{com}}{Mu} & 0 & -\frac{K_{com}}{M_u} & 0\end{bmatrix}\right\}\begin{bmatrix}x_1\\ x_2\\ x_3\\ x_4\end{bmatrix} + \begin{bmatrix}0\\ 0\\ 0\\ \frac{K_t}{M_u}\end{bmatrix}[x_0]$$

$$\dot{x} = \{A' + A''\}x + B\cdot x_0$$

4.2. Mathematical modelling of Semi-Active suspension with Variable Spring Stiffness and Variable Stiffness Spring

The mathematical model of this system figure 7 is complex that involves two nonlinear sub systems. The damping coefficient is a function of current, and the spring stiffness is a function of volume. Their values can be represented as: $C_s = C_s(i) \quad and \quad K_{com} = K_{com}(v)$

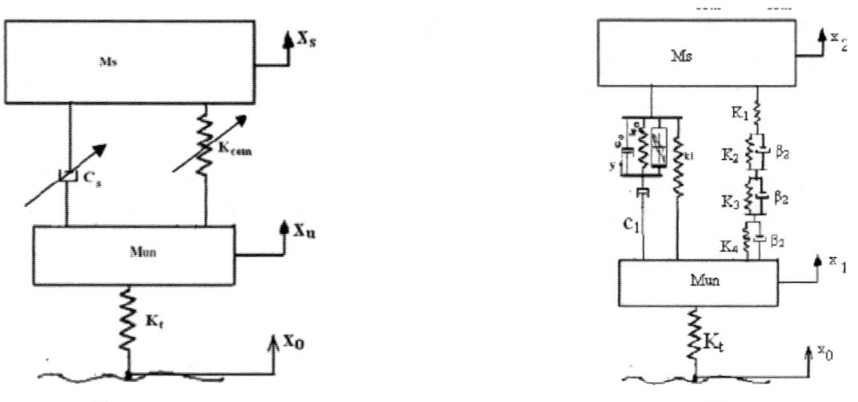

(a) (b)

Figure 7. Semi-active suspension system of variable damper & stiffness (a) physical model and (b) real model components.

The variable stiffness value $K_{com}(v)$ is determined using equation (21), Similarly the values of the variable damper Cs (i) is determined using equations (1 to 3) and (4 to 6). The equations of motion are:

$$M_s\ddot{x}_s + K_{com}(v)(x_s - x_u) + C_s(i)(\dot{x}_s - \dot{x}_u) = 0 \tag{27}$$

$$M_{un}\ddot{x}_u + K_{com}(v)(x_u - x_s) + C_s(i)(\dot{x}_u - \dot{x}_s) + K_t x_u = K_t x_0 \tag{28}$$

The state space representation of the system is:

$$\begin{bmatrix}\dot{x}_1\\ \dot{x}_2\\ \dot{x}_3\\ \dot{x}_4\end{bmatrix} = \begin{bmatrix}0 & 1 & 0 & 0\\ 0 & 0 & 0 & 0\\ 0 & 0 & 0 & 1\\ 0 & 0 & -\frac{K_t}{M_u} & 0\end{bmatrix}\begin{bmatrix}x_1\\ x_2\\ x_3\\ x_4\end{bmatrix} + \begin{bmatrix}0 & 0 & 0 & 0\\ -\frac{K_{com}(v)}{M_s} & 0 & \frac{K_{com}(v)}{M_s} & 0\\ 0 & 0 & 0 & 0\\ \frac{K_{com}(v)}{Mu} & 0 & -\frac{K_{com}(v)}{M_u} & 0\end{bmatrix}\begin{bmatrix}x_1\\ x_2\\ x_3\\ x_4\end{bmatrix} + \begin{bmatrix}0 & 0 & 0 & 0\\ 0 & -\frac{C_s(i)}{M_s} & 0 & \frac{C_s(i)}{M_s}\\ 0 & 0 & 0 & 0\\ 0 & \frac{C_s(i)}{M_u} & 0 & -\frac{C_s(i)}{M_u}\end{bmatrix}\begin{bmatrix}x_1\\ x_2\\ x_3\\ x_4\end{bmatrix} + \begin{bmatrix}0\\ 0\\ 0\\ \frac{K_t}{M_u}\end{bmatrix}[x_0] \tag{29}$$

$$\dot{x} = \{A' + A'' + A'''\}x + B \cdot x_0$$

4.3. Mathematical modelling of Semi-Active suspension with MR Damper

The analysis of semi-active suspension system with both variable stiffness and variable damping becomes a bit complex due to the existence of the two nonlinear variables Cs(i) and Kcom(v), especially this complexity made the development of the control law difficult. Hence in this section, the airspring stiffness is considered as a constant value represented as Ks shown in figure 8.

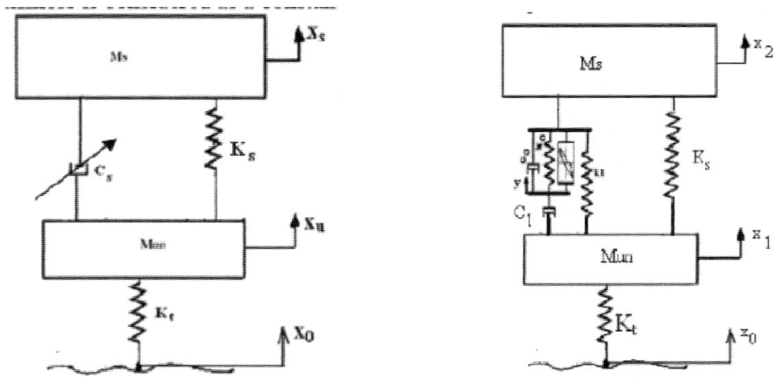

Figure 8. Semi-active suspension equipped with MR damper.

The equations of motion are:

$$M_s\ddot{x}_s + K_s(x_s - x_u) + C_s(i)(\dot{x}_s - \dot{x}_u) = 0 \quad (30)$$

$$M_{un}\ddot{x}_u + K_s(x_u - x_s) + C_s(i)(\dot{x}_u - \dot{x}_s) + K_t x_u = K_t x_0 \quad (31)$$

Accordingly, the state space representation of the system is modified to equation (32).

$$\begin{bmatrix}\dot{x}_1\\ \dot{x}_2\\ \dot{x}_3\\ \dot{x}_4\end{bmatrix} = \begin{bmatrix}0 & 1 & 0 & 0\\ -\frac{K_s}{M_s} & 0 & \frac{K_s}{M_s} & 0\\ 0 & 0 & 0 & 1\\ \frac{K_s}{Mu} & 0 & -\frac{K_t+K_s}{M_u} & 0\end{bmatrix}\begin{bmatrix}x_1\\ x_2\\ x_3\\ x_4\end{bmatrix} + \begin{bmatrix}0 & 0 & 0 & 0\\ 0 & -\frac{C_s(i)}{M_s} & 0 & \frac{C_s(i)}{M_s}\\ 0 & 0 & 0 & 0\\ 0 & \frac{C_s(i)}{M_u} & 0 & -\frac{C_s(i)}{M_u}\end{bmatrix}\begin{bmatrix}x_1\\ x_2\\ x_3\\ x_4\end{bmatrix} + \begin{bmatrix}0\\ 0\\ 0\\ \frac{K_t}{M_u}\end{bmatrix}\begin{bmatrix}x_0\end{bmatrix} \quad (32)$$

$$\dot{x} = \{A' + A'''\}x + B \cdot x$$

The model parameters used for this analysis are taken from Fox and Cebon [10], which are: Ms =1000 kg n = 2, γ = 500 N/m, υ= 613000 m-2, Mun = 100kg, A = 30.56, k_1 = 540 N/m, K_t =138000N/m and K_s =56850N/m.

5. Control of semi-active suspension system

In this work three control laws have been applied to the semi-active suspension system equipped with MR damper. The simulation results of these laws (PID, Skyhook and Fuzzy logic controllers) have been analysed. From the simulation results it has been observed that the fuzzy logic law gives good result. This is due to: in classical control theory good model result depends on strength of differential equation. However, fuzzy logic control depends on heuristics and rules. Therefore, fuzzy logic controller has an advantage over classical controller when it is applied to complex systems. It can be developed with minimal knowledge about the system dynamics. Tsampardoukas [11] and Mehrdad N. [12] have considered a fuzzy logic controller design consisting of the following four steps:

1. Rule-base (a set of If-Then rules)
2. Inference, which emulates the expert's decision in interpreting and applying knowledge about how best to control the system.
3. Fuzzification, in which the crisp inputs are transformed to fuzzy values.
4. Defuzzification in which the fuzzy output is converted back to the crisp values

The fuzzy membership functions of input and output is:

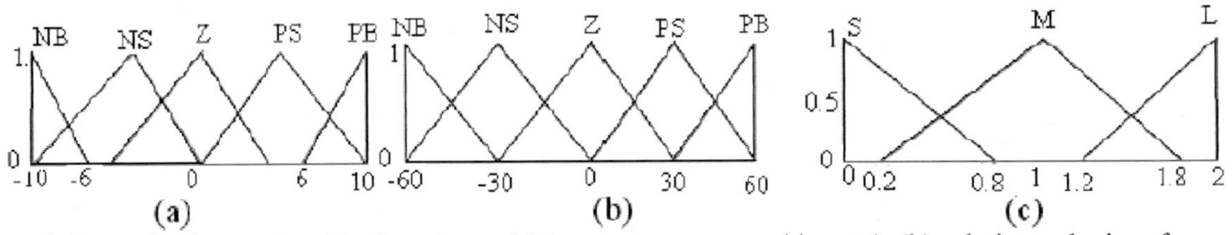

Figure 9. Fuzzy logic membership functions of (a) sprung mass speed input 1, (b) relative velocity of sprung mass to unspring mass input and (c) current output.

The fuzzy controller rules presented in table 2 will give the amount of current required continuously to reasonably damp the unexpected disturbances imposed to the system. This current results a damping coefficient Cs(i) according to the s-function developed as simulation subsystem of the main suspension system.

Table 2. Fuzzy rule out put.

Suspension speed	Relative speed				
	NB	NS	Z	PS	PB
PB	L	L	L	L	L
PS	L	M	S	M	L
Z	L	M	S	M	L
NS	L	M	S	M	L
NB	L	L	L	L	L

6. Simulation result

The two semi-active suspension systems: one with variable air spring stiffness and the other with both variable damper and variable stiffness airspring described have been analyzed for their time response to step input and transmissibility response. The semi-active suspension system with fixed airspring stiffness and MR damper has been analyzed with three control laws to a step input.

6.1. Variable Stiffness Airspring

The time response of the semi-active suspension system with variable airspring stiffness is given in figure 11. The response shows using a stiff spring in the system leads to an oscillatory response with higher initial overshoots, on the other hand low airspring stiffness results in small overshoot, reduced response time, with some steady state error. But the suspension system with medium airspring stiffness gives zero steady state error, small response time and larger overshoot.

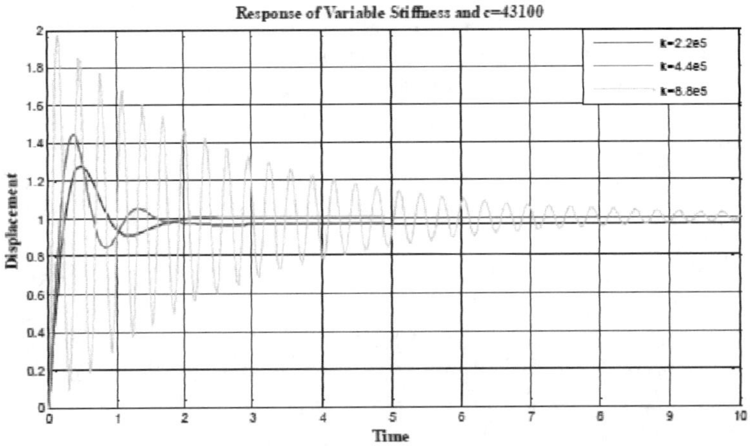

Figure 10. System responses for different stiffness values subjected to a step input excitation.

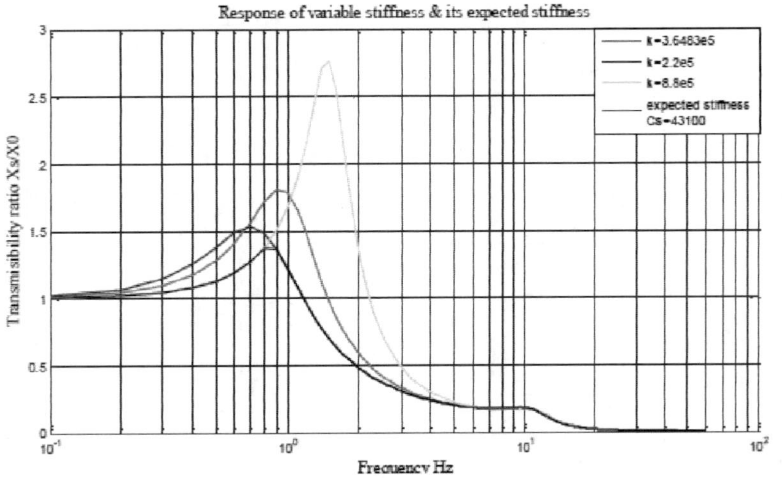

Figure11. Effects of different stiffness in vibration isolation, and expected response improvements for designing variable stiffness air spring.

The transmissibility of the semi-active suspension system with variable stiffness is given in figure11. From the transmissibility curve: the airspring with high stiffness gives small vibration transmissibility in the low frequency range; the spring with small stiffness gives good vibration isolation in the medium frequency area. However, airspring with any stiffness value has nearly the same behaviour to respond against vibration in the high frequency range. Hence, the new design approach helps to have variable stiffness airspring responding to any disturbance frequency as shown by a black colour in figure 11.

6.2. Variable Stiffness and Variable Damper Results

The time response for semi-active suspension system with variable damper and variable stiffness is shown in figure 12. The result is similar in its nature to the results of figure 10 except here the overshoot and number of oscillation before steady state have increased.

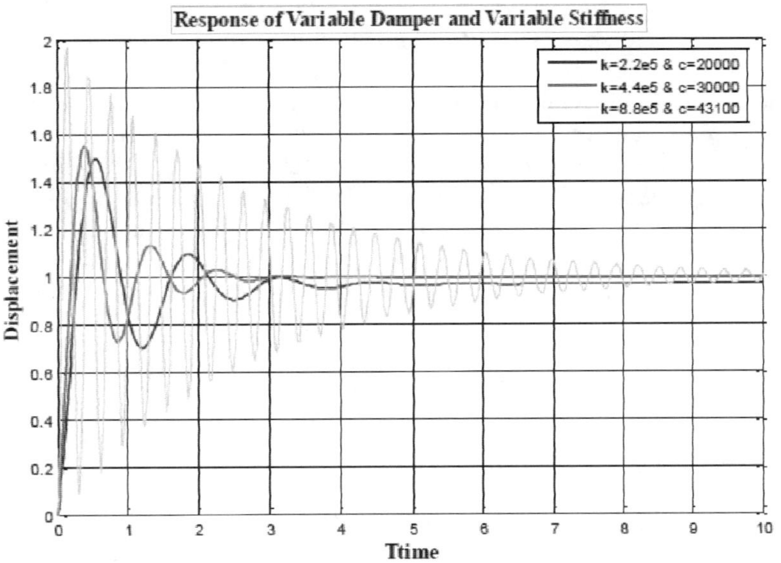

Figure 12. System responses for variable stiffness and variable damping to a step input excitation.

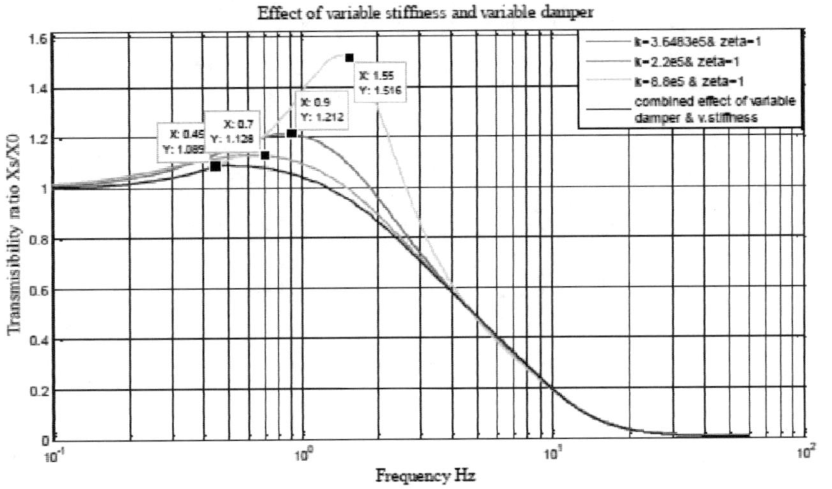

Figure 13. Transmissibility curve of variable damper and variable stiffness semi-active suspension system

One of the most important advantages of a semi-active suspension system is its capacity to shift the resonance frequency to the left. Secondly the resonance amplitude of vibration decreases substantially to a prominent value. In the low frequency area, when frequency is lower than 0.5 Hz, effective vibration isolation will be obtained by using high stiffness - high damping coefficient in the suspension system. In the medium frequency area where frequency is between 0.5 and 10 Hz the combined arrangement of low stiffness and high damping coefficient give better vibration isolation. Whereas the vibration isolation in the high frequency area, where frequency is greater than 10 Hz, almost all types of stiffness have equal effect of vibration isolation, however, a smaller damping coefficient results in good vibration isolation. Thus, introducing a variable stiffness airspring to the semi-active suspension system improves the ride comfort of vehicles as shown in figure13 drawn by a black colour.

6.3. Semi-Active Suspension System with MR Damper

Here the MR damper model parameters are simulated using three control rules (skyhook, fuzzy and PID). Both the fuzzy logic and skyhook control laws use an s-function for iterated analysis of MR damper. The time response of the three control laws are compared in figures 14 to 18. The steady state response of PID and fuzzy logic controllers are nearly the same except PID controller gives small response time with no oscillation after the first overshoot, but it has a bigger overshoot. The skyhook controller is accompanied by a maximum overshoot and oscillatory response with slow damping. Of the three control laws, the fuzzy logic controller has relatively good response both in terms of small overshoot, few unwanted oscillations after the first overshoot, and quick damping of oscillation.

IOP Conf. Series: Materials Science and Engineering **100** (2015) 012020 doi:10.1088/1757-899X/100/1/012020

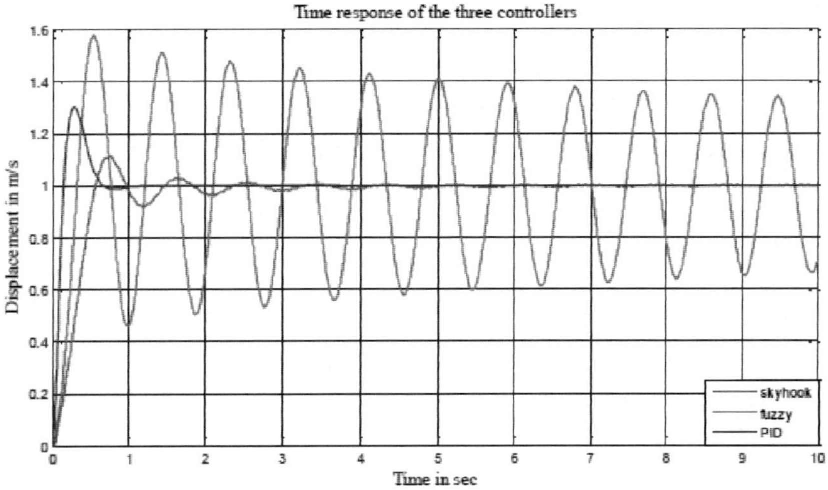

Figure14. Sprung mass displacement response of PID, Fuzzy and skyhook controller for a step input excitation.

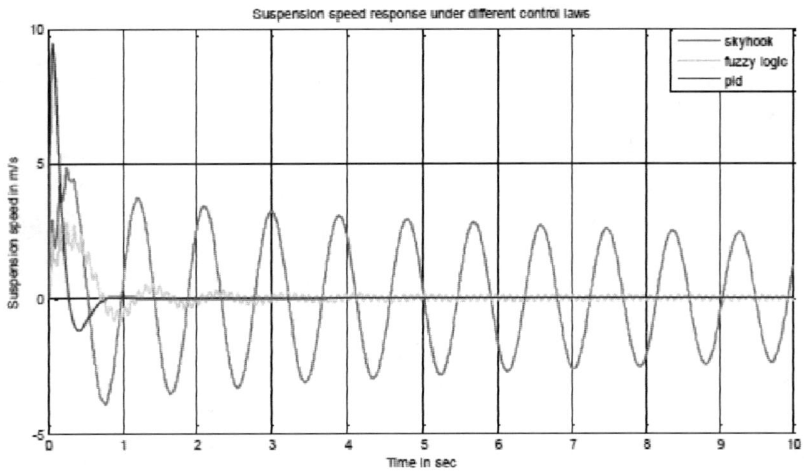

Figure15. Suspension speed under PID, Fuzzy and skyhook controller to a step input.

The sprung mass velocity response of the semi-active suspension quarter car model under these control laws is presented in figure 15. Both fuzzy and PID response shows the system returns to its original state after being disturbed. However, fuzzy controller has reduced overshoot. From the value of variable damping coefficient shown in figure 17, the damping coefficient value for PID controller is the reverse to the requirements in reality. Both the sky-hook and fuzzy logic controller response looks reasonable.

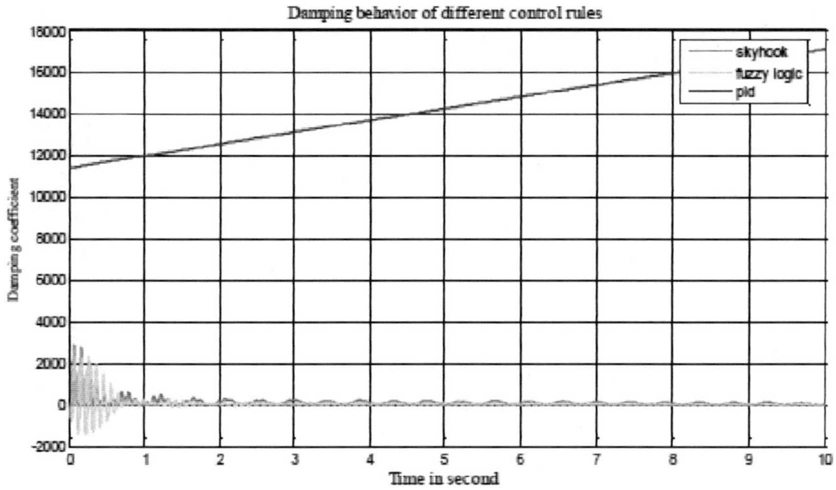

Figure16. Damping ratio response of PID, Fuzzy and Skyhook controller to step input.

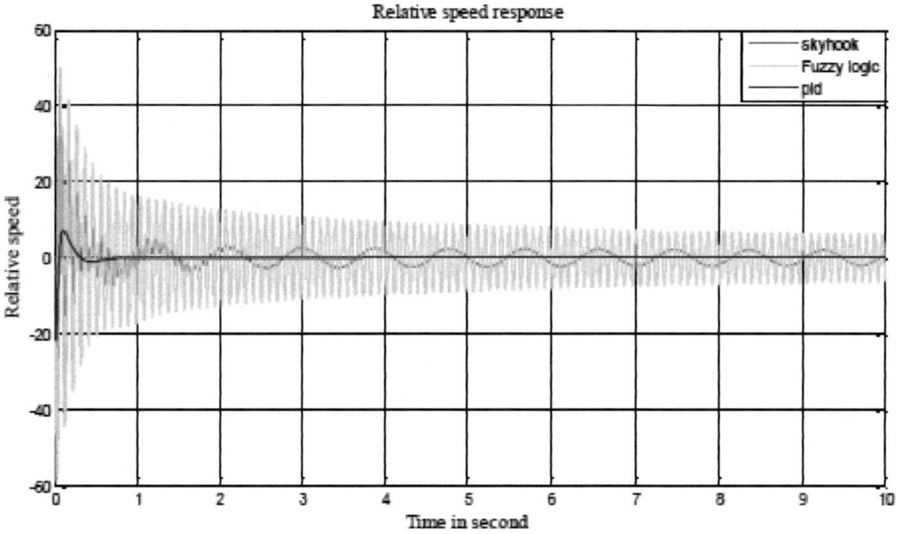

Figure17. Relative velocity of sprung mass to unspring mass when a step excitation is applied.

From the relative velocity of sprung mass to unspring mass response shown in figure17, the unspring mass of the system is highly unstable for the fuzzy logic controller, which is due to high stiffness of the tire spring. The current required by the MR damper is shown in Figure 18. The skyhook controller is acting as on/off switch. But the real requirement of current by the damper is a continuous process. This continuous requirement and continuous supply of current is only achieved through fuzzy controller.

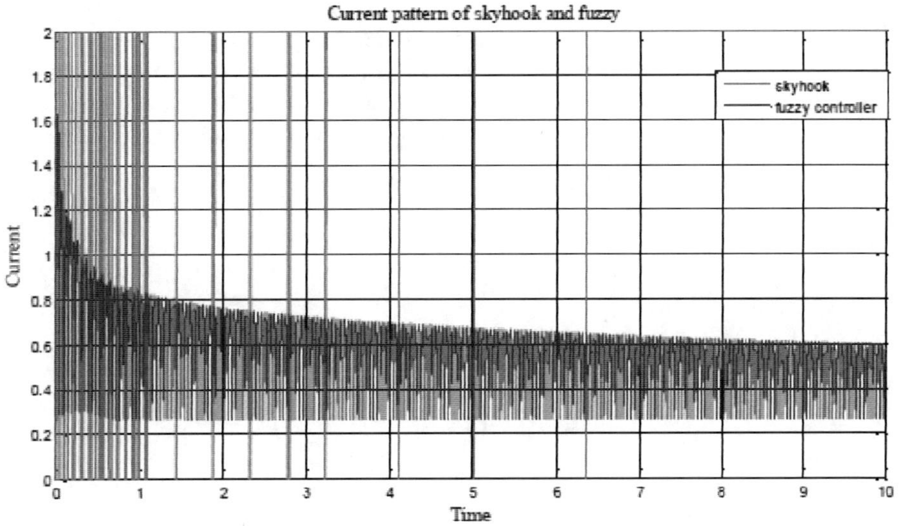

Figure 18. Current demand of skyhook and MR damper.

Looking the overall performance of semi-active suspension system with fuzzy controller shown in figures 15 to 18, it is the best control law suitable for semi-active suspension system equipped with variable damper (MR) that requires a continuous small current supply for its best performance

7. Conclusion

Looking at the conflicting requirements of suspension for comfort and handling; a new conceptual design of variable stiffness airspring is developed to facilitate independent control of ride height and airspring stiffness. The brain of semi-active suspension system is laid on the robustness of its control laws; three control laws have been implemented to analyse the mathematical model of semi-active suspension system with different stiffness and damper arrangement. The fuzzy logic control law and skyhook control law are designed to work with the s-function code for MR damper iterative analysis. The numerical simulation results indicate the fuzzy logic control has got a good reputation for best ride behaviour. The performance of semi-active suspension system with MR damper and fuzzy logic controller is optimum.

References

[1] Guglielmino E, Sireteanu T, Stammers C W, Ghita G, Giuclea M 2008 *Semi-active suspension control: improved vehicle ride and road friendliness* Springer Science & Business Media)
[2] Hamersma H A, Els P S 2014 *Journal of Terramechanics* **56** 91-101
[3] Qazi A J, Farooqui U A, Khan A, Khan M T, Mazhar F, Fiaz A 2013 *AASRI Procedia* **4** 160-6
[4] Qazi A J, Khan A, Khan M T, Noor S 2013 *AASRI Procedia* **4** 154-9
[5] Holtz M W, Van Niekerk J L 2010 *Journal of sound and Vibration* **329** 4354-66
[6] Giuclea M, Sireteanu T, Stancioiu D, Stammers C W 2004 *Proceedings of the Institution of Mechanical Engineers, Part I: Journal of Systems and Control Engineering* **218** 569-81
[7] Deo H V, Suh N P, Axiomatic design of customizable automotive suspension. Anais do ICAD2004-the 3rd International Conference on Axiomatic Design; 2004.
[8] Deo H, Suh N P, Pneumatic suspension system with independent control of damping, stiffness and ride-height. Proceedings of ICAD2006, 4th International Conference on Axiomatic Design, Firenze; 2006.
[9] Bensehop P 1995
[10] Fox M, Roebuck R, Cebon D 2007 *International Journal of Heavy Vehicle Systems* **14** 254-70

[11] Tsampardoukas G, Stammers C, Guglielmino E 2008 *Proceedings of the Institution of Mechanical Engineers, Part D: Journal of Automobile Engineering* **222** 325-52
[12] Khajavi M N, Abdollahi V, Comparison between optimized passive vehicle suspension system and semi active fuzzy logic controlled suspension system regarding ride and handling. Proceedings of world academy of science, engineering and technology; 2007.

IOP Conf. Series: Materials Science and Engineering **100** (2015) 012021 doi:10.1088/1757-899X/100/1/012021

Preparing side charging of PCM storage: theoretical and experimental investigation

A H Tesfay[1], F Y Hagos[2], K G Yohannes[3], O J Nydal[1] and M B Kahsay[3]
[1]Department of Energy and Process Engineering, Norwegian University of Science and Technology, 7491 Trondheim, Norway.
[2]Faculty of Mechanical Engineering, Universiti Malaysia Pahang, 26600 Pekan, Pahang, Malysia.
[3]Department of Mechanical Engineering, Ethiopian Institute of Technology-Mekelle, Mekelle University, P. O. Box 231 Mekelle, Ethiopia.

E-mail: kibrom2000@gmail.com

Abstract. In Ethiopia, there is an abundant source of solar energy that is estimated to 5.3 kWh/m^2/day. However, more than 90% of the society uses biomass as a main source of energy for cooking due to lack of technologies to convert this energy. Replacing these cooking activities by using renewable energy resources decreases pollution and reduces deforestation significantly. Using the solar energy in day time has no problem. For night time however, the system needs some kind of back-up system to make the daytime solar energy available. This back-up should have high-density energy storage and constant working temperature to perform a specific application. Latent heat storage using phase change materials (PCM) is one way of storing thermal energy. In the current study, a latent heat storage that uses a PCM material is used to store the solar energy aimed at utilizing solar energy for cooking Injera, main staple bread in Ethiopia. The PCM is a mixture of 60% $NaNO_3$ and 40% KNO_3 that are known as solar salts. The storage has a welded parallel aluminum fins with a gap of 40 mm in between to enhance the thermal conductivity during the charging-discharging process of the storage. The fins are extruded outside of the storage container to enable a side charging technique for the PCM. A prototype was developed with a solar salt of 17.5 kg and is tested for charging-discharging. The numerical simulation done on ANSYS and experimental results show an agreement and the system registered a 41.6% efficiency.

1. Introduction

Energy storage technologies have been in use for different purposes. Although, electro- chemical energy storage technologies are matured; majority of thermal energy technologies are still in their research stage. Since recently, there is a growing interest of solar thermal utilization. As a result, solar thermal storages got global attention. The two widely used heat storages in solar thermal applications are sensible and latent heat storages. These storages are used in different applications such as in drying, space heating, cooking, power generation. The temperature change in sensible heat storage comes as a result of the absorption of solar radiation by the material. On the other hand, latent heat is the energy absorbed and released by a substance during its phase change from solid to liquid, liquid to gas and vice versa.

Even though many developing countries are situated in in the solar belt (between 40° North to 40° South) that is endowed with huge solar energy potential, they are desperately in need of energy. Ethiopia, like other developing countries, is affected by shortage of energy. Consequently, biomass becomes the source of energy as the expense of deforestation and environmental impacts. The northern part of the country for example, has an average solar energy potential of more than 6.5 kWh/m^2/day during the dry season and 4.634 kWh/m^2/day during the rainy season [1, 2]. However, more than 90% of the region's community still depend on biomass for primary energy supply. Biomass took the lead in primary energy supply in most developing countries and majority of this primary energy is used for cooking. Cooking in biomass based energy affects children's' school time, woman and girls health, and contributes to greenhouse gas emission, land degradation and deforestation.

The interventions of solar cookers assure safe and clean energy utilization. However, their acceptance remains limited due to their longer cooking time and dependence to the presence of the sun. The mismatch of the availability and demand for solar energy can be bridged by using thermal batteries. Naturally, cooking requires nearly isothermal heat supply. Therefore, latent heat storage using phase change materials (PCM) are ideal candidate in this regard. Although PCM materials are good solar energy batteries, their charging-discharging process is difficult due to their poor thermal conductivity nature. This paper, deals with charging-discharging of nitrate based thermal storage (mixture of 60% $NaNO_3$ and 40% KNO_3). This mixture has two phase change points, a solid-solid phase change at 110°C and a solid-liquid phase change at 222°C. The material absorbs solar thermal to undergo these phase changes and release the heat in the reverse process during the application. The liquid-solid phase change state gives nearly isothermal heat supply suitable for cooking.

Phase change materials are good potentials to store solar energy, however, their charging-discharging techniques are premature for large-scale usage. Since recently many researches are undergoing on thermal properties of PCMs, heat transfer enhancement and design configurations of the heat storage for various end uses [3]. Cooking is one of the common applications particularly in developing countries, where it took the lion share of households energy share. Although many developing countries are found in the solar belt, there are only few solar cookers introduced in this region. The adoption and extension of solar cookers depends on its affordability, cooking speed, versatility, safety and cooking period. Although PCM materials have the potential to improve the utilization of solar energy, its technological status is at concept development and research level [4]. Some solar cookers with PCM storage enable cooking of family lunch while charging the storage and the stored heat has abled to cook dinner and the next day breakfast [5]. For example, many direct and indirect solar cookers with heat storage are installed in India and some African countries for family and community purpose [6]. Nitrate salts are suitable for high temperature isothermal applications. However, they have low thermal conductivity like other PCMs, which affects their charging-discharging process. For example, DLR test showed its usage for solar thermal power plants and process industry in the range of 2 to 100 kW at a melting temperature of 142°C and 222°C [7].

Some studies mainly focused on thermo physical properties of potential PCMs for high temperature latent heat storage [8]. Thermal reliability and stability of PCMs are the two most important factors to assure their latent heat storage. Manish and Jyotirmay have identified some of the most reliable PCMs for particular applications [9]. The solar salt PCM has a density of 1800 and 1700 kg/m^3 in its solid and liquid state, respectively. This density difference raises safety concerns and system challenge on how the storage charging-discharging technique was followed. Researchers by the Norwegian Science and Technology University (NTNU) have applied three different techniques to charge and discharge this PCM. While Foong used direct illumination of storage using double reflector collector, Maxim and Asfafaw used indirect charging by immersing the salt container in a hot oil bath and an aluminium block with salt cavities and steam channels, respectively [10-12]. These researches were focused to realize the use of stored solar energy for baking Ethiopia's Injera. Besides, the later author has able to bake Injera in a steam based solar stove [13]. Although, the literature shows that Injera baking is possible in the temperature range of 180 - 220°C [14], Asfafaw et al. [12] have shown Injera baking is possible in the range of 130 - 150°C. The charging time of PCMs can be improved by the thermal performance enhancement of solar collectors [15]. In addition, when a PCM is charged by using (heat transfer fluid) HTF, the inlet temperature of the HTF has greater influence over its mass flow rate in reducing the charging time [16].

The interest of this paper lies on the design and experimentation of PCM storage capable of storing 4-5 kWh thermal energy when fully charged (up to 250°C) by a side charging and top discharging technique. The outcome of this study will be implemented for solar thermal application using fixed focus offset parabolic dish concentrators. This charging technique will realize and simplify thermal system designs required for high temperature purpose.

.

2. Methodology and materials

Both theoretical and experimental methods are used to investigate the charging-discharging of the PCM Storage. The experiment is conducted in Mekelle University, Ethiopia. The geographical coordinate of the experimental location is 13°28.694' N latitude and 39°29.244' E longitude with 2,208 m above sea level [1, 2, 17]. ANSYS is used for the numerical analysis. In the experiment, about twelve K-type thermocouples are used to record continuous charging-discharging temperature of the storage at different positions as shown in figure 1 and 2.

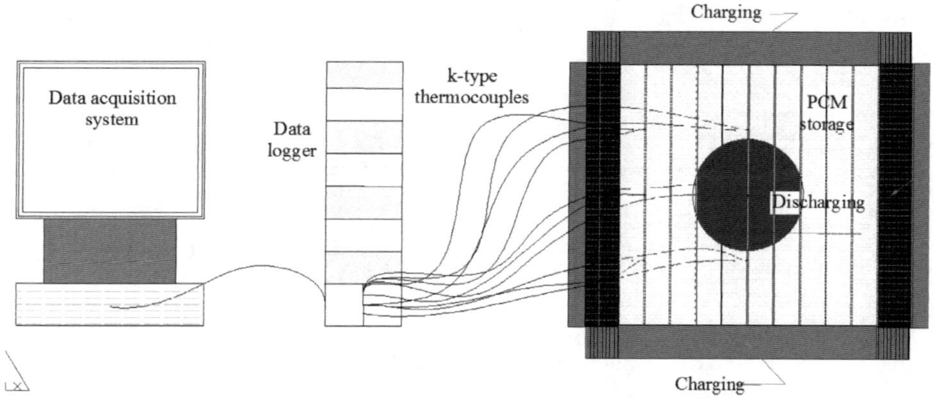

Figure 1. Schematic drawing of the system set up.

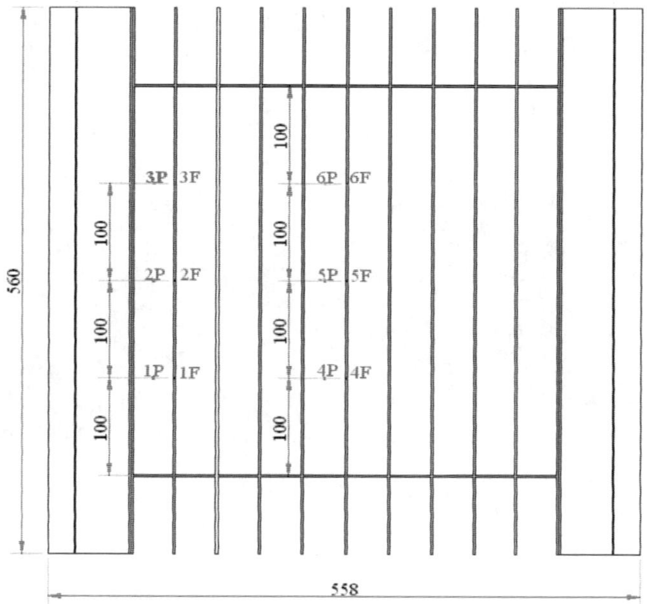

Figure 2. Position of thermocouples.

The storage has symmetric feature on which the thermocouples are connected to twelve points. Six of them were attached to the PCM and the remaining six attached to the fins. Heat was supplied to both sides of the storage and the extruded fins conduct this heat to the PCM. The temperature development inside the storage was sensed by the thermocouples and continuously recorded every second using National Instrument's data logger and Lab view software. Lab view gives continuous digital data and their graphical representation that help to understand how the charging-discharging process is undergoing. The simulated heat used in this study was obtained from charcoal.

3. System modeling

3.1. Theoretical modelling

Generally, heat flows from a high temperature zone towards a lower temperature zone. In the current analysis, only conduction and radiation heat transfers are considered. The conduction and radiation heats can be calculated by using Fourier's law and Stefan–Boltzmann law as given in equation (1) and (2)

$$q=-kA\ dt/ \quad (1)$$

$$q=\varepsilon\sigma A(T_s^4-T^4) \quad (2)$$

where q is the net rate of thermal energy, A is the radiating area, σ is the Stefan-Boltzmann constant (σ = 5.67x10-8 W/m^{2}K^4), ε is the emissivity coefficient, T_s is surface temperature and T is the temperature of the surrounding.

The rule of conservation of energy is used to model the physical problem and analysing the heat transfer problems

$$E_{in} - E_{out} + E_{gen} = E_{stored} \quad (3)$$

where, E_{in} and E_{out} represent the amount of energy crossing into and out of the surfaces of a system, E_{gen}, represents the rate of the conversion of energy from electrical, E_{stored} is the energy stored.

3.2. Fin design

The fins in the current study are made from aluminium sheet of 2 mm thickness with extruded features. Consecutive fins have a gap of 38 mm, which is be filled by the PCM material. Figure 3 shows the arrangement of the fins.

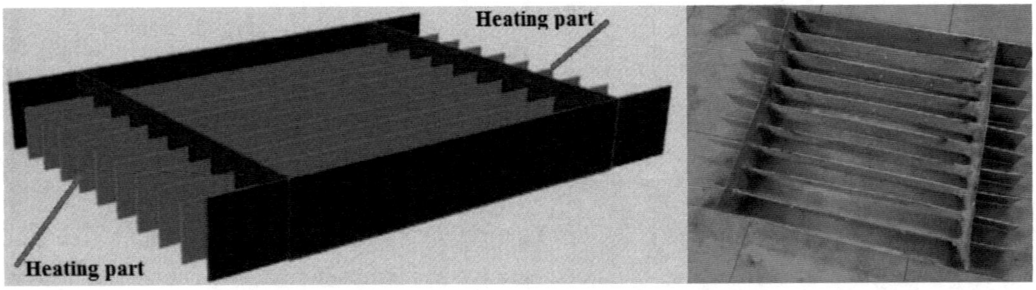

Figure 3. Aluminium fins inserted and welded to the storage.

3.3. Quantifying the amount of PCM

The PCM's melting temperature is 222°C and it has 108 kJ/kg heat of fusion. The amount of PCM is determined based on the average household energy requirement to bake one time Injera need (about 25 Injeras). The heat required to bake a single Injera is 239.5 kJ [18]. Assuming the efficiency of the storage to be 33.8%, then the amount of energy needed is 17,714.5 kJ and then the mass of PCM is found 17.5 kg from equation (4).

$$Q=\int_{T_i}^{T_f} mC_p\,dT \quad (4)$$

where m is the mass of PCM, dT is the temperature difference,

$$C_p(kJ/kg) = \begin{cases} 0.75 & T < 110^0C \\ 4.2 & 110^0C \le T \le 120^0C \\ 1.4 & 120^0C < T < 210^0C \\ 12 & 210^0C \le T \le 220^0C \\ 1.6 & T > 220^0C. \end{cases}$$

3.4. Modelling

The thermal behaviour of the system is simulated using ANSYS by formulating the linear thermal conduction problem using an element of plane 55 with temperature as a single degree of freedom variable at each node in the mesh and with the material properties model expressed deeply with temperature change to solve the nonlinear solution. Since transient thermal analysis uses the same elements as of the steady state thermal analysis. The nonlinear transient thermal analysis of the storage is simulated using ANSYS calculating the phase changing properties in two ways; the first one is by assigning the specific heat of the PCM at different temperatures and the second one is by giving enthalpy change through time. The phase changing properties of the solar salt is predicted by assigning specific heat of PCM versus temperature. The energy stored or released during phase change is considered by defining the enthalpy of the material as a function of temperature. Thermal conductivity, specific heat and density must be specified in thermal transient analysis. These inputs can be constant or temperature dependent. The material properties of each component are provided in the data base at different temperatures to make the analysis more accurate and are assigned from the database to the corresponding drawing of materials in the model. This transient analysis is arranged to give output with time steps until the steady state so that it can easily be compared with experimental values. The initial temperature of the PCM was assumed as 288 K for the transient analysis and heat begins to flow into the storage as soon as charging starts and after a period of time the temperature distribution throughout the PCM becomes steady. The element used in this analysis is Plane55 as shown in figure 4, an element description for a 2-D steady state or transient thermal analysis including phase change as shown in figure 4 and figure 5. The element can be used as a plane or ring element with a single degree of freedom, temperature at each node.

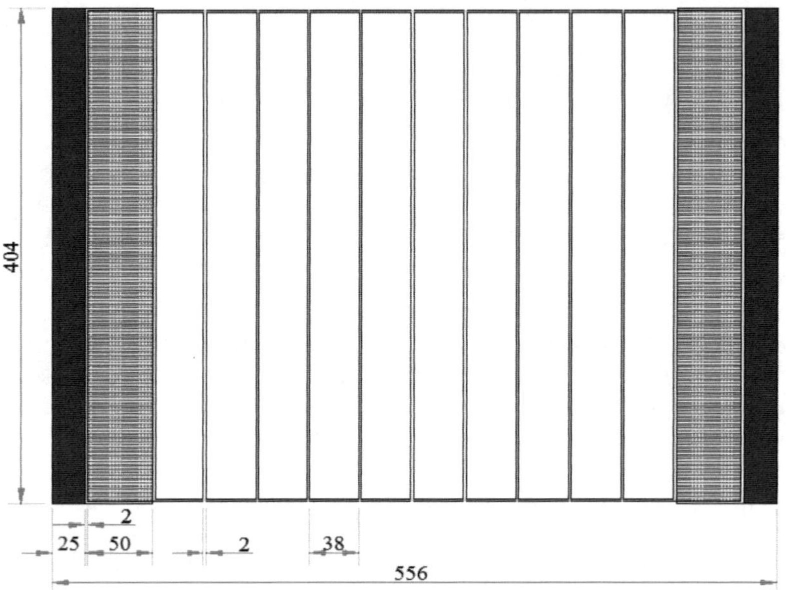

Figure 4. 2-D modeling for ANSYS charging analysis.

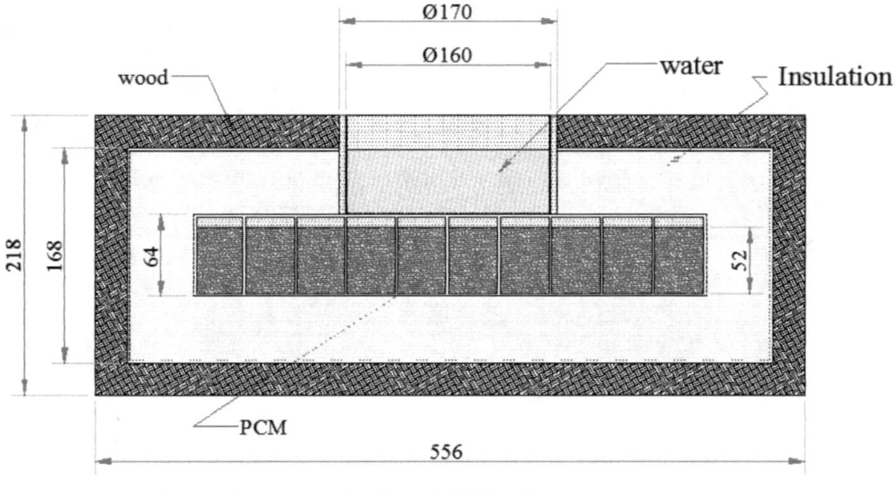

Figure 5. 2-D model for ANSYS discharging analysis.

4. Result and discussions

4.1. ANSYS results

Figure 6 shows the temperature development of the PCM during charging at a constant input temperature of 350°C. An optimized spacing between the fins gives best time in conserving the stored energy and helps to completely melt the PCM by fixing the charging temperature and time.

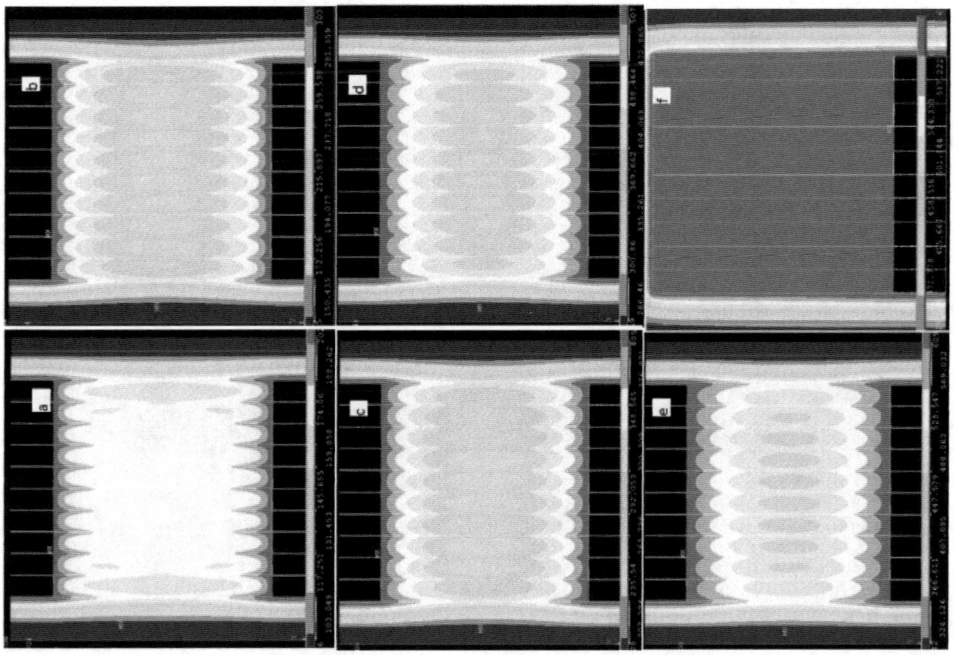

Figure 6. Results of ANSYS temperature contour during fin spacing design, (a) temperature contours at 2 hours of charging, (b) at 3 hours, (c) at 4 hours, (d) at 5 hours, (e) at 6 hours and (f) at steady state position.

The ANSYS result shows side charging is possible for solar collectors that can give an average temperature of 350°C at their receiver or charging sides of the storage for a consecutive eight hours. For less than 350°C input temperature to the storage charging side either it should get for more than 8 hours or there will not happen complete melting of the PCM. Optimizing the gap among the heat exchanger fins give the storage maximum efficiency by charging fully at the available time of solar energy and store for the whole night. If the gap is reduced to less than 40 mm the storage will be able to fully charged soon but the heat loss from the storage will increase due to high thermal conductivity of aluminium fins. During charging, it is expected to gain slightly higher temperature in the fins than the PCM. The PCM has zero slopes at about 110°C and 222°C where a solid-solid and solid-liquid phase change happens. Figure 7 shows results of ANSYS simulations while the storage is charging.

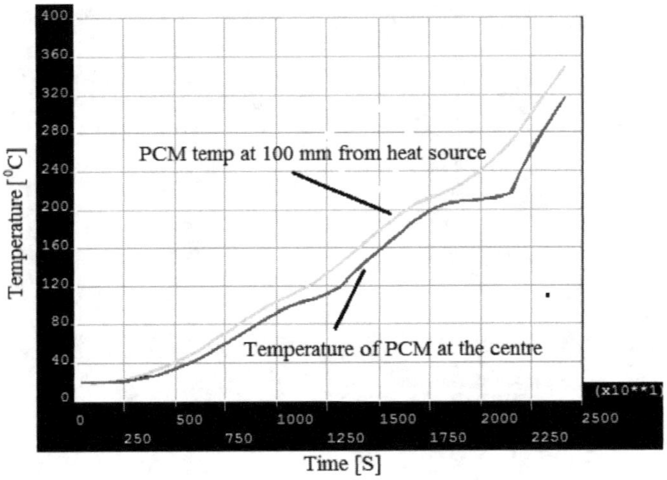

Figure 7. Temperature trend of PCM at the centre of the storage and 100 mm from the heat source.

Discharging of the stored heat is made using two cooking pots, which are aluminium and stainless steel pots. When the aluminium pot took 3,381 seconds to boil water to the temperature of 82.6°C, the stainless steel pot took 40,891 seconds to reach water temperature of 81.3°C. This difference happened due to the thermal conductivity of aluminium which is approximately 20 times greater than stainless steel. Figures 8 and 9 show the ANSYS simulation of water boiling on the two pots.

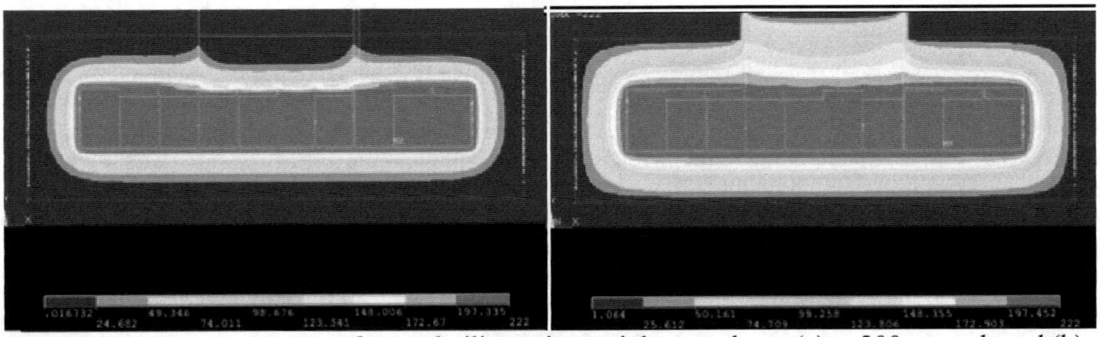

Figure 8. Temperature contour of water boiling using stainless steel pot, (a) at 300 seconds and (b) at 7200 seconds.

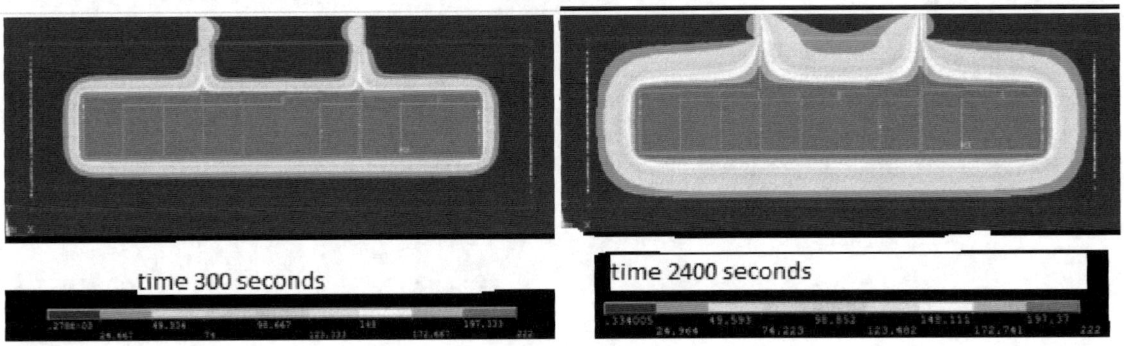

Figure 9. Temperature contour of water boiling using Aluminium pot, (a) at 300 seconds and (b) at 2400 seconds.

4.2. Experimental results

The experiment is carried out having 7 and 10.5 kg of KNO_3 and $NaNO_3$, respectively. The storage is a rectangular box with 400 mm length, 400 mm width and 60 mm height. The box has an allowance for expanding the PCM during melting. The storage can be considered as symmetric and representative temperature distribution can be taken from the first from left or right side and fifth fins at a distance of 100 mm from the charging side and 100 mm among the thermocouples. The thermocouples read temperatures of the fin and PCM. In order to prime a complete mixing of the two salt mixtures, the melting was performed using a digital electric furnace. After melting and mixing of the two salts pouring to the storage container is performed as shown in figure 10.

The temperature variation of the storage during charging is given in figure 11 and 12. The readings of 1F and 3F, 1P and 3P, 6P and 4P, and 4F and 6F, as given in figure 2, are expected to show the same temperature profile. In the ANSYS analysis, the same temperature contour is observed for these pairs. However, the temperature distribution of the heat source in the experimental investigation is not uniform. During ANSYS analysis a very small amount of temperature deviation is plotted between 1F and 4F, 1P and 4P, 3P and 6P, and 3F and 6F; even though these pairs are at the same distance from the heat source and are reading of the same material the slight difference comes due to the effect of insulation material.

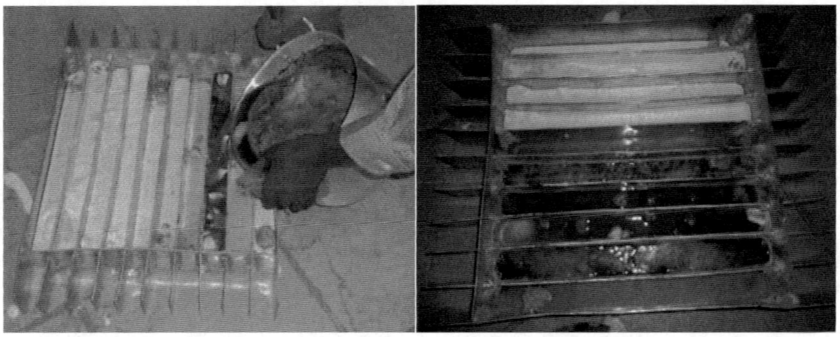

Figure 10. PCMs melted and pouring.

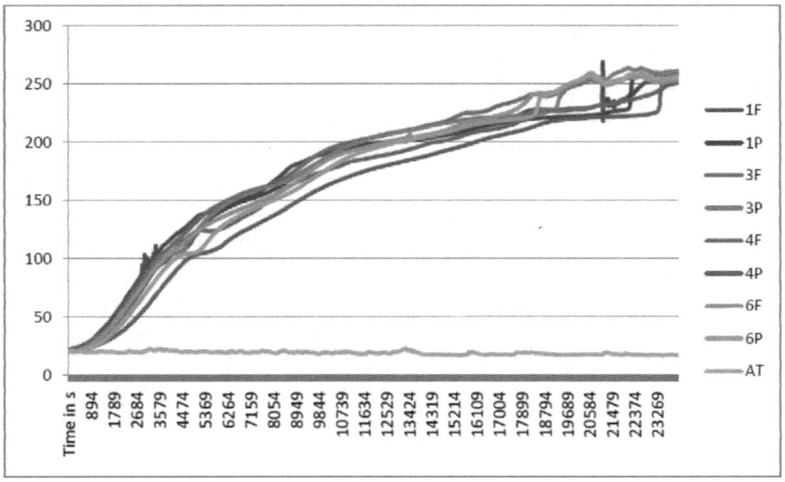

Figure 11. Charging temperature distribution 100 mm from the heat source.

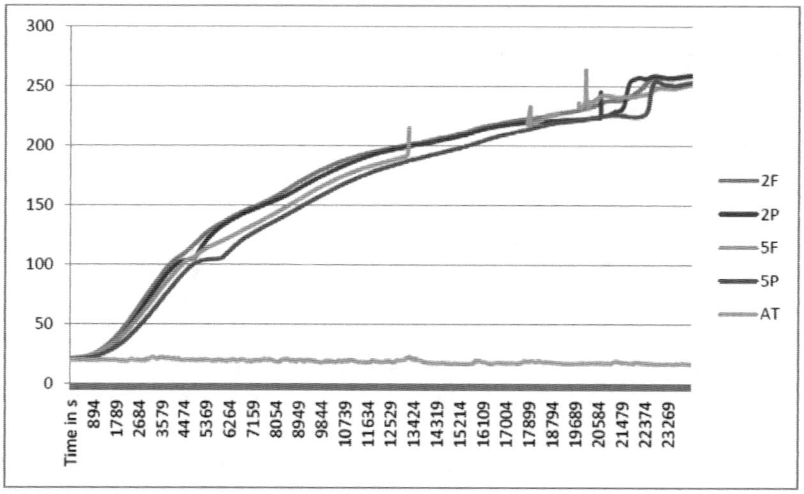

Figure 12. Temperature distributions at the centre of the storage during charging at a heat source of 490°C.

In the experimental study, the solid-solid phase transition temperature has happened in the range of 104 to 109°C and solid-liquid phase transition in the range of 217 to 228°C. The temperature variation from 228°C onward shows a sharp increase in its temperature. The maximum difference in temperature reading between the average readings of fins and PCM at 100 mm from heat source is 15°C and the average difference is calculated to be 7°C. Similarly, the maximum difference in average temperature reading between the fins and PCM at the centre or 200 mm away from charging side is 14°C and the average difference is calculated to be 5°C. The difference between average readings registered at PCM 100 mm away from heat source, PCM at the centre is 6°C, and the maximum temperature difference recorded is 11°C. On the other hand, the difference in the average reading between fins at 100 mm away from heat source and fins at the centre registered a maximum temperature of 17°C and the average difference is calculated to be 7°C. In all charging process the temperature of fins is greater than the PCM at the same position from the charging side.

4.3. Discharging tests

During charging, there is a considerable temperature variation throughout the storage until the phase transition is completed. After this state, the temperature increases almost at the same rate in all nodes. During discharging the storage temperature decreases at constant rate and at very small difference between fin and PCM temperature. Figure 13 shows a uniform declination of the storage temperature. The temperature of water initially has increased but at about 32,123 seconds and 32,905 seconds, the cover of the pot is opened twice and it started to maintain a constant temperature. At 39,730 seconds, the water is replaced by cold tab water and at 45,665 seconds, the temperature raised to 82.94°C. The thermal degradation of the container is less than 10°C per hour. The storage is tested for thermal degradation without load and a temperature of 85-95°C is recorded after 12 hours of disconnection from heat source. Heat loss is directly proportional to the external surface area of the storage. Better heat conservation is expected if it is made in full size as the surface area to volume ratio decreases. If the storage is fully charged during day time, its temperature in the next early morning will be about 90°C which makes the storage applicable for the whole night. The thermal efficiency of the system from the experimental work is also calculated to be 41.62%.

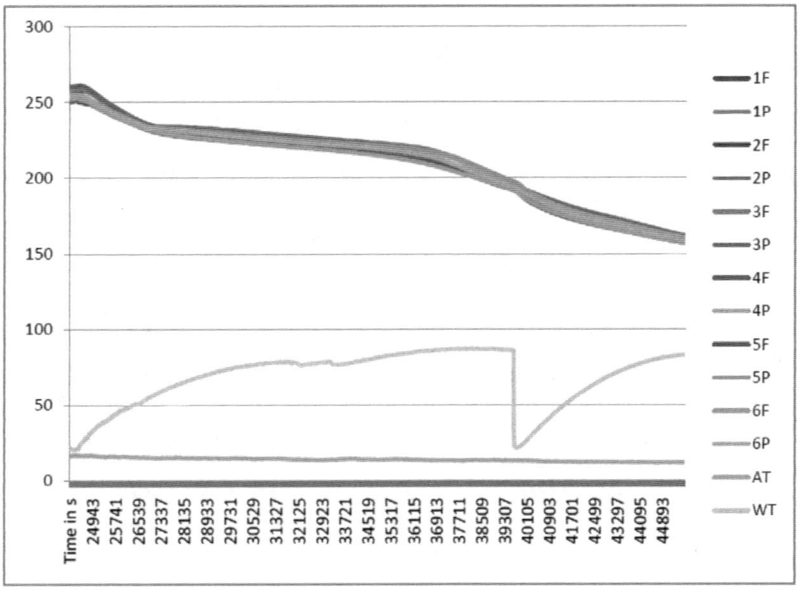

Figure 13. Discharge temperature distributions of the storage during water boiling.

5. Conclusions

In the current study, a solar-salt with a chemical composition of 60% $NaNO_3$ and 40% KNO_3 is used to store solar thermal energy. The PCM melts at 222°C, which is suitable for cooking and baking applications. The charging and discharging process is investigated experimentally and numerically. Twelve K-type thermocouples were used to record continuous charging-discharging temperature of the storage at different positions. Six of them were attached to the PCM and the remaining six attached to the fins as the storage has symmetric features. The PCM charging time depends on the level of input steady heat supply (input temperature), when the input temperature changes from 350 to 490°C the charging time has reduced dramatically. The charging time difference observed in this study comes from the different natures of heat supply. The heat supply in the numerical work considers an ideal steady heat input, however, the charcoal heat was difficult to regulate. The practicality of side charging has many advantages to utilize solar energy efficiently. Hence, combination of this design concept with offset parabola will give side charging and top discharging simultaneously, which is an ideal feature for cooking and baking applications.

Acknowledgments

The authors would like to acknowledge the support from Mekelle University, Department of Energy and Process Engineering and Norwegian University of Science and Technology (NTNU).

References

[1] Mahmud A M, Kahsay M B, Hailesilasie A, Hagos F Y, Gebray P, Kelele H K, et al. 2014 *Energy Procedia* **57** 1266-74
[2] Mahmud A M, Kahsay M B, Hailesilasie A, Hagos F Y, Gebray P, Kelele H K, et al. Solar Potential in Geba Catchment. International Congress: Livelihood 2013; Mekelle, Ethiopia2013.
[3] Kenisarin M, Mahkamov K 2007 *Renewable and Sustainable Energy Reviews* **11** 1913-65
[4] Muthusivagami R, Velraj R, Sethumadhavan R 2010 *Renewable and Sustainable Energy Reviews* **14** 691-701

[5] Lecuona A, Nogueira J-I, Ventas R, Legrand M 2013 *Applied Energy* **111** 1136-46
[6] Schwarzer K, Da Silva M E V 2003 *Solar Energy* **75** 35-41
[7] Laing D, Bauer T, Steinmann W, Lehmann D, Advanced High Temperature Latent Heat Storage System—Design and Test Results. Proceedings of the 11th International Conference on Thermal Energy Storage—Effstock; 2009.
[8] Cárdenas B, León N 2013 *Renewable and Sustainable Energy Reviews* **27** 724-37
[9] Rathod M K, Banerjee J 2013 *Renewable and Sustainable Energy Reviews* **18** 246-58
[10] Foong C W, Nydal O J, Løvseth J 2011 *Applied Thermal Engineering* **31** 1807-15
[11] Mussard M, Nydal O J 2013 *Solar Energy* **95** 144-54
[12] Tesfay A H, Kahsay M B, Nydal O J 2014 *Energy Procedia* **57** 1603-12
[13] Tesfay A H, Kahsay M B, Nydal O J 2014 *Energy Procedia* **57** 2946-55
[14] Hassen A A, Amibe D A
[15] Sharma S, Iwata T, Kitano H, Sagara K 2005 *Solar Energy* **78** 416-26
[16] Al-Abidi A A, Mat S, Sopian K, Sulaiman M, Mohammad A T 2014 *Energy and Buildings* **68** 33-41
[17] Kahsay M B, Mahmud A M, Hagos F Y, Kiros H, Tesfay A H, Gebray P, et al. 2013 *Wind Engineering* **37** 333-46
[18] Gebregziabher A. Gasification based Injera baking stove. Mekelle University: Mekelle University; 2012.

Reliability and availability analysis for robot subsystem in automotive assembly plant: a case study

A F Fudzin[1] and M A A Majid[1]
[1]Faculty of Mechanical Engineering, Universiti Teknologi PETRONAS, Perak, Malaysia

E-mail: fauzifudzin@gmail.com

Abstract. The automotive assembly plant in a manufacturing environment consists of conveying systems and robots. Robots with high reliability will ensure no interruption during production. This study is to analyze the individual robot reliability compared to reliability of robots subsystem in series configuration. Availability was computed based on individual robots breakdown data. Failures due to robots breakdown often occurred during the operations. Actual maintenance data for a period of seven years were used for the analysis. Incorporation of failures rate and mean time between failures yield the reliability computation with the assumption of constant failure rate. Result from the analysis based on 5000 operating hours indicated reliability of series configuration of robots in a subsystem decreased to 2.8% in comparison to 38% reliability of the individual robot with the lowest reliability. The calculated lowest availability of the robots is 99.41%. The robot with the lowest reliability and availability should be considered for replacement.

1. Introduction

Sequences of robotic system in production line increase the automation level of the industry, hence helps the industry to achieve high production volume. The automation in the automotive industry in fact helps the industry to improve in term of safety, quality, and standardization as well as to achieve production efficiency. However, failures and breakdown of robotic system is inevitable and give high impact to the production throughput.

The usage of robotic systems is to achieve high production throughput in all production lines such as stamping shop, body assembly shop, painting shop and assembly shop. Each production line is equipped with several robots, equipment, conveyors, manipulators and material handling equipment. In general, the stamping parts will transfer to body shop and flow from one process to another process via automatic conveyer systems and welded together by spot welder robotic systems. The body in white (BIW) then fed to painting shop for painting process and sprayed by a cluster of robotic systems. Finally, the painted bodies are fed to assembly shop for assembly processes via automatic conveyor system as shown in figure 1. There are 156 units of robots operated in this particular automotive plant. The robot failures data were kept as maintenance data which was well documented for a tracebility purposes.

Robots failures often occurred in industry which contributed to the downtime and production shut off. In the automotive industries with high level of automation, the failures are recorded as maintenance data. The robot failures data in many literatures are acquired from laboratory experiments, not from actual maintenance data. As stated by Vineyard, Amoako-Gyampah [1], most research studies on machine failures and maintenance decisions have assumed some failures distribution such as exponential or weibull and most of the data were provided by the suppliers of the equipment.

Relationship between the failures data and production data during warranty periods could be used for quality improvements for manufacturing industry. This approach was adopted by Jeon and Sohn [2] who used warranty data to activate an early warning for the design errors and highlight faults in manufacturing process.

Based on the current literature, most of the study analyzed the robots reliability as a unit which data was collected in the laboratory environment. However, this study was carried out for robots in the actual manufacturing environment in the automotive assembly plant. Simon, Javad [3] used maintenance data and daily reports data of conveyor systems in mining industry for a case study of reliability and maintainability. From the data the reliability and availability of the conveyors were computed.

A robot which performs without failure for a specified period of time under specified operating conditions is considered having high reliability. The discipline of reliability engineering basically is a study of the causes, distribution and prediction of failures [4]. Robotic systems are commonly used in automotive assembly line. It is important robot should have high reliability in order for production continuity. The lineups of robots exist in both series configuration and parallel configuration.

Study done by Tsarouhas, Varzakas [5] on the mobile robot reliability and a serial configuration of components in the mobile robots showed that the robot reliabilities are low as well as the entire system reliability. Sun, Xi [6] analyzed the reliability of machine-level reliability and system-level reliability for a manufacturing system in a series-parallel configuration of equipment for manufacturing plant. Result from the case study indicates the system reliability for serial-parallel hybrid system showed the highest reliability.

Reliability and availability co-exist in the sense of reporting for equipment maintainability issue. Availability of an equipment or a robot is measured as a factor of its reliability. Availability is defined as the probability that a system is operating satisfactorily at any random point in time t, when subject to a sequence of up and down cycles which constitute an alternating renewal process [7].

Performance of robot in a subsystem will affect the availability. Therefore, the performance of each individual robot in subsystem should be analyzed in order to determine how each robot affect the reliability and availability performance for the whole production line in automotive assembly plant.

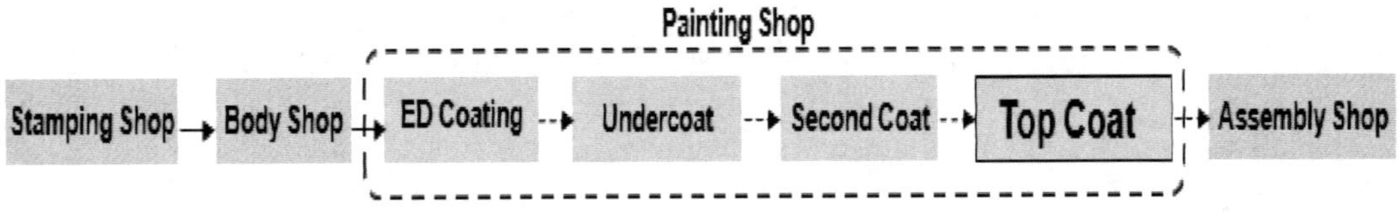

Figure 1. Schematic Process Flow in Automotive Plant.

The objective of the present study is to apply the reliability model to evaluate the reliability and availability of an automotive robotic system. The failure rate, λ, mean time to repair and mean time between failures were used in the analysis. The case study is on automotive assembly plant's robots at top coat line of painting shop. The case study used for comparison of individual robot and for the entire robots in the subsystem consists of 8 robots in the top coat system. The processes involved in top coat line are base coat and clear coat. The 8 robots in top coat line are arranged in the configuration 4 robots in left hand side and 4 robots in a right hand side as shown in figure 2. All robots must be fully operational and in the case of one of the robots fail to perform to function the whole top coat system will be shut off.

The failures data of robots, generally recorded maintenance data in top coat production line were acquired for the period of seven years. Result of the failures analysis is expected to assist painting production line decision maker to justify the decisions on the robot life cycle.

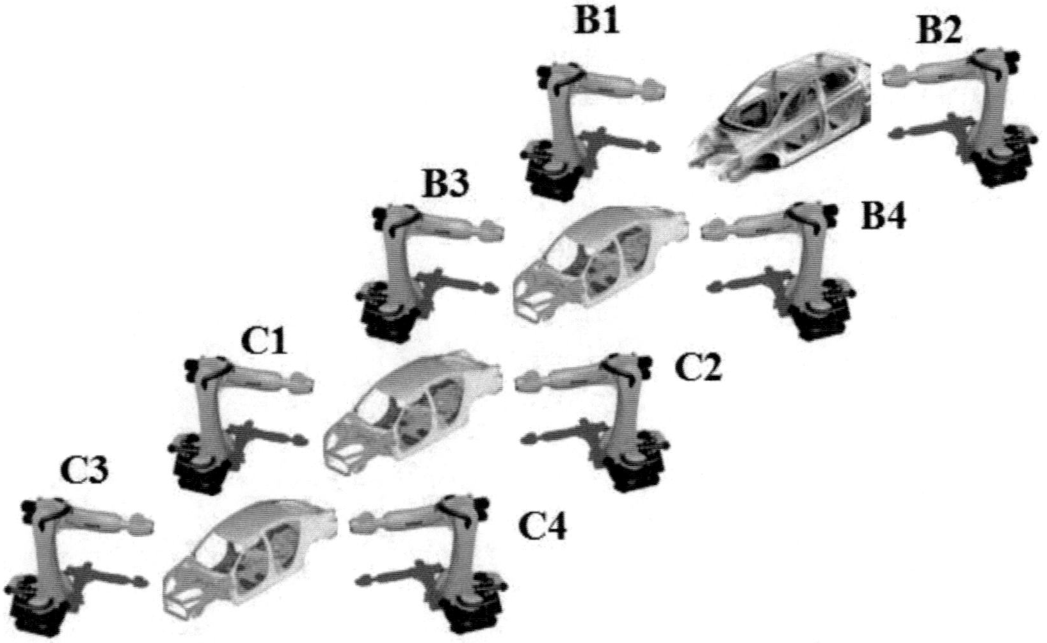

Figure 2. Typical view the top coat robot arrangement.

2. Mathematical Formulation

All 8 robots are configured as a series which represents reliability model in series configuration for the top coat line [7]. In the case of one robot fails to perform its function it will result to failure to the whole system.The individual reliability of robot was calculated as shown in equation (1).

$$R(t) = e^{-\lambda t} \tag{1}$$

Thus, the reliability of robot in top coat line is expressed in series configuration as shown in equation (2).

$$R_s(t) = R_1(t) \cdot R_2(t) \cdot R_3(t) \cdot \ldots \cdot R_n(t) = \prod_{i=1}^{n} R_i(t) \tag{2}$$

where, $R_1(t)$, $R_2(t)$......$R_n(t)$ are reliabilities of each individual robot respectively. As the assumption is constant failure rate, λ for all robots, which means that the exponential distribution for the reliability functions, then, is as shown in equation (3).

$$R_s(t) = e^{-\lambda_1 t} \cdot e^{-\lambda_2 t} \cdot \ldots \cdot e^{-\lambda_n t} = exp\left[-\sum_{i=1}^{n} \lambda_i t\right] = exp[-\lambda t] \tag{3}$$

As for the MTTR and MTBF calculations, the equations are shown in equation (4) and equation (5) respectively.

$$MTBF = \frac{T(t)}{r} \quad (4)$$

where, T(t) is the total operating time, and r is the number of failures.

$$MTTR = \frac{Production\ Time\ (min)}{Number\ of\ Failures} \quad (5)$$

Hence, the availability equation is shown in equation (6).

$$Availability = \frac{MTBF}{MTBF + MTTR} \times 100\% \quad (6)$$

3. Results and Discussion

Analysis on reliability of each robot and reliability of robots in a subsystem were computed in top coat line. The actual failures maintenance data were utilized which was acquired for a period of seven years as a case study and as a basis of computation of individual robot reliability and series robot reliability.

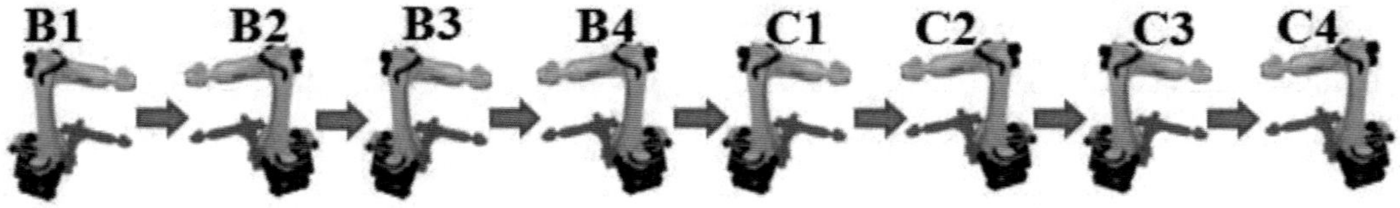

Figure 3. Schematic Block Diagram of Series Configuration of Top Coat Robots.

There are two groups of robots involved in the analysis; there are B1, B2, B3 and B4 are the base coat robots and C1, C2, C3 and C4 are clear coat robots. Four robots in each side left and right are configured in series as a modeling which illustrated in figure 3. In the case of failures occurred in one of the robot either left or right robot in top coat line, the whole production in top coat line will shut off.

The formulations of reliability for each individual robot were calculated. Table 1 shows the result after computation based on equation (1) and (2). The robot B3 is having the reliability 0.38; Robot B2, C1, C3 and C4 having the reliability 0.62; while robot B1, B4 and C2 having the reliability 0.79. It is clearly shown, robot B3 having the lowest reliability in comparison with others robots.

Table 1. Calculation Results of Reliability.

Robots	B1	B2	B3	B4	C1	C2	C3	C4
Reliability, R(t)	0.79	0.62	0.38	0.79	0.62	0.79	0.62	0.62
Reliability, Rs(t)					0.028			

The graph in figure 4(a) to 4(h) shows the results of reliability analysis for each individual robot, B1, B2, B3, B4, C1, C2, C3 and C4 respectively. The plot shows, the individual robots are having a decreasing reliability over the longer operating hours.

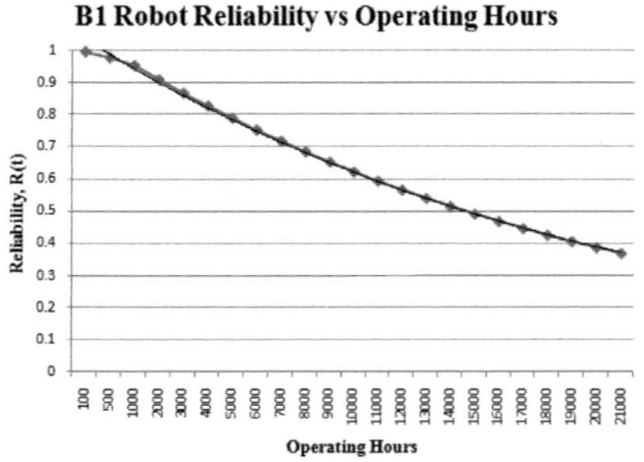
B1 Robot Reliability vs Operating Hours
Reliability, R(t)
0 0.1 0.2 0.3 0.4 0.5 0.6 0.7 0.8 0.9 1
100 500 1000 2000 3000 4000 5000 6000 7000 8000 9000 10000 11000 12000 13000 14000 15000 16000 17000 18000 19000 20000 21000
Operating Hours

(a)

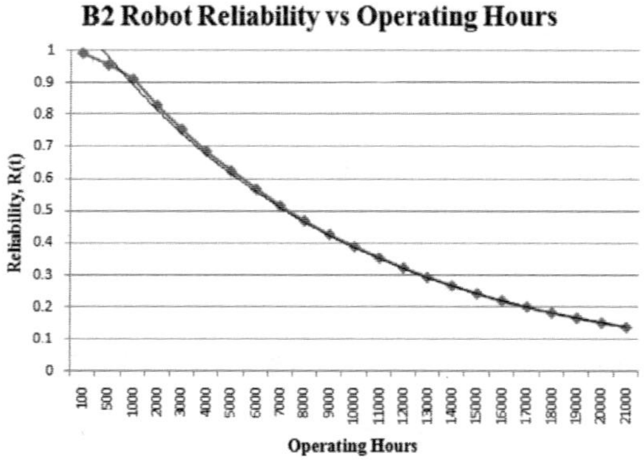
B2 Robot Reliability vs Operating Hours
Reliability, R(t)
0 0.1 0.2 0.3 0.4 0.5 0.6 0.7 0.8 0.9 1
100 500 1000 2000 3000 4000 5000 6000 7000 8000 9000 10000 11000 12000 13000 14000 15000 16000 17000 18000 19000 20000 21000
Operating Hours

(b)

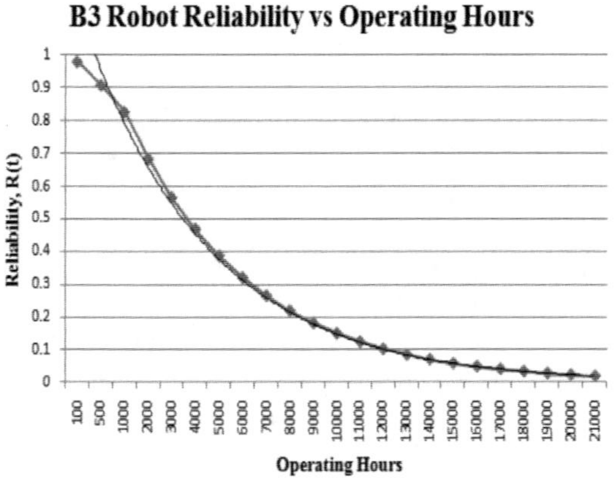
B3 Robot Reliability vs Operating Hours
Reliability, R(t)
0 0.1 0.2 0.3 0.4 0.5 0.6 0.7 0.8 0.9 1
100 500 1000 2000 3000 4000 5000 6000 7000 8000 9000 10000 11000 12000 13000 14000 15000 16000 17000 18000 19000 20000 21000
Operating Hours

(c)

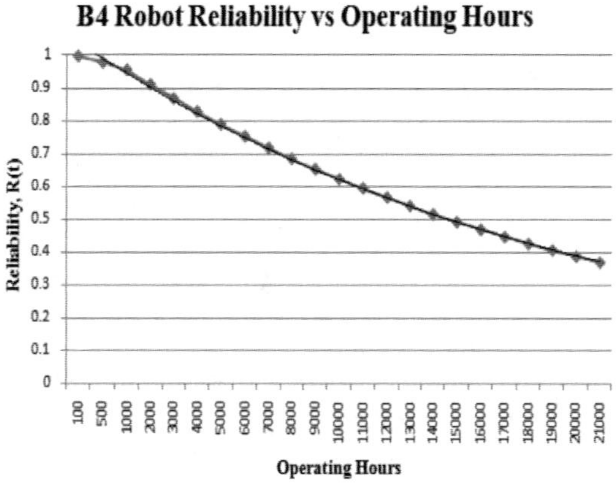
B4 Robot Reliability vs Operating Hours
Reliability, R(t)
0 0.1 0.2 0.3 0.4 0.5 0.6 0.7 0.8 0.9 1
100 500 1000 2000 3000 4000 5000 6000 7000 8000 9000 10000 11000 12000 13000 14000 15000 16000 17000 18000 19000 20000 21000
Operating Hours

(d)

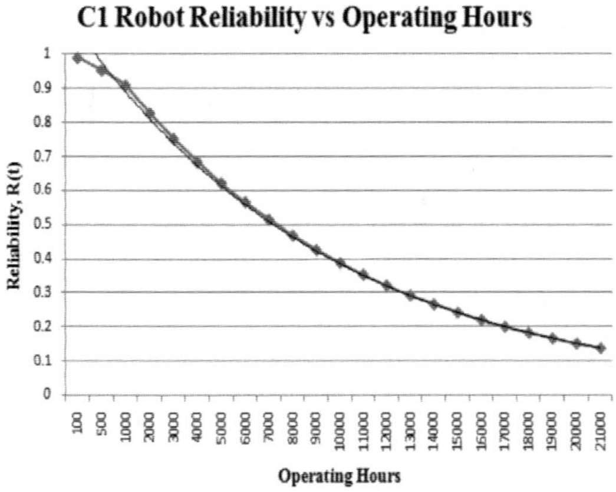
C1 Robot Reliability vs Operating Hours
Reliability, R(t)
0 0.1 0.2 0.3 0.4 0.5 0.6 0.7 0.8 0.9 1
100 500 1000 2000 3000 4000 5000 6000 7000 8000 9000 10000 11000 12000 13000 14000 15000 16000 17000 18000 19000 20000 21000
Operating Hours

(e)

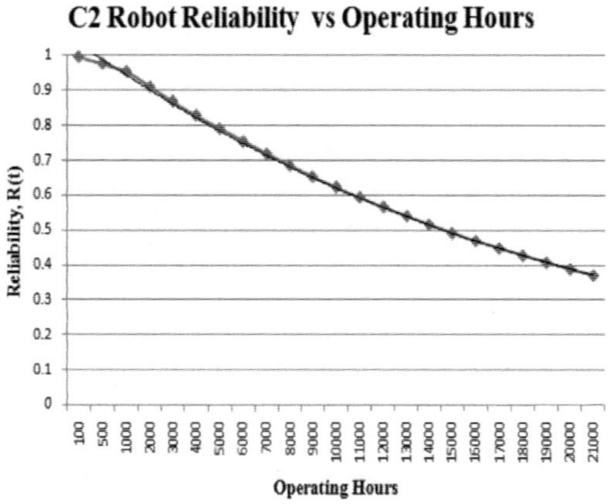
C2 Robot Reliability vs Operating Hours
Reliability, R(t)
0 0.1 0.2 0.3 0.4 0.5 0.6 0.7 0.8 0.9 1
100 500 1000 2000 3000 4000 5000 6000 7000 8000 9000 10000 11000 12000 13000 14000 15000 16000 17000 18000 19000 20000 21000
Operating Hours

(f)

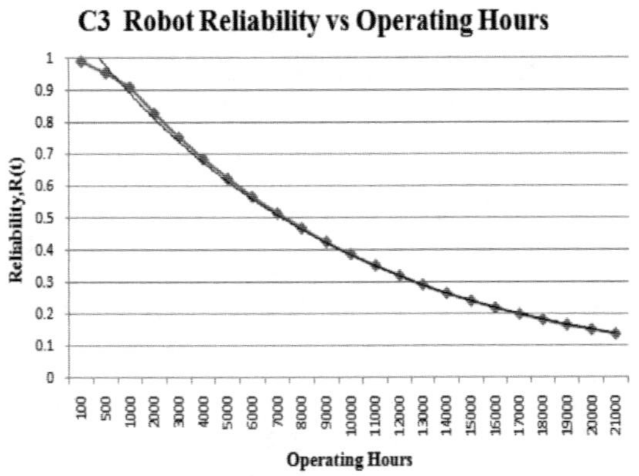

(g)

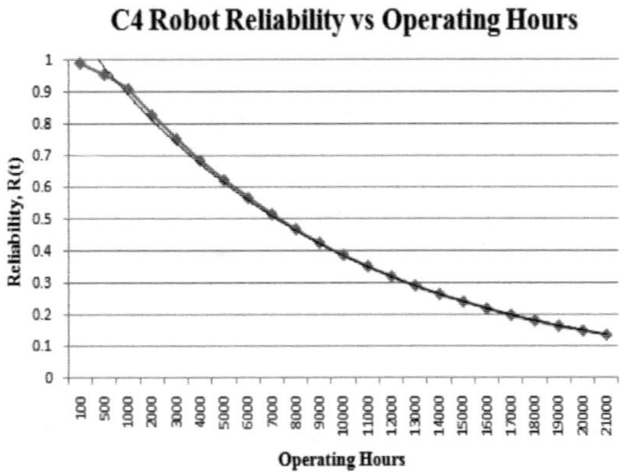

(h)

Figure 4. Individual Robot Reliability for B1, B2, B3, B4, C1, C2, C3 and C4 vs Operating Hours Graphs.

3.1. Analysis on Subsystem Robots Reliability

As for the subsystem reliability, Rs(t), the reliability is 0.028 at the 5000 operating hours. Based on the results of reliability analysis as drawn in figure 5, the behaviour of robot subsystem reliability, Rs(t) is decreasing compared to individual robot reliability, R(t) in series configuration.

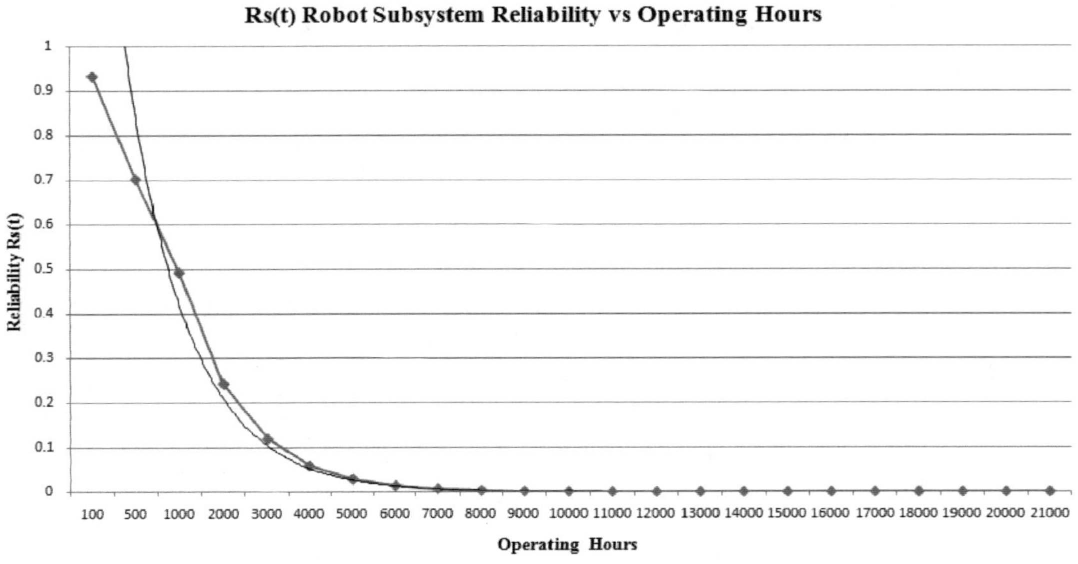

Figure 5. Robot Subsystem Reliability vs Operating Hours Graph.

3.2. Analysis on Robots Availability

Evaluation of availability analysis for each robot station is shown in figure 6. The availability were calculated using equation (4), equation (5) and equation (6). The availability of the robots ranging from 99.9993% to 99.4069%. Robot B3 has the lowest availability in comparison with others robots with the availability value is 99.4069%.

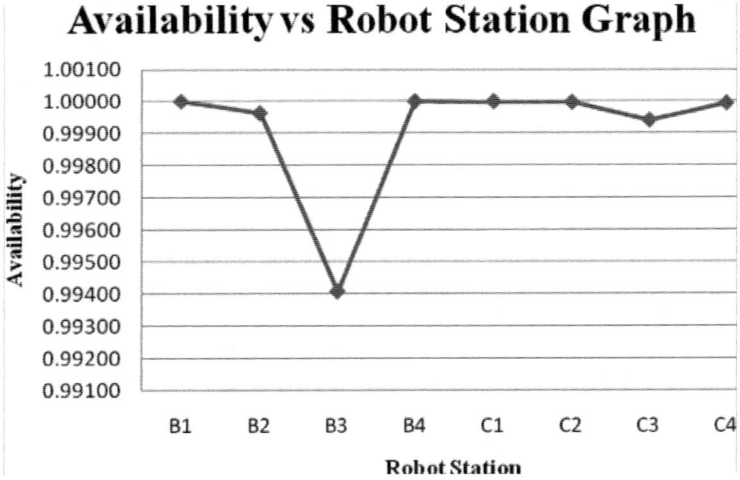

Figure 6. Availability vs Robot Station Graph.

4. Conclusion

Based on 5000 operating hours, the reliability of the robots B1, B4 and C2 are 79% respectively, while robots B2, C1, C3 and C4 are 62% respectively. The robot B3 having the lowest reliability is 38%. Further analyses were carried out in series configuration of robots in a system reliability decreased to 2.8% at the same operating hours. It is proven that the system reliability decreased further compared to reliability of the individual robot of the system. The availability analysis was carried out for all individual robots. The range of availability value is 99.9993% to 99.4069%. Robot B3 which is the third robot in the subsystem has the lowest availability value of 99.4069%.

It is recommended, immediate attention to the third robot is required. The plant management is recommended to consider immediate attention to the third robot on the replacement strategy. Proper maintenance schedule must be established for the second, fifth, seventh and eight robots.

Acknowledgement

The author would like to acknowledge to the Universiti Teknologi PETRONAS for support of this study.

References

[1] Vineyard M, Amoako-Gyampah K, Meredith J R 1999 *European journal of operational research* **116** 139-55

[2] Jeon J, Sohn S Y 2015 *Reliability Engineering & System Safety* **133** 176-83

[3] Simon F, Javad B, Abbas B 2014 *International Journal of Mining Science and Technology* **24** 587-91

[4] Dieter G E, Dieter G E, Schmidt L C 2012 *Engineering Design* **5** McGraw-Hill Higher Education)

[5] Tsarouhas P H, Varzakas T H, Arvanitoyannis I S 2009 *Journal of food engineering* **91** 250-9

[6] Sun J-w, Xi L-f, Du S-c, Ju B 2008 *International Journal of Production Economics* **114** 149-64

[7] Dept. of Defense United States 1998 *Electronic Reliability Design Handbook* **1** (Pennsylvania State University: Department of Defense)

Application of artificial neural network for prediction of marine diesel engine performance

C W Mohd Noor[1], R Mamat[1], G Najafi[2], W B Wan Nik[3] and M Fadhil[1]
[1]Faculty of Mechanical Engineering, Universiti Malaysia Pahang, 26600 Pekan, Pahang, Malaysia
[2]Tarbiat Modares University, P.O. Box 14115-111, Tehran, Iran
[3]School of Ocean Engineering, Universiti Malaysia Terengganu, Malaysia

E-mail: che.wan@umt.edu.my

Abstract. This study deals with an artificial neural network (ANN) modelling of a marine diesel engine to predict the brake power, output torque, brake specific fuel consumption, brake thermal efficiency and volumetric efficiency. The input data for network training was gathered from engine laboratory testing running at various engine speed. The prediction model was developed based on standard back-propagation Levenberg-Marquardt training algorithm. The performance of the model was validated by comparing the prediction data sets with the measured experiment data. Results showed that the ANN model provided good agreement with the experimental data with high accuracy.

1. Introduction

Simulation approaches such as artificial neural networks (ANN) model are widely used by researchers to solve a variety of problems in science and engineering, particularly for some areas where the conventional modelling methods fail [1]. An ANN model can accommodate multiple input variables to predict multiple output variables. The prediction by a well-trained ANN is normally much faster than the conventional simulation programs or mathematical models as no lengthy iterative calculations are needed to solve differential equations using numerical methods but the selection of an appropriate neural network topology is important in terms of model accuracy and model simplicity. On the other hand diesel engine laboratory performance testing are time consuming and very costly which include budget expenditure for the engine test set up. As an alternative method, diesel engine testing simulation can also be conducted to simulate the actual condition of the engine performance with different inputs and outputs. The application of ANN approach in automotive engineering could be progressed a very impressive rate in recent years. Some researchers used ANN to predict internal combustion engine characteristics. The predictive ability of an ANN results from the training on experimental data and then validation by independent data. ANN approach has been used by Xu, Xie [2], in forecasting engine systems reliability, Yuanwang, Meilin [3], analysed the effect of cetane number on exhaust emissions from engine, Korres, Anastopoulos [4], predict diesel lubricity, Lucas, Duran [5], setup ANN model for diesel particulate emission, Hafner et al.[6], for diesel engine control design, Shayler and Goodman [7], adapted ANN in automotive engine management systems and Tan [8], model the intake manifold and throttle body processes in an automotive engine. Najafi, Ghobadian [9] reported that artificial neural network (ANN) is among some of the most well-known methods to predict the actual combustion period and can be used for the experimental data validation and optimization. Mustafa Canakci and Erdil [10] and Arcaklioglu et al. [11] demonstrated that ANN can be an optimum tool for predicting the performance and exhaust emissions. Shanmugam and P.V. [12] has modelled his ANN concept to predict the performance and exhaust emissions of the single cylinder diesel engine operating with hybrid fuel under different load conditions. His report discovered that the use of the developed ANN model has proven to predict the performance and exhaust emissions of the diesel engine with a low root mean square error and a range between 0.975 and 0.999 for the correlation coefficient accurately. Wang, Zhang [13] used Ann radial basis function technique to predict emission parameters from marine diesel engine. Other ANN modelling and prediction works that related to the engine performance and exhaust emissions of diesel engine with biodiesel are Sharon H, Jayaprakash [14], Patil RC and S.U. [15], Sarala R and Rajendran [16] and Kökkülünk, Akdoğan [17]. According to these papers, it is important to have accurate experimental test data in order to give excellent and reliable ANN prediction results. In the existing literatures also stated that the ANN is a powerful modelling tool that has the ability to identify complex relationships between input–output data. However, there is still less investigation on marine diesel engine appears to have been published in the literature to date. Therefore, the present work intends to investigate the applicability of ANN method to predicting the marine diesel engine performance parameters. In this study, the prediction results were validated with the actual laboratory testing data.

2. Experimental Setup

2.1. Marine engine testing

In this research, the experiments were performed on a Cummins NT855 marine diesel engine with 4 stroke-Inline-6 cylinder. The full engine setup and details of engine specification are illustrated in figure 1(a) and Table 1 respectively. The eddy-current dynamometer SAJ SE-250 was attached to the engine to provide different load condition as illustrated in figure 1 (b). The operating conditions of the engine and the dynamometer brake are characterized by speed and torque. Engine speed is controlled by the throttle engine position lever. The dynamometer sizing is chosen to cover the full-load engine operation for the whole range of engine speeds as the characteristic parts of engine curve must fall within the characteristic curve of dynamometer. The maximum output power for this dynamometer is 250 kW.

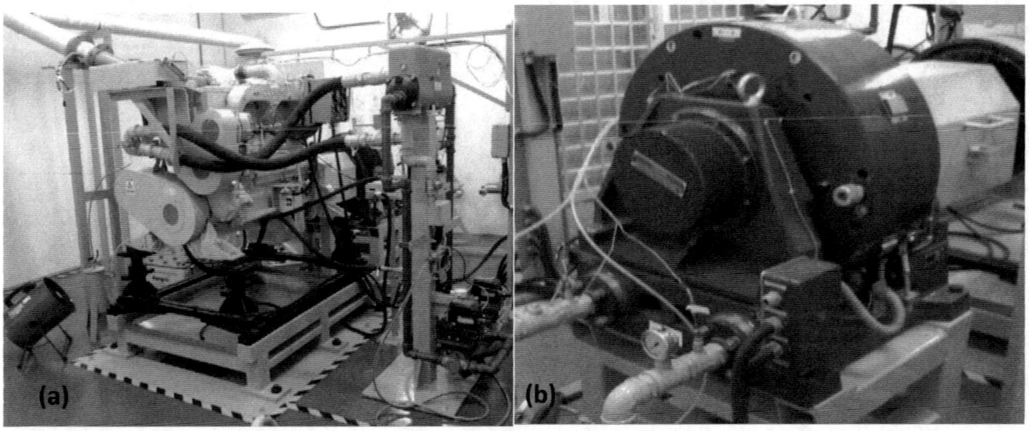

Figure 1. (a) Cummin NT855 marine diesel engine and (b) Eddy Current dynamometer SE-250.

Fuel consumption rate was measured by KOBOLD positive displacement type flow-meter. Air consumption was measured using an air flow-meter. Two separate fuel tanks were fitted to the engine and these contained diesel and the biodiesel blends. The engine control unit (ECU) used in this engine was REO-DCA control system with assisted of Samsung HDMI user interface as shown in figure 2(a). ECU function is to control the quantity of fuel, injection timing, ignition timing and engine speed by receiving signals from sensors. These sensors are oxygen sensor, knock sensor, manifold air pressure sensor, intake air temperature sensor, throttle position sensor, water temperature sensor and engine speed sensor. A multi-point fuel injection (MPFI) system with top-feed injectors is used to inject the fuel into the combustion chamber. The ignition system was semi-static distributor less ignition (DLI). A schematic diagram of the whole experimental setup is shown in figure 2(b). The marine diesel engine was fuelled by B-5 biodiesel. The engine speeds, brake power, torque and fuel consumption were measured, while the brake specific fuel consumption (bsfc), brake thermal efficiency and volumetric efficiency was computed later.

Table 1: Marine diesel engine specification.

Brand	Cummins NT855
Engine type	4 cycle-Inline-6 cylinder
Firing order	1-5-3-6-2-4
Bore x stroke (mm)	139 x 152
Displacement volume (litre)	14
Compression ratio	14.5
Maximum torque (N.m)	1068
Maximum power (kW)	201
Maximum speed (rpm)	2100
Cooling system	Water cooling

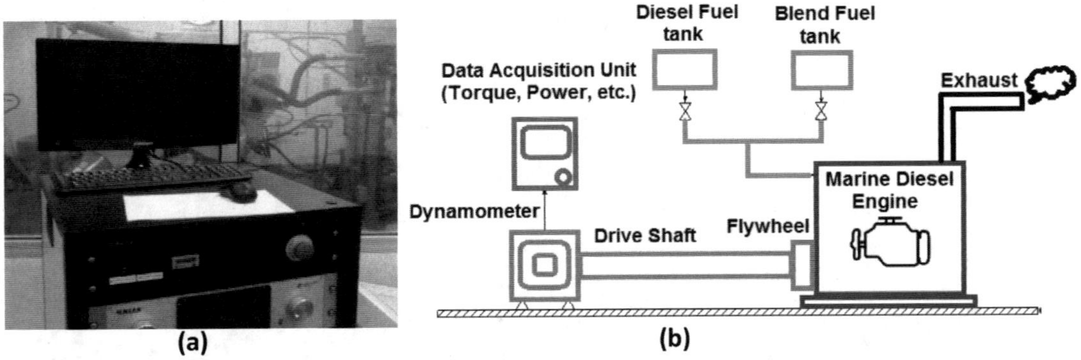

Figure 2. (a)REO-DCA data acquisition unit and (b) Schematic diagram of the experimental setup.

2.2. Neural network modelling

The development and the training of the network model were carried out using Neural Network Toolbox of MATLAB software. The ANN modeling comprises of two phases: the first phase is to train the network model, while the second phase is to validate the network model with new data which were not used for training. The flow of neural network modeling procedure is shown in figure 3. Choosing the optimum network architecture is one of the challenging steps in neural network modeling. Figure 4(a) shows the neural network architecture with back-propagation neural network model (BPNN) employed in this study. The network has three layers: input layer, hidden layer and output layer. As there are one input and five outputs, the numbers of neurons in the input and output layer had to be set to 1 and 5, respectively. In many ANN applications, the back propagation architecture with one hidden layer is enough. In order to find an optimal architecture, different numbers of neurons in the hidden layer were considered and prediction error for each network was calculated. The BPNN is based on the error correction learning rule. The operation of the neural network model can be divided into two main steps: forward computing and backward learning. In the forward computing, the input patterns applied to the neurons of the first layer are just a stimulus to the network. In fact, there is no computation in the input layer. As illustrated in figure 4(b), each neuron in the hidden layer determines a net input value based on all its input connections. These nodes are connected to each other so that the value of one node will affect the value of another. The relative influence that one node has on another one is specified by the “weight” that is assigned to each connection.

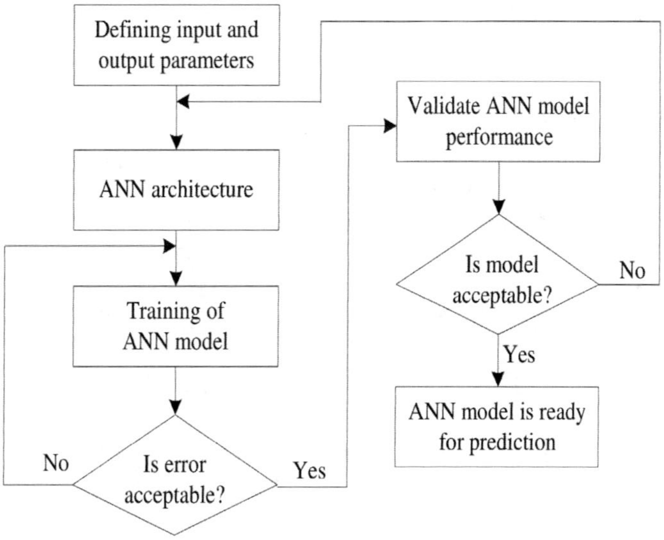

Figure 3. Neural network modeling procedure.

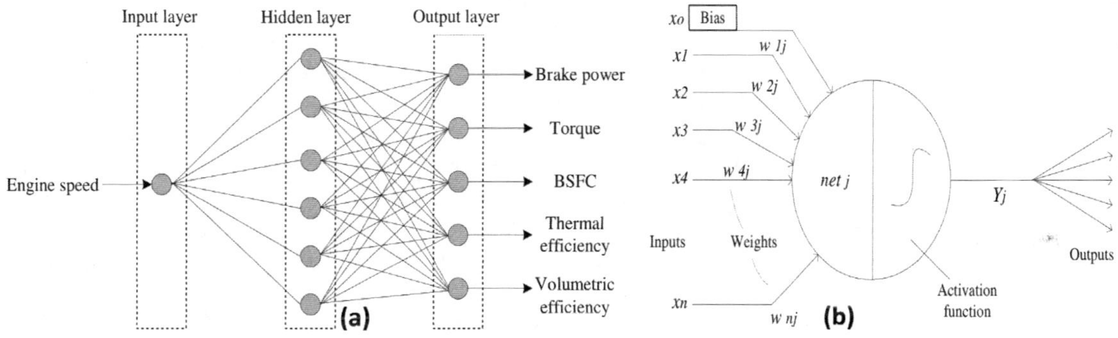

Figure 4. (a) Neural network architecture and (b) Architecture of an individual neuron.

The net input is calculated by summing the input values multiplied by their corresponding weight. Once the net input is calculated, it is converted to an activation value. The weight on the connection from the ith neuron in the forward layer to the jth neuron is indicated as w_{ij}. The output value of neuron j is computed by the following equation:

$$net_j = \sum_{i=0}^{n} w_{ij} x_j + x_0, \quad Y_j = f_{act}(net_j) \tag{1}$$

Where net$j$ is the linear combination of each of the x_i values multiplied by w_{ij}, x_0 is a constant known as the bias, n is the number of inputs to the jth neuron, and f_{act} is the activation of neuron j. In this study, the hidden layer with log-sigmoid (S-shaped curves) activation function is used for the prediction of weld bead geometry. The log-sigmoid activation function is given below:

$$Y_j = \frac{1}{1+\exp(-net_j)} \tag{2}$$

In backward learning, the generated output of the network is compared to the desired output, and an error is computed for each output neuron. The error vector E between desired values and the output value of the network is defined as:

$$E = \sum_j E_j = \sum_j \frac{1}{2}\left(T_j - Y_j\right)^2 \tag{3}$$

Where Y_j is the output value of the jth output neuron and T_j is the desired value of the jth output neuron.

Errors are then transmitted backward from the output layer to each neuron in the forward layer. The process is repeated layer by layer. Connection weights are updated by each neuron to cause the network to converge. The network was trained with the Levenberg-Marquardt training algorithm. This training algorithm was chosen due to its high accuracy in similar function approximation. The adjustments of weights and biases are done according to transfer following function:

$$\Delta w_{ij} = -\left(J^T J + \mu I\right)^{-1} J^T E \tag{4}$$

Where J is Jacobian matrix of derivation of each error, μ is a scalar and E is error function.

The input and output parameters data sets could not be trained by neural network in their original form due to the wide range of values among them. In order to become feasible neurons, all the values in the input neurons had to be preprocessed by normalizing and transformed within the range of -1and +1. The normalized value (X_i) for each raw input and output data set (d_i) was calculated as follows:

$$X_i = \frac{2}{d_{max} - d_{min}}\left(d_i - d_{min}\right) - 1 \tag{5}$$

Where d_{max} and d_{min} are the maximum and minimum values of the raw data, respectively.

The back-propagation neural networks model were trained using data sets from laboratory experiment results. They are randomly divided into tree data sets which 70% of data were used for training, 15% for testing and remaining 15% for validation. The neural network configuration for training was created and formulated according to specification given in table 2. In order to identify the optimum network architecture, it is essential to determine the number of neurons in the hidden layer. Therefore, the number of neurons was chosen from 4 to 28 neurons in a hidden layer. The accuracy of the network was evaluated by the mean squared error (MSE) and the coefficient of multiple determinations, R^2. As can be seen in table 3, the MSE in the training process is not directly related to increasing number of neurons. It can be noted that where there is a small number of neurons in the hidden layer, the output training performance of network is not satisfactory. However, the increasing number of neurons beyond 22 has no-significant improvement to the performance of the networks. It can be conclude that the network with 22 neurons in hidden layer shows give better results of minimum MSE and higher R^2 value.

Table 2. Neural network configuration for the training.

Parameter	**Specification**
No. of neurons in input layer	1
No. of neurons in hidden layer	4 to 28
No. of neurons in output layer	5
Training function	Levenberg-Marquardt (trainlm)
Performance function	Mean square error (MSE)
Activation function	Log-sigmoid
Performance goal	1.0 x 10^{-3}
Normalized range	-1 to 1

Table 3. Training performance of different network architecture.

Network architecture	MSE	R^2
1-16-5	0.00352	0.95989
1-18-5	0.00383	0.95435
1-20-5	0.00773	0.95094
1-22-5	0.00153	0.96222
1-24-5	0.00416	0.95232
1-26-5	0.01453	0.92544
1-28-5	0.02499	0.91357

3. Results and discussion

3.1. Experimental results

Engine performance testing was carried out by using Cummins NT855 Marine diesel engine with B-5 biodiesel as a fuel. Figure 5(a) and 5(b) show the variation of brake power and output torque at different engine speed respectively. The brake power is increased steadily with the increasing of speed until it reaches maximum point of 170 kW at 1700 rpm. On the other hand the torque value declined after speed of 1400 rpm. The maximum output torque of 1030 Nm was produced at about 1200-1400 rpm. Figure 5(c) shows the nature of fuel consumption (BSFC) of marine engine with varying speeds. Fuel properties such as density, viscosity and calorific value are normally will influence BSFC value. Optimum fuel consumption was observed at speed of 1200 rpm. Brake thermal efficiency appraises how efficiency an engine can transform the supplied fuel energy into useful work. Most of the supplied fuel will be lost as heat with the engine cooling water, lubricating oil and exhaust gas. As display in figure 5(d) the maximum brake thermal efficiency was around 55% at engine speed of 1000 rpm. The volumetric efficiency specifies the quantity of inflow air mass remaining in the cylinder relative to the theoretical air mass. Figure 5(e) shows the volumetric efficiency increase with increases of engine speed.

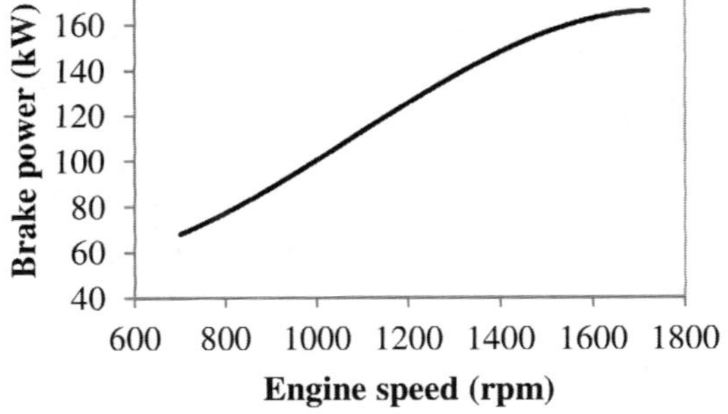

(a)

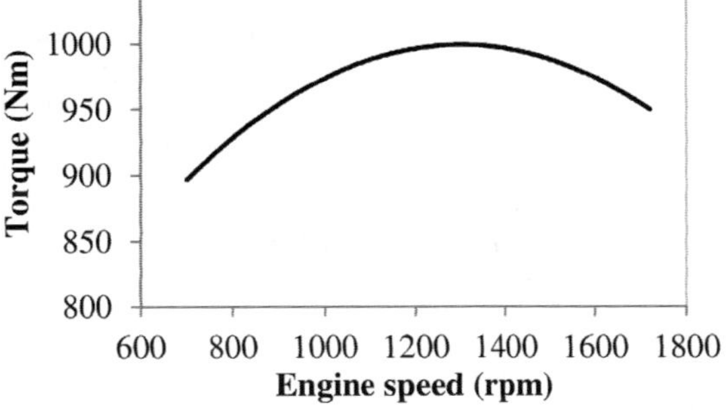

(b)

IOP Conf. Series: Materials Science and Engineering **100** (2015) 012023 doi:10.1088/1757-899X/100/1/012023

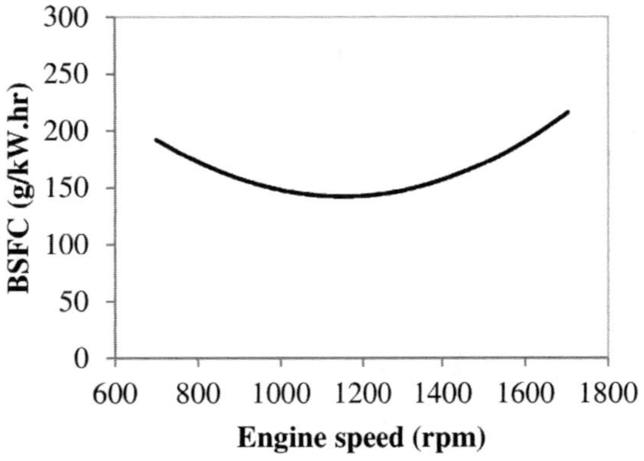

(c)

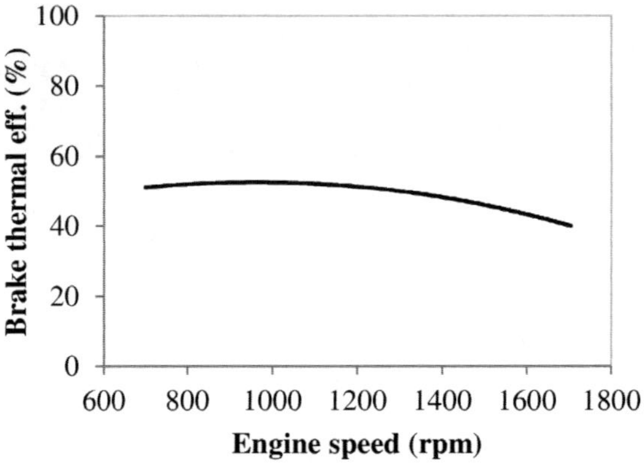

(d)

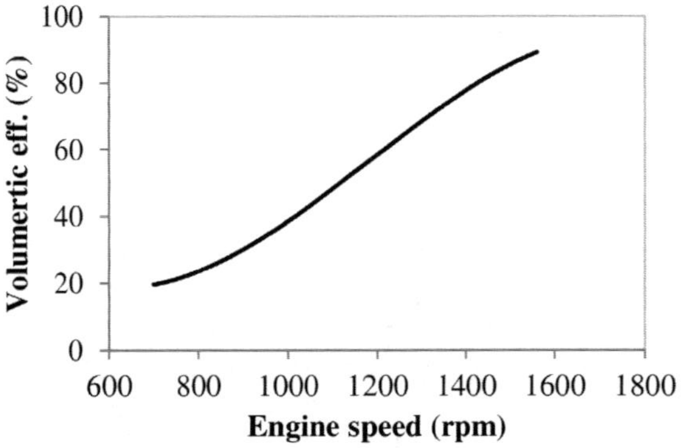

(e)

Figure 5. Experimental results of (a) brake power, (b) torque, (c) bsfc, (d) thermal efficiency and (e) volumetric efficiency.

3.2. ANN prediction results

The prediction of marine engine performance was carried out by using ANN with back-propagation model with Levenberg-Marquardt training algorithm. The number of neuron in each layer can be determined by the complexity of the problem and data sets. In this study, the network was decided to consist of 22 neurons in the hidden layer. The criterion of MSE and R^2 was selected to evaluate networks with optimum solution. A regression analysis between the network output and the corresponding targets was performed in order to investigate the network response in more detail. The results showed that the constructed model was sufficient to predicting marine diesel engine performance at various engine speeds. The ANN predicted outputs versus experimental values are indicated in figure 6. The ANN predictions for the (a) brake power, (b) torque, (c) brake specific fuel consumption, (d) brake thermal efficiency and (e) volumetric efficiency yield a correlation coefficient (R) of 0.9984, 0.9714, 0.9130, 0.9456 and 0.9976, respectively. This indicated high correlation between predicted model and experiment data.

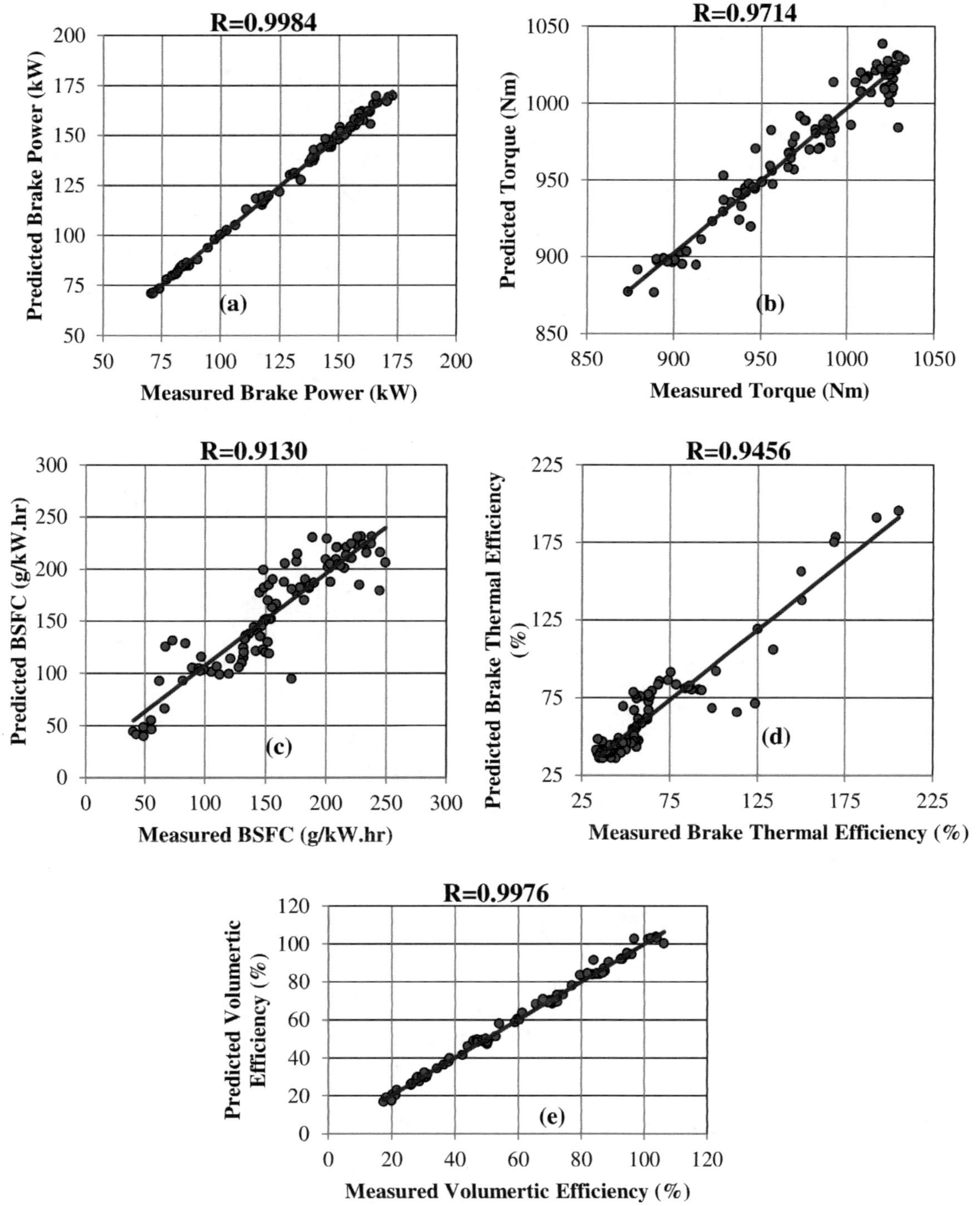

Figure 6. The ANN prediction for the (a) brake power, (b) torque, (c) bsfc, (d) thermal efficiency and (e) volumetric efficiency versus experimental values.

A comparative presentation of the error during model testing by ANN and experimental results is shown in figure 7. The test pattern consists of 5 pairs of results data set. It was observed that the ANN model can predict marine engine performance accurately with close agreement to actual experiment data.

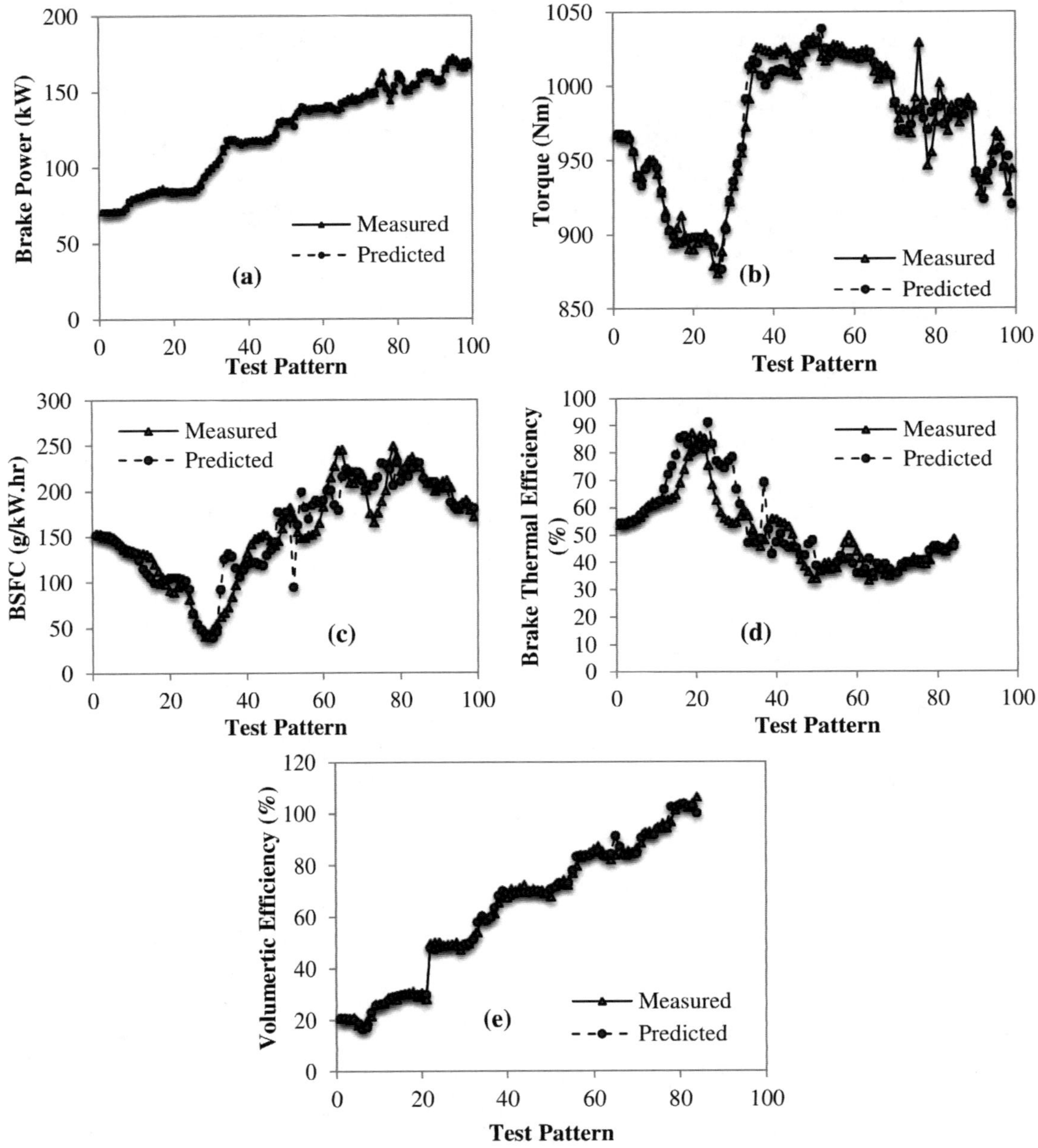

Figure 7. Comparisons of experimental results and the ANN predictions for the (a) brake power, (b) torque, (c) bsfc, (d) thermal efficiency and (e) volumetric efficiency for various test patterns.

4. Conclusion

An artificial neural model was successfully developed to predict the Cummins NT855 marine diesel engine performance. The back propagation ANN network with Levenberg-Marquardt training algorithm was used to train the data from engine laboratory testing. The main conclusions obtained in this study are as follows:

(i) The prediction result of the neural network model which has 22 neurons in hidden layer was found to be in good agreement with the experimental data.
(ii) The distribution of data points for neural network model almost similar and close to the actual experimental data with correlation coefficient (R) in range in range of 0.9 - 1.0. This indicated the developed neural network model is capable of making the prediction with reasonable accuracy.

(iii) Neural network is a powerful tool and is easy to use in complex or non-linear problems. The developed neural network models are reliable and able to predict marine diesel engine performance parameter with reasonable accuracy.

Acknowledgements
The authors wish to express their appreciation to the Department of Marine Technology, Universiti Malaysia Terengganu for providing facilities and support to carry out the research work

References

[1] Ghobadian B, Rahimi H, Nikbakht A M, Najafi G, Yusaf T F 2009 *Renewable Energy* **34** 976-82
[2] Xu S L H K, Xie M, Tang L C 2003 *Appl Soft Comput* **2** 255-68
[3] Yuanwang C X D, Meilin Z, Dong X 2003 *Fuel* **81** 1963-70
[4] Korres G B D M, Anastopoulos G, Lois E, Alexandridis A, Sarimveis H 2002 *Fuel* **81** 1243-50
[5] Lucas M L A, Duran M, Carmona M 2001 *FUel* **4** 548-93
[6] Hafner R I M, Schuler M, Nelles O 2000 *Control Eng Pr* **8** 1211-21
[7] Shayler T M P J, Goodman M 2000 *Eng Appl Artif Intell* 147-51
[8] Tan M S Y 2000 *Neurocomputing* **30** 129-42
[9] Najafi G, Ghobadian B, Tavakoli T, Buttsworth D R, Yusaf T F, Faizollahnejad M 2009 *Appl Energy,* **86** 630-9
[10] Mustafa Canakci E A, Erdil A 2009 *Appl Energy,* **86** 630-9
[11] CI A l E 2005 *Appl Energy,* **80** 11-22
[12] Shanmugam M A, P.V. S 2011 *Energy Sources, Part A: Recov, Utiliz, Environ Effects," Environ* **33** 1440-50
[13] Wang M, Zhang J, Zhang S, Ma Q 2010 *2010 Second WRI Global Congress on Intelligent Systems* **3** 184-8
[14] Sharon H K K, Jayaprakash R, Sundaresan A 2012 *Fuel* **99** 197-203
[15] Patil RC P H, S.U. P 2012 *Spec Issue Int J Electron Commun Soft Comput Sci Eng* 248-53
[16] Sarala R S B, Rajendran D M 2011 *Int J Res Environ Sci Technol* **2** 48-53
[17] Kökkülünk G, Akdoğan E, Ayhan V 2013 *Turkish J Electr Eng Comput Sci* **21** 2141-52

An investigation of urban boundary layer towards achieving similarity criteria in a short wind tunnel

A W M Yazid[1], **N A Che Sidik**[1], **S Mansor**[2], **A B Abdul Rahman**[3], **I S Ishak**[2], and **N Dahalan**[2]

[1]Department of Thermo-Fluids, Faculty of Mechanical Engineering, Universiti Teknologi Malaysia, 81310 Skudai, Johor Bahru, Malaysia

[2]Department of Aeronautics, Automotive & Ocean Engineering, Faculty of Mechanical Engineering, Universiti Teknologi Malaysia, 81310 Skudai, Johor Bahru, Malaysia

[3]Aeronautics Laboratory, Faculty of Mechanical Engineering, Universiti Teknologi Malaysia, 81310 Skudai, Johor Bahru, Malaysia

E-mail: azwadi@fkm.utm.my

Abstract. An experimental investigation was conducted in an effort to artificially thicken the boundary layer which has the characteristics of an urban boundary layer in a short wind tunnel. Turbulence grid was developed and then systematically tested to quantitatively investigate the formation of a thick and uniform boundary layer over location of interest. Investigated parameters were the mean velocity and turbulence intensity to assess the similarity of the simulated boundary layer. The measurement analysis shows that the current design provides the desired boundary layer profile under fully rough conditions at just 1.5 m downstream of the test section inlet, which allows for the building models to be placed over turntable. However, the roughness length value is high which reflects the determination of the model scale while the turbulence intensities levels is low which is not desirable characteristics of an urban boundary layer. The results presented here may facilitate future improvements in the design of the turbulence grid and measurement techniques.

1. Introduction

Vertical profile of atmospheric boundary layer is relatively different over specific location such as complex terrain (i.e., hill, mountain), offshore (i.e., coastal regions, sea) and built-up area (i.e. suburbs, city center) [1]. In particular, the urban boundary layer is essentially different from the boundary layer created over complex terrain or offshore due to the thickening of boundary layer as a result of vertical exchanges of momentum, heat and moisture amongst roughness elements. Figure 1 shows the schematic drawings of vertical structure of urban boundary layer, which indicates some of the layers and the terminology used throughout this paper. The layer extended from the ground ($z = 0$) up to the mean height of the roughness elements, zH is called urban canopy layer (UCL) [2]. In this layer, the wind flow is horizontally inhomogeneous as the processes of airflow and energy exchange are controlled by microscale, site-specific characteristics and processes [3, 4]. The effects of individual roughness elements on wind flow are discernible in the roughness sublayer (RSL) which extends from the ground up to height between 2zH in densely built areas (closely spaced), to about 5zH in low density areas [3-5]. Above the roughness sublayer is the inertial sublayer (IS) also known as constant flux layer, where the wind flow is horizontally homogeneous due to the constant turbulent fluxes with height [4]. Both RSL and IS are jointly addresses as the surface layer (SL), which extends up to an elevation approximately 10% of the total depth of the urban boundary layer [6]. The top layer in urban boundary layer is the outer layer (OL) also known as mixed layer or Ekman layer, in which the combined effects from pressure gradient force and Coriolis force resulting from convection process and surface friction due to the Earth's rotation respectively, becomes important [1]. For a detail descriptions about urban boundary layer, a comprehensive review can be found in [7].

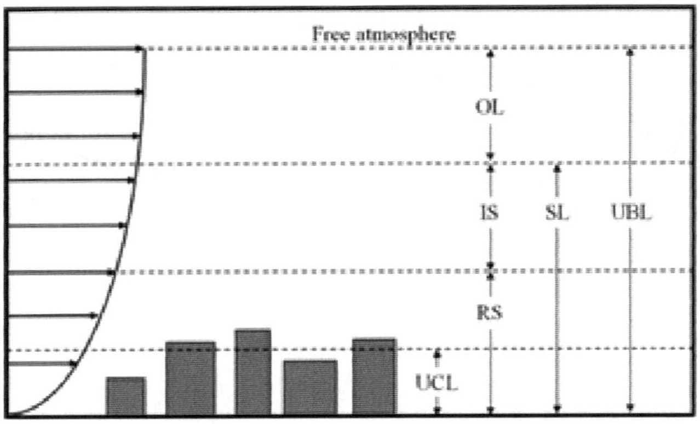

Figure 1. Schematic of the vertical structure of the urban boundary layer.

In general, there are two physical methods of studying the wind characteristic over Earth's surface (i.e. man-made structures, hills, forests, sea) namely the field measurement and reduced-scale measurement. Field measurement provides the most realistic results of parameter under investigations whereby it encompasses the original size of the structures or locations area with real atmospheric conditions. However, field measurements is not controllable due to the variable meteorological conditions and the data is limited due to the limitations of sensor and instrumentation capacities [8]. Besides, the experiment period is usually long and is costly to perform [9, 10]. Reduced-scale experiment via wind tunnel is an alternative to the physical experiment, which requires a special wind tunnel to simulate atmospheric boundary layer. The boundary conditions in wind tunnel measurements can be accurately controlled and the repeatability of the measurement is high [11]. One of the major drawback associated with wind tunnel measurement is the requirements of fulfilling the similarity criteria with that of full scale conditions [12], which can limit the extent and the range of problems that can be studied [8].The boundary layer wind tunnel must have capabilities for creating flows that simulate the basic characteristics of natural wind at a site in order to obtain wind-effect data representative of full scale conditions [13]. The test section in boundary layer wind tunnel is generally long enough (at least 6 m [14, 15] to more than 20 m [16, 17], to allow the development of a thick turbulent layer along the tunnel floor. The method of generating a thick boundary layer simulation in a wind tunnel is hugely depends on the length size of the test section and the target to achieve specific vertical wind profile and turbulence levels [18]. For a short wind tunnel, the method includes the combination of a few vortex generators align perpendicular to the freestream along the start of the test section, a barrier, and a fetch of roughness elements afterwards which will increase turbulence levels near the floor [13]. As the boundary layer thickness increases along the tunnel over roughness elements and model, the cross sectional area for the free stream becoming smaller, resulting in higher free stream velocity and lower static pressure as the flow propagate downstream. An adjustable ceiling height is needed to compensate for the loss of cross section area and thus ensure that there is no streamwise pressure difference. Turntable on the test section floor provide additional feature which allows for variety of incoming wind direction towards area of interest.

The Universiti Teknologi Malaysia, low speed tunnel (UTM-LST) is a research facility mainly used for aerodynamics testing of ground vehicles [19], aeronautics [20, 21] and marines transportation [22]. Industrial cooperation for the aerodynamics testing of an offshore drilling and production platform model immersed in marine boundary layer has been attempted before [23]. Past research on the application of UTM-LST on building aerodynamics research over urban area is limited, whereby uniform inlet wind profile was assumed [24]. On the other hand, the vertical profile of flow over Earth's surface is non-uniform and the profile is greatly influence by roughness parameters and atmospheric conditions as explained at the beginning of the section. Conventional methods in simulating the urban boundary layer such as combination of vortex generators and roughness elements in UTM-LST may not be practical as the ceiling of the test section is non-adjustable and the turntable is located 1.5 m from the inlet as figure 2(a). Thus, the motivation of this study was to check the feasibility of an initial design of turbulence grid developed by Aeronautics Laboratory group, UTM to simulate urban boundary layer flow within a short distance. The investigation will look at the variation of vertical wind profile and the similarity of the simulated mean wind velocity and turbulent intensity in the wind tunnel with that of full scale. The present simulation of the urban boundary layer is limited to be under neutral atmospheric conditions only, in which the conditions does not affected by thermal conditions.

2. Methodology

2.1. Similarity Considerations

A guideline on the similarity criteria for wind tunnel experimentation in the area of buildings aerodynamics covered by ASCE 1999 [13] is referred. There are two types of equations that represent the vertical profile of atmospheric boundary layer namely the power law and the logarithmic law. The power law is valid for any value of height, z within the atmospheric boundary layer [25]. Besides, the power law is often used instead of logarithmic law when other factors are not available such as surface roughness or atmospheric stability [26]. Meanwhile, the logarithmic law is only valid in the IS, which extends from a heights of 2 to 5 times the mean building height to approximately 10% of the boundary layer depth [27]. The power law is given as in equation (1). Consequently, the vertical mean velocity profile generated must be similar to the one observed in full scale where the relationship is given in equation (2) [13].

$$\frac{U(z)}{U_{ref}} = \left(\frac{z}{z_{ref}} \right)^{\alpha} \tag{1}$$

$$\alpha_{\mathrm{M}} = \alpha_{\mathrm{N}} \tag{2}$$

where U_{ref} is reference velocity at reference height, z_{ref}, α is the exponent constant of the profile and subscript M and N represents the model and nature respectively. The mean velocity profile within IS is expressed by a logarithmic law as given in equation (3):

$$\frac{U(z)}{U_*} = \frac{1}{k} \ln \left(\frac{z}{z_0} \right) \tag{3}$$

The logarithmic equations provides unknown values which are friction velocity, U_* and roughness length, z_0. By fitting the data of simulated boundary layer with equation (3), the value of both friction velocity and roughness length can simultaneously be determined. Table 1 shows the range of exponential and roughness length value over two typical terrain types of urban area.

Table 1. Characteristics of urban boundary layer over selected terrains.

Terrain type	[a]Power law exponent, α	[b]Roughness length, z_0 (m)
Suburbs	0.20-0.24	0.4-0.7
City centers	0.28-0.4	0.7-1.5

[a]from Plate, 1995 [26]
[b]from Wieringa, 1993 [28]

One of the important criterion that define an atmospheric boundary layer over a fully rough surface in wind tunnel is the roughness Reynolds number, Re* and is given as

$$\mathrm{Re}_* = \frac{U_* \times z_0}{v} \tag{4}$$

where v is the kinematic viscosity of working fluid. The roughness Reynolds number will determine whether the flow near the floor is affected by viscous effects. Uehara et al., 2003 [29] suggested a critical roughness Reynolds number of 5, regardless of model height. In addition to that, the ratio between the simulated roughness length in the wind tunnel and the nature will represent the model scale, M where the relationship is given in equation (5) [30].

$$\frac{z_{0,\mathrm{M}}}{z_{0,\mathrm{N}}} = \mathrm{M} \tag{5}$$

ASCE, 1999 [31] expressed a relationship of stream wise turbulence intensity profile with roughness length as given in equation (6). Besides the mean velocity profile and the roughness length, the turbulence intensity profile of the model must likewise be similar to the one observed in nature as given in equation (7) [31].

$$I_u(z) = \frac{1}{\ln\left(z / z_0\right)} \tag{6}$$

$$I_{\mathrm{u,M}} = I_{\mathrm{u,N}} \tag{7}$$

It is worth mentioned that a plane surface may have vertical displacement height value between zero and average height of roughness elements for flows over very rough surface (i.e. forests, buildings) and is highly dependent on the areal density of roughness elements. Due to that reason, the calculation of equation (1), (3) and (6) will have to be modified, where the displacement height must be deduct from the height elevation written as *z-d* instead of only *z*. However, for wind flow profile over smooth surface or flow past through isolated structure, $d = 0$ can be assumed [32, 33].

2.2. Experimental Set-up

The experiments in this study is carried out in a closed-loop and low speed wind tunnel (UTM-LST) at the Aeronautics Laboratory, Universiti Teknologi Malaysia. This wind tunnel was able to deliver a maximum air speed of 80 m/s inside the test section. The size of test section was 2.0 meters wide, 1.5 meters height, and 5.5 meters long. In order to develop the urban boundary layer, a turbulence grid made of a bar and rods are located at the entrance of the test section with horizontal alignment (figure 2(a) and (b). The size of the rods ranges from 1 cm to 2.8 cm in diameter with mostly is 2 cm in diameter and was displaced to each other according to the target vertical velocity profile as figure 3(a) which will be explain later in the next section. To reduce vibration on the turbulence grid, caused by high freestream wind speed, 2 rods separated 40 from each other were located at the center of the turbulence grid with vertical alignment. The location of in situ calibration and the subsequent measurements are located 1.5 m from the inlet. The measurement of urban boundary layer were made along vertical direction at three spanwise locations figure 3(b). Due to the size of the probe support and probe support holder, the measurement can only be made at the lowest elevation of 4 cm from the ground.

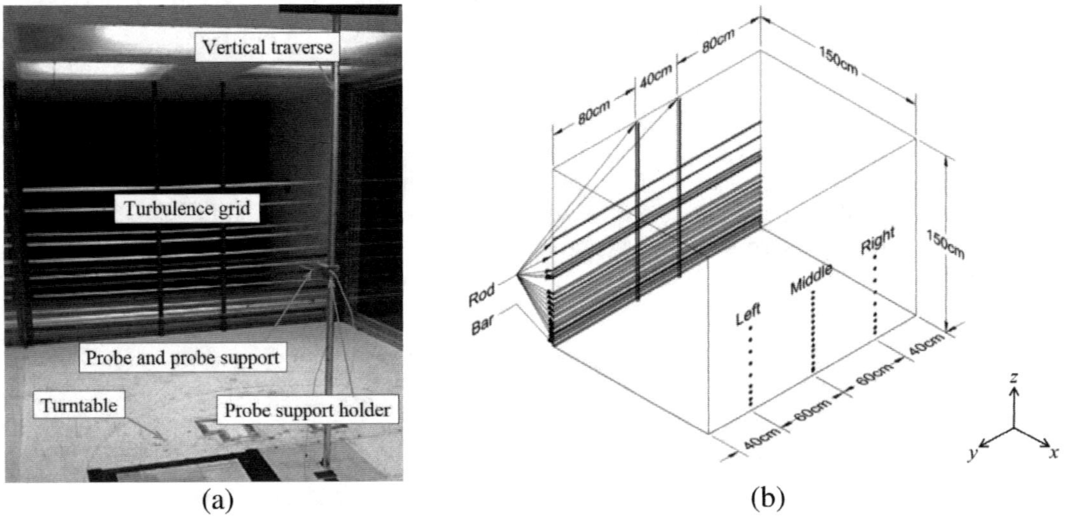

Figure 2. The installation of the turbulence grid and the measurement equipment in UTM-LST (a) overview of the experimental layout and (b) dimensions schematic of the experimental layout.

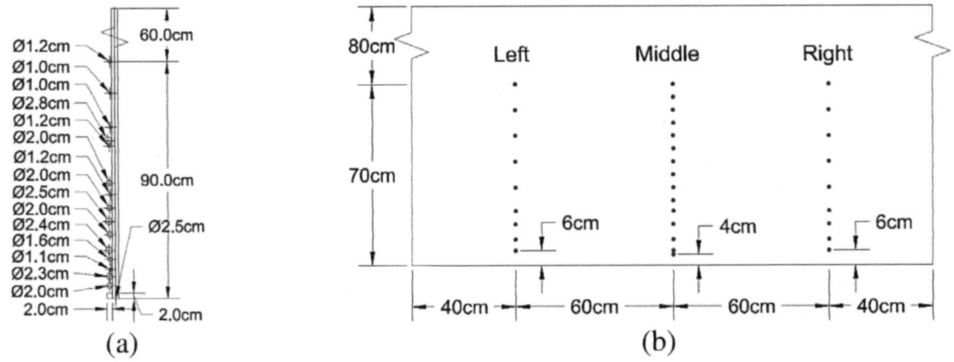

Figure 3. Schematic drawing of (a) turbulence grid from side view and (b) location of the measurement points looking upstream in UTM-LST.

2.3. Measurement Technique

The mean velocity and turbulence was measured using a constant temperature, 1-D hot wire anemometer (HWA) of Dantec 55P11 in conjunction with a 90C10 constant temperature anemometer, where both are connected into StreamLine 90N10 frame. First, the HWA is calibrated in situ facing the tunnel free stream in the middle of the wind tunnel test section against a pitot-static tube as a reference reading with 11 number of calibration points ranging from 2 to 32 m/s. Data was collected with a sample rate of 1 kHz for 10 seconds and a calibration curve for the HWA was obtained with polynomial curve fitting as in equation (8) where Ecorr is the corrected output voltage constant while C0, C1, C2, C3 and C4 are calibration constants.

$$U = C_0 + C_1 E_{corr} + C_2 E_{corr}^2 + C_3 E_{corr}^3 + C_4 E_{corr}^4 \quad (8)$$

Two main flow characteristics were simulated simultaneously, namely the mean streamwise velocity and the streamwise turbulence intensity. By simulating these 2 main characteristics, the similarity laws of wind tunnel studies, which are described in previous section, are expected to be fulfilled.

3. Result and discussion

Throughout this study, a reference height of z = 0.1 m from the floor tunnel with approximate velocity speed of 10 m/s was taken as a reference in order to ensure that the speed measured between 0.04 m up to 0.1 m or more are essentially higher than the lower limit of calibrated wind speed (2 m/s). Subsequently, the gaps between the rods are adjusted in the attempt to match with that of a typical velocity profile over city center (α = 0.3). The streamwise velocity profiles at each locations are approximated using power law to check the similarity of power law exponent value of 0.3. The profiles and power law trend lines are plotted in figure 4 and shows that the values obtained are close to 0.3, represented by the power over x-axis (z/zref).

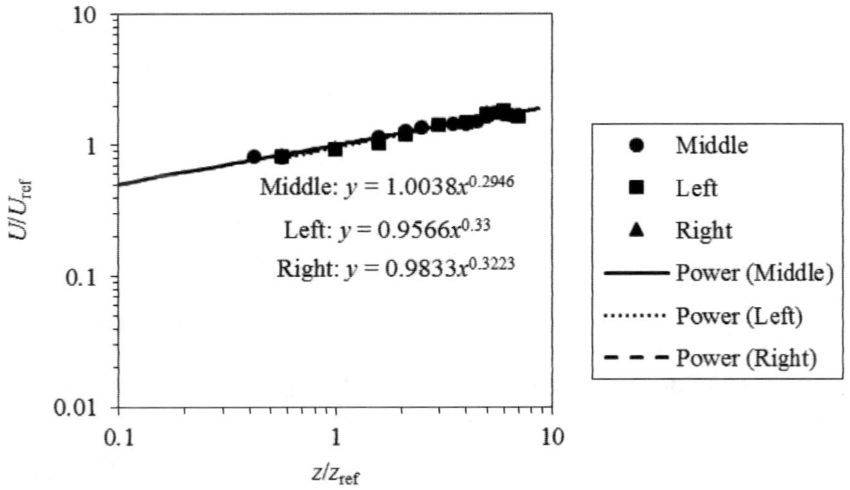

Figure 4. Vertical mean streamwise velocity profile at 3 locations.

Alongside the mean streamwise velocity profile and streamwise turbulence intensity were also measured along the vertical at three locations as figure 2(b), 1.5 m from the turbulence grid. Both the mean streamwise velocity and the turbulence intensity are plotted in logarithmic trends as shown in figure 5(a) and 5(b) respectively. The mean velocity profiles show linear log profile for a height between 158.4 mm up to 400 mm. The results of streamwise turbulence intensity also coincide with the mean velocity where the values are scattered for height below z = 158.4 mm and then reduced up to z = 400 mm before increased again for a height of more than 400 mm. The scattered values below 158.4 mm may indicate a region of roughness sublayer (RS), in which previous wind tunnel studies have shown that the streamwise turbulence intensities and Reynolds stress value are scattered below the roughness elements (made of sharp edge blocks, placed in staggered or in line configuration) up to twice the mean height of the roughness elements [34, 35]. Thus, the turbulence grid used in present study was able to produce the roughness sublayer without a fetch of roughness elements. A region of inertial layer was observed for height more than 158.4 mm where the flow is horizontally homogeneous. In other words, the profiles are linearly increased or decreased with height. At height of more than 400 mm, the profiles shows irregularity especially for the mean streamwise velocity and is at increasing trends for longitudinal turbulent intensity due to a few huge gaps between specific rods of the turbulence grid (The first gap is between 445 mm up to 572 mm, the second gap is between 657 mm up to 784 mm and the third gap is 784 to 900 mm). These large gaps may have caused the effects of an individual rod to take place rather than small gaps. Nevertheless, the region at height more than 400 mm is not really concerned with the characteristics of the simulated boundary layer in the present study as our intention is to look at the profile near the ground (in particular, up to a height which consists partial region of IS) which reflects the similarity characteristics of the simulated boundary layer with that of actual boundary layer.

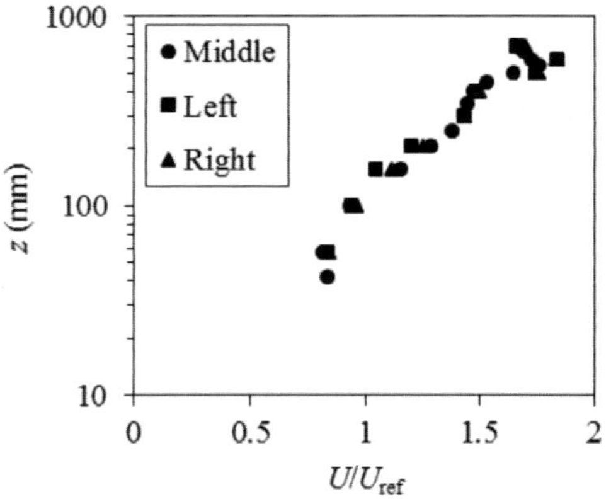

(a)

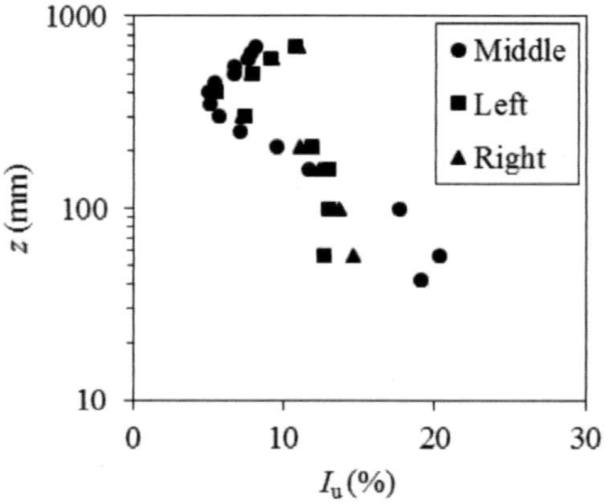

(b)

Figure 5. Vertical profiles at three locations of (a) mean streamwise velocity and (b) longitudinal turbulent intensity.

By further examining the mean streamwise velocity within the simulated IS, the roughness length and consequently the friction velocity of each profile were obtained through logarithmic law relationship. Such relationship is plotted in figure 6(a) and the roughness length was deduced from the line fitted to the mean velocity profile in logarithmic form which intercept at zero velocity value which are: Middle: z0 = 6.07 mm; Left: z0 = 19.83 mm; Right: z0 = 11.49 mm. The friction velocity of each profiles is determine by substituting the predetermined roughness length values into the logarithmic law equation which then yields the following friction velocity: Middle: U* = 1.44 m/s; Left: U* = 2.03 m/s; Right: U* = 1.72 m/s. By taking the average of these three profiles of roughness length and friction velocity, the average roughness length is 12.463 mm while the average friction velocity is 1.73 m/s.

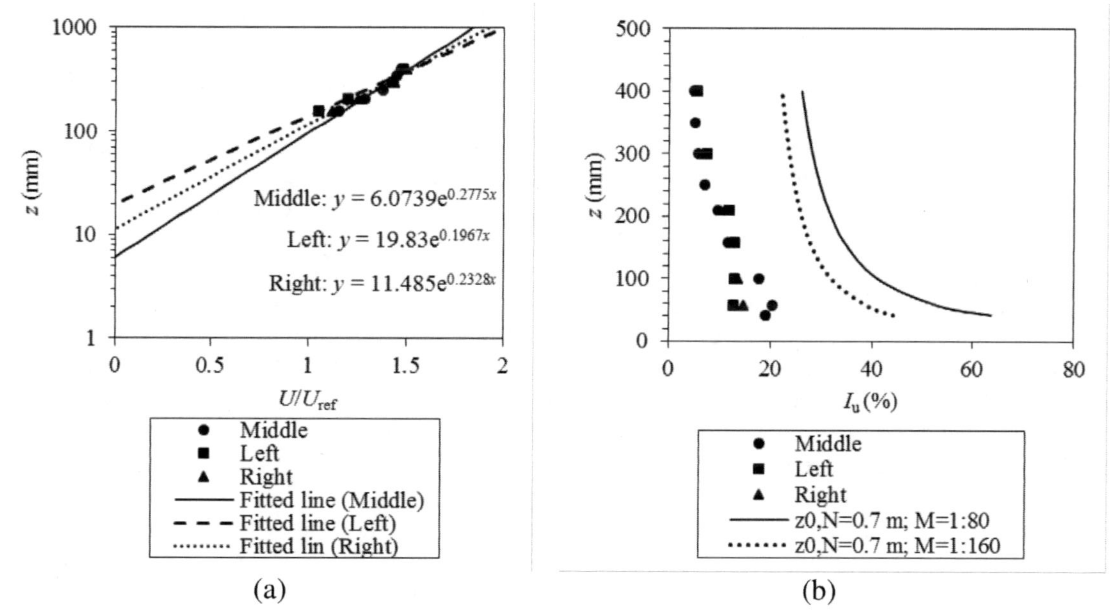

Figure 6. Vertical profiles at three locations a) mean streamwise velocity and b) longitudinal turbulent intensity.

Based on the results of the simulated flow described above with the following experimental conditions at 27 °C and pressure at 105 kPa, the average roughness Reynolds number can be calculated as

$$\mathrm{Re}_{*,\mathrm{M}} = \frac{u_{*,\mathrm{M}} \times z_{0,\mathrm{M}}}{\nu} = \frac{1.73\,\mathrm{m/s} \times 0.0125\,\mathrm{m}}{1.576 \times 10^{-5}\,\mathrm{m^2/s}} = 1372\,\mathrm{m/s}$$

Hence, the simulated boundary layer was fully under rough condition. By taking the roughness length in nature of 1 m representing the mid-range of city center [29], the ratio between the z0,M and z0,N produce the average model scale of the present study as

$$\frac{z_{0,\mathrm{M}}}{z_{0,\mathrm{N}}} = \frac{0.0125\,\mathrm{m}}{1\,\mathrm{m}} \approx \frac{1}{80}$$

With such model scale and the expected model height to be at maximum of only half the height of RS (≈ 80 mm) [3, 5], the simulated boundary layer could only be possibly immersed over an actual building height of 6.4 m. The 6.4 m building height is a typical building height found in suburban residential area and is not a building height found in city center where it was typically found to be more than 10 m high [35]. It is worth noting that the roughness length at the center and lateral side of the wind tunnel are varied, where the least roughness length is located at the center while higher roughness length are at lateral sides. Such variations has direct implication in the determination of model scale, M. If the roughness length could be maintained at values which is same with that at the middle of the test section, the model scale will be 1:160. At 1:160 model scale, a model height of 80 mm is equivalent to an actual building of 13 m high, which is well representative of an average buildings height in city center. Uehara et al., 2003 [29] obtained consistent roughness length value regardless of freestream tunnel speed. Hence, the variations of roughness length values in the present study suggest that there is lack of vertical rods align at lateral side of turbulence grid and may have contributed to the inconsistency.

The streamwise turbulence intensity of the present study and the theoretical streamwise turbulence intensity profile at the scale of 1:80 with typical minimum roughness length of city center (0.7 m [30]) is plotted in figure 6(b). Another theoretical streamwise turbulence intensity profile but at the scale of 1:160 was also included. From figure 6(b), the profiles of the whole streamwise turbulence intensity profile of the simulated boundary layer has failed to reproduce within the minimal value of streamwise turbulence intensity representing a full scale profile over central city. In particular, the simulated turbulence intensity at height less than 100 mm is about 70% to 80% lower than what it should be like in nature at 1:80 scale. Even if the model scale is at 1:160, there are still deficiencies of about 50% to 68%. The comparisons were made with the lower limit of city center category and the deficiency will be even more at higher roughness length. The lack of turbulence profile may be caused by the high free stream velocity of wind tunnel where previous wind tunnel study by Uehara et al., 2003 [29] shows the same trend. The reason of using higher freestream velocity in current study is to allow the measurement to be made within the calibrated speed of the HWA but at the cost of turbulence intensity deficit. However, the deficit of turbulence intensity reported by Uehara et al. 2003 [29] was not significantly reduced at higher speed which suggest that a fetch of roughness elements on the floor must have properly maintained the levels of turbulence intensity even at higher freestream speed. Recent studies have shown the important of having a combination of various passive devices such as screen, spires, and roughness elements to produce an acceptable boundary layer profiles within a short wind tunnel [36, 37]. Therefore, using only turbulence grid may not be enough to produce high turbulence intensities near the ground and an artificial element to augment the turbulence intensity levels near the tunnel floor must be considered.

4. Conclusions

A turbulence grid has been developed and tested to control the simulation of urban boundary layer development under neutral atmospheric condition in UTM-LST within a short distance. Results show that a good vertical mean velocity profile representing an urban boundary layer flow over city center with fully rough condition can be produced in UTM-LST. However, the following statements related to the deficiency in the present study can be deduced as follows:

i. The roughness length simulated in this study is rather high, which affects the determination of the model scale. The reason is due to the influence from the lack of vertical rods at lateral sides of turbulence grid.
ii. There is much deficiencies of streamwise turbulence intensity in the whole simulated urban boundary layer. The reasons may be due to the high freestream velocity of the wind tunnel and the lack of physical interactions near the tunnel floor to enhance the turbulence production.

The following actions can be taken to improve measurement techniques and enhance flow quality:

i. Inclusion of more vertical rods to reduce and maintain the roughness length values along spanwise of the test section inlet.
ii. Calibrate the HWA at lower wind tunnel speed to check whether the turbulence could be enhance at lower tunnel freestream speed. X-probe HWA can also be used to determine the similarity of lateral and vertical turbulence intensity and also to complement the turbulent intensity obtained with Reynolds stress.
iii. A fetch roughness elements may help to enhance turbulence levels near the floor. However, careful consideration must be taken especially in determining its size, due to the wind tunnel constriction of non-adjustable ceilings.

Based on the deficiencies observed in the present study, future works should focus on the enhancement of current design and measurement technique for quality assurance of future study of flow through building models or transport phenomena experiments immersed in urban boundary layer.

Acknowledgements
Authors would like to express their very great appreciation to the UTM-LST technicians for their support during this study.

References

[1] Emeis S 2012 *Wind Energy Meteorology: Atmospheric Physics for Wind Power Generation* Springer Science & Business Media)
[2] Rotach M W 1999 *Atmospheric Environment* **33** 4001-8
[3] Roth M 2000 *Quarterly Journal of the Royal Meteorological Society* **126** 941-90
[4] Arnfield A J 2003 *International journal of climatology* **23** 1-26
[5] Raupach M, Antonia R, Rajagopalan S 1991 *Applied Mechanics Reviews* **44** 1-25
[6] Arya S P 1988 *San Diego*
[7] Barlow J F 2014 *Urban Climate* **10** 216-40
[8] Blocken B 2014 *Journal of Wind Engineering and Industrial Aerodynamics* **129** 69-102
[9] Rotach M, Vogt R, Bernhofer C, Batchvarova E, Christen A, Clappier A, et al. 2005 *Theoretical and Applied Climatology* **81** 231-61
[10] Grimmond C 2006 *Theoretical and Applied Climatology* **84** 3-22
[11] van Hooff T, Blocken B 2012 *Journal of Wind Engineering and Industrial Aerodynamics* **104** 330-41
[12] Snyder W H 1972 *Boundary-Layer Meteorology* **3** 113-34
[13] Cermak J E, Isyumov N, Wind tunnel studies of buildings and structures1999: American Society of Civil Engineers.
[14] Zaki S A, Hagishima A, Tanimoto J, Mohammad A F, Razak A 2014 *J Eng Sci Technol* **9** 176-90
[15] Shojaee S, Uzol O, Kurç Ö 2014 *International Journal of Environmental Science and Technology* **11** 59-68
[16] Cermak J E 2003 *Journal of wind engineering and industrial aerodynamics* **91** 355-70
[17] Uehara K, Murakami S, Oikawa S, Wakamatsu S 2000 *Atmospheric Environment* **34** 1553-62
[18] Durgin F, Isyumov N, Cermak J, Davenport A, Irwin P, Peterka J, et al. 1996 *JOURNAL OF AEROSPACE ENGINEERING* **9** 19-36
[19] Rahman A, Riza M. Numerical simulation and wind tunnel measurements of lateral aerodynamic characteristics on simplified automotive model: Universiti Teknologi Malaysia, Faculty of Mechanical Engineering; 2010.
[20] Ishak I S, Mansor S, Mat Lazim T 2008
[21] Zakaria I A, Mustaffa M R, Mohamed W, Najmi W A, Mamat A M I, Steady-State Potential Energy Recovery Modeling of an Open Cathode PEM Fuel Cell Vehicle. Applied Mechanics and Materials; 2014: Trans Tech Publ.
[22] Tofa M M, Maimun A, Ahmed Y M, Jamei S, Priyanto A 2014 *The Scientific World Journal* **2014**
[23] Wiriadidjaja S, Hasim F, Mansor S, Asrar W, Rafie A S M, Abdullah E J, Subsonic Wind Tunnels in Malaysia: A Review. Applied Mechanics and Materials; 2012: Trans Tech Publ.
[24] Beladi B. Wind flow analysis around a four-building cluster: Universiti Teknologi Malaysia, Faculty of Mechanical Engineering; 2011.
[25] Xu Y-L 2013 *Wind effects on cable-supported bridges* John Wiley & Sons)
[26] Aly A M 2014 *Building and Environment* **75** 206-21
[27] Drew D R, Barlow J F, Lane S E 2013 *Journal of Wind Engineering and Industrial Aerodynamics* **121** 98-105
[28] Plate E 1995 *Windtechnologische Gesellschaft WTG, Karlsruhe (ISBN 3-928909-02-9)*
[29] Uehara K, Wakamatsu S, Ooka R 2003 *Boundary-layer meteorology* **107** 353-70
[30] Wiernga J 1993 *Boundary-Layer Meteorology* **63** 323-63

[31] ASCE. Wind Tunnel Studies of Buildings and Structures, ASCE Manuals and Reports on Engineering Practice No. 67. Reston, Virginia: ASCE, 1999.
[32] Kovar-Panskus A, Moulinneuf L, Savory E, Abdelqari A, Sini J-F, Rosant J-M, et al. 2002 *Water, Air and Soil Pollution: Focus* **2** 555-71
[33] Ikhwan M, Ruck B 2006 *Journal of wind engineering and industrial aerodynamics* **94** 745-65
[34] Benson J. Boundary-layer response to a change in surface roughness: The University of Reading; 2005.
[35] Ricciardelli F, Polimeno S 2006 *Journal of Wind Engineering and Industrial Aerodynamics* **94** 815-32
[36] Grimmond C, Oke T R 1999 *Journal of applied meteorology* **38** 1262-92
[37] Pires L B M, de Paula I B, Fisch G, Gielow R, da Mota Girardi R 2013 *Journal of Aerospace Technology and Management* **5** 305-14

IOP Conf. Series: Materials Science and Engineering **100** (2015) 012025 doi:10.1088/1757-899X/100/1/012025

Simulation analysis of the oil/water structure in the downhole with presence of hydrocyclone separator

I E Jamil[1] and H H Al-Kayiem[1]
[1]Mechanical Engineering Department, Universiti Teknologi Petronas, 32610 Bandar Seri Iskandar, Perak, Malaysia.

E-mail: hussain_kayiem@petronas.com.my

Abstract. The maturing oil fields with increasing water production can pose a challenge in terms of produced water handling and disposal issues. This paper presents the modelling and simulation procedure of the two-phase flow of water/oil in a downhole using ANSYS-FLUENT 14 software. The developed procedure successfully simulated the production zone and the interaction of the two fluids in a natural environment where the reservoir pressure is the sole driving force. The results show significant difference of volume distribution in the flows with different oil content. The model can become an essential tool to assist in prediction of the behavior of oil/water mixture flow in the wellbores, and to serve in designing downhole oil/water separators.

1. Introduction

The largest volume waste stream associated with oil and gas production is produced water [1, 2]. Virtually, water production in mature field is unavoidable. Oil is commonly accompanied by an underlying aquifer. As production rate is increasing, oil water contact (OWC) also increases until water breakthrough into the wellbore. The unavoidable premature water breakthrough or water coning could happen due to high pressure drawdown around the wellbore and high mobility of water than oil, where water viscosity much lesser than the oil, and water relative permeability much greater than oil. The consequences of increased water production translates to high operating expenses and produced water disposal costs, increased in environmental risk associated with large volumes of produced water, additional requirement for produced water lifting and significantly reduced surface processing facilities for oil. Water production can also limit the well production life via fines migrations and high water production inhibits oil production to production tubing, hence causing low recovery at oil layer [3-5].

Treatment and disposal of produced water represent significant costs for operators. A new technology, down-hole oil/water separators (DOWS), has been developed to reduce the cost of handling produced water [1, 2]. DOWS is a hydrocyclone-based system for downhole separation of produced oil and water and subsequent disposal of the produced water by reinjection within the same wellbore using electric submersible pump [6]. However, concurrent with the project presented in this paper, is the exploitation of downhole separation *without* the use of electric submersible pump. Theoretically such concept requires relative positioning of the zones of a high pressure production layers on top and a low pressure water zone below to accomplish hydrocyclone separation.

In order for better prediction of the behavior of the flow entering the hydrocyclone separator, we take a step back to study the behavior of the oil and water flow inside the wellbore. In this investigation, computational fluid dynamics (CFD) is adopted to obtain quantitative description of the oil and water flow behavior from the reservoir into the wellbore. The wellbore is modelled in 3 dimensions using ANSYS-Fluent 14 software, while the fluid properties were taken from Bayan offshore field in Miri.

2. Model formulation of the downhole zone

The design of the wellbore is depicted from real well. As seen in figure 1, production in the well occurs in the upper zone and dumping in the lower zone [1]. The wellbore model is broken down into three parts: the upper zone (production zone) with high pressure, the lower zone (dumping zone) with low pressure, and the liquid-liquid hydrocyclone separation zone. The investigation presented in this paper focuses on the upper zone.

The wall of the zone is perforated with 12 shots per foot for the inflow of production fluid into the well. Table 1 and figure 2 show the design and parameters of the upper zone model of the wellbore.

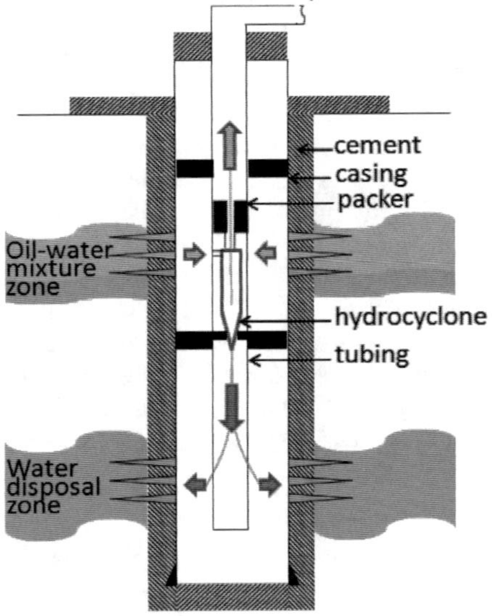

Figure 1. The wellbore model.

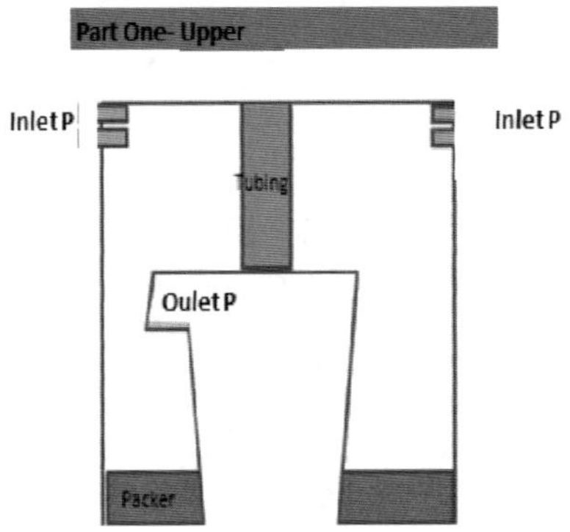

Figure 2. Model sketch of the upper zone.

Table 1. Parameters of upper zone design.

Parameter	Value
Casing OD	9 5/8" (0.244m)
Casing ID	8.535" (0.217m)
Casing wall thickness	0.545" (0.014m)
Casing length (packer to packer)	50' (15.240m)
Inner tubing OD	3.5" (0.089m)
Perforation	12 shots per foot, 120 degrees phasing

3. Numerical implementation

In ANSYS-FLUENT, 5 stages of simulation are need to be followed which are geometry, mesh, setup, solution and results. The order of the simulation is as such, so it is a top to down sequence. First, a 3 dimensional (3-D) model of the production zone is created using ANSYS Design Modeler. With the dimensions stated in table 1, the 2-D outline of the model's vertical cross-section is first sketched then revolved into a 3-D solid model. The perforated holes on the walls and the flow outlet are included later, drawn at a separate plane then extruded and cut out from the solid wellbore model.

Table 2. First mesh setting.

Details of Mesh	Description
Sizing	
Use Advanced Size Function	Off
Relevance Centre	Coarse
Element Size	Default
Initial Size Seed	Active Assembly
Smoothing	Low
Transition	Slow
Span Angle Centre	Coarse
Minimum Edge Length	3.1919e-002m
Inflation	
Use Automatic Inflation	Program Controlled
Inflation Option	First Aspect Ratio
First Aspect Ratio	5
Maximum Layers	5
Growth Rate	1.2
View Advanced Options	No
Assembly Meshing	
Method	None

Figure 3. Mesh model of the top part of the wellbore.

Figure 3 shows the completed 3-D model for the wellbore. The next step leading to simulation is the meshing of the model as shown in Figure 4. In mesh, the inlets (perforated holes) and outlets (exit into separator) of the model as well as the sizing, smoothing, inflation and all other mesh options are set. Table 2 shows the first mesh setting used to create mesh model. Other elements unspecified are using the default setting of the program. The statistics for this mesh model is 349, 690 nodes and 1, 832, 214 elements.

4. Mesh quality

Before going straight with the first mesh model to generate solution, there is great importance in carrying out the mesh independency study to check the quality of the mesh. We may gain a benefit in terms of computing time if the number of elements can be reduced but still maintaining the quality of the mesh. A high-quality mesh with lesser number of elements would require less computing time while giving equally good results. The changes made to the second mesh model are shown in table 3. The statistics for the second mesh is 253, 615 nodes and 1, 338, 868 elements.

Three primary factors to evaluate mesh quality analyzed here which are skewness, aspect ratio and orthogonal quality (figures 4-6). This second mesh model is used in the next stage of the simulation which is setting up the boundary conditions. The preliminary results of these are shared in the results section of this paper.

Table 3. Second mesh setting.

Details of "Mesh"	Description
Sizing	
Use Advanced Size Function	On: Curvature
Relevance Center	Coarse
Initial Size Seed	Active Assembly
Smoothing	Medium
Transition	Slow
Span Angle Center	Fine
□ Curvature Normal Angle	Default (18.0 °)
□ Min Size	Default (6.1304e-003 m)
□ Max Face Size	Default (0.613040 m)
□ Max Size	Default (1.22610 m)
□ Growth Rate	Default (1.20)
Minimum Edge Length	3.1919e-002 m
Inflation	
Use Automatic Inflation	None
Inflation Option	Smooth Transition
□ Transition Ratio	0.272
□ Maximum Layers	5
□ Growth Rate	1.2
Inflation Algorithm	Pre
View Advanced Option	No

5. Skewness

Skewness is one of the primary quality measures for the cells in a mesh. Skewness determines how close to ideal (i.e., equilateral or equiangular) a face or cell is. Skewness is valued from 0 to 1 and is computed directly based on the mesh built. 0 being the best and 1 being the worst quality of mesh elements. The skewness distribution of the present model is shown in figure 4. The average skewness of this initial model is 0.24, which falls under excellent category. Hence this model is acceptable.

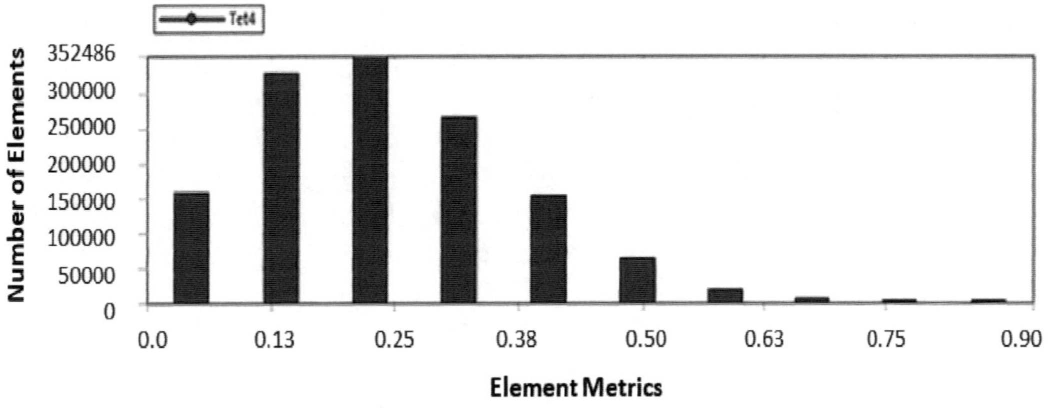

Figure 4. Mesh statistics – skewness.

6. Orthogonal quality

Orthogonal quality helps determine the quality of the mesh itself. The worst cells will have an orthogonal quality closer to 0 and the best cells will have an orthogonal quality closer to 1. The minimum orthogonal quality for all types of cells should be more than 0.01, with an average value that is significantly higher. The average orthogonal quality calculated for the second mesh for the present model is 0.85, as in figure 5; hence it is very good.

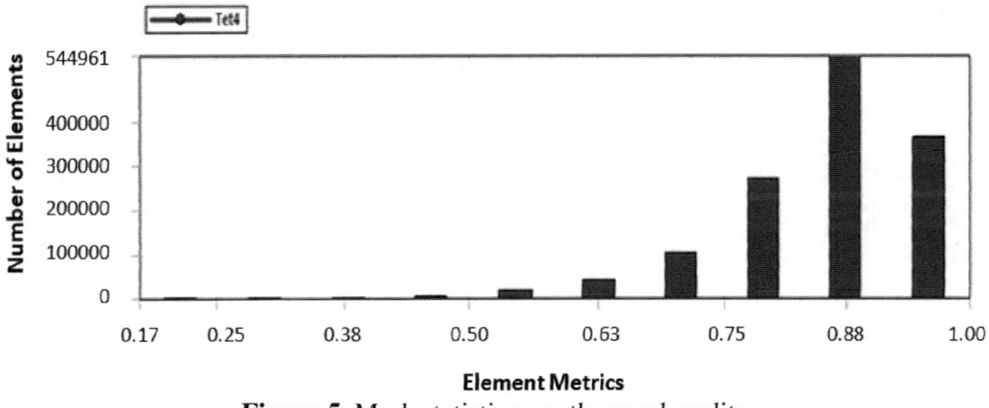

Figure 5. Mesh statistics – orthogonal quality.

7. Orthogonal quality

The aspect ratio is a measure of the stretching of a cell. It is computed as the ratio of the maximum value to the minimum value of any of the following distances: the normal distances between the cell centroid and face centroids, and the distances between the cell centroid and nodes. Generally, it is best to avoid sudden and large changes in cell aspect ratios in areas where the flow field exhibit large changes or strong gradients. Analysis of the mesh in the present work in terms of aspect ratio is shown in figure 6.

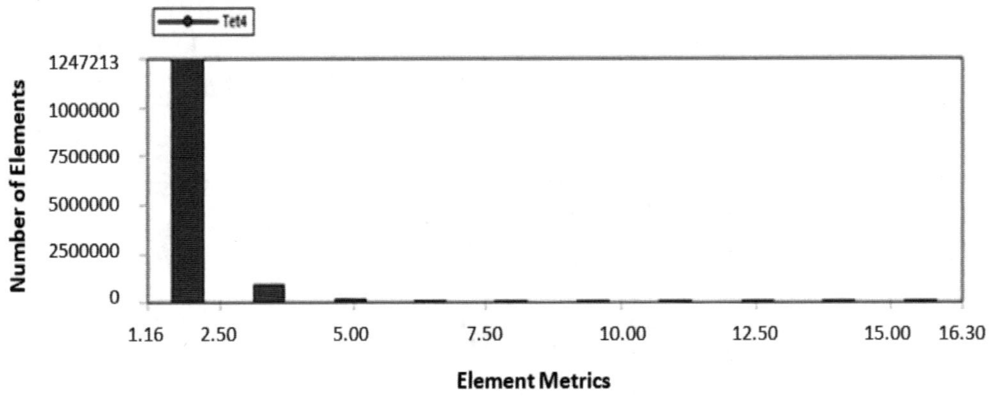

Figure 6. Mesh statistics – aspect ratio.

8. Boundary conditions

The boundary conditions for the setup is taken from the Bayan field in Miri. Table 4 shows the input into the numerical solution.

Table 4. Data for the simulation.

Parameter	Value
Flow Properties	
Inlet flowrate	3.036 x 10^{-4} m^3/s
Inlet pressure	12,776 kPa
Outlet relative pressure	620.53 kPa
Turbulence intensity at inlet	10%
Gravity	(0, -9.81, 0) m/s^2
Fluid Input	
Density Water	1000 kg/m^3
Viscosity Water	0.00097kg/m-s
Density Hydrocarbon	840 kg/m^3
Viscosity Hydrocarbon	0.00045 kg/m-s
Variation of Oil Content	10%, 30%, 50%, 70%, and 90%

9. Results and discussion

The solution is computed for 5 different cases, the variation being the oil content in the flow (10%, 30%, 50%, 70%, and 90%). Figure 8 shows the density distribution at the middle section of the simulation zone i.e. at the perforated holes and at the bottom of the simulation zone, near the inlets to the separator. We verify the simulation results by comparing the densities of the mixture in the flow obtained from the simulation against the mathematical formula. Equation (1) is used to calculate the mixture density ρ where c_{oil} is the volume fraction of oil in the mixture, ρ_{oil} is the density of oil and ρ_{water} is the density of water. The error from this calculation is shown in table 5.

$$\rho = c_{oil}\rho_{oil} + (1 - c_{oil})\rho_{water} \tag{1}$$

Table 5. Verification of density.

Oil Content	Calculated Density	Average Density from Simulation	Error
10%	984.00	985.88	0.2%
30%	952.00	961.58	1.0%
50%	920.00	936.69	1.8%
70%	888.00	900.64	1.4%
90%	856.00	876.40	2.4%

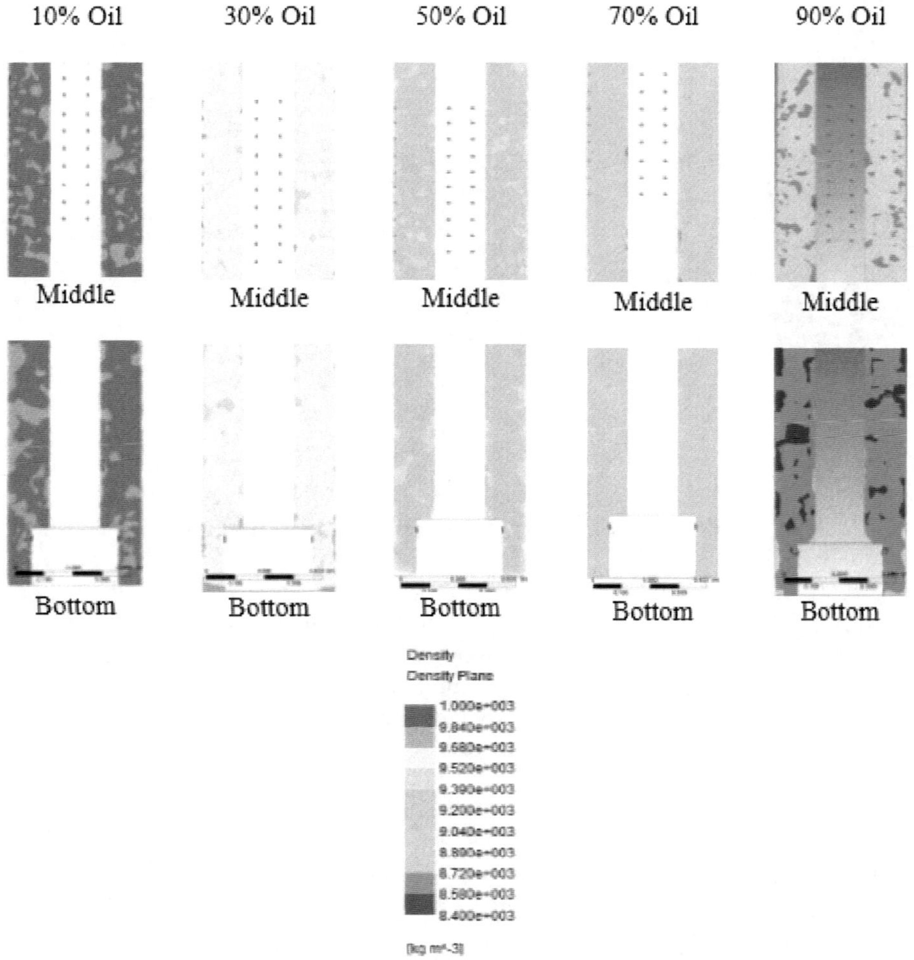

Figure 7. The predicted density distribution of oil inside the wellbore at different oil/water percentage in the production zone.

Visualization of the oil/water mixture, as predicted by the simulation, is shown in figure 7 in terms of density distribution. Dark red color means 100% oil. Two locations in the upper zone of the well, which is the production zone, identified as the middle and bottom. At 10%/90% oil/water production, the simulation in the upper location of the upper zone reveals that there is no spots with pure oil or pure water. But the mixture structure is changing at the bottom of the upper zone, where it is clear that lower density portions are produced. Same behavior could be noticed for the case of 90% /10% oil/water mixture. At the middle, the mixture is more homogenous compared to the bottom. This is due to the high turbulence and jet interaction from the perforating holes in the middle region.

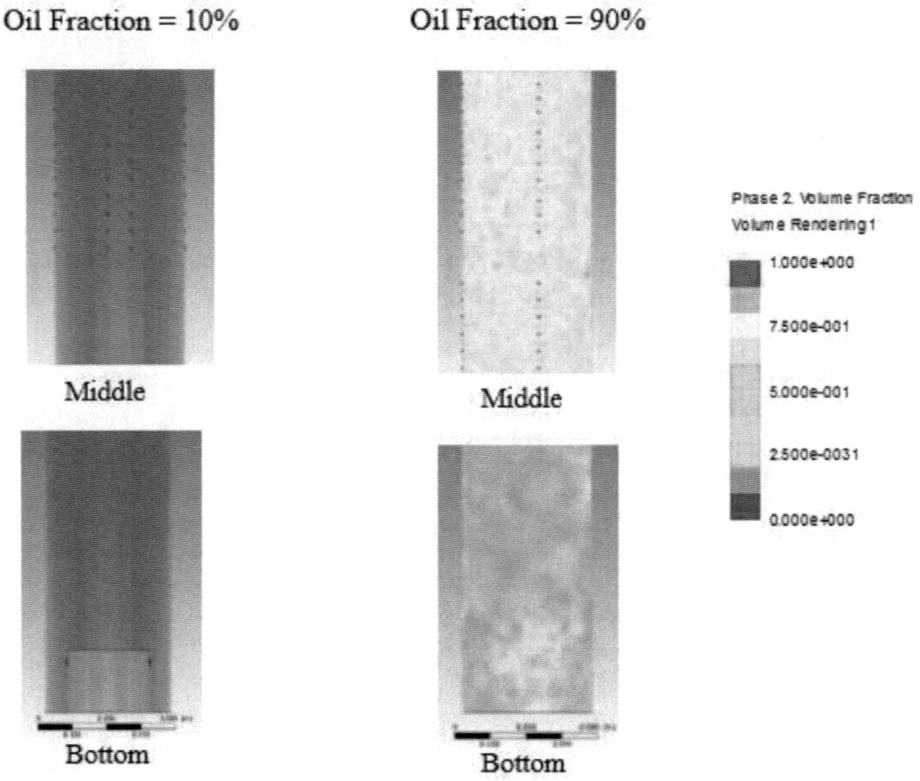

Figure 8. Density distribution at the middle and bottom sections.

Figure 8 shows the predicted volume fraction of oil at cases of 10%/90% oil/water and 90%/10% oil/water mixtures. In the case of 10%/90% oil/water, significant difference of oil distribution between the middle section where flow enters the wellbore and the bottom section before the hydrocyclone separator inlets. The even color in the bottom section depicts that the flow is more settled. In the case of 90%/10% oil/water, it is obvious that the distribution of oil in entire wellbore length is even.

10. Conclusions

A wellbore with production in the upper zone is designed and modeled by Computational Fluid Dynamics using ANSYS-Fluent 14. The model can become an essential tool to better predict the behavior of oil/water mixture flow inside the wellbores, and to serve in designing downhole separators. The simulation results revealed that the mixture is more even in the upper portion of the production zone where the jet interaction from the perforation holes induce high turbulence which enhance the even distribution of the fluids droplets.

Acknowledgements
The authors acknowledge Universiti Teknologi Petronas for the financial and technical support to produce this paper under the research grant YUTP – FRG 015-3AA-A73.

IOP Conf. Series: Materials Science and Engineering **100** (2015) 012025 doi:10.1088/1757-899X/100/1/012025

References

[1] Bowers B, Brownlee R, Schrenkel P, Development of a Downhole Oilwater Separation and Reinjection System for Offshore Application. Offshore Technology Conference; 1998: Offshore Technology Conference.

[2] Matthews C, Chachula R, Peachey B, Solanki S, Application of downhole oil/water separation systems in the alliance field. SPE Health, Safety and Environment in Oil and Gas Exploration and Production Conference; 1996: Society of Petroleum Engineers.

[3] Aziz K 2001 *SPE Reservoir Evaluation & Engineering* **4** 327-35

[4] Flores J G, Chen X T, Brill J P, Characterization of oil-water flow patterns in vertical and deviated wells. SPE Annual Technical Conference and Exhibition; 1997: Society of Petroleum Engineers.

[5] Ahmed M M, Ayoub M A 2013 *Journal of Applied Sciences* **14** 3162-71

[6] Amini S, Mowla D, Golkar M, Esmaeilzadeh F 2012 *Chemical Engineering Research and Design* **90** 2186-95

Investigation of mechanical field weakening of axial flux permanent magnet motor

M Syaifuddin Mohd[1,3], A Rashid A Aziz[1,3] and M Syafiq Mohd[2]
[1]Mechanical Engineering Department, Universiti Teknologi PETRONAS, Bandar Seri Iskandar, Malaysia
[2]Electrical and Electronics Engineering Department, Universiti Teknologi PETRONAS, Bandar Seri Iskandar, Malaysia
[3]Centre for Automotive Research and Electric Mobility (CAREM), Universiti Teknologi PETRONAS, Bandar Seri Iskandar, Malaysia

E-mail: syaifuddin@petronas.com.my

Abstract. An investigation of axial flux permanent magnet motor (AFPM) characteristics was conducted with a proposed mechanical field weakening control mechanisms (by means of stator-rotor force manipulation) on the motor through modeling and experimentation. By varying the air gap between at least two bistable positions, the peak torque and top speed of the motor can be extended. The motor high efficiency region can also be extended to cover greater part of the motor operating points. An analytical model of the motor had been developed to study the correlation between the total attraction force (between the rotor and the stator) and the operating parameters of the motor. The test results shows that the motor output complies with the prediction of the research hypothesis and it is likely that a spring locking mechanism can be built to dynamically adjust the air gap of the motor to increase the operating range and could be applied in electric drivetrain applications to improve overall efficiency of electric and hybrid electric vehicles.

1. Introduction

Axial flux permanent magnet motor (AFPM) inherently has one of the highest specific power due to torque generation from the inner radius to outer radius of the magnets on the rotor face as compared with radial flux permanent magnet motor (RFPM), where torque is only generated at the outer radius of the motor where the magnets are located. However, the permanent flux linkage between the rotor and stator causes a fixed relationship between winding current and motor torque. This "torque constant", kT, also causes a fixed top speed for a PM motor. The higher the kT, or torque per amp, the lower the top speed of the motor [1].

Various methods of getting around this property have been developed in radial flux PM motors. Some of these methods allow for increasing the top speed of the motor by 2 or 3 times what would otherwise be achievable. Unfortunately, this is achieved at a severe cost in terms of loss and motor efficiency. This is because kT reduction is done purely by using winding current that has been offset the maximum phase current amplitude which would deliver maximum torque output from the motor shaft [2].

The axial flux motor on the other hand is different from other types of PM motors in that the same effect of varying kT can be achieved without introducing additional losses because it can be accomplished by the virtue of altering the geometry of the motor. One of the reasons the torque constant, kT is called a constant is because it is an inherent property due to the geometry of the motor. Although there are many different contributors to kT, the one, which is of the main interest is the air gap, or the distance between the magnet rotor and the stator. The size of the air gap determines the electromagnetic coupling between the rotor and the stator and therefore the value of kT. By simply moving either the rotor or the stator axially relative to the other, kT can be adjusted.

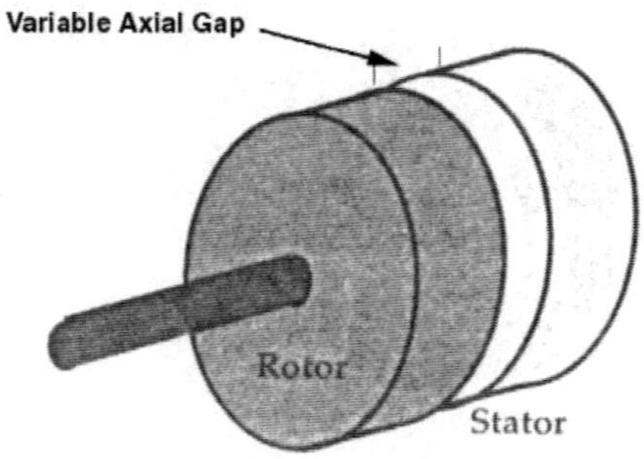

Figure 1. Axial Flux Permanent Magnet Motor Topology.

J. Jermakian et. al. introduced a pioneering work on testing and modeling of variable air gap axial flux permanent magnet motor and discussed on the operating characteristics of the motor at different air gaps and the effects on operating speeds and torque outputs [3-6]. Later, J. Kern et. al. Oh, Kern [7] from Argonne National Lab continue with modeling and simulation work using the same type of motor topology and study the motor operation in all four quadrants. It is determined that the main advantage of adjusting the flux is that the motor torque-speed characteristics can better match the changing load especially in applications which requires variable operating load and speed such as an electric vehicle. The challenge in implementing an electric machine with these qualities is to develop a control strategy that takes advantage of the available efficiency improvements without using excessive energy to mechanically adjust the air gap and reduce the potential energy savings [7].

There are several existing patents [8-10], patent applications [11, 12] and publications [2, 13, 14] on different ways to achieve mechanical field weakening. All of the methods described employ the use of servo motors or some other mechanical devices, which need to be powered externally to the motor and will add to the system mass and costs. At present, the proposed method of manipulating the rotor-stator attractive force to vary the air gap distance appears to be novel in comparison to the prior work on mechanical field weakening of a PM motor.

2. Research methodology

An analytical model of an axial flux PM motor had been developed and simulated in a Matlab Simulink environment. The peak torque values from simulations had been validated with peak torque output of an actual motor [3] with the same design parameters and configurations. A model of a Belleville spring mechanism is also included in the simulation to reduce the electromagnetic axial force between the rotor and the stator.

In addition, a quasi-static 2-dimensional finite element analysis (FEA) model was created by modeling the motor as multiple layers from inside radius, Ri to the outside radius, Ro. A periodic type boundary condition as shown in Figure 3 is utilized since the magnetic field repeat periodically at every pole pitch at no-load operation and every double pole pitch under-load conditions [15]. The main results from the finite element analysis are the flux distribution and the back emf waveform. The back emf waveform shape is then utilized to refine the dynamic model of the motor in the Matlab Simulink environment. Figure 2 shows the validation of the analytical model as compared with an actual motor torque output data.

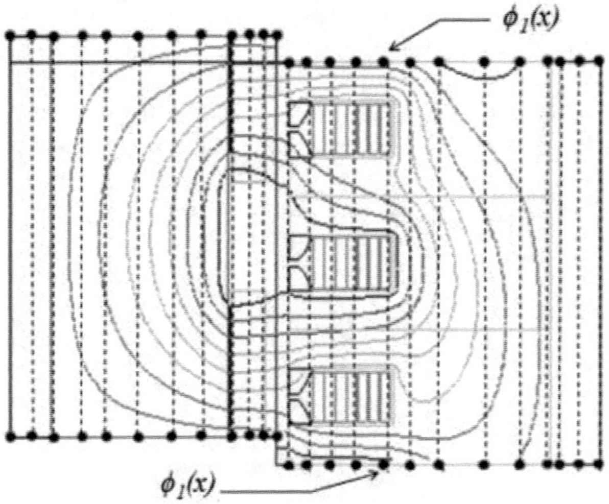

Figure 2. FEA electromagnetic simulation of a single pole pair with periodic boundary condition applied.

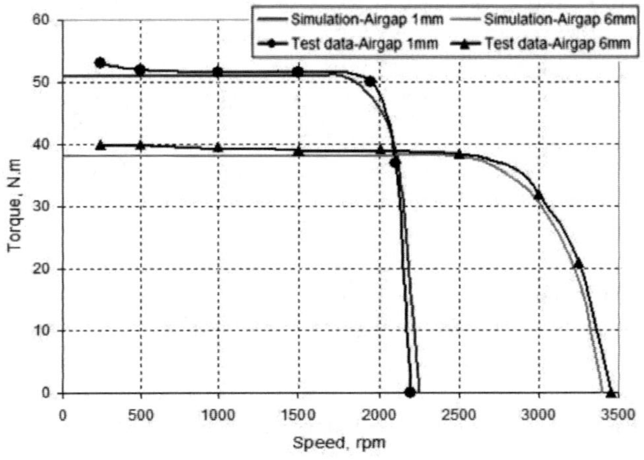

Figure 3. Peak torque values – Simulation & Test Data.

Figure 4 shows the torque output of one of the motors as a function of phase angle offsets. As the phase angle is advanced or retarded, the maximum amount of torque generated is reduced while the rotor-stator attraction force is either respectively decreased or increased. The loss in torque production due to the phase current offset is directly converted into these additional rotor-stator forces. Based on these results, it is possible to determine the total rotor-stator attraction force as a function of the three main parameters; size of air gap, phase current offset and amplitude of phase current.

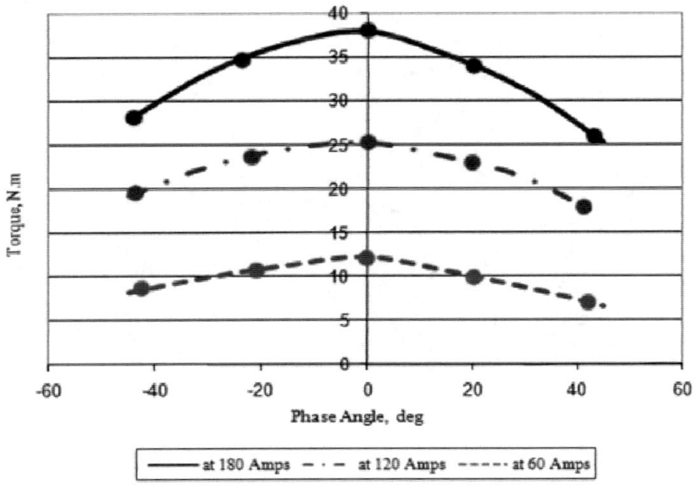

Figure 4. Torque as a function of phase current offset and amplitude.

An analytical model of the rotor-stator attraction force was developed based on the test data. However, the derivation of the equations for the analytical model of the electromagnetic force is too extensive to be covered in this paper, thus only the reduced final equations (for the given motor dimensions and design parameters) are shown below:

- Electromagnetic attraction force of rotor and stator [16], Fr-s:

$$F_{r\text{-}s} = \int B_G(x)^2/(2\mu_o)\ .dA \quad (1)$$

where:

BG – air gap flux density, μ_o - permeability of air
A – rotor magnet area, x – air gap distance

- Attraction force due to induced magnetic field, $F(I,\theta,x)$:

$$F(I,\theta,x) = (-4kT(x).\ I.\ \theta 2)\ /\ (Reff.\pi 2) \quad (2)$$

where:

kT – torque constant, I – phase current magnitude
θ – phase current offset, $Reff$ – effective radius of rotor

-Total attraction force, *Ftotal:*

$$Ftotal = Fr\text{-}s + [-sign(\theta)] \cdot F(I,\theta,x) \quad (3)$$

The resultant equation for total attraction force is a function of 3 variables; magnitude of phase current, phase current offset and air gap distance. A typical four dimension sliced plot is shown in figure 5.

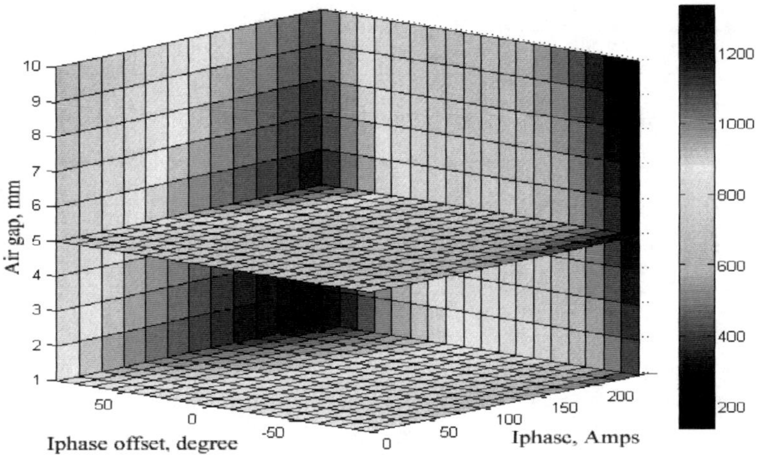

Figure 5. 4D plot of total attraction force (in N).

3. Results and discussion

The resultant four-dimensional matrix will be used in the dynamic model of the system. Based on the calculated total attraction force as a function of the three variables, it is possible to design a bistable air gap control mechanism by manipulation of the internal forces and insertion of a Belleville spring mechanism to reduce the total attraction force.

The mechanism of the proposed system can be explained using the force profiles in figure 6 above as following:

1. As the phase current is advanced to -30°, the total attraction force is reduced to less than 840 N becoming lower than the repulsion force from the Belleville spring mechanism.
2. Hence, the rotor is pushed away from the stator as Fspring>Ftotal until the maximum air gap distance is achieved. The motor is now capable of delivering torque at higher speed at a larger air gap.
3. To return the rotor back to the minimum air gap distance, the phase current is changed back to 0° (or even several additional degrees to increase the attraction force). Now the total attraction force is becoming larger than the repulsion force from the spring mechanism.
4. Therefore, the rotor is pulled back to the minimum air gap distance as Fspring<Ftotal.

In the actual operation of the motor, a locking mechanism might be required to lock the rotor at the minimum or maximum air gaps. The effects of cogging torque and torque ripples on the proposed mechanism will also be investigated later using a dynamic model of the system.

The dynamic simulation of motor-controller-load is performed in Matlab Simulink environment. The model takes into account the actual shape of the back EMF and the shape of the current waveform — sinusoidal or trapezoidal — injected into the windings [17] and includes a model of the motor drive subsystem to simulate the control of the phase current timing. The main outputs of the simulations are the torque output of the motor, rotational speed of the motor and the variation in the rotor-stator attraction force as the functions of air gap length and phase current offsets.

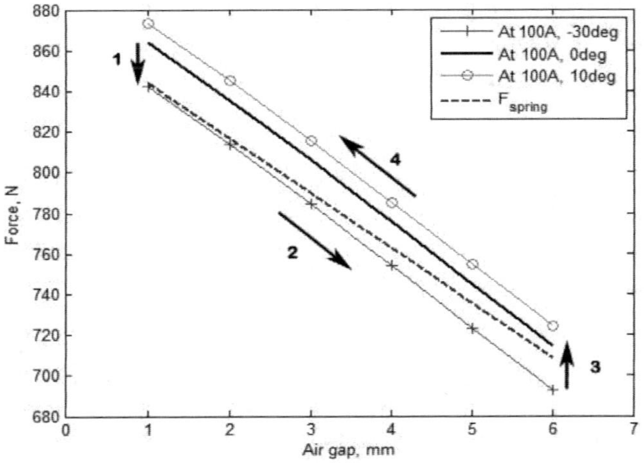

Figure 6. Rotor-stator axial forces at I = 100 amps.

4. Scaled down experimentation

A custom made test bench (as shown in figure 7) for a scaled down 700 Watt motor was fabricated with two load cells to measure i. the axial attraction force between rotor and the stator and ii. torque output from the motor. Two operating parameters of the motor can be physically adjusted: i. air gap size by 1 mm increment by inserting spacers on the rotor assembly and ii. phase current offset by 10 electrical degrees increment of the hall sensor position. Figure 8 shows the construction of the rotor and stator coils.

Figure 7. Custom test bench set-up to measure torque and axial force.

(a) (b)

Figure 8. (a) Stator coils and (b) Rotor assembly.

Preliminary results in figure 9 suggest that the shape of the torque output profiles comply with the model prediction. However, the peak torque of the 700 Watt motor is 'flat' between 30 degrees and -30 degrees phase angle offset compared with the sinusoidal shape of the torque output of a typical motor as shown in figure 7. This can be explained by the six-step motor drive scheme used for the 700 Watt motor compared with the sinusoidal drive scheme used for the motor in the larger motor.

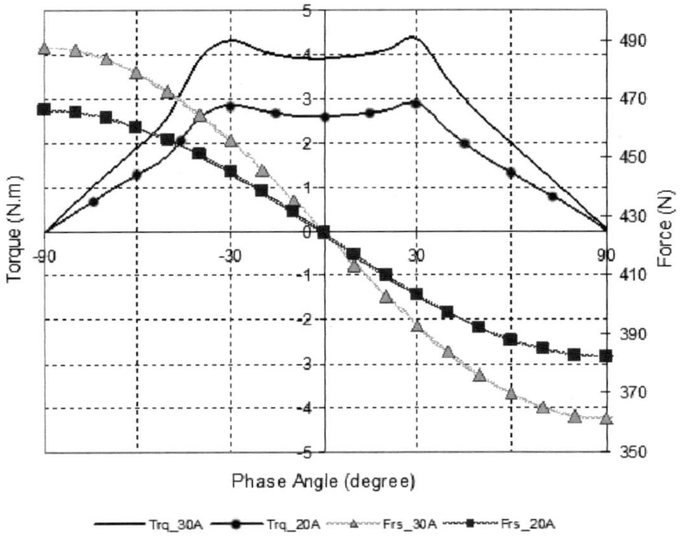

Figure 9. Motor torque and F_rs versus phase angle offset.

As predicted, the attraction force reduced by advancing the phase current and it reaches the minimum value at phase angle offset of 90 degrees. The electromagnetic force vector increases in the axial direction and reduces the rotor-stator attraction force as the phase current is advanced from 0 degree to 90 degree. As a result, the motor torque output is also reduced as the electromagnetic force vector decreases in the tangential direction to the rotor disc.

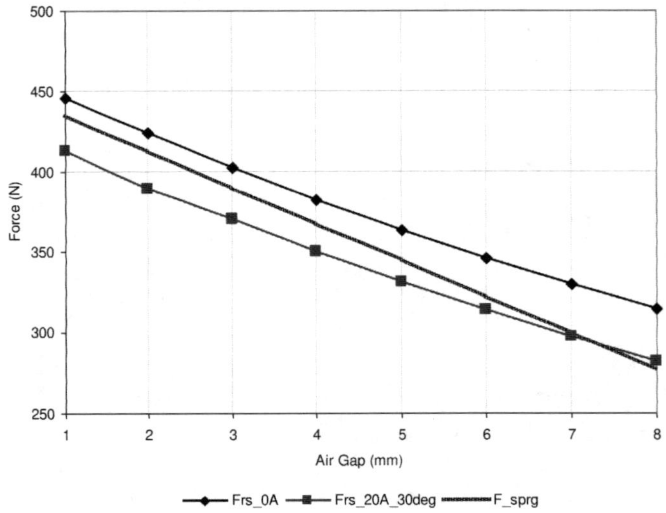

Figure 10. Rotor-stator attraction force versus air gap size.

Figure 10 shows the rotor-stator attraction force results versus the increment in air gap size from 1 mm to 8 mm. As the phase current of 20 Amps is advanced by 30 degree, the rotor-stator attraction force decreases in linear manner as it is a function of torque constant, kT which reduces linearly as a function of air gap size. A pre-compressed spring force profile shown in the same figure suggests that it is possible to dynamically vary the air gap using the method suggested in figure 5.

5. Conclusion

This research is investigating into the feasibility of manipulating the rotor-stator attractive force in order to mechanically introduced field weakening to axial flux PM motor. An analytical model of the axial flux PM motor had been developed and simulated in Matlab Simulink environment. Based on the simulated outputs of the rotor-stator attraction force profiles, it is feasible to design and control a bistable air gap switching mechanism. The model can be further improved and enhanced in order to study the system stability and to determine an appropriate control algorithm when the system is operating under dynamic transitional loads expected in actual applications. A custom made test bench (and motor controller) for a 700 W motor was built specifically to characterize the motor as a function of phase current, air gap size and phase current offset.

Based on current results, the motor output complies with the prediction of the research hypothesis and it is likely that a spring locking mechanism can be built to dynamically adjust the air gap of the motor to increase the operating range (higher torque at minimum air gap and higher top speed at maximum air gap). Additional sets of experiments can be conducted to further validate the model, to test transient response of the proposed system.

Acknowledgement
This work is partly funded by an STIRF (Short Term Internal Research) grant from Universiti Teknologi PETRONAS.

References

[1] Hanselman D C 2003 *Brushless permanent magnet motor design* (umaine.edu: The Writers' Collective)
[2] Lipo T, Aydin M, Field weakening of permanent magnet machines: Design approaches. EPE Power Electronics and Motion Control Conference (EPE-PEMC 04); 2004.
[3] Jermakian J, Mohd M-S, Motevalli V. Testing and modeling of variable airgap axial flux brushless DC motor. SAE Technical Paper, 2001 0148-7191.
[4] NGM. NGM Drive System's Superior Performance Demonstrated in Solar Car Races. http://www.ngmcorp.com: New Generation Motors Corporation 1999.
[5] Patterson D, Recent advances in the design and construction of axial flux permanent magnet machines. Proceedings, IEEE Australia Summit; 1996; Darwin.
[6] Pinkham M. Chinese NdFeB Magnet Presence Grows Along with Demand. High Tech Materials Newsletter. 2002.
[7] Oh S C, Kern J, Bohn T, Rousseau A, Pasquier M. Axial flux variable gap motor: application in vehicle systems. SAE Technical Paper, 2002.
[8] Jermakian J B, Crain S G, Knudtson C D, Piacesi R F. Electric motor with active hysteresis-based control of winding currents and/or having an efficient stator winding arrangement and/or adjustable air gap. United States Patent Office: US Patents; 2000.
[9] Kim S-m. Motor having variable air gap. United States Patent Office: US Patent No. 5,770,908; 1998.
[10] Maslov B A, Soghomonian Z. Rotary permanent magnet electric motor with varying air gap between interfacing stator and rotor elements. United States Patents Office: U.S. Patent No. 6,727,630; 2004.
[11] Yagi H, Matsueda H, Saitou M. Wheel driving apparatus and electric vehicle including the same. Unites States Patents Office: US Patent Application No. US2009/0212728 A1; 2009.

[12] Ai X, Mularcik B, Knepper R. Electric motor with field weakening. United States Patents Office: US7960888 B2; 2011.

[13] Cirani M, Thelin P, Sadarangani C, Design procedure for an AFPM machine for field-weakening applications. Proceedings of the Nordic Workshop on Power and Industrial Electronics, NORpie; 2002 August 2002; Stockholm, Sweden.

[14] Chalmers B J, Akmeşe R, Musaba L, Design and field-weakening performance of permanent-magnet/reluctance motor with two-part rotor. IEE Proceedings-Electric Power Applications; 1998 2nd of March 1998.

[15] Hameyer K, Belmans R 1999 *Numerical modelling and design of electrical machines and devices* (Computational Mechanics: WIT Press)

[16] Gieras J F, Wang R-J, Kamper M J 2008 *Axial flux permanent magnet brushless machines* **1** (Springer Science & Business Media: Springer)

[17] Ong C-M 1998 *Dynamic simulation of electric machinery: using MATLAB/SIMULINK* **5** (Upper Saddle River, NJ: Prentice hall PTR)

Effect of wear on the burst strength of l-80 steel casing

S Irawan[1], A M. Bharadwaj[1], B Temesgen[1], S Karuppanan[2] and M Z B Abdullah[2]
[1]Department of Petroleum Engineering, Universiti Teknologi PETRONAS, 32610 Bandar Seri Iskandar, Perak, Malaysia.
[2]Department of Mechanical Engineering, Universiti Teknologi PETRONAS, 32610 Bandar Seri Iskandar, Perak, Malaysia.

Email: abhilashbharadwaj19@gmail.com

Abstract. Casing wear has recently become one of the areas of research interest in the oil and gas industry especially in extended reach well drilling. The burst strength of a worn out casing is one of the significantly affected mechanical properties and is yet an area where less research is done The most commonly used equations to calculate the resulting burst strength after wear are Barlow, the initial yield burst, the full yield burst and the rupture burst equations. The objective of this study was to estimate casing burst strength after wear through Finite Element Analysis (FEA). It included calculation and comparison of the different theoretical bursts pressures with the simulation results along with effect of different wear shapes on L-80 casing material. The von Misses stress was used in the estimation of the burst pressure. The result obtained shows that the casing burst strength decreases as the wear percentage increases. Moreover, the burst strength value of the casing obtained from the FEA has a higher value compared to the theoretical burst strength values. Casing with crescent shaped wear give the highest burst strength value when simulated under nonlinear analysis.

1. Introduction

The deep wells or extended reach wells have seen an increasing problem of casing wear [1]. These wells often follow highly-deviated, horizontal, and multilateral well path trajectories leading to the drill string coming in contact with the casing leading to wear. Wear is the removal of material from a surface as a result of dynamic conditions [2]. It can be caused due to the rotational effect of the drill string or due to the contact force in a dogleg section when a directional drilling is conducted. In the years prior to 1980s, casing wear was not considered as a big problem in oil and gas industry [3]. However, recently more emphasis has been given on the investigation and monitoring of casing wear given that wear has significant effect on the strength of the casing. This is achieved by considering the various loads, specifically the burst and collapse loads, which the casing needs to resist during its life time. The burst strength of a casing is the ability of a casing to resist the internal pressure exerted on the casing before failure [4]. If a thorough analysis of the burst strength of casing is not performed, well control problem such as casing leakage, kick or blow out might occur. Thus, the resulting burst strength analysis is significant to overcome such undesirable scenarios and ensure good casing design.

This study estimates the casing burst strength after wear using Finite Element Analysis (FEA). Findings were compared with the theoretical values obtained from equations in literature. The scope of this work mainly focused on analyzing the mathematical equations developed to compute the allowable internal pressure for a worn out casing and to generate various models using FEA to estimate the burst strength after wear. The initial, full and rupture yield strengths are estimated and compared with each other. The shape of the wear, as a factor impacting the magnitude of burst strength, is also studied in this work.

2. Literature Review

Casing wear has a significant impact in the performance of a casing in the life of a well especially for future plans in operations such as artificial lift. The oil and gas industry allocates additional investment per year on additional well thickness to allow for wear [3]. One of the main responsibilities of a well engineer is to design a casing that has the strength to withstand the various forces it may face in its life time. The three important mechanical properties that are used to describe the strength of a casing are collapse, burst and tensile strength. The customary practice to estimate the burst strength of a casing is by using the American Petroleum Institute (API) standard equation which is also known as Barlow's equation as equation (1). Barlow's equation related the internal pressure with the tensile strength of the pipe and its dimensions. In addition to Barlow's equation, there are three other casing burst strength estimation equations which Wu and Zhang [1] had briefly discussed in their paper. The initial yield burst equation as equation (2) calculates the burst pressure as the casing yields at the inner diameter before reaching the entire wall thickness. The full yield burst equation as equation (3) is related to the pressure as casing yield throughout its entire wall thickness and the casing rupture burst equation as equation (4) refers to the pressure where ductile failure of the casing takes place [1].

Bradley [5] performed a theoretical analysis to determine the effects of wear on the burst strength of casing and showed that the API methods for determining burst resistance may result in burst values that have very low probabilities of failure.

Song, Bowen [6] conducted a study focusing on the burst strength of a casing after wear. An analytical solution for the hoop stress of worn casing was developed by dividing the entire worn casing into three shapes that are mirror to one another. This superimposition principle was used to obtain the induced hoop stress of the worn casing. Other studies were conducted to show assumptions of slotted ring in a casing wall can be used to create a more simplified casing wear models [1]. The research conducted by Wu and Zhang [1] reflects the relationship between casing wear, hoop stress and burst strength. They performed FEA modeling to study the effect of internal pressure on the hoop stress. They observed the casing is deformed in to an oval shape when exposed to an internal pressure loading and zero external pressure [1].

Researchers studied the wear depth caused by the contact pressure applied to the inner wall of casing. They utilized different sizes of drill string to find the wear depth as a function of time [7]. Field studies have revealed the different parameters that affect the intensity of wear. The most common parameters are side loads, dogleg severity, chemical composition of drilling mud, ability of drill pipe to cause wear, resistance of casing to wear, rotation time and revolution per minute [8]. In directional wells, the rotating tool joint is forced by the drill string against the inner wall of the casing for a longer period of time. As a result, it grinds against the casing wall, creating material erosion in both the rotating tool and casing surfaces i.e. wear. The decrease in the thickness of a casing wall affects the geometry and load distribution on the casing.

Mark Haning, James Doherty [9] performed FEA modeling of an eccentrically worn casing to determine the burst capacity of a worn out casing. They analyzed the various casing burst strength equations i.e. API burst capacity equation, rupture burst strength equation and Klever Stewart's burst capacity. Their research concluded as Barlow equation is more stringent compared to the rest of the existing burst strength equations.

3. Methodology

3.1. Material Properties

L-80 steel is used as the casing material in this project. The reasons for choosing L-80 casing are: (a) it is suitable for sour drilling environment, (b) it is widely available and (c) it is suitable for effective steam injection in shallow wells. The mechanical and physical properties of the casing material are summarized in table 1.

Table 1. Properties of Experimental.

Parameter	Value
Material Grade	L-80 Steel
Length, L (mm)	2000
Nominal Outer Diameter, OD (mm)	244.475
Nominal Wall Thickness, t (mm)	11.9888
API Minimum yield strength (MPa)	552
API Minimum Tensile strength (MPa)	655
Poisson ratio	0.3
Modulus of elasticity (MPa)	200,000

3.2. Theoretical Solution

The four equations used to estimate the theoretical values of burst strength of a worn out casing are mentioned in equation (1) – (4).

$$P_{API} = \frac{1.75\sigma_y t}{D} \tag{1}$$

$$P_{API} = \frac{1.75\sigma_y}{\sqrt{3}}\frac{2t}{D}\left(1 - \frac{t}{D}\right) \tag{2}$$

$$P_{API} = \frac{1.75\sigma_y}{\sqrt{3}}\frac{2t}{D}\left(1 + \frac{t}{D}\right) \tag{3}$$

$$P_{API} = \frac{1.75\sigma_{ult} t}{D-t} \tag{4}$$

where, P_{API} is the burst strength of the casing under API standards, σ_y is the yield strength of the casing material, t is the casing wall thickness, D is the casing outer diameter, σ_{ult} is the ultimate tensile strength of the casing material.

3.3. Finite Element Modeling

The model involves a casing pipe of the required dimension and placing the defector wear accurately. Since the actual shape of a wear is irregular, three different wear models are used to represent the idealized model i.e. to represent the wear shape using regular geometric shapes. These wear models have varying depth and shape.

Case 1- Rectangular shaped wear

The first case represents a casing with a rectangular shaped wear as shown in figure 1.

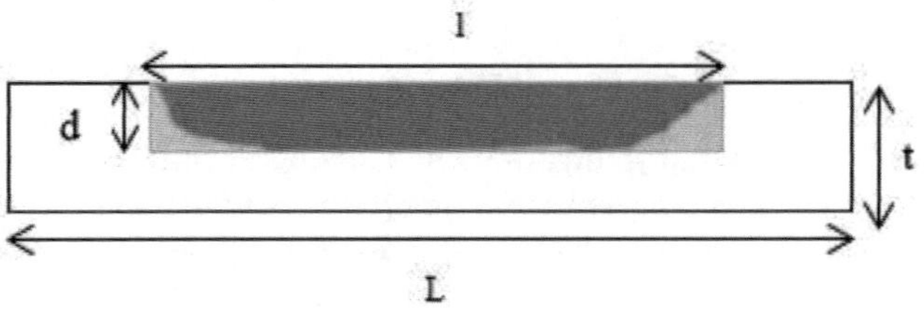

Figure 1. Geometry of rectangular wear.

The dimensional parameters for modelling case 1 are shown in table 2.

Table 2. Dimensions for the models in case 1.

Wear depth (%)	Wear depth, d (mm)	Wear length, l (mm)
20	2.39776	200
40	4.79552	200
60	7.19328	200
80	9.59104	200

Case 2 **-** Crescent shaped wear

The second case is developed by assuming that a crescent shaped wear is formed. The geometry of such wear is represented by figure 2.

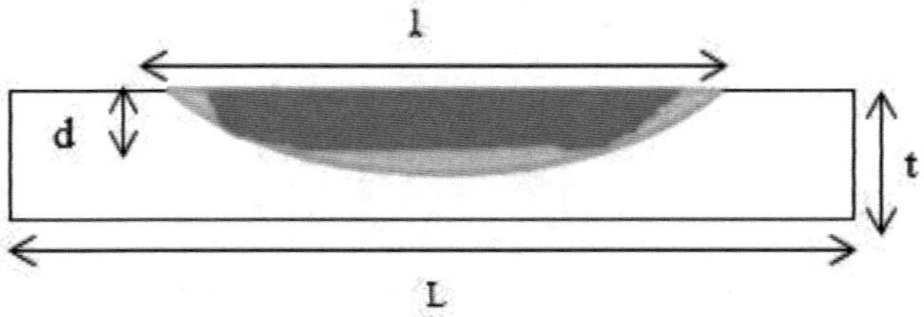

Figure 2. Geometry of cresent wear.

The depth and length of the wear case is presented in table 3.

Table 3. Dimensions for the models of case 2.

Wear depth (%)	Wear depth, d (mm)	Wear length , l (mm)
20	2.39776	200
40	4.79552	200
60	7.19328	200
80	9.59104	200

***Case 3*-** Multiple wear

The third geometric model considers the possibility of multiple wear located on the casing. Figure 3 illustrates this case.

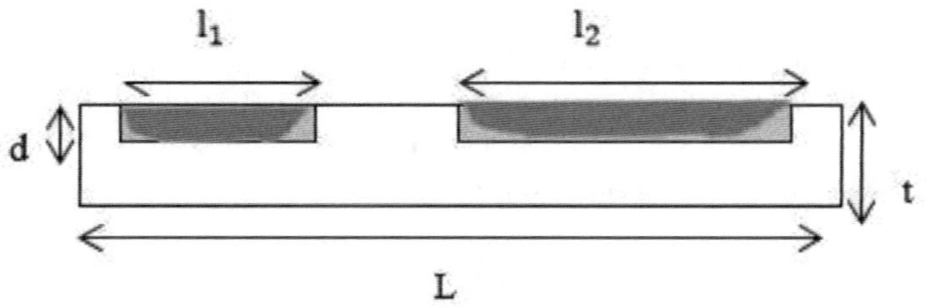

Figure 3. Geometry of case 3.

The data tabulated in table 4 show the wear depth in percentage as well as the length for major and minor wear on the casing.

Table 4. Dimensions for the models of case 3.

Wear depth (%)	Wear depth, d (mm)	Minor Wear length l_1 (mm)	Major Wear length l_2 (mm)
20	2.39776	100	200
40	4.79552	100	300
60	7.19328	100	400
80	9.59104	100	500

Hexahedron mesh is used for the Finite Element Analysis. Figure 4 represents the FE model along with the meshing and a couple of boundary conditions. The meshing properties for each model described in above are presented in table 5.

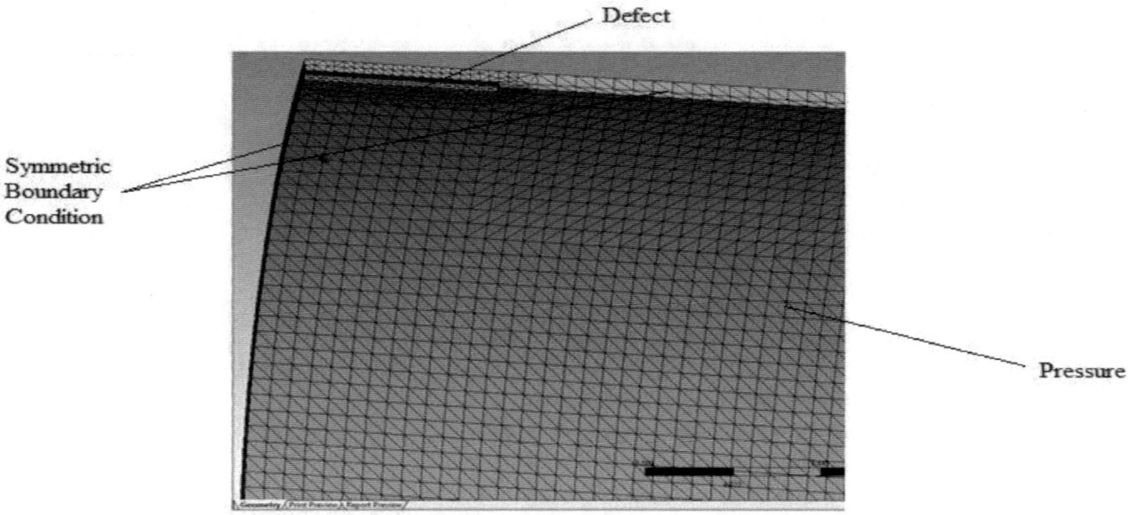

Figure 4. Finite Element model of the defect and boundary conditions.

3.4. Boundary conditions and Load applications

A symmetric boundary condition is applied on three edges of the casing model in order to be able to simulate a quarter of the casing pipe. Also at the end faces of the model all degrees of freedom are fixed. The major loads applied are the internal pressure and axial load on the model. The internal pressure applied is increased until the casing equivalent Von Mises stress equals to the minimum tensile strength of the casing i.e. when the casing bursts. An axial load is applied at the ends of the pipe to represent the closed end of the pipe during a burst test. A displacement vector equivalent to zero is applied at the end of the pipes to constrain the pipe from moving.

Table 5. Mesh Property.

Model	Wear Depth (%)	Number of nodes	Number of Elements
Case 1	20	23566	11843
	40	26194	13751
	60	26497	13619
	80	26828	13790
Case 2	20	27022	14053
	40	26491	13674
	60	26295	13481
	80	27386	14128
Case 3	20	16684	8410
	40	18566	9636
	60	21428	11706
	80	16686	8799

4. Result and discussion

4.1. Type of analysis

Both linear and nonlinear analyses are considered in this work. A linear analysis demonstrates a direct relationship between stress and strain. On the other hand nonlinear analysis allows for a nonlinear relationship between stress and strain beyond the yielding point and it takes into account the effect of temperature on material properties. The stress-strain curve for L-80 steel casing is shown below in figure 5.

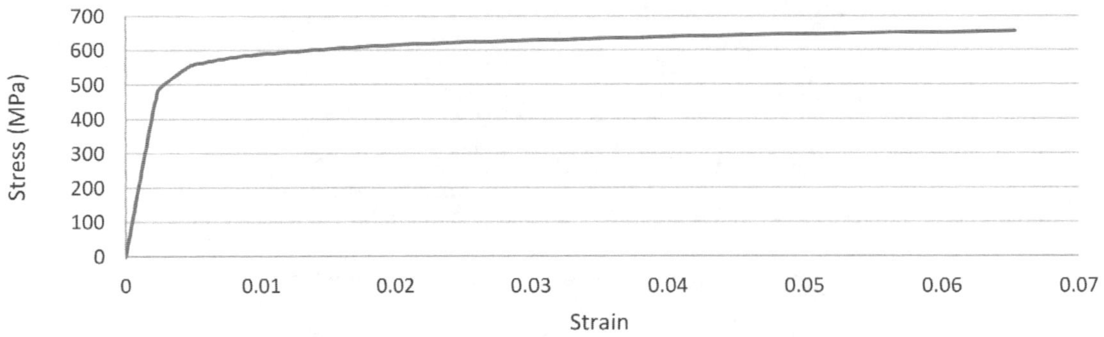

Figure 5. Stress-Strain Curve for L-80 steel casing.

The models are simulated with increasing internal pressure loading (P) until the Von Mises Stress, $\sigma_{VonMises}$ of the entire nodes ligament values is equal to the specific yield strength of the casing, i.e. 552 MPa. Figure 6 shows a sample output result of how the casing with a 40% wear depth percentage under nonlinear analysis looks like when exposed to an internal pressure.

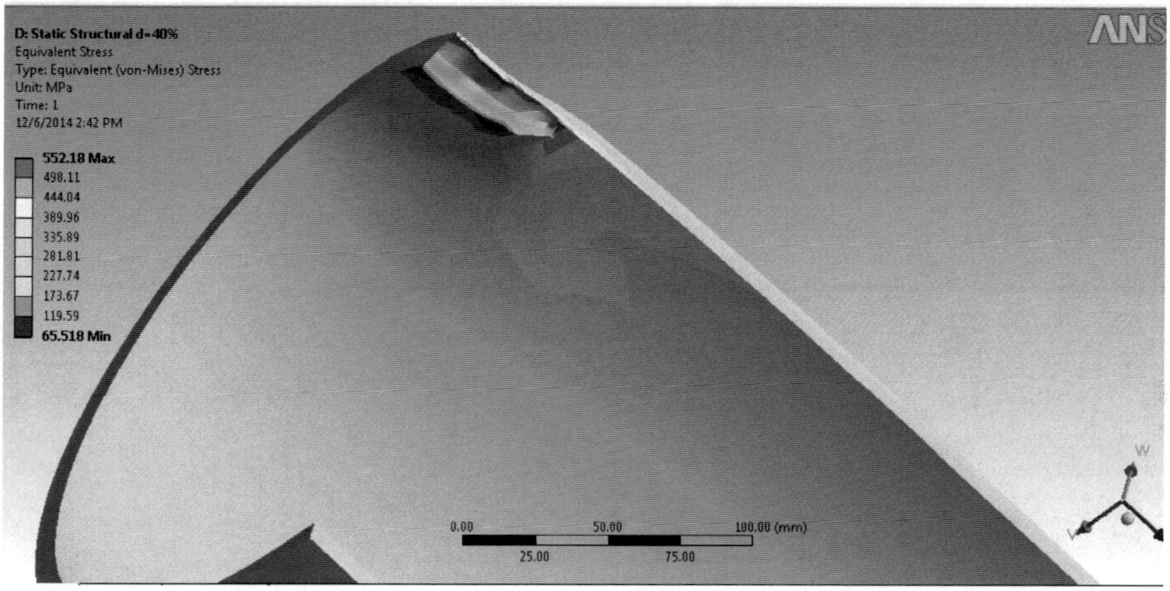

Figure 6. Von Mises stress distribution for non-linear analysis of 40% crescent shaped wear depth.

The theoretical values plotted in figure 7 shows that Barlow's equation gives the lowest value of burst strength whereas the rupture burst equation gives the highest value amongst the four theoretical equations. This is reasonable considering that the rupture strength model assumes burst after the casing string has completely failed or ruptured. Therefore, Barlow's equation should be used in the casing design. As the burst strength is extremely crucial in well operations, the equation providing the lowest failure point is always noteworthy. The burst strength decreases as the wear depth percentage increases which is naturally expected [1]. It is also seen that all theoretical results tend to converge with increase in wear depth. This is because all the theoretical models depend on the t/D ratio. At higher wear depth, the ratio approaches zero and hence, the values begin to converge.

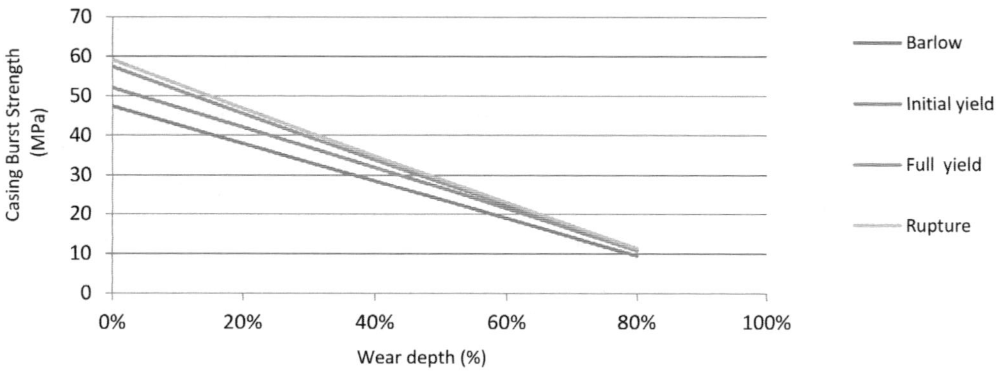

Figure 7. Theoretical casing burst strength values.

Figure 8 shows the graphical representation of the linear and non-linear FE analysis of a worn out casing with varying wear depth and shape respectively. The values for linear analysis of the casing shows similar trend as the analytical solution i.e. the burst strength values decrease in a similar trend with increase in wear depth percentage showing that the FE analysis validates the theoretical analysis. However, the linear and the theoretical results underestimate the burst strength of the casing material. This can be seen through the result for the nonlinear analysis which provides a higher burst strength value. The flattened trend between 40% wear depth and 60% wear depth as seen in figure 8 indicates that the material is able to absorb some of the stresses developed. This is related with the nonlinear response of the L-80 steel. Therefore, non-linear analysis is important to have a realistic estimate of the burst strength. Different shapes have different impact from the analysis and it is seen that the casing bearing crescent shaped wear has higher burst strength than rectangle or multiple wear. This is primarily due to the lower volume of material being lost through wear in crescent shape as compared to rectangular. Larger volume of wear results in lower burst strength and less than 10 MPa difference in burst strength is noted for the given wear geometry mentioned above.

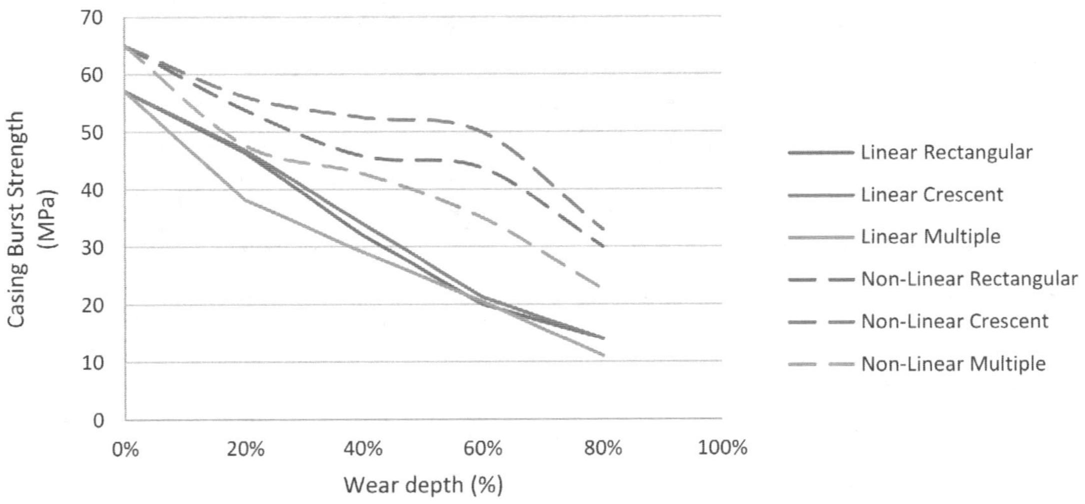

Figure 8. Linear and Non-Linear FEA results for casing burst strength with wear.

The result from nonlinear analysis for a crescent shaped wear is selected to be the most representative of the actual burst strength condition for a worn out casing. The reason is that it takes into consideration the nonlinear material property of steel and the shape of the tool joint causing the wear. The values of burst strength obtained using this case are the highest among the other two cases. Barlow's equation gives a smaller burst pressure value compared to the result from FEA as shown in figure 9. This comparison is crucial for the drilling engineers to estimate the onset of casing failure post Barlow's burst strength estimation. It will enable the engineers to take the necessary steps in managing the pressures inside the wellbore.

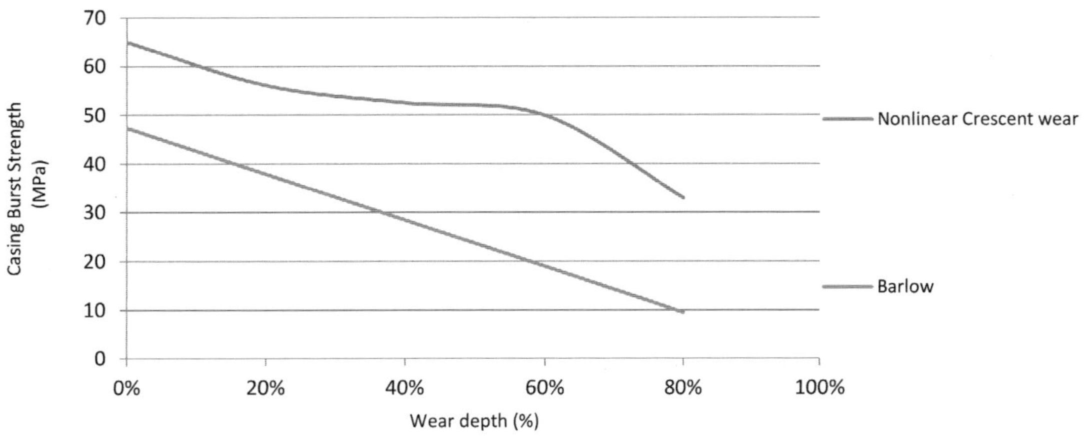

Figure 9. Comparison between Barlow's burst strength and nonlinear FEA for a crescent shaped wear.

5. Conclusion

The comparison between the existing theoretical burst strength results for L-80 steel are shown in this paper. It is seen that the rupture burst equation provides the highest burst strength while the Barlow's equation provides the lowest value of burst strength. This suggests that using Barlow's equation for initial prediction of casing burst strength is a better option as casing failure would occur around 10 MPa from that value. Linear and non-linear finite element analysis results for different wear shapes and depths are obtained and compared. Non-linear finite element results provide more realistic burst strength as compared to the linear or theoretical values. The non-linear analysis of the crescent shape wear seems to best predict the general type of wear in the extended reach wells. It is also seen that casing with crescent shape wear tends to have higher burst strength as compared to rectangular wear. The results of non-linear crescent shaped casing wear (highest value) are plotted with the results of Barlow's equation (lowest value) to obtain an indication of onset of casing failure. With this result drilling engineers can perform accurate pressure management in the wellbore and prevent mishaps such as kick, blow out or casing leakage from occurring. A future study on buckling and bending stresses at the worn section can help in estimating the life of the casing.

Acknowledgements

The authors would like to thank Universiti Teknologi PETRONAS for providing facilities and resources to conduct this research.

References

[1] Wu J, Zhang M, Casing burst strength after casing wear. SPE Production Operations Symposium; 2005: Society of Petroleum Engineers.
[2] Srivastava S *Developments in Lubricant Technology* 87-101
[3] White J P, Dawson R 1987 *SPE Drilling Engineering* **2** 56-62
[4] Azar J J, Samuel G R 2007 *Drilling engineering* PennWell Books)
[5] Bradley W 1976 *Journal of Manufacturing Science and Engineering* **98** 686-94
[6] Song J, Bowen J, Klementich F, The internal pressure capacity of crescent-shaped wear casing. SPE/IADC Drilling Conference; 1992: Society of Petroleum Engineers.
[7] Shen Z, Beck F E, Ling K 2014 *Journal of Petroleum Engineering* **2014**

[8] Haberer J. New solutions may ease hardbanding controversy. Drilling contractor. 2000.
[9] Mark Haning, James Doherty, House A, 3D Non-Linear FEA to Determine Burst and Collapse Capacity of Eccentrically Worn Casing. CEED Seminar Proceedings 2012; Crawley: University of Western Australia

Impact of the electric compressor for automotive air conditioning system on fuel consumption and performance analysis

A A Zulkifli[1], A A Dahlan[1], A H Zulkifli[1], H Nasution[2], A A Aziz[1], M R M. Perang[1], H. M. Jamil[1] and M N Misseri[1]
[1]Automotive Development Centre, Universiti Teknologi Malaysia, Skudai, 81310 Johor Bahru, Johor, Malaysia
[2]Department of Mechanical Engineering, Faculty of Industrial Technology, Universitas Bung Hatta, Padang 25132, Sumatera Barat, Indonesia

Email: henry@fkm.utm.my

Abstract. Air conditioning system is the biggest auxiliary load in a vehicle where the compressor consumed the largest. Problem with conventional compressor is the cooling capacity cannot be control directly to fulfill the demand of thermal load inside vehicle cabin. This study is conducted experimentally to analyze the difference of fuel usage and air conditioning performance between conventional compressor and electric compressor of the air conditioning system in automobile. The electric compressor is powered by the car battery in non-electric vehicle which the alternator will recharge the battery. The car is setup on a roller dynamometer and the vehicle speed is varied at 0, 30, 60, 90 and 110 km/h at cabin temperature of 25°C and internal heat load of 100 and 400 Watt. The results shows electric compressor has better fuel consumption and coefficient of performance compared to the conventional compressor.

1. Introduction

The belt-driven compressor is the mechanical type of the compressor and this is a concept of the conventional air conditioning (AC) system in automobile. The compressor is connected to the engine crankshaft through belting system. The compressor works by using the energy from the engine. Therefore workload is applied to the engine due to the use of this type of compressor in AC system [1-13]. This might reduce the performance of the engine. To overcome such problem, changing the power source for the AC system to the electrically driven will greatly benefit in reducing the fuel consumption and also increases the performance of the engine [1, 2, 5-8].

Fuel consumption of a vehicle can be affected by up to 20% for AC system [1]. This number can be reduced by lowering the energy consumed by the compressor. This study proposed to retrofit an electric compressor to the current conventional AC system in conventional vehicle.

This system is proposed so that the compressor speed is independent to the engine crankshaft speed, thus better energy efficiency and temperature control inside vehicle cabin [2]. The electric compressor is powered by direct current (dc) so that it does not need a power inverter that alternate current compressor would require. Electric compressor is easy to maintain and install to a compact system and offers low energy consumption [2]. Previous study of such as the compressor that creates a mobile heating, ventilation and air conditioning (HVAC) system for operation with the engine off. It is a solution to disable idle stop to reduce fuel consumption by temporarily switching off the engine into a sleep mode when the vehicle is at a stop [4]. Another study developed small, lightweight and high-efficiency electric compressors. The performance of the dc compressor for automobile AC system shows that the electric compressor gives better performance than that the conventional compressors despite the cabin temperature distribution and cooling effect are a letdown by the dc compressor [6]. Guyonvarch et al. [8] carried out fuel consumption for conventional and electrically driven AC compressor. However, the comparison is only possible to a limited extent. This is due to the AC system was not specifically designed for the 12V onboard voltage system from the conventional system. The research studied 42V electric AC system for low emission, architecture, comfort and safety of the next-generation vehicles.

The objective of this paper is to investigate the coefficient of performance (COP), cabin temperature fuel usage of electric compressor retrofitted into conventional vehicle air conditioning system. The results are compared to belt-driven compressor.

2. Experimental Set Up

Experimental approach is used for this study. A compact hatchback vehicle with 1.3 liter gasoline engine is retrofitted with the electric compressor. The electric compressor is powered by the car battery which is charged by the alternator. Variable parameters are shown in table 1 and the schematic diagram for electric compressor is shown in figure 1.

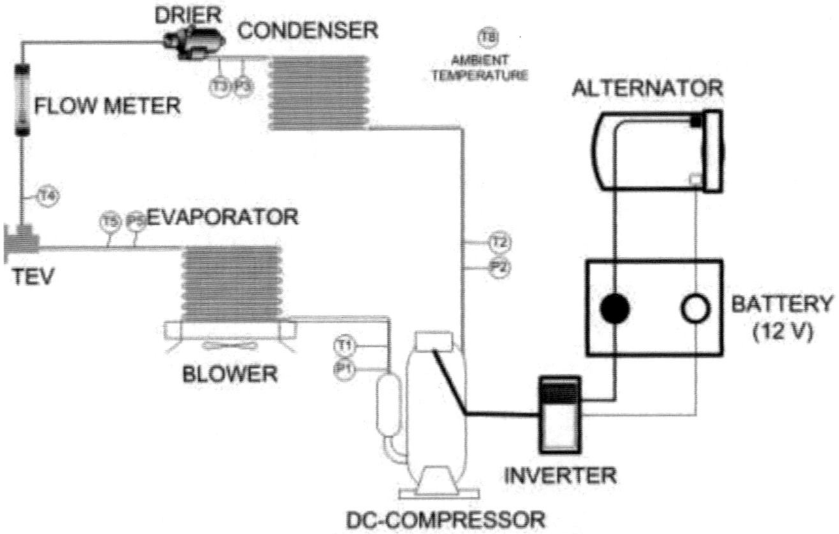

Figure 1. Schematic diagram for electric compressor vehicle air conditioning system.

Table 1. Variable parameters.

Parameter	Range of variation
Engine speed (km/h)	0, 30, 60, 90, 110
Internal Load (Watt)	100, 400

The temperature, pressure and mass flow rate of the system is taken at the points shown in figure 1. The temperature is taken using thermocouples connected to the temperature data logger. The pressure and mass flow rate of the air conditioning system is taken manually using pressure gauge and flow meter, respectively.

The vehicle is setup on roller dynamometer to emulate the flat road condition. The internal heat load is to represent the heat coming from passengers inside vehicle cabin which is one passenger emits 100 W of heat. The refrigerant used is HFC-R134a. Figure 2 shows the vehicle on the roller dynamometer and the air conditioning control setting.

The electric compressor is working at maximum speed 2500 rpm. The control system for both electric compressor and conventional compressor is on/off controller placed in the evaporator. All tests are started at 30°C. Fuel flow meter is used to measure the usage of fuel through the one hour test period.

(a) (b)

Figure 2. (a) Vehicle on roller dynamometer and (b) Temperature and blower setting.

3. Mathematical Modelling

An efficient AC system is the one that manage to remove heat greatly with less energy needed. The performance of AC system can be expressed in term of coefficient of performance, COP. In term of the refrigerating effect, Q_e and work input, $W_{net,in}$, the COP can be expressed as:

$$\text{COP} = \frac{\text{Desired Output}}{\text{Required Input}} = \frac{Q_e}{W_{net,in}} = \frac{h_1 - h_4}{h_2 - h_1} \tag{1}$$

where h_1, h_2 (kJ/kg) are the enthalpy at the compressor inlet and outlet respectively, h_4 (kJ/kg) is the enthalpy at the evaporator inlet.

The formula used to find the percentage of fuel consumption reduction is:

$$\%\ \text{Fuel Consumption Reduction} = \frac{Fuel_{DC} - Fuel_{Conventional}}{Fuel_{Conventional}} X\ 100 \tag{2}$$

4. Results and Discussion

Figure 3 (a) and (b) show the fuel consumption against vehicle speed of conventional compressor and electric compressor. The electric compressor consumes less fuel compared to conventional compressor up to 18%. This is because there is fewer loads to the vehicle come from the compressor and better energy efficiency. Increase internal heat load would cause fuel increment to both of the system.

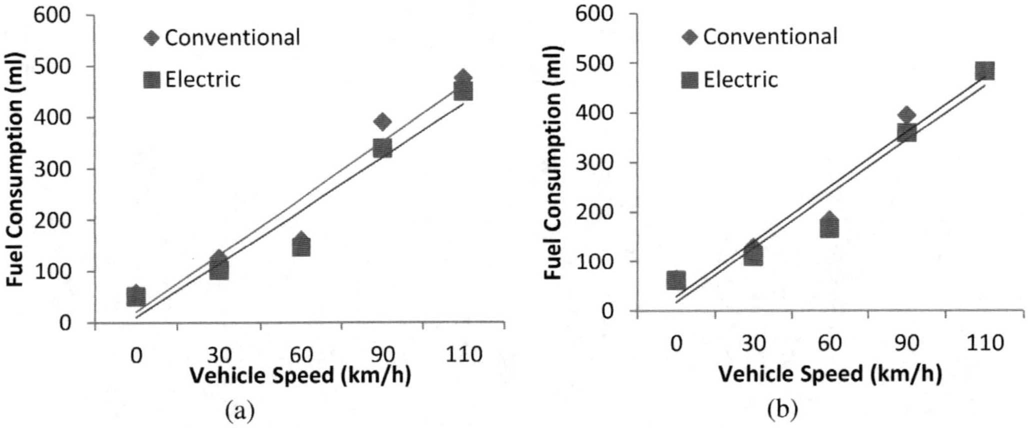

Figure 3. Fuel consumption against vehicle speed;
(a) 100 W internal heat load and (b) 400 W internal heat load.

Figure 4 (a) and (b) show the temperature responses against vehicle speed. The temperature of electric compressor is always at the temperature setting nevertheless the vehicle speed. This is because the electric compressor is independent from the engine speed. Conventional compressor on the other hand is dependent to the vehicle speed causing the compressor to work at higher speed and hence, the temperature cannot be controlled efficiently. This causes waste of energy and hence higher energy consumption.

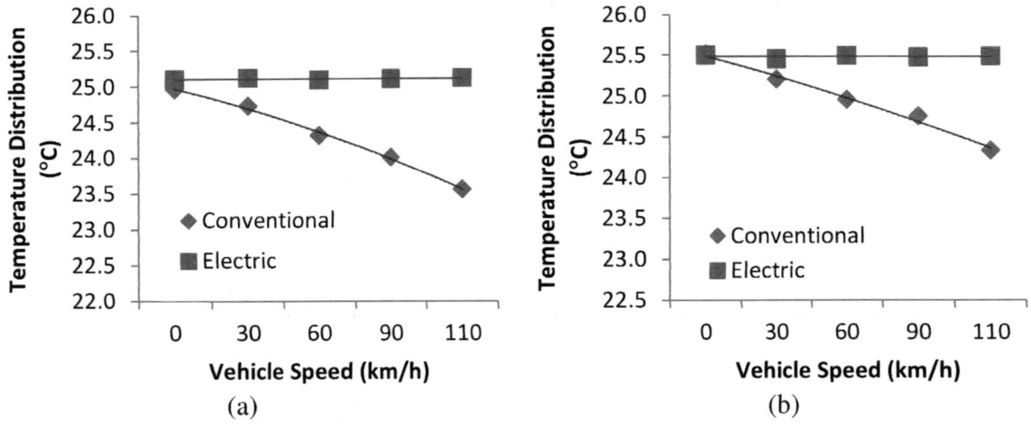

Figure 4. Temperature distributions against vehicle speed.
(a) 100 W internal heat load and (b) 400 W internal heat load.

Figure 5 (a) and (b) show the COP of the systems against vehicle speed. Electric compressor shows better COP because of less energy needed to compress the refrigerant of the system. The COP for 400 W of internal heat is higher compared to 100 W. COP for conventional compressor decreasing as the vehicle speed increasing because the compressor consumes higher energy as the compressor speed also increasing.

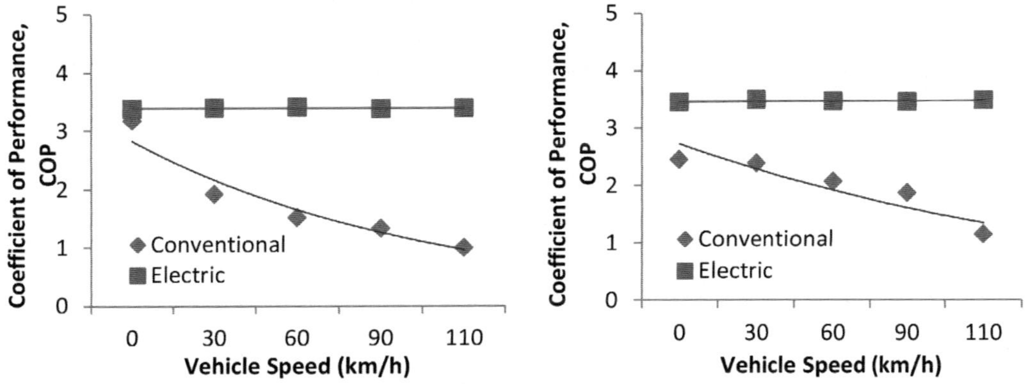

Figure 5. Vehicle speed against coefficient of performance.
(a) 100 W internal heat load and (b) 400 W internal heat load.

5. Conclusions

This research presents the cabin temperature responses by the electrically driven compressor is more accurate to the temperature set point even though the belt-driven compressor temperature distribution is lower. The compressor performance of electrically driven compressor is higher than the belt-driven compressor about 15 to 54%. The experiment shown that the fuel reduction percentage by the electrically driven compressor to the belt-driven compressor about 5 to 14%.

We can conclude that using the direct current compressor has high possibility in reducing the load of engine system and will result in the higher coefficient of performance and lowering the fuel consumption. This can help in fuel saving of the vehicle.

Acknowledgements

The research was supported financially by Universiti Teknologi Malaysia: Knowledge Transfer Program (KTP) No. R.J130000.7809.4L509 and Automotive Development Center, Ministry of Education (MOE) Malaysia. Their guidance and assistance are gratefully acknowledged.

References

[1] Farrington R, Rugh J, Impact of vehicle air-conditioning on fuel economy, tailpipe emissions, and electric vehicle range. Earth technologies forum; 2000.
[2] Kim S, Lee G, Lee J, Hong J 2010 *International Journal of Automotive Technology* **11** 277-82
[3] Bharathan D, Chaney L, Farrington R B, Lustbader J, Keyser M, Rugh J 2007 *An Overview of Vehicle Test and Analysis from NREL's A/C Fuel Use Reduction Research* National Renewable Energy Laboratory)
[4] Martin G, Galloway C S, Collier S J, Diaz G C. Development of a mobile HVAC system for operation with engine off. SAE Technical Paper, 2004.
[5] Makino M, Ogawa N, Abe Y, Fujiwara Y. Automotive air-conditioning electrically driven compressor. SAE Technical Paper, 2003.
[6] Henry N, Yamani M, Sumeru K, Performance Study of DC Compressor for Automotive Air Conditioning System. Advanced Materials Research; 2013: Trans Tech Publ.
[7] Maraisa C, Holdack-Janssena H, van Niekerkb T 2010
[8] Guyonvarch G, Aloup C, Petitjean C, De Savasse A D M. 42 V Electric Air Conditioning Systems (EA/CS) for Low Emissions, Architecture, Comfort and Safety of Next Generation Vehicles. SAE Technical Paper, 2001.
[9] Heydari A, Jani S. Entropy-minimized optimization of an automotive air conditioning and HVAC system. SAE Technical Paper, 2001.

[10] Tian C, Li X 2005 *Energy conversion and management* **46** 2718-38
[11] Tian C, Xu H, Zhang L, Li X 2009 *Applied Thermal Engineering* **29** 2824-31
[12] Ratts E B, Brown J S 2000 *Applied Thermal Engineering* **20** 1039-58
[13] Kaynaklı Ö, Horuz I 2003 *International communications in heat and mass transfer* **30** 273-84

Performance investigation of a waste heat driven pressurized adsorption refrigeration cycle

K Habib[1]
Department of Mechanical Engineering, Universiti Teknologi PETRONAS, Bandar Seri Iskandar, 31750 Tronoh, Perak, Malaysia

E-mail: khairul.habib@petronas.com.my

Abstract. This article presents performance investigation of a waste heat driven two bed pressurised adsorption refrigeration system. In this study, highly porous activated carbon (AC) of type Maxsorb III has been selected as adsorbent while n-butane, R-134a, R410a, R507a and carbon dioxide (CO2) are chosen as refrigerants. All the five refrigerants work at above atmospheric pressure. Among the five pairs studied, the best pairs will be identified which will be used to provide sufficient cooling capacity for a driving heat source temperature above 60°C. Results indicate that for a driving source temperature above 60°C, AC-R410a pair provides highest cooling capacity while AC-CO2 pairs works better when the heat source temperature falls below 60°C.

1. Introduction

One of the most pestering environmental problems that the refrigeration industry is intrigued is ozone depletion. Thermally driven sorption based cooling systems are gradually emerging as environmentally friendly alternatives to conventional vapor compression based refrigeration cycles. Since the detection of ozone holes in the stratosphere, caused by the CFCs (chlorofluorocarbons) and HCFCs (hydro- chlorofluorocarbons) of vapor compression based coolers, interest on adsorption cooling systems have increased a lot [1]. The attractive features of these systems are; they have the ability to operate by low temperature driving heat sources typically below 100°C, thus makes it is possible to use solar energy and industrial waste heat as the driving heat sources. The construction of these systems is simpler, as there are almost no moving parts, and they require low electricity usage. Furthermore, it is possible to operate these systems with heat source temperature as low as 50°C if multi-stage regeneration is performed and an improvement in the waste heat recovery efficiency can be attained by using multi-bed scheme [2].

Thermally powered adsorption cooling systems could be operated in both partial vacuum and pressurized conditions. Of the adsorbent-refrigerant pairs working at sub-atmospheric conditions are silica gel-water [2], zeolite-water [3], activated carbon-methanol [4], activated carbon fiber-ethanol [5] pairs etc. However, the main drawback of vacuum based adsorption cooling cycles is that the leakages will drastically reduce its performance. In addition, the high vacuum operation demands advanced technology and special designs. Several other studies have been conducted to investigate the possibility of using different types of adsorbent-refrigerant pairs for pressurized adsorption cooling systems [6-10]. For the above mentioned perspectives, the main objective of the study is to analyze the performance of a two bed adsorption cooling cycles at pressurized conditions employing highly porous activated carbon of type Maxsorb III as adsorbent paired with five different refrigerants namely n-butane, R-134a, R410a, R507a and carbon dioxide (CO_2) . The performance of the best pair will be investigated to analyze cooling capacity and COP variations by varying regeneration, cooling and chilled water temperatures.

2. Working principle of adsorption chiller

In the present study, a two bed adsorption cooling cycle has been studied. The schematic diagram of the adsorption cycle is shown in figure 1. The chiller is composed of four heat exchangers namely two adsorber/desorber heat exchangers, an evaporator and a condenser. Four refrigerant valves are used for the operation four pre-determined processes namely, adsorption, pre-heating, desorption and pre-cooling. As can be seen from figure 1, refrigerant evaporates from the evaporator and adsorbed by the adsorber bed through refrigerant valve V4. At the same time desorbed refrigerant goes to the condenser through refrigerant valve V2 which is known as the desorption-condensation process. The desorber bed is powered by solar thermal heat which is the only heat source of the studied system. The details of the working principle of adsorption chiller have been described in detail elsewhere [1].

3. Adsorbent-refrigerant pair

In the present study, activated carbon of type Maxsorb III has been selected as adsorbent while n-butane, R-134a, R410a, R507a and carbon dioxide (CO2) are chosen as refrigerants. Table 1 lists the thermo-physical properties of some of the activated carbon used as adsorbents. It is observable from the table 1 that Maxsorb III has the highest surface area and pore volume compared with the others. Figure 2 shows the SEM photograph of Maxsorb III.

Table 1. Thermophysical properties of various activated carbons.

Adsorbent	BET Surface area (m^2/g)	Pore volume (cc/g)	Source
ACF (A-20)	1900	1.028	[7]
Granular AC	1150	0.43	[8]
AC of type Maxsorb III	3120	1.7	[9]

4. Mathematical modeling

4.1. Adsorption isotherms

Dubinin-Astakhov (D-A) model is used to estimate the equilibrium uptake of all the pairs.

$$W = W_0 \exp\left[-\left\{\frac{RT}{E}\ln\left(\frac{p_s}{p}\right)\right\}^n\right] \tag{1}$$

with W = xVa , where x is the mass concentration (kg-kg-1) and Va is the adsorbed phase specific volume. Equation (1) is used to evaluate the adsorption uptake of AC-n-butane, AC-R134a, AC-R507a and AC-R410a pairs. For AC-CO_2 pair, equation (2) can be written without the adsorbed phase volume which is in the following form,

$$x = x_0 \exp\left[-\left\{\frac{RT}{E}\ln\left(\frac{p_s}{p}\right)\right\}^n\right] \tag{2}$$

In the case of D-A, equation (2), the adsorbed phase volume is calculated as [11],

$$V_a = V_b \exp\left[\alpha\left(T_{des} - T_b\right)\right] \tag{3}$$

where, $\alpha = \dfrac{\ln\left(\dfrac{b}{V_b}\right)}{T_c - T_b}$ and b is the Van der Waal's volume ($= RT_c/8P_c$)

The values of Tc, Pc and Tb of n-butane, R134a, R410a, R507a and CO2 can be found elsewhere [12]. Table 2 lists the adsorption characteristics of the various pairs studied.

In addition, the D-A model has to be modified to accommodate the pseudo-vapour pressure of carbon dioxide since its critical temperature is as low as 304.12K. Hence, the saturation pressure is given by [11],

$$P_s = \left(\frac{T}{T_c}\right)^2 P_c \tag{4}$$

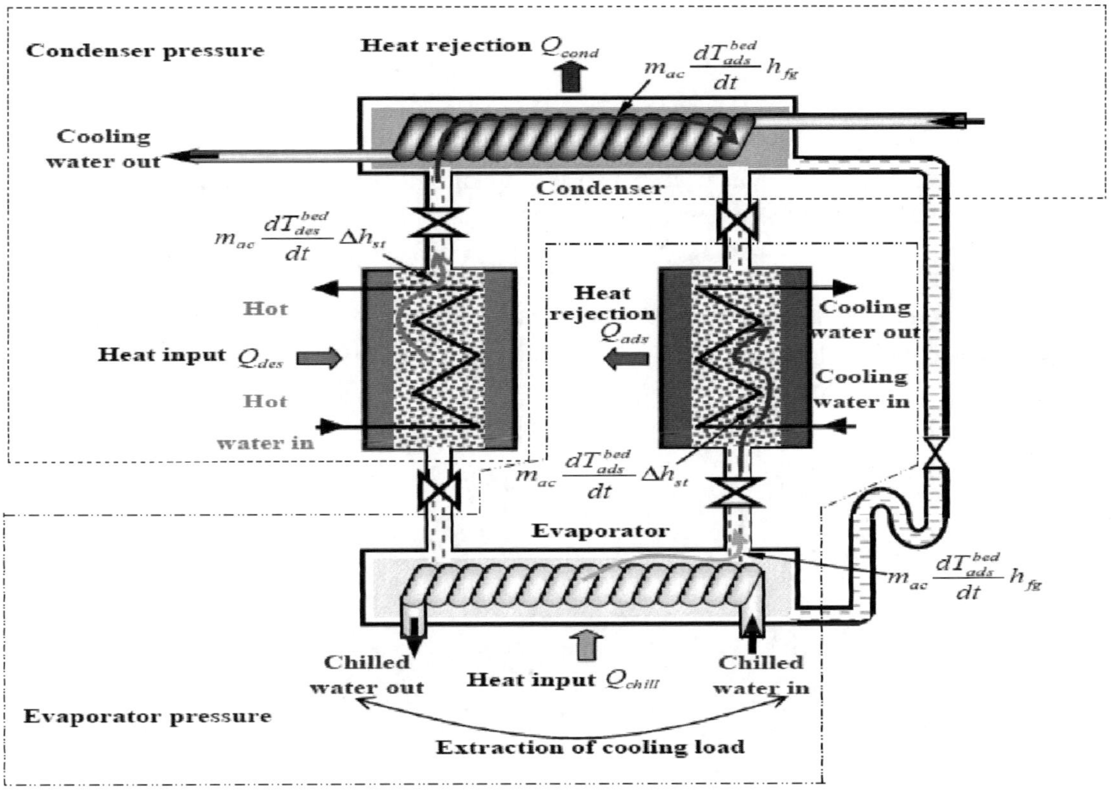

Figure 1. Schematic diagram of two bed adsorption chiller.

Table 2. Adsorption characteristics of five various parameters on activated carbon.

Refrigerant	R134a [13]	R410a [13]	*n*-butane [14]	R507a [13]	CO_2 [10]
W_0 (m^3/kg)	2.22	2.07	0.8	2.05	1.67
E (kJ/kg)	71.87	72.37	300	76.34	94.502
n	1.29	1.36	1.05	1.34	1.18

4.2. Heat of adsorption

The heat of adsorption at a constant concentration can be expressed as:

$$\Delta h_{st} = -\frac{R\partial(\ln P)}{\partial\left(\frac{1}{T}\right)} \tag{5}$$

The specific cooling capacity (SCC) is achieved at the evaporator and is given by,

$$SCC = (x_{\max} - x_{\min})\left[h_{fg} - \int_{T_{eva}}^{T_{con}} C_{p,ref}\,dT\right] \tag{6}$$

where xmax indicates the amount of refrigerant uptake on AC as a function of adsorption temperature and evaporator pressure, xmin denotes the amount of refrigerant uptake on AC during desorption at the desorption temperature and condenser pressure, hfg represents the latent heat of evaporation at the cooling load temperature and Cp shows the specific heat capacity of refrigerant.

The coefficient of performance (COP) of the adsorption cooling cycle can be defined as:

$$COP = {SCC}/{Q_h} \tag{7}$$

where Q_h is the summation of heat added to the adsorbent, Q_{adb} and the heat added to the refrigerant, Q_{ref}, as represented below:

$$Q_h = Q_{adb} + Q_{ref} \tag{8}$$

The heat added to the adsorbent and refrigerant are further derived as shown in equation (9) and equation (10) respectively:

$$Q_{adb} = \int_{T_{e,a}}^{T_{e,d}} C_{p_{adb}}\,dT \tag{9}$$

$$Q_{ref} = x_{\max}\int_{T_{e,a}}^{T_{s,d}} xC_{p_{ref}}\,dT + \int_{T_{s,d}}^{T_{e,d}} xC_{p,ref}\,dT + \int_{x_{\max}}^{x_{\min}} \Delta h_{st}\,dx \tag{10}$$

Figure 2. SEM photograph of Maxsorb III.

5. Results and discussion

5.1. Ideal cooling cycle

Using the isotherm data, the conceptual ideal adsorption cooling systems of AC-R134a, AC-R507A, AC-R410a, AC-n butane and AC-CO_2 are shown in figure 3. One can understand from figure 3 that for the same operating temperature conditions, the difference between the maximum and minimum uptake, Δx = xmax – xmin, of AC-R410a cycle is higher than that of other cycles. The Δx values for AC-R410a cycle is 0.4 kg/kg, while for AC- n butane cycle the Δx value is 0.08 kg/kg.

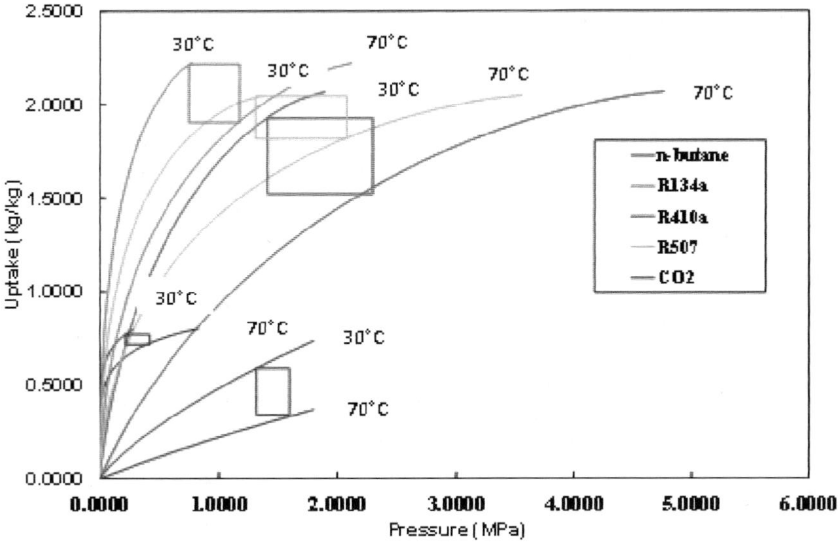

Figure 3. Adsorption isotherms of AC-n butane, AC-R134a, AC-R410a, AC-R507a and AC-CO_2 pairs.

5.2. Heat transfer fluids

Figure 4 shows the effect of regeneration temperature on the specific cooling effect (SCC) for all the pairs. From figure 4, it is apparent that the increasing regeneration temperature yields in a linear increase in the SCC for all five various adsorbent-refrigerant pairs. Maximum SCC value of 54 kJ/kg is achieved by AC-R410a pair at when the regeneration temperature is 80°C. For a temperature range of 60°C to 80°C, the AC-R410a pair shows the highest SCC followed by AC-R410a, AC-CO_2, AC-R134a, AC-R507a and lastly AC-n-butane. AC-n-butane pair shows the lowest SCC value of 9 kJ/kg at driving source temperature of 60°C.

Figure 5 illustrates the effect of regeneration temperature on COP for all the pairs. It is observed that the COP value of the refrigeration cycles increase sharply with increasing regeneration temperature up to 75°C. However, it can be seen that COP exhibits an asymptotic behaviour when the regeneration temperature approaches 80°C. This is due to the higher heat intake requirement when the difference between heat sink temperature and heat source temperature grows significantly large. From the figure, it is obvious that the AC -R410a pair provides the highest COP for regeneration temperature range of 70°C to 80°C. However, for a relatively lower regeneration temperature, the COP value drops lower than that of AC -CO_2 pair. For regeneration temperature lower than 70°C, the COP of AC -CO_2 pair is 0.1 compared 0.095 of AC -R410a pair.

Figure 6 shows the effects of cooling water temperature on the SCC of all the pairs. As opposed to the case of regeneration temperature and evaporator temperature, the increase in cooling water temperature results in a linear decrease in the SCC of the refrigeration cycle, as shown in figure 6. AC -R410a pair provides the highest SCC value among all assorted adsorbent-refrigerant pairs for a temperature range of 10°C to 30°C.

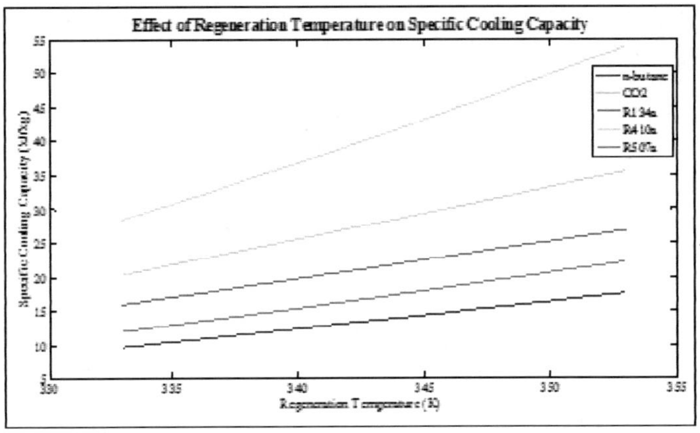

Figure 4. Effect of regeneration temperature on specific cooling capacity for all the pairs.

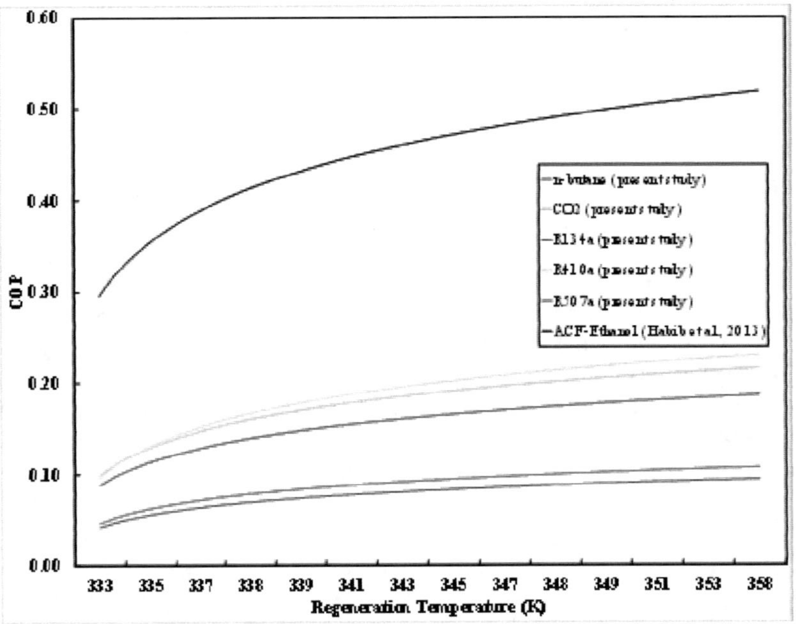

Figure 5. Effects of regeneration temperature on COP.

IOP Conf. Series: Materials Science and Engineering **100** (2015) 012029 doi:10.1088/1757-899X/100/1/012029

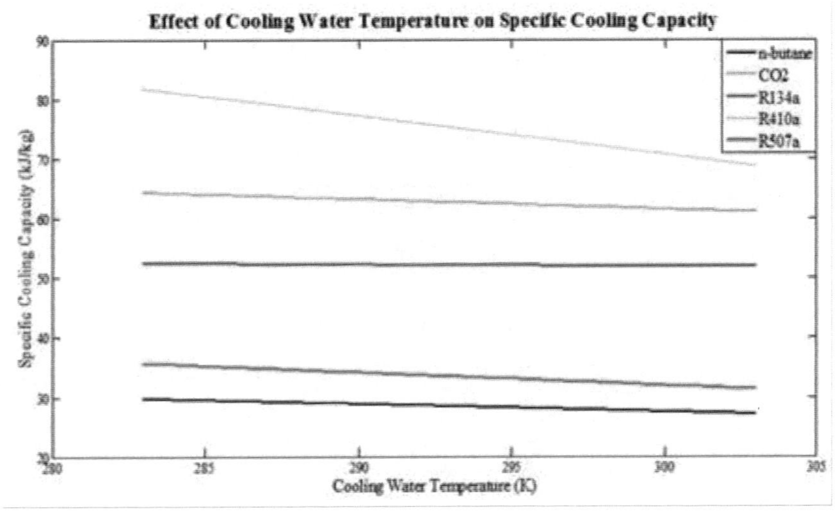

Figure 6. Effects of cooling water temperature on specific cooling capacity.

As for the case of COP, it is observed from figure 7 that the COP value of the refrigeration system decreases with increasing cooling water temperature but the decrement is insignificant as Tcw approaches 20°C. It is observable from figure 7 that AC - R410a pair provides the highest COP for cooling water temperature of 10°C to 30°C.

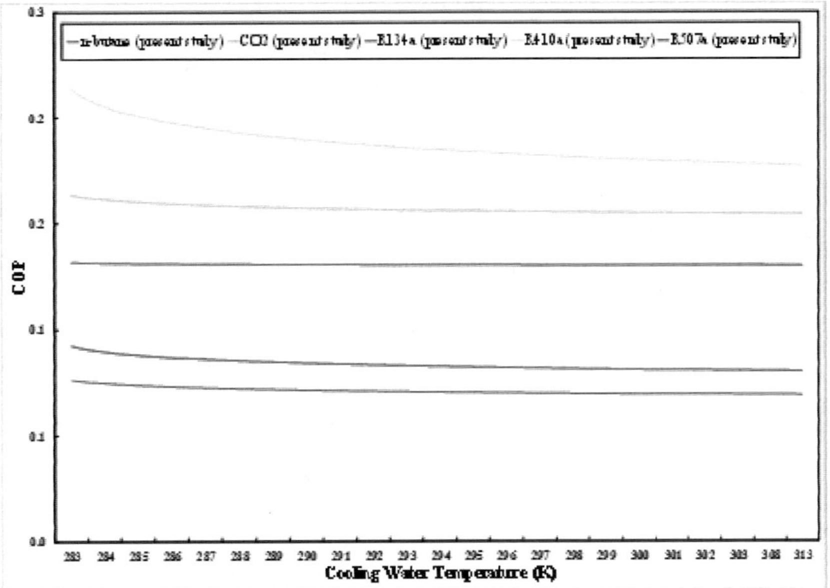

Figure 7. Effects of cooling water inlet temperatures on COP.

6. Conclusion

The aim of the present study is to analyse the performance of pressurized adsorption refrigeration cycle using activated carbon of type Maxsorb III as adsorbent and five various refrigerants. The main conclusions are as follows:

i. AC-R410a pair exhibits the highest adsorption uptake compared to the other pairs.

ii. AC-R410a pair shows higher values of cooling capacity with the variation of regeneration temperature of 70°C to 80°C and evaporator temperature range of 5°C to 20°C.
iii. When the regeneration temperature and evaporator temperature drop below 70°C and 5°C respectively, the pair of AC-CO_2 proves to be a better option.
iv. As for the case of varying cooling water temperature, the pair of AC-R410a provides the best cooling capacity and COP value, for cooling water temperature ranging from 10°C to 30°C.
v. The present study may provide useful information in designing pressurized based adsorption cooling cycles driven by solar energy or waste heat.

Acknowledgement
The authors of current work wish to thank Universiti Teknologi PETRONAS for their facilities and financial support.

References

[1] Wang D C, Wang Y J, Zhang J P, Tian X L, Wu J Y 2008 *Energy Conversion and Management* **49** 1063-73
[2] Habib K, Choudhury B, Chatterjee P K, Saha B B 2013 *Energy* **63** 133-41
[3] Myat A, Kim Choon N, Thu K, Kim Y-D 2013 *Applied Energy* **102** 582-90
[4] Pons M, Guilleminot J 1986 *Journal of solar energy engineering* **108** 332-7
[5] El-Sharkawy I I, Kuwahara K, Saha B B, Koyama S, Ng K C 2006 *Applied Thermal Engineering* **26** 859-65
[6] Loh W S, Saha B B, Chakraborty A, Ng K C, Chun W G 2010 *Journal of Thermal Science and Technology* **5** 252-65
[7] Saha B B, Chakraborty A, Koyama S, Yoon S-H, Mochida I, Kumja M, et al. 2008 *International Journal of Heat and Mass Transfer* **51** 1582-9
[8] Askalany A A, Saha B B, Ismail I M 2014 *Applied Thermal Engineering* **72** 237-43
[9] Saha B B, El-Sharkawy I I, Habib K, Koyama S, Srinivasan K 2008 *Journal of Chemical & Engineering Data* **53** 1872-6
[10] Saha B B, Jribi S, Koyama S, El-Sharkawy I I 2011 *Journal of Chemical & Engineering Data* **56** 1974-81
[11] Saha B B, Koyama S, El-Sharkawy I I, Habib K, Srinivasan K, Dutta P 2007 *Journal of Chemical & Engineering Data* **52** 2419-24
[12] REFPROP 2002 *Reference fluid thermodynamics and transport properties* **23** (NIST: Standard Reference Data Bases)

IOP Conf. Series: Materials Science and Engineering **100** (2015) 012030 doi:10.1088/1757-899X/100/1/012030

The effect of substantive parameters on the efficiency of Archimedes screw microhydro power: a review

A Nurul Suraya[1], N M M Ammar[1] and J Ummu Kulthum[1]
[1]Faculty of Mechanical Engineering, Universiti Malaysia Pahang, 26600 Pekan, Pahang, Malaysia.

Email: surayanurul@gmail.com

Abstract. Due to the increasing demands in electricity and decreasing in fossil fuels sources, then hydropower are being developed and most of the project involve with large dams construction and may cause the people live surrounding to be flooded. Hence, the construction of Archimedes screw runner blade turbine can be the best option to generate energy without cause too much environmental impact like a fish friendly turbine and low maintenance cost. This review focus on the parameters that affected the performance of turbine. Besides the review also presented a methodologies based on parameters studied by previous researcher and from that review, it has been found that the uses of external parameter have an effect on the internal parameter in terms of efficiency. In further investigation, another parameter such as gap leakage, velocity of water, blade thickness should be considered to investigate the relationship of efficiency.

1. Introduction

Renewable energy has become one of the important medium for future needs as demand for fossil fuels keep increasing progressively [1]. Renewable energy sources that are commonly used by people nowadays to solve diminishing of fossil fuels include as solar, geothermal, biomass, biofuel wind, water and fuel cells. All these sources of energy will not diminished as it will always be renewed naturally and have high potential to produce a clean environment and reduce the greenhouse gas emission as well as produce a little of waste products that can give unscrupulous effect to earth.

About 71% of this earth is covered by water. So that means the innovation on generating the power from water sources can be expand rapidly. There are several ways on how to generate energy from the water source. For example, energy can be created by using the tidal waves, ocean waves and hydropower. The hydropower can be classified into four categories in terms of power output, such as large, small, mini, micro and Pico hydropower. Table 1 below shows the classification of hydropower plant. Large hydropower can achieved about more than 1000kW of electricity, small hydropower range is of 500kW to 1000kW, mini hydropower have capacity between 100kW to 500kW and micro hydropower installed that can have output power between 5kW to 100kW [2, 3]. Micro hydropower power output can be used in rural areas specifically for small or medium houses.

Table 1. Classification of Hydropower plant [2, 3].

Classification of Hydropower	Power Output
Large	More than 1000kW
Small	Installation range between of 500 to 1000kW
Mini	Power output of 100 to 500kW
Micro	Installation range of 5kW to 100kW

As the electricity demand increasing year by year, the hydropower is one of the best alternative way to generate energy and have many advantages in many aspects which is highly stable, economical, accomplished of generating reliable energy needed, minimal civil works, suitable on low head scheme. Classification of Hydropower can also be describes in terms of head difference and flow rate required. Figure 1 below explains the graph of head, H (m) difference against the flow rate, Q (m^3/s) for each hydropower [4, 5].

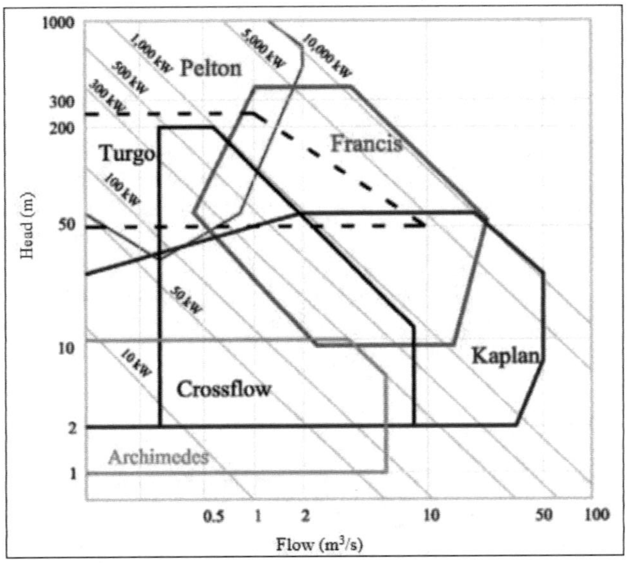

Figure 1. Classification of hydropower for head vs. flow rate [4, 5].

In this study, the main purpose of the research is to determine the suitable hydropower to install in rural areas, specifically the remote areas [6]. The study on scope of the area is more focus in the rural region in Malaysia. Since the opportunity of exploring the low head hydropower for generation power output in Malaysia have been given slight responsiveness, the introduction of Archimedes micro hydro screw could be a way to response to the assessment. The Archimedes micro hydro power assists the rural topology with its small head range and lower flow rate. The benefit of selecting Archimedes screw runner also is that environment impact has very small impression on the efficiency of the turbine. Although the output electricity only can achieve about less than 100kW, it may become one of best ways in generating electricity to small house and small resorts at the rural areas.

The aspect of generating power with small flow rate and head arrangement is highly suitable to install Archimedes Screw micro hydro power in rural areas, specifically the remote area. Water sources such as river, pond, and water fall and river stream condition area in generous in remote area [7]. According to figure 1 above, the Archimedes screw Micro Hydro Power head difference ranges from 1 to 10 meter in height while the flow rate required is approximately between speeds of 0.5 m^3/s to 6 m^3/s. Flow rate on the current flow rate on River Pahang in Malaysia can generate to the flow rate of that amount and remote area can produce heads as required [8].

Each turbine may produce different efficiencies if the topology of the area and climate change. The Archimedes screw is also similar. It can be describe as a gravity machine [9].It is comparable to waterways turbine but it can produce higher efficiency as presented in figure 2 below compared to a water wheel and a low speed radial turbine for smaller and moderate flow rate. In fact, hydro screw can sustain its efficiency at high rates. Even without the disturbances from torrential river waterway[9]. As for environmental effect, the Archimedes screw is much proven friendly toward marine animals. The slow rotational speed as well as the gap between the blades produces small hydrodynamic forces, and creates minimal contact with moving obstacles[9].

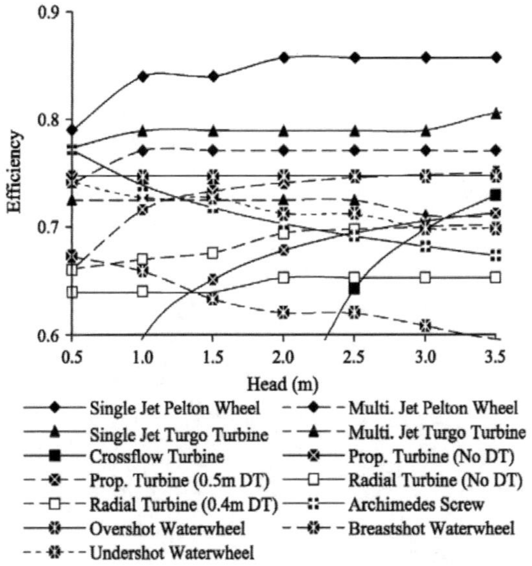

Figure 2. Comparison of low head turbine plots in Effeciency vs Head (m) [5, 10].

Thus, the development of renewable energy in contact of micro hydro power is one of the important source of power generation, but there are such problems that disturb the performance of turbine and at the same time effect the quality of power that can be created by turbine and it is important to study the suitable related parameter based on current situation in order to achieve some improvement in energy production.

2. Controlling parameter

The benefit of improving the design of Archimedes screw runner is to improve its efficiencies and its power output. However, there are many criteria displayed in figure 3 below that are needed to develop better efficiencies. These criteria are the external parameter (the outer radius of the blade R0, length of the blade L, Head difference H, inflow depth d0 and slope of the penstock, α) and internal parameters (hub diameter Dh, inner radius Ri, number of blades N, volume of bucket V and the pitch of the blades P). The outer parameters are the controlled parameters of the Archimedes screw runner and the flow rate of the water. The internal parameters are the required characters to optimize the enactment of the screw [11]. In this study, the parameter of the research is to emphasis on the study of the outer diameter of the penstock Do, the slope α, head difference H and the length, L as shown in figure 3(a) and 3(b) to control the flow of water thus allow the optimization on design of Archimedes screw runner for micro hydro power. This outer parameter can determine the relationship of internal parameter on the effect of efficiencies.

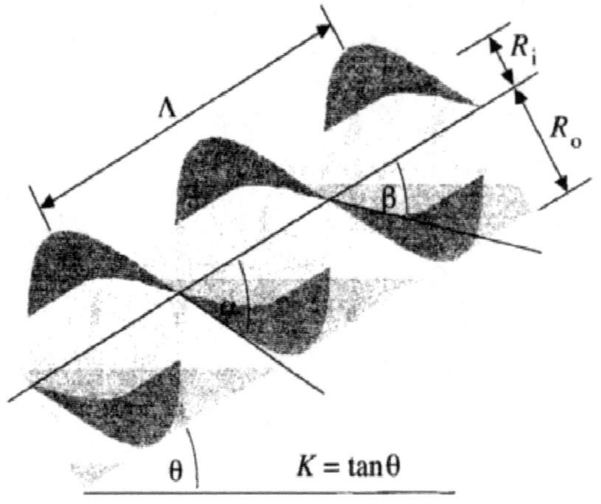

Figure 3(a). Blade design of Archimedes screw [11].

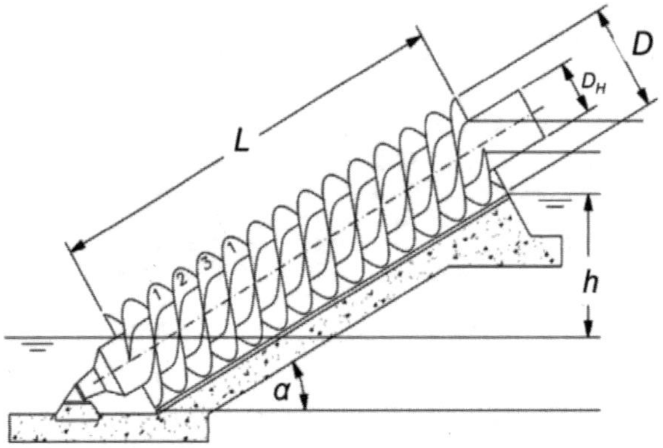

Figure 3(b). Side elevation of screw blade [12].

Head difference exists as the level of water is differ between top and bottom reservoir. The water flow from the top reservoir and forming a bucket which make the blade rotate along the length of the Archimedes screw blade [13] while the inclination of blade has contribute to the formation of hydrostatics force where the downstream pressure that acting on blade is smaller compared to the upstream level [12]. Because of the hydrostatic force, the blade is rotating (which contribute to kinetic energy) and make turbine turns generator rotor. Lastly, the electrical energy is generated.

3. Parameter evaluation

Several methods of analytical were conducted by controlling the head difference, H, length of the blade, L and the slope of the penstock, α [11, 12, 14, 15] . Some researcher conducted the analytical method by abandoning the slope of the penstock to determine the efficiency of the blade [12]. The first parameter that commonly used as independent parameter is slope inclination angle. Muller and Senior, Yulistiyanto et al., and Stergiopoulou and Kalkani have been considered the effect of slope angle towards the performance of the Archimedes screw runner blade turbine. All of the researchers possess a dissimilar approach on how to collect the data relating to efficiencies. Müller and Senior [12] proposed that the efficiency of the blade is not consistent with the slope angle . However, Yulistiyanto et al., determined the increasing slope angle is increase with the efficiency, but have optimum efficiency occurred at 350 angle and after 350 angle, efficiency starts to decline [16]. William David Lubitz [17] obtained the results equally with Muller which is lower slope angle is better in performance. Stergiopoulou and Kalkani compared angle between 220, 330 and 440 angle of slope and study on the behavior of hydrodynamics performance and only shows the preliminary results of simulation [18]. Lubitz et al., and Muller have an identical findings where lower angle have a better efficiencies while Yulistiyanto et al., is vice versa.

In head difference aspects, there are two researchers which focus on this parameter. Both of the researchers use a dissimilar methodology. Stergiopoulou and Kalkani have preliminary results which study more on the effected head in terms of hydrodynamics behavior than the efficiency of turbine [18]. On the other hand, Hendra and Indriani consider the head difference with three values of head and compare between these 450mm, 550mm and 650mm on the prototype [19]. The experimental results from Hendra and Indriani formularize that head is directly proportional with the power output. These can be believed that the head difference plays an important role in the turbine's performance.

Additional parameter that leads to the variation performance of the turbine is the number of turns and the outer diameter of the blade. Based on the study made by Rorres, Muller and Senior, Stergiopoulou and Kalkani and Raza et al., study on the number of turns but in different methods as summaries in table 2. Rorres conclude in terms of volume of water in cycle which increases with the number of blades. Muller identified that efficiency dependable with number of turns while Stergiopoulou and Kalkani studied the behavior of hydrodynamic using one or more blades with various values of flow input. Both Muller and Senior and Raza et al. have the similar results in number of turns, so it can be summaries that to have better efficiencies, then the number of turns should rise depending on the site.

Another parameters which have an important role as controlling parameter is the length of the blade. Rorres, Muller and Senior, Yulistiyanto et al., Raza et al., Stergiopoulou and Kalkani, Lubitz et al., and Ammar et al. control the length as constant parameter as the length did not contribute most in feature of efficiency.

Consequently, by considering the above parameters such as slope angle, length of the blade, head difference, number of turns and diameter of blade would not be improved if the associated problem is not identified including the climatic change, the prototype cost such as material structure and material joining, installation cost, the suitability of the site and etc.

Table 2. Summary of parameter from previous researchers.

Researcher	Method of analysis	Control parameter	Independent parameter	Results/ Findings
Rorres (2000) [11]	Analytical	1.Diameter 2.Length,L	1.Number of blade, N	1. Volume of bucket depend on slope angle. 2. Volume of water in cycle increase with number of blade.
G. Müller and J. Senior (2009) [12]	Analytical	1.Length, L, 2.Outer diameter, 3.Head, H,	1. Inflow depth over radius, d1/R 2. Slope Angle,α 3. Number of turns, N.	1. Inflow depth over radius increase, efficiency, η increase. 2. Angle increase, efficiency, η decrease. 3. Number of turns increase, efficiency, η increase.
Yulistiyanto et al.,(2012) [16]	Experimenta l	1.Length, L, 2.Outer diameter, D0 3.Head, H, 4. Pitch diameter, P	1. Slope angle, α 2. Water discharge, Q	1. At angle 25°-40°, efficiency increase, at 41°-50°, start to decrease, have maximum efficiency and power at 35° angle. 2. Water discharge increase, then efficiency and power increase too.
Raza et al., (2013) [20]	Numerical	1.Head, H, 2.Length, L, 3.Outer diameter,D0 4.Head, H, 5. Pitch diameter,P	1. Water discharge, Q 2.Outer radius screw 3.Blade portion length of screw	1. Simulate to obtain electrical power output.

Table 2. Continued.

Researcher	Method of analysis	Control parameter	Independent parameter	Results/ Findings
Strgiopoulou and Kalkani (2013)	Computational	1. Length ,L 2.Diameter	1. Slope angle, α 2.Number of rotation,N 3.Water discharge,Q 4.Head,H	1. Inlet pressure was set to zero under homogenous Neumann condition.
Lubitz (2014) [13]	Experimental	-	1. Rotational speed	1. All friction losses at entry and exit effect are neglected. 2. Reducing the speed does not minimize the gap leakage.
Lubitz et al., (2014) [17]	Analytical and experimental	1. Pitch diameter, P 2.Outer diameter, D0 3.Length ,L 4. Head, H 5.Number of rotation, N	1.Fill level (full, under full and overfull) 2.Gap leakage between flights 3. Slope angle, α 4.Pitch ratio 5.Water discharge, Q	1. All friction losses at entry and exit effect are neglected. 2. Gap leakage decreases the efficiency. 3. Gap leakage decreases the rotational speed.
Ammar et al.,(2014) [21]	Computational	1.Length, L, 2.Outer diameter, D0 3.Head, H 4. Pitch diameter, P	1. Water discharge, Q 2. Velocity streamline	1. Obtain pressure contour and velocity contour to determine particle impact prediction for Archimedes screw design.

Table 2. Continued.

Researcher	Method of analysis	Control parameter	Independent parameter	Results/ Findings
E. Fiardi (2014) [14]	Experimental comparison	Slope Angle,α Length, L Head, H, Outer diameter, D0	-	1. Efficiency for analytical is 50% and power output is 0.236W. 2. Efficiency of experiment is 41% and power output is 0.098W.
Hendra and A.Indriani (2014)[19]	Experimental	1. Pitch diameter, P 2.Outer diameter, D0 3.Length, L	1.Head, H 2. Water discharge, Q	1. Flow rate increase, rotation increase. 2. Head increase as the rotation increase.

4. Conclusions

The Archimedes screw runner blade turbine is one of most guaranteed alternative solution to overcome the lacking of fossil fuels source and environmental impacts. A suitable Archimedes screw runner blade turbine efficiency is depending on the function of screw parameters. Therefore, this review is focused on parameters study by comparing the effect of parameters in aspect of efficiency. The uses of external parameter have an effect on the internal parameter and the external parameter data can be collected for the prototype. Besides, the effect of climate change, material selection and cost installation also should be considered to build an Archimedes screw runner blade turbine with good performance and have benefits to the people especially to those people who lives in rural areas.

Acknowledgements

The authors would like to be obliged to University Malaysia Pahang for providing laboratory facilities and financial assistance under project no. RDU140348.

References

[1] Kandpal J B, Madan M 1995 *Renewable Energy* **6** 159-60
[2] Mohibullah M, Radzi A M, Hakim M I A, Basic design aspects of micro hydro power plant and its potential development in Malaysia. Power and Energy Conference, 2004 PECon 2004 Proceedings National; 2004 29-30 Nov. 2004.
[3] Keawsuntia Y 2014 *Frontiers of Manufacturing and Design Science IV* 496-500
[4] Fraenkel P P O, Bokalders V, Harvey A, Brown A, Edwards R. 1999 *Micro-hydro power: a guide for development workers* IT Publications)
[5] Elbatran A H, Yaakob O B, Ahmed Y M, Shabara H M 2015 *Renewable and Sustainable Energy Reviews* **43** 40-50
[6] Date A, Akbarzadeh A 2009 *Renewable Energy* **34** 409-15
[7] Tushar K. Ghosh M A P 2011 *Energy Resources and Systems* **2** Springer)

[8] Muhamad Barzani Gasim M E T, Mushrifah Idris, Pan Ian Lun, M. K.A. Kamarudin, A. A. Nor Azlina, Mazlin Mokhtar and S. A. Sharifah Mastura 2013 *American Journal of Applied Sciences* **10** 42-57
[9] Barelli L, Liucci L, Ottaviano A, Valigi D 2013 *Energy* **58** 695-706
[10] Williamson S J, Stark B H, Booker J D 2014 *Renewable Energy* **61** 43-50
[11] Rorres C 2000 *Journal of Hydraulic Engineering* **126** 72-80
[12] Müller G, Senior J 2009 *Journal of Hydraulic Research* **47** 666-9
[13] Lubitz W D 2014 *Proceedings of The Canadian Society for Mechanical Engineering International Congress 2014* 1-6
[14] Fiardi E 2014 *Journal of Ocean Mechanical and Aerospace Science and Engineering-* **5** 1-12
[15] Brada K 1999 **14** 52-6
[16] Bambang Yulistiyanto Y H, and Lisdiyanti 2012 *Dinamika TEKNIK SIPIL* **12** 1-5
[17] William David Lubitz M L a S S 2014 *Journal of Hydraulic Engineering* **140** 04014050
[18] Alkistis Stergiopoulou E K 2013 *International Journal of Engineering Research & Technology* **2** 193-9
[19] Hendra H a I, Anizar 2014 *International Conference on Mechanical Design, Manufacture and Automation Engineering*
[20] Ali Raza M S, Yasir Saleem 2013 *International Journal of Engineering Research & Technology* **2** 2471-7
[21] Mutasim M A N A, N.S. ; Adam, A.A.A., Prediction of Particle Impact on an Archimedes Screw Runner Blade for Micro Hydro Turbine. 4th international conference (ICME 2013); 2014; Bangi-Putrajaya, Malaysia: Trans Tech Publications , Durnten-Zurich.

IOP Conf. Series: Materials Science and Engineering **100** (2015) 012031 doi:10.1088/1757-899X/100/1/012031

A passive cooling system of residential and commercial buildings in summer or hot season

M M Rahman[1], M Mashud[2], C M Chu[1], M S bin Misaran[1], M Sarker[2], S Kumaresen[1]
[1]Materials and Mineral Research Unit, Faculty of Engineering, Universiti Malaysia Sabah (UMS) Kota Kinabalu
Sabah-88400, Malaysia
[2]Department of Mechanical Engineering, Khulna University of Engineering & Technology, Khulna-9203, Bangladesh

Email: mizanur@ums.edu.my

Abstract. The increasing number of high rise buildings may contribute to lack of natural ventilation in modern buildings. Generally, fans and air conditioning are used in the modern building for cooling and air ventilation. Most of the energy in tropical regions are consumed by heating, cooling and ventilation appliances. Therefore, solar power appliances for cooling, heating and ventilation will be a suitable option for saving energy from the household sector. A modified-structure building is designed and constructed with solar chimney to enhance ventilation rate that increases cooling performance and ensure thermal comfort. An evaporative cooler is introduced with a newly designed room to enhance the temperature reduction capacity. The room temperature is compared with a non-modified room as well as with ambient temperature. The results show that passive cooling system with evaporative cooler was able to reduce temperature by 5°C compared to the ambient temperature and about 2°C to 3°C below the reference room temperature.

1. Introduction

The demands of electrical energy in developing countries depend on the economic development as well as the standard of living. Consequently, electricity is viewed as the key indicator of the economic development of the country. The generation of commercial electricity depends on the natural resources of that country generated mainly from natural gas and hydropower [1], however the reserve of the natural gas in Bangladesh is likely to deplete by the year 2020 [2].

In Bangladesh, about 40 per cent of the total energy used in the urban areas is consumed by the residential and commercial building and about 50-60 per cent of this energy is used for cooling, heating, ventilating and space conditioning in residential and commercial buildings [3]. Therefore, during hot climate or in summer when the outside temperature is high, effective passive cooling system for commercial and residential building will be a viable option. It can reduce the electricity consumption as well as extending the availability of natural resources. Bangladesh is situated between 20.30° and 26.38°

IOP Conf. Series: Materials Science and Engineering **100** (2015) 012031 doi:10.1088/1757-899X/100/1/012031

north latitude and 88.04° and 92.44° east of GMT which provides this country with abundant solar energy. The average daylight or sunlight hour throughout the year is more than 10 hours and average solar radiation varies from 4 to 6.5 kWh/m^2-day. It indicates a good prospect for solar power ventilation system or passive cooling system for commercial and residential building in Bangladesh [1, 2]. In this study, the space cooling is defined as the rate of energy removed from the air space in order to maintain a set lower temperature than the surrounding atmospheric temperature. During summer or hot climate when the ambient temperature is high, poor ventilation and cooling system can change the environment of the residential and the commercial building in the cities and causes thermal discomfort. In addition, electricity blackout is very common in the cities of Bangladesh consequently worsening the situation [1-4].

At present all modern commercial and residential buildings are furnished with electrical appliances mainly used for cooling and ventilation, resulting in excessive energy consumption during summer. Therefore, an efficient passive cooling system coupled with a solar chimney system can minimize this energy used for the commercial and residential buildings in urban areas of Bangladesh. There are some well-established methods available for space heating by using solar energy which is direct radiation, trombe wall, transparent insulation etc. [5]. On the other hand very limited number of studies is available on space cooling by using solar energy. This is due to the skepticism on the use of night ventilation, shading, evaporative cooling etc. The earliest methods of space cooling during dry season was discussed by Bahadori in 1978 [6] and the summary of the state-of-the-art of passive cooling systems was provided by Givoni in 1991 [7]. Study also shows that in India larger parts of the central and northern regions including New Delhi experience mixed climates such as hot – humid, hot – dry and cold climate systems. Attempt to implement passive cooling in the city buildings in New Delhi was made in 1990's but limited literature is found related to this study [8]. Ventilation by using natural draft system is one of the passive cooling techniques initially used for space cooling [9, 10].

Introduction of modern science and technology has significant changes in the controlling of indoor environment of commercial and residential buildings. Passive cooling system is also closely related to thermal comfort of occupants of the building. Although some of the passive cooling techniques do not help to remove the cooling load significantly it does extend the tolerance of humans to achieve thermal comfort in a given space. Thermal comfort for the building can also be achieved by controlling the climate of the building, changing the site design, introducing thermal insulation, changing the behavior of the occupant, preventing the heat gains from internal sources. A hybrid system, by a combination of natural convection cooling and cooling system with evaporative cooler can prevent heating of the building as well as enhance the cooling efficiency to improve thermal comfort conditions in buildings [11-15]. In this paper a solar chimney-driven passive cooling system capable of enhancing ventilation above normal natural draft system is discussed for providing thermal comfort conditions inside a building throughout the summer season in hot and humid climates.

2. Methodology

Figure 1 shows the model of the passive solar ventilation and cooling system consisting of a solar chimney or solar air heater or simply a mechanical ventilator system. The model is developed in the laboratory of the Mechanical Engineering Department at Khulna University of Engineering and Technology (KUET) Bangladesh. In the solar passive cooling model (figure 1), solar air heater with chimney is used to enhance the suction of the hot air from the space during day time. At bottom, a water tank is used as evaporative cooler. A circular copper tube is used as inlet duct which is completely immersed in the water tank. The whole system worked as a heat exchanger. Thermal energy was transferred from the hot air to the cold water. The evaporative cooler was insulated to protect heat transfer from ambient air and was shaded to protect radiation absorption.

3. Design

3.1. Modification of Room for Thermal Load Reduction

The physical dimension ($long \times wide \times high$) of the existing single room was approximately$150\ cm \times 120\ cm \times 90\ cm$. The longer side of the room was aligned to the east-west direction. Thirteen centimeter wide bricks were used to build surrounding walls with a reinforced concrete slab of 10 cm thickness. A wooden door of about 60 cm^2 face area was used in the north facing wall. After completing the construction of the room, ambient and room temperatures were measured and found to differ by 5^0C.

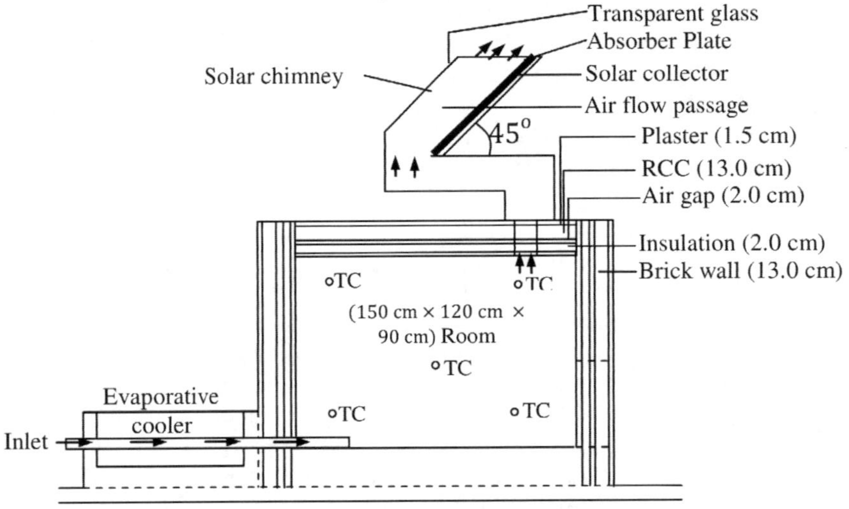

Figure 1. Cross-sectional View of Solar Passive Model.

The above mentioned passive cooling system used a natural draft solar chimney which had a solar heat collector and a transparent glass. In the solar chimney, the solar energy was absorbed by the solar collector and transferred to the adjacent air. The stack effect caused air flow through the solar chimney. Air flow rate in a normal solar chimney may reduce due to cross wind and flow reversal [16]. To overcome this problem, the chimney was installed on a base opposite to the air flow direction. Solar chimney was also placed at 45° to the horizontal axis, to receive maximum sunlight during day time.

3.2. Solar chimney

In the solar passive cooling system, the solar chimney was fabricated with mild steel sheet. The dimensions of the solar chimney were 130 cm × 120 cm (length × width). Glass wool insulation was used at the bottom and at the sides. The solar chimney was installed to face north, where the open loop system was set with a tilt of 45° from the horizontal, having an air flow passage above the absorber plate.

3.3. Evaporative cooler

An insulated water tank was used as evaporative cooler. A 7.5 cm diameter copper pipe was immersed in the water and was used as the inlet of the solar passive model room. Water of the tank exposed to the ambient air evaporated and cooled. Warm fresh air passing through the copper tube transferred heat to the water and air and water became cool compared to ambient air, subsequently this cooled air would reduce the room temperature.

IOP Conf. Series: Materials Science and Engineering **100** (2015) 012031 doi:10.1088/1757-899X/100/1/012031

4. Performance analysis

An ordinary room with the same physical dimensions as the solar passive model room and without solar chimney is selected as the reference room. The room temperature was collected from 21st to 25th May, 2014 and presented in the figure 2. It was found that, the solar chimney passive cooling model room was able to maintain the temperature at about 4 °C to 5 °C below the outside ambient temperature and at about 2 °C to 3 °C below the reference room temperature.

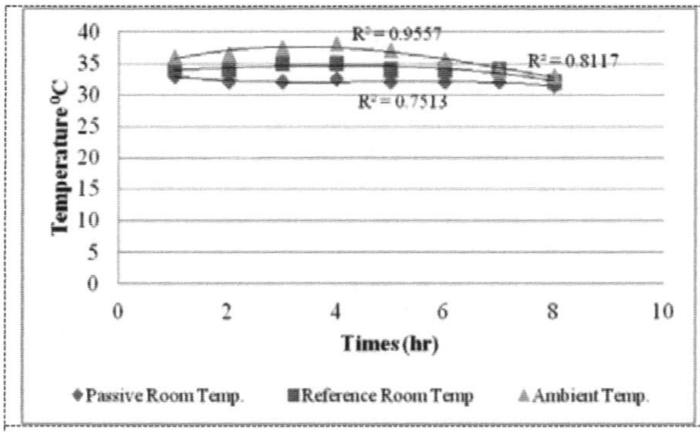

Figure 2. (a) 21st May, 2014

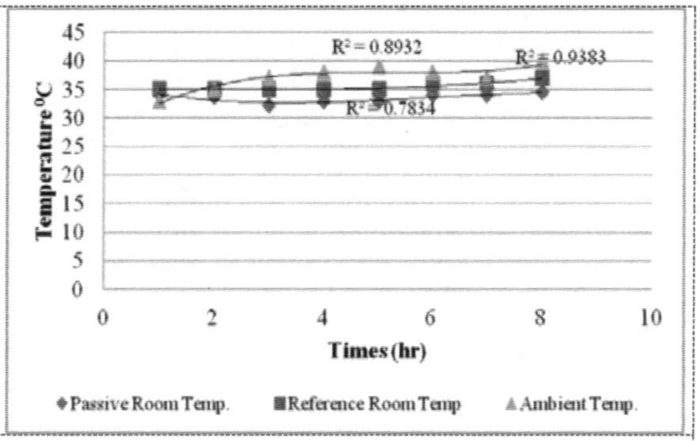

Figure 2. b) 22nd May, 2014.

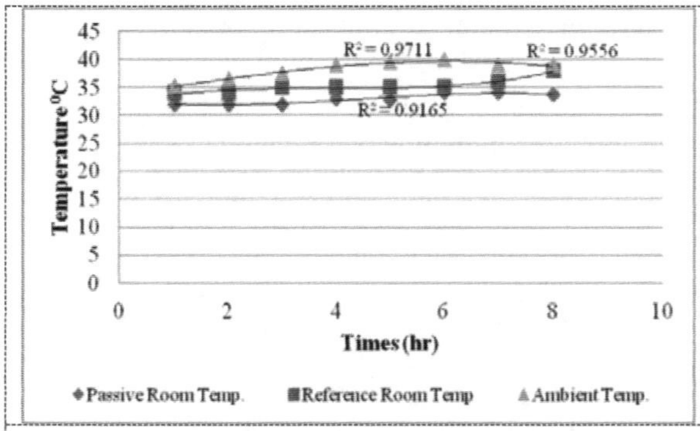

Figure 2. (c) 23rd May, 2014.

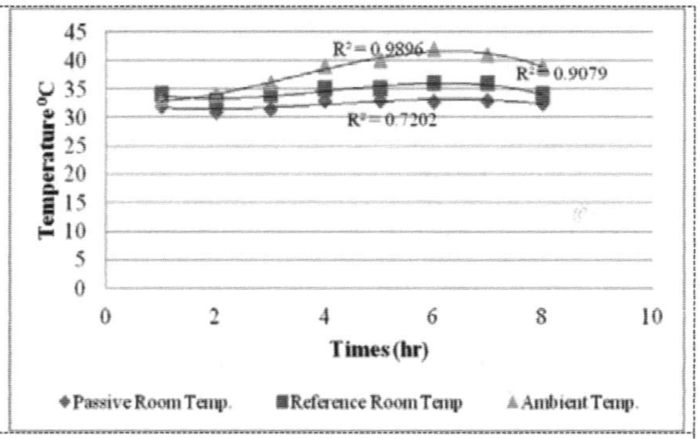

Figure 2. (d) 24th May, 2014.

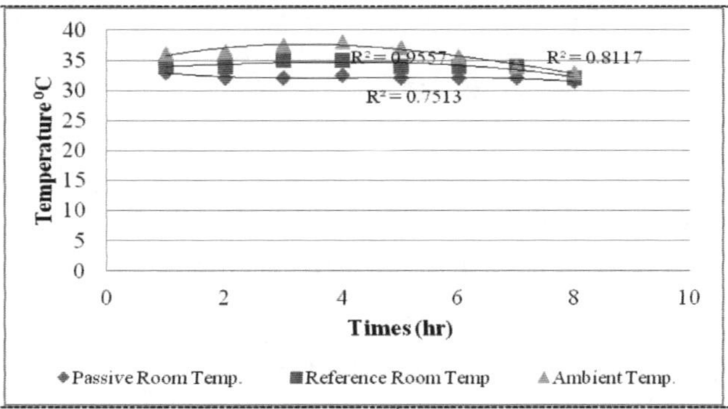

Figure 2. (e) 25th May, 2014

Figure 2. Variation in Temperature between Passive Room, Reference Room and Ambient Atmosphere.

Cheikh and Bouchair in 2008 [17] also developed a passive cooling system for house by introducing a passive cooling roof and were able to lower the temperature about 6°C to 10°C than ambient. The passive cooling roof was fabricated with steel structure, rocks and a water bed which was costly in comparison with brick structure. Raman *et al.* in 2001 [8] also used solar chimney with evaporative cooling system and found that during hot – dry months (April – July), the experimental room temperature was 2°C -3°C less than the room without passive cooling system. Typical ambient temperature during these times is varied from 30°C to 42°C. The comfortable room temperature was about 22°C therefore it cannot be concluded that the above described passive cooling system performances were satisfactory but that it could reduce indoor room temperature significantly.

Cubic polynomial curves were obtained for the passive cooling room, reference room and ambient temperature by using least-squares regression analysis from Microsoft Excel. All curves had good fit with data points. Among all the curves, the passive cooling room temperature curve shows weak relation with data. This is because solar chimney and evaporative cooler might have enhanced ventilation and cooling rate therefore time had very little effect on passive cooling room temperature. The data also indicate that the passive cooling room temperature was significantly below the reference room temperature.

5. Discussion
The main components of solar passive system are solar chimney, evaporative cooler and additional wall insulation described in this paper. The solar passive system showed good performance for providing thermal comfort during hot climate. It also had the ability to reduce the air conditioning load of buildings during summer and able to reduce the extensive use of electrical cooling appliances. Solar passive cooling system performance data were taken for an empty room. It must be borne in mind that human activity, appliances, metabolism etc. may alter the performance of solar passive system. In addition the performance of solar chimney may decrease due to back flow, cross wind etc. that may change also the performance of solar passive system.

6. Conclusion
A passive solar system for cooling, de-humidification and ventilation during hot environment has been tested on an experimental room with modified structure. The investigation has confirmed the effectiveness of passive cooling system in comparison to a reference room. The results show that cooling inside buildings can be enhanced by changing the design of the building to include a solar chimney for natural draft ventilation. By installing solar chimney, evaporative cooler and wall insulation, it is estimated to increase the cost by about 10 per cent of the cost of a conventional room. Therefore, the described passive cooling system with effective modified chimney seems to have very good potential value in terms of economy and also technological point of view.

References
[1] Anam K, Bustam H A 2011 *Cyber Journals: Multidisciplinary Journals in Science and Technology, Journal of Selected Areas in Renewable and Sustainable Energy (JRSE)* **September Edition**
[2] Rahman M S, Saha S K, Khan M R H, Habiba U, Hosse S M 2013 *GJRE-F: Electrical and Electronic Engineering* **13**
[3] Roodman D M, Lenssen N K, Peterson J A 1995 *A building revolution: how ecology and health concerns are transforming construction* Worldwatch Institute Washington, DC)
[4] Cheikh H B, Bouchair A 2004 *Renewable Energy* **29** 1877-86
[5] *Passive Solar Design Handbook* Van Nostrand Reinhold, New York.)
[6] Bahadori M N 1978 *Sci Am;(United States)* **238**

[7] Givoni B 1991 *Energy and Buildings* **17** 177-99
[8] Raman P, Mande S, Kishore V 2001 *Solar Energy* **70** 319-29
[9] Fischer S K, Creswick F. Energy-use impact of chlorofluorocarbon alternatives. Oak Ridge National Lab., TN (USA), 1989.
[10] Van der Maas J, Roulet C 1991 *ASHRAE Technical Data Bulletin* **7** 32-40
[11] H. A M, M. M, M. A M, R. R H M. Passive Cooling System of a Building: A New Approach. International Conference and Exhibition on Sustainable Energy and Advanced Materials; October 3-4, 2011; Solo, Indonesia2011. p. 88-94.
[12] Handbook-Fundamentals A 2009 *Inc, NE Atlanta, GA* **30329**
[13] Santamouris M, Asimakopoulos D 1996 *Passive cooling of buildings* Earthscan)
[14] Khedari J, Boonsri B, Hirunlabh J 2000 *Energy and buildings* **32** 89-93
[15] Lal S, Kaushik S, Bhargava P 2013 *International Journal of Development and Sustainability* **2**
[16] Chu C C-M, Chu R K-H, Rahman M 2012 *Advanced Computational Methods and Experiments in Heat Transfer XII, WIT Transactions on Engineering Sciences* **75** 73-82
[17] Cheikh H B, Bouchair A 2008 *The Open Fuels & Energy Science Journal* **1**

The combustion behavior of diesel/CNG mixtures in a constant volume combustion chamber

Firmansyah[1], A R A Aziz[1] and M R Heikal[1]
[1]Center for Automotive Research and Energy Management, Universiti Teknologi Petronas 31750, Bandar seri iskandar, Perak, Malaysia

E-mail: firmansyah@petronas.com.my

Abstract. The stringent emissions and needs to increase fuel efficiency makes controlled auto-ignition (CAI) based combustion an attractive alternative for the new combustion system. However, the combustion control is the main obstacles in its development. Reactivity controlled compression ignition (RCCI) that employs two fuels with significantly different in reactivity proven to be able to control the combustion. The RCCI concept applied in a constant volume chamber fueled with direct injected diesel and compressed natural gas (CNG) was tested. The mixture composition is varied from 0 – 100% diesel/CNG at lambda 1 with main data collection is pressure profile and combustion images. The results show that diesel-CNG mixture significantly shows better combustion compared to diesel only. It is found that CNG is delaying the diesel combustion and at the same time assisting in diesel distribution inside the chamber. This combination creates a multipoint ignition of diesel throughout the chamber that generates very fast heat release rate and higher maximum pressure. Furthermore, lighter yellow color of the flame indicates lower soot production in compared with diesel combustion.

1. Introduction

Stringent emissions regulations and a significant increase of fuel price highly affecting the growth rate of automotive technology development with the objective of reducing fuel consumption and improving engine efficiency. Controlled auto-ignition (CAI) based combustion system such as homogeneous charge compression ignition (HCCI), stratified charge combustion ignition (SCCI), and premixed charge compression ignition (PCCI) are the recent engine development with high efficiency, low emissions and fuel consumption.

In the effort of finding the effective method of controlling the combustion process, dual fuel or later known as reactivity charge compression ignition (RCCI) is found to be able to tailor the combustion process and phasing [1]. RCCI is a combination of the two fuels with significant difference in reactivity, high reactivity and low reactivity fuels. Various fuel combinations had been tested such as diesel-methane [2], diesel-gasoline [3, 4] and heptane-methane [5].

Originally, it was started with the combination of diesel and CNG where diesel fuel is used as pilot igniter for CNG. While, CNG selection as the combination of diesel is due to its low carbon ratio as well as its high octane number. It was proven that this combination can reduce the unburned hydrocarbon and nitric oxide emissions. Furthermore, it was stated that the methane addition significantly affect the intensity and spreading of the flame during the combustion, especially with the suitable in-cylinder bulk flow motion. Furthermore, diesel-CNG combination is able to significantly reduce the soot and NOx emissions [6].

The interaction between methane and diesel was described by Carlucci, Laforgia [7] and Carlucci, De Risi [8]. Methane addition in the intake port is significantly affect the intensity and spreading of the flame during the combustion process of the dual fuel thus increase the combustion efficiency. As the results, hydrocarbon and NOx emission generated from this fuel combination is significantly lower compared to diesel. These earlier research have proven the ability of dual fuel RCCI fuel combination method in increasing the performance and reducing the emission of the diesel engine. However, the fundamental interactions between the two fuels have not been explained. It is necessary to improve the understanding on the interaction between the two fuels in order to improve the efficiency as well as opening the possibilities for other types of fuel combination.

The earlier diesel-CNG RCCI introduced CNG at the intake manifold while diesel is directly injected into the combustion chamber. This method is found to be advantageous because this setup does not require significant modification to the engine. However, it gives a limitation on the control parameter to the diesel injection parameters such as injection pressure and multiple injection system. Previous research stated that the combustion results from the dual fuel combination also affected by the bulk in-cylinder flow motion [7]. It is known that the cylinder bulk motion have significant effect on the fuel mixing process which also maybe the cause of the variation in the combustion process of the dual fuel.

In order to gain more control in fuel delivery method and fuel-air mixing variation, dual-direct injection method is proposed in this research. The freedom to control the injection strategies for both fuels, will give further advantages in shaping the combustion process of RCCI and further improve the combustion performance as well as the operating range of the CAI based engine. This study focuses on the combustion properties of direct injected diesel, and directly injected CNG mixture in the constant volume chamber (CVC) at elevated temperature. This method isolates the mixture distribution affecting parameters to the injection strategy which excluding the bulk in-cylinder flow. Furthermore, the mixture composition effect to the combustion output of the mixture will be discussed further.

2. Experimental Setup

The study performs extended experimental investigation on the auto-ignition behavior of dual fuel (Diesel-CNG) in a constant volume combustion chamber (Figure 1). A probe heater is used to increase the temperature inside the combustion chamber. Heater temperature is set at 800°C in order to get a stable auto-ignition from the mixture of combustion chamber pressure equal to atmospheric. Furthermore, oxygen with purity 99.5% is used as the replacement of air in order to reduce the complexity of the reaction and to increase the auto-ignition occurrence probabilities.

IOP Conf. Series: Materials Science and Engineering **100** (2015) 012032 doi:10.1088/1757-899X/100/1/012032

Figure 1. Constant volume chamber.

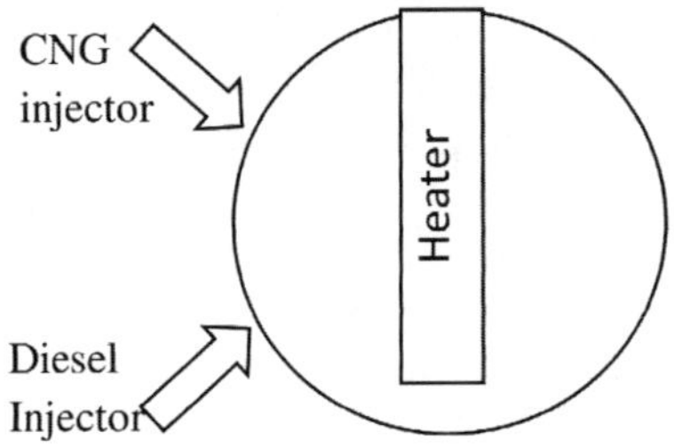

Figure 2. Fuel injector and heater arrangement in the CVC.

Table 1. Injector delivery rate for each fuel.

Fuels	Injector delivery rate (g/s)
Diesel	4.2 @ 3 bar
CNG	7.2 @ 7.5 bar

In the experiment, three fuel mixtures was tested, Diesel-CNG and heptane-CNG. The composition was varied from 10-100% Liquid fuel/CNG. The combustion data for these mixtures were obtained for lambda 1.

The injector used is Siemens Deka 4 with 3 bars injection pressure and 4.3 g/s delivery rate (manufacturer specification and fueled with gasoline). Due to density variations of the fuels, the calibration process is carried out in the ambient condition for each file to measure the actual delivery rate of the injector. The calibration results are shown in

Figure 2. Fuel injector and heater arrangement in the CVC.

Table 1. The CNG injector was using low-pressure CNG injector from orbital with 7.2 g/s at injection pressure 7.5 bars. The injectors were placed 90° relative to each other and both are directly injected into the combustion chamber (

Figure 2).

Lambda was generated by calculating the stoichiometric reaction between the fuels and oxygen.

$$x\ C_{10}H_{22} + (1-x)CH_4 + a(O_2 + 3.76N_2) \leftrightarrow bCO_2 + c\ H_2O$$

Lambda resulted from the above equation for each mixture is depicted in Figure 3. It is ranges from 17.8 to 16.6.

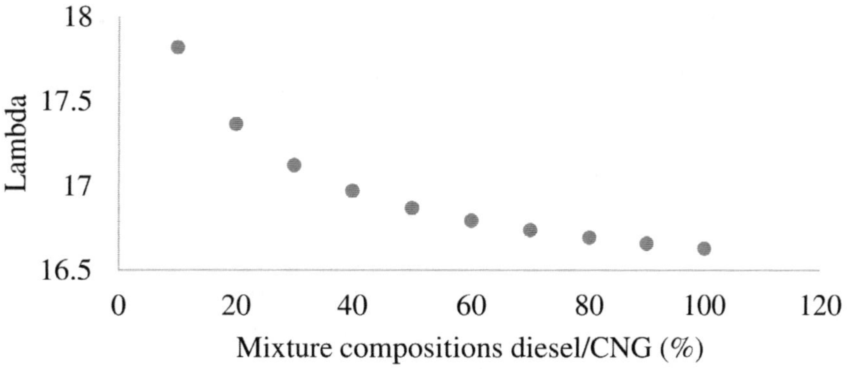

Figure 3. Lambda for various mixture compositions.

The required fuel amount at constant volume (0.000492 m^3) was determined and translated to the injector opening by dividing the total fuel required with the injector delivery rate.

3. Results And Discussion

Various compositions of diesel and CNG ranges from 0 to 100% diesel/CNG percentage were tested. It is found that there is a lowest limit for diesel/CNG composition percentage in order for the combustion to occur. 50% diesel/CNG shows the lowest composition limits for the mixture to combust in this experiment as shown in Figure **4**. The inexistence of combustion for lower diesel percentage may be caused by the dominant CNG properties that have very low reactivity with high auto-ignition temperature and very high octane number that suppressed the diesel combustion.

This diesel-CNG trend has a similar pattern to the simulation done by Zoldak, Sobiesiak [1] that shows increasing maximum pressure with additional CNG in the mixture. The main reason for this behavior is the unique interaction between diesel and CNG. Diesel combustion categorized as diffusive where the combustion occurs during the injection period. CNG, on the other hand, have very low reactivity that makes it difficult to combust.

Diesel as the source of ignition for CNG creates a favorable condition for CNG to combust yet CNG is suppressing the combustion of diesel to the point where diesel droplet has been distributed all over the chamber. The phenomena happen in such ways that distributed source of ignition occurs inside the chamber to ignite CNG and starts the flame propagation process similar to the spark ignition combustion.

This interaction also caused faster combustion rate with a shorter duration as shown in the heat release graph in Figure 5 where the maximum pressure increased from 0.02 kJ/s to 0.1 kJ/s, and the combustion duration reduced from 1.5 s to 0.9 s. Furthermore, Figure 5 also shows the significance of CNG in diesel combustion. 10% CNG composition in the mixture is able to improve the HRR twice of the single diesel combustion. However, diesel on the other hand shows second pressure peak at 90 ms after the combustion starts. This is caused by the combustion of unburned diesel at the center of the chamber as shown in Figure 6 as the yellow bright point while the earlier flame shows darker flame that indicates the final stage of the combustion process.

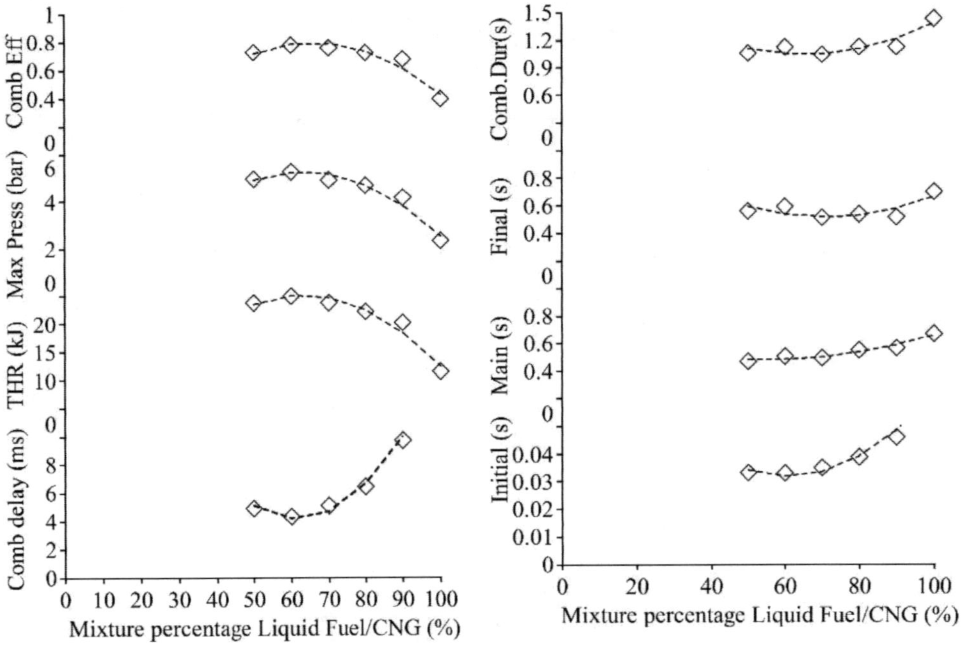

Figure 4. Combustion properties of Diesel-CNG at lambda 1.

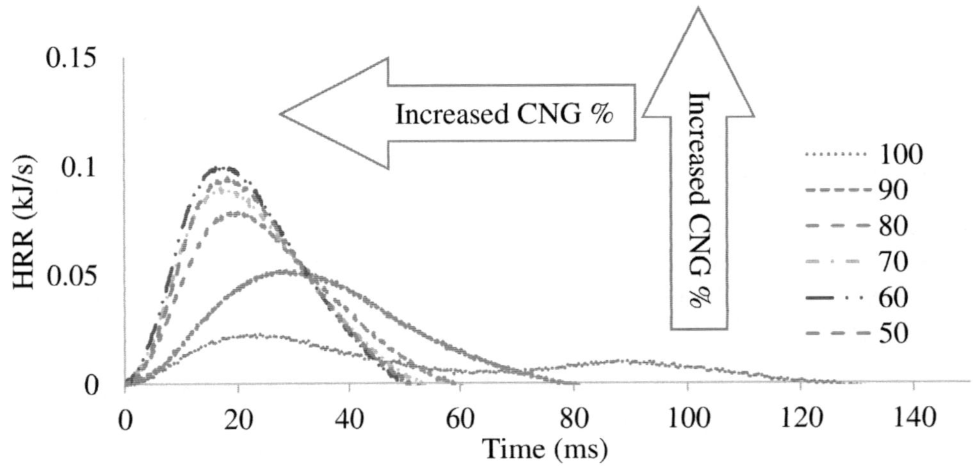

Figure 5. Heat release rate profile for various mixture of diesel-CNG combustion at lambda 1.

However, the combustion limitation occurs at the point where diesel combustion could not create the suitable environment for CNG to combust which is shown by the decreasing rate of the combustion at 50-50 diesel-CNG composition.

Figure 6. Combustion image at 90ms after start of combustion.

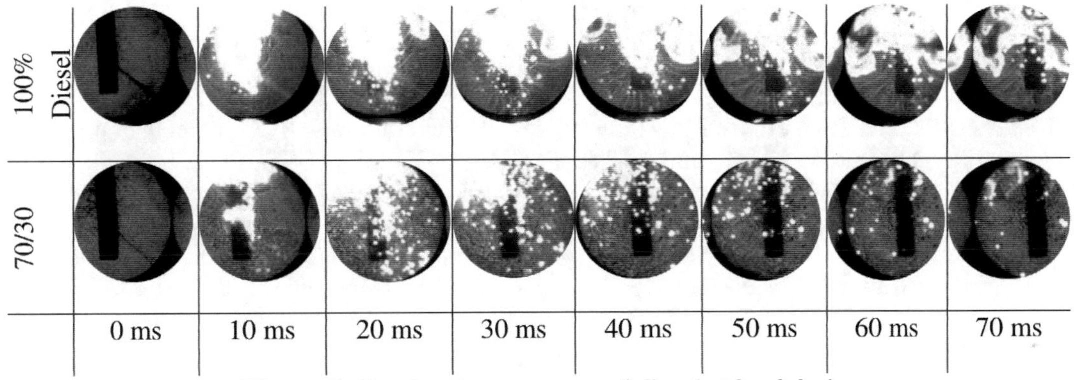

Figure 7. Combustion sequence of diesel at lambda 1.

Figure 7 also shows that CNG injection creates distributed hotspot along the diesel injection axis. It is marked by the bright spots along the axis. These scattered hotspots could increase the heat release rate of the mixture up to four times faster and significantly shorter duration compared to the single fuel (diesel). In addition, the flame at the final stage of the combustion for 100% diesel shows darker yellow color that indicates potentially high soot concentration in the exhaust. The DCNG mixture, on the other hand, shows bright yellow flame during the combustion process and subtle percentage of the darker flame on the final stage of the combustion. This behavior align with the previous literature involving diesel and CNG mixture that stated that diesel-CNG mixture combination is able to drastically reduce the soot emission in diesel engine.

4. Conclusions

In this paper, diesel and CNG mixture combustion behavior is analyzed. The addition of CNG in diesel can increase the combustion output of the mixture. It generates higher maximum pressure, faster heat release rate, and shorter combustion duration. The minimum mixture composition is at 50% diesel/CNG that is caused by the low reactivity of CNG that suppresses the diesel combustion.

The main reason for the increased combustion output is the multi ignition point generated with CNG injection. CNG is suppressing the diesel combustion and delayed the timing until the diesel is distributed throughout the chamber and create a multipoint ignition.

The flame image shows that the combustion of diesel-CNG mixture is better than diesel combustion shown by bright yellow flame throughout the combustion process while diesel shows darker yellow flame color. The flame color also indicates the probabilities of soot generation where darker flame has larger probabilities to generate more soot emissions. Further experiment on the engine need to be done in order to verify the combustion characteristics of diesel-CNG mixture combination. Furthermore, dual fuel direct injection system concept need to be tested in the actual engine in order to prove its capabilities in controlling CAI based engine technology.

Acknowledgements

The authors would like to be obliged to Center for Automotive Research and Energy Management (CAREM) Universiti Teknologi PETRONAS in providing the experimental facilities.

References

[1] Zoldak P, Sobiesiak A, Bergin M, Wickman D D. Computational Study of Reactivity Controlled Compression Ignition (RCCI) Combustion in a Heavy-Duty Diesel Engine Using Natural Gas. SAE Technical Paper, 2014 0148-7191.

[2] Doosje E, Willems F, Baert R. Experimental demonstration of RCCI in heavy-duty engines using diesel and natural gas. SAE Technical Paper, 2014 0148-7191.

[3] Hanson R, Kokjohn S, Splitter D, Reitz R D 2011 *SAE International Journal of Engines* **4** 394-411

[4] Splitter D, Reitz R D, Hanson R. High efficiency, low emissions RCCI combustion by use of a fuel additive. SAE Technical Paper, 2010.

[5] Rahimi A, Fatehifar E, Saray R K 2010 *Proceedings of the Institution of Mechanical Engineers, Part D: Journal of Automobile Engineering* **224** 1141-59

[6] Zhang Y, Kong S-C, Reitz R D. Modeling and simulation of a dual fuel (diesel/natural gas) engine with multidimensional CFD. SAE Technical paper, 2003 0148-7191.

[7] Carlucci A P, Laforgia D, Saracino R, Toto G 2011 *Energy Conversion and Management* **52** 3004-17

[8] Carlucci A, De Risi A, Laforgia D, Naccarato F 2008 *Energy* **33** 256-63

Investigation of mechanical properties for open cellular structure CoCrMo alloy fabricated by selective laser melting process

A Azidin[1], Z A M Taib[1], W S W Harun[2*], S A Che Ghani[2], M F Faisae[2], M A Omar[3], H Ramli[3]

[1]Institute of Postgraduate Studies, Universiti Malaysia Pahang, Lebuhraya Tun Razak, 26300 Gambang, Kuantan, Pahang.

[2]Faculty of Mechanical Engineering, Universiti Malaysia Pahang, 26600 Pekan, Pahang.

[3]Kolej Kemahiran Tinggi MARA Kuantan, KM.8, Jalan Gambang, 25150 Kuantan, Pahang.

[*]Email: sharuzi@ump.edu.my

Abstract. Orthodontic implants have been a major focus through mechanical and biological performance in advance to fabricate shape of complex anatomical. Designing the part with a complex mechanism is one of the challenging process and addition to achieve the balance and desired mechanical performance brought to the right manufacture technique to fabricate. Metal additive manufacturing (MAM) is brought forward to the newest fabrication technology in this field. In this study, selective laser melting (SLM) process was utilized on a medical grade cobalt-chrome molybdenum (CoCrMo) alloy. The work has focused on mechanical properties of the CoCrMo open cellular structures samples with 60 %, 70 %, and 80 % designed volume porosity that could potentially emulate the properties of human bone. It was observed that hardness values decreased as the soaking time increases except for bottom face. For compression test, 60 % designed volume porosity demonstrated highest ultimate compressive strength compared to 70 % and 80 %.

1. Introduction

Dental implant consists of parts such as crown, abutment and bridge where is restored in the interior or posterior region of human jaw or skull. There are two factors that must take into consideration for the purpose to produce the dental implants product that are mechanical performance and also the surface properties of the finished product [1]. Selective laser melting (SLM) process has been proven best to produce a dental implant as a ready used product with better stress distribution and also the primary stability [2]. SLM has been widely used in the application of medical implants manufacturing as this process can produce complex geometry that impossible to fabricate using conventional manufacturing processes [3].

Cobalt chromium molybdenum (CoCrMo) has been widely utilized in orthodontic application nowadays due to its high corrosion resistance and biocompatibility. The biocompatibility of CoCrMo is due to exhibition of spontaneous formation of a passive oxide film layer and thus results a barrier for metal ion release [4]. A study on CoCrMo has been performed on the human body and it has been proven to be a preferred alloy for the application of the metal-on-metal contact [5]. From EOS datasheet, this material possess excellent characteristics as it able to stands to Young modulus of 200 – 210 GPa, tensile strength at 1150 – 1400 MPa and yield strength of 880 – 980 MPa [6].

The size of pore in a finished part plays an important role in biomedical applications. In medical term, osseointegration is significantly in strengthening the internal fixation by allowing the bone ingrowth into the implant in order to enhance the functionality and longevity of the dental implants [7]. The pores distribution and the shape of the pore must emulate natural porous structure like sandstone that is mimicking the human bone as they exhibit the fractal characteristics [8]. In the human nature, there are complex ways to determine the pore-structure completely as it has a character of abnormity, uncertainty, and intangibility. Accuracy description of the pore structure has led to a scientific issue and experimental method cannot be available for the natural porous structure [9].

Cellular structures of natural orthodontic such as cellular biomaterials device comes from either two dimensional unit cells or also three dimensional unit cells. Honeycomb is an example of two dimensional unit cells and tetradecahedron and cubic is example of three dimensional unit cells [10]. The mechanical properties of the specimen are based on the type of unit cell and its dimensions [11].

An idea of introducing porous structure in biomedical application is to minimize the stress shielding effect that occurred between the bones and also implant due to mismatch of stiffness characteristics. Therefore, this study is conducted to define the mechanical properties of open cellular structure CoCrMo alloy sample with different designed volume porosity that could mimic the bone.

2. Experimental Set Up

2.1 Sample Design

Three (3) groups of samples were prepared based on the designed volume porosity namely 60 %, 70 % and 80 % that were being designed by aid of Solid Works® 2013 software. Wall thickness of open cellular structures samples were fixed at 0.80 mm with pore sizes in square shape ranging between 1.20 and 1.80 mm within 15 mm cubic. Detailed specifications of individual CAD models are shown in Table 1. The models were exported from CAD software in STL file format into Materialise Magics 17.01 software. The models were then positioned on the build platform (substrate) and sliced files were generated to enable physical sample fabrication process by SLM followed. Typical CAD model of a sample is shown in figure 1.

Table 1. Dimensions and volume based porosity of open cellular structure samples.

Sample Group	Strut Thickness (mm)	Pore Size (mm)	Porosity (%)
1	0.80	1.20	60
2	0.75	1.60	70
3	0.50	1.80	80

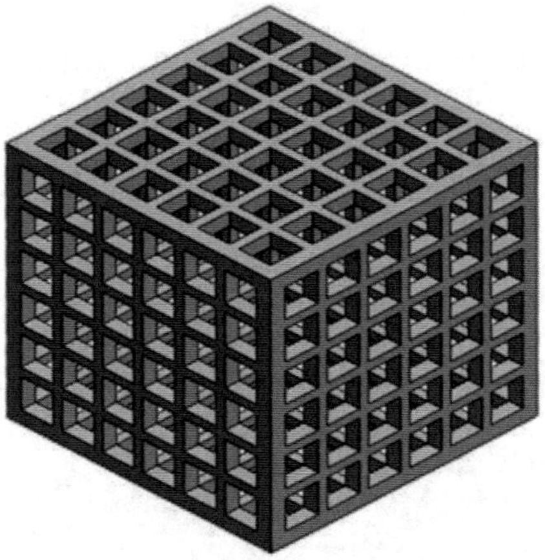

Figure 1. The CAD model of sample with designed volume porosity 80 %.

2.2 Fabrication Process

The material used for this study was EOS CobaltChromeMP1^{TM} supplied by EOS GmbH, Germany. This powder consist mainly element of Co, Cr, and Mo equipped with fine spherical morphology and average particle size of 25 μm. Figure 2 shows the shape of each particle of the powder that was obtained by scanning electron microscope (SEM). The chemical composition of the powder is summarizes in table 2.

Table 2. Chemical composition of CoCrMo alloy.

Element	Mass (%)
Co	60-65
Cr	26-30
Mo	5-7
Si	1.0
Mn	1.0
Fe	0.75
C	0.16
Ni	0.10

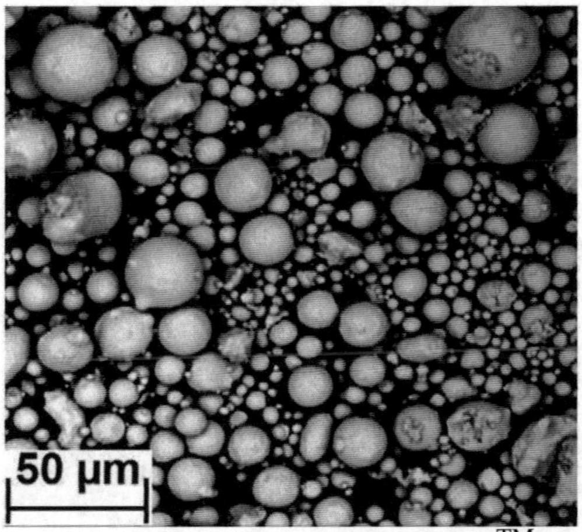

Figure 2. SEM image of EOS Cobalt Chrome MP1^{TM} powder particles.

All of the samples were fabricated by EOSINT M280 machine that was operated on PSW 3.06 software. The powder was manually sieved into the dispenser platform to avoid any powder that agglomerate to each other. Substrate dimension was 250 × 250 × 36 mm and made from tool steel plates. Table 2 summarizes the parameter that is set for the machine during fabrication process. The fabrication chamber was filled by nitrogen gas with less than 1.3 % oxygen present. Figure 3 shows the virtual process inside the fabrication chamber.

Table 3. Setting parameters of SLM process.

Process parameters	Parameters Value
Layer Thickness	20 µm
Laser Power	195 W
Scanning Space	0.1 mm
Laser Scan Speed	800 mm/s

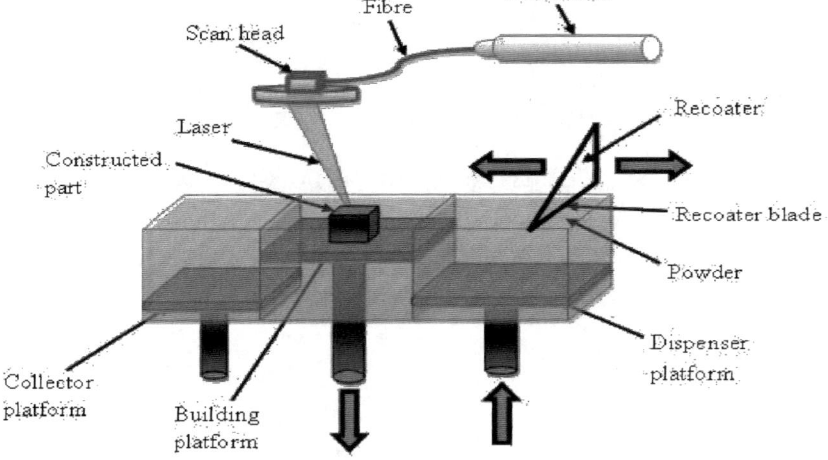

Figure 3. Schematic diagram of SLM process.

2.3 Heat Treatment

Samples from each group was subjected to stress relieve process for 8 and 20 hour of soaking time at 250 ^{o}C and pressure at -0.1 bar. The samples were then oven-cooled to room temperature. Figure 4 shows the oven used for the stress relieves process.

Figure 4. Vacuum oven for stress relieve.

2.4 Hardness

Hardness test was conducted on the sample from each group. The measurement involved samples with and without stress relief process. Each measurement was subjected to five (5) indentations that were on top, bottom, inner, and side face. The setting for hardness measurement was 10 seconds contact for the indentation with 500 gram applied load.

2.5 Compression

Uniaxial compression test was conducted in this study using Shimadzu materials testing machine that possess the capability of 100 kN maximum applied load. The test was performed to determine the effect of different soaking time on mechanical properties of the sample. The tests were performed with stroke of 8 mm and test speed of 0.5 mm/min. The samples were being tested until failure or until the maximum load capacity of the machine was reached. The stress strain relationship for each individual component was calculated from the real time force versus displacement data obtained from the test machine.

3. Results and Discussion

3.1 Hardness

Figure 5 concludes the result for hardness testing on all surfaces using Vickers hardness test. We can clearly observed that hardness values for top, inner, and side faces shows decrease trend as the soaking time increased. At high temperature, the grain size will be increased and deformation occurred which will promote recrystallization process accompanies with softening due to formation of new grains free of dislocations. Recrystallization resulted in reduction of strength and hardness of a material and a simultaneous increase in the ductility. Nonetheless, bottom face has shown distinctive trend, where instead of decreasing, the hardness values on this face keep increasing as the soaking time increased. The phenomenon was due to presence of heat affective zone (HAZ) as a result from electro discharge machining (EDM) wire-cut process to detach samples from the substrate. The rapid rate of cooling during the cutting process favours the formation of hard and brittle martensite in all the sub zones of the HAZ. The presence of martensite in the HAZ results in a very high hardness value for the heat affected zone. Figure 6 shows the effect of HAZ on bottom face of the sample.

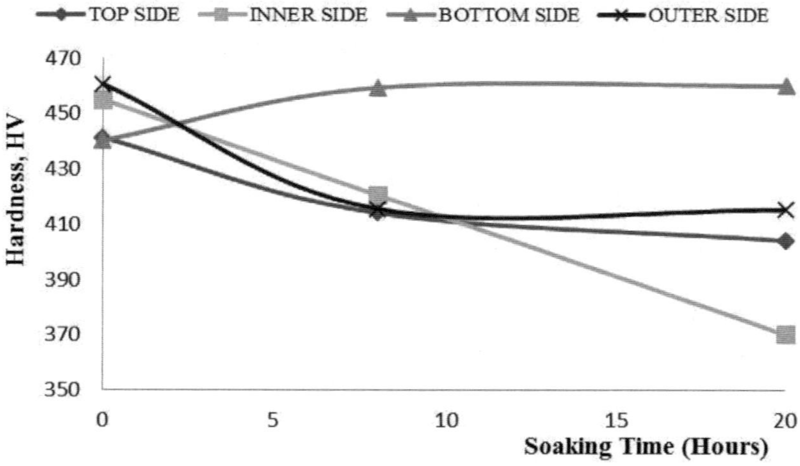

Figure 5. Hardness values for different surfaces.

Figure 6. Heat affective zone (HAZ) on the sample.

IOP Conf. Series: Materials Science and Engineering **100** (2015) 012033 doi:10.1088/1757-899X/100/1/012033

3.2 Compression Test

Compression test data for all samples are presented in figure 7 to figure 9. As expected, 60 % designed volume porosity possess the highest value of ultimate compressive strength compared to 70 % and 80 % in all conditions. Samples with higher porosity will show lower mechanical characteristics. In general, the stress-strain curves can be divided into three (3) distinct regions. First, a quasi-elastic linear increase of the stress-strain curve, which is controlled by cell-wall bending and cell-face stretching. Next region is a deformation "plateau" characterized by a small or vanishing slope of the stress-strain curve with nearly constant flow stress to a large strain. This region is associated with cell collapse. The third region involved of rapidly increasing stress, where the cell walls touch each other and the material is densified.

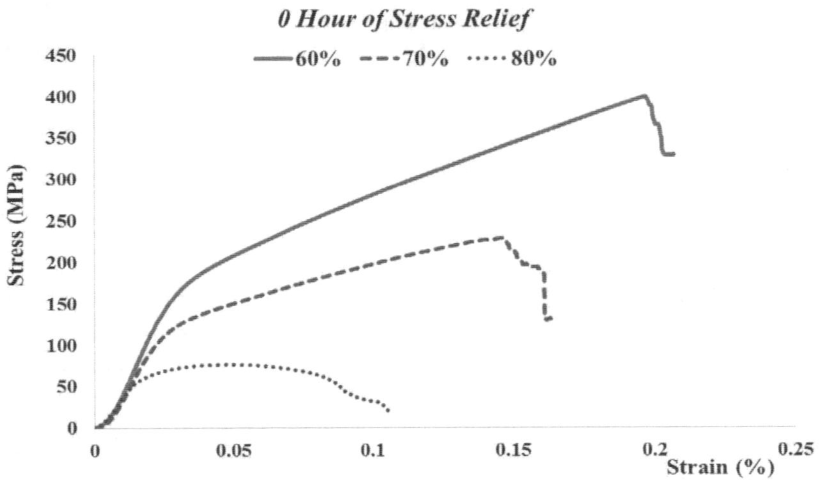

Figure 7. Without stress relief process.

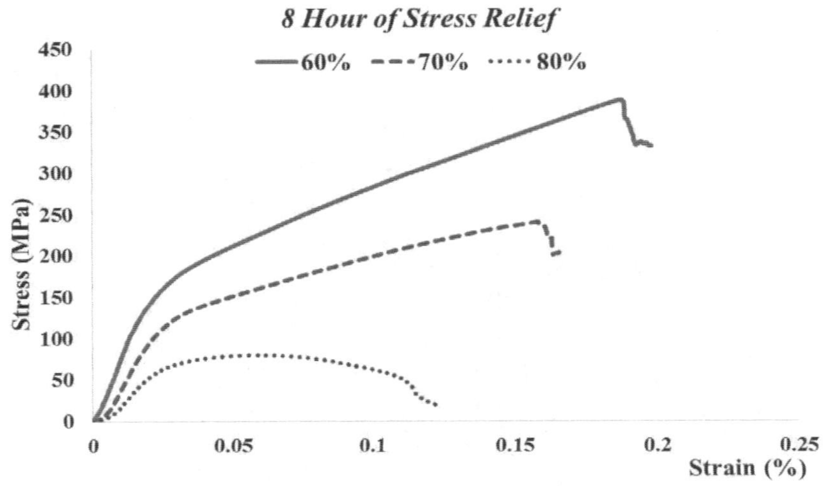

Figure 8. 8 hour of stress relief process.

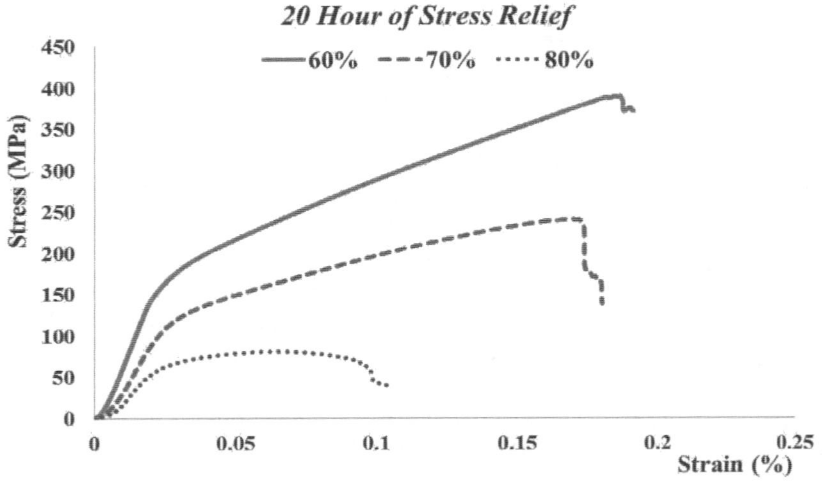

Figure 9. 20 hour of stress relief process.

Table 4 summarizes the ultimate compressive strength along with its strain values in bracket at different condition of soaking time. As mentioned earlier, 60 % designed volume porosity demonstrated the highest ultimate compressive strength and strains for all conditions compared to 70 % and 80 %. In general, prolong the soaking time to 20 hour resulted in improvement of ultimate compressive strength for all samples. It indicates the effectiveness of stress relief process regarding to its function which is to eliminate any residual stress that contained in the samples. However, low-temperature stress relief seem does not give any significant effect to the compressive strength. Therefore, a study will be conducted in future to investigate the differences in compressive strength values at higher temperature of stress relief.

Table 4. Ultimate compressive strength for different soaking time.

Porosity (%)	0 Hour	8 Hour	20 Hour
60	398.49 MPa (0.1967 %)	387.66 MPa (0.1876 %)	389.62 MPa (0.1846 %)
70	228.52 MPa (0.1458 %)	240.14 MPa (0.1583 %)	241.42 MPa (0.1688 %)
80	76.42 MPa (0.0486 %)	79.10 MPa (0.0594 %)	81.00 MPa (0.0649 %)

4. Conclusions

The effects of different designed volume porosity; 60 %, 70 %, and 80 % on mechanical properties were successfully identified. Hardness values showed decrease trend as soaking time increased for all surfaces except for bottom face. Samples with higher porosity demonstrated lower mechanical properties. 60 % designed volume porosity exhibit the highest ultimate compressive strength compared to 70 % and 80 %. Meanwhile, higher soaking time seem to give positive impact to the mechanical properties of the samples.

Acknowledgements

The authors would like to be obliged to Universiti Malaysia Pahang for providing laboratory facilities and financial assistance under project no. RDU140354, and thank to Kolej Kemahiran Tinggi Mara Kuantan for technical support

References

[1] Jamshidinia M, Wang L, Tong W, Kovacevic R 2014 *Journal of Materials Processing Technology* **214** 1728-39

[2] Chen J, Zhang Z, Chen X, Zhang C, Zhang G, Xu Z 2014 *The Journal of Prosthetic Dentistry* **112** 1088-95.e1

[3] van Noort R 2012 *Dental Materials* **28** 3-12

[4] Rodrigues W C, Broilo L R, Schaeffer L, Knörnschild G, Espinoza F R M 2011 *Powder Technology* **206** 233-8

[5] Xin X Z, Chen J, Xiang N, Gong Y, Wei B 2014 *Dental Materials* **30** 263-70

[6] EOS 2011 *Electro Optical System (Ed)*

[7] Lin H-Y, Bowers B, Wolan J T, Cai Z, Bumgardner J D 2008 *Dental Materials* **24** 378-85

[8] Tang H P, Wang J Z, Zhu J L, Ao Q B, Wang J Y, Yang B J, et al. 2012 *Powder Technology* **217** 383-7

[9] Tang H, Zhu J, Xi Z, Di X, Wang J, Ao Q 2010 *Science China Technological Sciences* **53** 348-51

[10] Li X, Wang C, Zhang W, Li Y 2010 *Rapid Prototyping Journal* **16** 44-9

[11] Parthasarathy J, Starly B, Raman S, Christensen A 2010 *Journal of the mechanical behavior of biomedical materials* **3** 249-59

IOP Conf. Series: Materials Science and Engineering **100** (2015) 012034 doi:10.1088/1757-899X/100/1/012034

Development of a KT driving cycle for UMT PHEV powertrain

W H Atiq[1], M N Haezah[1], J S Norbakyah[1] and A R Salisa[1]
[1]School of Ocean Engineering, Universiti Malaysia Terengganu,
21030 Kuala Terengganu, Terengganu, Malaysia

E-mail: salisa@umt.edu.my

Abstract. Driving cycles were identified as one of the core sources that contribute to develop the powertrain for vehicle. Plug-in hybrid electric vehicles (PHEVs) are the future transport for next generation. Compared to conventional internal combustion engine vehicle, hybrid and electric vehicle can improve fuel economy and reduce green house gases. This paper describes a development of Kuala Terengganu driving cycle for Universiti Malaysia Terengganu PHEV. Car speed-time data along the two selected fixed route is obtained by using on-board technique which is Global Positioning System, GPS. The developed driving cycle contains a 1050s speed time series, with a distance of 2.17 km, and an average and a maximum speed of 20.67 km/h and 61.47 km/h, respectively. The results obtained from this analysis are within reasonable range and satisfactory.

1. Introduction

The significant growth of world population had led to an increased use of transportation resulting in pollution and other serious environmental problem. The urge for more efficient and greener vehicle is greatly on demand for modern cities. Proactive measures should be taken to minimize these emissions. Nowadays, the automotive industry move their sight from producing internal combustion engine (ICE) vehicles to more efficient and eco-friendly hybrid electric vehicles. Plug-in hybrid electric vehicle (PHEV) is a vehicle which uses two or more kinds of propulsion. This paper introduces a method of developing Kuala Terengganu (KT) driving cycle for Universiti Malaysia Terengganu (UMT) PHEV. Figure 1 shows a schematic ilustration of the UMT PHEV powertrain. The UMT PHEV has only one electric machine (EM) which functions as either a motor or generator at a time and powered by batteries and ultracapacitors packs. Driving cycle for car had been developed in several parts of the world including America, Europe, Australia and Asia [1]. The purpose of a driving cycle is to simulate actual driving characteristics on the road, in order to measure the vehicle exhaust emission and fuel consumption [2]. Actual driving cycle, are the cycle derived from the movement of test vehicle on the road under real traffic conditions [3]. The driving cycle for different country have various set of parameters.

2. Driving cycle development

Basically, developing a driving cycle includes of recording the driving condition using one or several instrumented vehicles, which are usually driven for normal purposes [4]. Data will be recorded as the image of the traffic conditions. Then, all data will be analyzed in order to subsequently describe or characterized this condition. To produce a driving cycle, the research tackles three major tasks, which are data collection, route selection and cycle construction. In this paper, on-board measurement technique will be conducted to collect vehicle speed data. The data are obtained using on-board measurement techniques, which is global positioning system (GPS) to collect car speed-time data along the selected route. The flowchart for KT driving cycle is illustrated in figure 2.

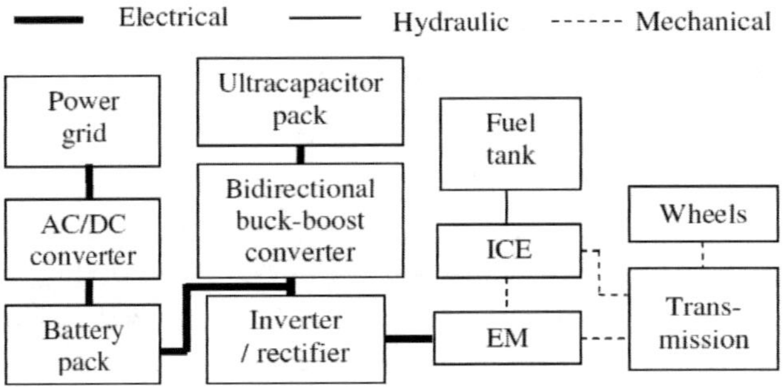

Figure 1. Schematic illustration of UMT PHEV powertrain.

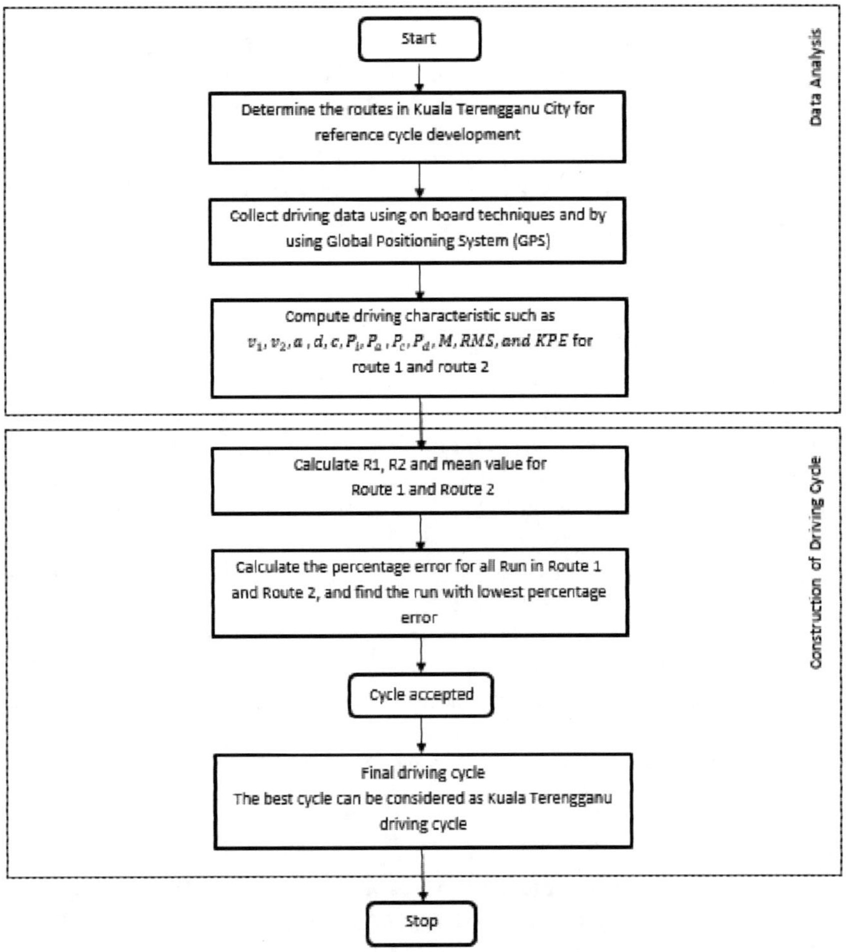

Figure 2. Flowchart for KT driving cycle development.

Two different routes were selected in KT city are shown in figures 3 and 4, due to the level of congestion. The 12 sets of relevant variable parameters were adopted in the development of a car's driving cycle construction. Table 1 lists the variable used in a KT driving cycle development.

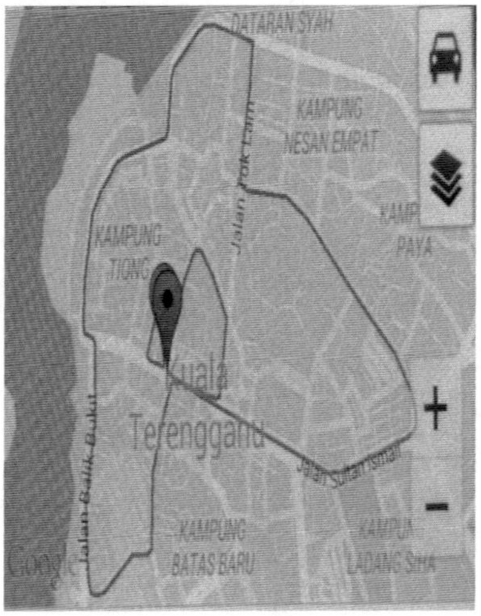

Figure 3. Selected route 1.

Figure 4. Selected route 2.

Table 1. Variables used in KT driving cycle construction [4].

No	Variable	Unit	Formula
1	Average speed,	km/h	$v\_avg = 3.6\frac{dist}{T_{total}}$
2	Average running speed, v_2	km/h	$v\_run = 3.6\frac{dist}{T_{drive}}$
3	Average acceleration of all acceleration phases, a	m/s^2	$a\_avg = \left(\sum_{i=1}^{n}\begin{cases}1\,(a_i>0)\\0\text{ (else)}\end{cases}\right)^{-1}\sum_{i=1}^{n}\begin{matrix}1 & (a_i>0)\\0 & \text{(else)}\end{matrix}$
4	Average deceleration of all deceleration phases, d	m/s^2	$d\_avg = \left(\sum_{i=1}^{n}\begin{cases}1\,(a_i<0)\\0\text{ (else)}\end{cases}\right)^{-1}\sum_{i=1}^{n}\begin{matrix}1 & (a_i<0)\\0 & \text{(else)}\end{matrix}$
5	Mean length of a driving period, c	s	-
6	Time proportion of idling, P_i	%	$\%drive = \frac{T_{drive}}{T_{total}}$
7	Time proportion of acceleration, P_a	%	$\%acc = \frac{T_{acc}}{T_{total}}$
8	Time proportion of cruise, P_c	%	$\%cruise = \frac{T_{cruise}}{T_{total}}$

9	Time proportion of deceleration, P_d	%	$\%dec = \frac{T_{dec}}{T_{total}}$
10	Average number of acceleration-deceleration changes within one driving period, M		-
11	Root mean square of acceleration, RMS	m/s^2	$RMS = \sqrt{\frac{1}{T_{total}}\sum_{i=1}^{n} a_i^2}$
12	Acceleration energy per kilometer, PKE	m/s^2	$PKE = \frac{1}{dist}\sum_{i=2}^{n}\begin{cases} v_i^2 - v_{i-1}^2 & (v_i > v_{i-1}) \\ & (else) \end{cases}$

After the route is selected, the test run is handled to compute all the 12 parameters, which are average speed, average running speed, average acceleration of all acceleration phases, average deceleration of all deceleration phases, time proportion of idling, time proportion of acceleration, time proportion of cruise, time proportion of deceleration, average number of acceleration-deceleration changes within one driving period, root mean square of acceleration and acceleration energy per kilometer. After all 40 runs are computed, the mean value and the percentage error for both road are calculated.

3. Results and discussions

Speed-time data was collected by using GPS within two selected road. The data was collected at a peak hour ranging from 5:00 - 7:00 p.m for 40 days on November and December 2014. The assessment parameters derived from the test run are shown in tables 2 and 3. There were 40 tests run for both selected routes. There are significant differences between runs 1 - 20 and 21- 40. The road pattern is different in KT city and 20 assessments for each selected KT road is calculated and the mean value is listed in table 4. After the mean value is calculated, percentage difference relative to target summary statistics is calculated and the result is illustrated in table 5. Minimum percentage error of the assessment parameters of grouped runs is shown in table 6 and the KT driving cycle is selected based on the lowest percentage error. The lowest percentage error for run 20 in route 1 is selected as the best KT driving cycle. The driving pattern shows that the recommended KT driving cycle, as shown in figure 5 has an average speed of 20.67 km/h and the running speed is 23.54 km/h with idling. The maximum speed for KT driving cycle is 61.47 km/h. The rate of acceleration and deceleration is 0.39. The mean length of driving period is low, which is 131.25. It shows that the route is full of obstacle and the car has to stop for several times. The root mean square for this run is 0.55 m/s^2 and 0.32 m/s^2. The best KT driving cycle is constructed as presented in figure 5.

Table 2. Analysis data for route 1.

Criterion	v_1 (km/h)	v_2 (km/h)	a (m/s^2)	d (m/s^2)	c (s)	P_i (%)	P_c (%)	p_a (%)	P_d (%)	M	RMS (m/s^2)	KPE (m/s^2)
j	1	2	3	4	5	6	7	8	9	10	11	12
Run 1	21.09	27.58	0.71	0.67	52.31	23.54	4.02	35.35	37.16	351.37	1.03	0.59
Run 2	21.61	23.03	0.34	0.37	167.67	6.16	8.45	44.38	41.09	446.41	0.45	0.28
Run …	…	…	…	…	…	…	…	…	…	…	…	…
Run …	…	…	…	…	…	…	…	…	…	…	…	…

Run …	…	…	…	…	…	…	…	…	…	…	…	…
Run 19	20.96	22.21	0.35	0.40	144.43	5.64	2.67	49.01	42.77	495.43	0.48	0.33
Run 20	20.67	23.54	0.39	0.39	131.25	12.19	8.29	39.75	39.85	417.40	0.55	0.32

Table 3. Analysis data for route 2.

Criterion	v_1 (km/h)	v_2 (km/h)	a (m/s^2)	d (m/s^2)	c (s)	P_i (%)	P_c (%)	p_a (%)	P_d (%)	M	RMS (m/s^2)	KPE (m/s^2)
j	1	2	3	4	5	6	7	8	9	10	11	12
Run 21	25.81	27.17	0.32	0.36	168.00	5.00	11.67	44.34	44.34	372.39	0.45	0.24
Run 22	20.66	22.31	0.28	0.32	174.00	7.38	11.97	39.98	37.68	449.38	0.40	0.22
Run …	…	…	…	…	…	…	…	…	…	…	…	…
Run …	…	…	…	…	…	…	…	…	…	…	…	…
Run …	…	…	…	…	…	…	…	…	…	…	…	…
Run 39	19.06	21.34	0.24	0.33	79	10.67	0.36	51.86	37.92	573.38	0.45	0.27
Run 40	18.61	20.79	0.28	0.33	70.12	10.52	0.27	48.88	40.95	548.41	0.44	0.25

Table 4. Mean values of the assessment parameters of grouped runs.

Criterion	v_1 (km/h)	v_2 (km/h)	a (m/s^2)	d (m/s^2)	c (s)	P_i (%)	P_c (%)	p_a (%)	P_d (%)	M	RMS (m/s^2)	KPE (m/s^2)
j	1	2	3	4	5	6	7	8	9	10	11	12
Route 1 (run 1 – run 20)	20.48	23.97	0.49	0.49	121.23	14.78	9.48	38.34	37.91	399.28	0.67	0.40
Route 2 (run 21 – run 40	20.26	21.91	0.28	0.33	124.80	7.61	10.53	44.34	37.64	477.97	0.42	0.23
Mean Value	20.37	22.94	0.39	0.41	123.02	11.20	10.00	41.34	37.77	438.63	0.55	0.32

Table 5. Percentage difference relative to target summary statistics.

Criterion	V_1 (km/h)	V_2 (km/h)	a (m/s^2)	d (m/s^2)	c (s)	P_i (%)	P_c (%)	P_a (%)	P_d (%)	M	RMS (m/s^2)	KPE (m/s^2)	Total Error
j	1	2	3	4	5	6	7	8	9	10	11	12	
Mean Value	20.37	22.94	0.39	0.41	123.02	11.20	10.00	41.34	37.77	438.63	0.55	0.32	
Run 1	3.53	20.23	82.05	63.41	57.48	110.18	59.80	14.49	1.62	19.89	87.27	84.37	666.11
Run 2	3.53	0.39	12.82	9.76	36.29	45.00	15.50	7.35	8.79	1.77	18.18	12.50	172.85
Run …	…	…	…	…	…	…	…	…	…	…	…	…	…
Run	…	…	…	…	…	…	…	…	…	…	…	…	…

IOP Conf. Series: Materials Science and Engineering **100** (2015) 012034 doi:10.1088/1757-899X/100/1/012034

… Run …	…	…	…	…	…	…	…	…	…	…	…	…	…
Run 19	2.90	3.18	10.26	2.44	17.40	49.64	73.30	18.55	13.24	12.95	12.73	3.13	219.72
Run 20	1.47	2.62	0	4.88	6.69	8.84	17.10	3.85	5.51	4.84	0	0	55.80
Run …	…	…	…	…	…	…	…	…	…	…	…	…	…
Run …	…	…	…	…	…	…	…	…	…	…	…	…	…
Run …	…	…	…	…	…	…	…	…	…	…	…	…	…
Run 39	6.43	6.97	38.46	19.51	35.78	4.73	96.40	25.45	0.40	30.72	18.18	15.62	298.65
Run 40	8.64	9.37	28.21	19.51	43.00	6.07	97.30	18.24	8.42	25.03	20.00	21.88	305.67

Table 6. Minimum percentage error of the assessment parameters of grouped runs.

Criterion	v_1 (km/h)	v_2 (km/h)	a (m/s^2)	d (m/s^2)	c (s)	P_i (%)	P_c (%)	p_a (%)	P_d (%)	M	RMS (m/s^2)	KPE (m/s^2)	Total Error
J	1	2	3	4	5	6	7	8	9	10	11	12	
Run 2	3.53	0.39	12.82	9.76	36.29	45.00	15.50	7.35	8.79	1.77	18.18	12.50	172.85
Run 4	1.62	0.83	2.56	2.44	7.30	19.64	84.70	10.11	6.14	10.55	3.64	3.13	148.96
Run 11	0.59	2.48	5.13	7.32	8.93	24.11	27.20	8.30	8.53	7.35	1.82	6.25	108.01
Run 12	2.75	2.27	7.69	2.44	3.00	3.93	87.90	3.92	17.21	5.20	3.64	18.75	158.70
Run 18	0.98	1.87	0	4.88	6.69	7.32	78.10	10.23	11.04	9.07	0	18.75	148.93
Run 20	1.47	2.62	0	4.88	6.69	8.84	17.10	3.85	5.51	4.84	0	0	55.80
Run 22	1.42	2.75	28.21	21.95	41.44	34.11	19.70	3.29	0.24	2.45	27.27	31.25	214.08
Run 23	6.97	7.24	25.64	21.95	17.53	2.41	11.10	0.19	2.65	4.96	21.82	34.38	156.84
Run 27	3.68	2.88	25.64	12.20	18.94	6.61	33.10	0.19	11.60	3.58	16.36	28.13	162.91
Run 34	3.93	5.97	28.21	14.63	17.90	17.23	21.50	6.34	8.18	11.34	29.09	28.13	192.45

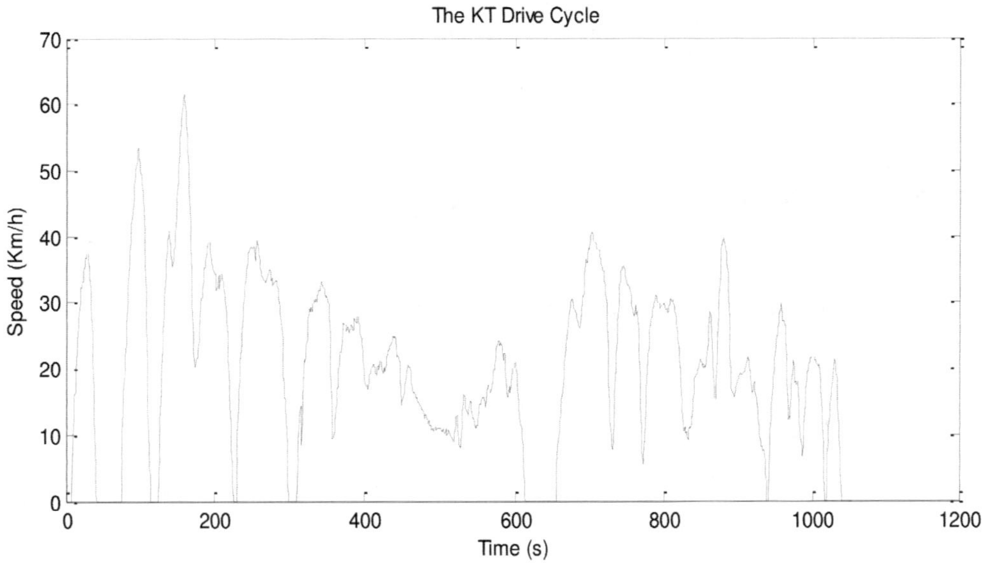

Figure 5. The recomended KT driving cycle.

4. Conclusions

Based on the results of this study, it can be concluded that the proposed method is possible to generate a recomended KT driving cycle that can be used for UMT PHEV powertrain, in order to measure fuel economy and emissions.

Ackowledgements

The authors would like to be obliged to Universiti Malaysia Terengganu and Ministry of Education Malaysia for providing financial assistance under project no. 59353.

References

[1] Hung W, Tong H, Lee C, Ha K, Pao L 2007 *Transportation Research Part D: Transport and Environment* **12** 115-28
[2] Tong H Y, Hung, W.T., & Cheung, C.S 1999 *Atmospheric Environment* **33** 2323-35
[3] Kamble S H, Mathew T V, Sharma G 2009 *Transportation Research Part D: Transport and Environment* **14** 132-40
[4] BARLOW T J, Latham S, McCrae I, Boulter P 2009 *A reference book of driving cycles for use in the measurement of road vehicle emissions*

Power requirements for PHERB powertrain

J S Norbakyah[1], **W H Atiq**[1] **and A R Salisa**[1]
[1]School of Ocean Engineering, Universiti Malaysia Terengganu,
21030 Kuala Terengganu, Terengganu, Malaysia

E-mail: salisa@umt.edu.my

Abstract. Boats are considered as favourite maritime transportation designed for recreation activities, fishing and surveillance purposes. However, in tropical developed countries, boats are employed for different applications such as passenger and goods transportation. In this paper, the power requirements for a proposed plug-in hybrid electric recreational boat (PHERB) powertrain is determined using a steady state velocity and the Kuala Terengganu river driving cycle according to the boat parameters, specifications and performance requirements. The boat power requirements can be used to size the main components for PHERB powertrain. The results obtained from this analysis are within reasonable range and satisfactory.

1. Introduction

There are several energy resources which are used widely in transportation, such as petroleum, coal, and natural gases. Currently, fuel from the petroleum sources is the major power and almost drained due to transportation consumptions [1]. The strain pertaining fuel availability and the increasing need of power in vehicle due to modern lifestyles, is resulting in a need for social and technological revolution that is capable of bringing a new, stable and durable balance between consumption and environmental concerns [2]. Boat propulsion is significantly responsible for environmental problems such as the greenhouse effect, acid rain and air pollution related to the emission issues [3-5].

The purpose of this paper is to introduce the plug-in hybrid electric recreational boat (PHERB) in order to reduce the fuel consumption and emissions for marine transportation by Figure 1 shows a schematic illustration of the series-parallel PHERB powertrain. The PHERB powertrain has only one electric machine (EM) to function as either electric generator or motor in different time intervals specified by a special developed energy management strategy that control the power flow according to desired operating mode. In order to improve the drive performance and energy efficiency, the PHERB uses a hybrid energy storage system (ESS) combining batteries and ultracapacitors, which can work together effectively. The size of the internal combustion engine (ICE) can be reduced since it is operated during certain operation mode only.

In this paper, a theoretical research of power requirement for PHERB powertrain was studied using a steady state velocity and a Kuala Terengganu (KT) river driving cycle. The calculation of power requirements for PHERB powertrain are based on the boat parameters, specifications and performance requirements.

Published under licence by IOP Publishing Ltd

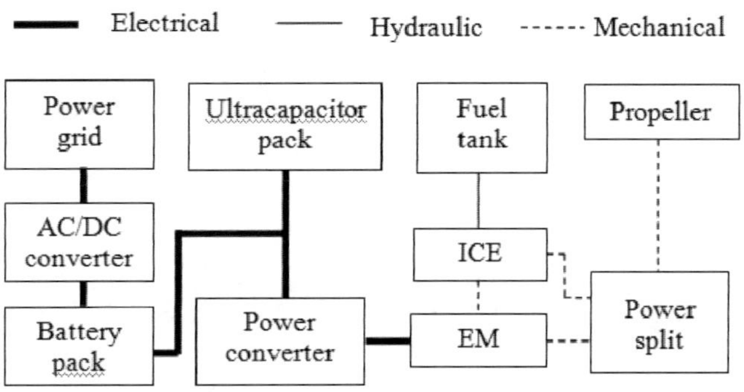

Figure 1. Schematic illustration of series-parallel PHERB powertrain.

2. Theoretical equations

To calculate the propulsion power for PHERB powertrain, the resistance and the total propulsive efficiency have to be determined with the highest possible accuracy. As empirical methods are normally used for these calculations, it is worthwhile to at least to know the accuracy of the different elements in the calculation procedures such that the propulsive power can be predicted in combination with an estimate of the uncertainty of the result.

2.1. Resistance

The resistance of a boat at a given speed is the force required to tow the boat at that speed in smooth water for the boat velocity, v, and power propulsion, P_P in the low velocity range. When cruising, a boat encounter frictional resistance, C_F , incremental resistance, C_A, air resistance, C_{AA}, and residual resistance, C_R. At low velocity, frictional resistance, C_F is dominant. The total resistance coefficient, C_T of boat can be defined by

$$C_T = C_F + C_A + C_{AA} + C_R = \frac{R_T}{0.5 \rho V^2 S} \tag{1}$$

where R_T is the boat resistance, ρ the density of water, S the wetted surface and v the speed.

The wetted surface, S is normally calculated by hydrostatic programs. However for a quick and fairly accurate estimation of the wetted surface many different methods and formulas exist based on only boat main dimensions.

$$S = 1.025 \times L_{PP} \left(C_B B + 1.7T \right) \tag{2}$$

where L_{pp} is the length between perpendicular, C_B the block coefficient, B the breadth and T the traught.

The frictional resistance coefficient is defined by

$$C_F = \frac{0.075}{\left(\log R_n - 2 \right)^2} = \frac{R_F}{0.5 \rho V^2 S} \tag{3}$$

where the frictional resistance, is sum of tangential stresses along the wetted surface in the direction of the motion. R_n is the Reynolds number.

$$R_n = \frac{v \times L_{wt}}{\upsilon} \tag{4}$$

where υ is the kinematic viscosity of water, L_{wt} the length hull in waterline and v the boat speed.

The total resistance, R_T is defined as,

$$R_T = 0.5\rho C_T V^2 S \tag{5}$$

where ρ is the density of water, S the wetted surface, v the speed and C_T total resistance coefficient

2.2. Propulsive efficiency

Total propulsive efficiency can be defined as

$$\eta_T = \eta_H + \eta_O + \eta_R + \eta_S \tag{6}$$

where η_H is the hull efficiency, η_O the propeller in open water condition, η_R the relative rotative efficiency and η_S the transmission efficiency. In order to determined th propulsion power, the effective power can calculated using equation (7)

$$P_E = R_T \times V \tag{7}$$

where R_T is the total resistance and v the boat speed.

Power requirement for propulsion can be estimated , P_P using equation (8)

$$P_P = P_E \times \eta_T \tag{8}$$

where P_E is the effective power and η_T the total propulsive efficiency.

3. Results and discussions

Based on the parameters, specifications and performance requirements of PHERB powertrain in table 1, the power requirement of the boat can be determined. Figure 2 shows the selected route for KT river driving cycle. As in figure 3, the KT drive cycle lasts for 2414 s covering a distance of 12.09 km with an average speed of 18.02 km/h and maximum speed of 22 km/h.

Table 1. The parameters, specifications and performance requirements of PHERB powertrain

Parameters and Specifications		
Variable	**Description**	**Dimension**
L	Lenght overall	12.4 m
L_{WT}	Lenght at waterline	11.0 m
B	Breath	1.8 m
T	Draught	0.64 m
L_{PP}	Length between perpendicular	10.67 m
ρ	Density of water	1000 kgm^{-3}
W	Displacement	395 kg
Performance Requirements		
Maximum speed		30 km/h
EV range		10 km

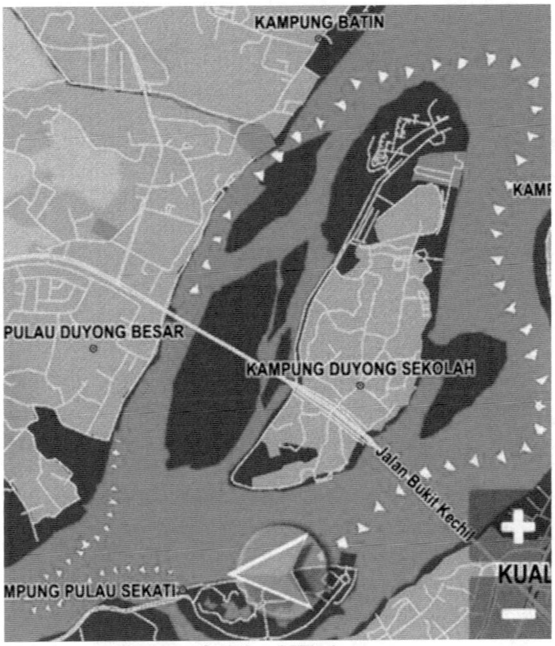

Figure 2. The KT river route.

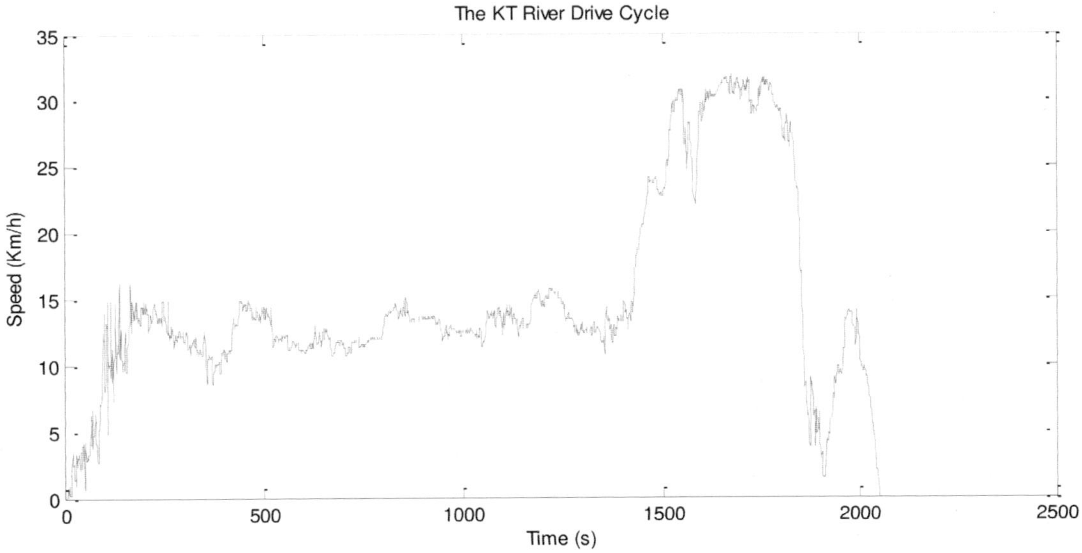

Figure 3. The KT river driving cycle.

The relationship between speed and power requirement of the PHERB powertrain are shown in figures 4 and 5. The drivetrain power required for PHERB is determined using equation (8). During the steady state velocity, when speed is higher, the power requirement is higher too. The maximum speed of 40 km/h needed 17 kW of power requirement and for the minimum speed of 5 km/h, only 0.4 kW power is required. The drivetrain power requirements are different during different speed for KT river driving cycle. The size and capacity of the main components, which are EM, ESS and ICE can be selected based on the drivetrain power requirement.

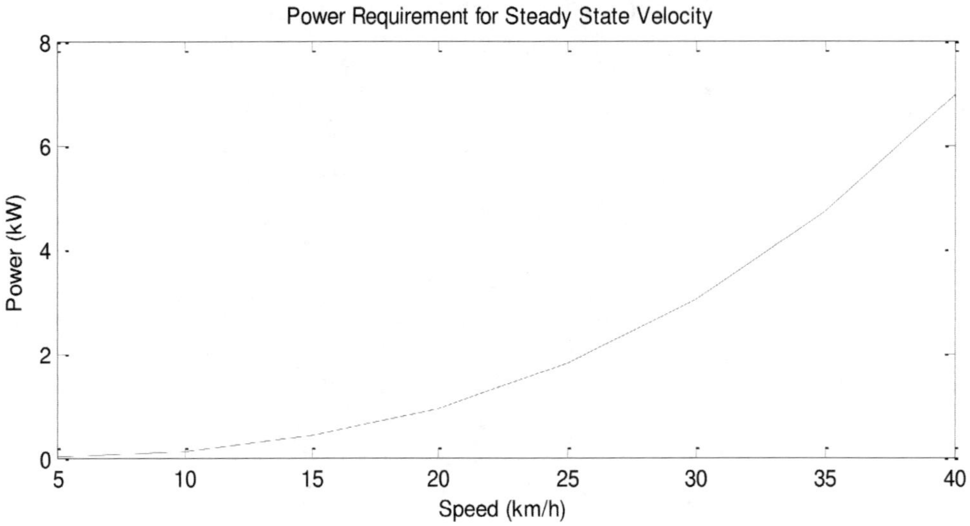

Figure 4. The PHERB power requirement for steady state velocity.

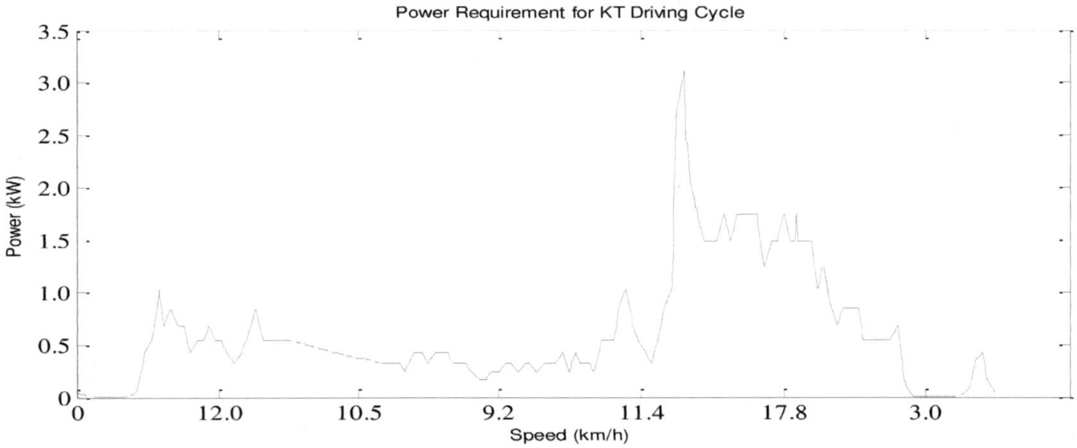

Figure 5. The PHERB power requirement for KT river driving cycle.

4. Conclusions

Based on the results of this study, it can be concluded that the proposed method is possible to generate a recomended KT driving cycle that can be used for UMT PHEV powertrain, in order to measure fuel economy and emissions.

Ackowledgements

The authors would like to be obliged to Universiti Malaysia Terengganu and Ministry of Education Malaysia for providing financial assistance under project no. 59353.

IOP Conf. Series: Materials Science and Engineering **100** (2015) 012035 doi:10.1088/1757-899X/100/1/012035

References

[1] Freire T, Sousa D M, Branco P J C, Aspects of modeling an electric boat propulsion system. Computational Technologies in Electrical and Electronics Engineering (SIBIRCON), 2010 IEEE Region 8 International Conference on; 2010: IEEE.

[2] Minami S, Hanada T, Matsuda N, Ishizu K, Nishi J, Fujiwara T 2013 *Journal of Asian Electric Vehicles* **11** 1653-7

[3] Corredor L, Baracaldo L, Jaramillo L, Gutiérrez J, Jiménez D. A comprehensive energy analysis of a hybrid motorization for small/medium boats. International Conference on Renewable Energies and Power Quality; 28-30 March; Santiago de Compostela, Spain2012. p. 1-6.

[4] Luttenberger L R A I, Šestan A, Vladimir N, Integrated power systems in small passenger ships. PlugBoat 2013 World Electric & Hybrid Boat Summit; 2013 10-11 October 2013; Nice, France.

[5] Nóbrega J, Dan T, Rubanenco I. Electric Propulsion Applied for Research Vessel. International Conference on Renewable Energies and Power Quality; 20-22 March; Bilbao, Spain2013. p. 1-6.

A conceptual design of main components sizing for UMT PHEV powertrain

M N Haezah[1]**, J S Norbakyah**[1]**, W H Atiq**[1] **and A R Salisa**[1]
[1]School of Ocean Engineering, Universiti Malaysia Terengganu,
21300 Kuala Terengganu, Terengganu, Malaysia

E-mail: salisa@umt.edu.my

Abstract. This paper presents a conceptual design of main components sizing for Universiti Malaysia Terengganu plug-in hybrid electric vehicle (UMT PHEV) powertrain. In the design of hybrid vehicles, it is important to identify a proper component sizes. Component sizing significantly affects vehicle performance, fuel economy and emissions. The proposed UMT PHEV has only one electric machine (EM) which functions as either a motor or generator at a time and using batteries and ultracapacitors as an energy storage system (ESS). In this work, firstly, energy and power requirements based on parameters, specifications and performance requirements of vehicle are calculated. Then, the parameters for internal combustion engine, EM and ESS are selected based on the developed Kuala Terengganu drive cycle. The results obtained from this analysis are within reasonable range and satisfactory.

1. Introduction

Along with the common goal of reducing fuel consumption and resulting emissions compared to classical vehicles with internal combustion engine (ICE) for vehicles, the hybrid electric vehicles (HEVs) are increasingly popular as the most promising solution to energy crisis and urban air pollution [1]. In order to obtain a better fuel economy and vehicle performance, the design and sizing of its powertrain components are important [2, 3]. This paper introduces a method for the powertrain component sizing of a series-parallel Universiti Malaysia Terengganu plug-in hybrid electric vehicle (UMT PHEV) [4, 5]. In the design of HEV, it is important to identify proper component sizes [6-8]. This study focuses on the impact of real driving cycle, most daily trips demand. Figure 1 shows schematic illustration of the proposed series-parallel UMT PHEV powertrain.

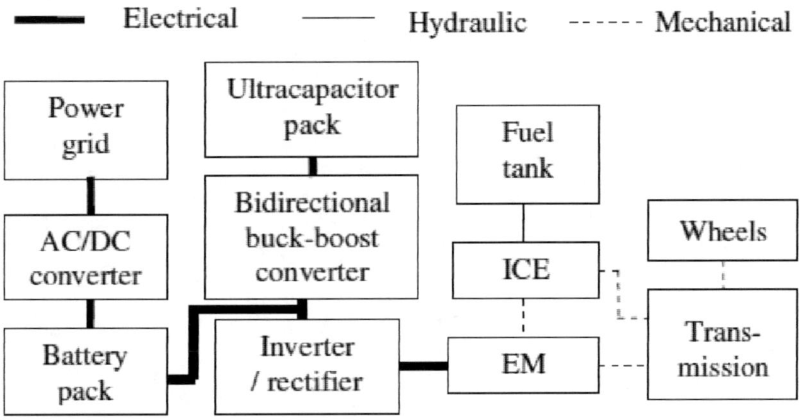

Figure 1. Schematic illustration of series-parallel UMT PHEV powertrain.

2. UMT PHEV parameters, specifications and power requirements

To develop the UMT PHEV powertrain, calculations of the vehicle energy and power requirements are based on the parameter, specifications and performance requirements. Table 1 lists the typical parameters, specifications and performance requirements of a vehicle of this type.

Table 1. UMT PHEV parameters, specifications and performance requirements.

Parameters and specifications	
Vehicle configuration	Series-parallel
Vehicle class	Midsize
Vehicle mass	1360 kg
Aerodynamic drag coefficient	0.26
Coefficient of rolling resistance	0.01
Frontal area	2.57 m^2
Wheel radius	0.282 m
Air density	1.2 kg/m^3
Gravitational acceleration	9.81 m/s^2
Performance requirements	
Maximum speed	Over 80 km/h
EV range	54 km

The power required, P_{req} for a vehicle as shown in figure 2 is calculated as

$$P_{req} = P_{aero} + P_{roii} + P_{grade} + P_{accel} \quad (1)$$

where P_{aero} is a power to overcome aerodynamic drag which is the present by the standard aerodynamic equation that includes the density of air (ρ), the drag coefficient of the vehicle (C_d), the frontal area of the vehicle (A) and the velocity(v). P_{roii} is a power to overcome rolling resistance which the product of the vehicle mass (m), the gravitational acceleration due to gravity (g), the rolling resistance (C_{rr}), the grade angle of the road measured from the horizontal plane (θ) and velocity. P_{grade} is a power for ascending which is product of the vehicle mass, the gravitational acceleration due to gravity, the sine of the grade angle and velocity. Finally, P_{accel} is a power for acceleration is merely product of the mass of the vehicle, acceleration (a) and v is velocity.

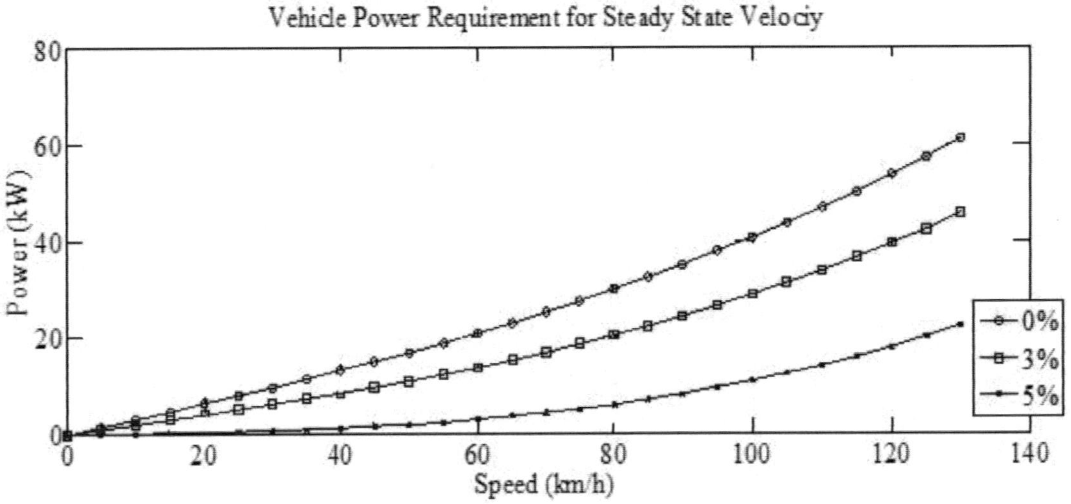

Figure 2. Vehicle power requirement for steady state velocity.

3. Main components sizing

Based on the vehicle power requirements for steady state velocity, the main components of the UMT PHEV powertrain were sized.

3.1. Electric Machine (EM)

The power requirement of the electric propulsion motor is determined by the maximum speed and the maximum gradient at this speed. The maximum gradient is 5 %. The designed maximum speed is assumed as 130 km/h. All calculations are undertaken with maximum mass. To achieve 130 km/h with 5 % gradient, the propulsion motor power requirement is:

PEM (5 %, 130 km/h) = 61.27 kW

Motor size and cost may be reduced if the speed demand is relaxed. If the vehicle is designed to run at 90 km/h it will still get the requirements, but allowing for a smaller propulsion:

PEM, continuous = PEM (5 %, 100 km/h) = 40.75 kW

3.2. Internal Combustion Engine (ICE)

The ICE requirements are determined by the average power requirements in the series UMT PHEV concept. Cruising at 110 km/h, the maximum velocity is assumed to define the average power in the worst case scenario. The continuous ICE output power requirement is:

PICE, continuous = PEM (0 %, 110 km/h) = 14.18 kW

The electric output is 14 kW with an estimated efficiency of 85 %, the mechanical input power has to be 20 kW. This is the minimum continuous ICE power requirement.

PICE, continuous = 20 kW

IOP Conf. Series: Materials Science and Engineering **100** (2015) 012036 doi:10.1088/1757-899X/100/1/012036

3.3. Energy Storage System (ESS)

There are two main energy storage requirements, which are an available energy and a maximum power. The available energy should be sufficient 50 km/h in pure electric driving mode. The average velocity is about 50 km/h. In a simplified calculation, an average of 50 km/h is assumed. This is to take into account that the average speed is based on a higher speed plateau but with frequent starts and stops. The motor power to propel the boat at 50 km/h is:

$$PEM\ (0\ \%,\ 50\ km/h) = 2.01\ kW$$

Assuming an overall drivetrain efficiency of about 60 %, the required battery storage capacity is at least:

$$Ebattery,\ min = (50\ km/\ 50\ km/h)\ x\ (2.01\ kW\ /\ 0.6) = 3.35\ kWh$$

The battery power should be sufficient to boost the propulsion motor to its highest power. Maximum motor power is 1.5 times continuous motor power.

$$Pbattery,\ max = 1.5\ x\ PEM,\ continuous\ -\ PICE\ ,\ continuous = 46.95\ kW$$

In order to achieve full performance, a maximum discharge of 3C (3 times the rated capacity) was assumed. The battery storage capacity is determined by the requirement, provided it also meets the criteria for pure electric range:

$$Ebattery, = Pbattery,\ max\ /\ 3\ x\ h = 15.65\ kWh$$

4. Selected components parameters and specifications

The selected main components of UMT PHEV powertrain for steady state velocity, which are EM, ICE and ESS based on each component specifications and requirements during the sizing process are list in table 2.

Table 2. Main components of the UMT PHEV powertrain.

Component	Specifications
ICE	1.5L, 57kW @ 5000rpm
EM	50kW AC induction motor
Battery	Li, 10kWh, 28Ah

5. Results and discussions

The analysis on the influence of different drive cycles on the EM, ICE and ESS components that make up the overall structure is carried on the UMT PHEV powertrain using Kuala Terengganu (KT) drive cycle, which is shown in figure 3. The KT dive cycle end at 1049 s covering a distance of km with an average speed of 20.67 km/h and maximum speed of 61.47 km/h.

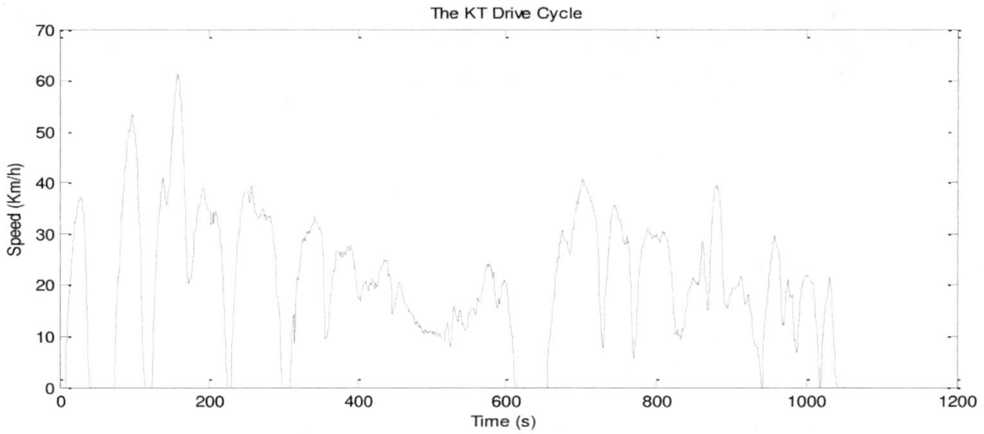

Figure 3. The KT drive cycle.

The main components sizing for KT drive cycle are listed in Table 3 based on the UMT PHEV power requirements as illustrated in figure 4.

Table 3. Components Sizing For KT Drive Cycle.

EM	
PEM (62 km/h)	62.7 kW
PEM , continuous = PEM (41 km/h)	40 kW
ICE	
PICE , continuous = PEM (30 km/h)	31 kW
PICE , continuous	40 kW
ESS	
PEM (20 km/h)	20 kW
EESS, min	33.3 kWh
PESS, max	13.5 kW
EESS	40.5 kWh

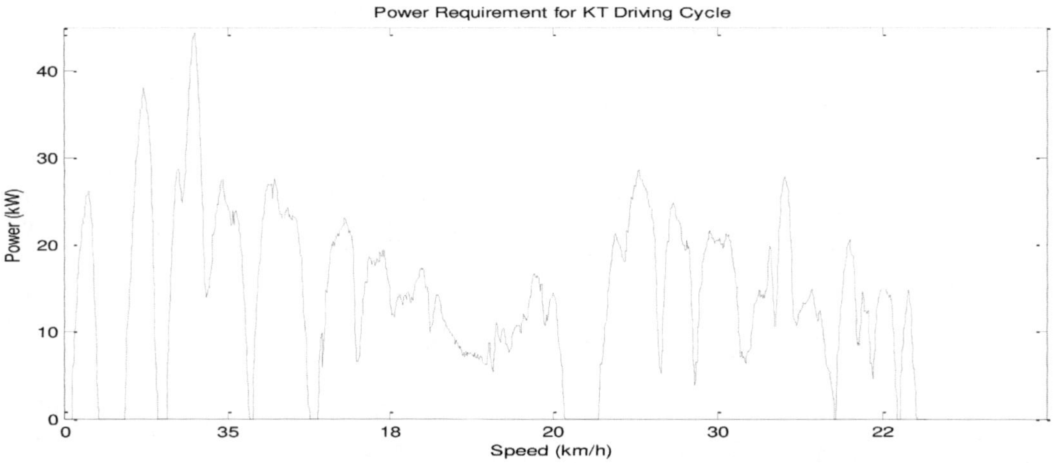

Figure 4. UMT PHEV power requirement for KT driving cycle.

6. Conclusions

To design an optimal power management vehicle in terms of all electric drive performance and energy efficiency is sizing and selecting the UMT PHEV powertrain main components is the most important task. According to vehicle parameters, specifications and performance requirements, the results of an individual component that make up the overall structure of the UMT PHEV powertrain are within reasonable and expected range using the different drive cycles.

Acknowledgements

The authors would like to be obliged to Fundamental Research Grant Scheme and Universiti Malaysia Terengganu for the financial support of this work.

References

[1] Ravey A, Roche R, Blunier B, Miraoui A, Combined optimal sizing and energy management of hybrid electric vehicles. Transportation Electrification Conference and Expo (ITEC), 2012 IEEE; 2012: IEEE.
[2] Weinstock R, Krein P T, White R, Optimal sizing and selection of hybrid electric vehicle components. Power Electronics Specialists Conference, 1993 PESC'93 Record, 24th Annual IEEE; 1993: IEEE.
[3] Pourabdollah M, Murgovski N, Grauers A, Egardt B 2013 *Vehicular Technology, IEEE Trans. on* **62** 2469-80
[4] Abdul Rahman S, Zhang N, Zhu J 2008
[5] Hegazy O, Van Mierlo J, Particle swarm optimization for optimal powertrain component sizing and design of fuel cell hybrid electric vehicle. Optimization of Electrical and Electronic Equipment (OPTIM), 2010 12th International Conference on; 2010: IEEE.
[6] Zhang L, Dorrell D G, Genetic Algorithm based optimal component sizing for an electric vehicle. Industrial Electronics Society, IECON 2013-39th Annual Conference of the IEEE; 2013: IEEE.
[7] Zhang X, Peng H, Sun J, A near-optimal power management strategy for rapid component sizing of power split hybrid vehicles with multiple operating modes. American Control Conference (ACC), 2013; 2013: IEEE.
[8] Liao Y, Quail A M, Component sizing of traction motor in hybrid powertrains. Vehicle Power and Propulsion Conference (VPPC), 2011 IEEE; 2011: IEEE.

Fatigue strength of woven kenaf fiber reinforced composites

A E Ismail[1] **and M A Che Abdul Aziz**[1]
[1]Faculty of Mechanical and Manufacturing Engineering, Universiti Tun Hussein Onn Malaysia, Batu Pahat, 86400 Johor, Malaysia

Email: emran@uthm.edu.my

Abstract. Nowadays, green composites provide alternative to synthetic fibers for non-bearing and load-bearing applications. According to literature review, lack of information is available on the fatigue performances especially when the woven fiber is used instead of randomly oriented fibers. In order to overcome this problem, this paper investigates the fatigue strength of different fiber orientations and number of layers of woven kenaf fiber reinforced composites. Four types of fiber orientations are used namely 0^0, 15^0, 30^0 and 45^0. Additionally, two numbers of layers are also considered. It is revealed that the fatigue life has no strong relationship with the fiber orientations. For identical fiber orientations, the fatigue life can be predicted considerably using the normalized stress. However as expected, the fatigue life enhancement occur when the number of layer is increased.

1. Introduction

It is an obvious fact that composite failure has distinct failure mechanisms compared with metals generally dominated by plastic deformation [1]. While, the toughening mechanisms during failure progressive of composite materials revealed several basic processes such as matric cracking, delamination, fiber breakage and interfacial debonding [2-7]. Fatigue strength of synthetic fiber can be found easily [8]. For example Wicaksono and Chai [9] reviewed fatigue behavior of different volume fractions of woven and non-woven glass fibers. It is found that the woven fiber resulted lower fatigue life compared with random oriented fiber reinforced composites.

Due to higher environmental awareness and sustainability aspects among researchers and industrialists, they recently utilized natural fiber to reinforce the polymeric composites. Shahzad et al. [10] studied the fatigue performances of hemp fiber reinforced composites where the fibers are randomly oriented. They also treated the fibers chemically. They found that the fiber treated chemically produced better fatigue capabilities. On the other hand, Vasconcellos et al. [11] weaved the hemp fiber to reinforce epoxy composites consisted of seven layers of hemp mats. The results exhibited that higher fatigue life can be obtained if the fiber is configured using $\pm 45^0$ instead of $0/90^0$.

Liang et al. [12] investigated the flax/epoxy composites under fatigue loading. They lay-up the fibers as follows 0^0, 90^0, $0^0/90^0$ and $\pm 45^0$. Their work also lead to conclude that $\pm 45^0$ composites produced higher fatigue performances. Kalam et al. [13] used empty fruit bunch reinforced composites with different fiber volume fractions. In their work, the fibers are aligned uni-axially. The results obtained showing that increasing fiber fractions have significantly reduced the fatigue life. Abdullah et al. [14] investigated the fatigue performance of uni-directional kenaf fiber. However, their results inconsistence with Kalam et al. [13] where they found that fatigue life increased when fiber volume fractions are increased.

From literature review, there is a noticeable lack of information in the investigations of woven kenaf yarn fiber reinforced composites when subjected to fatigue loadings. Therefore, this work presents on the effect of fiber orientations of woven kenaf fibers. Additionally, the effects of two numbers of layers are also considered. The fatigue life is thus correlated with the fiber orientations and the number of layers before the empirical mathematical model to predict the fatigue life is formulated.

2. Experimental set up

2.1. Fiber and composite preparations

The as-received kenaf fiber (1 mm in diameter) is in the form of yarn as in figure 1(a). This kenaf yarn is then weaved to form twill woven-type mats as in figure 1(b). It is consisted of two sets of yarns warp and weft which is interlaced at 90^0 relative to each other to produce a balanced twill weave where the number of warp and weft is 4 x 4 (4 yarns in warp and 4 yarn in weft directions per $10mm^2$). Woven kenaf mats are produced manually using in-house machine as in figure 2(a) and the kenaf mat produced is shown in figure 2(b).

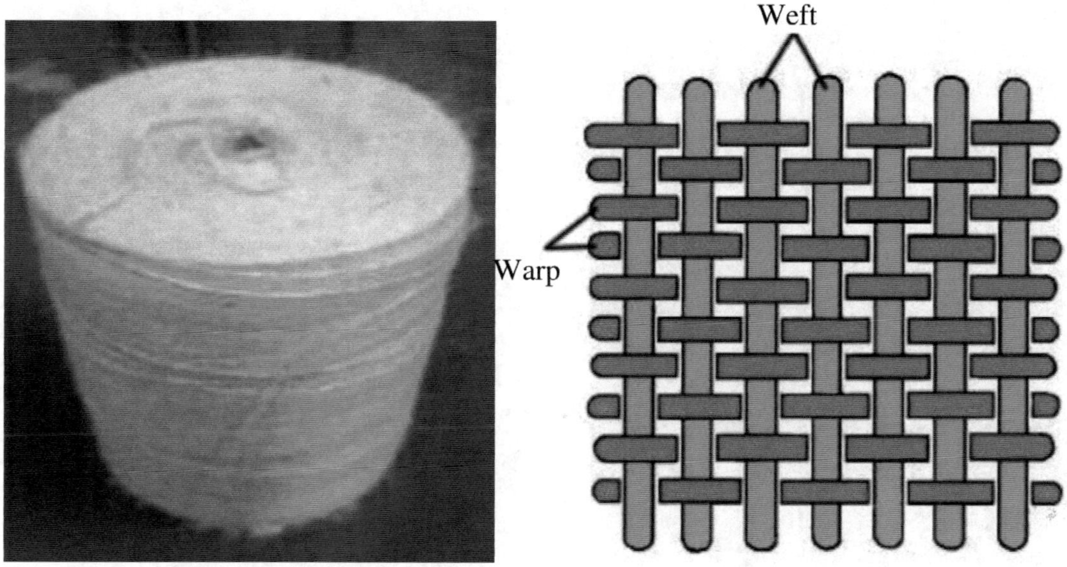

(a) (b)

Figure 1. (a) As-received kenaf yarn fiber and (b) Twill-type woven mat.

Two important parameters are considered namely fiber orientations and number of layers. Four fiber orientations are used 0^0, 15^0, 13^0 and 45^0. While single and double mat layers are produced. For double layers, the angle configurations as follows: $0^0/0^0$, $15^0/0^0$, $30^0/0^0$ and $45^0/0^0$ as in figure 3(a). The mats are then placed into a rectangular steel mold (figure 3(b)) before polyester resin is poured and compressed in order to squeeze the excessive resin out as in figure 4.

2.2. Experimental Procedure

The hardened plates are shaped according to the standard geometry suggested by ASTM D3039. Fatigue test (figure 5) is performed in an ambient temperature and table 1 lists the testing parameters used in this study. Displacement-controlled fatigue test is used to examine the fatigue performances of the composites. In order to analyze the cause of failure for the composites, a scanning electron microscope (SEM) is used.

(a) (b)

Figure 2. (a) In-house weaving machine and (b) Finished woven type fiber mats.

(a) (b)

Figure 3. (a) Fiber mat configuration and (b) Fiber mat placement in a mold.

(a) (b)

Figure 4. (a) Fiber mats and resin is under compression and (b) The hardened composite plates.

Figure 5. Universal tensile test machine with a high temperature chamber.

Table 1. Fatigue testing parameters

Parameter	Condition				
Loading	Tension-tension fatigue loading				
Stress ratio	0.5				
Stress level	$0.9\sigma_{ult}$	$0.8\sigma_{ult}$	$0.7\sigma_{ult}$	$0.6\sigma_{ult}$	$0.5\sigma_{ult}$
Frequency	5Hz				
Temperature	25^0C				

*σ_{ult} is an ultimate tensile strength.

3. Results and discussion

Figure 6 reveals the curves of stress against normalized fatigue life for single and double layers of woven kenaf fiber reinforced composites, respectively. Four fiber orientations and their combinations are considered. As expected, the fiber which is aligned parallel with the principal loading axis produced higher fatigue life compared with other type of composites. For single layer composites, the distributions of fatigue life are not significant where it is seemed that the life accumulated into a single fatigue life curve.

Figure 6(a) shows the fatigue life behavior of single layered woven kenaf reinforced composites. It is indicated that 0^0 oriented fibers produced higher fatigue life compared with other types of composites. For all type of composites except 0^0 fiber orientation, the fatigue curves seemed not to affect significantly where the curves are almost similar to each other. However, 30^0 fiber oriented composites show a slight reduction on the fatigue life. While figure 6(b) on the other hand, showing the behavior of fatigue life of the double layered fiber reinforced composites. In general, the curves are relatively straight and the distributions of the results are quite high especially for 0^0 fiber orientations. For other type of composites, the fatigue curves are almost identical with those in figure 6(a). However, 300 fiber oriented composites have shifted downward obviously showing that the composite unable to resist the fatigue damage longer.

Figure 7 shows the curves of fatigue life distributions against the normalized stress for single and double layered composites. As expected, the fatigue life degraded as the normalized stresses increased for all composite conditions. The behavior of fatigue life against normalized stress shown in figures 7(a) and 7(b) are almost identical. However, for the double layered composites, the curves reveal a slightly drop showing the composite degradations are relatively higher than that single layered composites due to the brittle effect.

Figure 8 shows the fracture mechanisms of two fiber orientations 0^0 and 15^0 for double fiber layers. For the composites which have the fibers aligned parallel with the loading axis, the most failure mechanisms found are pull-out fiber. It is responsible to produce better fatigue life. While, for the fibers which are aligned obliquely, lack of push-out mechanisms are observed instead of the fiber yarns have been peeled-off resulted lower fatigue life.

Tables 2 and 3 tabulated the mathematical model used to predict the fatigue life of single and double layered composites, respectively. These equations are formulated when the axis of fatigue life is transformed into logarithmic values against normalized stresses. Then, the regression technique is used to determine the mathematical representations with high degree of accuracies (R^2 > 0.99). These models can be used to predict the fatigue life of the composites as long as within the specified range of normalized stresses.

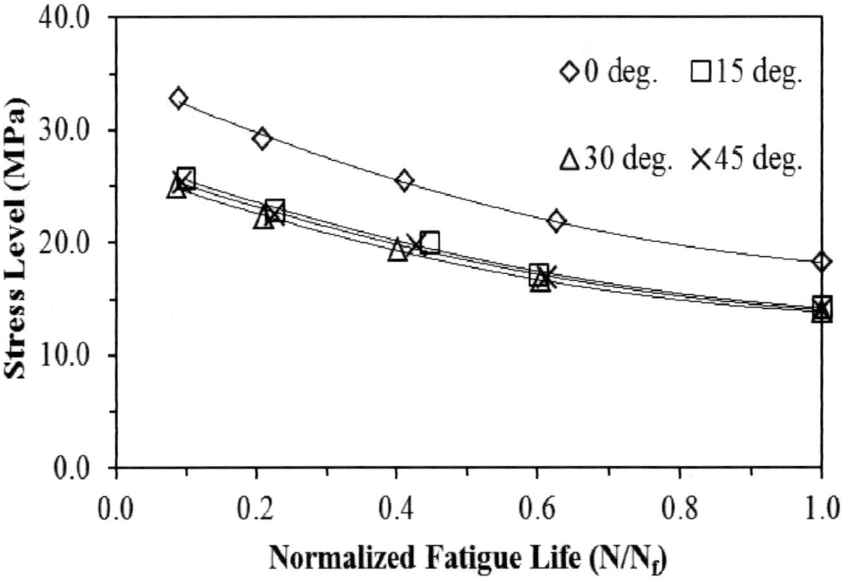

(a)

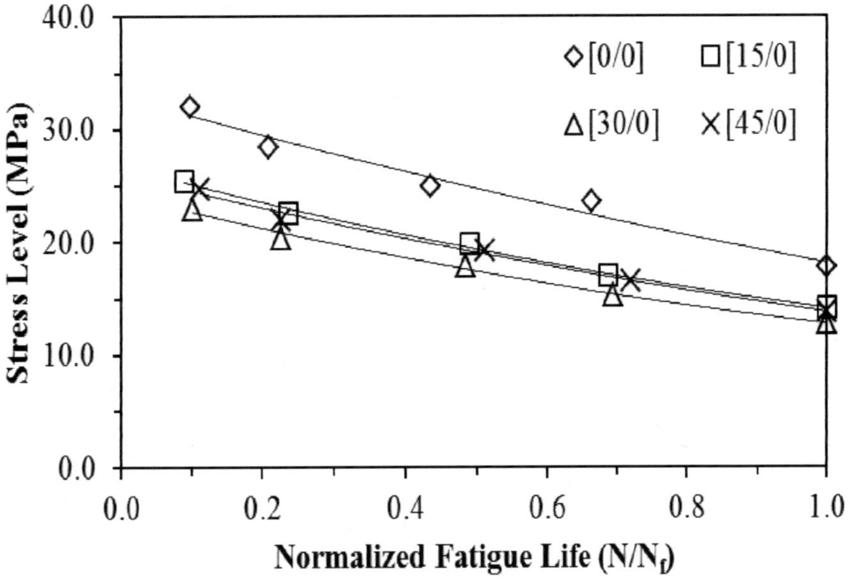

(b)

Figure 6. Fatigue life of woven kenaf fiber reinforced composites, (a) single and (b) double layers.

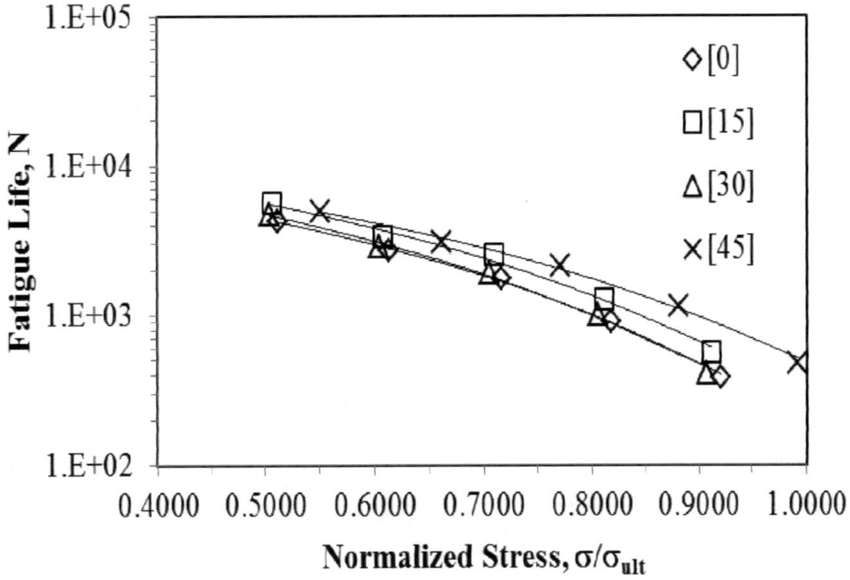

(a)

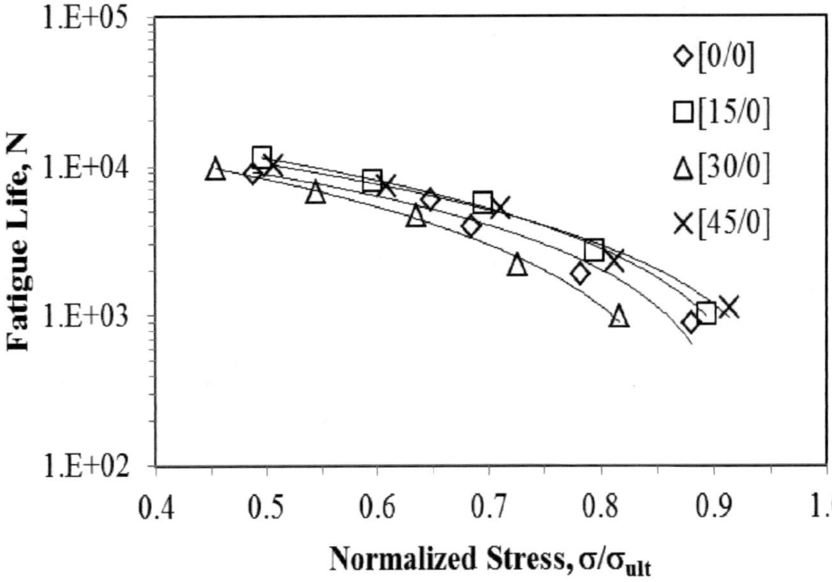

(b)

Figure 7. Fatigue life plotted against normalized stress level for various composites, (a) single and (b) double layers of composites.

Table 2. Mathematical model to predict fatigue life for single layered composites.

Types of Composites	Fatigue Life, N	R Square
[0^0]	$N = 15127\left(\frac{\sigma}{\sigma_{ult}}\right)^2 - 31021\left(\frac{\sigma}{\sigma_{ult}}\right) + 16139$	$R^2 = 0.9977$
[15^0]	$N = 18552\left(\frac{\sigma}{\sigma_{ult}}\right)^2 - 38379\left(\frac{\sigma}{\sigma_{ult}}\right) + 20177$	$R^2 = 0.9890$
[30^0]	$N = 18375\left(\frac{\sigma}{\sigma_{ult}}\right)^2 - 36421\left(\frac{\sigma}{\sigma_{ult}}\right) + 18368$	$R^2 = 0.9955$
[45^0]	$N = 14370\left(\frac{\sigma}{\sigma_{ult}}\right)^2 - 32176\left(\frac{\sigma}{\sigma_{ult}}\right) + 18293$	$R^2 = 0.9940$

*σ is an applied stress level and σ_{ult} is an ultimate tensile strength.

IOP Conf. Series: Materials Science and Engineering **100** (2015) 012037 doi:10.1088/1757-899X/100/1/012037

Table 3. Mathematical model to predict fatigue life for double layered composites.

Types of Composites	Fatigue Life, N	R Square
$[0^0/0^0]$	$N = 14344\left(\frac{\sigma}{\sigma_{ult}}\right)^2 - 41303\left(\frac{\sigma}{\sigma_{ult}}\right) + 25881$	$R^2 = 0.9729$
$[15^0/0^0]$	$N = 22993\left(\frac{\sigma}{\sigma_{ult}}\right)^2 - 58231\left(\frac{\sigma}{\sigma_{ult}}\right) + 34661$	$R^2 = 0.9969$
$[30^0/0^0]$	$N = 26368\left(\frac{\sigma}{\sigma_{ult}}\right)^2 - 57663\left(\frac{\sigma}{\sigma_{ult}}\right) + 30409$	$R^2 = 0.9967$
$[45^0/0^0]$	$N = 17626\left(\frac{\sigma}{\sigma_{ult}}\right)^2 - 47620\left(\frac{\sigma}{\sigma_{ult}}\right) + 29758$	$R^2 = 0.9939$

* σ is an applied stress level and σ_{ult} is an ultimate tensile strength.

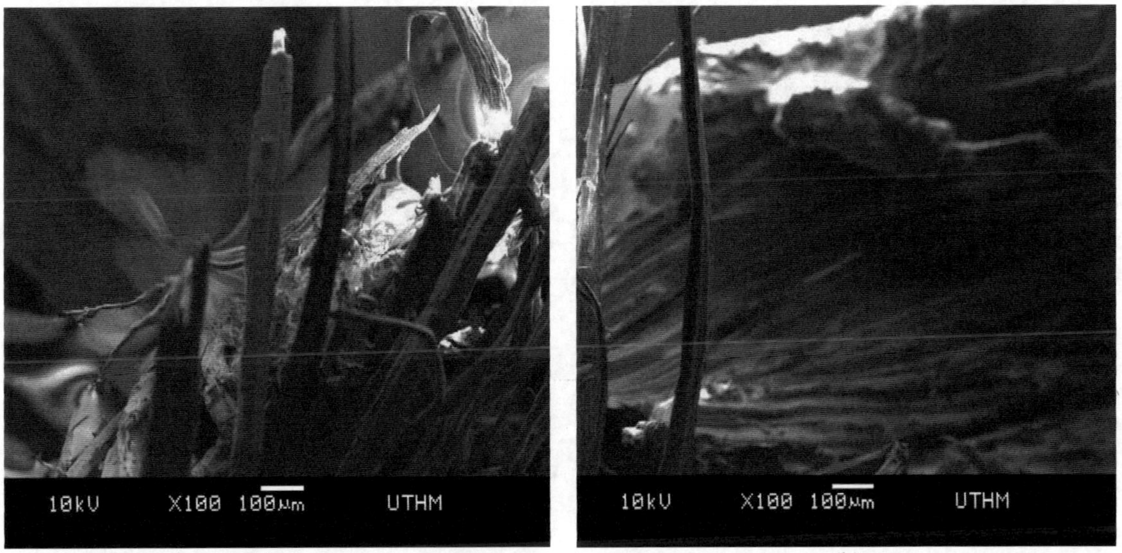

(a) (b)

Figure 8. Fracture mechanisms of woven kenaf fiber reinforced composites of different fiber orientations (a) 0^0 and (b) 15^0.

4. Conclusion

In this paper, an attempt is to investigate the fatigue performances of woven kenaf yarn fiber reinforced composites. Two important parameters are considered such as fiber orientations and number of layers. After fatigue tests completed, an empirical formulation for each composite condition is developed using a simple regression technique. It can be concluded that higher fatigue life is produced when the fiber is aligned parallel with the principal loading axis. This behavior contradictorily agreed with the previous works where better fatigue life can be obtained when the fiber configuration $\pm 45^0$. The reliable empirical models are also produced and can be used as long as the normalized stress level within the range of 0.5 to 0.9.

Acknowledgements
The authors would like to be obliged to Universiti Tun Hussein Onn Malaysia for providing laboratory facilities and financial assistance for this work.

References

[1] Mohanty A, Misra M, Drzal L, Selke S, Harte B, Hinrichsen G. Natural Fibers, Biopolymers and Biocomposites: An Introduction. Natural Fibers, Biopolymers and Biocomposites. United States of America, Taylor & Francis Group, CRC Press; 2005.
[2] Akil H, Omar M, Mazuki A, Safiee S, Ishak Z, Bakar A A 2011 *Materials & Design* **32** 4107-21
[3] Al Emran Ismail M, Awang K, Sa'at M H, Tensile Strength of Natural Fiber Reinforced Polyester Composite. AIP Conf Proc; 2007.
[4] Ismail A E, Zainulabidin M, Roslan M N, Mohd Tobi A, Nor M, Hisyamudin N, Effect of Velocity on the Impact Resistance of Woven Jute Fiber Reinforced Composites. Applied Mechanics and Materials; 2014: Trans Tech Publ.
[5] Roslan M N, Ismail A E, Hashim M Y, Zainulabidin M H, Khalid S N A, Modelling analysis on mechanical damage of kenaf reinforced composite plates under oblique impact loadings. Applied Mechanics and Materials; 2014: Trans Tech Publ.
[6] Ismail A E, Hassan M, Low Velocity Impact on Woven Kenaf Fiber Reinforced Composites. Applied Mechanics and Materials; 2014: Trans Tech Publ.
[7] Gassan J 2002 *Composites part A: applied science and manufacturing* **33** 369-74
[8] Wambua P, Ivens J, Verpoest I 2003 *composites science and technology* **63** 1259-64
[9] Wicaksono S, Chai G B 2013 *Proceedings of the Institution of Mechanical Engineers, Part L: Journal of Materials Design and Applications* **227** 179-95
[10] Shahzad A, Isaac D 2014 *Polymer Composites* **35** 1926-34
[11] Belaadi A, Bezazi A, Maache M, Scarpa F 2014 *Procedia Engineering* **74** 325-8
[12] Liang S, Gning P-B, Guillaumat L 2014 *International Journal of Fatigue* **63** 36-45
[13] Kalam A, Sahari B, Khalid Y, Wong S 2005 *Composite structures* **71** 34-44
[14] Abdullah A H, Alias S K, Jenal N, Abdan K, Ali A 2012 *Engineering Journal* **16** 105-14

IOP Conf. Series: Materials Science and Engineering **100** (2015) 012038 doi:10.1088/1757-899X/100/1/012038

The effect of welding parameters on surface quality of AA6351 aluminium alloy

S Yacob[1], M A MAli[2], Q Ahsan[2] N Ariffin[1], R Ali[1], A Arshad[1], M I A Wahab[1], S A Ismail[1], N S M Roji[1], W B W Din[1], M H Zakaria[1], A Abdullah[1], M I Yusof[1], K Z Kamarulzaman[1], A Mahyuddin[1], M N Hamzah[1] and R Roslan[1]
[1]Faculty of Quality Engineering, University Kuala Lumpur Malaysian Institute of Industrial Technology (MITEC),81750, Bandar Seri Alam,Johor,Malaysia
[2]Faculty of Manufacturing Engineering, Universiti Teknikal Malaysia Melaka, 76100,Durian Tunggal, Melaka, Malaysia

Email: suhailay@unikl.edu.my

Abstract. In the present work, the effects of gas metal arc welding-cold metal transfer (GMAW-CMT) parameters on surface roughness are experimentally assessed. The purpose of this study is to develop a better understanding of the effects of welding speed, material thickness and contact tip to work distance on the surface roughness. Experiments are conducted using single pass gas metal arc welding-cold metal transfer (GMAW-CMT) welding technique to join the material. The material used in this experiment was AA6351 aluminum alloy with the thickness of 5mm and 6mm. A Mahr Marsuft XR 20 machine was used to measure the average roughness (Ra) of AA6351 joints. The main and interaction effect analysis was carried out to identify process parameters that affect the surface roughness. The results show that all the input process parameters affect the surface roughness of AA6351 joints. Additionally, the average roughness (Ra) results also show a decreasing trend with increased of welding speed. It is proven that gas metal arc welding-cold metal transfer (GMAW-CMT)welding process has been successful in term of providing weld joint of good surface quality for AA6351 based on the low value surface roughness condition obtained in this setup. The outcome of this experimental shall be valuable for future fabrication process in order to obtained high good quality weld.

1. Introduction

The quality of the joints depends on the processing parameters. Surface finish is very crucial characteristic to be considered in assessing the performance of fabrication process. The main objective of modern industries is to manufacture low cost, high quality products in short time [1]. Many authors [2-4] have agreed that the welding parameters can affect the weld bead characteristics. It is proved that it is vital to know how the welding parameters responses affect the surface roughness.

Welding is one of the most important and versatile means of fabrication available to industry [5] .Gas metal arc welding cold metal transfer (GMAW-CMT) process is part of the fusion welding and it has been increasingly appreciated in automotive industry. The demand to improve process performance in automotive had led to the innovation of conventional arc welding method to develop GMAW-CMT welding technique. The usage of GMAW-CMT method exhibits advantages as it generate high productivity, high deposition rate ,low heat input, low cost and controllable spatter.

Most of researchers were satisfied with the performance of GMAW-CMT. Cold metal transfer welding of aluminum alloy to zinc coated steel[6], titanium to copper [7], cladding of nickel base inconel 718 [8] , aluminum to galvanized mild steel [9],magnesium to aluminum [10] and aluminum to aluminum [11].

Many specific properties of aluminum alloys including light weight and good structural strength enable them to be applied for structural parts. The demand of aircraft and automotive industries for lightweight materials is met by aluminum alloys [12]. Aluminum 6351 was chosen as the main material, because of its good mechanical properties and widespread usage including automation industry, structural work and shipbuilding [13].There is no single work concern on the surface roughness of AA6351 using GMAW-CMT method. Therefore, this study is aimed at determining the effect of welding speed, material thickness and contact tip to work distance on the surface roughness of AA6351 joints.

2. Experimental set up

The AA6351 material is supplied by NovaScientific Resource (M) Sdn.Bhd. The aluminum alloy 6351 was in plate shape having two dimensions of the raw material in this sample preparation which is 150mmx 53mmx 5 mm and 150mmx 53 mm x 6mm as presented in figure 1. The plates of AA6351 were prepared using plasma cutting machine by supplier.AA6351 consists of aluminum, iron, magnesium, manganese, zinc, titanium, silicon and copper. Experiments were conducted using a GMAW-CMT welding machine as presented in figure 2. ER4043 with a diameter of 1.2mm was used as filler metal. The GMAW-CMT process was carried using single pass layer, fixture application and forehand technique. An argon shielding gas with a flow rate of 15L/min was used in the welding process. The welding parameters used during the GMAW-CMT processs as shown in Table 1.Three input parameters with two levels were selected and full factorial technique was utilized to drive the experimental planning process. Totally 16 conditions with different welding speed, material thickness and contact tip to work distance have been shown in table 2. After welding process, a Mahr Marsuft XR 20 machine was used to evaluate the surface roughness of the GMAW-CMT weld by measuring average roughness (Ra). A Mahr Marsuft XR 20 surface roughness machines is a direct measurement method with static workstation. The surface roughness measurement was performed on the centre of weld zone area.The cut-off length was fixed at 0.8mm and the evaluation length measured was 4mm with sampling length 5.The measuring speed 0.1 mm/s was setup in the experiment for all measurements. All measurements were run in a surface laboratory room at UNIKL MITEC Pasir Gudang, Johor with temperature 25°C.

Figure 1. Raw material AA6351.

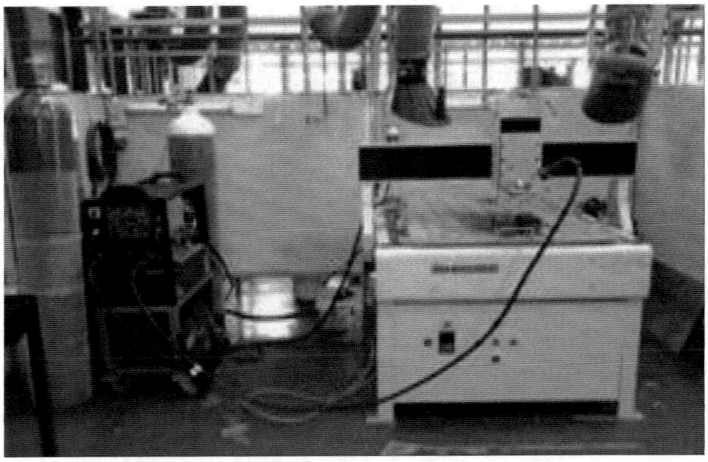

Figure 2. GMAW-CMT welding machine.

Table 1. Welding process parameters and their levels.

Process parameters	Levels	
	Level 1	Level 2
Material thickness (mm)	5	6
Welding speed (mm/min)	500	600
Contact tip to work distance(mm)	10	15

3. Results and discussion

The surface roughness performance in this work was observed based on the quality characteristic is the smaller the better. The collected data were analyzed using design of experiment analysis in MINITAB 16 software with average roughness as dependent variable and welding speed, contact tip to work distance and material thickness as independent variables. The results of surface roughness are shown in table 2.As experiment conducted had proved, GMAW-CMT welding process for AA6351 was observed show more great average roughness, Ra values. Information obtained from the experiments designated that none of the welding condition produced the same value of surface roughness .It also showed major differences between largest value of surface roughness and the smallest value of surface roughness. However, all trials successfully attained average roughness, Ra below than seven (Ra< 6).This means that they are categorized as smooth weld bead surface. The main effect response graph of weld bead surface roughness is shown in figure 3. The average roughness, Ra value shows a decreasing trend with increased of welding speeds. Welding speed has greatest effect on weld bead surface roughness of AA6351. So, it can be stated that lowest surface roughness can be achieved when an increase in welding speed and material thickness. This finding meets an agreement with the theoretical information stated that increased welding speed does not result rough bead. Figure 4 shows the outline for interaction plot of GMAW-CMT welding parameters on the surface roughness. It was observed no interaction effects between welding speed and contact tip to work distance.

Table 2. Results of surface roughness.

Trial	Average roughness, Ra (µm)
1	5.0569
2	3.9546
3	3.4118
4	1.971
5	2.2029
6	5.9394
7	3.7081
8	3.2535
9	2.1689
10	3.0413
11	4.085
12	2.5745
13	5.3105
14	4.7378
15	2.5095
16	2.5046

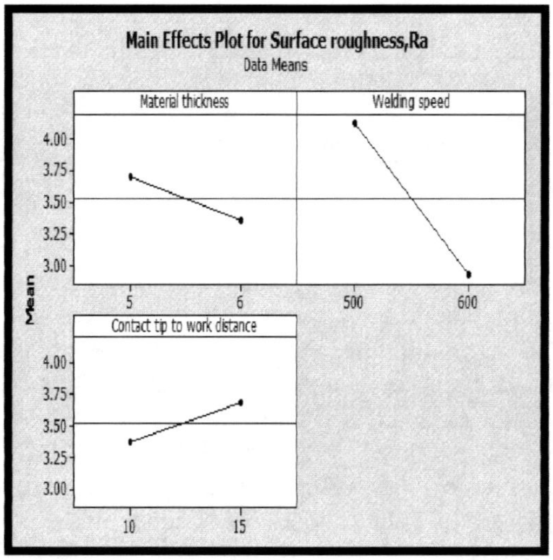

Figure 3. GMAW-CMT welding machine

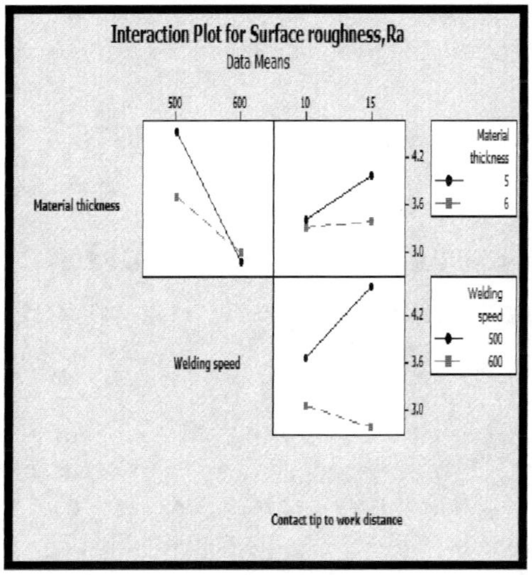

Figure 4. Interaction effect of surface roughness.

4. Conclusion

It can be concluded that increasing welding speed and material thickness decreased the surface roughness whereas increasing the contact tip to work increased the surface roughness. The welding speed, contact tip to work distance and material thickness has significant effect on the surface roughness of AA6351 welded joints. In additional, it can be stated that lowest surface roughness can be achieved when an increase in welding speed and material thickness which material thickness substitute to welding current and welding voltage.

References

[1] Hossain S J, Ahmad N 2012 *Journal of Mechanical Engineering Research* **4** 112-29

[2] S R Meshram N S P 2013 *International Journal of Engineering, Business and Enterprise Application (IJEBEA)* 46-52

[3] Wagiman A, Wahab M S, Mohid Z, Mamat A 2013 *Applied Mechanics and Materials* **465** 1370-4

[4] Ahmad R, Bakar M 2011 *Materials & Design* **32** 5120-6

[5] Hussain A K, Lateef A, Javed M, Pramesh T 2010 *International Journal of App. Engineering Research* **1** 518-27

[6] Yang S, Zhang J, Lian J, Lei Y 2013 *Materials & Design* **49** 602-12

[7] Cao R, Feng Z, Chen J 2014 *Materials & Design* **53** 192-201

[8] Ola O, Doern F 2014 *Materials & Design* **57** 51-9

[9] Cao R, Huang Q, Chen J, Wang P-C 2014 *Journal of Alloys and Compounds* **585** 622-32

[10] Cao R, Wen B, Chen J, Wang P-C 2013 *Materials Science and Engineering: A* **560** 256-66

[11] Gungor B, Kaluc E, Taban E, Aydin S 2014 *Materials & Design* **54** 207-11

[12] Palanivel R, Mathews P K, Murugan N, Dinaharan I 2012 *Materials & Design* **40** 7-16

[13] Kishore K, Krishna P G, Veladri K, Kumar G K 2010 *Int. Journal of Applied Engineering Research* **5** 393-9

Design and development of split-parallel through-the road retrofit hybrid electric vehicle with in-wheel motors

S A Zulkifli[1,4], M Syaifuddin Mohd[2,4], M Maharun[2,4], N S A Bakar[2,3], S Idris[2,3], S H Samsudin[2,3], Firmansyah[2,4], J J Adz[1], M Misbahulmunir[2], E Z Z Abidin[2,4], M Syafiq Mohd[1], N Saad[1] and A R A Aziz[2]
[1]Electric and Electronics Engineering Department
[2] Mechanical Engineering Department
[3] Design and Prototyping Centre (DPC)
[4] Centre for Automotive Research and Electric Mobility (CAREM) Universiti Teknologi PETRONAS, Bandar Seri Iskandar, Malaysia

E-mail: saifulazrin_mz@petronas.com.my

Abstract. One configuration of the hybrid electric vehicle (HEV) is the split-axle parallel hybrid, in which an internal combustion engine (ICE) and an electric motor provide propulsion power to different axles. A particular sub-type of the split-parallel hybrid does not have the electric motor installed on board the vehicle; instead, two electric motors are placed in the hubs of the non-driven wheels, called 'hub motor' or 'in-wheel motor' (IWM). Since propulsion power from the ICE and IWM is coupled through the vehicle itself, its wheels and the road on which it moves, this particular configuration is termed 'through-the-road' (TTR) hybrid. TTR configuration enables existing ICE-powered vehicles to be retrofitted into an HEV with minimal physical modification. This work describes design of a retrofit-conversion TTR-IWM hybrid vehicle – its sub-systems and development work. Operating modes and power flow of the TTR hybrid, its torque coupling and resultant traction profiles are initially discussed.

1. Introduction

An HEV draws its propulsion power from an ICE and an electric motor, which also operates as a generator to re-charge the onboard battery pack. Depending on connectivity of the power sources to the load (wheels), an HEV has several possible configurations, which include series hybrid, parallel hybrid and combined series parallel hybrid. A parallel hybrid vehicle typically has the motor-generator on board the chassis. Alternatively, the electric motor can actually be part of the wheel itself, called an in-wheel motor (IWM).

In such vehicle, the ICE operates on one drive-shaft (e.g. front) while the in-wheel motors are fitted into the other pair of wheels (e.g. rear), resulting in a split-axle parallel architecture (figure 1) [1]. There is no dedicated mechanical device coupling the two propulsion sources. Instead, the sources are coupled through the road, giving rise to the term TTR hybrid. This paper describes specific operating modes, power flow, torque coupling and resultant traction profiles of the split-parallel TTR hybrid. It also explains sub-systems and development work for conversion of conventional vehicle into a TTR hybrid vehicle. The parallel HEV drive-train is a combination of two separate power-trains, each having an energy source/reservoir and an energy converter. The first power-train consists of the fuel tank (source) and the combustion engine (converter), while the second is made up of the battery bank (source) (or any electrical energy storage device) and the electric motor-generator (converter). As illustrated in figure 2, while power flow in the former is unidirectional, the latter allows for energy to be dispensed (electric motoring) and replenished (battery charging through electric generation and re-generative braking). In general, this results in nine possible operating modes for a parallel HEV. In a split-axle parallel hybrid vehicle with in-wheel motors, it is not possible for the ICE to directly charge the battery bank through the electric generator since the two sources are not coupled together [2].

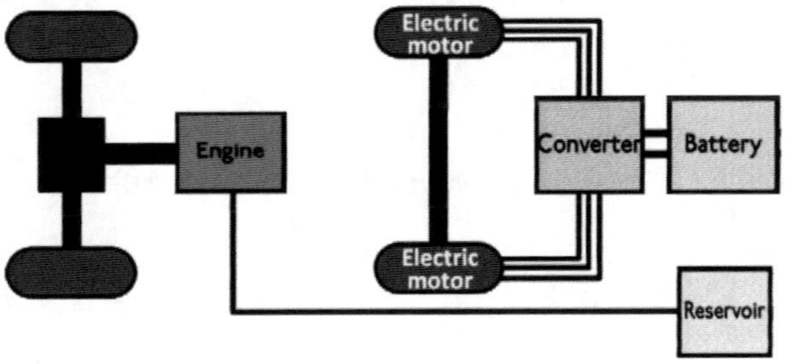

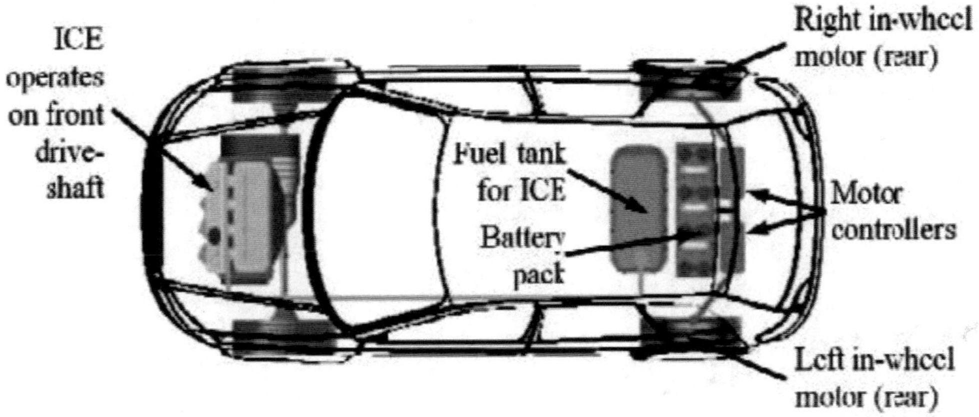

Figure 1. Split-parallel through-the-road hybrid electric vehicle configuration.

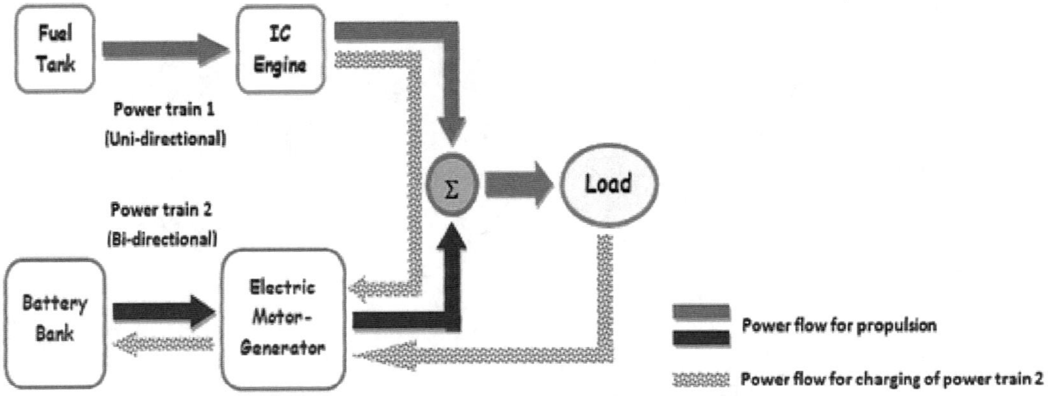

Figure 2. Power flow paths of a parallel hybrid electric vehicle.

Rather the batteries can be charged only when the vehicle is moving – when the wheels and in-wheel motors are turning and operating as generators. Charging cannot occur when the vehicle is stationary. In a TTR hybrid, the ICE does not deliver power to the battery bank directly but through the load (vehicle mass) – when the vehicle is propelled by the ICE and power is delivered to the battery through the vehicle's chassis, wheels and road coupling [1, 2].

Figure 3(a) shows configuration of the split-parallel hybrid vehicle, clearly showing the two power sources acting independently on different drive axles-there is no mechanical coupler as in the single axle parallel hybrid vehicle (shown in figure 3(b) and 3(c)). These results in only five possible operating conditions for the TTR-IWM hybrid, determined by the possible combination of energy flow paths, as depicted in figure 4:

1. *Mode A*:
   Similar to Mode 1, only the ICE is providing propulsion power to the vehicle. The in-wheel motors are not operating.
2. *Mode B*:
   Similar to Mode 2, only the motors are providing power to the vehicle. The ICE is idling.
3. *Mode C*:
   Similar to Mode 3, both the ICE and in-wheel motors are providing propulsion power to the vehicle.
4. *Mode D*:
   Similar to Mode 4, the in-wheel generator, to electrical power charge the battery bank, during regenerative braking.
5. *Mode E*:
   Similar to Mode 9, while the ICE is propelling the vehicle, the in-wheel generator converts the vehicle's energy to charge the battery bank.

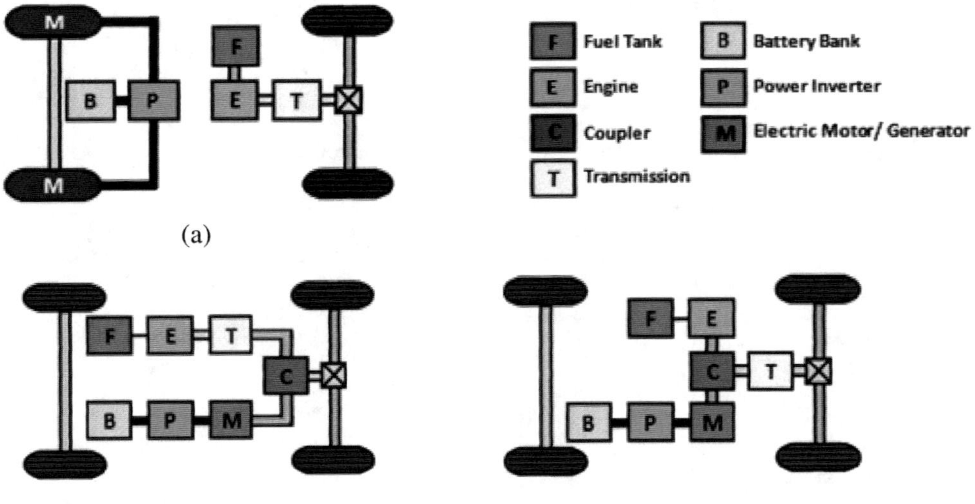

Figure 3. Drive-train configuration: (a) TTR-IWM split-parallel hybrid, (b) single-axle post-transmission coupled parallel hybrid and (c) single-axle, pre-transmission coupled parallel hybrid

IOP Conf. Series: Materials Science and Engineering **100** (2015) 012039 doi:10.1088/1757-899X/100/1/012039

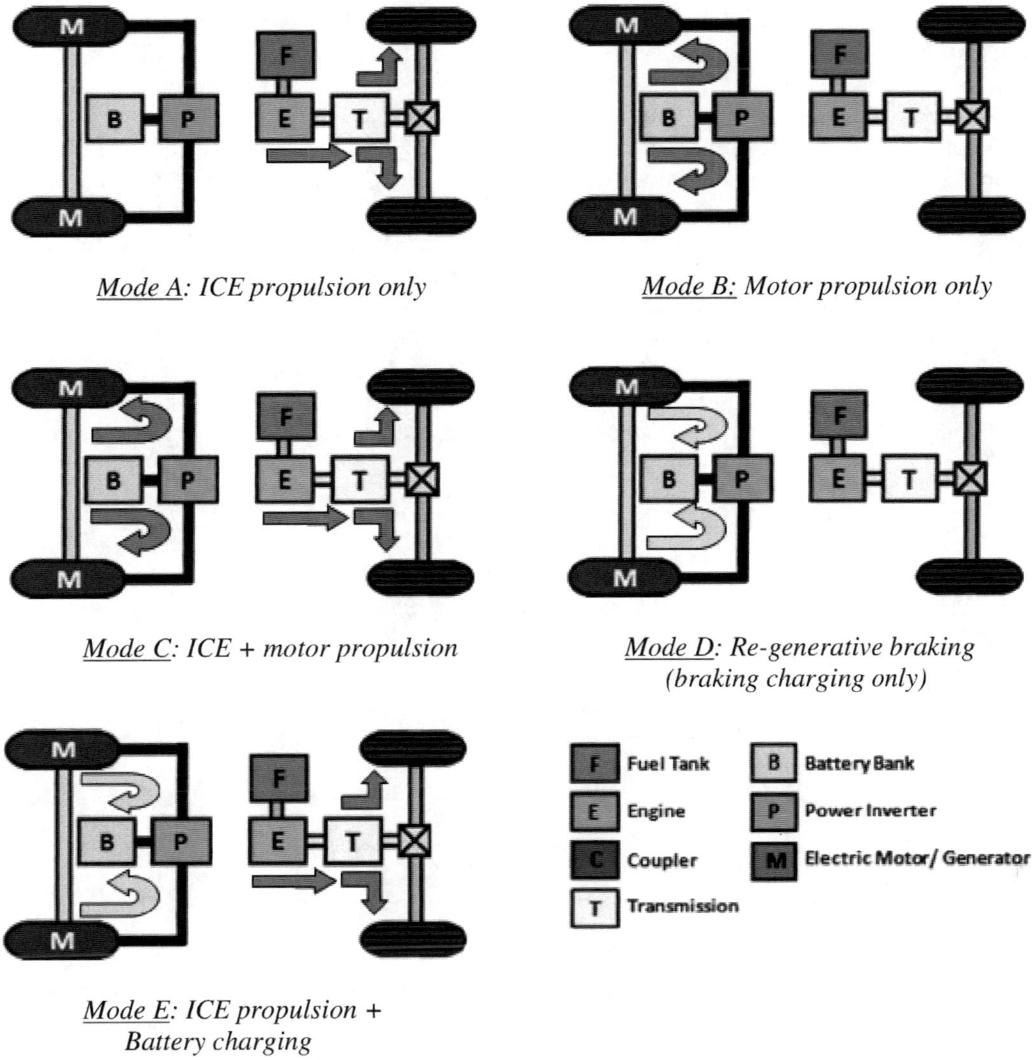

Figure 4. Five operating modes of a TTR-IWM hybrid vehicle.

2. Torque coupling and resultant tractive profiles

A parallel hybrid vehicle has a mechanical-coupled drive train-it has some form of mechanical coupling device such as pulley, gear or chain assembly, which adds together mechanical powers from the two power sources. Due to power conservation, in a torque-coupled drive train, the torques from both sources adds up to give the final torque, while speed of the power sources cannot be individually controlled [2-4].

The split-parallel configuration is a torque-coupled vehicle, with torque contributions from the ICE and electric motor added up through the vehicle chassis, wheel and road-there is no dedicated coupling device. For the in-wheel motor hybrid vehicle, the electric motors care in the rear wheels; thus, the motor speed (rear wheel speed) is equal to the speed of the front wheels.

For the TTR hybrid with in-wheel motors, the ICE is connected to a multi-gear transmission (existing) while there is no transmission for the motor (similar to a directly-coupled motor). The torques from both sources combines to give a final traction profile. The resultant tractive profiles of a TTR-IWM hybrid vehicle are shown in figure 5(a). As mentioned above, the traction profiles are similar to device-coupled parallel hybrid vehicle with single-axle, post-transmission coupling as shown in figure 5(b) [4].

3. Retrofit-conversion TTR-IWM hybrid vehicle

The TTR split-parallel configuration enables an existing ICE-driven vehicle to be converted into a hybrid electric vehicle with minimal modification-by replacing the otherwise non-driven wheels with motors (IWM) and powering them with a battery bank located in the vehicle's trunk. Along with controllers and an energy management system controller (EMS) that intelligently manages the vehicle's energy and power flow, they form a TTR-IWM hybrid vehicle system [1, 5].

One such retrofit project is ongoing at Universiti Teknologi *PETRONAS* (UTP) in Bandar Sri Iskandar, Malaysia, to convert an existing conventional-drivetrain P*ERODUA* MyVI into a TTR-IWM hybrid vehicle. The prototype shall be used for validation of TTR hybrid dynamic simulation model and investigation of different control strategies to achieve optimum fuel saving benefits. A retrofit-conversation TTR hybrid vehicle consists of several sub-systems, as explained below, with reference to the TTR hybrid vehicle development project at UTP.

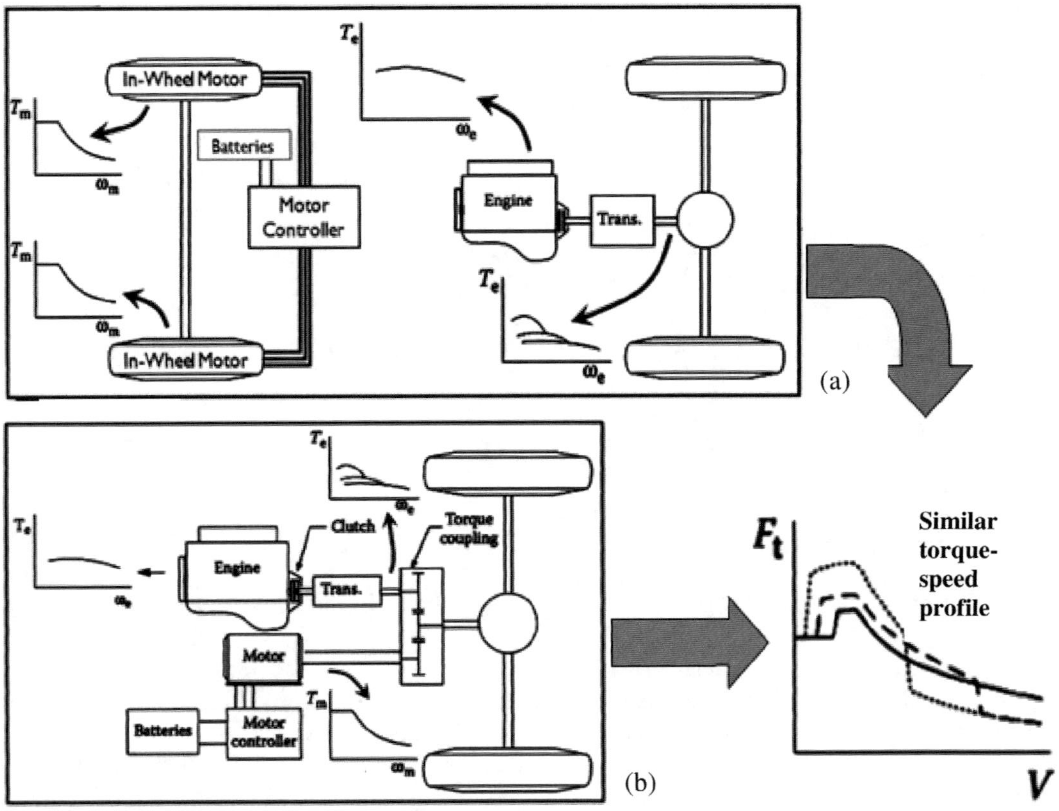

Figure 5. Resultant torque-speed profiles: (a) TTR-IWM hybrid and (b) single-axle, parallel hybrid with post-transmission, single-gear motor and multi-geared ICE.

In-wheel motors used in this project are sourced from Kelly Motor Controls (figure 6). Installation of the IWM system involves design and fabrication of a special adapter for attachment of the in-wheel motors into the rear axle (figure 6). The work consists of attaching two in-wheel motors for a split-axle hybrid drive train configuration and replacing the original drum brake system with a disc brake (figure 7).

The existing drum brake system needs to be replaced as it would have taken up a large amount of space at the center and will result in extending the wheels further outwards from the vehicle body. The disc brake system is preferable as it occupies less space, since the brake calliper mounting is located outside the central area and the disc itself has a much thinner profile. Braking performance of the rear tires is expected to be equal or better than the replaced drum brake system since it combines both mechanical braking and electrical re-generative braking (see separate section below).

Figure 6. In-wheel motor(left) and motor adapter (right) for attachment of IWM to the rear axle.

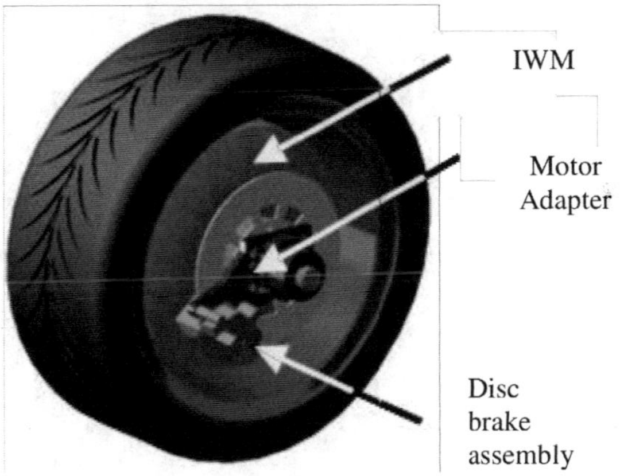

Figure 7. Motor adapter (left) and disc brake assembly for IWM (right)

This sub-subject consists of installation and configuration of motor controller hardware and power circuitry for the two in-wheel motors. The motor controller used in this project is sourced from the same manufacturer of the in-wheel motors: *Kelly Motor Controls* (figure 8). The work requires programming and commissioning of the controller, tuning for bench-top testing of motor speed control and interfacing to other parts of the hybrid system (figure 9).

If the existing vehicle has a mechanical-cable driven throttle valve body for the engine's air-fuel intake, it can be replaced with an electric throttle body (ETB) to have electronic throttle-by-wire for vehicle. In this project, an ETB model *BOSCH DV-E5* is tested for the electronic throttle (figure 10) [6]. The work requires configuration of the interface and power circuitry for the electronic throttle (figure 11), programming and configuration of the throttle controller and simulation and control tuning for the throttle position control.

Figure 8. In-wheel motor (left) and motor controller (right) from *Kelly Controls.*

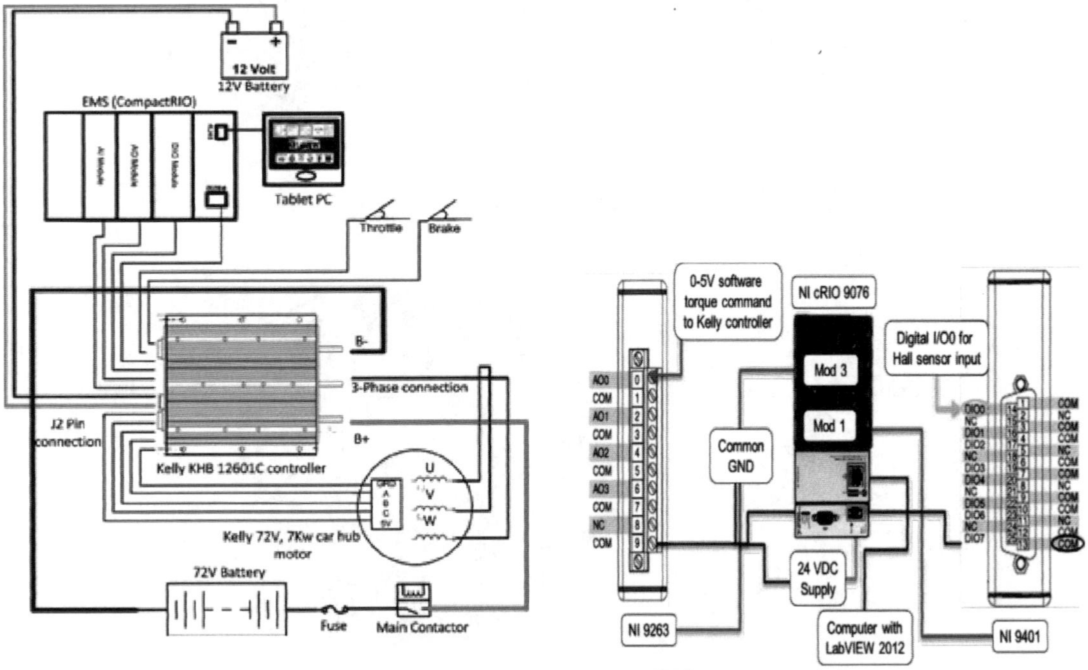

Figure 9. In-wheel motor system (IWM) connectivity.

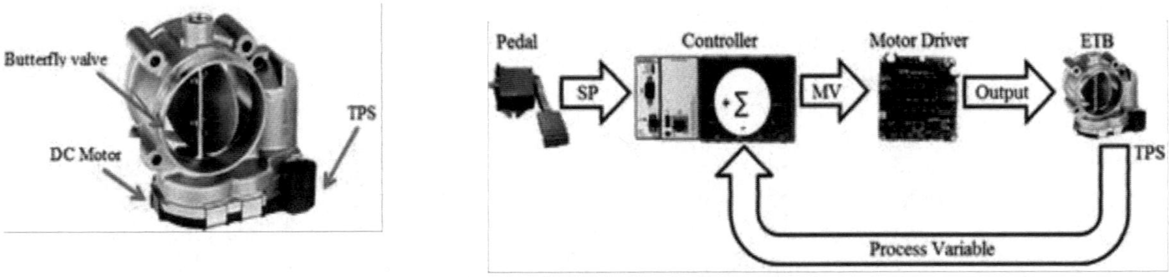

Figure 10. Electronic throttle body (*BOSCH DV-E5)* and throttle position control.

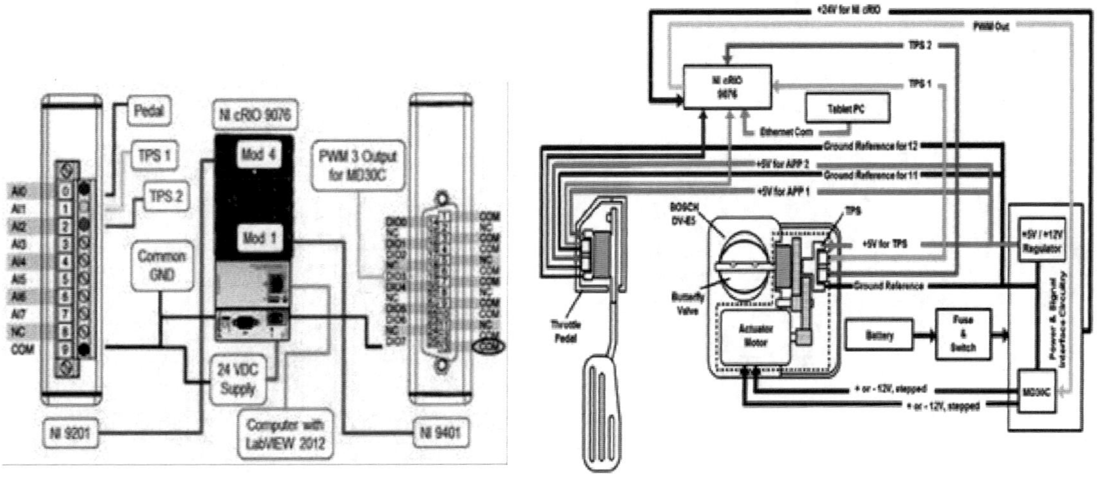

Figure 11. Electronic throttle-by-wire system (TBW) connectivity.

4. Energy management system (EMS)

The energy management and supervisory control system (EMS) is the heart of the retrofit hybrid system. In this project, the EMS is developed on National Instrument's CompactRIO embedded controller and LabVIEW Real-Time software platform (figure 12). The EMS functions as the primary vehicle controller whose objective is to achieve optimal operation of the hybrid propulsion system-reduced fuel consumption and self-sustaining on-board energy storage, since the vehicle is a non-plug in hybrid [7, 8].

The EMS reads the inputs of throttle position, vehicle speed, engine rpm and battery state of charge (SOC). Based on a certain control algorithm, it then determines the two sources operation status and power distribution and charging of the battery pack [8, 9]. A graphical driver interface is implemented on a tablet PC with a TCP/IP connection to the EMS controller. A LabVIEW-based interface appears on a dynamically-controllable HTML page hosted by the web server function of the CompactRIO, enabling the driver to monitor and override control of the hybrid vehicle. Figure 13 shows connectivity of the EMS to other sub-systems of the retrofit hybrid vehicle, showing the name, cable type and signal type of each input or output (I/O) parameter of the system.

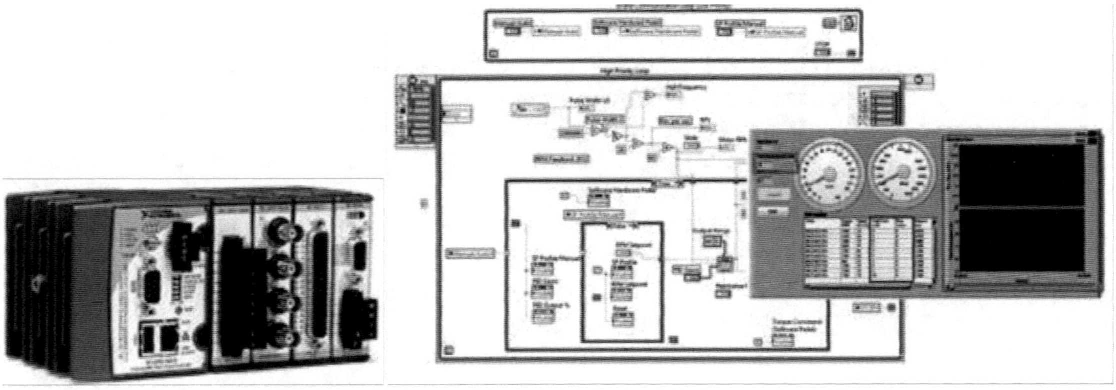

Figure 12. *National Instruments CompactRIO* controller and *LabVIEW Real-Time* software.

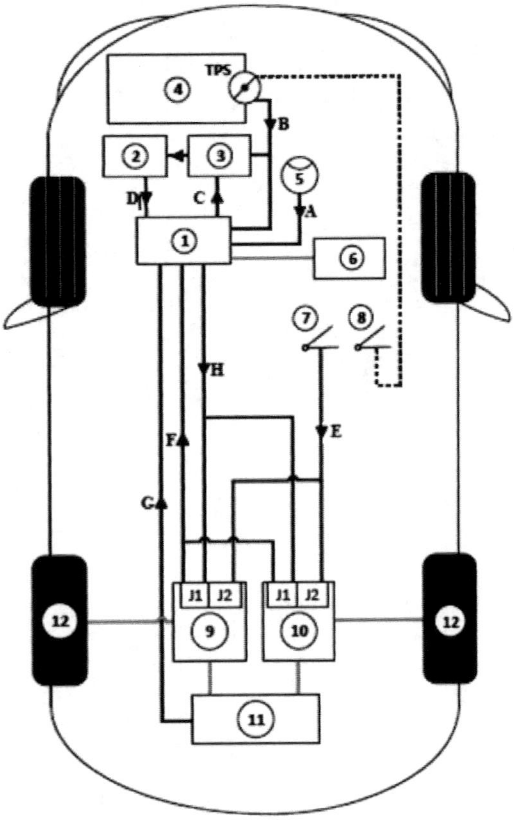

	I/O Parameter	Cable Type	Signal Type
A	Fuel Flow	2-core shielded	Digital (pulse)
B	Throttle Position Sensor (TPS)	2-core shielded	Analog
C	Enhanced Throttle Position Sensor (E-TPS)	2-core shielded	Analog
D	ECU Parameters (engine rpm, vehicle speed, etc.)	4-core shielded	Digital (pulse)
E	Brake Pedal Position	3-core shielded	Analog
F	Motor Parameters (motor rpm, status, current, etc.)	4-core shielded	CANBus
G	Battery State-of-Charge (SoC)	2-core shielded	Analog
H	Motor Torque Command	4-core shielded	Analog

Mechanical Link — Signal Cable
CAT-5 Ethernet Cable — HV DC Power Cable

Figure 13. EMS layout and connectivity to other sub-systems and I/O parameter details.

5. Re-generative braking system (RBS)

R-generative braking is a major energy saving mechanism for a hybrid vehicle, since the vehicle's kinetic energy which would otherwise be lost during braking is captured and converted to electrical energy-by operating the in-wheel motor as electric generator-to be saved in the onboard battery pack for later use. In a retrofit hybrid vehicle, a mechanism is required to incorporate electrical re-generative braking onto the existing mechanical braking system, which is retained in the vehicle for safety and maximum braking performance.

The system is designed as follows: when the pedal is depressed, only re-gen braking occurs initially (no mechanical braking) up to a certain stroke of the pedal. Beyond this pre-set position, mechanical braking is applied along with re-gen braking, until the end of the stroke. This ensures that more braking energy is re-captured through re-gen, before mechanical braking takes place. This requires modification of the existing brake pedal (figure 14).

The main criterion in re-design of the brake pedal is to ensure its ergonomic position and pedal travel while in operation. By maintaining as much of the original design as possible, the mechanism is incorporated without major change to the brake pedal's initial position. The strategy of the integrated braking mechanism consists as following (figure 15):

1. Applying re-gen braking to harvest electrical energy form the IWM for the first 30% of the pedal travel and
2. To continue with mechanical braking using standard OEM braking mechanism for the remaining 70% of the travel.

If the vehicle is required to stop abruptly, such as in an emergency situation, the EOM brake will still be fully functional. Additionally, actual amount of re-gen braking energy depends on the battery pack's state-of-charge (SOC)-if the battery is almost full, very little re-gen is possible and thus braking that can be applied to the vehicle also becomes reduced, which affects effectiveness of braking.

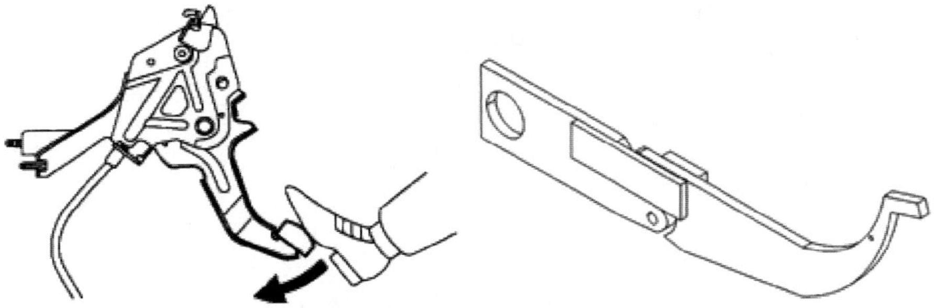

Figure 14. Brake pedal adaptation to incorporate re-generative braking.

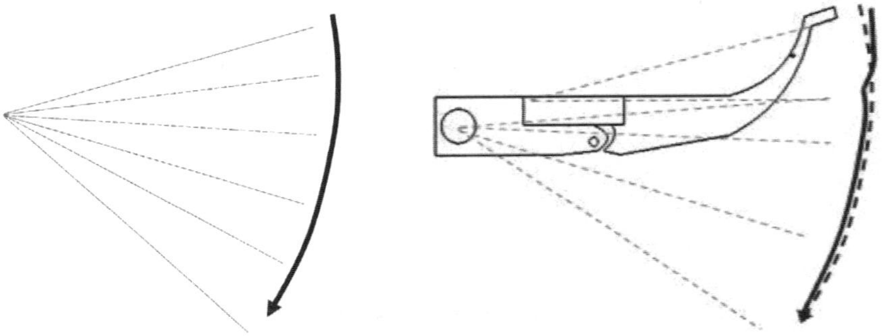

Figure 15. Travel pattern of brake pedal: original OEM brake pedal (left) and modified brake pedal to incorporate re-gen braking (right)-a free-travel zone is created to retard mechanical braking, whereby only re-gen braking is activated in this zone.

Thus mechanical braking remains a significant braking component for the retrofit vehicle, to ensure safety and maximum braking performance. Apart from pedal modification to enable delay in mechanical braking, the system requires installation of a brake pedal position sensor and an electronic interface circuit for the brake sensor's connection to the motor controller and EMS.

6. Thermal management system (TMS)

The vehicle IWM system is subject to high temperatures. This sub-project consists of design and fabrication of a thermal management system and a fire-retardant enclosure for the motor controller and battery pack, which are located in the vehicle's trunk (rear boot space-figure 16). The thermal system's function is to maintain the batteries and motor controller at their optimum operating temperature, which is important not only for safety objectives but also to prolong the batteries life and performance.

Figure 16 shows the rear view of the vehicle which indicated two main areas with different temperature profile. Interior cabin space will have lower temperature due to the vehicle's air condition system. It cooler air is blown into the rear boot space via fans to cool the battery pack and motor controller located within. A basic model for temperature profile in the battery pack is as equation (1).

$$C_c T_{c,i} = I^2 R_{in} + \frac{T_{s,i} - T_{c,i}}{R_c} , \tag{1}$$

where C_c is lumped heat capacity, $T_{c,i}$ is the cell core temperature, I is current applied to the cell, R_{in} the internal electric resistance of the cell, $T_{s,i}$ is the cell surface temperature and R_c is the thermal resistance between the core and surface of the cell.

An active thermal management system is built based on this equation to predict the battery core temperature. For the retrofit vehicle, an air-cooled system with two fans is selected for the cooling system since it requires less modification in the vehicle's interior and it can maintain of the battery cells and motor controller for temperature monitoring. The readings are fed to a closed-loop temperature control system, in which a temperature controller (CAREL) will adjust the speed of one of the fans to keep the system temperature about a set point.

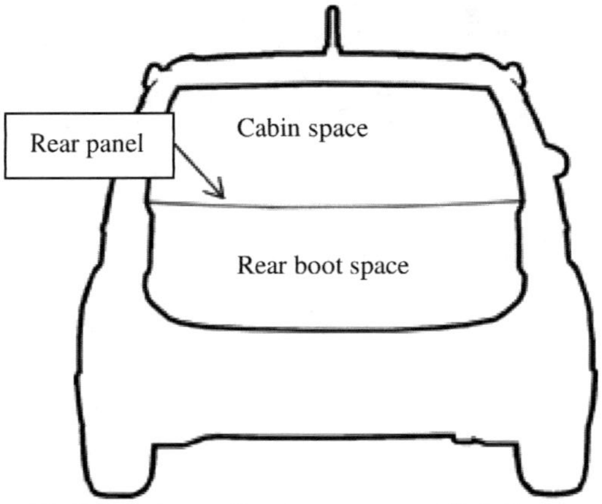

Figure 16. Rear view of prototype retrofit hybrid vehicle.

The driver also has the option to manually activate further cooling in the event of over-temperature, by turning on the second fan. The thermal management system's control schematic is shown if figure 17.

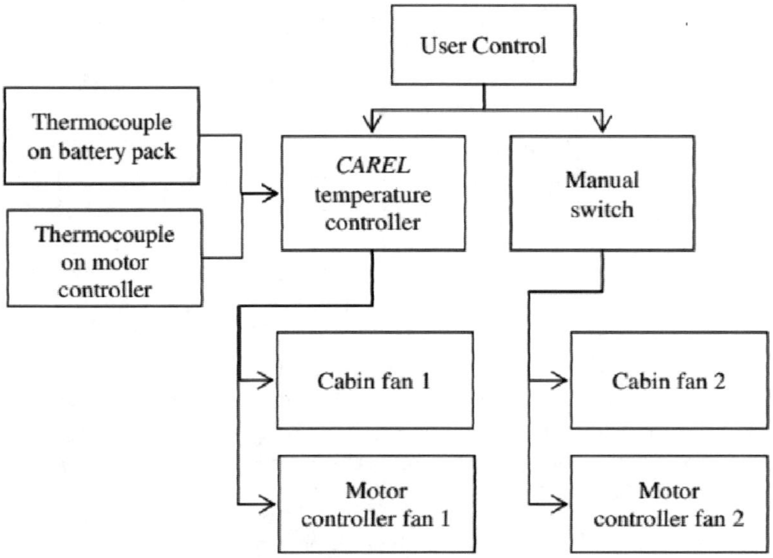

Figure 17. Schematic diagram of thermal management system.

Figure 18 indicates air-flow direction in the two areas, showing hot air being blown out of the rear boot compartment via a separate exhaust fan. Thus in this design, cooler air from the cabin plays an important role to cool down the vital system components. This is further illustrated in figure 19, which also shows a vertical cross section of the system.

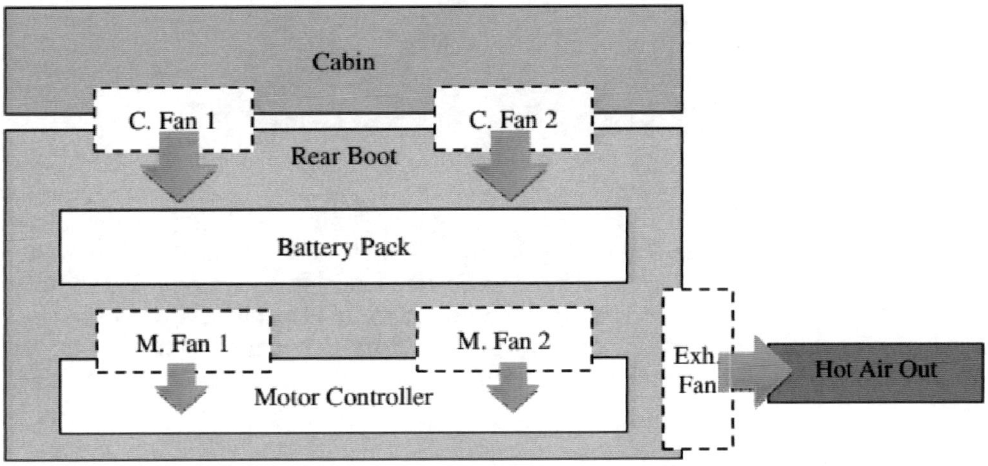

Figure 18. Component position and direction of air flow (vertical cross-section of system).

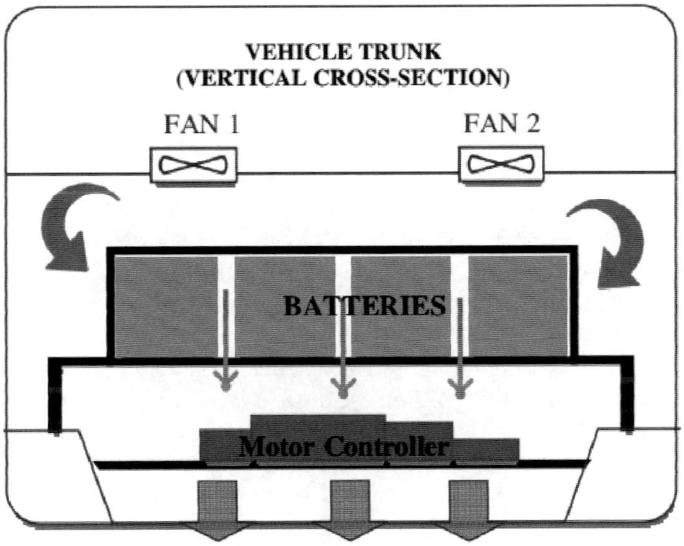

Figure 19. Thermal management system for battery pack and motor controller (vertical cross-section of vehicle trunk)

An in-house designed fire-retardant material (figure 20) is used as coating for and enclosure which houses the battery pack and motor controller to prevent flames from spreading to the vehicle cabin in the event of a fire. This coating suppresses flame by releasing CO_2 when it is in contact with the flame. It can withstand surrounding temperatures of up to 1000°

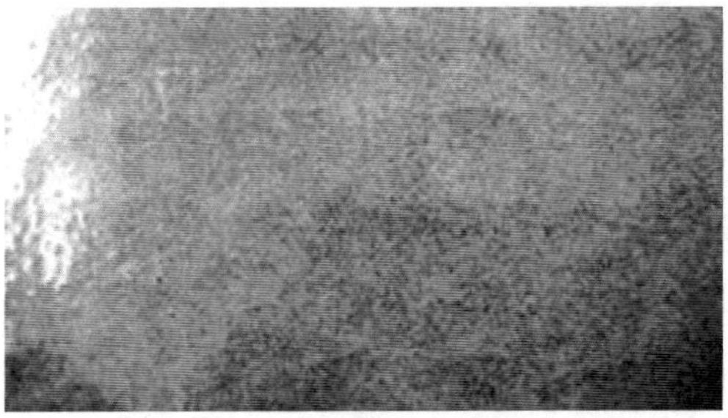

Figure 20. Fire-retardant coating for enclosure housing battery pack and motor controller.

7. Conclusion

A split-axle parallel HEV configuration enables existing ICE-powered vehicles to be retrofitted into a hybrid electric vehicle with minimal physical modification. This paper describes specific operating modes, power flow torque coupling and torque profiles of the split-parallel, through-the-road hybrid vehicle. Design and development of sub-systems for a TTR hybrid vehicle with in-wheel motors are provided as part of a retrofit-conversion hybrid electric vehicle project.

Acknowledgments

This work is partly funded by an ERGS grant (Exploratory Research Grant Scheme) from the Ministry of Higher Education (MoHE), Malaysia.

References

[1] Zulkifli S, Saad N, Mohd S 2012 *Power Engineering and Optimization Conference (PEDCO) Melaka, Malaysia, 2012 Ieee International* IEEE) p. 11-6
[2] Ehsani M, Gao Y, Emadi A 2009 *Modern electric, hybrid electric, and fuel cell vehicles: fundamentals, theory, and design* CRC press)
[3] Chan C-C, Bouscayrol A, Chen K 2010 *Vehicular Technology, IEEE Transactions on* **59** 589-98
[4] Zulkifli S A, Mohd S, Saad N B, Aziz A R A 2014 *Applied Mechanics and Materials* vol 663 Trans Tech Publ) p. 498-503
[5] Young M, Molen G M, Oglesby D, Crawford K, Walp K, Lewis R, et al. 2007 *Vehicle Power and Propulsion Conference, 2007 VPPC 2007 IEEE* IEEE) p. 511-8
[6] Zulkifli S A, Asirvadam V S, Saad N 2014 *Intelligent and Advanced Systems (ICIAS), 2014 5th International Conference on* IEEE) p. 1-6
[7] Fallahi N, Niasar A H 2014 *Intelligent Systems in Electrical Engineering* **4**
[8] Asaei B, Habibidoost M 2010 *Vehicle Power and Propulsion Conference (VPPC), 2010 IEEE* IEEE) p. 1-5
[9] Sciarretta A, Guzzella L 2007 *Control systems, IEEE* **27** 60-70

IOP Conf. Series: Materials Science and Engineering **100** (2015) 012040 doi:10.1088/1757-899X/100/1/012040

Review of magnetorheological fluids and nanofluids thermal behaviour

M S A Rahim[1] and I Ismail[1]

[1]Faculty of Manufacturing Engineering, Universiti Malaysia Pahang, 26600 Pekan, Pahang, Malaysia

Email: sallehrahim@icloud.com

Abstract. Magnetorheological fluids are smart materials with the capability to reversibly change from liquid to near solid state under the presence of external magnetic fields. The interest in magnetorheological fluid as a heat transfer medium is due to the possibility of controlling its flow and heat transfer process through external magnetic field. This review highlights on the recent developments of magnetorheological fluid and nanofluid in the field of thermal behaviour. Special emphasis is paid to the understanding of their thermal conductivity property by several parameters which include particle volume fraction, shape and size of particles, materials of particles and base fluid, and magnetic field. Reports indicate that the increase with particle volume fraction and magnetic field strength may increase the thermal conductivity of magnetorheological fluid. From the theory and experimental given, relationship between these parameters and thermal conductivity can be found.

1. Introduction

Magnetorheological fluids are intelligent materials made of magnetic particles that have the ability to rapidly change their flow characteristics in the presence of an applied magnetic field. Some of the unique characteristics of magnetorheological fluids are adjustable viscosity, rapid response time and high dynamic flow strength [1]. However, the field-dependent thermal conductivity of magnetorheological fluids has received much less attention despite its importance in heat dissipation of active damping devices and potential new applications but this is not the case for nanofluids. Nanofluids on the other hand are two-phase media that consist of liquid and solid nanometer-scale particles. The interest to nanofluids relates to the fact that even low concentration of nanoparticles can significantly change its thermal behaviour. Practical application of nanofluids is connected with a possibility to create a high performance liquid heat carrier. By considering the fact that metallic solids possess better thermal properties than fluids, nanofluids contained with suspended solid particles are expected to exhibit significantly higher thermal properties than that of conventional heat transfer fluids [2].

This review mainly focuses on the most relevant progress during the recent years and further application in future regarding thermal behaviour of magnetorheological fluids. In particular, it deals with the most significant experimental results in magnetorheological fluids and nanofluids relating to thermal conductivity for heat transfer applications. The analysis is made to identify the influence of different factors on the expected results. It is expected that this review provides a framework for the future direction of studies on magnetorheological fluids to enhance its thermal properties.

2. Particle Volume Fraction

Solid suspension is a type of fluid consists of small solid particles for various purposes which some of the notably discovered suspension types are magnetorheological fluid, ferrofluid, electrorheological fluid, and nanofluid. Generally, these suspensions comprise 3 main elements; solid particles, base fluid and additives. Numerous studies have been conducted on solid particles with mostly by varying the particle volume fraction to enhance the suspension performance.

In magnetorheological fluid, it is found that thermal conductivity increases in the absence of magnetic field as particle volume fraction is increased. From figure 1, the thermal conductivity ratios were recorded at approximately 1.4 and 2.0 for 10 vol% and 20 vol% of iron based magnetorheological fluid, respectively. The measured thermal conductivities are normalized by the oil/grease base fluid conductivity of 0.1242 W/m K [3].

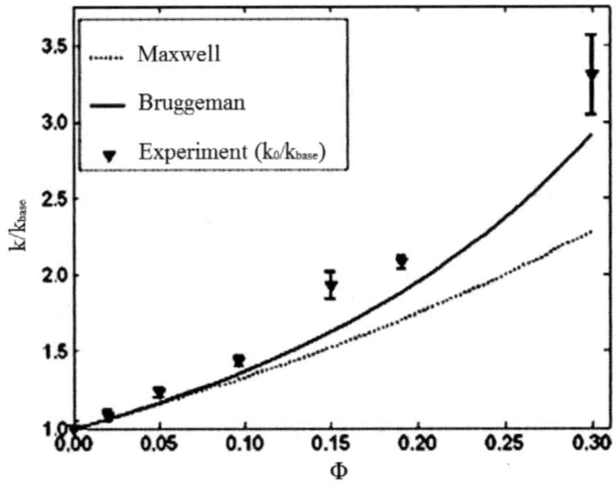

Figure 1. Measured zero-field thermal conductivity for varying particle volume fraction [3].

Additionally, in the presence of magnetic field, the thermal conductivity of magnetorheological fluid will increase when particle volume fraction and magnetic field strength are increased. At 2 mT of magnetic field strength, the thermal conductivity values for 20 vol% and 40 vol% of iron based magnetorheological fluids were recorded at approximately 0.42 W/m K and 0.46 W/m K, respectively. While at 15 mT of magnetic field strength, the thermal conductivity values increased again to approximately 0.49 W/m K and 0.54 W/m K, respectively [1]. However, the ratio of thermal conductivity in bulk magnetorheological fluids without magnetic field and with 290 kA/m of magnetic field is approximately 1.3. The experiment conducted to produce this finding was using bulk iron based magnetorheological fluids with particle volume fraction up to 33 % along the magnetic field direction. The result indicates that increasing particle volume fraction is insignificant on enhancing huge amount of thermal conductivity [4]. In other study, a relationship between thermal conductivity and particle volume fraction was reported to be in linear correlation. The linear correlation model as shown in equation (1) and (2), were applied in an experiment and the result shows that the model is able to evaluate the sedimentation stability of the magnetorheological fluids [5].

$$k_{eff} = 0.0245\Phi - 0.08 \tag{1}$$

$$k_{eff}/k_f = 0.167\Phi - 0.53 \tag{2}$$

where $10 \leq \Phi \leq 50$

k_{eff} is the magnetorheological fluid effective thermal conductivity,
k_f is the thermal conductivity of the base fluid,
Φ is the particle volume fraction

By comparison with magnetorheological fluid, it is also found that the nanofluid thermal conductivity increases with the increase of particle volume fraction in nanoparticles [6]. As shown in figure 2, the increase of up to 1.5 vol% of Al_2O_3 nanoparticles resulted in thermal conductivity increase at about 5 % [7].

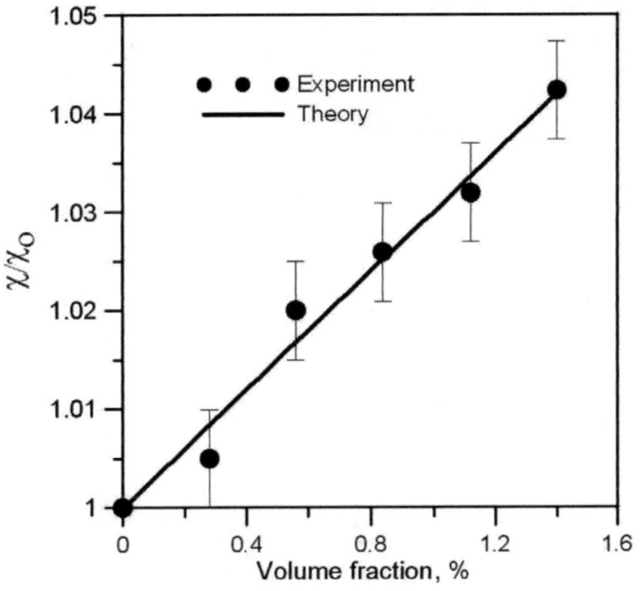

Figure 2. Relative thermal conductivity versus nanoparticles volume fraction [7].

3. Shape and Size of Particles

The magnetorheological particles are traditionally in spherical shape while typical particle sizes are in the range of micrometres [8]. The size distribution range of magnetorheological particles is 0.5-20 μm. Surface morphology of carbonyl iron particles is observed with a scanning electron microscope (SEM) and the images with size distributions are shown in figure 3 [9]. Although shape and size of particles were proven to affect thermal conductivity in nanofluids, no reports found any description in magnetorheological fluid [10]. Nevertheless, mechanism of shape and size of particles in affecting the thermal conductivity remains relevant in the case of magnetorheological fluid.

The enhancement of effective thermal conductivity due to particle shape effect is strongly reduced by interfacial effects proportional to the total surface area of particles. For example in nanofluid research, the thermal conductivities of Al_2O_3 nanofluids with different particle shapes measured at room temperature are presented in figure 4 as a function of nanoparticle volume fraction [10]. This result indicates spherical shape of particles produce least enhancement in thermal conductivity as in the magnetorheological fluid. Therefore, introduction of cylinder shaped particles or additive will improve the thermal conductivity properties of magnetorheological fluid.

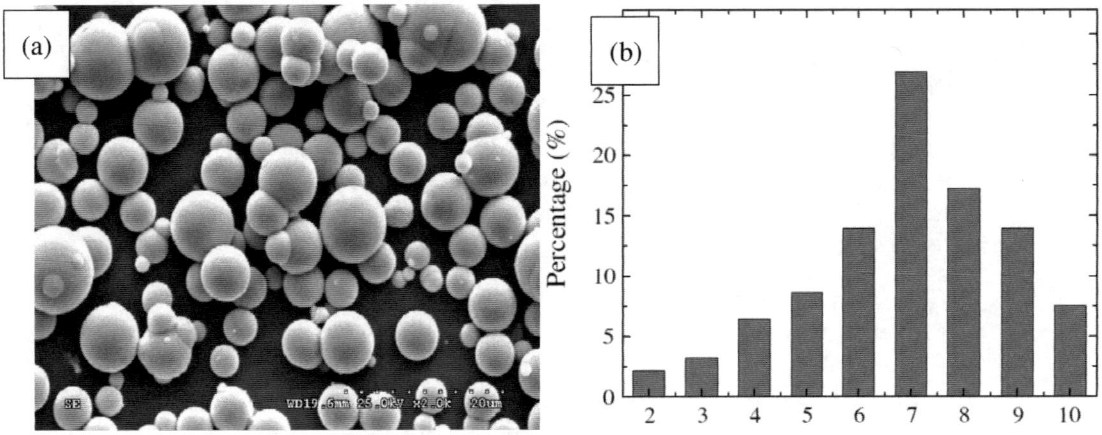

Figure 3. (a) SEM images and (b) size distributions of carbonyl iron particles [9].

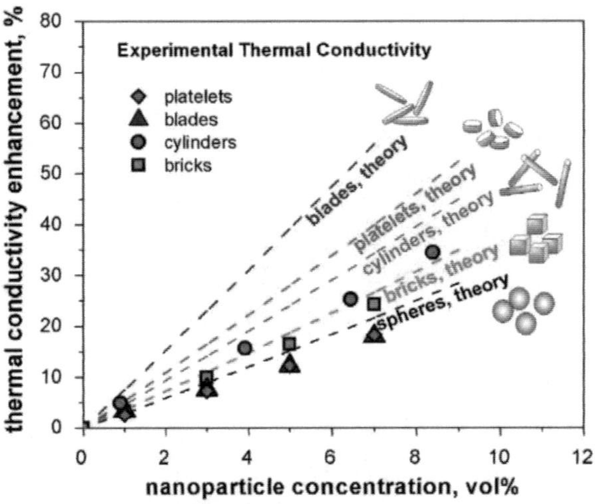

Figure 4. Experimentally measured thermal conductivity of Al_2O_3 nanofluids in ethylene glycol/H_2O for corresponding particle shapes [10].

Furthermore, it was found that the reduction of particle size with temperature rise enhanced the thermal conductivity ratio of a solid suspension. Taking nanofluid as an example, the thermal conductivity ratios could be enhanced by up to 14.7 % for 20 nm nanoparticles at 50 °C as can be seen in figure 5. The experiment used water based Al_2O_3 nanofluids with different nominal diameters of 20, 50, and 100 nm under varied temperatures [11].

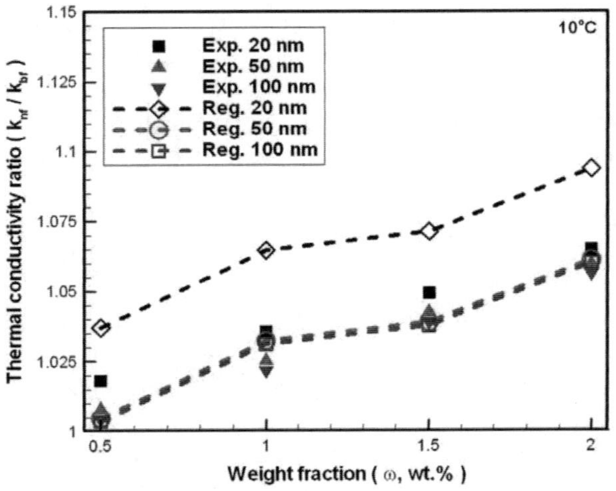

Figure 5. Dependence relationship between weight fraction and thermal conductivity ratio of Al_2O_3/water nanofluid with different particle sizes at 50 °C [11].

This is further evidenced by an experiment conducted on three different Al_2O_3 nanofluids with 11, 47 and 150 nm sized nanoparticles where the effective thermal conductivity is normalized by the base fluid thermal conductivity. The thermal conductivity enhancement at 70 °C is merely 5 % for 150 nm sized nanoparticles, but increased further to 10 % and 20 % for 47 nm and 11 nm sized, respectively [12]. Hence, it is proven that thermal conductivity will be improved by reducing particle size and can be applied to magnetorheological fluid.

4. Material of Particles

In practice, the most frequently used material of particles in magnetorheological fluid is carbonyl iron powder. It has several important features such as high magnetic permeability, low remnant magnetization, fine particle size and common availability. The iron content is 97–99 % and the surface morphology is typically spherical [9]. However, the lack of research on material of particles in enhancing the thermal conductivity of magnetorheological fluid has forced this review to look into nanofluids. At room temperature (300 K), the thermal conductivity of carbonyl iron powder is 80.4 W/m K [13]. Alternatively, the thermal conductivities of common materials of particles used in solid suspensions are summarized in table 1.

Table 1. Thermal conductivities of several solid suspension particles at 300 K [14].

Suspension particles	Thermal conductivity, k (W/m K)
CIP	80.4
TiO_2	8.04
CuO	17.65
Al_2O_3	39
Al	237
Cu	401

It is found that the thermal conductivity enhancement is higher for water based Cu nanofluids compared to water based Al_2O_3 nanofluids. The thermal conductivity ratios for Cu nanofluids and Al_2O_3 nanofluids are 1.12 and 1.10, respectively. This result is attributed to the higher thermal conductivity of Cu nanoparticles than Al_2O_3 nanoparticles [15]. Likewise, water based Al_2O_3 nanofluid showed a higher enhancement of thermal conductivity compared with TiO_2 nanofluids with respect to particle volume fraction. While there was only 9 % enhancement of thermal conductivity with TiO_2 nanoparticles, 11 % thermal conductivity enhancement was observed with Al_2O_3 nanoparticles, which is attributed to the higher thermal conductivity of Al_2O_3 against TiO_2 [16].

On the other hand, it is found that metallic particles are giving higher enhancements of thermal conductivity than oxide particles. For instance, by using ethylene glycol as the base fluid, 11 % and 17 % of thermal conductivity enhancement were recorded for the nanofluids with 150 nm and 45 nm sized Al_2O_3 nanoparticles, respectively. Whereas 27 % of enhancement was observed for nanofluid with 80 nm Al nanoparticles, metallic nanofluids thus proved to have higher enhancement of thermal conductivity than oxide nanofluids [17]. Despite of possessing lower thermal conductivity than metallic particles like Al in this case, it is expected that carbonyl iron particles to contribute a substantial enhancement of thermal conductivity in magnetorheological fluid. Either way, the metallic nanoparticles can also be as an additive so as to enhance the thermal properties of magnetorheological fluid.

5. Material of Base Fluid

The main element of a magnetorheological fluid is the base fluid which serves as a continuous insulating medium. It also has a strong impact on the fluid performances. Currently, a commonly used base fluid is the silicone oil due to its wide operating temperature range, strong anti-oxidation ability and some other important characteristics. As material of base fluid has yet to become a focal study in enhancing thermal conductivity of magnetorheological fluid, it is more sensible to look into research areas in nanofluids. Silicone oil has a thermal conductivity of 0.14 W/m K [18]. Meanwhile, the thermal conductivities of other common base fluids used in solid suspensions can be seen in table 2.

Table 2. Thermal conductivities of several base fluids at 300 K [14].

Base fluids	Thermal conductivity, k (W/m K)
Silicone oil	0.14
Deionized water	0.607
Ethylene glycol	0.255
Engine oil	0.145

When considering a theoretical model to predict the thermal conductivity of a solid–liquid mixture, the Maxwell model as shown in equation (3) can be used which is applicable to statistically homogeneous and low volume fraction of solid suspensions with randomly dispersed, uniformly sized and non-interacting spherical particles [19]. The equation (3) also shows that the thermal conductivity of base fluid has a very close relationship with the thermal conductivity of particles. This relationship as shown in equation (4) can be used to obtain the thermal conductivity enhancement or thermal conductivity ratio (without %) of a solid suspension [20].

$$\frac{k}{k_f} = \frac{k_p + 2k_f + 2\Phi(k_p - k_f)}{k_p + 2k_f - \Phi(k_p - k_f)} \quad (3)$$

where

k is the thermal conductivity,
k_f is the thermal conductivity of the base fluid,
k_p is the particle thermal conductivity,
Φ is the particle volume fraction

$$\text{Thermal conductivity enhancement, } \% = \frac{(k_p - k_f)}{k_f} \times 100 \quad (4)$$

Most of the studies have shown an increased thermal conductivity enhancement with decrease in thermal conductivity of the base fluid. For example in nanofluids, as can be observed in figure 6, dramatic improvement in thermal conductivity of graphene oxide nanosheets suspension is seen for a base fluid with lower thermal conductivity. This experiment on thermal conductivity of nanofluids comprises several base fluid materials like liquid paraffin (0.24 W/m K), propyl glycol (0.206 W/m K), ethylene glycol (0.255 W/m K) and distilled water (0.58 W/m K) [21]. Hence in theoretically, magnetorheological fluid would have a good thermal conductivity ratio as silicone oil has a very low thermal conductivity combined with the high thermal conductivity of carbonyl iron particles.

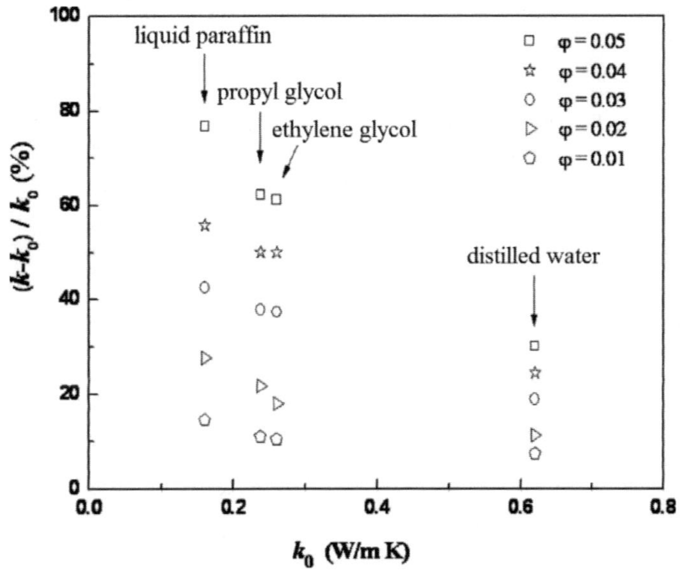

Figure 6. Thermal conductivity enhancement ratios as a function of the thermal conductivities of the base fluids with different volume fraction [21].

6. Strength and Direction of Magnetic Field

It is found that thermal conductivity of magnetorheological fluid increases with increasing magnetic field in the temperature intervals from 0 to 50 °C and from 50 to 100 °C. The substantial enhancement of the thermal conductivity is by almost 188 % at magnetic field of 15 mT as can be seen in table 3 [1].

Table 3. Thermal conductivity enhancement of magnetorheological fluids with respect to an applied magnetic field H = 15 mT [1].

	0-50 °C at 150 G		50-100 °C at 150 G	-20-0 °C at 150 G	
	ΔT = 35 K (%)	ΔT = 20 K (%)	ΔT = 15 K (%)	ΔT = 20 K (%)	ΔT = 15 K (%)
40SM	12	18	134	-42	-2
40SQ	9	11	50	-29	-8
20SM	14	3.9	188	-25	-6
20SQ	20	21	36	-7	0.2
5SM	29	32	78	-25	-4
5SQ	23	34	95	-31	-0.2

Direction of the magnetic field also has a strong influence on the thermal conductivity, with the maximum being achieved when the field was parallel to the direction of thermal gradient. As in figure 7, for parallel alignment of the magnetic field, thermal conductivity increases with increasing magnetic field strength while for perpendicular alignment, the thermal conductivity has an inverse relationship to the magnetic field strength [22].

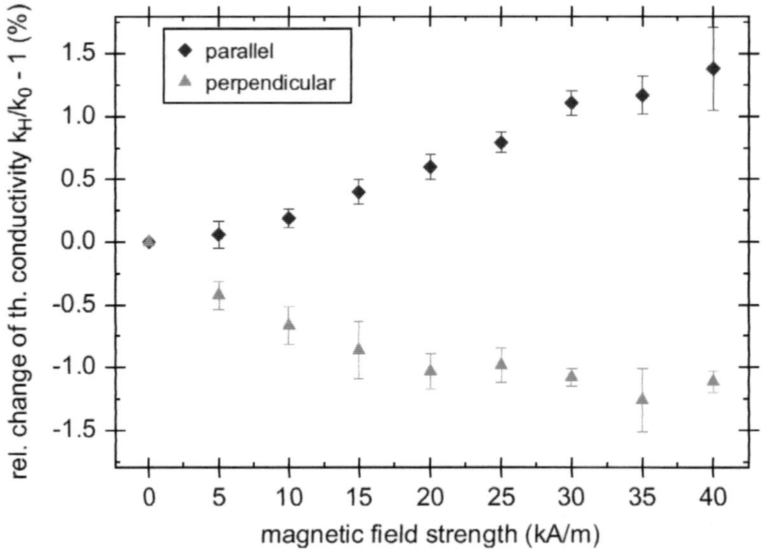

Figure 7. Relative change of thermal conductivity depending on magnetic field strength and orientation against heat flux direction [22].

In particular, the increase in thermal conductivity under external magnetic field is attributed to the effective conduction of heat through the chainlike structures formed under magnetic field when the dipolar interaction energy becomes greater than the thermal energy [23]. Additionally, hysteresis was found in which when the thermal conductivity is enhanced by the magnetic field, it remained high until the fluid is disturbed to break the particle chains. By referring to figure 8, once the suspension is magnetized to 202 kA/m, the thermal conductivity remains constant with decreasing field and only returns to its initial zero field value if the suspension is stirred [3].

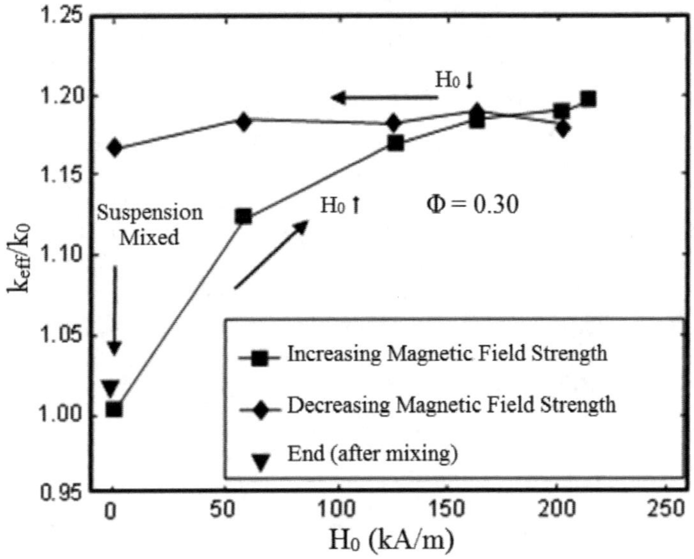

Figure 8. Thermal conductivity ratio of magnetorheological fluid with 0.30 vol% of iron depends on time history of the applied magnetic field [3].

7. Conclusions

A review of the studies conducted on magnetorheological fluids and nanofluids thermal behaviour with special attention on their thermal conductivity property was conducted. The increase in thermal conductivity has been studied according to variations in the following parameters which are particle volume fraction, shape and size of particles, material of particles, material of base fluid, and strength and direction of magnetic field. Although there are several studies on the enhancement of thermal conductivity of magnetorheological fluids from 0.42 W/m K to 0.54 W/m K, they are still room for improvement to enable optimization of the fluids to be used in practical applications. The development of this area is still challenged by many aspects ranging from the preparation and characterization to design and applications. The potential applications of magnetorheological fluids with better thermal properties would be like thermomagnetic convection and heat pipe. Further research must be related to understanding the thermal conductivity mechanisms enhancement. More detailed studies are necessary on improving magnetorheological fluids for heat transfer applications especially in the off and on-state conditions.

Acknowledgements

The authors would like to be obliged to Universiti Malaysia Pahang for providing facilities and financial assistance under project no. RDU131409.

References

[1] Yildirim G, Genc S 2013 *Smart Materials and Structures* **22** 1-8
[2] Philip J, Shima P D 2012 *Advances in Colloid and Interface Science* **183–184** 30-45
[3] Reinecke B N, Shan J W, Suabedissen K K, Cherkasova A S 2008 *J Appl Phys* **104** 1-7
[4] Cha G, Ju Y S, Ahuré L A, Wereley N M 2010 *J Appl Phys* **107** 09B505
[5] Cheng H, Hong W, Liu C 2014 *Journal of the Chinese Advanced Materials Society* **2** 130-7
[6] Xiao B, Yang Y, Chen L 2013 *Powder Technol* **239** 409-14

[7] Serebryakova M A, Dimov S V, Bardakhanov S P, Novopashin S A 2015 *International Journal of Heat and Mass Transfer* **83** 187-91
[8] de Vicente J, Klingenberg D J, Hidalgo-Alvarez R 2011 *Soft Matter* **7** 3701-10
[9] Wang D, Zi B, Zeng Y, Hou Y, Meng Q 2014 *J Mater Sci* **49** 8459-70
[10] Timofeeva E V, Routbort J L, Singh D 2009 *J Appl Phys* **106** 014304
[11] Teng T-P, Hung Y-H, Teng T-C, Mo H-E, Hsu H-G 2010 *Appl Therm Eng* **30** 2213-8
[12] Chon C H, Kihm K D, Lee S P, Choi S U 2005 *Applied Physics Letters* **87** 153107--3
[13] Perry D L, Phillips S L 1995 *Handbook of Inorganic Compounds* Taylor & Francis)
[14] Sattler K D 2010 *Handbook of Nanophysics: Nanoparticles and Quantum Dots* Taylor & Francis)
[15] Wang X-j, Zhu D-s, yang S 2009 *Chemical Physics Letters* **470** 107-11
[16] Palabiyik I, Musina Z, Witharana S, Ding Y 2011 *J Nanopart Res* **13** 5049-55
[17] Patel H, Sundararajan T, Das S 2010 *J Nanopart Res* **12** 1015-31
[18] Totten G E 2011 *Handbook of Hydraulic Fluid Technology, Second Edition* Taylor & Francis)
[19] Syam Sundar L, Singh M K, Sousa A C M 2013 *Int Commun Heat Mass* **44** 7-14
[20] Hadadian M, Goharshadi E K, Youssefi A 2014 *J Nanopart Res* **16** 1-17
[21] Yu W, Xie H, Chen W 2010 *J Appl Phys* **107** -
[22] Krichler M, Odenbach S 2013 *Journal of Magnetism and Magnetic Materials* **326** 85-90
[23] Philip J, Shima P, Raj B 2008 *Nanotechnology* **19** 305706

Performance of CO_2 enrich CNG in direct injection engine

Firmansyah[1], **W B Ayandotun**[1], **E Z Zainal** [1]**A, A R A Aziz**[1] **and M R Heika**[1]
[1]Centre for Automotive Research and Energy Management, Universiti Teknologi PETRONAS, 32610 Bandar Seri Iskandar, Perak, Malaysia.

E-mail: ezrann.zharif@gmail.com

Abstract. This paper investigates the potential of utilizing the undeveloped natural gas fields in Malaysia with high carbon dioxide (CO_2) content ranging from 28% to 87%. For this experiment, various CO_2 proportions by volume were added to pure natural gas as a way of simulating raw natural gas compositions in these fields. The experimental tests were carried out using a 4-stroke single cylinder spark ignition (SI) direct injection (DI) compressed natural gas (CNG) engine. The tests were carried out at 180° and 300° before top dead centre (BTDC) injection timing at 3000 rpm, to establish the effects on the engine performance. The results show that CO_2 is suppressing the combustion of CNG while on the other hand CNG combustion is causing CO_2 dissociation shown by decreasing CO_2 emission with the increase in CO_2 content. Results for 180° BTDC injection timing shows higher performance compared to 300° BTDC because of two possible reasons, higher volumetric efficiency and higher stratification level. The results also showed the possibility of increasing the CO_2 content by injection strategy.

1. Introduction

The search for alternative fuels for internal combustion engines with a vision to improve the engine fuel economy and reduce harmful exhaust emissions becomes necessary [1] as the energy needs in the transportation sector is growing drastically due to the increase in the number of vehicles. However for fossil fuels specifically crude oil has been the major source of fuel for transportation, power generation and other applications. Hence, the crude oil are depleting at alarming rates which leads to increase in fuel prices. Furthermore, the increased in consumption of these fuels are causing environmental pollution thereby causing global warming and health related issues. Therefore, dependency on fossil fuels (gasoline and diesel) calls for alternative energy sources for future energy needs. Moreover, with the increasing stringent emission standards continually legislated, alternative fuels are currently being used [2, 3] to tackle the negative resultant effects of environmental pollution such as unburned hydrocarbon (uHC), carbon monoxide (CO) and nitrogen oxide (NO_x) among others generated by all sectors. In lieu of this, extensive research studies have been carried out to determine alternative fuels that are best suited for both spark ignition (SI) and compression ignition engines; and other applications. For spark ignition engines, biogas and syngas [4, 5] have been studied and proved to enhance performance and reduce emissions under appropriate operating condition while for compression ignition engines, seed-oil bio fuels [5, 6] have been examined and found to give similar performance and emissions when compared with conventional diesel fuels, particularly when blended with diesel fuel or emulsified with ethanol or water.

Natural gas is amongst the fast growing component of the world's primary energy consumption because of its availability, abundance and adaptability to the gasoline and diesel engines. Malaysia is endowed with abundant natural gas reserves of 2400 billion cubic metres (84.76 Trillion cubic feet) and thus ranked the 15th largest in the world [7]. However 37 Tcf of these natural gas reserves are non-developed due to the presence of large quantities of CO_2 (from 28% to 87%), thus rendering it uneconomically valuable. CO_2 capturing from natural gas is seen as an option of utilizing these undeveloped resources. However, there are challenges associated with this technique such as high cost, advanced technology and processes corrosion and high energy demand [8, 9]. This study therefore was focused on increasing the CO_2 minimum requirement in the Malaysian natural gas composition from 0.57% to 40% as a way of simulating the natural gas field's composition and at the same time study the effect of high CO_2 content-natural gas on engine performance.

Carbon dioxide and hydrogen sulphide are part of the constituents of natural gas reservoirs that are present in large quantities [10]. It is majorly produced from combustion of coal or hydrocarbons, fermentation and decomposition of organic materials and breathing life forms. Carbon dioxide is chemically inert in nature and slightly heavier than air which has high specific heat capacity and would neither combust nor support combustion process.

The presence of high amount of CO_2 in a mixture of natural gas and carbon dioxide can cause lower heating value of the mixture compared to pure natural gas. This leads to reduction in burning velocity which eventually affects the performance of the engine [11]. Experiment conducted on a Ricardo engine using methane and CO_2 at different compression ratios, speeds and equivalent ratios showed that there is an improvement in NO_x emissions. However, lower cylinder pressure obtained led to the reduction in engine power and thermal efficiency and increase in the level of unburnt hydrocarbon. The results obtained are the same for different engines and operating conditions [4]. Furthermore, the presence of high CO_2 CNG may not affect much on the efficiency of the engine if the right amount of CO_2 is introduced into the mixture. Henham and Makkar [12] found that the efficiency of the mixture of natural gas and carbon dioxide is not much affected up to 37% natural gas substitution. It was reported also that introduction of carbon dioxide more than 40% into the mixture could lead to harsh and irregular running of an engine. Shrestha and Karim also found that the presence of diluents (carbon dioxide) beyond about 50% in the mixture could result to significant drop in the power output of the engine [13].

It is obvious from the results of the experimental studies conducted on engine using mixture of methane and large amount of CO_2 causes performance to dropped, which could be as a result of weak turbulence inside the combustion chamber caused by large specific heat capacity of the mixed fuel which absorbed more heat released and reduced the cylinder gas temperature [14]. In addition, the use of port-injection or carbureted fuel delivery could also be a contributing factor in reduction of power, volumetric efficiency and brake mean effective pressure [15].

2. Experimental Setup

In this research, the experimental work was carried out on a single-cylinder, 4-stroke SI DI-CNG engine. The detailed specifications of the engine are given in Table 1. The CO_2 used in the experiment had a purity of 99.8%. The schematic diagram of the experimental setup is shown in figure 1. All experimental studies were conducted at the Centre for Automotive Research and Energy Management (CAREM) located in the Department of Mechanical Engineering, Universiti Teknologi PETRONAS (UTP), Malaysia.

Table 1. Engine Specifications.

Engine Description	Specifications
Displacement volume	399.25 cm^3
Cylinder Bore	76 mm
Cylinder Stroke	88 mm
Compression Ratio	14:1
Exhaust Valve Closed	350° BTDC
Exhaust Valve Open	225° BTDC Inlet
Valve Open	12° BTDC Inlet
Valve Closed	132° BTDC
Dynamometer	Direct Current with maximum reading is 50 Nm

The operation of the engine used for the experimental work was managed and controlled by a PC-based data acquisition and control system. The in-cylinder pressure was measured using a water cooled Kistler 6061B piezoelectric pressure transducer. The fuel injection system was designed to allow two types of fuels (CNG and CO_2) to be injected at high injection pressure. The CNG and CO_2 were supplied through different gas inlets and directly injected into the combustion chamber at 18 bar using a synerject injector. Prior to the injection, the fuel were mix in a mixing chamber near the injector. The flow rate of carbon dioxide was measured using a CO_2 digital mass flow meter with an accuracy of ±1.0% of the full scale. The amount of the injected CO_2 was varied by adjusting the needle valve which in turn controlled the flow rate.

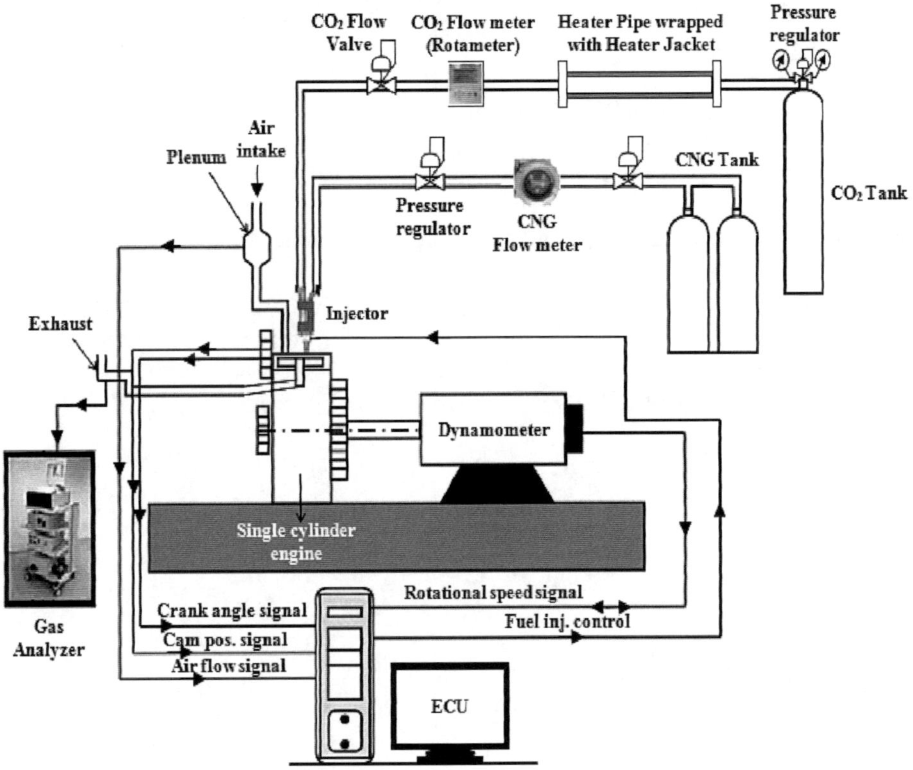

Figure 1. Experimental setup schematic diagram.

All tests and measurements were carried out on a CNG-DI engine with the injection timing set at 180° and 300° BTDC, the air-fuel ratio was set at stoichiometric ratio ($\lambda = 1$); while the ignition timing was adjusted to obtain the maximum brake torque. The CO_2 proportions (0%, 10%, 20%, and 30% by volume) were added to the CNG while the experiments were performed at full open throttle. Engine parameters such as injection timing, ignition timing, and air-fuel ratio were controlled by the ECU Remote Interface (ERI).

3. Results and Discussion

The effect of CO_2 content in natural gas to the engine performance at 3000 rpm can be seen in figure 2. It is observed that the torque and power output is decreasing as the CO_2 percentage increases. It can be explained by the fact that total energy content of the mixture is reducing as he CO_2 content increased. Furthermore, CO_2 is a stable compound with -393.5 kJ/mol heat formation in which it requires a significant amount of energy to dissociate its bonding. In the current process, there are high possibilities that CO_2 absorbs the energy from the combustion of CNG, which will suppress and slows down the combustion rate of the mixture.

Decreasing CO_2, CO, and NO_x emissions and increasing THC emission shown in figure 3 indicates that CO_2 addition is suppressing the combustion of CNG. Decreasing CO_2 may indicate the event of CO_2 dissociation and formation of HC especially for the CO_2 percentage above 10%. A significant amount of energy is absorbed by this process and cause torque and power drop.

The presence of CO_2 suppression can also be seen in the pressure and heat release rate data in figure 4. The heat rate gradient for CNG is higher compared to the one with CO_2 addition. It shows the combustion propagation is slower for high CO_2 content fuels. The maximum pressure and heat release also decreases for fuels with CO_2 addition due to the low energy content of the mixture.

The effect of CO_2 content to the engine performance, emission and combustion for different injection timing also shown in figure 2, figure 3 and figure 4. High performances were recorded for 180° BTDC for all CO_2 contents compared to 300° BTDC. The main reason is because of higher volumetric efficiency due to late injection.

Fuel injection starts during the intake process of 300° BTDC injection timing which will disrupt the air intake process to the chamber. Fuel will displace the air thus reducing total air intake to the cylinder which relates to reducing engine volumetric efficiency. This effect amplifies when CO2 was added since the total volume of gas injected increases the effort of maintaining the lambda. Longer injection duration will further reduce the engine volumetric efficiency and causing engine performance to decrease.

The decrease in cylinder peak pressure and heat release rate with increasing CO_2 content differs for a variation of injection timing. Performance drop for 180° BTDC is less compared to 300° BTDC especially for 30% CO_2 content. Fuel injected at 180° BTDC produces higher level of stratification, rich local air-fuel ratio, therefor suitable for the combustion regardless of CO_2 content in the mixture.

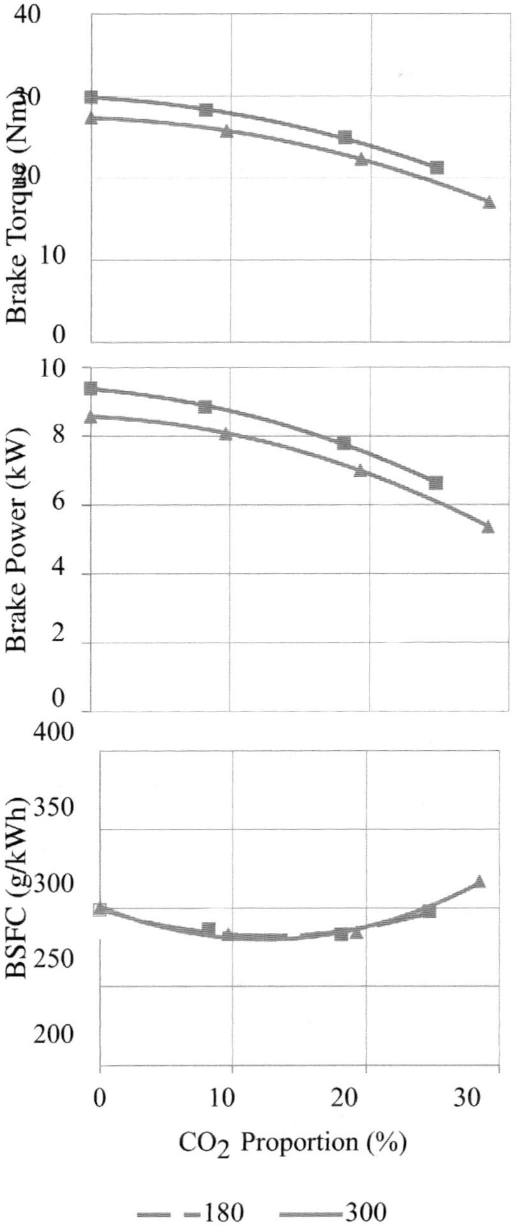

Figure 2. Engine performance for various CNG-CO_2 mixture composition.

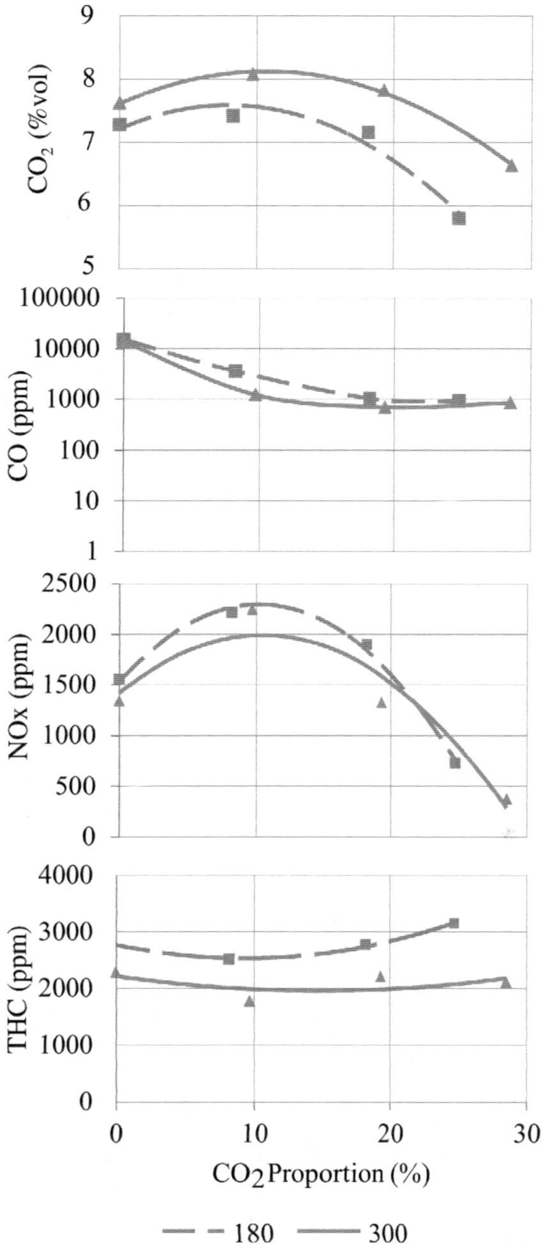

Figure 3. Engine emissions for various CNG-CO_2 mixture composition.

IOP Conf. Series: Materials Science and Engineering **100** (2015) 012041 doi:10.1088/1757-899X/100/1/012041

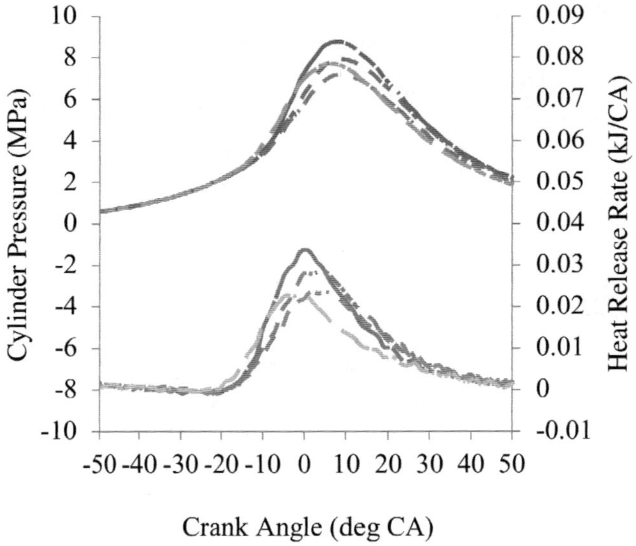

(a)

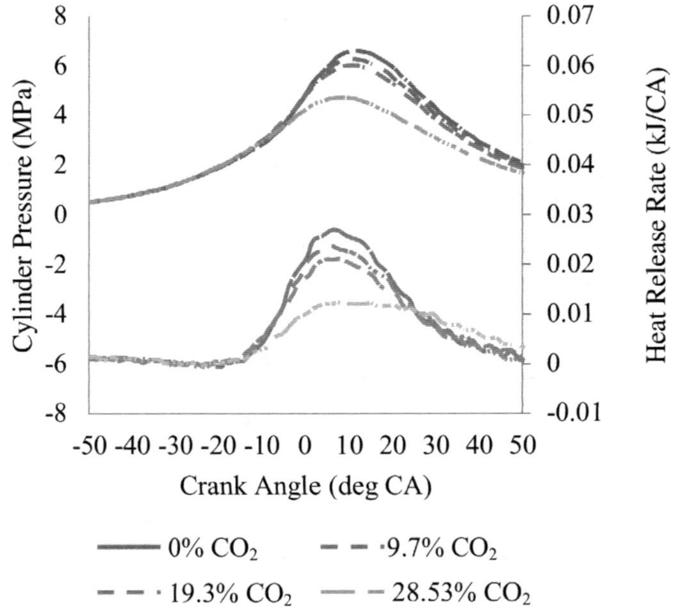

(b)

Figure 4. Cylinder pressure and heat release rate for various CO_2 composition at 3000 rpm with (a) 180° and (b) 300° BTDC injection timing.

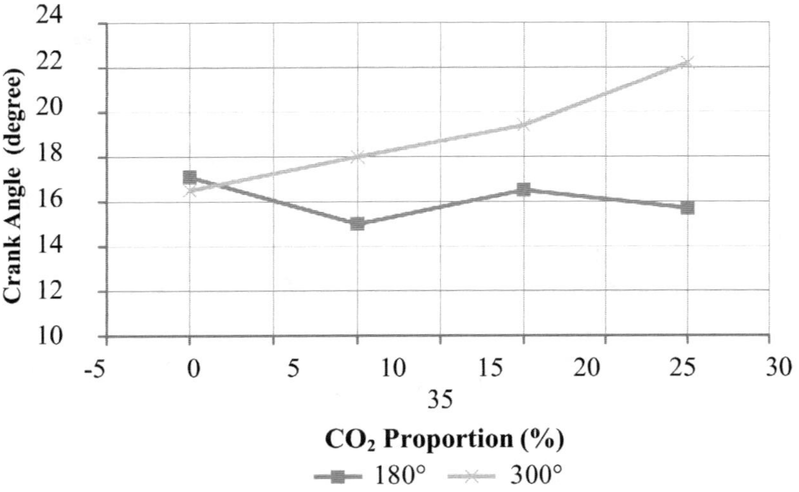

Figure 5. Effect of CO_2 percentage towards ignition delay for 180° and 300° BTDC injection timing at 3000 rpm.

Figure 5 shows two ignition delay profiles for 180° and 300° BTDC injection timing. The profile for 180° BTDC shows consistency throughout the test which indicates that fuel stratification is able to maintain the ignition delay similar to pure CNG and creates wider operating range window for CNG with high CO2 content. However for 300° BTDC shows an increase in ignition delay as CO_2 percentage increases up to 22.2° BTDC at 30% CO_2 content. The ignition delay profile for 300° will limit engine operations and add complexity in engine control which is not desirable.

4. Conclusions

In this paper, the effect of different CO_2 content in CNG while varying injection timing towards the performance, emission, and combustion were explained. It shows that CO_2 suppresses the combustion of CNG while on the other hand CNG combustion is causing CO_2 dissociation shown by decreasing CO_2 emission when combusted with higher CO_2 content. The performance significantly drops when the CO_2 content percentage is above 10%. Injection timing at 180° BTDC shows higher overall performance compared to 300° BTDC because of two possible reasons which are higher volumetric efficiency and higher stratification level. The results showed possibilities of CO_2 enrich CNG usage as fuel by implementing injection strategy.

Acknowledgement

The authors would like to acknowledge Ministry of Education of Malaysia (MoE) on the MyRA project funding entitled "CO_2 rich natural gas value chain program from wells to wealth: a green approach" (0153AB-J07) and Centre for Automotive Research and Energy Management (CAREM), Universiti Teknologi PETRONAS, Malaysia, in providing the experimental facilities.

References

[1] Hu E, Huang Z, Liu B, Zheng J, Gu X 2009 *International Journal of Hydrogen Energy* **34** 1035-44
[2] Semin R A B 2008 *Am J Eng Appl Sci* **1** 302-11
[3] Cho H M, He B-Q 2007 *Energy Conversion and Management* **48** 608-18
[4] Huang J, Crookes R 1998 *Fuel* **77** 1793-801

[5] Crookes R 2006 *Biomass and Bioenergy* **30** 461-8
[6] Gogoi T, Baruah D 2011 *Applied Energy* **88** 2713-25
[7] Ong H, Mahlia T, Masjuki H 2011 *Renewable and Sustainable Energy Reviews* **15** 639-47
[8] Azmi N, Hilmi M, Sabil K M, Removal of high CO2 content in natural gas by formation of gas hydrates as a potential solution for CO 2 gas emission. International Conference on Environment; 2010.
[9] Folger P, Carbon capture: a technology assessment2013: DTIC Document.
[10] Mohammed S E, Baharom M, Aziz A R A 2011 *International Journal of Hydrogen Energy* **36** 4029-37
[11] Bari S 1996 *Renewable energy* **9** 1007-10
[12] Henham A, Makkar M 1998 *Energy Conversion and Management* **39** 2001-9
[13] Shrestha S B, Karim G 2001 *Applied thermal engineering* **21** 331-42
[14] Das L, Gulati R, Gupta P K 2000 *International Journal of Hydrogen Energy* **25** 783-93
[15] Peng H, Cui Y, Shi L, Deng K 2008 *Energy* **33** 471-9

Parametric studies on tensile strength in joining AA6061-T6 and AA7075-T6 by gas metal arc welding process

M Ishak[1,2], N F M Noordin[1] and L H Shah[1]
[1]Faculty of Mechanical Engineering, Universiti Malaysia Pahang, 26600, Pekan, Pahang, Malaysia
[2]Automotive Engineering Centre, Universiti Malaysia Pahang, 26600, Pekan, Pahang, Malaysia

Email: mahadzir@ump.edu.my

Abstract. Proper selection of the welding parameters can result in better joining. In this study, the effects of various welding parameters on tensile strength in joining dissimilar aluminum alloys AA6061-T6 and AA7075-T6 were investigated. 2 mm thick samples of both base metals were welded by semi-automatic gas metal arc welding (GMAW) using filler wire ER5356. The welding current, arc voltage and welding speed were chosen as variables parameters. The strength of each specimen after the welding operations were tested and the effects of these parameters on tensile strength were identified by using Taguchi method. The range of parameter for welding current were chosen from 100 to 115 A, arc voltage from 17 to 20 V and welding speed from 2 to 5 mm/s. L16 orthogonal array was used to obtained 16 runs of experiments. It was found that the highest tensile strength (194.34 MPa) was obtained with the combination of a welding current of 115 A, welding voltage of 18 V and welding speed of 4 mm/s. Through analysis of variance (ANOVA), the welding voltage was the most effected parameter on tensile strength with percentage of contribution at 41.30 %.

1. Introduction

Gas metal arc welding (GMAW) process is a welding process which yields coalescence of metals by heating with a welding arc between a continuous filler metal electrode and the work piece. This method of welding is widely used in various industrial applications such as construction, piping and automotive sector [1-3]. The GMAW process is one of the most widely used in many industrial operations due to its comparatively easier handling, most cost effective and a variety of metals can be joined [4, 5]. Moreover, GMAW also offers the advantage of high deposition rate and high welding speed as well as deeper penetration because of high heat input [6, 7].

Arc welding of aluminum alloys with different grades will give problems due to the difference in thermal conductivity. The heat produced by the arc will flow more easily in the material with the larger thermal conductivity. This can result in lack of fusion of this material, at the same time an excessive melting at the material with lower thermal conductivity [6]. Luijendijk, have successfully studied about GMAW method on dissimilar aluminum alloys 5xxx series and 6xxx series [6]. Recently, Guo et al. been studied about joining dissimilar AA6061 and AA7075 by friction stir (FSW) [7]. However, the information about joining AA6061 and AA7075 by GMAW process is scarce.

Heat treatable AA6061-T6 (170 W/mK) and AA7075-T6 (130 W/mK) with its different thermal conductivity pose a challenge in obtaining good weld bead quality [8]. This is particularly evident when joining these dissimilar aluminum alloys together. The problems such as cracks, burn through and distortion is predicted to occur [9]. The input parameters such as welding current, welding voltage and welding speed play a very important factor in order to produce high quality of the joining [10]. Thus, this parameter should be estimated and their effect on the output joint quality should be investigated in order to achieve the optimum output and to mitigate the defects.

The aims of this study is to investigate the effect of process parameters such as welding current, welding voltage and welding speed on tensile strength on AA6061-T6 and AA7075-T6 by using the GMAW process. The highest tensile strength values of the processes parameter is obtained by analyzing the raw data for each of the specimen by using Taguchi method.

2. Experimental set up

For the experimental study, both materials AA6061 - T6 and AA7075 - T6 were prepared by cutting in to dimensions of 150 mm × 100 mm × 2 mm. The chemical compositions of both aluminum alloys and the filler are presented in table 1. Then, they were welded using a butt joint configuration with filler ER5356 by using a semi-automated GMAW machine model Dr Well DM-500 as shown in figure 1. The welding table is controlled with a touch screen welding program interface. All welded specimens were cut in to the standard dog bone shape using the American society for Testing of Materials (ASTM-E809) standards [11]. For each of experimental level, three tensile specimens were prepared. In this experimental study, the Taguchi method is used for optimization of process parameters of GMAW of dissimilar aluminum alloys. Taguchi approach is a robust design method that uses experimental design called orthogonal arrays (OAs). Taguchi method is to study a large number of decision variables with a small number of experiments [12, 13]. By using and understanding the Taguchi method, welding quality and experimentation process were developed and improved. The range of parameter of welding current, welding voltage and welding speed were obtained by several preliminary experiments. The parameters (factors) and its levels are shown in table 2. Table 3 shows the L16 orthogonal array (OA) which were selected for the analysis.

Table 1. Chemical composition of aluminum alloys and filler wire.

Alloy	Al	Si	Fe	Cu	Mn	Mg	Zn	Cr	Ni	Ti
AA6061-T6	97.30	0.79	0.43	0.30	0.03	0.86	0.01	0.18	<0.01	0.03
AA7075-T6	89.8	0.07	0.27	1.60	0.02	2.28	5.58	0.21	0.01	0.03
ER5356	Bal	0.03	0.15	0.01	0.14	4.83	0.01	0.11	---	0.09

Table 2. Process parameters and their levels.

Process Parameters	Levels of factors			
	1	2	3	4
Welding Current (A)	100	105	110	115
Welding Voltage (V)	17	18	19	20
Welding Speed (mm/s)	2	3	4	5

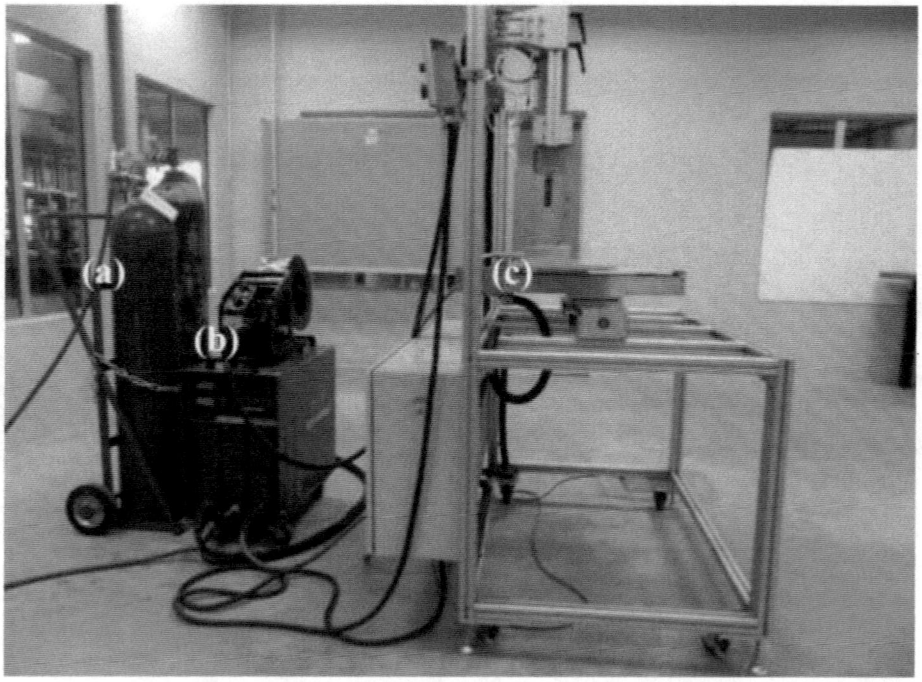

Figure 1. The welding machine and its apparatus used in experiments (a) welding gas, (b) wire feeder and (c) welding table.

Table 3. L16 Orthogonal Array.

Experimental run	Welding current (A)	Welding voltage (V)	Welding speed (mm/s)
1	100	17	2
2	100	18	3
3	100	19	4
4	100	20	5
5	105	17	3
6	105	18	2
7	105	19	5
8	105	20	4
9	110	17	4
10	110	18	5
11	110	19	2
12	110	20	3
13	115	17	5
14	115	18	4
15	115	19	3
16	115	20	2

3. Results and Discussion

Figure 2 shows some of the weld appearance for the different combinations of parameter. All samples show good quality joints with minimal weld defects.

Figure 2. Weld appearance of specimens (a) specimen 1, (b) specimen 6, (c) specimen 9 and (d) specimen 14.

3.1. Parameter optimization

In order to investigate the effect of process parameters of the GMAW process, the tensile strength of the welded specimens were observed. In the Taguchi method, a loss function is defined to calculate the deviation between the experimental value and the desired value. Usually, there are three categories of the quality characteristic in the analysis of the signal-to-noise ratio, that is, the-lower-the-better, the-higher-the-better, and nominal-the-better [14-17]. Analysis of mean and signal-to-noise (S/N) ratio for each experiment collectively can generate and suggest the optimum level of process parameters to obtain the highest tensile strength of the weld. Therefore, this investigation focused on the-higher-the-better optimization analysis. All response was calculated as a follows:

$$S/N = -10\log10\,[1/n \sum (1/y2)] \quad (1)$$

Where n is number of experiment, in this case is 16 and y is the experimental value of the quality characteristic.

Mean responses of raw data and S/N ratios of tensile strength for each parameter are calculated and tabulated in table 4. The effect of process parameters on tensile strength were identified by obtaining the main effects plot for mean S/N ratios which expressed are in figure 3. The graphs show the level effects of each welding parameter on tensile strength.

In figure 3 (a), the mean S/N ratio shows an upward trend from 100 A to 115 A with only a negligible decrease at 105 A, indicating an overall increase in weld strength with the increase of welding current. Meanwhile, not recognizable pattern can be seen in figure 3 (b), where the S/N value increase from 17 V to 18 V, and began to decrease at higher voltage value. The decrease in S/N ratio value at higher voltage is due to excessive heat given to the work piece causing the joint to receive an excessive penetration, consequently creating defects due to high penetration. On the other hand, insufficient penetration is occurred at low value of welding voltage, also causing a decrease in tensile strength. The welding speed effect on tensile strength is clearly plotted in figure 3 (c). It showed that, the tensile strength value is low at the slowest (2 mm/s) and fastest (5 mm/s) welding speed. The highest tensile strength is obtained at welding speed of 4 mm/s and slightly lower at welding speed of 3 mm/s.

The response for S/N ratios and mean results were tabulated in table 5. It can be seen that, welding voltage is the most important parameter in order to obtain the optimum joint quality. This is followed by the welding speed and the welding current, in that order.

Table 4. Data summary for tensile strength.

Experimental run	WC	WV	WS	Tensile strength	S/N ratio for TS
1	100	17	2	182.37	45.2191
2	100	18	3	186.92	45.4331
3	100	19	4	182.70	45.2348
4	100	20	5	179.45	45.0789
5	105	17	3	185.02	45.3444
6	105	18	2	184.45	45.3176
7	105	19	5	180.21	45.1156
8	105	20	4	180.51	45.1300
9	110	17	4	186.82	45.4285
10	110	18	5	184.67	45.3279
11	110	19	2	180.84	45.1459
12	110	20	3	184.80	45.3340
13	115	17	5	184.24	45.3077
14	115	18	4	194.34	45.7712
15	115	19	3	187.21	45.4466
16	115	20	2	181.34	45.1699

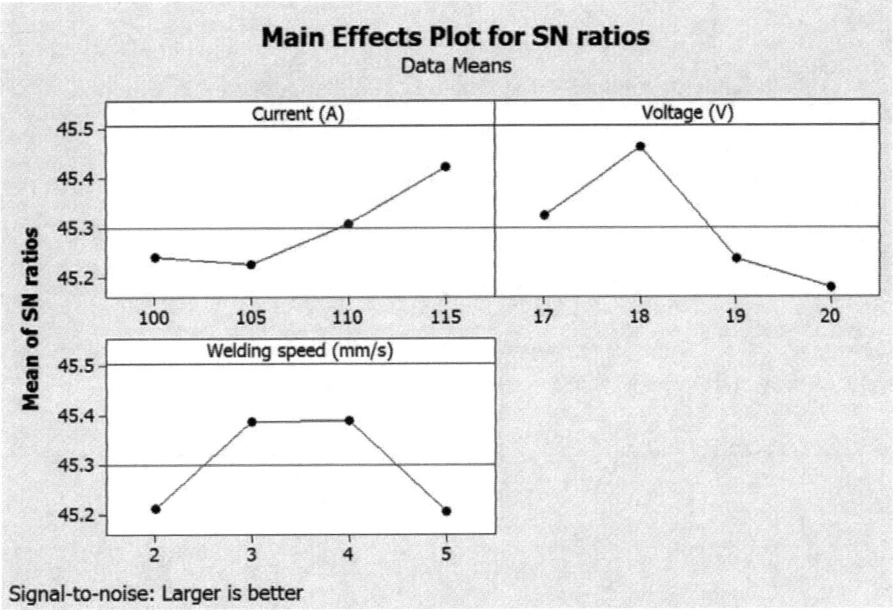

Figure 3. Main affects plots for means (tensile strength).

Table 5. Response for S/N ratios and mean.

Level	Response data S/N rations for Tensile Strength		
	WC	WV	WS
1	45.21	45.32	45.24
2	45.39	45.46	45.23
3	45.39	45.24	45.31
4	45.21	45.18	45.42
Delta	0.81	0.28	0.20
Rank	3	1	2

3.2. Analysis of Variance (ANOVA)

The effects of different welding parameters on tensile strength were obtained by ANOVA. The most influences parameter on output process can be identify by ANOVA [18]. In the analysis, the sum of squares and variance are calculated. The ANOVA result is tabulated in Table 6 showed the sum of square (S), variance, F-ratio (F), P-value (P) and percentage of contribution (%C) were calculated. The percentage of contribution can be calculated by using equation below:

$$\% \mathrm{C} = \frac{\mathrm{SSd}}{\mathrm{SSt}} \tag{2}$$

First, statistically, F-ratio named after Fisher [19], is used in order to find out the significant factors that affecting the welding process. In analysis, the larger the F value specifies that the variation of the process parameter makes a big effect on the quality in terms of tensile strength in welding process [18, 20, 21]. Welding voltage has the highest F value with 10.53 verified welding voltage is the most influenced factor as ranked as shown in table 5. Meanwhile, for P-value, it is to verify the significance of factors. In this case, at 90% confident level, P-value should be <0.1. In descending order, the percentage of factors contribution was identified with; welding voltage (41.30%), welding speed (29.12%) and welding current (21.73%) at residual error of 7.85%, welding voltage was proven to be dominant factor.

Table 6. Analysis of variance (ANOVA) for S/N ratios

Source	DF	Sum of square (S)	Variance	F-ratio (F)	P-value (P)	%C
Current (A)	3	0.09677	0.09677	5.54	0.037	21.73
Voltage (V)	3	0.18393	0.18393	10.53	0.008	41.30
Welding Speed (mm/s)	3	0.12970	0.12970	7.42	0.019	29.12
Residual error	6	0.03495	0.03495			7.85
Total	15	0.44534				100

S/N : Signal-to-noise; ANOVA: Analysis of variance

S= 0.07632, R-Sq=92.2% and R-Sq(Adj)=80.4%

4. Conclusion

In this paper, the effect of process parameter in GMAW dissimilar aluminum alloys sheets on tensile strength was successfully conducted. The conclusion that can be made based on the results obtained in the investigation, welding voltage is the most highly effective parameters that control the weld strength, whereas, welding current and welding speed are slightly less effective factors. The percentage of contribution of parameter is determined by using ANOVA. Through ANOVA, welding voltage was found to be dominant contribution with 41.30 %. The highest tensile strength obtained by Taguchi method in combining these three factors are experimental number 14 at welding current of 115 A, welding voltage of 18 V and welding speed of 4 mm/s with calculated highest tensile strength value of 194.34 MPa

Acknowledgement

The author would like to thank Mr. Rafiqul, Mr. Ihsan and Mr. Naqiudin for technical supports. Also,financial support through Fundamental Research Grant Scheme, project no. FRGS/1/2013/TKOI/UMP/02/2 is gratefully acknowledged

References

[1] Karadeniz E, Ozsarac U, Yildiz C 2007 *Materials & Design* **28** 649-56
[2] Wahab M A, Painter M 1997 *International journal of pressure vessels and piping* **73** 153-9
[3] Kim I-S, Son K, Yang Y, Yaragada P 2003 *International Journal of Machine Tools and Manufacture* **43** 763-9
[4] Sevim I, Hayat F, Kaya Y, Kahraman N, Şahin S 2013 *The International Journal of Advanced Manufacturing Technology* **66** 1825-34
[5] Mutombo K, Du Toit M 2011 *International Journal of Fatigue* **33** 1539-47
[6] Luijendijk T 2000 *Journal of Materials Processing Technology* **103** 29-35
[7] Guo J, Chen H, Sun C, Bi G, Sun Z, Wei J 2014 *Materials & Design* **56** 185-92
[8] Khodir S A, Shibayanagi T 2008 *Materials Science and Engineering: B* **148** 82-7
[9] Balasubramanian V, Ravisankar V, Madhusudhan Reddy G 2008 *Int. Journal of Fatigue* **30** 405-16
[10] Sapakal[1] S, Telsang M 2012
[11] American Society for Testing of Materials (ASTM E8-09) 2010.
[12] Kumar S K S, Kumar A 2012 *Proc. of the Institution of Mechanical Engineers, Part C: Journal of Mechanical Engineering Science* **227** 1150-63
[13] Bilici M K 2012 *Materials & Design* **35** 113-9
[14] Yang W p, Tarng Y 1998 *Journal of materials processing technology* **84** 122-9
[15] Kulkarni S S J S R, Ganjigatti J P. International Journal of Latest Trends in Engineering and Technology (IJLTET2014.
[16] Lee J, Um K 2000 *Journal of materials processing technology* **108** 106-13
[17] Kim I, Son J, Kim I, Kim J, Kim O 2003 *Journal of Materials Processing Technology* **136** 139-45
[18] Rawal M R, Inamdar K, Optimization of Spot Welding Process Parameters for CRCA Material using Taguchi Method. International Journal of Engineering Development and Research; 2014: IJEDR.
[19] Lindman H. Analysis of Variance in Experimental Design, 1992. Springer-Verlag, Berlin.
[20] Eşme U 2009 *Arabian Journal for Science and Engineering* **34** 519
[21] Anawa E, Olabi A-G 2008 *Optics & Laser Technology* **40** 379-88

Performance investigation of the S-Rotors

B A Bhayo[1], H H Al-Kayiem[1] and N Z Yahaya[2]
[1]Department of Mechanical Engineering, Universiti Teknologi Petronas, Bandar Seri Iskandar, 32610, Perak, Malaysia.
[2]Department of Electrical and Electronics Engineering, Universiti Teknologi Petronas, Bandar Seri Iskandar, 32610, Perak, Malaysia.

E-mail: bilawal.bhyao@yahoo.com

Abstract. This paper presents and discusses results from an experimental investigation of three models of wind S-rotors. Models 1 is modified from conventional Savonius rotor with a single stage and zero offsets zero overlaps; model 2 is three blade single stage wind rotor; and model 3 is double stage conventional Savonius rotor. The three models were designed, fabricated and characterized in terms of their coefficient of performance and dynamic torque coefficient. A special open wind simulator was designed for the test. The optimum parameters for the models were based on previous studies. The results showed that the model 1, model 2 and model 3 has the maximum power coefficient of 0.26, 0.17, and 0.21 at the correspondence tip speed ratio (TSR) of 0.42, 0.39 and 0.46, respectively. Model 1 is further optimized in terms of the aspect ratio resulting in improved power coefficient by 24%. The maximum dynamic torque coefficient of model 1, model 2 and model 3 was found as 0.81, 0.56 and 0.67 at the correspondence minimum TSR of 0.28, 0.21 and 0.17, respectively. It was noted that the all three models have high torque coefficient because the models were tested at higher applied torque on the rotors.

1. Introduction

Wind energy is a clean renewable energy freely available on the earth surface in the form of kinetic energy of air. However, wind technology is a mature technology to exploit the wind energy for electricity generation [1]. According to Betz (1966) [2], a maximum of about 59% of wind energy can be transformed into mechanical energy. In pursuit of generating electricity using the wind as a clean and renewable energy for remote rural communities, the Savonius rotor (S-rotor) can be helpful [3].

The S-rotor is a simple and cheaper vertical axis wind turbine (VAWT). It can be installed near to living places because of less noisy and can accept the wind from any direction to drive it. Although the S-rotor is an inherently low efficient wind turbine, but its self-starting characteristics at the low cut in speed due to high torque makes it more attractive for the stand-alone power generation system [4, 5]. Thus, the S-rotor can be used for a self-driven stand-alone power system with very low adverse environmental effects. Since S-rotor has a poor coefficient of performance inherently because of negative torque at the convex side. Therefore, to increase the torque difference at positive and negative aspect; the geometrical profile and parameters should be optimized.

Akwa et al. [5] carried out the review to discuss the different geometrical parameters to enhance the performance of S-rotor. From the discussion, it was observed that the increase in an aspect ratio of the rotor and the rotor with end plates can result in performance improvement. The S-rotor with two blades has higher coefficient of power (C_p) than three blades rotor. Morshed et al. [6] have analyzed the effect of overlap ratio and Reynolds number on the aerodynamic characteristics of the three blade S-rotor with blade angle 120° apart and no central shaft between the end plates. In their experimental study, it was observed that the better static torque coefficient is at an overlap ratio of 0.12. However C_p is higher for the model without overlap ratio at low Reynolds number. Mahmoud et al. [7] have carried out study to test different Savonius wind rotors in a subsonic wind tunnel at wind speed of 13 m/s. It was concluded that the rotor with two blades, two stages, without overlap and with end plates has higher mechanical power. However the mechanical power increases with the increase of aspect ratio.

Al-Kayiem and Ming [8] have experimentally investigated three different rotors and compared their performance in bounded flow, partially bounded flow and fully open flow. It was observed that the conventional S-rotor had higher power coefficient in the partially bounded flow. However Rotor 2, based on Kamoji and Kedare [9] and Rotor 3, a new design have maximum power and torque coefficient in the fully bounded flow. Kamoji et al. [9] have investigated the effect of geometrical parameters on single stage modified Savonius rotor in an open jet wind tunnel. It was found that the modified Savonius rotor (without shaft) has higher coefficient of power of 0.21 than that of conventional Savonius rotor (with and without shaft) and also with the modified Savonius rotor (with shaft). The optimum parameters for modified Savonius rotor were found as an overlap ratio zero, aspect ratio 0.7, blade arc angle 124° and blade shape factor 0.2. Hayashi et al. [10] carried out a study to minimize the negative torque coefficient, a three stage S-rotor with bucket stages angle 120° from each other was designed and tested in an open wind tunnel. During experimental testing, it was found that the three-stage rotor has an average and constant torque variation than that of one stage two-blade rotors. For the three stages rotor, the maximum power coefficient decreased by 25% relative to the single-stage rotor.

Menet and Bouraba [11] have designed and developed a small prototype of S-rotor for an off grid system. The optimum geometrical parameters were found as two-step rotor, with an end plate ratio of 1.1, aspect ratio of 4 and the overlap ratio of 0.15 to 0.3. Percival et al. [12] have carried out an experimental study to design and test S-rotor for domestic electricity generation. The optimum overlap ratio was found 0.2 after testing ten different configurations for overlap ratio in the range of 0.6 to 0.3 at a maximum wind speed of 10 m/s.

To improve the performance of S-rotor, this experimental study is carried out to investigate and compare the performance of three models. Also, the experimental comparative analysis of the designed wind rotor models with the conventional double stage Savonius rotor is carried out in this study. To do so, the models were designed, fabricated and experimentally tested in an open wind flow. The experiments were conducted to evaluate the performance coefficient and dynamic torque coefficient.

2. Experimental Expects

2.1. Design Concept of the Models

In this experimental analysis, three different wind rotor models were designed, fabricated and tested in an open wind flow. The main dimensions of wind rotor models are given in table 1. The models are named as Model 1, Model 2 and Model 3. All the models have an aspect ratio, AR (H/D) equal to 2, end plate ratio (Do/D) of 1.1 and shaft diameter (d) of 0.01 m. The geometrical parameters of simple Savonius rotor are taken optimum as suggested by [11].

Table 1. The main dimensions of the wind rotor models.

H	0.628 m	A	0.1897 m^2
D_O	0.302 m	AR	2
D	0.151 m	d	0.01 m

IOP Conf. Series: Materials Science and Engineering **100** (2015) 012043 doi:10.1088/1757-899X/100/1/012043

2.1.1. Model 1. Refers to figure 1(a), Model 1 is a single-stage rotor, with zero overlap ratio (e/D) between the blades, the optimum blade arc angle is 124° and the shape factor (p/q) is set to 0.2 as recommended by [9]. Model 1 was designed in such a way that its blade diameter should be 150 mm.

2.1.2. Model 2. Refers to figure 1(b), Model 2 is a three bladed, single-stage rotor. The curvature ratio of rotor radius to the depth, (R/Y) is set to 2.5. However the skew factor (X/R), which is the ratio of the maximum depth location to a radius of the rotor is set to 0.75 as recommended by [13].

2.1.3. Model 3. Refers to figure 1(c), Model 3 is a two blade, double stage simple Savonius rotor. The rotor is provided an overlap (e) between the blades with an overlap ratio of 0.2. The height to diameter ratio of the model 3 is same as of the model 1 and model 2. Since the rotor has two stages with phase shift angle of 90°, therefore it is expected to have optimum performance characteristics of the simple Savonius rotor [11, 14].

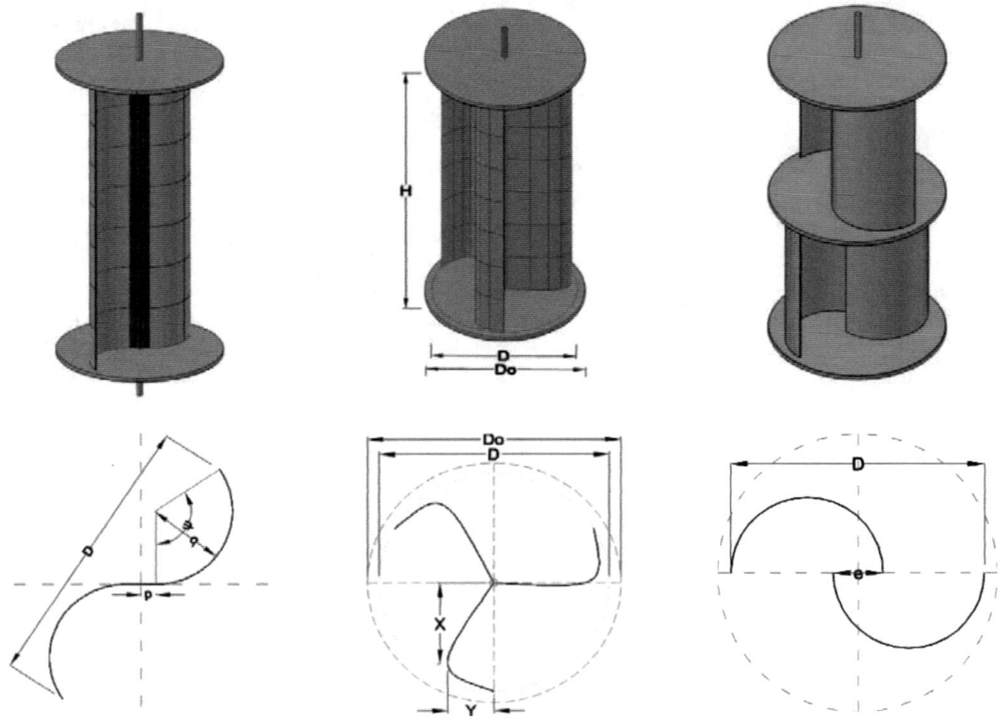

(a) Model 1 (b) Model 2 (c) Model 3

Figure 1. The schematic drawings of wind rotor models.

2.2. Fabrication of Models

The models blades were fabricated from 1 mm aluminum sheets because of its low weight and manufacturability. The 2 mm slot as per design was cut in 8 mm acrylic end plates, of 330 mm diameter, to fix the bended sheets easily. For bending the aluminum sheets to a particular profile, three wooden patterns for each model were made on five-axis CNC machine and the sheets were bended by using pattern and manual rolling machines. The rotors were assembled using super glue and bonding epoxies.

Two protruding shafts of mild steel at top and bottom with diameter 10 mm and height 150 mm were also assembled.

3. Experimental Setup

3.1. Open wind simulator

The experimental setups consist of two sections, section 1 and two as shown in figure 2. Section 1 was an open wind tunnel, where a big propeller fan with variable RPM drive was used to blow the wind at different velocities. The propeller fan has a power capacity of 750 watts and can blow air up to 32000 m^3/hr. The fan was placed at the inlet of wind tunnel section and wind was directed to the honeycomb of dimension 1 m x 1 m with each cell dimension of 0.2 m x 0.2 m to reduce the flow turbulence and swirl. At exit section of honeycomb flow convergent setup was attached with exit area of 0.8 m x 0.8 m to increase further the wind speed. Section 2 was a test rig to hold the models vertically for testing and data acquisition. The models extended shafts at ends were embedded to frictionless bearings to make the rotor rotate smoothly. While designing test rig, it was confirmed that the rotor end plates should also experience wind effect.

A brushed 24V, 135W DC generator was mounted with the setup to get the electrical power output. The timing belt and pulley system was also provided, to run the generator at a higher speed of about 3.75 times the wind rotor speed.

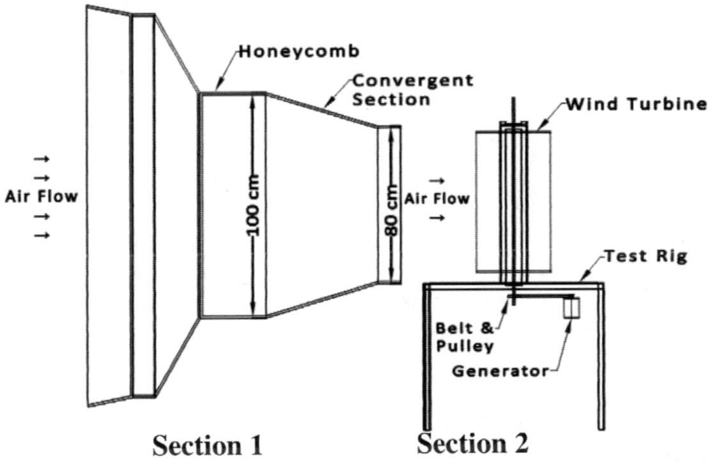

Figure 2. Schematic of the experimental setup for open wind simulation.

3.2. Experimental methodology

To test the models, the test rig was placed in such a way that the center of wind rotor models should be at 0.8 m downstream from the wind tunnel exit area. The propeller air fan was operated for variable wind velocities and the wind velocity was measured by using KIMO VT200 Vane Anemometer, which was held for 10 seconds at each of 6 different points and their mean was taken. The dynamic torque for the wind rotor models was calculated from the electric power output of a brushed 24 volts, 135 watts DC generator. The laser tachometer was also used to get wind rotor speed at a particular wind velocity. For every reading, the procedure was repeated four times to validate the readings. The variation of wind velocity from the concave side to convex side of the rotor is ±0.2 m/s.

4. Results and Discussion

To evaluate the performance of a wind rotor, the coefficient of power and torque should be determined. Hence, the electrical power output and rpm at a constant wind speed were measured. However, the dynamic torque was determined using mathematical relation of electrical power output and measured torque values. After the data had been acquired, the below provided mathematical formulations were used to identify performance parameters of the rotor.

Coefficient of Power,

$$C_p = \frac{P_{electrical}}{\frac{1}{2}\,\rho\, V^3\, A_{rotor}\, \eta_g} \tag{1}$$

Dynamic Torque,

$$T = P_{electrical} \;/\; \omega\, \eta_g \tag{2}$$

Dynamic Coefficient of Torque,

$$C_T = \frac{T}{\frac{1}{2}\rho\, V^2\, A_{rotor}\, R} \tag{3}$$

Tip Speed Ratio (λ),

$$TSR = Rotor\ Tip\ Speed \;/\; Wind\ Speed \tag{4}$$

or

$$\lambda = \omega\, R \,/\, V$$

Reynolds Number *(Re)*

$$Re = \rho\, V\, 2R \,/\, \mu \tag{5}$$

Where, ρ is the density of air (= 1.22 kg/m^3), V is the free stream wind velocity, A_{rotor} is the rotor area, R is the radius of rotor, ω is the angular velocity of the rotor, T is the torque generated by the rotor, η_g is the generator efficiency and μ is the air absolute viscosity (= 1.83×10^{-5} Pa·s at 20 °C). C_T, is the dynamic torque coefficient.

4.1. Power Coefficient

Figure 3 represents the variation of power coefficient with the variation of tip speed ratio (TSR) for the model 1, model 2 and model 3. The magnitude of the power coefficient (C_p) at the minimum, optimum and maximum TSR with the corresponding Reynolds number (*Re*) for all three models is shown in Table 2.

For model 1, the minimum TSR observed is 0.28 at which power coefficient is 0.23 below this TSR rotor goes to stop rotating. The maximum power coefficient for the model 1 is noticed as 0.26 at the TSR of 0.42. The power coefficient decreases with the further increase of TSR. The percentage increase in power coefficient of Model 1 is about 24% than that of modified Savonius rotor by [9], which might be because of the positive effect of increasing aspect ratio from 0.7 to 2. The rise in performance with an increase in aspect ratio was also reported by [15, 16].

Nevertheless, the modified Savonius rotor is further optimized as rotor with zero overlaps having end plate ratio of 1.1, Aspect ratio (H/D) of 2, blade arc angle (Ψ) of 124° and blade shape factor (p/q) of 0.2.

For model 2, the maximum power coefficient is recorded as 0.17 at the TSR of 0.39. The highest power coefficient of the model 2 was expected to be about double than the conventional Savonius rotor with same geometrical configuration [13]. However, the power coefficient of the model 2 from this experimental study is 1.13 times the power coefficient of the three bladed simple Savonius rotor having an aspect ratio of 1.59 as reported by [14]. Thus, the power coefficient of Model 2 is about 43% less than the expected. This might be because of fabrication of wind rotor in a different way than that of reported by [13].

For model 3, the power coefficient of the model at lowest possible TSR of 0.17 is recorded as 0.12. The power coefficient increases with the rise of TSR to 0.46, where the maximum power coefficient is 0.21. The further increases of TSR cause to decrease power coefficient and at the highest TSR of 0.56, the power coefficient is 0.19. The maximum power coefficient of the model 3, double stage simple Savonius rotor is greater about 31% than the maximum power coefficient of a single stage two-bladed wind rotor from the referred study [10]. The highest power coefficient of the model 3 is also greater about 24% than the model 2, and the maximum power coefficient is less about 19% than the model 1.

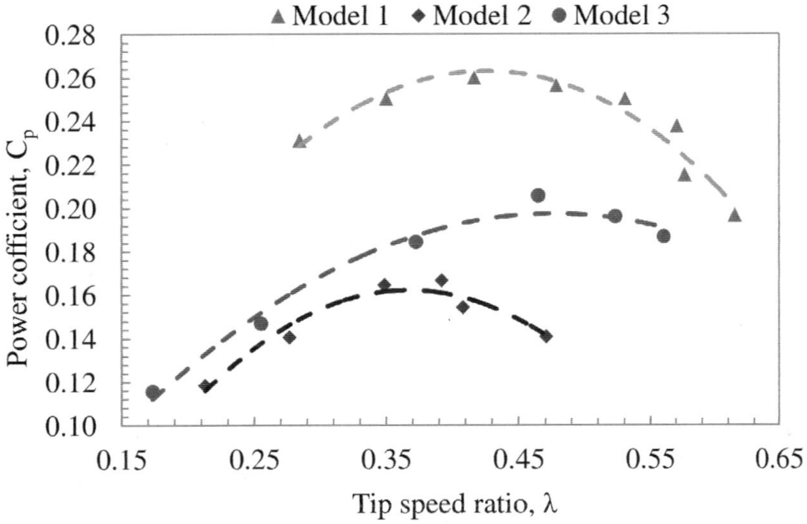

Figure 3. The power coefficient of the wind rotor models.

4.2. Dynamic Torque

Figure 4 represents the variation of dynamic torque coefficient with the variation of TSR for the model 1, model 2 and model 3. The magnitude of the dynamic torque coefficient (C_T) at the minimum, optimum and maximum tip speed ratio with the corresponding Reynolds number (*Re*) for all three models is shown in table 2.

For model 1, the maximum dynamic torque coefficient is 0.81 at the minimum TSR of 0.28, and the torque coefficient decreases with the increase of tip speed ratio. The minimum torque coefficient can be seen as 0.32 at the TSR of 0.62. As there are larger variations in torque coefficient with small variations in tip speed, which can be because of the testing of the wind rotor models at a higher torque.

For model 2, the maximum dynamic torque for the wind rotor model is 0.56 at the minimum TSR of 0.21 and the torque coefficient decreases with the increase of TSR. The minimum torque coefficient of the model is 0.3 at the TSR of 0.47. The maximum dynamic torque coefficient of the model 2 is less than the maximum dynamic torque coefficient of the model 1. A reason for less torque coefficient of model 2 is that the model 2 is a three bladed rotor, which produce comparative less torque than a two bladed rotor. Also, the three blades rotor rotate slower than the two blades rotor, which is due to more area facing toward the wind and have more negative torque positions. The increases of negative torque positions cause the decrease of net torque produced by the rotor.

For the model 3, a double stage simple Savonius rotor. The maximum dynamic torque coefficient is noted as 0.67 at the minimum TSR of 0.17, and the torque coefficient decreases with the increase of TSR. The minimum torque coefficient is 0.33 at the higher TSR of 0.56. The maximum torque coefficient of the model 3 is 16% less than the maximum torque coefficient of the model 1. Moreover, the maximum torque coefficient of Model 3 is about 20% greater than the model 2. A reason for decrease of torque coefficient of the model 3 (double stage simple Savonius rotor) from the model 1 (a two-bladed single-stage rotor) is that the model 3 has become bulky due to two stages and the wind rotor loses some of its energy in rotating idle stage. The decrease of torque coefficient for such model was also reported by [14]. However, a reason for increase of torque coefficient of the model 3 than the model 2 is that a two stage, two bladed rotors produce more net torque than three bladed single-stage rotor [14].

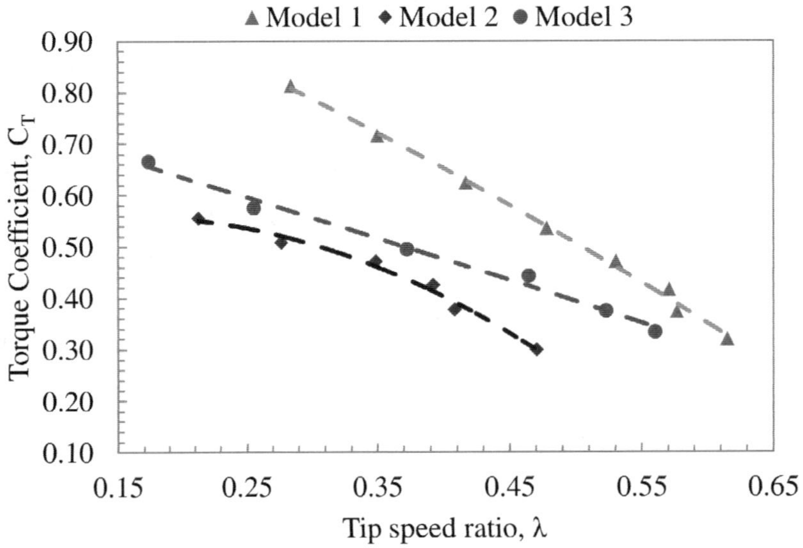

Figure 4. Dynamic torque coefficient of the wind rotor models.

Table 2. Power coefficient and torque coefficient at the correspondence minimum, optimum and maximum tip speed ratio

Model #	λ	*Re*	C_p	C_T
	0.28	90041	0.231	0.81
Model 1	0.42	104980	0.260	0.62
	0.62	149799	0.197	0.32
	0.21	104980	0.118	0.56
Model 2	0.39	129206	0.167	0.43
	0.47	149799	0.141	0.30
	0.17	104980	0.115	0.67
Model 3	0.46	129206	0.206	0.44
	0.56	149799	0.187	0.33

5. Conclusion

Three models of wind rotors have been tested in an open wind flow. The power coefficient of model 1 determined to be higher than the power coefficient of model 2 and model 3. The power coefficient of model 1 is also found to be higher about 24% than that of previously referred study of the same rotor because of the positive effect of changing the aspect ratio to 2. The maximum power coefficient for model 1, model 2 and model 3 is noted as 0.26, 0.17 and 0.21 at their respective tip speed ratio of 0.42, 0.39 and 0.46. The power coefficient of model 3 was observed higher than the power coefficient of the mode 2, a reason of this is model 3 is a two bladed, double stage rotor. Whereas, model 2 is a three bladed single-stage rotor therefore it can be concluded that the double stage two-bladed rotors have higher performance. However, the blade profile design also contributes to performance coefficient therefore it is needed to test the design of model 2 as a two bladed, double stage rotor.

The dynamic torque coefficient of all three models decreases with the increase of tip speed ratio. The maximum dynamic torque coefficient of the wind rotor models is at the minimum tip speed ratio. Dynamic torque coefficient of model 1, model 2 and model 3 is 0.81, 0.56 and 0.67 respectively at the correspondence minimum tip speed ratio of 0.28, 0.21 and 0.17. The models in this study have larger variations in torque coefficient with smaller variations in tip speed, which can be because of the testing of the wind rotor models at higher applied torque.

Acknowledgment
The authors acknowledge Universiti Teknologi PETRONAS (UTP) for supporting the work financially and technically under the International research agreement with University of Stavanger – Norway, grant URIF 22/2013.

References

[1] Panwar N L, Kaushik S C, Kothari S 2011 *Renewable and Sustainable Energy Reviews* **15** 1513-24
[2] Betz A 2014 *Introduction to the theory of flow machines* Elsevier)
[3] Borhanazad H, Mekhilef S, Saidur R, Boroumandjazi G 2013 *Renewable Energy* **59** 210-9
[4] Sathyajith M 2006 *Wind energy: fundamentals, resource analysis and economics* Springer Science & Business Media)
[5] Akwa J V, Vielmo H A, Petry A P 2012 *Renewable and Sustainable Energy Reviews* **16** 3054-64

[6] Morshed K N, Rahman M, Molina G, Ahmed M 2013 *International Journal of Energy and Environmental Engineering* **4** 1-14
[7] Mahmoud N, El-Haroun A, Wahba E, Nasef M 2012 *Alexandria Engineering Journal* **51** 19-25
[8] Al-Kayiem H H, Ming G J 2011 *World acade Sci, Eng and Tech* **60** 144-9
[9] Kamoji M A, Kedare S B, Prabhu S V 2009 *Applied Energy* **86** 1064-73
[10] Hayashi T, Li Y, Hara Y 2005 *JSME International Journal Series B* **48** 9-16
[11] Menet J L 2004 *Renewable Energy* **29** 1843-62
[12] Percival M, Leung P, Datta P, The development of a vertical turbine for domestic electricity generation. European Wind Energy Conference London; 2004.
[13] Vanderhye R A, Dexter M H, Aldrich A L, Rotsky B A, Hascup J R, inventorsThree bladed Savonius rotor. U.S. patent 7,314,346 B2. 2008.
[14] Saha U, Thotla S, Maity D 2008 *Journal of Wind Engineering and Industrial Aerodynamics* **96** 1359-75
[15] Blackwell B F, Sheldahl R F, Feltz L V 1977 *Wind tunnel performance data for two-and three-bucket Savonius rotors* Sandia Laboratories)
[16] Alexander A J, Holownia B P 1978 *Journal of Wind Engineering and Industrial Aerodynamics* **3** 343-51

IOP Conf. Series: Materials Science and Engineering **100** (2015) 012044 doi:10.1088/1757-899X/100/1/012044

Modeling of diesel/CNG mixing in a pre-injection chamber

H A Abdul-Wahhab[1], A R A Aziz[1], H H Al-Kayiem[1], and M S Nasif[1]
[1]Mechanical Engineering Department, Universiti Teknologi PETRONAS, 32610 Bandar Seri Iskandar, Perak, Malaysia

E-mail: abu_ameer5@yahoo.com

Abstract. Diesel engines performance can be improved by adding combustible gases to the liquid diesel. In this paper, the propagation of a two phase flow liquid-gas fuel mixture into a pre-mixer is investigated numerically by computational fluid dynamics simulation. CNG was injected into the diesel within a cylindrical conduit operates as pre-mixer. Four injection models of Diesel-CNG were simulated using ANSYS-FLUENT commercial software. Two CNG jet diameters were used of 1 and 2 mm and the diesel pipe diameter was 9 mm. Two configurations were considered for the gas injection. In the first the gas was injected from one side while for the second two side entries were used. The CNG to Diesel pressure ratio was varied between 1.5 and 3. The CNG to Diesel mass flow ratios were varied between 0.7 and 0.9. The results demonstrate that using double-sided injection increased the homogeneity of the mixture due to the swirl and acceleration of the mixture. Mass fraction, in both cases, was found to increase as the mixture flows towards the exit. As a result, this enhanced mixing is likely to lead to improvement in the combustion performance.

1. Introduction

Due to the large occurrence in a variety of natural and practical applications, multiphase turbulent mixtures have attracted significant engineering interest. These applications include liquid-gas fueled combustors, liquid fuel injected combustion engines, mixtures propellant combustion and fire suppression and control. All of these situations are involving dispersed phase species which considerably affects the thermochemical nature of the surrounding carrier phase. Therefore, mixing and evaporation processes in liquid fuel combustors have marked effects on ignition characteristics [1]. Since this area is an emerging one and the technology has not been disseminated to the scale of driving market, it is essential that specialized components, that require modification, need to be studied. Air-fuel ratio characteristics exert a large influence on exhaust emissions and fuel economy in internal combustion engine. With increasing demand for high fuel efficiency and low emissions, the need to supply the engine cylinders with a well-defined mixture under all circumstances has become more essential for better engine performance [2]. A liquid-gas fuels mixer is one of the important components in such application and it is identified that additional research work need to be carried out in establishing a design procedure for this application.

To ensure proper performance, the liquid-gas mixer should be reproducible and has unequivocal adjustment procedures. For that, a Diesel-CNG mixer has been simulated and this paper presents the simulation results at various design configurations and operational conditions. The simulation was carried out using that ANSYS-FLUENT version 14 software. The turbulence model used in this study is k-ε model which represents a compromise between computational time and precision, for subsonic internal flow [3-5].

2. Mixer design and computational modelling

The purpose of the pre-mixer is to achieve a proper homogenous two phase fluid of liquid Diesel with gas CNG before injection into the engine cylinder. The two phase mixing is very important in dual-fuel engine, as it will provide a combustible mixture of liquid-gas in terms of the required quantity and quality for efficient operation of the engine under all operation conditions. In this study, six designs of mixers labeled as SA1, SA2, DB1, DB2, SC1, and DD1 have been modeled and simulated. The design variables were different CNG jet diameters (1 mm and 2 mm) and the injection angle of CNG into the core of the mixer (55° and 90°). The 90° is selected as it provides the best tangential flow of the jet in the mixture. Then, we selected the 55° as it represents almost the midway between the perfect tangential and the normal (0°). We haven't select the 45° where the flow will create jet flow in the mixture with high turbulence without enhancement to the flow to swirl in the mixer body. The length of the mixer was selected arbitrarily as 60 mm. Table 1 shows the geometrical dimensions of each design.

Table 1. Mixers labeling and geometrical dimension.

Mixer label	Diesel pipe (dp) diameter (mm)	CNG jet diameter (dinj) (mm)	Length L (mm)	Injection angle β_{inj}
SA_1	9	1 (single)	60	90°
SA_2	9	2 (single)	60	90°
DB_1	9	1 (double)	60	90°
DB_2	9	2 (double)	60	90°
SC_1	9	1 (single)	60	55°
DD_1	9	1 (double)	60	55°

Figure 1 shows a model of single injection mixer (SA1), where the diesel and CNG inlets are identified clearly. The liquid diesel is injected axially while the gaseous CNG is injected tangentially.

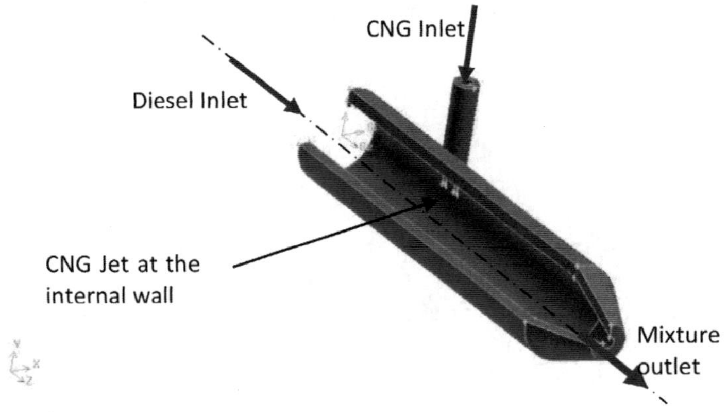

Figure 1. Single injection pre-mixer as constructed in software, model SA_1.

3. Governing equation

Turbulence consists of fluctuations in the swirl flow field in time and space and can have a significant effect on the characteristics of the flow. Turbulence occurs when the inertia forces in the fluid become significant compared to the viscous forces, and is characterized by a high Reynolds number [6] and [7].The k-ε model of turbulence is widely chosen for fluid flow analysis where k is the turbulence kinetic energy and is defined as the variance of the fluctuations in velocity and ε is the turbulence eddy dissipation (the rate at which the velocity fluctuations dissipate). To simulate the turbulence parameters and swirl analysis, a standard k-ε model was chosen with isothermal heat transfer condition at 300 K. The Solver uses this model with two new variables and the continuity equation as follows [8]:

the continuity equation:

$$\frac{\partial}{\partial z_j}\left(\rho \bar{u}_j\right) = 0 \tag{1}$$

the momentum equation:

$$\bar{u}_j \frac{\partial}{\partial z_j}\left(\rho \bar{u}_i\right) = B - \frac{\partial \bar{p}}{\partial z_i} + \frac{\partial}{\partial z_j}\left(\mu\left(\frac{\partial \bar{u}_j}{\partial z_i} + \frac{\partial \bar{u}_i}{\partial z_j}\right) - \rho \overline{u_i u_j}\right) \tag{2}$$

the energy equation:

$$\bar{u}_j \frac{\partial}{\partial z_j}\left(\rho c_p \bar{T}\right) = H + \frac{\partial}{\partial z_j}\left(\rho c_p \alpha \frac{\partial \bar{T}}{\partial z_j} - \rho c_p \overline{u_j T}\right) \tag{3}$$

the model with two new variables:

$$\bar{u}_j \frac{\partial}{\partial z_j}\left(\rho \bar{c}\right) = R + \frac{\partial}{\partial z_j}\left(\rho D \frac{\partial \bar{c}}{\partial z_j} - \rho \overline{u_j c}\right) \tag{4}$$

$$\rho \frac{\partial k}{\partial t} + \rho u_j k_{1,j} = \left(\mu + \frac{\mu_t}{\sigma_k} k_{1,j}\right)_{1,j} + G + B - \rho\varepsilon \tag{5}$$

$$\rho \frac{\partial \varepsilon}{\partial t} + \rho u_j \varepsilon_{1,j} = \left(\mu + \frac{\mu_t}{\sigma_\varepsilon} \varepsilon_{1,j}\right)_{1,j} + c_1 \frac{\varepsilon}{k} G + c_1(1 - c_3)\frac{\varepsilon}{k} B - c_2 \rho \frac{\varepsilon^2}{k} \tag{6}$$

where, j=1, 2, 3, which are refereeing to three axis.

4. Computational modelling

The computational fluid dynamics (CFD) is based on the numerical solutions of the fundamental governing equations of fluid dynamics namely the continuity, momentum, energy, species and turbulent equations, two phase flow [9]. The FLUENT software package was used to accomplish this job. This software flow solver is a finite volume, pressure based, fully implicit code solving the 3D Navier-Stokes equations governing fluid flow and associated physics considering incompressible fluids. The code is used for the modeling of a wide range of industrial problems involving fluid flow, heat transfer (including radiation), turbulence, mixing of chemical species, multi-step chemistry, two-phase flows, moving-rotating bodies and other complex physics. Figure 2 shows the injection angle and the mesh for cases under study.

The three-dimensional model of the mixer is developed by using the preprocessor CFD software. The number of cells used in this model (SA2) was 316,177. Mesh refinement investigation has been carried out to optimize the number of cells used. It has been found that increasing the number of cells to 377,697 and 411,045 has no effect on the results accuracy. Hence, 316,177 were selected to be the optimum number of cells that can be used in the simulation.

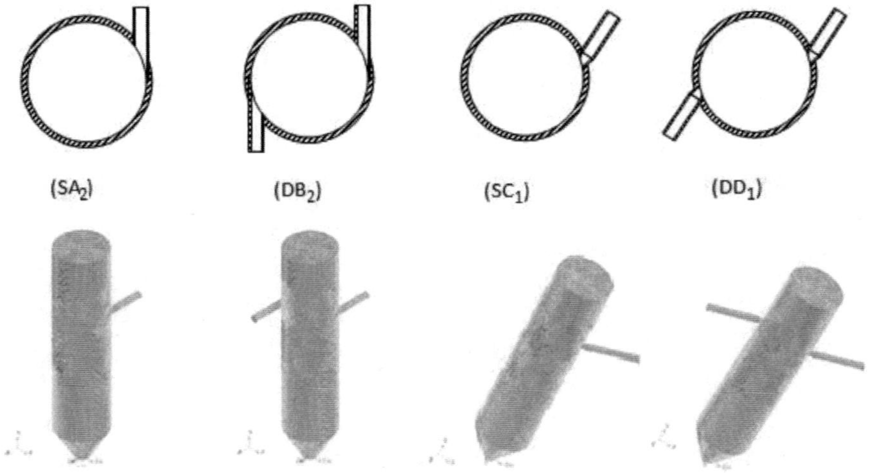

Figure 2. The injection angle and the generated mesh for cases under study.

4.1. Turbulent flow
This involves the use of a turbulence model, which generally requires the solution of additional transport equations and k-ε transport equation is used in this study. Three quantities; turbulent kinetic energy, k, dissipation rate, D and length scale, L are very important in specifying the turbulence characteristics at the inlet in two phase flow cases. If k and D are specified, the value for L will be ignored. It is sometimes more convenient to provide a lengths scale instead of a value for the dissipation rate. The length scale that would be used for an internal flow is usually the inlet diameter or pipe length.

4.2. Mixing flow without reaction (Two Phase Flow)
This requires the solution of additional equations to two phase flow for mixture fractions or species mass fraction. In this simulation, it is assumed that there is no reaction between Diesel and natural gas. Table 2 Properties of Diesel Fuel and CNG.

Table 2 Properties of Diesel Fuel and CNG.

Parameter	Diesel Fuel	CNG
Density (kg/m3) (20 °C)	830	0.72
Viscosity (Cst) (20 °C)	6	$7.8*10^{-6}$
LCV (Mj/kg)	43.8	45.8
Octane Number	50	125
Cloud point (°C)	-9	-27
Flash point (°C)	70	220

4.3. Diesel Inlet boundary
The condition needed to be specified at inlet is the pressure. For this condition, the flow solver determines the pressure applied to each face of the boundary using the velocity specification, the mass flow rate, the temperature and the fluid density. Table 3 initial conditions for the simulation.

4.4. CNG inlet boundary
Static pressure inlet boundary condition was used at the CNG inlet with ratio of 1.5 to 3 from Diesel pressure inlet. This higher gas pressure than the liquid pressure is the practical case to prevent back flow in the gas injector. This range of Gas-to-liquid pressure is reasonable to simulate the mixing phenomena. The mass mixing ratio of CNG to Diesel is selected as 0.7 and 0.9. This mixing ration is

the common practice in the steady of the gas to liquid mixing for fuel application, as reported by [3, 5, 10].

Table 3. the list of initial conditions for the simulation.

Parameter	Initial Condition
Diesel inlet pressure (bar)	2
CNG inlet pressure (bar)	3 to 6
Injection temperature (K)	300
Wall temperature (K)	300
Mass CNG/Diesel ratio	0.7 to 0.9

5. Result and discussion

The mixing efficiency is very important to determine the quality of combustion. A high combustion quality will produce low exhaust gas emission components. In dual fuel engines, the mixer should be designed to be able to produce lean air-fuel mixture in the cylinder, as the diesel engine is a compression ignition which is usually produce lean air/fuel mixture [11]. In this work, the idea of pre-mixing CNG-Diesel was used to increase the air quantity inside the cylinder more than if a mixture of CNG-Air is injected in a diesel fuelled engine which may result is a rich mixture. An incomplete combustion will produce high amounts of CO, which is hazardous to the environment. Moreover, the excessive air will lead to high NOx emission. The mixing quality for the 6 injection models was investigated using CFD. The simulation was done for the same diesel inlet pressure and same outlet pressure of the real engine geometry. Comparing the six mixers, it can be seen that the mixing is more homogeneous for the double injection mixer with jet diameter 1 mm and injection angle 55°. CNG mass fraction in all mixers is shown in figures 3 and 4. Figure 3 shows the results due to the change of the injection angle from 90° to 55° and decreasing the jet diameter from 2 mm to 1 mm. The results demonstrate that the mixing is more homogeneous for the single, 1 mm mixer compared to single 2 mm mixer. While for the double injection case, shown in figure 4, shows that inlet with double injection side is smoother than single injection side because of the increased swirl and turbulence.

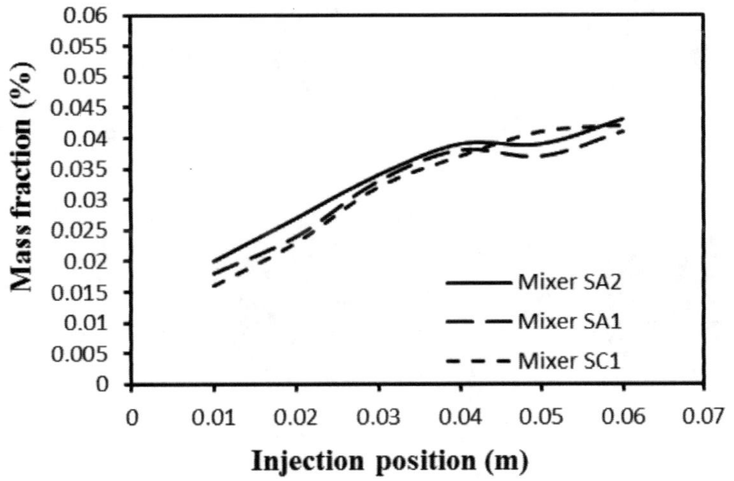

Figure 3. CNG mass fraction (X) in single injection mixers SA_2, SA_1 and SC_1.

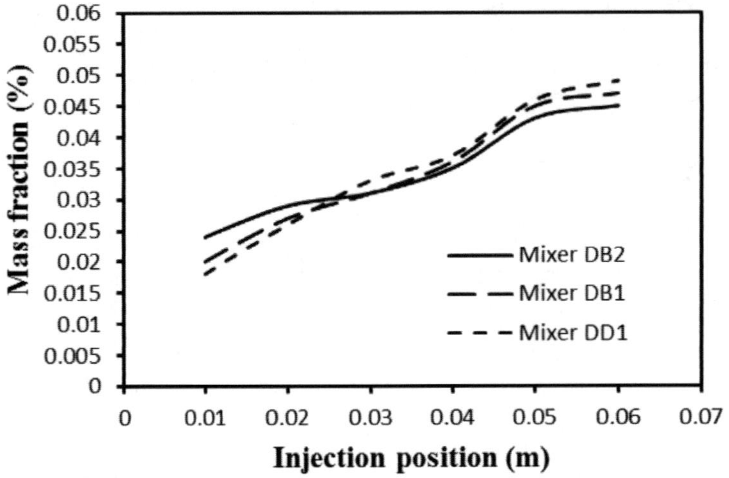

Figure 4. CNG mass fraction (X) in double injection mixers DB_2, DB_1 and DD_1.

The contours in figure 5 shows the variation of mixing pressure in cases of one inlet and two inlets at 90° and 55° injection angle for each case. If the CNG flow rates increase at certain points there is a tendency of the flow to attain an annular type flow regime, which is beneficial to obtain dispersed bubble and droplet populations. It is found that when CNG injection, in both cases of injection angles, that the injection at 55° impose the swirl to be born far from the wall. In contrast, the 90° injection enhances the swirl and initiates the swirl to initiate near the wall. This will lead to proper dispersion of the CNG in the Diesel fuel.

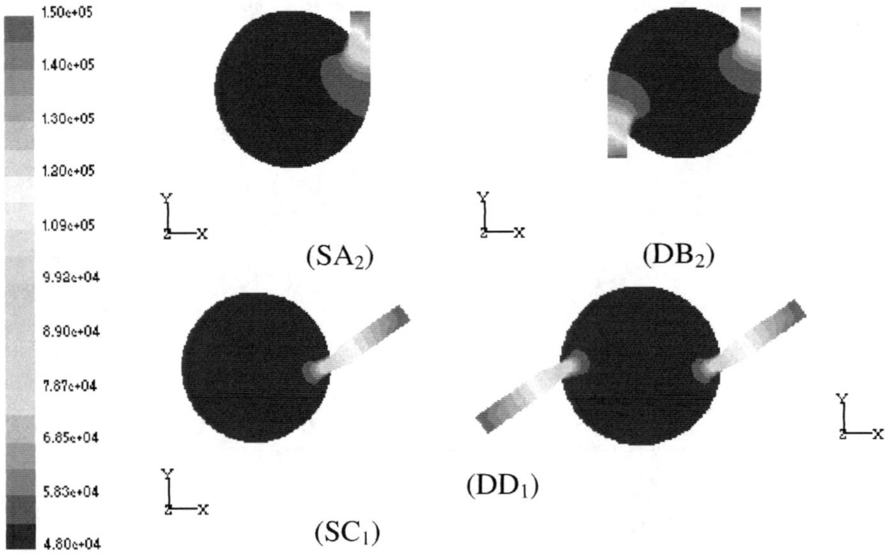

Figure 5. Contours of the total pressure (Pa) for cases under study.

Figures 6 and 7 show the contours of the velocities in the mixer for 2 mm and 1 mm CNG injector diameters, respectively. The simulation results for 2 mm single injection side, shown in figure 6, is misleading where the values and intervals of streamlines are not the same. Rotation in the plane seems stronger and more extended for the 2 mm double injection side, since much shorter intervals of stream function values have been employed, in order to be better detected. Thus, what could be seen with the green and blue areas of figure 6 at SA2 are only weak vortexes that consequently encompass large areas. Figure 7 The flow predicted by the simulation in the case of the mixer with double injection at 1 mm diameter and injection angle 55°, actually rotates much more vigorously. It is clearly exhibiting vortexes with concentrated rings and of a stronger inner structure. Finally, it should be noted that the prediction of mixer DD1 is almost axisymmetric, which is in agreement with the boundary conditions, barring the surely minor detail of the primary swirl. In contrast, the results of mixer SC1 showed an asymmetric flow, especially near the throat, in a manner that cannot be related easily to the geometry.

On the other hand, it can be noticed from the simulation results in these figures that the bubbles coalescence and break-up are continuous processes in the transport phenomena of the two-phase flows in the mixer. If the liquid flow rate increases, the associated liquid turbulent dissipation is noticed to increase. This increase is attributed to the break up mechanism of the bubbles due to severe turbulent splitting. As a result of the predominant splitting action, the relatively large bubbles in the flow will divide into smaller almost spherical bubbles. With a decrease in the volume of the bubble, the interfacial tension dominates maintaining the spherical bubble shape. Under such extreme conditions, dispersed bubbles or homogeneous flow patterns would exist, even at high void fractions. Thus, highly turbulent dissipation and breaking is the key mechanism responsible for flow pattern transition in turbulent two-phase flows. From the contours of relative velocity angle shown in figures 6 and 7, it can be noticed that this phenomenon increases when mixing by using double side with 1mm diameter nozzle and injection angle 55° because of the increase in turbulence due to the increased swirl.

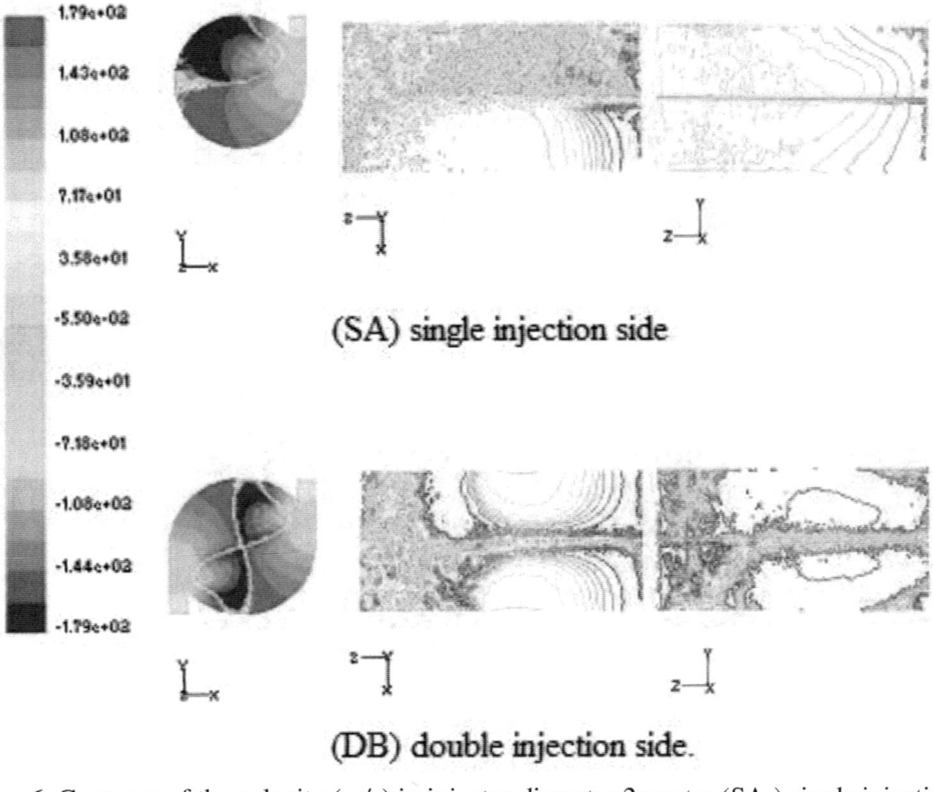

Figure 6. Contours of the velocity (m/s) in injector diameter 2mm to: (SA_2) single injection side, (DB_2) double injection side.

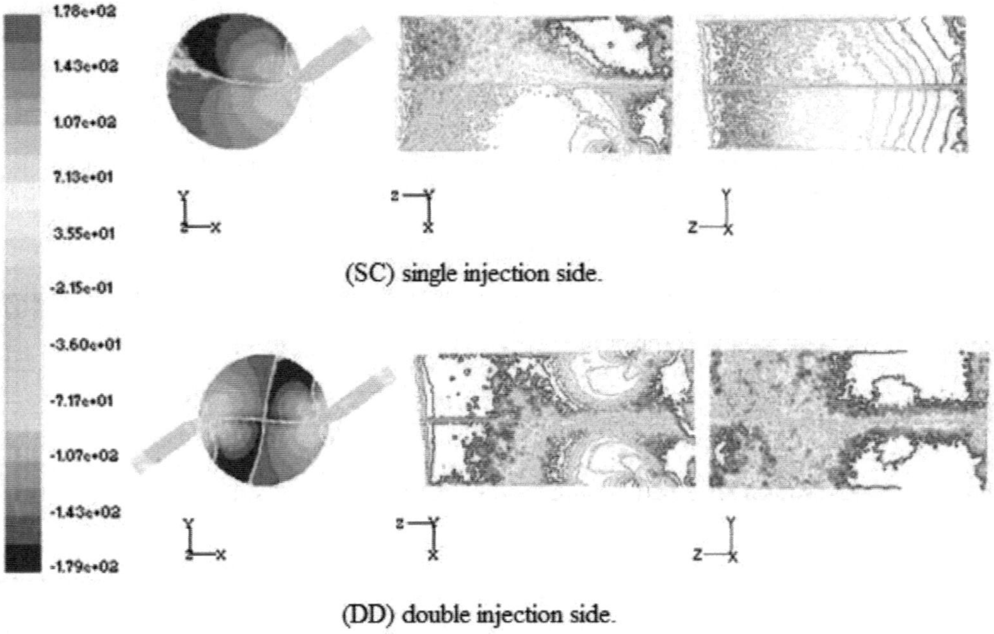

Figure 7. Contours of the velocity (m/s) in injector diameter 1mm to: (SC_1) single injection side, (DD_1) double injection side.

Figure 8 shows the validation of the simulation procedure, where a comparison was made with the experimental results of Rankin, et al. [10]. They carried out the experiments of co-axial CNG injector at mass flow ratio 0.7 with double injection mixer (DD1) at 1 mm injector diameter with mass flow ratio of 0.7 and 0.9. The figure compares our simulation results with their experimental results in terms of the axial velocity of the mixture.

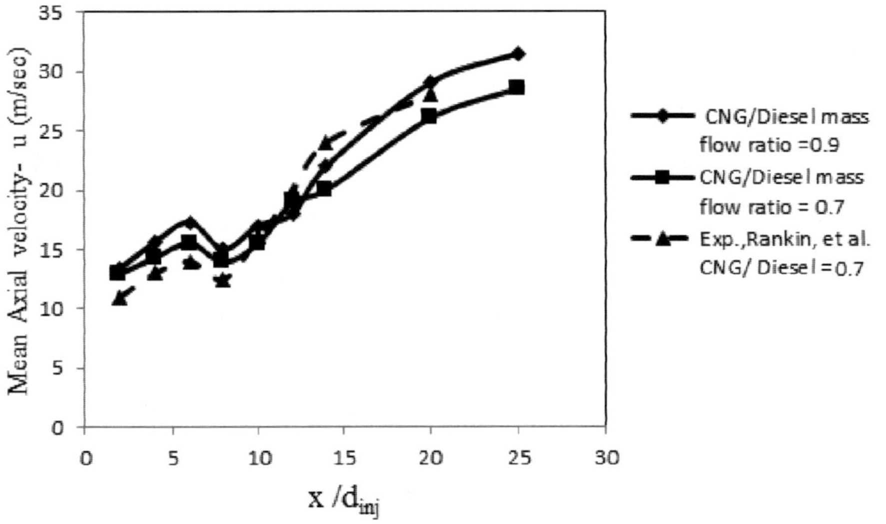

Figure 8. Comparing the results with Rankin, et al. [10].

6. Conclusions

A liquid-gas mixer for Diesel-CNG dual-fuel stationary engine was studied using 3D CFD analysis. Mixers with single and double injections and different jet diameters and injection angles were simulated. It was found that the double-sided gas injection mixer gave better mixing performance as compared to the single injection mixer. Also, the smaller jet diameter produced better mixing due to the increased jet velocity and enhanced turbulence. A high quality mixer should provide the very optimum environment for the engine operating condition.

Acknowledgments
The authors acknowledge Universiti Teknologi PETRONAS for providing the facilities existing in center for automotive research and energy management (CAREM). The main author expresses his thankful remarks to the Iraqi government for the financial support to pursue his PhD.

References

[1] Boivin M, Simonin O, Squires K D 1998 *J Fluid Mech* **375** 235–63
[2] Devarajan R 2008 *International Conference on Association of Science and Technology for Development Power and Energy Systems* (Langkawi) (Langkawi Power and Energy Systems) p. 359-62
[3] Yusaf T, Yusoff M Z 1995 *Proceedings of the International Conference and Exhibition and Natural Gas Vehicles*
[4] Smith K, Arellano L, Jerry I, Tustaniwsky J 2006 University of California, San Diego Department of Mechanical Engineering, USA)
[5] Mardani H 2004 Universiti Teknologi Malaysia) 42-7
[6] Sutar P 2003 *IRF-International Conf. Organized by Institute of Research* (Deemed University, Pune)
[7] Raghunathan B D, Kenny R G 1997 *International Congress and Exposition* (Detroit, Michigan)
[8] Xu B Y, Furuyama M 1997 *Technical Notes, JSAE Review* **18** 57-82
[9] Rahim A R 2008 *Malaysian Technical Universities Conference on Engineering and Technology* (Perlis) vol 2 99-104
[10] Rankin B A, Guilden Becher D R, Gore J P 2012 *Spring Technical Meeting of the Central States Section of the Combustion Institute* 22–4
[11] Ganesan V. Internal Combustion Engine. Internal Combustion Engine Second edition: McGraw-Hill Book Company; 2004.

Conceptual design of semi-automatic wheelbarrow to overcome ergonomics problems among palm oil plantation workers

N S M Nawil[1], B M Deros[1], M N A Rahman[1], E H Sukadarin[2], N Nordin[3], S B M Tamrin[4], S A Bakar[5], M L Norzan[5]

[1]Department of Mechanical and Materials Engineering, Faculty of Engineering and Built Environment, Universiti Kebangsaan Malaysia

[2]Occupational Safety and Health Program, Faculty of Technology Engineering Universiti Malaysia Pahang

[3]Department of Technology Management, School of Technology Management and Logistic Universiti Utara Malaysia

[4]Department of Community Health, Faculty of Medicine and Health Sciences, Universiti Putra Malaysia

[5]School of Mechatronic Engineering, Universiti Malaysia Perlis (UniMAP)

E-mail: nswanimn@gmail.com

Abstract. An ergonomics problem is one of the main issues faced by palm oil plantation workers especially during harvesting and collecting of fresh fruit bunches (FFB). Intensive manual handling and labor activities involved have been associated with high prevalence of musculoskeletal disorders (MSDs) among palm oil plantation workers. New and safe technology on machines and equipment in palm oil plantation are very important in order to help workers reduce risks and injuries while working. The aim of this research is to improve the design of a wheelbarrow, which is suitable for workers and a small size oil palm plantation. The wheelbarrow design was drawn using CATIA ergonomic features. The characteristic of ergonomics assessment is performed by comparing the existing design of wheelbarrow. Conceptual design was developed based on the problems that have been reported by workers. From the analysis of the problem, finally have resulting concept design the ergonomic quality of semi-automatic wheelbarrow with safe and suitable used for palm oil plantation workers.

1. Introduction

Malaysia has a rapid economic growth and able to meet global demand of palm oil products [1-3]. This is an important industry and has emerged as a major agricultural activity in Malaysia [3, 4]. However, since 2006 palm oil production in Malaysia had showed a decline compared to previous years, which has brought down Malaysia to become world second top producer of palm oil after Indonesia [1]. There were many problems that need to be resolved immediately in order to increase the productivity and profit. Among the influential factors are majority of plantation workers are still using manual tools, workplace environment, and worker's abilities in using high technological equipment, interrelation human–machine problems and management programs [5, 6]. This is because, traditional methods to collect palm oil fruits are still been widely practiced in palm oil plantations. This approach

IOP Conf. Series: Materials Science and Engineering **100** (2015) 012045 doi:10.1088/1757-899X/100/1/012045

is no longer efficient, wasting time and the source of work related musculoskeletal disorders (WMSDs) problems to the workers. Therefore, to stay competitive in global market, the palm oil industry needs to improve the productivity of fresh fruit bunches (FFB) through technological improvement and innovation of harvesting and handling equipment [6, 7]. For example, uses of mechanized harvesting equipment and material handling device that can help workers work safely, efficiently and smoothly.

Up till now, there are various harvesting and collecting machines have been developed either by researchers, industrial or agriculture machine manufacturers such as; tractor-mounted grabber, CantasTM, pruner and harvester machinery [6, 8, 9]. However, not all these new technologies were implemented successfully. Presently, the new technology and machineries are still facing low level acceptance and limited application of harvesting technology in palm oil industry [5, 10]. The low acceptance or rejection was due to several factors such as; machine tool vibration effects on the body, the machine is too large, difficult to use, unskilled workers due to lack of training [11]. Thus, the workers still preferred to use manual handling compared to new existing harvesting technologies. This finding is in line with the study result by Abdullah & Samah [12] which focused on factors impinging farmers' to use new agriculture technology.

2. Manual tools in palm oil plantation

In the long run, the continued use of manual tools will negatively affect the health of workers. Manual plantation workers need to perform many movements and lifting heavy loads every day. Table 1 shows the daily work activities by a palm oil plantation worker. Starting with harvesting, workers will cut FFB by using a chisel for shorter and a sickle for taller palm oils trees. Having done that, FFB collector will pick up the detached FFBs on the ground by using either a hook or metal pole to pierce and load them onto a wheelbarrow. When the wheelbarrow is full, the FFB collector will push it to the roadside along the main truck route and unload it at a collection point. Later, trucks will stop at each collection point along the main route to load all the FFBs to be sent to the palm oil mill.

Table 1. Daily activity of palm oil workers.

Activities in palm oil plantation	Description of the activities
Harvesting	Harvesters performed the FFB harvesting activity using a chisel or sickle; cutting the ripe palm oil fruit.
Collecting and Loading	Collector and loader using a skewer of sharp metal to load the FFB in a wheelbarrow. The loader then pushes the fully loaded cart with FFB to the nearest collection station.
Time of pruning jobs (every evening or weekend)	Pruning (cutting the palm oil frond at the base of the trunk), Gather, wedding and spraying using herbicides.

From the table 1, the manual plantation workers are exposed to ergonomics problems during performing job tasks in the palm oil plantation that include manual handling of FFBs, repetitive upper extremity motions and awkward postures. The palm oil industry shows many employees are reported to be at risk for work-related musculoskeletal disorders (WMSDs) [7]. Among the workers in palm oil plantations, collectors and loaders have a high risk of experiencing MSD pain. This is because, palm oil workers had to bend their body to pick the FFBs on the ground and put them in the wheelbarrow. Workers need to use a lot of energy to lift heavy FFBs and push fully loaded wheelbarrow from one tree to another tree. This finding is supported previous study by Nawi et al. [13] which had identified risks of working postures in the palm oil plantation based on REBA (Rapid Entire Body Assessment) developed by Hignett and McAtamney [14]. From the results, it is clear that the risk of experiencing MSDs is high for FFB collectors and loaders. The purpose of this study is to design an ergonomically wheelbarrow, which is intended to be used by palm oil workers. Therefore, this study is focused on palm oil collectors and loaders, which discussed on development of the semi-automated wheelbarrow specially dedicated for palm oil workers.

3. Issues of manual wheelbarrow

Wheelbarrow or wheel loaders are working machines used to lighten the load to carry in many different fields and situations. The use of manual wheelbarrow gives advantages, in terms of the cheap price and low maintenance costs. However, it results in many shortcomings, particularly to the palm oil plantation workers. Based on analysis of the observations and the results of preliminary study by [13], several key issues have been identified. Among the issues that arise from the use of manual wheelbarrow is:

Among the issues that arise from the use of manual wheelbarrow are as follows:

a) Its size is too small and cannot be used to carry FFBs in large quantities. It can only carry for 5 to 6 bunches of FFBs weighing 10-20 kg each at a time, which is not efficient to be used especially for large plantations. Workers had to work repeatedly. It takes a long time and may reduce the number of FFBs daily collections.

b) As shown in figure 1, workers have to manually push the fully load wheelbarrow with FFBs through the hilly and bushy undergrowth by using a lot of effort. The workers had to do it repeatedly for many times every day, from collected area to its deposition area.

This situation shows that the collectors and loaders are exposed to the risk of having musculoskeletal injury due to their high work demands. Due to the small size of wheelbarrow, large-sized FFBs always fall off the wheelbarrow. Workers had to repeat the work of collecting the fallen fruit and push it back the on the cart. This practice will increase the risk of accidents in the workplace. Therefore, the safety of the collectors and loaders is also an important issue here. A lot of time is wasted in collecting back the fallen FFBs and eventually decreasing the overall productivity.

Figure 1. Palm oil workers pushing manually of fully loaded wheelbarrow.

4. Methodology

Challenges faced by collectors and loaders in these situations have been identified and new approaches to solve the problems need to be considered. To overcome these problems, mechanize equipment has been suggested as solutions to improve productivity and make it easier during performing the work tasks compared to being performed manually [6, 15]. It could be achieved through the introduction of efficient and economical mechanical technology with an ergonomic interface. There are many existing machineries and tools, which have been developed, but they are too expensive to buy and high maintenance costs. It is not efficient to small-scale palm oil plantation owner. Besides, large and heavy machines have not suitable for use at palm oil plantations which have hilly landscape.

Several steps are carried out to create improvement designs on existing wheelbarrow. The steps were; analysing current problems, identify product attributes and impact of the equipment to be designed on workers performance. It was performed by conducting observation at palm oil plantation, interviewing palm oil workers and analyses on the existing design. Several aspects need to be taken in account to ensure the newly designed machine or tool can be easily utilized and commercialized to market. Therefore, the required innovation should have the ergonomics characteristics, such as: good human and machine interface with work environment as shown in figure 2. The main features of semi-automatic wheelbarrow to be developed for the improvement of traditional wheelbarrow are: lower cost that can be afforded not only for large plantation owner but also for small holders, easy to use without the need for a lot of training, can be used even without requiring a lot of energy, easy to access into the hillside, no or low vibration to avoid the long term health effects, easy to take care and maintenance.

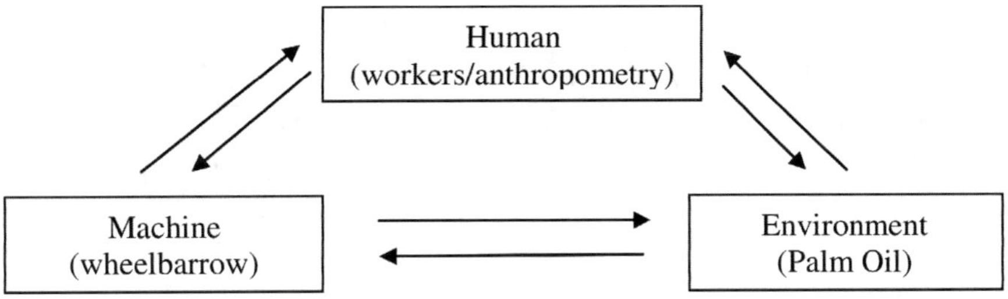

Figure 2. Interrelation of ergonomics relationship.

Finally, the product criteria specification were determined. There are multiple criteria which need to consider in designing new model of wheelbarrow. These criteria cover technical, ergonomic, environment and economical consideration. Table 2 shows the evaluation criteria and parameter of product design.

Table 2. Evaluation criteria and parameter of product design.

Evaluation Criteria	Parameter
Function	Overall geometry, capacity
Safety	Operational, human, environmental
Quality	Quality assurance, quality control, reliability
Manufacturing	Production and purchase of components
Timing	Schedule in designing, development and production
Economic	Cost in resources, designing and manufacturing
Ergonomic	User needs, ergonomic design
Ecological	Environment impact, sustainability, material selection
Aesthetic	Customer (workers) appeal, future expectation
Life-Cycle	Operation, Maintenance, Disposal
Legal/Ethical	Regulation, ethics, intellectual property

5. Result and discussion

New design and innovation of the semi-automatic wheelbarrow is very important to overcome the inadequacy of the traditional manual wheelbarrow. From the data collected and previous researches, the authors have proposed a conceptual design of a semi-automatic wheelbarrow as shown in figure 3. Critical analyses were done to satisfy palm oil plantation owners, small holders and workers. This semi-automatic wheelbarrow was chosen based on the following factors: lower costs, affordable by small holder of palm oil plantation; it require little training for workers; low maintenance cost to the owners; and easy to control in the hilly plantation environment.

In designing a product, ergonomic aspects such as working posture, energy, conditions of environment, safety and usability need to be taken into account. These features were emphasized in the design of this semi-automatic wheelbarrow. A few improvements have been made on the design of semi-automatic wheelbarrow. Among new design innovations that have been done are: using a power motor to reduce the burden of manual worker pushing the wheelbarrow; increase the number of wheels from two to three to get a better balance; increase the size or storage capacity to avoid the collected FFBs fall to the ground and cart that can be opened at the front to facilitate palm oil plantation workers placing and lifting the FFBs.

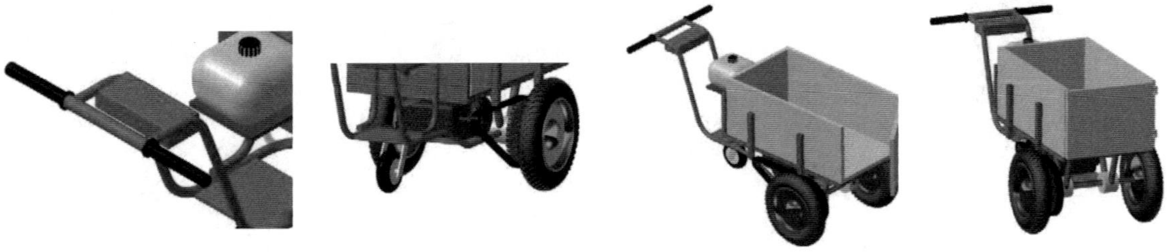

Figure 3. Conceptual design of a semi-automatic wheelbarrow.

There were several significant differences between the traditional manual wheelbarrow and the new design of a semi-automatic wheelbarrow, such as: wheelbarrow wheels, power motor, hand grip, cart size, maintainability, improve safety and lower cost. A comparison between manual and semi-automatic wheelbarrow is shown in table 3.

Table 3. Comparison between manual and semi-automatic wheelbarrow.

Characteristics	Manual Wheelbarrow	Semi-automatic wheelbarrow	Differences
Cart size	Small	Medium	The cart size or load area of semi-automatic wheelbarrow is larger than the traditional wheelbarrow. Therefore, workers will have no problems of loading the FFBs and it would not fall off to the ground.
Hand grip	Normal hand grip	Rubbery elastic hand grip	Semi-automatic wheelbarrow also has a firmer, longer and rubbery elastic hand grip, which can reduce hand pain.
Maintainability	Low maintenance cost	Low maintenance cost	Semi-automatic wheelbarrow using low cost materials and spare parts are easily available.
Safety	High risk of injury	Low risk injury	Semi-automatic wheelbarrow will eliminate the need to lift and push which is present in manual wheelbarrow.
Cost	Cheap	Affordable cost	Semi-automatic wheelbarrow is still considered at affordable costs to palm oil small holders.

Power motor	No power motor	Small power motor	Addition of small power motors on handcarts of semi-automatic wheelbarrow. The power motors will help workers to reduce burden from using a lot of energy while pushing a fully loaded wheelbarrow.
Wheels	1 medium size wheel	2 medium size and 1 small size of wheels	Semi-automatic wheelbarrow adding more wheels from one commonly used to three wheels. Additional wheels will strengthen the equilibrium of the carts and can be handled more easily even to accommodate heavy loads of FFBs.
Environment	Not suitable for hilly landscape	Suitable for hilly landscape	Semi-automatic wheelbarrow is easy to use in the hilly plantation because does not require a lot of energy to push a fully loaded of wheelbarrow.

The innovation of semi-automatic wheelbarrow for palm oil collectors was designed to solve their current problems. The palm oil collectors and loaders can get many benefits such as lighten the load and reduce the risk of injury to workers from the use of this semi-automatic wheelbarrow. The two most significant improvements are in the use of power motor and two extra wheels to stabilize the semi-automatic wheelbarrow as shown in figure 4. The heavy manual task is taken over by the power motor, which powers the semi-automatic wheelbarrow. Palm oil collectors and loaders only need to hold and steer the semi-automatic wheelbarrow. This situation will help the palm oil collectors and loaders to reduce energy consumption by eliminating the need to push heavy load and thus shall reduce work related muscular skeletal disorders (WMSDs) risks and injuries. Furthermore, this new semi-automatic wheelbarrow could help collectors and loaders to collect more FFBs from one tree to the next tree more efficiently and subsequently it will increase work productivity. The manufacturing process is also very practical because it does not require expensive machines or materials that are difficult to obtain.

(a) (b)

Figure 4. Design of semi-automatic wheelbarrow (a) before and (b) after.

6. Conclusion

Technology and innovation appears to provide more benefits to users. Improved technology and operation will exhibit higher productivity for Malaysian palm oil production. However, technology alone will not succeed without the support of other factors such as financial and knowledge capability, human-technology relationship, training and others. Thus, the innovation of the proposed design semi-automatic wheelbarrow considers ergonomic relationship between the human, machines and the environment. Compared with manual wheelbarrow, this new design is much better and provides many benefits to palm oil plantation workers. Application of ergonomic relationship in the design of semi-automatic wheelbarrow will help collectors and loaders in palm oil industry in providing the following positive impacts, such as: increasing productivity, reducing energy and workload, low maintenance cost and reducing risk of collectors and loaders' injury. Palm oil industry should pay more attention to the principles and application of ergonomics rules on occupational safety and health of manual workers.

Acknowledgement
The authors would like to thank all researchers from Universiti Malaysia Perlis, Universiti Putra Malaysia and Universiti Kebangsaan Malaysia for this integrated ergonomic approach project. This research was funded by AgriOsh Sciencefund (06-10-04-SF1539) grant.

References
[1] Liwang T, Daryanto A, Gumbira-Sáid E, Nuryartono N 2009 *XVI Int. oil palm conference and expopalma Fedepalma, Colômbia*
[2] Gan P Y, Li Z 2012 *Institute of Ergonomics Economics Japan*
[3] Panapanaan V, Helin T, Kujanpää M, Soukka R, Heinimö J, Linnanen L. Sustainability of palm oil production and opportunities for fiish technology and know-how transfer. Faculty of Technology, Lappeenranta University of Technology: Lappeenranta University of Technology, 2009.
[4] Basiron Y 2007 *European Journal of Lipid Science and Technology* **109** 289-95
[5] Ishak W, Ismail W, Bardaie M Z 1997 *Pertanika J Soc Sci Humanit* **5**
[6] Jelani A R, Hitam A, Jamak J, Noor M, Gono Y, Ariffin O 2008 *J Oil Palm Res* **20**
[7] Ng Y G, Shamsul Bahri M T, Irwan Syah M Y, Mori I, Hashim Z 2013 *Journal of Occupational Health* **55** 405-14
[8] Jayaselan H A J, Ismail W I W 2010 *Int J Agric Biol Eng* **3**
[9] Ismi Intara Y, Mayulu H, Radite P A S 2013 *Physical and Mechanical Properties of Palm Oil Frond and Stem Bunch for Developing Pruner and Harvester Machinery Design* **4**
[10] Duraj E W V, Miles J A, Meyers J M, Faucett J A, Janowitz I L, Tarter M E, et al., Harvesting aids for reducing ergonomics risk factors in wine grape hand harvesting. Proceedings of the American Society of Agricultural Engineers Meeting; 2000.
[11] Jelani A R, Shuib A R, Hitam A, Jamak J, Noor M M, Hand-Held Mechanical Cutter. MPOB Information Series; 2003.
[12] Abdullah F A, Samah B A 2013 *Asian Soc Sci* **9** 120-4
[13] Nawi N S M, B. M. Deros, Nordin N 2013 *Adv Eng Forum* **10** 122-7
[14] Hignett S, McAtamney L 2000 *Applied Ergonomics* **31** 201-5
[15] Abd Rahim S, Abdul Razak J, Salmah J, Mohd Solah D, Mohd Ramdhan K, Ahmad H, et al. *Development of a machine for harvesting tall palms*

IOP Conf. Series: Materials Science and Engineering **100** (2015) 012046 doi:10.1088/1757-899X/100/1/012046

Aerobic rice mechanization: techniques for crop establishment

K M Khusairy[1], H Ayob[3], C S Chan[1], M I Mohamed Fauzi[1], Z O Mohamad Fakhrul[2], G S M Shahril Shah[1], O Azlan[1], M A Rasad[1], A M Hashim[1], Z Arshad[2], E Ibrahim E[2], M N Saifulizan[1]

[1]Mechanisation and Automation Research Center, MARDI Seberang Perai 13200 Kepala Batas, Pulau Pinang, Malaysia

[2]Mechanisation and Automation Research Center, MARDI Alor Setar 05400 Alor Setar, Kedah, Malaysia

[3]Mechanisation and Automation Research Center, MARDI HQ 43400 Serdang, Selangor, Malaysia

E-mail: mkhusairy@mardi.gov.my

Abstract. Rice being the staple food crops, hundreds of land races in it makes the diversity of rice crops. Aerobic rice production was introduced which requires much less water input to safeguard and sustain the rice production and conserve water due to decreasing water resources, climatic changes and competition from urban and industrial users. Mechanization system plays an important role for the success of aerobic rice cultivation. All farming activities for aerobic rice production are run on aerobic soil conditions. Row seeder mechanization system is developed to replace conventional seeding technique on the aerobic rice field. It is targeted for small and the large scale aerobic rice farmers. The aero - seeder machine is used for the small scale aerobic rice field, while the accord - seeder is used for the large scale aerobic rice field. The use of this mechanization machine can eliminate the tedious and inaccurate seeding operations reduce labour costs and increases work rate. The machine is easy to operate and it can increase crop establishment rate. It reduce missing hill, increasing planting and crop with high yield can be produce. This machine is designed for low costs maintenance and it is easy to dismantle and assemble during maintenance and it is safe to be used.

1. Introduction

To achieve higher self-sufficiency level (SSL) required for food security rice yield need to be increased rapidly by increasing the rice production areas. Major challenges to meet the objectives are short of natural resources such as suitable lowland, insufficient water supplies throughout growth period and lack of labour force. Rice is one of the few crops cultivated under flooded conditions. The process of cultivation of rice requires huge amount of water as well as long working hours. According to an estimate it requires approximately 5000 litres of water to cultivate 1 kg of rice [1, 2].

Conventionally, there are two methods to cultivate rice i.e. sowing method and menugal method. In sowing method, the farmers and/or labourers sow the seeds manually on a prepared ground for sowing. This method requires lanes for manual sowing and needs 200 hrs/ha of man/labour hours. Further, said method is only suitable for small-scale area and cultivation of rice in a large-scale area i.e. more than 1 ha is not possible with this method [3].

The menugal method is an improved method as compared with the sowing method. However, alike sowing method it requires a large amount of water as well as substantial amount of man/labour hours to cultivate rice. In this method, farmers and/or labourers sow the seeds into the soil by using a dibble and requires 92.30 hrs/ ha of man/labour hours for an area of 1 ha. Though this method may be used to cultivate a large-scale area, however, due to substantial amount of time and high dependency on skilled labourers this method is not suitable for the same [4].

However, such aforesaid technologies/methods possess following limitations and/or disadvantages. First and the foremost requirement of conventional method is the need of highly skilled workers for sowing seeds which leads to high production cost. Further, the conventional methods are tedious and require very hard work. The conventional methods require a huge amount of water for cultivating rice. Water being a limited natural resource, the conventional methods over - exploit the ground water. In addition, the conventional methods are time consuming and incapable of carrying out rice cultivation in a large-scale area i.e. more than 1 ha. Also, sowing of seeds is not accurate by the conventional methods which hamper the final quantum of crop produced at the time of harvesting.

A lot of research and development has taken place in this regard for having an alternate method of cultivating rice. Aerobic rice crop cultivation is one of the improved methods [2] known wherein cultivation of rice is carried out under aerobic situation. In the aerobic crop cultivation method, the farmers are not required to grow rice sapling separately in a nursery and it also eliminates the need of ploughing the land. The farmers, by using the seed drill pulled by tractor or bulls, can sow the seeds directly in the land. Further, by using the aerobic crop cultivation method, paddy can be sown with good spacing in between and prevented from developing any disease as well as the yield as compared with conventional method. The aerobic crop cultivation method requires less labour, less human intervention and less quantity of seeds, therefore is economical as compared with the conventional methods. However, existing aerobic cultivation systems/methods include number of limitations and shortcomings.

Accordingly, there remains a need in the prior arts to have an improved method of cultivation of crop especially rice with an aero - seeder machine which enables the cultivation of crops in a sustainable manner which in turn increases crop establishment rate. Further, it can reduce the amount of land which remains left out during manual cultivation of crops and increases yield of the crop. Aerobic rice planting is fundamentally a different approach of rice cultivation proposed by MARDI to overcome this issue where a high yielding rice variety is grown in non-puddled and non-saturated fields with supplementary irrigation and high external inputs [5]. Mechanization system plays an important role for the success of aerobic rice production. The development of machine that efficient to help aerobic rice farmers is needed to achieve the mission of the aerobic rice production in this country. A modern agricultural practice emphasizes the ability of farmers to meet that demand of the consumers of rice. The development of mechanization system of row seeding machine is one way to increase the efficiency of the farm work, while reducing dependence on labour and also to create a method of rice cultivation more systematically. Row seeding mechanization systems for aerobic rice have been developed by MARDI to help maximize the profitability of the farmers. This will ensure the completeness of seeds cultivation operations to be more accurate in terms to get a better crop yield and quality result [6].

2. Materials and Methods-Accord-seeder

Embodiments of the present invention aim to provide an aero - seeder machine which replaces the conventional seeding techniques of cultivation of rice. The proposed machine is suitable for small as well as large scale area. Further, aero - seeder machine eliminates the tedious and inaccurate seeding operations carried out by labourers manually. The aero - seeder machine is easy to operate and it can increase crop establishment rate. Also it reduces missing hill and increases cultivation, thereby crop with high yield can be produced. Embodiments of the present invention are made possible in this way:

An aero - seeder machine comprises a frame, a plurality of sliding shaft mounted on the frame, a plurality of seed hopper mounted on the plurality of sliding shaft and configured to carry a plurality of seed, a plurality of metering device mounted on the plurality of sliding shaft and configured to attach at a bottom end of the plurality of seed hopper, a plurality of seed hose cone mounted on the plurality of sliding shaft and configured to couple at the bottom end of the plurality of seed hopper, a plurality of drop hose configured to couple at a bottom end of the plurality of seed hose cone, a plurality of furrow opener attached with the frame and configured to construct a plurality of row seed lane on a ground and a plurality of wheel coupled with a proximal end of the frame. Where in the frame is attached with a two wheeler tractor. The plurality of seed hose cone drops the plurality of seed in the plurality of drop hose and the plurality of metering device regulates the amount of plurality of seed flows from the bottom end of the plurality of seed hopper. Thereafter, the plurality of drop hose drops the plurality of seed inside the plurality of row seed lane which is having depth in the range of 0.03 m to 0.05 m and the plurality of wheel closes the plurality of row seed lane.

Figure 1. Aero-seeder.

As compared to the existing technologies, the proposed machine is economical as it has low maintenance costs, easy to dismantle and assemble during maintenance, reduces labour costs and increases work rate.

Figure 2. Accord seeder.

Accord seeder machine is use for the large scale area for aerobic rice. The machine is attached to the four wheel tractor and pneumatic systems used for dropping the seeds. The central metering device accurately measures any desired volume of seed from 2 kg /ha to 200 kg/ha. For the sowing of aerobic rice, the metering device can be infinitely adjusted to fine seed/micrometering by means of a spindle, without any need for tools.

The central, totally enclosed cell wheel of the metering device accurately measures the required volume of seed and discharges it into the venturi cone where it is mixed with the air stream and then conveyed through the diffusor tube and the seed delivery hoses to the coulters. Distribution installed in a protected position in the seed hopper. Advantages using of this machine is the machine easy to be adjusted, the hopper is placed close to the tractor and have a wide breadth of the sower which is up to 1.5 - 3.0 meters for sowing seeds in the large-scale area.

Table 1. Specifications and functions of the row seeder.

Aero – seeder	Accord - seeder
Equipped 6 row seeder hose	Equipped 9 row seeder hose
Seed rate 80-150kg/ha	Seed rate 60-150kg/ha
Work rate 3-4 hour/ha	Work rate 1-1.5 jam/ha
Work rate/day (2 ha/day)	Work rate/day (6-7 ha/day)
Attached to 2 wheel tractor (power tiller)	Attached to 4 wheel tractor 40hp above
Hopper capacity 6kg/tank (machine equipped with 6 hopper)	Hopper capacity 150-200kg/tank
Seed drop by gravity flow	Seed drop by pneumatic air flow system

3. Results And Discussions

Conventional sowing method and the menugal method is only suitable for small-scale area. This method is not practical for a large-scale area of more than 1 ha. This is because the conventional method requires a lane for manual sowing and it requires the work of 200 hours hour/ha , while the method of mech - tugal required working hour 92.3 hours/ha for an area of 1 ha. This method is very detrimental to the time and high dependence for labour required on the large scale area.

Compared to mechanization system of row seeder, time and dependence on labour can be saved and reduced by more than 50 %. For the use of aero - seeder machine, sowing the seeds for the area of 1 ha take time to work 3-4 hours / ha , while the use of accord – seeder machine, time work for an area of 1 ha area is 1-1.5 hours / ha . Therefore, the economic impact of the use mechanization system of row seeder for aerobic rice farming is more economical. In addition besides reducing dependence on labour, it also enhances work efficiency. Mechanized system of row seeding also enhances aerobic rice seed bed for good crop growth and uniformly planting line.

Figure 3. Mechanization row seeder boost the crop performance and uniform plant growth.

4. Conclusions

Aerobic rice mechanization system is urgently needed to ensure that agricultural activities can be cultivated intensively and effectively and provide returns to the farmers and make a meaningful contribution to national food production.

The potential of growing aerobic rice is good when it can be grown as upland crop while saves water. Thus, in areas where water is relatively scarce than land, the best way to maximize total rice production is by growing aerobic rice. However, the level of mechanization operations should be as good as wetland land rice cultivation if not better. This is to ensure farming activities can be carried out without depending heavily on labour force and water can be supplied to the field in the right time, right amount and right place. By using row seeder for Aerobic rice production, labour dependency problems can be solved and working time can be reduced compared with the work of conventional sowing. Therefore, it is recommended to use the mechanization of the row seeder to increase rice yield besides improving the quality of aerobic rice production and farmers work quality.

Acknowledgements

The authors would like to be obliged to express my sincere appreciation and thanks to especially Mechanization and Automation Research Centre (MA 02), Central Rice Research and Industrial Crops (RIC) and the parties directly involved with their cooperation and assistance in making this study.

References

[1] Sariam O, Chan C S, Ismail C H, Zainuddin B H, Alicia J, Mohd Fitri M, et al. Manual Penanaman Padi Aerob. MARDI. 2013.

[2] Sariam O, Zainudin P M D H, Chan C S, Azmi M, Rosniyana A, Badrulhadza A. Padi Aerob Untuk Mengatasi Masalah Kekurangan Air. Penerbit UTM Press. 2014.

[3] Abu Hassan D, Sariam O, Saleh B, K. M K, Mohamed Fauzi I, Hashim A, et al. Poster Penabur Berbaris Aero : Mekanisasi Penanaman dalam pengeluaran Padi Aerobik. MARDI.

[4] Mohd Khusairy K, Ayob H, Chan C S, Mohamed Fauzi M I, Mohamad Fakhrul Zaman O 2014 *Penerbit UTM Press* 33-5

[5] Bouman B A M, Lampayan R M, Tuong T P. Water Management in Irrigated Rice: Coping with Water Scarcity. International Rice Research Institute. 2007.

[6] Sariam O, Afliza R. Padi Aerob–Penanaman Padi Tanpa Air Bertakung. Penerbit UTM Press. 2005.

Analysis on the grinding quality of palm oil fibers by using combined grinding equipment

H L Gan[1], L M Gan[1] and H C Law[1]
[1]Faculty of Mechanical Engineering, Universiti Malaysia Pahang, 26600 Pekan, Pahang, Malaysia

E-mail: ming@ump.edu.my

Abstract. As known, Malaysia is the second largest palm oil producer worldwide after Indonesia, therefore indicating the abundance of its wastes within the country. The plantation would be seen to increase to at least 5.2 million ha by 2020, and the waste generation would be 50-70 times the plantation. However, the efficiency of bulk density is reduced. This is one of the main reasons of the initiation of this size reduction/ grinding research. With appropriate parameters, grinding will be seen to be helping in enhancing the inter-particle bindings, subsequently increasing the quality of final products. This paper focuses on the grinding quality involving palm oil wastes by using the Scanning Electron Microscope (SEM). The samples would first be ground to powder at varying grinding speed and finally got the randomly chosen particles measured to obtain the size range. The grinding speed was manipulated from 15 Hz to 40 Hz. From the data obtained, it was found the particles fineness increased with increasing grinding speed. In general, the size ranged from 45 μm to about 600 μm, where the finest was recorded at the speed of 40 Hz. It was also found that the binding was not so encouraging at very low speeds. Therefore, the optimum grinding speed for oil palm residues lied in the range of 25 Hz to 30 Hz. However, there were still limitations to be overcome if the accuracy of the image clarity is to be enhanced.

1. Introduction

As to produce a good quality of densified product, grinding process is necessary. Grinding is the process designated to reduce particle size prior to subsequent processing. The rationale of size reduction is none other than to transform biomass materials into a form which optimizes handling, storage, transportation as well as conversion [1]. Some of the known parameters to have direct influence on the quality of grinded product include moisture content, type of equipment utilised, as well as rotary speed (rpm) [2].From studies, biomass has been proven as one of the most promising alternative energies in Malaysia due to its sustainability and environmental friendliness [3]. Then, it was found that the largest biomass source in Malaysia is oil palm since Malaysia is the second largest palm oil producer worldwide after Indonesia [4, 5] This is simply because 14.9 % out of 32.9 million ha land available in Malaysia is used for agricultural activities, where 34.16 % is oil palm [6]. As summarized by [7], there were 400 ha of oil palm plantation back in 1920, then rose to 4.06 million ha in 2009, and expected to expand to 5.2 million ha by 2020. A hectare of oil palm plantation is capable of producing 50-70 tonnes of waste and thus, electricity generation would be able to be generated from those wastes. The resulting yearly energy potential is projected to be at an amount of 800 GWh [8] while the capacity potential would be 1300 MW [9].However, the quality of grinding has to be preserved in order to maximise the efficiency of conversion into fuel source. For instance, the ground biomass powder could be combusted to produce syngas, also known as producer gas. This type of gas results in a better efficiency of 30-40 % when used in internal combustion engines, as compared to the conventional combustion Rankine cycle biomass power plants which have the efficiency of at most 25 % [10].Therefore, this paper will be aiming at determining the functionality of a grinding machine through verification of Scanning Electron Microscope (SEM) outcomes. Its magnifications would enable the extent of internal bonding of lignin for instance within the particles to be determined [11]. The higher the extent of the bindings, the better the resulting strength and other properties would be. The bonding could be achieved through a good extent of particle fineness besides uniformness, at appropriate moisture content. A good grinding facility will have to be able to produce particles with desired uniform shape and size.

2. Experimental set up

In this paper, the material involved is palm oil stem and palm oil fibers (empty fruit bunches-EFB).The standard operating procedures cover two major parts, grinding and scanning.

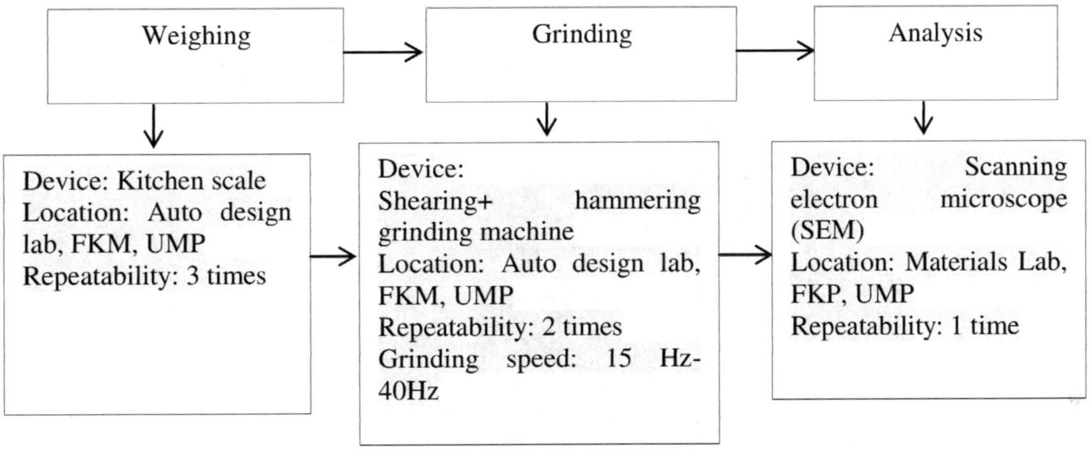

Figure 1. SOP for palm oil biomass.

Firstly, each sample would be weighed using a kitchen scale for three times, at the precision of 1 g. The average weight was then calculated and recorded in a datasheet. The weighing session was necessary to determine the load to be involved in the subsequent grinding process. The load included in the experiment was 5 g each.This particular process was performed using the combined grinding (shearing + hammering) equipment available in the Auto Design Laboratory, Faculty of Mechanical Engineering. In the process, samples would be ground to varying particle sizes through the manipulation of grinding speed, controlled through an inverter. The speeds used in the palm oil waste samples ranged from 15 Hz to 40 Hz, at an interval of 5 Hz. Each of the grinding speeds was repeated twice to allow better results consistency. The samples were first fed into the sampling pipe, then pushed to the blades by using the push piston. After shearing, the samples were allowed to be hammered into finer particles for 30 s. Then, the collection was conducted. All the resulting particles at the specimen collector were placed into either a plastic bag or a container. The machine would be cleaned after each session of grinding to avoid contamination.

3. Analysis

After grinding, the samples were sent for SEM analysis. The analysis focused on the particle sizes randomly chosen at each grinding speed. For the SEM analysis, the preliminary studies involving the palm oil stems were conducted in Central Lab located at Universiti Malaysia Pahang Gambang campus. On the other hand, the SEM for the EFB fibers ground at different grinding speeds were conducted in the Engineering Materials Laboratory, Faculty of Manufacturing Engineering, after the facility was made available in the Pekan campus. As there were limitations to the facility in terms of material conductivity, each sample could only be conducted once. However, the multiple images captured for each sample shall be adequate for analysis. Table 1 shows the equipment used for each process where their respective purpose was highlighted.

Table 1. Equipment used for each process with their purpose.

Process	Equipment	Figure	Purpose
Weighing	Kitchen scale (SF-400)	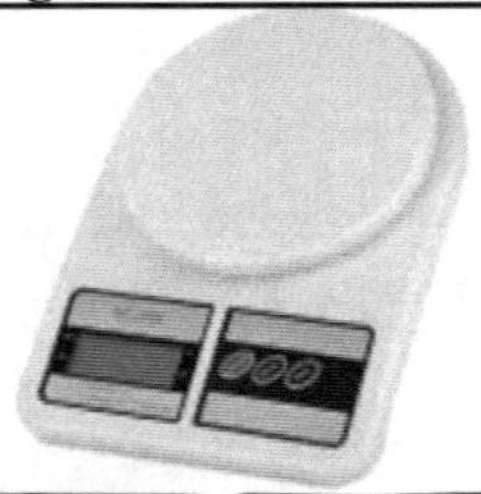	Measuring the weight of samples.
Grinding	Combined grinding machine		Reducing size of samples at manipulated grinding speed and grinding time.
Scanning	Scanning electron microscope (Phenome)		Determining the particle size and extent of bonding at higher magnifications.

4. Results and discussions

After a specimen was grinded, it would be essential to observe the fineness of the powder or ground specimen. It's simply because too fine or coarse powders would in turn affect the binding effects among the particles and subsequently the strength and durability of the products. Before the analysis on size measurements were performed, the preliminary studies to observe the internal structure were first conducted. In preliminary studies, the material involved was the palm oil stem biomass. However, the specific parameters were not taken into consideration; therefore no relationships could be derived yet. Anyhow, the internal structures especially multiple display of surface structure could be observed clearly and acted as the stepping stone for further analysis. Table 2 portrays the SEM outcomes together with the original ground specimens.

Table 2. Preliminary studies for SEM.

Specimen	**Palm oil stems**		
Sample	**Original figure**	**Magnified figure**	**Observation**
1	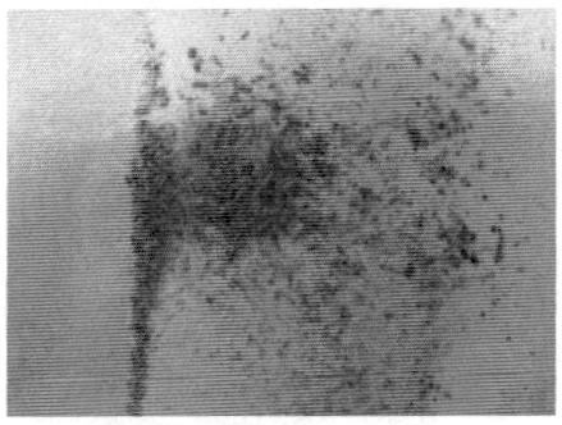	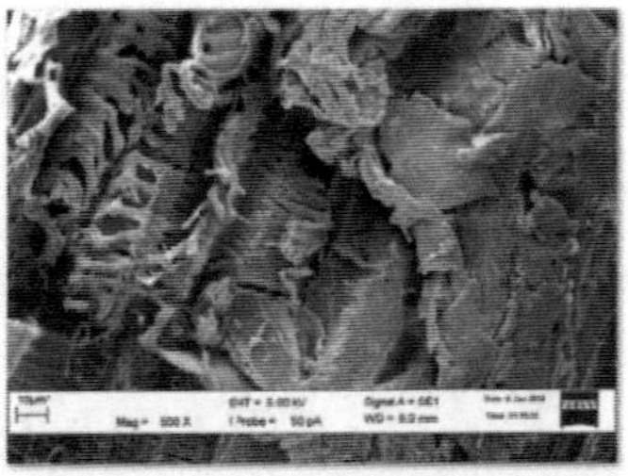	The figure shows the structure similar to the shape of DNA, however, without twisting. This would be one of the features preferred as to encourage better binding.
2		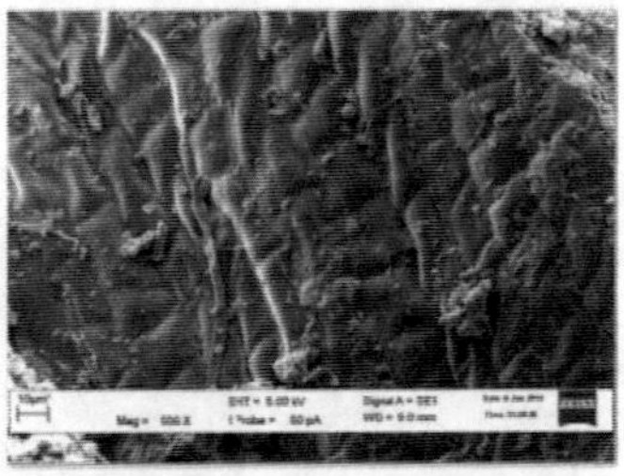	The figure portrays the uneven surface structure of the particles, captured at 500x of magnification power.
3		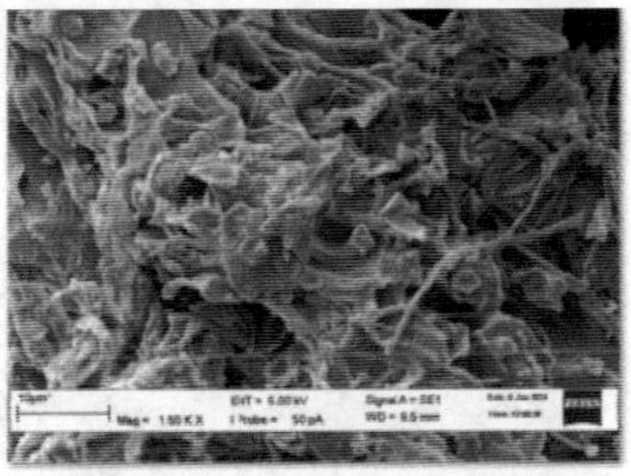	The figure displays the 1500x magnified view of the internal structure of particles with 'thorny' feature for binding.
4		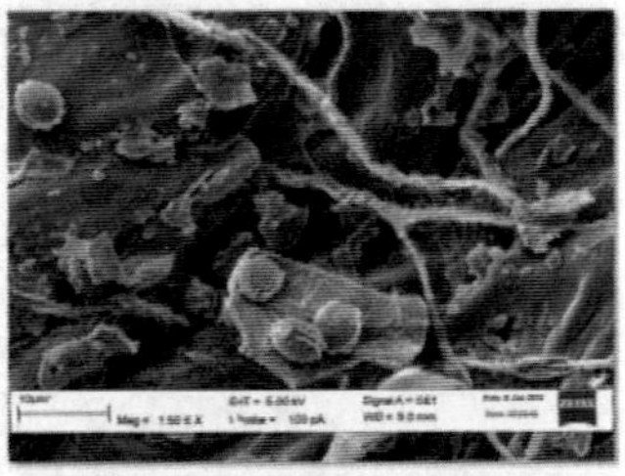	At 1500x magnification, the 'thorny' feature contained in a prtticle could be visible and the feature would aid in the binding between particles.
5		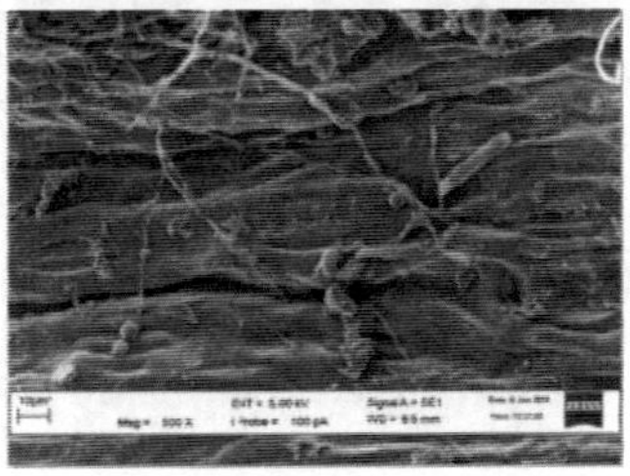	The figure depicts the binding between particles at 500x magnification.

By using the preliminary studies as guidance, it was found that manipulating grinding parameters would bring about different properties, as the binding is closely related to particle size. Therefore, the grinding speed was manipulated, as confirmed by studies in order to figure out the relationship of speed with particle size. The other parameters, i.e. grinding time and load were kept constant at 30 s and 5 g, respectively. For these parameters, only EFB was involved. The outcomes from the subsequent SEM analysis are depicted clearly in table 3.

Table 3. SEM analysis for EFB at different grinding speeds.

Specimen	**Empty fruit bunches (EFB) fibers**			
Grinding speed (Hz)	**Before grinding**	**After grinding**	**Size (µm approx.)**	**Magnified figure**
15	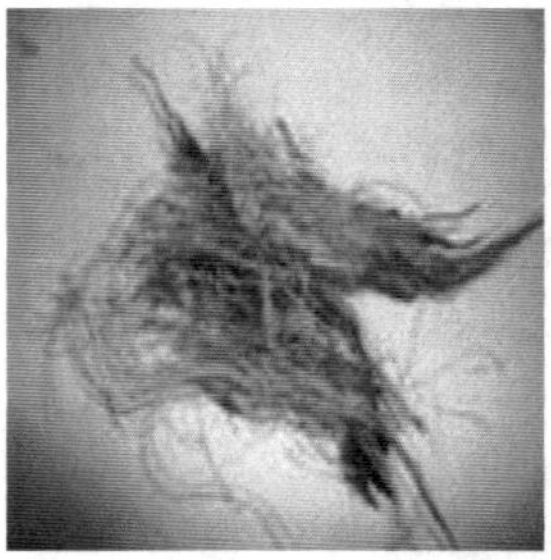		98-480	
20	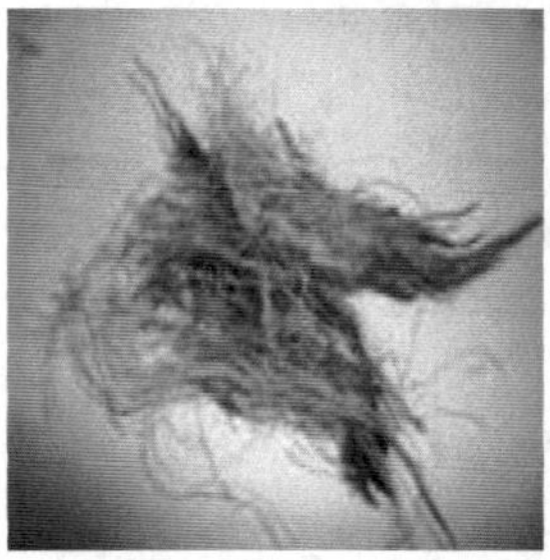		54-565	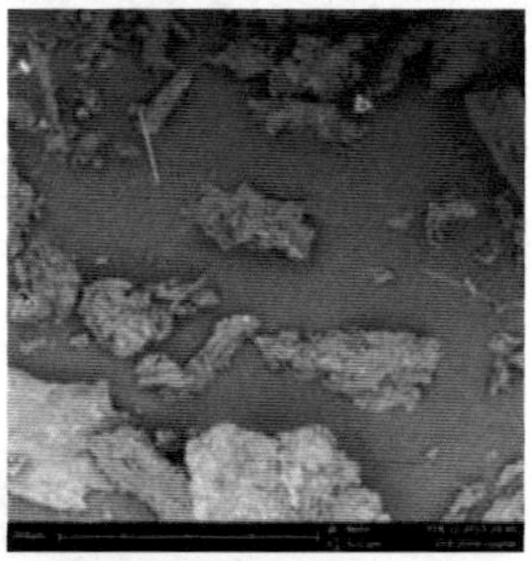
25	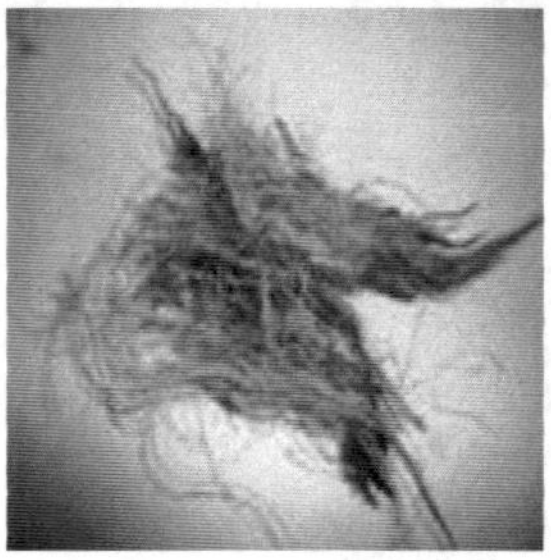		44-400	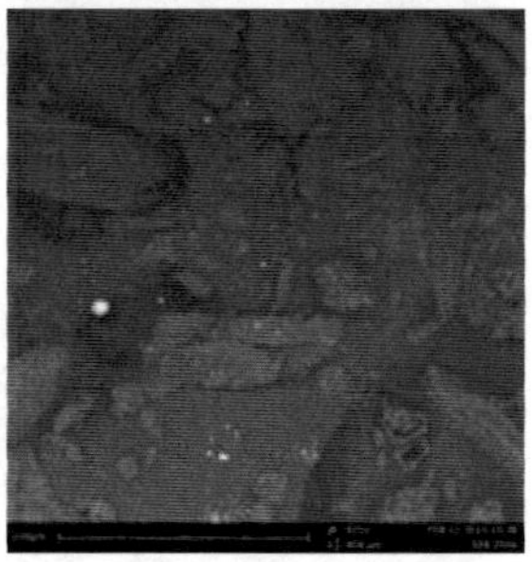

30	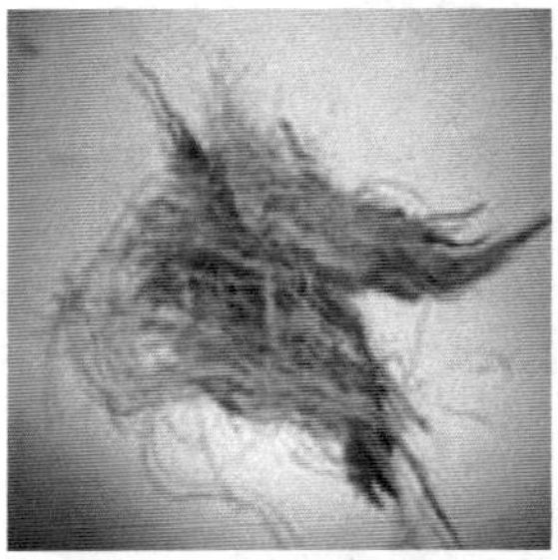		77-340	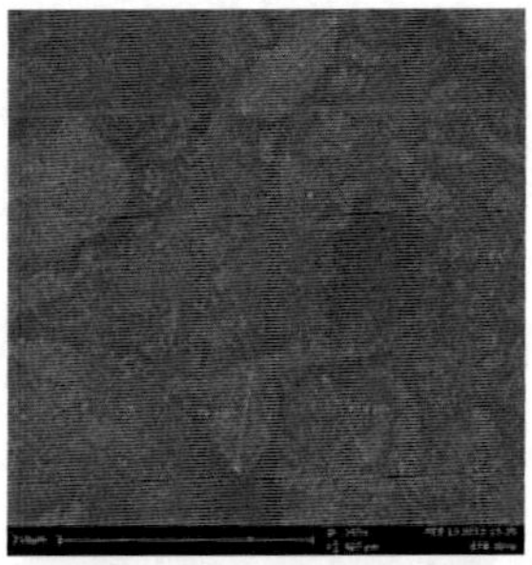
35	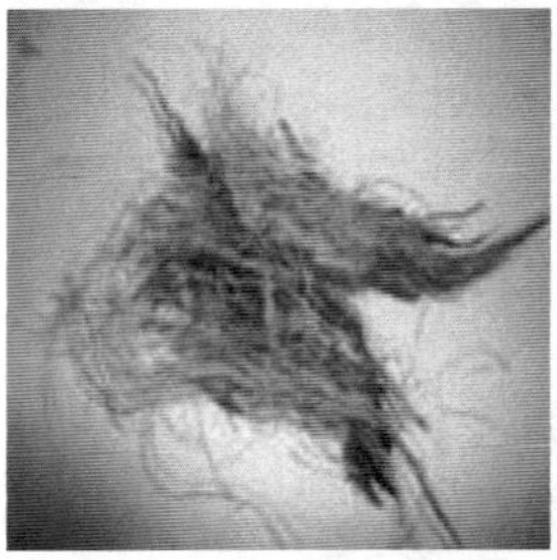		52-331	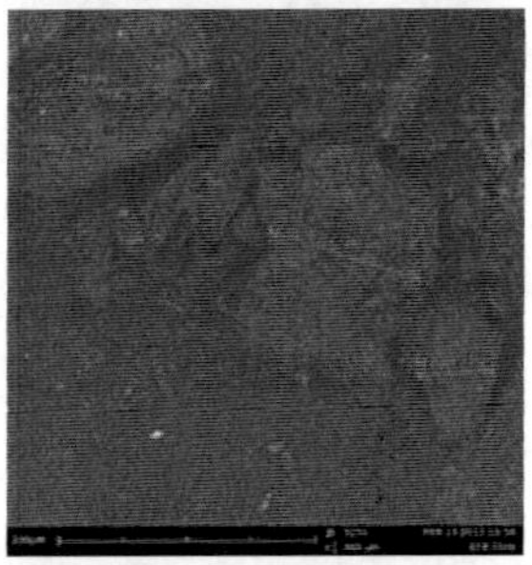
40	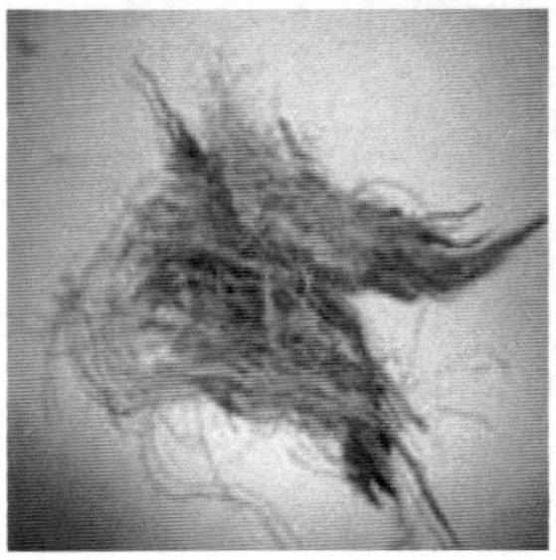		90-214	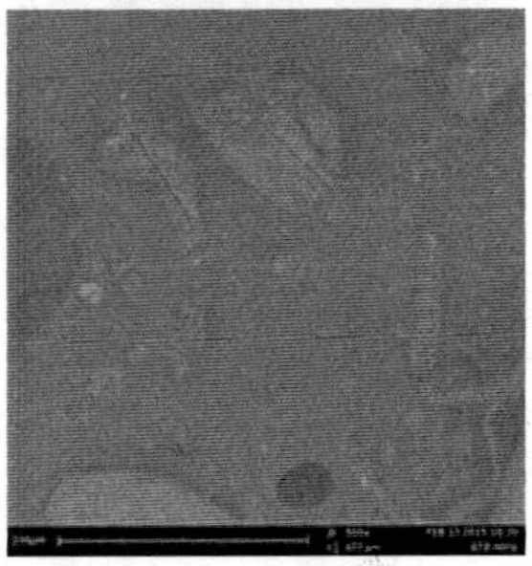

As seen in table 3, grinding at different grinding speeds resulted in a variety of particle size. The size of the particles was measured using Scanning Electron Microscope. At high resolution and magnification, the bonding between the particles could be verified. From the analysis, the relationship between the grinding speed and particle size could be derived. The trend was that as the grinding speed increased, the particle fineness also experienced increment, implying the production of finer particles. The relationship was that the particle was proportional with the grinding speed, as shown in equation (1).

$$\text{Particle size} \ \alpha \ \text{Grinding speed} \qquad (1)$$

In general, the particle sizes ranged from 45 μm to about 600 μm, at the grinding speed ranging from 15 Hz to 40 Hz. The lowest size of about 45 μm was achieved at the grinding speed of 25 Hz, in the case of EFB. Each of the grinding speed was conducted twice and the SEM once due to machine limitation.

Figure 2. A magnified photo to show binding of particles.

Figure 3. Binding performance at 20Hz.

As depicted in figures 2 and 3, the bonding between particles was shown clearly. Of course, from studies, the better the binding, the better the quality of the final products formed. From a lower magnification, the bonding looked encouraging even at low speeds. However, when the magnifications were further increased, it was proven the bindings were not so encouraging at low grinding speeds, as proven in Figure 3. The speeds identified were 15 Hz and 20 Hz. Thus, it could be concluded the optimum grinding speed to produce a good quality of oil palm wastes powder specifically for EFB would be at 25-35 Hz. However, as the tested specimens were chosen at random and due to the limitation of the equipment used for analysis, the accuracy was reduced. This was due to only clearly seen images were captured and measured, indicating forced negligence on ultra-fine particles which were normally not visible clearly even at high resolutions.

5. Conclusions

From the study, it was found that the newly designed grinding machine with strong cutting blades equipped with the functions of shearing and hammering, helped in producing fine particles with features for good binding. The 'thorny' feature was verified through SEM, as shown in the result section. As expected, the particle fineness increased with increasing grinding speed, manipulated at an interval of 5 Hz. The particle sizes ranged from 45 μm to about 600 μm, where the coarsest particle was recorded at the grinding speed of 20Hz and the finest at 40Hz. However, there were recorded limitations of the device which subsequently reduced the results accuracy. The bindings at 15Hz and

20 Hz were recorded to be below than par when they were broken, subsequently reducing the quality of final product formed, in terms of strength in particular. In future, analysis involving SEM shall be applied with appropriate coating as to increase the accuracy and clarity of images. Otherwise, the particle size analysis shall be performed using some other device that could result in high accuracy outcomes. Also, in future studies, the experimentations would involve more variety of samples which are abundant within the country such as rice residues as well as food residues. Then, the relationship between the grinding parameters with resulting particle size could be explored further.

Acknowledgements
The authors would like to be obliged to Universiti Malaysia Pahang for providing laboratory facilities and financial assistance under the project no RDU131401.

References
[1] Miao Z, Grift T E, Ting K 2014
[2] Adapa P, Tabil L, Schoenau G 2011 *biomass and bioenergy* **35** 549-61
[3] Gross R, Leach M, Bauen A 2003 *Environment International* **29** 105-22
[4] Sulaiman F, Abdullah N, Gerhauser H, Shariff A 2011 *Biomass and bioenergy* **35** 3775-86
[5] Umar M S, Jennings P, Urmee T 2013 *Renewable Energy* **60** 107-15
[6] Ong H, Mahlia T, Masjuki H 2011 *Renewable and Sustainable Energy Reviews* **15** 639-47
[7] Ahmad S, Ab Kadir M Z A, Shafie S 2011 *Renewable and Sustainable Energy Reviews* **15** 897-904
[8] Bazmi A A, Zahedi G, Hashim H 2011 *Renewable and Sustainable Energy Reviews* **15** 574-83
[9] Oh T H, Pang S Y, Chua S C 2010 *Renewable and Sustainable Energy Reviews* **14** 1241-52
[10] San Miguel G, Dominguez M, Hernandez M, Sanz-Perez F 2012 *biomass and bioenergy* **47** 134-44
[11] Wengenmayr R, Bührke T 2011 *Renewable energy: Sustainable energy concepts for the future* John Wiley & Sons)

Combustion and emissions characteristics of a compression ignition engine fueled with n-butanol blends

I M Yusri[1], R Mamat[1], O M Ali[1], A Aziz[1], M K Akasyah[1], M K Kamarulzaman[1], C K Ihsan[1], H M Mahmadul[1], and S M Rosdi[1]
[1]Faculty of Mechanical Engineering, Automotive Engineering Centre
Universiti Malaysia Pahang, 26600 Pekan, Pahang, Malaysia

Email: rizalman@ump.edu.my

Abstract. The use of biomass based renewable fuel, n-butanol blends for compression ignition (CI) engine has attracted wide attention due to its superior properties such as better miscibility, higher energy content, and cetane number. In this present study the use of n-butanol 10% blends (Bu10) with diesel fuel has been tested using 4-cylinder, 4-stroke common rail direct injection CI engine to investigate the combustion and emissions of the blended fuels. Based on the tested engine at BMEP=3.5Bar Bu10 fuel indicates lower first and second peak pressure by 5.4% and 2.4% for engine speed 1000rpm and 4.4% and 2.1% for engine speed 2500rpm compared to diesel fuel respectively. Percentage reduction relative to diesel fuel at engine speeds 1000rpm and 2500rpm for Bu10: Exhaust temperature was 7.5% and 5.2% respectively; Nitrogen oxides (NO_x) 73.4% and 11.3% respectively.

1. Introduction

Compression Ignition (CI) engine is a well-known internal combustion engine available in the present day. Generally CI engine is producing higher thermal efficiency compared to spark ignition (SI) engine because of higher compression ratio of the engine and the carbon content of the fuel itself [1]. Unfortunately the pollution emitted by the CI engine usually producing higher nitrogen oxides (NO_x) and soot. In order to meet the stringent emissions regulations, increasing energy demand and depletion of non-renewable fuels the present worldwide research is directed to search for alternatives fuel; alcohol and biodiesel for CI engine.

Alcohol fuels such as methanol (CH_3OH), ethanol (C_2H_5OH), and butanol (C_4H_9OH) can be used with diesel fuels in various percentage blends for CI engine as a clean alternative fuel source. Low percentages of alcohol; 5%, 10% and 15% in diesel fuel blends does not require any modifications to the engine [2]. Study on alcohol fuels blended with standard diesel fuels has been studied extensively on CI engines to observe the engine performance and emissions. However the use of n-butanol fuel is still not widely explored by the researchers.

Butanol is produce by fermentation of biomass; algae, corn and plant materials that contain cellulose. There are four of butanol isomers namely normal butanol, $CH_3CH_2CH_2CH_2OH$ (n-butanol), secondary butanol $CH_3CH_2CHOHCH_3$ (2-butanol), isobutanol $(CH_3)_2CH_2CHOH$ (i-butanol), and ter-butanol $(CH_3)_3COH$ (t-butanol). Each structure of butanol has the same formula and amount of heat of energy. Despite their similarity, they have different solubility properties [3].

Using butanol diesel fuel blends in diesel engines and its effects on engine performance and exhaust emissions have been investigated in several literatures. Yao et al. [4] investigated the effects of butanol ratios (5%, 10% and 15%) by volume in diesel blends on six cylinders diesel engine equipped with common rail injection system. Throughout the results increasing butanol blends reflected to reduction of CO and soot emissions while little increased in BSFC. Rakopoulos et al. [5-7] performend the experimental tests on single-cylinder, compression–ignition, direct injection, naturally aspirated diesel engine. Based on the data, with addition of n-butanol (8, 16 and 24%, by vol.) to diesel fuel increased the BSFC, BTE and HC emissions while significantly decreased CO, NO_x and soot.

Presently there are limited numbers of study on combustion and emission characteristics using n-butanol as an alternative fuel. Thus an effort has been done to investigate the use of n-butanol blends on the water-cooled engine fitted with a high pressure direct fuel injection system from common rail equipped with turbochargers and exhaust gas recirculation (EGR). The engine was tested at engine speed 1000rpm and 2500rpm with single brake mean effective pressure (BMEP) level 3.5Bar.

2. Experimental set up

2.1. Fuel Properties

Compared to the other alcohol kinds, n-butanol has more advantages than methanol and ethanol as fuel substitutions for CI engine. Butanol has a lesser auto-ignition temperature than methanol and ethanol. Thus, butanol can be ignited easier when combusted in the combustion chamber. Moreover Butanol has also a higher cetane number, therefore more suitable fuel blends than ethanol and methanol for diesel fuel. Energy content of the butanol is the highest among the alcohol family thus it released more energy per unit mass. The physical and chemical properties of butanol indicate that it is capable to seize the limitations from low carbon alcohols which are methanol and ethanol [8].

Table 1. Physicochemical properties of butanol and diesel fuels.

Property	Diesel Fuel	Butanol
Research octane number (RON)	15-25	96
Cetane No.	40-55	25
Energy content (Lower heating value) (MJ/Kg)	42.8	33.1
Heat of vaporization (MJ/Kg)	44.8	36.6
Density at 20 °C (g/ml)	0.829	0.8098
Flash Point (°C)	74	35
Auto ignition temperature (°C)	235	397

2.2. Engine setup

The experimental test setup was conducted on a 4-cylinder, 4-stroke CI engine. The engine was a water-cooled, fitted with a high pressure direct fuel injection system from common rail and equipped with turbochargers and EGR. Commercial Diesel fuel produced by Petronas was used as the based fuel and will be referred to as "Diesel". Apart of the base fuel, 10% of n-butanol blended with diesel fuel were tested and referred to as "Bu10". The engine was operated at engine speeds (1000rpm and 2500rpm) and constant BMEP level 3.5Bar. One of the four engine cylinders was attached with a Kistler water cooled piezoelectric transducer (Type 6041A) to measure the in-cylinder pressure of the engine. The pressure transducers were synchronized with kistler cam crank angle encoder type 2713B1 attached to the end crank shaft and the reading is measured by Dewe-5000. The brake torque of the engine was measured with an eddy-current dynamometer model ECB-200F SR No.617 from Dynalec Controls. The emissions of the engine are measured by KANE gas analyzer. Figure 1 shows the schematic diagram of the experimental setup. The specifications of the engine are based on Table 2.

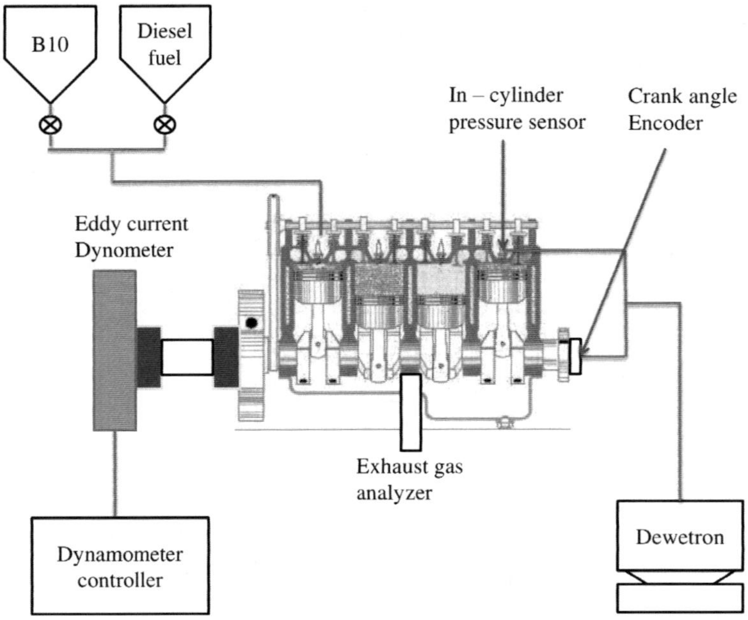

Figure 1: Experimental diagram.

Table 2: Engine specifications.

Engine model	Isuzu 4JJ1
Type	Inline 4 – cylinder
Injection system	Common rail direct injection
Bore x stroke	95.4mm x 104.9mm
Displacement	3.0L
Compression ratio	17.5 to 1
Max power at 2500 rpm	61kW
Max torque at 1800 rpm	280Nm

3. Results and discussion

3.1. Combustion characteristics

The in-cylinder pressure profile of a 4-cylinder, 4-stroke common rail direct injection CI engine are presented. Figure 2 depicts the combustion profile at engine speed 1000rpm with constant BMEP=3.5Bar. The graph denoted as the scale graph of in-cylinder pressure in the range -60° to 60° CA. The circle indicates the specified area of the analysis at peak combustion.

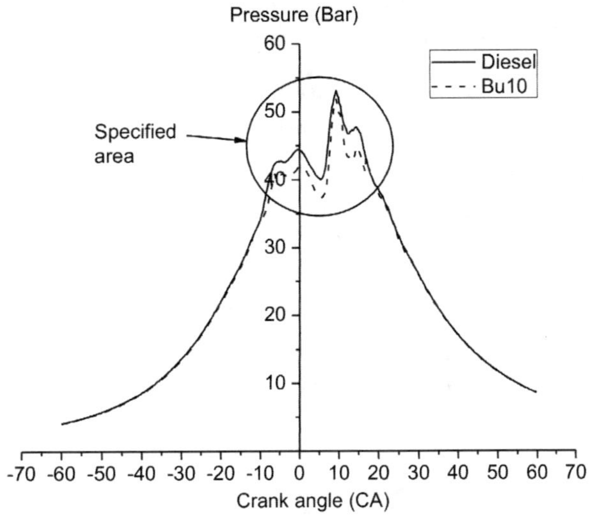

Figure 2. In-cylinder pressure at engine speed 1000rpm.

Figure 3 shows the specified graph with smaller scale in the range of -20° to 20°CA. Based on the graph two stage of in-cylinder peak pressure can be observed under 1000rpm engine speed at constant BMEP=3.5Bar. The resulted peak is reflected by the pilot and main injection strategy of the engine behavior. During first stage, the peak in-cylinder pressure decreased as the torque increased to a high load conditions. Bu10 indicates lower first and second peak in-cylinder pressure by 5.4% and 2.4% respectively compared to diesel fuel. This phenomenon is due to the lower auto ignition of the fuel properties [9]. The cetane number of the blended fuel decreases as 10% of n-butanol was mixed with diesel fuel. Thus, less fuel combusted at the first and second stage of combustion when more n-butanol is blended hence reflected to lower heat release for both stages.

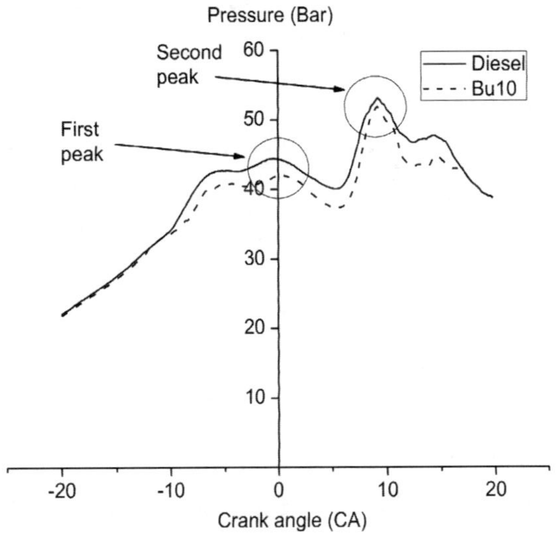

Figure 3. Focus area of peak pressure at engine speed 1000rpm.

Figure 4 shows the in-cylinder pressure at engine speed 2500 with constant BMEP=3.5Bar. The graph indicates scale of combustion profile in the range of -60° to 60° CA. The circle indicates the specified area of the analysis at peak combustion. The in-cylinder pressure is directly proportionally to the engine speed, thus increase of in-cylinder peak pressure can be observed.

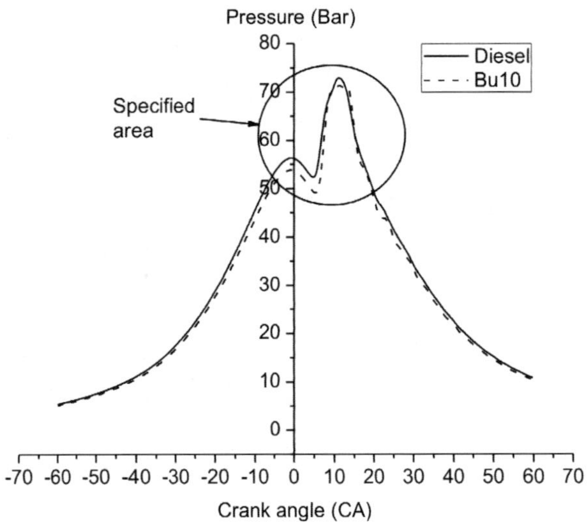

Figure 4: In-cylinder pressure at engine speed 2500rpm.

Figure 5 denoted as the smaller scale of peak combustion in the range of -20 to 20° CA. The combustion profile shows similar trend of injection strategy. Bu10 indicates 4.4% and 2.1% of reduction for both first and second peak in-cylinder pressure respectively.

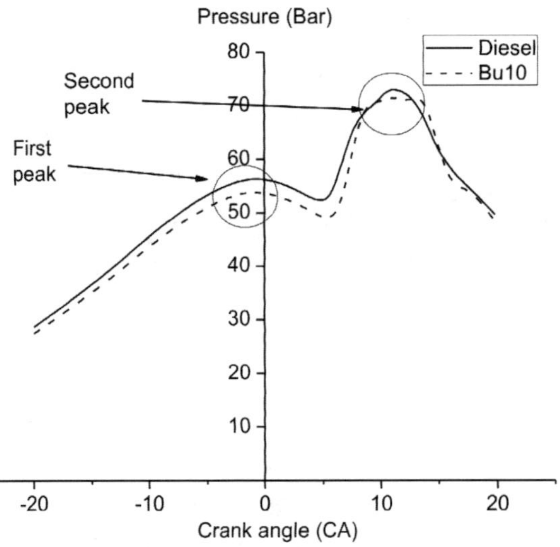

Figure 5: Focus area of peak pressure at engine speed 2500rpm.

3.2. Emissions characteristics

Figure 6 shows the effect of n-butanol/diesel fuel blends on exhaust temperature at engine speeds 1000rpm and 2500rpm with constant BMEP=3.5Bar. It was observed that n-butanol/diesel fuel blends resulted to lower exhaust temperature than diesel fuel by 7.5% and 5.2% at engine speeds 1000rpm and 2500rpm respectively. This is due to the lower energy content and the cetane number of n-butanol/diesel fuel blends [9, 10].

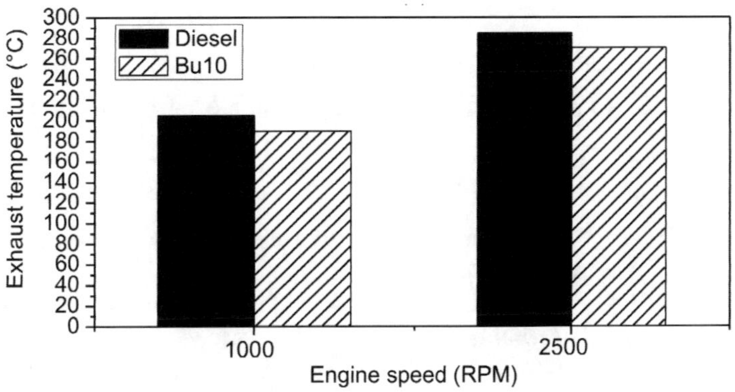

Figure 6: Exhaust temperature at engine speed 1000 and 2500rpm.

Figure 7 shows NO_x emissions at engine speeds 1000rpm and 2500rpm with constant BMEP=3.5Bar. It was observed that NO_x emissions decreased at engine speeds 1000rpm and 2500rpm by 73.4% and 11.3% than diesel fuel respectively. The emissions of NO_x strongly related to in-cylinder temperature during combustion. The mixture of n-butanol/diesel fuel blends lead to lower combustion temperature due to lower heating value and oxygen content of n-butanol fuel properties [11, 12].

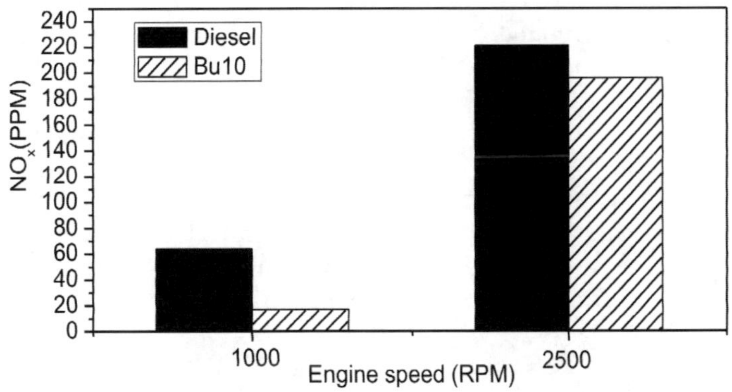

Figure 7: NO_x emissions at engine speed 1000 and 2500rpm.

4. Conclusion

As for the conclusion, the influences of 10% n-butanol blend with diesel fuel on combustion and emissions characteristics were investigated under two different engine speeds (1000rpm and 2500rpm) with constant BMEP=3.5Bar. The main results can be summarized as follows.

(i) Combustion characteristics of n-butanol/diesel fuel blends indicates lower first and second peak pressure by 5.4% and 2.4% for engine speed 1000rpm and 4.4% and 2.1% for engine speed 2500rpm compared to diesel fuel respectively.

(ii) Exhaust temperature of n-butanol/diesel fuel blends are reduced significantly by 7.5% and 11.3% for both engine speeds 1000rpm and 2500rpm compared to diesel fuel respectively.

(iii) Reduction of NO_x emissions using n-butanol/diesel fuel blends by 73.4% and 11.3% for both engine speeds 1000rpm and 2500rpm compared to diesel fuel respectively.

Acknowledgements

Appreciation and acknowledgement to the Ministry of Higher Education (KPT) for providing author the scholarship under My Brain 15 scheme and financial support from Universiti Malaysia Pahang grant.

References

[1] Karabektas M and Hosoz M Performance and emission characteristics of a diesel engine using isobutanol–diesel fuel blends 2009 *Renewable Energy* **34** 1554-1559

[2] Van Stam J, Kronberg B, Golubkov I, and Hull A Alternative Fuel for a Standard Diesel Engine 2006 *International Journal of Engine Research,* **7** 51-63

[3] Doğan O The influence of n-butanol / diesel fuel blends utilization on a small diesel engine performance and emissions 2011 *Fuel* **90** 2467-2472

[4] Yao M, Wang H, Zheng Z, and Yue Y Experimental study of n-butanol additive and multi-injection on HD diesel engine performance and emissions 2010 *Fuel* **89** 2191-2201,.

[5] Rakopoulos DC, Rakopoulos CD, Giakoumis EG, Papagiannakis RG, and Kyritsis DC Influence of properties of various common bio-fuels on the combustion and emission characteristics of high-speed DI (direct injection) diesel engine: Vegetable oil, bio-diesel, ethanol, n-butanol, diethyl ether 2014 *Energy* **73** 354-366,

[6] Rakopoulos DC, Rakopoulos CD, Giakoumis EG, Dimaratos AM, and Kyritsis DC Effects of butanol–diesel fuel blends on the performance and emissions of a high-speed DI diesel engine 2010 *Energy Conversion and Management* **51** 1989-1997

[7] Rakopoulos DC, Rakopoulos CD, Giakoumis EG, and Dimaratos AM Characteristics of performance and emissions in high-speed direct injection diesel engine fueled with diethyl ether / diesel fuel blends 2012 *Energy* **43** 214-224

[8] Kumar S, Cho JH, Park J, and Moon I Advances in diesel–alcohol blends and their effects on the performance and emissions of diesel engines 2013 *Renewable and Sustainable Energy Reviews* **22** 46-72

[9] Chen G, Yu W, Li Q, and Huang Z Effects of n - Butanol Addition on the Performance and Emissions of a Turbocharged Common Rail Diesel Engine 2012 *SAE International*

[10] Valentino G, Iannuzzi S, and Corcione FE Experimental Investigation on the Combustion and Emissions of a Light Duty Diesel Engine Fuelled with Butanol-Diesel Blend 2013 *SAE International*

[11] Siwale L, Kristóf L, Adam T, Bereczky A, Mbarawa M, Penninger A, *et al.*, "Combustion and emission characteristics of n-butanol / diesel fuel blend in a turbo-charged compression ignition engine 2013 *Fuel* **107** 409-418

[12] Siwale L, Kristóf L, Bereczky A, Mbarawa M, and Kolesnikov A Performance, combustion and emission characteristics of n-butanol additive in methanol–gasoline blend fired in a naturally-aspirated spark ignition engine 2014 *Fuel Processing Technology* **118** 318-326

[13] Ismail MY, Alimin AJ, and Osman SA Mono-Gas Fuelled Engine Performance and Emissions Simulation Using GT-Power 2013 *Applied Mechanics and Materials* **465-466** 125-129

[14] Chen G, Shen Y, Zhang Q, Yao M, Zheng Z, and Liu H Experimental study on combustion and emission characteristics of a diesel engine fueled with 2,5-dimethylfuran–diesel, n-butanol–diesel and gasoline–diesel blends 2013 *Energy* **54** 333-342

[15] Chen Z, Wu Z, Liu J, and Lee C Combustion and emissions characteristics of high n-butanol / diesel ratio blend in a heavy-duty diesel engine and EGR impact 2014 *Energy Conversion and Management* **78** 787-795

[16] Lujaji F, Kristóf L, Bereczky A, and Mbarawa M, "Experimental investigation of fuel properties, engine performance, combustion and emissions of blends containing croton oil, butanol, and diesel on a CI engine 2011 *Fuel* **90** 505-510
[17] Rakopoulos CD, Rakopoulos DC, Giakoumis EG, and Kyritsis DC The combustion of n-butanol / diesel fuel blends and its cyclic variability in a direct injection diesel engine 2011 *Proceedings of the Institution of Mechanical Engineers, Part A: Journal of Power and Energy* **225** 289-308

The influence of the equivalent hydraulic diameter on the pressure drop prediction of annular test section

A H H Al-Kayiem[1] and M A Ibrahim[1]
[1]Mechanical Engineering Department, Universiti Teknologi PETRONAS, 32610 Bandar Seri Iskandar, Perak, Malaysia

E-mail: hussain_kayiem@petronas.com.my

Abstract. The flow behaviour and the pressure drop throughout an annular flow test section was investigated in order to evaluate and justify the reliability of experimental flow loop for wax deposition studies. The specific objective of the present paper is to assess and highlight the influence of the equivalent diameter method on the analysis of the hydrodynamic behaviour of the flow and the pressure drop throughout the annular test section. The test section has annular shape of 3 m length with three flow passages, namely; outer thermal control jacket, oil annular flow and inner pipe flow of a coolant. The oil annular flow has internal and external diameters of 0.0422 m and 0.0801 m, respectively. Oil was re-circulated in the annular passage while a cold water-glycol mixture was re-circulated in the inner pipe counter currently to the oil flow. The experiments were carried out at oil Reynolds number range of 2000 to 17000, covering laminar, transition and turbulent flow regimes. Four different methods of equivalent diameter of the annulus have been considered in this hydraulic analysis. The correction factor model for frictional pressure drop was also considered in the investigations. All methods addressed the high deviation of the prediction from the experimental data, which justified the need of a suitable pressure prediction correlation for the annular test section. The conventional hydraulic diameter method is a convenient substitute for characterizing physical dimension of a non-circular duct, and it leads to fairly good correlation between turbulent fluid flow and heat transfer characteristic of annular ducts.

1. Introduction

A commonly accepted approach to extend the pipe flow equations to annular geometry is by modification of the diameter. The equations that are used in the pipe flow to calculate the friction factor (FF) and Reynolds number (Re) are being used in annular flow by replacing the diameter, as the characteristic geometry in pipe flow equations, by the effective diameter. Different effective diameter definitions have been proposed in the literature both for Newtonian and non-Newtonian fluids. Re and FF relationship is unique to a given effective diameter definition, De. Different values of frictional pressure drop might be obtained according to the De used in characteristic geometry. Several studies have also been proposed to modify the Re definition with shape factor while continuing to use the classical approach of equivalent diameter. Since the results from these studies have significant deviations from each other, the selection of appropriate correlation for the respective fluid and flow regime becomes very important for the accurate estimation of pressure losses in annuli Lawn and Elliott [1].

Dosunmu and Shah [2] developed a turbulent friction factor correlation that includes wall roughness for straight pipe and annular flow. A systematic approach is adopted in which pipe and annular pressures loss data for two fluids (xanthan gum and guar gum) are used. The first pressure loss data with xanthan gum have been used for correlation development while the second set of data with guar gum have been used for the correlation evaluation. They recommendation that the hydraulic diameter definition of equivalent diameter for an eccentric annulus provided the best annular definition for friction pressure prediction.

In addition to the hydraulic diameter customarily used and accepted as "equivalent" diameter for the annular concentric ducts, there are many other equivalent diameter correlations suggested in the literature. In the past, annuli have been evaluated on the basis of hydraulic diameter as equivalent diameter, but this may not always be the best way to represent the dimension for flow in an annulus. Lawn and Elliott [1], and Jones and Leung [3] recommended the need of further studies to explore the influence of different equivalent diameters in the annular flows. Anifowoshe and Osisanya [4] tested seven different equivalent diameters definitions to estimate their effect on the pressure losses in wellbore hydraulics. The predicted results have been compared with experimental data to select the best model to fit the pressure losses prediction of power low models. The concluded that the pressure losses estimation is significantly affected by the equivalent diameter model. They suggested that the hydraulic diameter definition provided the best estimation of the pressure losses for the power law fluid in laminar flow conditions.

The objective of the present paper is to demonstrate and discuss the pressure drop results, which are predicted theoretically and measured experimentally, for the annulus oil flow using various equivalent diameter models. This effort is part of a hydraulic characterization and correction of unjustified flow loop for wax deposition studies, and to introduce a correction method to adjust the pressure drop prediction. In addition to the commonly used hydraulic diameter model, another three models have been checked, and the pressure drop correction factor model. The investigated region covers all the flow regimes, laminar, transition, and turbulent with Re ranging from 2000 to 17000. The results assisted in understanding the influencing of the equivalent diameter on the prediction of the pressure drop in the deposition section.

2. Experimental apparatus

The experimental set-up used for this study, shown in figure 1, is a high pressure/high temperature (HPHT) flow loop, developed for wax deposition research of waxy crudes. The main test section has an annular flow shape of 3 m long. The test section is fully jacketed with hot glycol-water mixture re-circulated in the heating system of the HPHT flow loop from the hot fluid reservoir. While a cold glycol-water mixture is re-circulated in the central pipe and counter-currently to the oil flowing in the annular space. The aim of the cold mixture is to provide a cold conditions for the flowing oil to impose the waxing deposition.

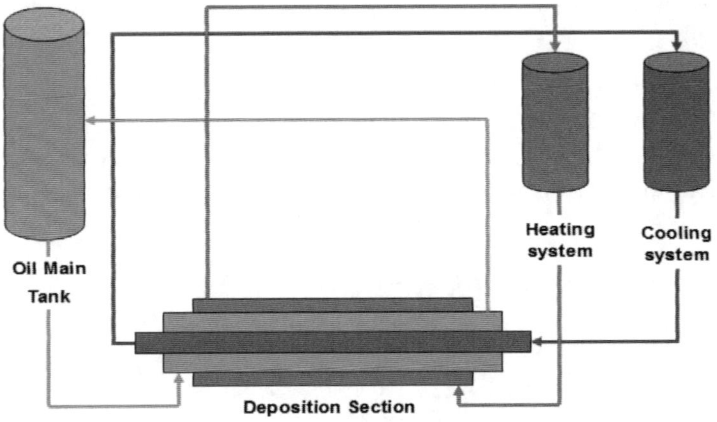

Figure 1. Schematic of the wax deposition test loop in the HPHT flow loop.

The oil is re-circulated through the experimental flow loop by 3″ × 3 stage gear pump. The flowrate and density of oil are measured by MICROMOTION flow meter installed downstream the pump. The oil enters and leaves the test section vertically from 2 inch pipe.

A Rosemount differential pressure transducer is installed across the oil section to measure the pressure drop. The inlet of the oil section is equipped by pressure transmitter. The heating jacket temperature, oil temperature and cold glycol-water mixture temperature are monitored at the inlet and outlet of the test section by type-T (copper-constantan) thermocouple probes.

3. Equivalent diameter approaches for non-circular pipe flow

Prediction of Re, FF, and the pressure loses, Δp may differ according to the use of different De. In this contest, several studies have also been proposed to modify the Re definition shape factor while continuing to use the classical approach of equivalent diameter. Since the results from these studies have significant deviations from each other, the selection of appropriate correlation for the respective fluid and flow regime becomes very important for accurate estimation of pressure losses in annuli, Lawn and Elliott [1].

2.1. The Hydraulic Diameter Method, Dh

The common representation of diameter for flow in the non-circular conduits is the hydraulic diameter defined as:

$$D_h = \frac{4A}{P} \tag{1}$$

where A is the flow area and P is the perimeter. For annular configuration, D_h becomes,

$$D_h = d_2 - d_1$$

In addition to the hydraulic diameter customarily given by equation (1) and accepted as "equivalent" diameter for the annular concentric ducts [5], there are many equivalent diameter correlations also suggested in the literature. These correlations have been checked against the hydraulic diameter to extend the criteria of the pressure drop characterization. By defining the inner and outer diameters of the annulus by d_1 and d_2, respectively, the three methods defining the annulus equivalent diameter are as below.

2.1.1. The Petroleum Diameter Method, De, p. One of the suggested ways of calculating the pressure drop of a non-circular, concentric- annular flow cross-section named "The Petroleum Engineering Method" uses equation (2) for evaluating an equivalent calculation diameter to replace the circular pipe diameter Bertuzzi F [6].

$$D_{e,p} = \sqrt[5]{(d_2 + d_1)^2 \times (d_2 - d_1)^3} \tag{2}$$

2.1.2. The Equivalent Area Diameter Method, De, a. Equation (3), simply, is assuming an equivalent diameter of a circular pipe having the same flow area. The "equivalent area" diameter for annulus is calculated as:

$$D_{e,a} = \left(d_2^2 - d_1^2\right)^{1/2} \tag{3}$$

2.1.3. The Equivalent Geometry Diameter Method, De, g. Another criteria used to obtain an equivalent circular diameter from comparing the geometry in the pressure-loss equations for pipe flow and concentric annulus flow of Newtonian fluid terms is the geometry terms is the "equivalent geometry" diameter given by Bertuzzi F [6]:

$$D_{e,g} = \sqrt{d_2^2 + d_1^2 - \frac{d_2^2 - d_1^2}{\ln(d_2 / d_1)}} \tag{4}$$

2.2. The "Correction Factor" Model, knon-c

Another method for predicting the frictional pressure drop for flow in a concentric annular conduits (long pipe stabilized flow assumption) suggests the use of a correction coefficient for non-circular passages, $k_{non\text{-}c}$. This factor is to be separately evaluated for laminar and turbulent flow conditions and applied to the frictional pressure drop calculations with the aid of a circular pipe standard correlations but having a diameter equal to the hydraulic diameter, ($D_h = d_2 - d_1$) of annular conduit Idelchik and Fried [7] . For a rounded annular tube, the correction factor, which is a function of the diameter ratio, can be found for ($Re \leq 2000$) from:

$$k_{non-c,lam} = \frac{1 - \left(\frac{d_1}{d_2}\right)^2}{1 + \left(\frac{d_1}{d_2}\right)^2 + \left[1 - \left(\frac{d_1}{d_2}\right)^2\right] \Big/ \ln\left(\frac{d_1}{d_2}\right)^2} \tag{5}$$

Equation (6), obtained by Jones & Leung (3), is also proposed to predict the correction factor for the laminar flow in the annular passage, as:

$$k_{non-c,lam} = \frac{(d_2 - d_1)^2 \cdot (d_2 - d_1)^2}{(d_2^4 - d_1^4) - (d_2^2 - d_1^2)^2 / \ln(d_1 / d_2)} \tag{6}$$

Both equation (5) and (6) are not a function of Re, they are only function of the diameters of the annular flow conduits.

In the case of turbulent flow, $k_{non\text{-}c}$ depends only slightly on the diameter ratio and lies in the range 1.0 – 1.07. The correction factor of such a tube can be calculated from the following formula.

$$k_{non-c,turb} = f\left(\frac{0.02 d_1}{d_2} + 0.98\right)\left(\frac{1}{f} - 0.27\frac{d_1}{d_2} + 0.1\right) \tag{7}$$

where, d_1 and d_2 are the inner and outer diameters of the annulus passage.

According to the handbook of Hydraulic Resistance [7], the actual frictional pressure drop in a stabilized "long pipe assumption" flow in an annular conduits should be calculated as a product of a correction coefficient and the frictional pressure drop calculated for the hydraulic diameter of the annular conduits.

IOP Conf. Series: Materials Science and Engineering **100** (2015) 012049 doi:10.1088/1757-899X/100/1/012049

4. Results and discussion

In order to decide the appropriate model of equivalent diameter, the predicted pressure drop in the test section is compared to the measured pressure drop. A Rosemount differential pressure transducer is installed across the oil section to monitor the pressure drop. The pressure drop is predicted using Darcy equation assuming Newtonian, incompressible and steady flow in the annulus,

$$\Delta P = f \frac{L}{D_e} \frac{\rho U^2}{2} \tag{8}$$

Darcy friction factor, f is estimated using Churchill's equation. The Churchill equation combines expressions for friction factor in both, the laminar and the turbulent flow regimes. It is accurate to within the error of the data used to construct the Moody diagram. This equation also provides an estimate for the transition region. The Churchill equation shows very good agreement with the Darcy equation for laminar flow. In the turbulent regime a difference of around 0.5-2% is observed between the Churchill equation and the Colebrook equation, Co [8].

$$\left.\begin{aligned} f &= 8\left[\left(\frac{8}{\mathrm{Re}}\right)^{12} + \frac{1}{(A+B)^{1.5}}\right]^{1/12} \\ A &= \left[2.45\ln\frac{1}{\left(\frac{7}{\mathrm{Re}}\right)^{0.9} + 0.27\frac{e}{D_e}}\right]^{16} \\ B &= \left(\frac{37530}{\mathrm{Re}}\right)^{16} \end{aligned}\right\} \tag{9}$$

Re is based on the relevant equivalent diameter, and e is the surface roughness. Knowing that the annulus inner and outer diameters are 0.0422 and 0.0801 m, respectively, the values of the equivalent diameters, D_e are obtained from the four models using the set of equation (1) to (4). The calculated values of the equivalent diameters are shown in table 1.

4.1. Results of the Pressure Drop using Equivalent Diameter Models

Figure 2 displays the measured and predicted pressure drop across the annulus versus Reynolds number. The actual Re calculated from the measurement in the annulus is ranging from 3000 to 9400. The predicted ranges of Re are extended to 17000 to examine the capacity of each model in simulating the turbulence regime. The results of pressure drop obtained by using the various equivalent diameter models are compared with the measured results. The equivalent diameter has significant effect on the turbulent flow regime length and the pressure drop. Different trends of the pressure drop could be observed for different equivalent diameters used in the prediction. In case of the equivalent area method, which has the lowest values of the pressure drop estimation, the turbulence region is over estimated compared to the realistic estimation. The length of the turbulent flow regime decrease with the increase of the value of the equivalent diameter. For this point, care must be taken in the selection of the appropriate equivalent diameter which can give a good representation for the flow phenomenon in the annular passage. As it is clear in the figure 2, all the models are underestimated the pressure drop compared to the measured values. In fact, deeper investigation has demonstrated that the models are correct if the losses at the inlet and outlet of the annulus test section are taken into consideration. This will be elaborated further in the coming sections.

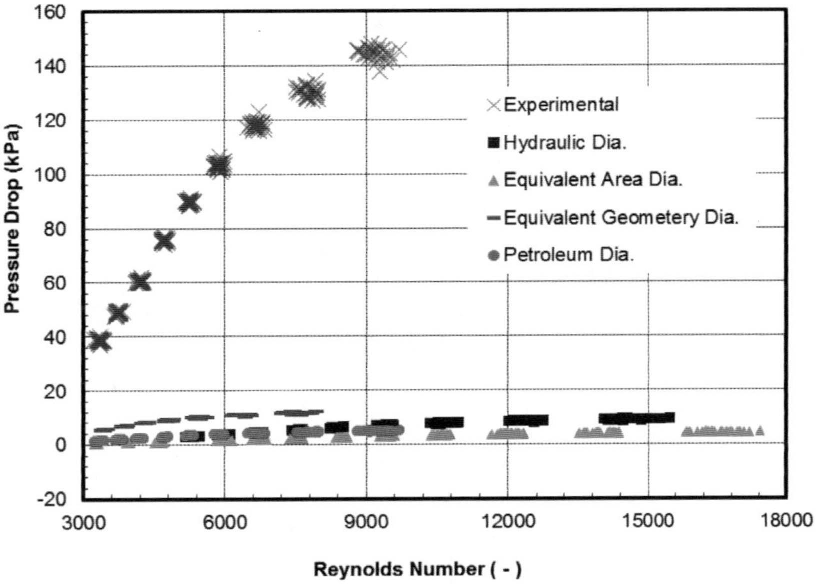

Figure 2. The effect of the various equivalent diameter models on the prediction of the pressure drop at various turbulent regimes.

Figure 3 shows the regression analysis for the calculated and measured pressure drop using different equivalent diameter models. The "equivalent area" model shows the biggest deviation of 96% between the measured and calculated pressure drop, and the values of the measured pressure drop are 31 times the calculated values with standard error of 3.8. The "equivalent geometry" model showed the smallest deviation of 91% and the measured pressure drop is 11 times the calculated pressure drop with standard error of 3.99.

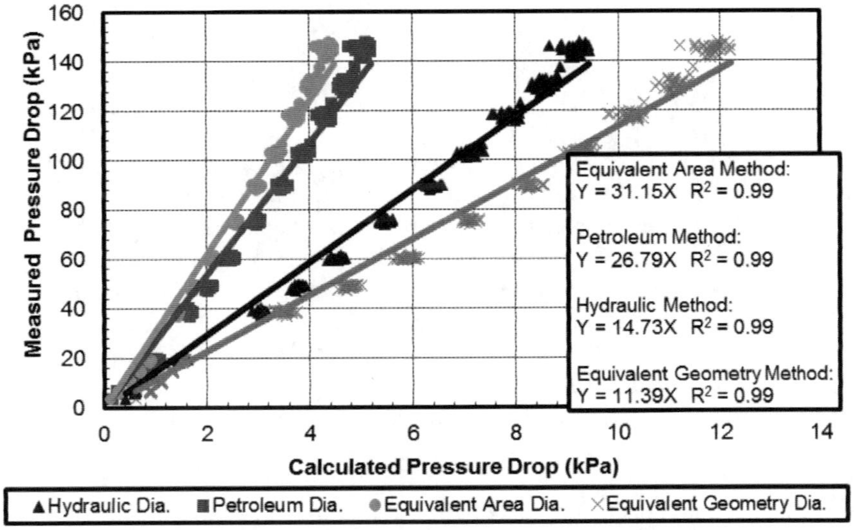

Figure 3. Calculated versus measured pressure drop and linear regressions; the effect of various equivalent diameter models used for the annulus.

4.2. Results of the Pressure Drop using Correction Factor Model

For the deposition apparatus under consideration in this study, the correction factor for laminar flow, $k_{non\text{-}c,\ lam}$, according to equation (8), is equal to 1.49. The same value is obtained by using equation (9). For turbulent flow, the correction coefficient, $k_{non\text{-}c,\ turb}$, is dependent on Re number. This dependency is found to be week for Re ranging from 2000 to 10000, $k_{non\text{-}c,\ turb} = 1.058 - 1.060$. This is because of the presence of the transition zone and low turbulence level in this range of Re. Figure 4 illustrates the correction factor effect on the calculation of the pressure drop. The use of Jones & Leung (3) correction factor reduces the root mean square error between the measured and calculated values by 2%, while the standard deviation from the measured values remains the same about 44.33 with or without J-L correction. Therefore the correction factor produces insignificant (positive) effect for improving the calculation routine. However, the correction factor method should be considered for "polishing" calculation results when the hydraulic diameter approximation is used, (3, 5, 7).

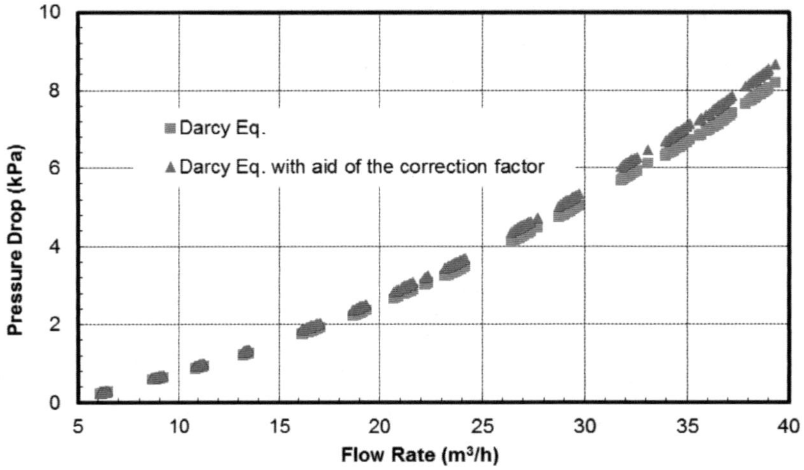

Figure 4. The effect of the correction factor on the calculation of the pressure drop.

Although the equivalent geometry model, equation (4), slightly reduces the discrepancy between the measured and calculated pressure drop results compared to other equivalent diameters, the hydraulic diameter will be used as standard for the calculations and further analysis. This decision is justified by the following reasons:

- The hydraulic diameter is a convenient substitute for the characteristic physical dimension of a non-circular duct, and it leads to fairly good correlation between turbulent fluid flow and heat transfer characteristic of circular and noncircular ducts. The hydraulic diameter is also used for ducts involving laminar flow to provide a consistent basis of comparison with turbulent flow results (1, 3, 4). However, for laminar flow itself, this quantity does not lead to satisfactory correlations between circular and noncircular ducts, Idelchik et al. (7).
- Using a general correlation for the equivalent length in not advised as it demonstrated that the equivalent length is not sufficient for accurate description of the observed behavior. No existing correlation or method has yet found with general acceptance Jones and Leung (3).

5. Conclusion

Annular flow test section, designed originally for wax deposition studies, is investigated experimentally and theoretically to introduce acceptable hydraulic simulation model. Four models, and also the pressure drop correction factor model, have been investigated. The followings are concluded.

- The equivalent diameter models have significant influence on the estimation results of the pressure drop in annuli.

- The four models of the equivalent diameters studied in the present work show different trend in estimating the turbulence regions in the annulus.
- The equivalent geometry diameter model under-estimate the turbulence level, while the other models over-estimated the turbulence level.
- The hydraulic diameter is reasonably estimated the turbulence level in the annulus.

However, the theoretically obtained results of the frictional pressure losses have shown large deviation compared to the measurement results. Accordingly, a correction method is necessary to reduce the mathematical simulation error of the frictional pressure losses of oil flow in the annular test section in PRSB HPHT flow loop.

Acknowledgement
The authors acknowledge Universiti Teknologi PETRONAS, (UTP) for the financial support to carry out the project. In particular, the second author expresses his appreciation to UTP for sponsoring his study under the GA scheme. Thanks are due to PRSB/PETRONAS for providing the HPHT flow loop apparatus and for the logistic support to carry out the experimental measurement.

References
[1] Lawn C, Elliott C 1972 *Journal of Mechanical Engineering Science* **14** 195-204
[2] Dosunmu I T, Shah S N 2013 *Journal of Petroleum Science and Engineering* **109** 80-6
[3] Jones O C, Leung J 1981 *Journal of Fluids Engineering* **103** 615-23
[4] Anifowoshe O L, Osisanya S O, The Effect of Equivalent Diameter Definitions on Frictional Pressure Loss Estimation in an Annulus With Pipe Rotation. SPE Deepwater Drilling and Completions Conference; 2012: Society of Petroleum Engineers.
[5] Davidson P 2015 *Turbulence: an introduction for scientists and engineers* Oxford University Press, USA)
[6] Bertuzzi F F M J, Poettmann F H and Thomas L K Wellbore Wellbore Hydraulics Petroleum Engineering Handbook. Richardson TX USA: Society of Petroleum Engineers 1987.
[7] Idelchik I E, Fried E 1986
[8] Co C 1988 *Flow of fluids through valves, fittings, and pipe* Crane Co.)

Design consideration for design a flat and ring plastics part using Solidworks software

M A M Amran[1], K M Faizal[1], M S Salleh[1], M A Sulaiman[1], and E Mohamad[1]
[1]Faculty of Manufacturing Engineering, Universiti Teknikal Malaysia Melaka, 76100 Durian Tunggal, Melaka, Malaysia

E-mail: mohdamran@utem.edu.my

Abstract. Various considerations on design of plastic injection moulded parts were applied in initial stage to prevent any defects of end products. Therefore, the objective of this project is to design the plastic injection moulded part by taking consideration on several factors such as draft angle, corner radius and location of gate. In this project, flat plastic part, ring plastic part, core inserts for flat and ring plastic part were designed using SolidWorks software. The plastic part was drawn in sketching mode then the 3D modeling of solid part was generated using various commands. Considerations of plastic part such as draft angle and corner radius with location of gate was considered in the design stage. Finally, it was successfully designed the two plastic parts with their respectively insert by using SolidWorks software. The flat plastic part and ring plastic part were designed for the purpose for future researches for study the weld lines, meld lines, air trapped and geometrical size of the product. Thus, by designing the flat plastic part and ring plastic part having core insert on each part, the completed mould design of two plate mould can be considered. This is because, plastic injection parts are needed to be designed properly in order to neglect any defect when the mould was made.

1. Introduction

Over recent years, there are many types of Computer Aided Design (CAD) software available around the world. These softwares lead the imagination and creativity of engineers to interpret and deliver their good ideas using this kind of medium. Then, others people can understand well what and how they want to execute their ideas. CAD softwares are used to increase the productivity of the designer, improve the quality of design, improve communications through documentation, and to create a database for manufacturing [1]. CAD software outputs are often in the form of electronic files for print, machining, or other manufacturing operations [2].

In mechanical design it is known as mechanical Design Automation (MDA) or Computer-Aided Drafting (CAD), which includes the process of creating a technical drawing with the use of computer software [3]. CAD softwares for mechanical design use either vector-based graphics to describe the objects of traditional drafting, or may also produce raster graphics showing the overall appearance of designed objects. However, it involves more than just shapes. As in the manual drafting of technical and engineering drawings, the output of CAD must convey information, such as materials, processes, dimensions, and tolerances. There are various types of CAD softwares available in the market nowadays such as Solidworks, AutoCAD, CATIA, Pro Engineering, etc., not to mention with various capability to analyze the after design product.

In designing and producing the plastic injection molding parts, it is hard by only using the conventional method and by not using Computer Aided Design (CAD) softwares. However, creating or designing the part is not enough, injection molding parts are affected by many factor such as corner design, draft angles, the location of the edge gate and those must be consider in order to get the best design without neglecting the main purpose of the part itself. Sharp corner will block the resin flow when molding, which will become a cause of the flow marks. Also, the stress will concentrate on the sharp corner part, so it likely to be the cause of deterioration in strength such as notch effect and residual strain. In this project, two plastic parts has been selected for design. Selections of two types of these plastic parts are based on the wall thickness, shape, and its suitability in producing a plastic injection mold for use in the future research. In future research, plastic parts that are designed in this project, are taken into consideration the plastic flow that can cause product defects such as weld lines, meld lines, air trapped and geometrical size of the product.

2. Methodology

2.1. Experimental Procedure

There are several steps involving plastics part design from sketching until 3D modeling in order to achieve the finish completed 3D solid modeling. Figure 1 shows the flow chart in order to design the plastic part. In this process, Solidworks 2013 software was used to design the flat and ring plastic part.

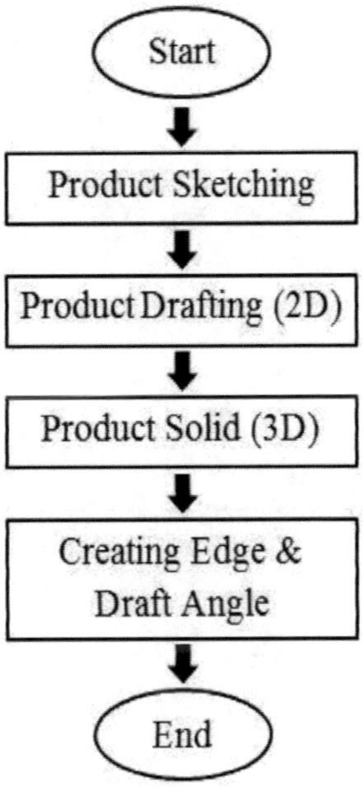

Figure 1. Flow chart to design 3D solid plastic part.

2.1.1. 2D Drawing. The processes started by having the product sketching, in this stage, the idea on the basic structure of the part were drawn on a piece of paper in order to overview the part. The second process was the product drafting processing, this was made by using the CAD software to generate the Two Dimensional (2D) drawing as shown in figure 2.

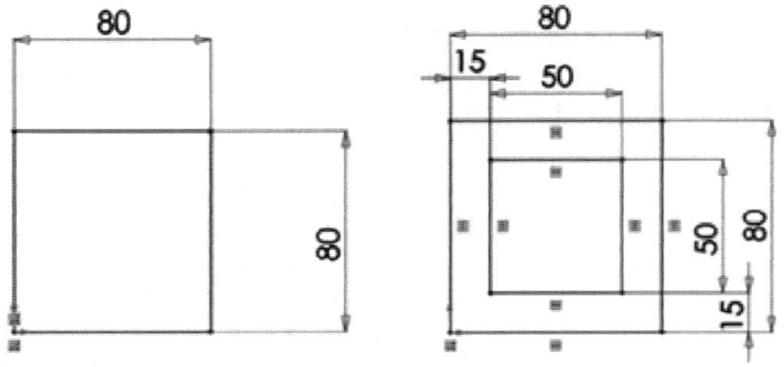

Figure 2. 2D drafting of flat and ring part.

2.1.2. 3D Solid Modeling. In this stage, all drawings must be given dimension in order to constrain the shape from changing. The third step in designing the plastic part was extruding, in other CAD software this step also called padding, and this was done in order to create Three Dimensional Solid Modeling (3D Modeling). Figure 3 shows the extruded phase. In the extruded command, the height or thickness of the part should be decided through recommendation wall thickness of the plastic part.

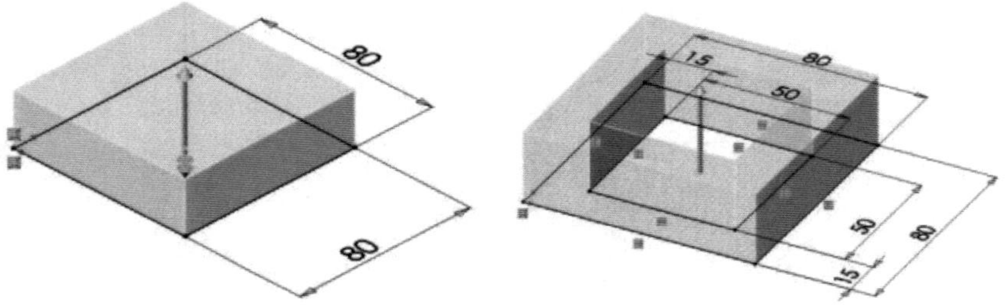

Figure 3. Extruded process for flat and ring part.

The next step in order to made the plastic part easily removal from the mould, it needed to create the draft angle and filleting at the selected edge around the corner of the plastic part. Figure 4 shows the drafting step takes place where the draft surface and normal surface that is need to be chosen.

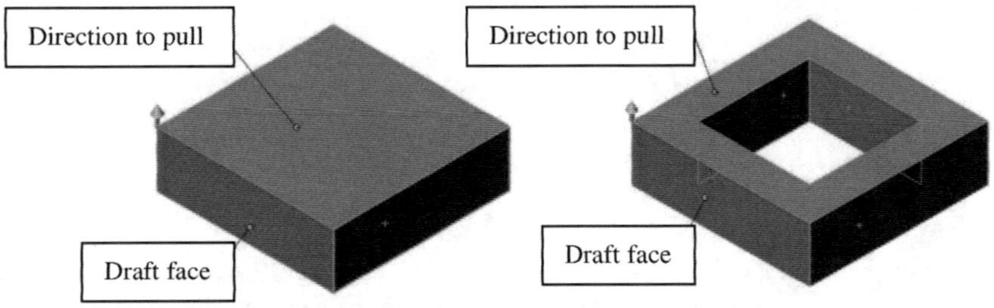

Figure 4. Draft surface and normal surface in drafting.

The purple surface was the face that the product dimension will not change and the draft surfaces were the blue color. Each face has to be selected separately and the draft angle can be dimensioned to the desired value. Figure 5 shows the cross section of the part with 5° draft angles were indicated by arrow at wall face of plastic parts.

Figure 5. Draft angle 5° at the flat and ring part.

The next step was filleting the desired edge or in some CAD software it was called fillet radius, figure 6 shows the filleting steps with 5mm radius at four corners on the flat plastic parts and eight corner on the ring plastic parts. Fillet can be made on the desired edge and in this case of plastic part, the edge in figure 6 shows the selected edge by the yellow mark.

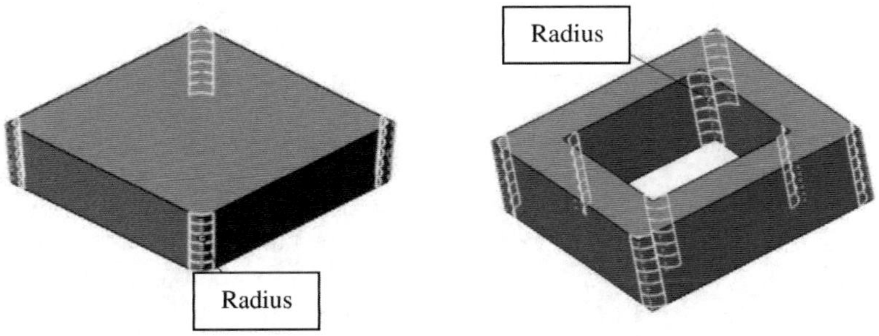

Figure 6. Fillet on the desired edge on flat and ring part.

2.1.3. Completed Part 3D Solid Modeling. Finally, the designs for both plastic parts were completed. Figure 7 shows the completed flat and ring plastic part. Location of gate on the completed plastic parts will be decided when the structure of plastic injection mould was decided. There two types of plastic injection mould, where two plate mold used side gate system and three plate mould used pin point gate system.

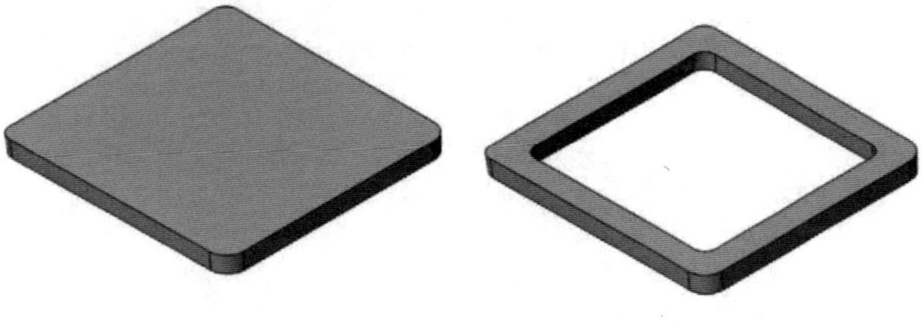

Figure 7. Flat plastic part and ring plastic part.

3. Steps for Modeling Core Insert for Flat Plastic Part

Starting from 2D drawing, there are several command and steps such as 2D drafting, following by extruding to make the main solid body and also some other command such as drafting and filleting that is need to be done in order to complete the 3D solid modeling for certain part that is designed.

There are many types of command in SolidWorks that is not covered in this paper however it is still be used in order to model other part of the mould. Figure 8 shows the core insert that is work as cavity in the plastic injection mould and table 1 shows the steps and command that is used in SolidWorks.

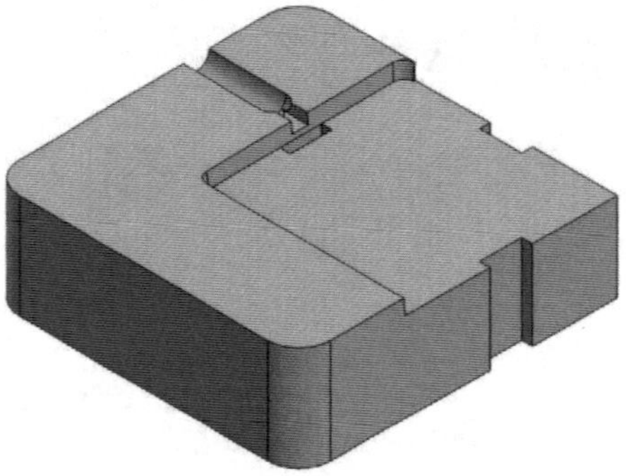

Figure 8. 3D model of core insert for flat plastic part.

Table 1. Steps to design core insert.

	Command	Descriptions
Step 1	Sketch	2D drawing for base (60x60 mm)
Step 2	Extrude	Create 3D solid modeling (20 mm)
Step 3	Sketch	2D drawing on top plane for cavity
Step 4	Extrude cut	Cut the area for cavity draft angle 5° (40x40x3 mm)
Step 5	Sketch	2D drawing on cavity plane for flat ejector pin (8x2 mm)
Step 6	Extrude cut	Cut the area for flat ejector pin
Step 7	Sketch	2D drawing on top plane for runner (2.5x15 mm)
Step 8	Cut revolve	Cut the area for runner in semi-cylindrical
Step 9	Sketch	2D drawing for gate on top plane (3x5x2 mm)
Step 10	Extrude cut	Cut the area for gate
Step 13	Fillet	Filleting the edge of cavity area (R2 mm)
Step 14	Fillet	Filleting the edge of main body (R6 mm)

4. Steps for Modeling Core Insert Ring for Ring Plastic Part

The other part is the core insert ring where ring plastic part is made using this insert. Figure 9 shows the core insert ring while table 2 shows the steps for designing the core insert ring.

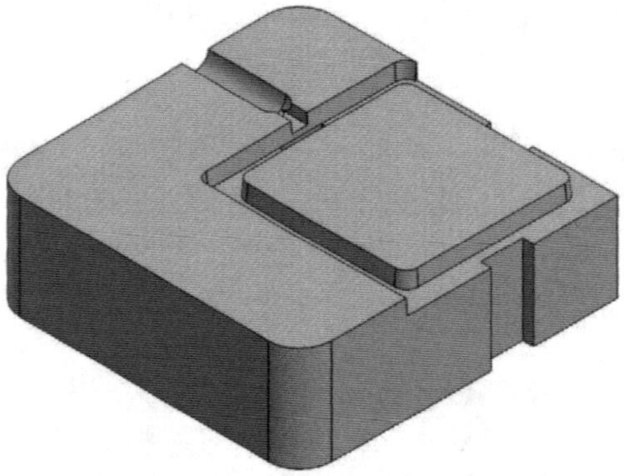

Figure 9. 3D model of core insert ring for ring plastic part.

Table 2. Steps to design core insert ring.

	Command	Descriptions
Step 1	Sketch	2D drawing for base (60x60 mm)
Step 2	Extrude	Create 3D solid modeling (20 mm)
Step 3	Sketch	2D drawing on top plane for cavity
Step 4	Extrude cut	Cut the area for cavity draft angle 5° (40x40x3 mm)
Step 5	Sketch	2D drawing on cavity plane for flat ejector pin (8x2 mm)
Step 6	Extrude cut	Cut the area for flat ejector pin
Step 7	Sketch	2D drawing on top plane for runner (2.5x15 mm)
Step 8	Cut revolve	Cut the area for runner in semi-cylindrical
Step 9	Sketch	2D drawing for gate on top plane (3x5x2 mm)
Step 10	Extrude cut	Cut the area for gate
Step 11	Sketch	2D drawing on cavity plane for ring shape
Step 12	Extrude	Create 3D solid modelling with draft angle 5° (32x32x3 mm)
Step 13	Fillet	Filleting the edge of cavity area (R2 mm)
Step 14	Fillet	Filleting the edge of main body (R6 mm)

Noted that several steps are same because the difference between this two core insert only on step 11 and step 12. The steps for designing the part by using CAD Software are not constraints to these following steps in the table. The designer may choose any command that is appropriate to what they want to do first. However the first two steps are must because it is the basic to form the main 3D solid before modifying it to the desired design.

5. Discussion

Good design is important for any manufactured product but for plastics it is absolutely vital. There are so many considerations to be look of in order to get good quality finish product. However, according to the two main plastic parts that introduced, the consideration about the good and bad corner design, the stress concentration factor and also the wall thickness are to be look.

5.1. Corner Design

Walls usually meet at right angles, at the corners of a box for example. Where the box walls meet the base, the angle will generally be slightly more than 90° because of a draft angle on the walls. However, this can causes two difficulties. The first difficulty is that the increase in thickness at the corner breaks the rule of uniform wall thickness. The maximum thickness at a sharp corner is about 1.4 times the nominal wall thickness shows in figure 10. The result is a longer cooling time accompanied by a risk of sink marks and warping due to differential shrinkage [4].

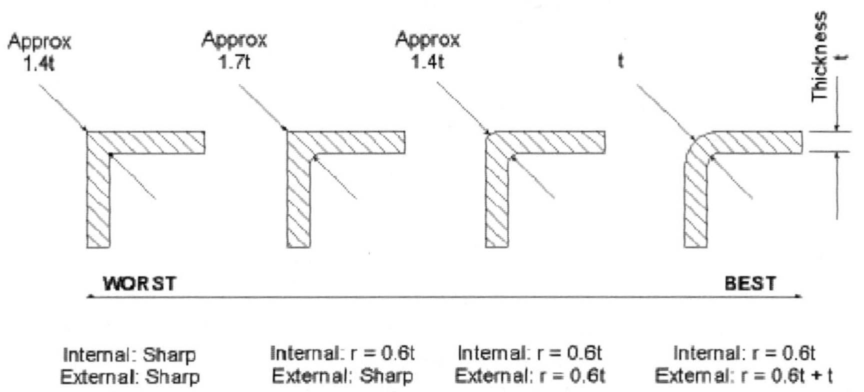

Figure 10. Good and bad corner design.

The other difficulty is even more serious. Sharp corners concentrate stress and greatly increase the risk of the part failing in service.

Plastics are said to be notch-sensitive because of their marked tendency to break at sharp corners. This happens because the stress concentration at the corner is sufficient to initiate a microscopic crack which spreads right through the wall to cause total failure of the part. Sharp internal corners and notches are the single most common cause of mechanical failure in moulded parts [5]. There is some solution for this problems, first is to avoid sharp internal corners. Second is to make sure that internal radii should be at least 0.5 and preferably 0.6 to 0.75 times the wall thickness. On the other hand is to keep corner wall thickness as close as possible to the nominal wall thickness. Ideally, external radii should be equal to the internal radii plus the wall thickness [6].

By properly designed corners will make a big difference to the quality, strength and dimensional accuracy of a moulding parts such as smooth curved corners will help plastic flow in the mould by reducing pressure drops in the cavity and minimising flow-front break-up.

5.2. Ribs

When the normal wall thickness is not stiff enough or strong enough to stand up to service conditions, the part should be strengthened by adding ribs rather than making the whole wall thicker. The principle is the same with the 'I' and 'T' beam where they are almost as rigid as solid beams but are only a fraction of the weight and cost.

Ribs are used to improve the rigidity of a plastics part without increasing the wall thickness so much that it becomes unsuitable for injection moulding. Ribs are important in the design of plastics parts because they allow us to make a component rigid without making it too thick. The usual rule is to make the gap at least twice the nominal wall thickness and preferably three times or more [7]. The consideration is that, these ribs must not be too close together. This is because the gap between the ribs is produced by an upstanding core in the mould. If this core is too thin it becomes very difficult to cool and there may also be a shrinkage effect that will cause ejection problems. Figure 11 shows example of ribs on the plastic parts.

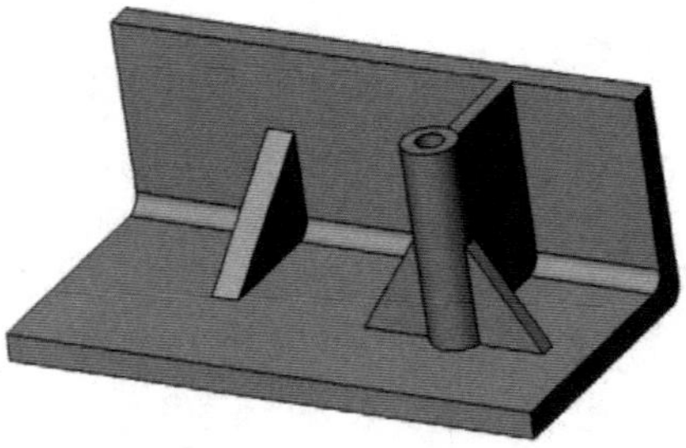

Figure 11. Example of ribs on the plastic parts.

5.3. Wall Thickness

Moulded plastics do not lend themselves to solid forms. The reason is, plastics are processed with heat but they are poor conductors of heat. This means that thick sections take a very long time to cool and so are costly to make. The problems posed by shrinkage are equally severe. During cooling, plastics undergo a volume reduction. In thick sections, this either causes the surface of the part to cave in to form an unsightly sink mark, or produces an internal void. Furthermore, plastics materials are expensive, it is only high-speed production methods and net-shape forming that makes mouldings viable. Thick sections waste material and are simply uneconomic.

Solid shapes usually do the job well in wood or metal usually transformed to a 'shell' form in plastics. This is done by hollowing out or 'coring' thick parts so the thin walls part can be made however, in order to achieve a strong part by only thin wall impossible, it must be reinforce by ribs. In practice, there must be some variation in thickness to accommodate and fulfil functions or aesthetics. It is very important to keep this variation to a minimum. Figure 12 shows the gradual transition between thick and thin wall sections by considering reducing the defects. A plastics part with thickness variations will experience differing rates of cooling and shrinkage. The result is likely to be a part that is warped and distorted, where close tolerances become impossible to hold on the other hand, variations in thickness are unavoidable [8].

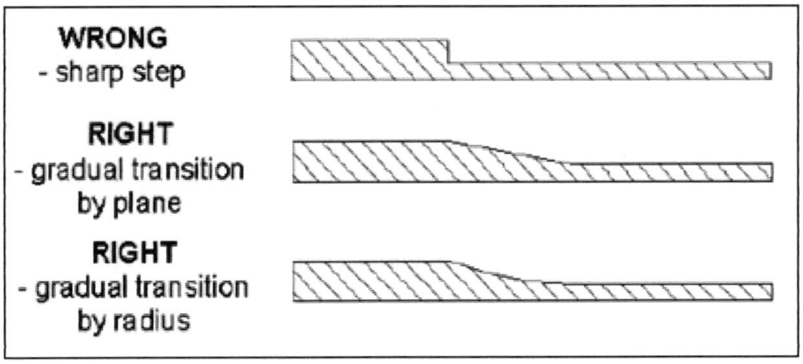

Figure 12. Gradual transitions between thick and thin sections.

6. Conclusion

The design consideration for designing a flat and ring plastic part using SolidWorks software was carried out. This project was successfully design the flat plastic part, ring plastic part, core insert for flat plastic part and ring plastic part using SolidWorks software. Flat plastic part was added external draft angle at the four edges meanwhile the ring plastic part was added to internal and external draft angle with both angle of 5°. The edge gate location was considered for flat plastic part and ring plastic part. By taking the consideration to design the plastic part, the new completed of two plate injection mould will be proposed.

Acknowledgements

The authors would like to thank Faculty of Manufacturing Engineering, Universiti Teknikal Malaysia Melaka for providing facilities for this research to be conducted successfully. This research is funded by the Ministry of Higher Education through Fundamental Research Grant Scheme (FRGS/1/2014/TK01/FKP/02/F00224).

References

[1] Rao K M, Narayan K L 2008 *Computer Aided Design and Manufacturing* PHI Learning)
[2] Farin G E, Hoschek J, Kim M-S 2002 *Handbook of computer aided geometric design* Elsevier)
[3] Madsen D A, Madsen D P, Turpin J L 2012 *Engineering drawing & design* (Clifton Park, NY: Delmar, Cengage Learning)
[4] Suzuki M, Ali M A M, Okamoto K, Taniike T, Terano M, Yamaguchi M 2009 *Advances in Polymer Technology* **28** 185-91
[5] Amran M, Izamshah R, Hadzley M, Shahir M, Amri M, Sanusi M, et al. 2015 *Applied Mechanics and Materials* vol 695 Trans Tech Publ) p. 709-12

[6] Amran M A, Hadzley M, Amri S, Izamshah R, Hassan A, Samsi S, et al. 2007 *International Conference on Advancement of Materials and Nanotechnology:(ICAMN—2007)* (Langkawi Island, Malaysia) vol 1217 AIP Publishing) p. 309-13
[7] Amran M, Salmah S, Faiz A, Izamshah R, Hadzley M, Manshoor B, et al. 2015 *Applied Mechanics and Materials* vol 699 Trans Tech Publ) p. 20-5
[8] Amran M, Salmah S, Zaki M, Izamshah R, Hadzley M, Kasim M S, et al. 2014 *Advanced Materials Research* vol 903 Trans Tech Publ) p. 61-6

Size effects in thin sheet metal forming

T A Chang[1], A R Razali[1], N A I Zainudin[1] and W L Yap[1]
[1]Faculty of Mechanical Engineering, Universiti Malaysia Pahang, 26600 Pekan, Pahang, Malaysia

E-mail: cta_wen@hotmail.com

Abstract. Negligible factors in bulk materials, such as grain-size effects, have proven inappropriate to be neglected for micro-forming processes. Studies had shown that material behaviour varies greatly with the increasing of the scale in the micro-forming world. Therefore, in every micro-forming-related process, especially in micro-stamping, studies and analyses of each material used for the process have to be considered as indispensable in order to be able to understand their behaviour and to be able to correlate their behaviour with the process. Uniaxial tensile-testing experiments have been carried out to determine the strip's properties, behaviour and its correlation with the feeding process in micro-stamping/micro-sheet-forming application. Based on the results of the uniaxial tensile-test experiments conducted, the flow stress was found to decrease with the decrease of the strip thickness and vice versa, due to the size/scale effect. A surface model was used to explain the findings.

1. Introduction

Desire for better quality of life, good health and high working efficiency has been one of major drives to the innovation of many products hence, new products models were invented, for instance netbook, handheld computers and cars, emerging of new products such as smart mobile phones, MP3 player, ultra-thin flat-screen displays, as well as new medical instruments/implants. Dramatic changes of the global, economic development maps and demand for global market of the said micro-products during last 15 years have significantly influenced how the manufacturing is organized and implemented. It was estimated that micro-products industry through evolvement of miniaturization concept would bring more than 1.5-3.5billion dollars trade within five years [1-4].

Nevertheless, differ from conventional manufacturing process which is seen more mature and stable, theory of micro-manufacturing process is seen immature enough for industrial transformation or even 24/7 application. A lot of issues have been addressed by researchers around the globe covering the manufacturing methods and processes [5-13]. Not only issues on the machineries and tooling aspect to be tackled and dealt accordingly, challenging issues such as material behaviors also have to be well-versed in order to guarantee total success.

When feature size of a part is reduced smaller than 1mm, a so-called size effect came up and made the knowledge in terms of empirical and analytical know-how of conventional forming process cannot be directly applied in micro-forming world. A lot of efforts has been done and proved that material behavior in micro-scale is different from macro-scale. The purpose of this research is to study and understand the influence of the size effects on material behavior. Then the material behavior is discussed with a view to the micro-stamping material handling process.

2. Equipment and Materials

2.1. Tensile Testing Machine
Universal testing machine made by Instron was used to measure the material's mechanical properties. The machine has various ranges of working tension and compression load; 0-2.5, 0-5, 0-10, 0-25, 0-50kN with resolution of 1N. Maximum actuator stroke is at ±50mm with actuation speed of up to 5000mm/min. The screw grips measures 60 mm wide and 55 mm deep and can hold material as thick up to 10mm and as thin as micron range thickness.

2.2. Electron Backscattered-Diffraction (EBSD)
Efforts made elsewhere [14-16] have revealed that the use of EBSD is proven as an excellent tool for quantitative metallographic. The effort also has revealed that capability of EBSD analysis which is not only limited to grain size determination, but also there are a number of important microstructural parameters may be determined through the analysis which is not obtainable from any conventional light microscopy.

EBSD analysis also may reveal more number of grain count compared to conventional light microscopy methods. This in turn leads to better microstructural analysis range that covers analysis from the smallest to the biggest grain size. Consequently, this leads to better and more accurate grain size measurement, deviation and aspect ratio.

2.3. Tensile Testing Specimen
Tension test first require the preparation of a test specimen typically from sheet coil. The specimens were prepared according to ASTM E8M specification and cut by using wire cut process. This cutting process is favored in order to avoid burry edges which may have potential to initiate crack during testing. Three types of specimens were used, 50μm and 100μm thick carbon steel and 50μm stainless steel strips and prepared with the similar size to one another.

3. Procedures
Uniaxial tensile test is the simplest and most convenient method and is used to determine material properties for particular temperature and strain-rate specifications. Tensile tests are mostly used to test the materials which are to be sued in mechanical structures in which the materials deflect elastically or undergo low levels of plastic deformation. Mechanical properties analyses on thin sheet metal had received vast studies worldwide covering some popular engineering material [4, 17]. 15 specimens of each strip (50μm and 100μm thick carbon steel and 50μm thick stainless steel strips) were tested and stress-strain curve for each strip was plotted and discussed.

Apart of determination of mechanical properties through uniaxial testing which was conducted on each strip, measurement of grain structure is also important because grain size does have an influence on the material mechanical properties [17]. Therefore electron backscattered-diffraction analysis was conducted on each strip with a view to determine grain orientation, boundaries, shapes and sizes. The EBSD system used was an Oxford Instruments/HKL Nordlys detector. The polished samples were mounted at 70degrees to the horizontal and were examined in a Zeiss Supra 40 FEGSEM at 20kV with a beam current of about 2nA.

4. Results
Brittle materials such as 50μm and 100μm thick carbon-steels do not have a yield point, and do not strain-harden. This means that both ultimate and breaking strength of each material are the same. The stress–strain curves for both the thin and thick carbon-steel strips are shown in figure 1(a) and figure 1(b). Typical brittle materials such as these carbon-steel strips show little plastic deformation. One of the characteristics of the failure of carbon-steel strips failure is that the two broken parts can be reassembled to produce the same shape as that of the original component.

From the figures, it was obvious that the flow stress–strain curves vary greatly with the sheet thickness. For carbon-steel strips, with the decrease of the thickness, flow-stress decreases. The testing also revealed that less deformation was observed where none of the tested carbon-steel strips had significant plasticity deformation, hence no region of obvious plastic deformation was observed. The strips, however, were believed to fracture during elastic deformation, which was indicated by an almost linear stress–strain curve.

Differently to the case for stainless-steel strip (SS50) in figure 1(c), significant plastic deformation was observed in the given recorded stress–strain curves, as shown in figure 1(c) the strip was seen to be more ductile and possibly plastically-deformed before fracture. However, less force was required to result in fracture and failure. Table 1 shows the calculated modulus of elasticity for all of the strips tested.

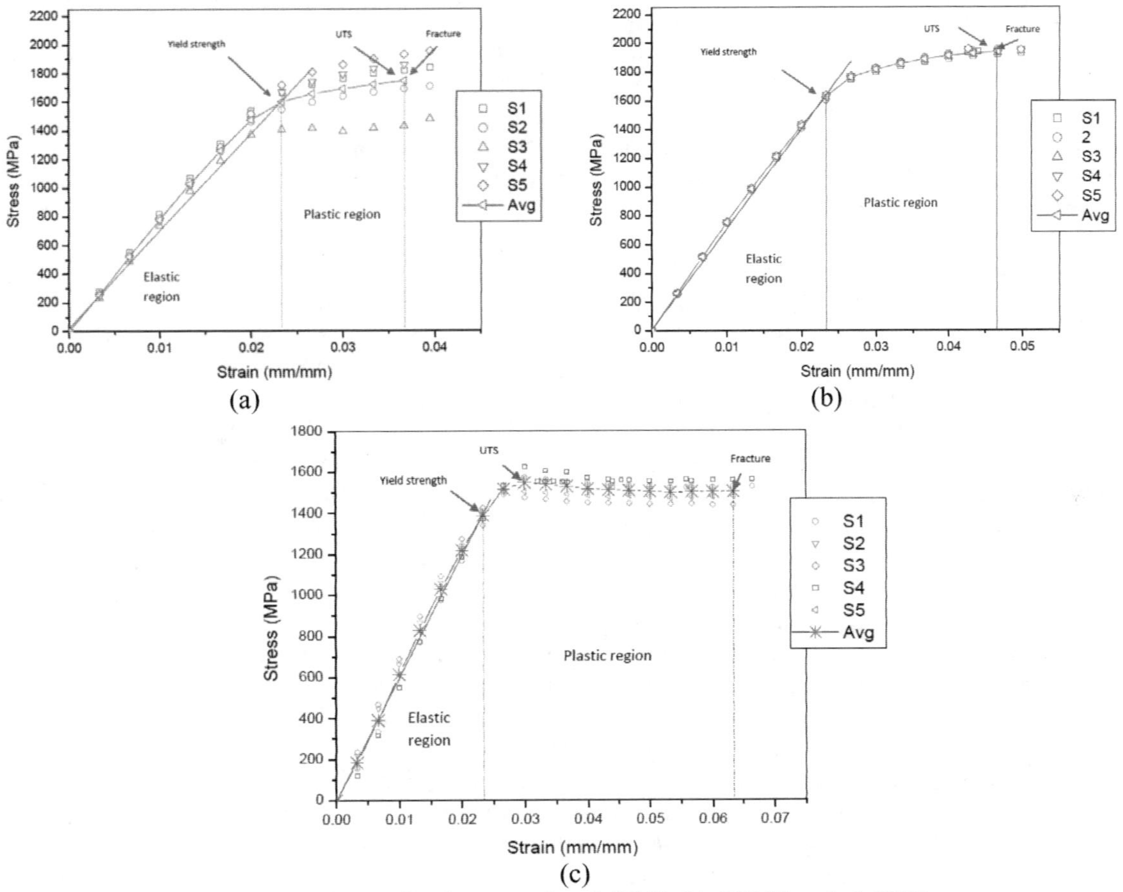

Figure 1. Stress-Strain curve for a) CS50, b) CS100 and c) SS50.

4.1. EBSD Results

Inspection by conventional light-microscopy at 500x magnification as illustrate in figure 2(a) revealed a bluish martensite region as well as cementite content in the 50μm carbon-steel strip (CS50) due to the heat-treatment process. Quenching of the strip has resulted in some of the carbon content being formed into iron carbide. Iron carbide, also known as cementite (white spots), usually measures less than 1μm and increases the mechanical properties of the strip. Figure 2(b) shows grains coloured by size with the darker/red coloured being the larger grain size for 50μm carbon-steel strip. Although no specific shape may be used to define the grain, the grain was defined as areas enclosed by boundaries. A step-size of 0.05μm was used to create a reasonably clear resolution for accurate indexing of the microstructure. This step-size covers an image-mapping area of 30x35μm. Based on the analysis of the mapped image, there were 3404 counted grains. The smallest and largest measured grains were 0.0564μm and 5.2757μm, with an average of 0.4028μm. The numbers of small grains were seen to be dominating the microstructural pattern and no specific and uniform grain shapes were observed. This suggests that the previous deformed-grain (due to hot-rolling) might be reformed/reshaped due to the tempering process. A large aspect ratio was observed also, where each grain on average has an aspect ratio of 2.2288 and a maximum of 12.2060. A large and slender aspect ratio may cause the microstructure to become weak and less resistant to external forces.

Figure 3(a) shows an image of a photomicrograph of a 100μm thick carbon-steel strip (CS100) under light microscopy at 500x magnification. This technique may be found useful as a fast way to reveal the type of process that has been undergone by the strip material. Although this technique was unable to reveal greater numbers of grains than EBSD was capable of, the technique has proven able to identify irregularity-content of the strip. White spot on the image was identified as cementite, which may cause an increase of the strength and brittleness of the strip. The presence of cementite on this strip was observed to be twice as much as for the CS50 strip.

An EBSD microstructure-mapping image of CS100 strip is shown in Figure 3(b). A similar scanning-step as that for CS50 was used to observe the microstructure. A similar number of pixels were used to ensure that a broad and wide image would be captured, using the same total mapped area of 30x35μm. 6438 grains was successfully mapped with a grain-size average of 0.2661μm, the smallest and largest grains being 0.0564μm and 3.2762μm, respectively. The number of grains revealed was found to be twice many when compared with the number for CS50. This indicates that the CS100 microstructure was very dense and also consists of many tiny grain-sizes, smaller than those for CS50. This was confirmed by the average grain-size obtained throughout the examination, which revealed that CS100 grains were 34% smaller than CS50 grains. An almost similar maximum aspect ratio as that for CS50 was found for CS100, which was a 12.3760. However, the average aspect ratio for CS100 was found to be slightly smaller than for the thinner carbon-steel strip, which was 1.8331. Due to the smaller aspect ratio, the grains might be able to stay close to each other, resulting in a very dense grain pattern.

Mapping of 50μm thick stainless-steel strip (SS50) using conventional light-microscopy at 500x magnifications in figure 4(a) shows many grains, which were heavily deformed due to the hot-rolling process. Nevertheless, the image from this technique was seen to be unable to show important microstructural information such as grain boundaries and clear distribution of neighbouring grains. Due to the larger grain size that may be obtained from a stainless-steel specimen, a large scanning step may be used to map and index its microstructure. 5618 grains were successfully mapped and indexed in figure (4b) with an average grain size of 0.7698μm, which was larger than that for the tested carbon-steel strips. The minimum and maximum indexed grain-size was 0.2820μm and 20.2870μm, respectively. Due to the grains being heavily deformed, a large part of the mapped area could not be indexed, hence a large number of blanks were observed on the image. In order to obtain a clear view of grain boundaries and orientation, the mapped image was split into two images. Both of these images represent the indexed grain size, which is $<1\mu m$ and $>1\mu m$. The analysis also reveals that SS50 has many larger grains compared to carbon-steel and has the greatest grain aspect-ratio of 19.7060. This has resulted in a lesser grain density, which may be affecting its mechanical properties (table 1).

IOP Conf. Series: Materials Science and Engineering **100** (2015) 012051 doi:10.1088/1757-899X/100/1/012051

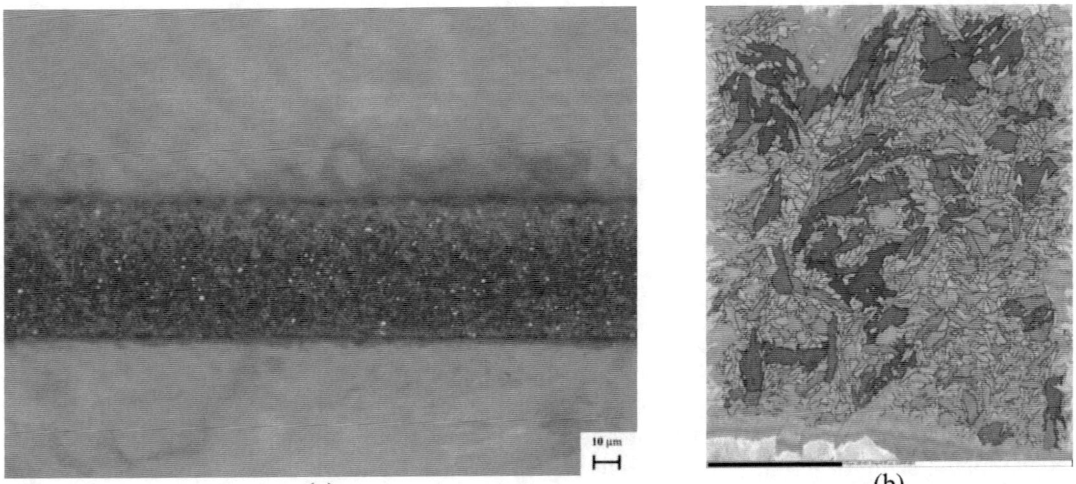

(a) (b)

Figure 2. a) CS50 at 500x magnifications, showing iron carbide/cementite (white colour) and b) EBSD microstructure mapping image of CS50 strip.

(a) (b)

Figure 3. a) Light-microscopy image of CS100 strip at 500x magnification with cementite shown (white colour) and b) EBSD microstructure-mapping image of CS100 strip.

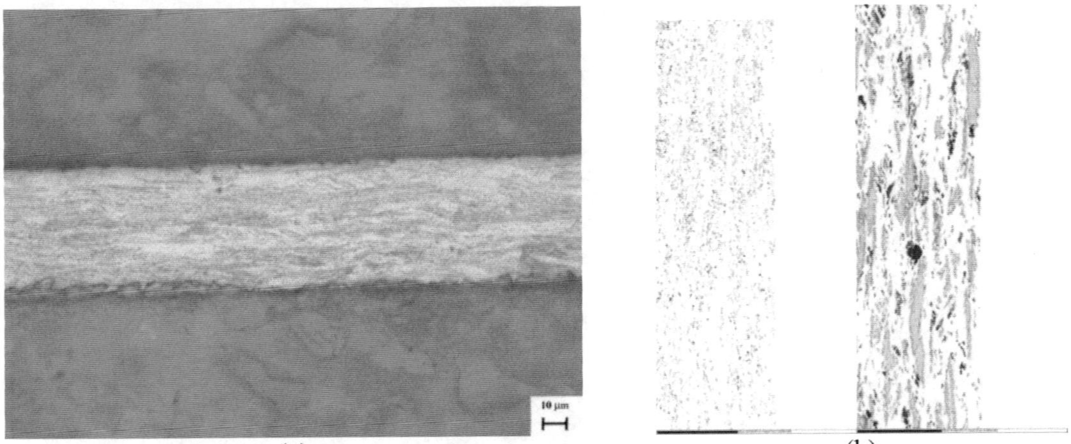

(a) (b)

Figure 4. a) Light-microscopy image of SS50 strip at 500x magnification and b) EBSD microstructure-mapping image of SS50.

5. Discussions

5.1. Size Effect and Material Stress–Strain Relation

Studies had demonstrated that the flow stress of thin sheet-metal decreases compared to that of its bulk material due to the size effect, and decreases proportionally with miniaturization [17, 18]. Commonly, there are two categories of the size effect that can be investigated: the grain-size effect and the feature-size effect.

Many investigations have been conducted elsewhere [19, 20] and have shown that sheet metals of different grain sizes show different mechanical properties. The grain-size effect was already known to follow the Hall–Petch equation, which topic has been elaborated upon elsewhere [21]. This equation simply states that a material with a greater grain size has less strength compared to one which has a smaller grain size. This effect is purely reliant on the average grain-size of the material and is mostly dominant in macro-scale material. Although the tested strips were very thin, due to their relatively large width and length, they can still be considered as macro-scale material. According to the EBSD results conducted on all of the strips, the stainless-steel strip (SS50) were found to have a greater grain size, followed by that of the 50μm-thick carbon-steel strips (CS50), and finally that of the 100μm-thick carbon-steel strips (CS100). Good agreement of the results achieved between those for EBSD and those for tensile testing have established and confirmed the effect of varying grain size. The least flow-stress was observed with SS50, which in turn has the larger grain size, this being followed by CS50 and finally CS100. This is because when a material of greater grain size is deformed, dislocations are easier to move compared to the situation with a smaller grain size. Less force is required and a lesser number of grains have made a larger grain-size material prone to fail a lot more quickly than would a smaller grain-size material.

Moreover, feature size also has a significant effect on a material's flow stress. The decrease of flow stress with the decrease of sheet thickness may be explained by the so-called surface-model, as depicted in figure 5. This model was used to describe the correlation between flow stress and miniaturization and has been popularly validated elsewhere [4, 22, 23]. According to the model, the grains located at the free surface are less restricted than are the internal grains. This in turn leads to the surface grains being subjected to less hardening, and having a lower resistance against deformation, because dislocations move through the grains during deformation and pile-up at grain boundaries, but not at the free surface. With the decrease of strip thickness, the share of surface grains increases, hence lower flow stress. On the other hand, thicker sheet, in this case 100μm carbon-steel strip, the share of surface grains decreases and this in turns leads to a relatively higher flow stress compared to that for thinner strip.

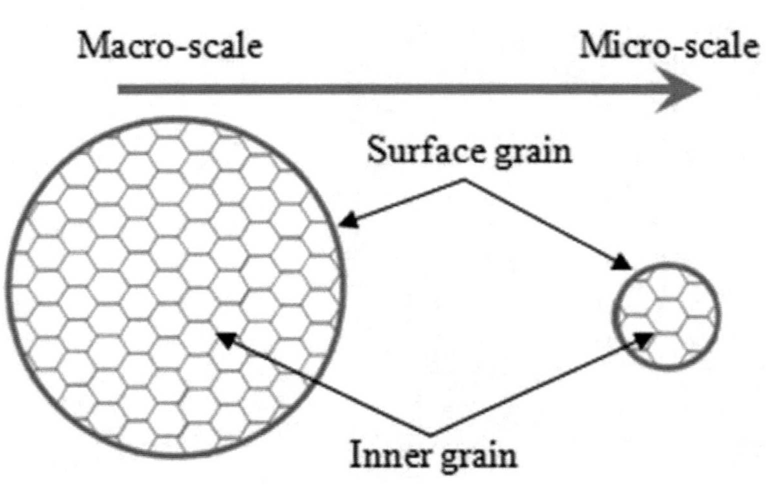

Figure 5. Grain distribution in a material section of different material/feature size.

Table 1. EBSD and Young's Modulus of the tested strips.

	CS50	CS100	SS50
Count grain, N	3404	6438	5618
Mean (μm)	0.4028	0.2661	0.7698
Minimum (μm)	0.0564	0.0564	0.2820
Maximum (μm)	5.2757	3.2762	20.2870
Standard deviation (μm)	0.3980	0.3072	1.0114
Grain aspect ratio			
Mean	2.2288	1.8331	2.0039
Minimum	1.0000	1.0000	1.0000
Maximum	12.2060	12.3760	19.7060
Standard deviation	1.0546	0.8834	1.2024
Young's Modulus (GPa)	69	70	60

6. Conclusions

Based on the experiments conducted, the following conclusions are drawn:

i. Tested thicker carbon steel strip has the smallest grain size, followed by the tested thinner carbon steel strip and finally stainless steel strip.
ii. The thick carbon-steel strip has the greatest modulus of elasticity (stiffness) followed by the thinner carbon-steel strip and finally the stainless-steel strip.
iii. It may be concluded that thicker carbon steel strip has the stiffer characteristic compared to the thinner carbon and stainless steel strip.

References

[1] Zhao J, Qin Y, Razali A, Yip A L K, Fei Z, Hot-embossing of polymeric micro-tubes. 7th International Conference on MicroManufacturing; 2012.
[2] Razali A R, Qin Y 2013 *Procedia Engineering* **53** 665-72
[3] Qin Y 2006 *Journal of Materials Processing Technology* **177** 8-18
[4] Lai X, Peng L, Hu P, Lan S, Ni J 2008 *Computational Materials Science* **43** 1003-9
[5] Razali A, Qin Y, Harrison C, Brockett A, Investigation of feeding devices and development of design considerations for a new feeder for micro-sheet forming. 6th International Conference on Manufacturing Research, ICMR 08; 2008.
[6] Razali A, Qin Y, Zhao J, Harrison C, Smith R 2011 *Journal of Manufacturing Sc. & Engineering* **133** 061025
[7] Qin Y, Brockett A, Ma Y, Razali A, Zhao J, Harrison C, et al. 2010 *The International Journal of Advanced Manufacturing Technology* **47** 821-37
[8] Qin Y, Ma Y, Harrison C, Brockett A, Zhou M, Zhao J, et al. 2008 *International Journal of Material Forming* **1** 475-8
[9] Razali A, Qin Y, Harrison C, Zhao J 2009 *Journal of Advanced Manufacturing Technology*
[10] Zhao J, Brockett A, Razali A, Qin Y, Harrison C, Ma Y 2010 *Steel Research International* **81** 1185-8
[11] Qin Y, Razali A R, Zhou M, Zhao J, Harrison C, Nawang W, et al., Dynamic Characteristics of a Micro-Sheet-Forming Machine System. Key Engineering Materials; 2012: Trans Tech Publ.

[12] Jie Z, Yip A L K, Yi Q, Razali A, Fei Z-c 2012 *Transactions of Nonferrous Metals Society of China* **22** s214-s21
[13] Law A R J Z F, Yi Q 2011 *Journal of Plasticity Engineering* **6** 015
[14] Mingard K, Roebuck B, Bennett E, Gee M, Nordenstrom H, Sweetman G, et al. 2009 *International Journal of Refractory Metals and Hard Materials* **27** 213-23
[15] Wagner F, Allain-Bonasso N, Berbenni S 2011 *Materials Characterization* **62** 681-3
[16] Klocke F, Hensgen L, Klink A, Mayer J, Schwedt A 2014 *Procedia CIRP* **13** 237-42
[17] Peng L, Hu P, Lai X, Mei D, Ni J 2009 *Materials & Design* **30** 783-90
[18] Anand D, Kumar D R 2014 *Procedia Materials Science* **6** 154-60
[19] Dong X, Hong X, Chen F, Sang B, Yu W, Zhang X 2014 *Materials & Design* **64** 400-6
[20] Kim J, Golle R, Hoffmann H 2010 *Materials Science and Engineering: A* **527** 7220-4
[21] Mahabunphachai S, Koç M 2008 *International Journal of Machine Tools and Manufacture* **48** 1014-29
[22] Peng L, Liu F, Ni J, Lai X 2007 *Materials & design* **28** 1731-6
[23] Vollertsen F, Niehoff H S, Hu Z 2006 *International Journal of Machine Tools and Manufacture* **46** 1172-9

IOP Conf. Series: Materials Science and Engineering **100** (2015) 012052 doi:10.1088/1757-899X/100/1/012052

Numerical simulation of rock cutting using 2D AUTODYN

D E Woldemichael[1,2], A M Abdul Rani[1,2], T A Lemma[1] and K Altaf[1]
[1]Mechanical Engineering Department, Universiti Teknologi PETRONAS, 32610 Bandar Seri Iskandar, Perak, Malaysia
[2]MOR-Enhanced Oil Recovery, Universiti Teknologi PETRONAS, 32610 Bandar Seri Iskandar, Perak, Malaysia

E-mail: dereje.woldemichael@petronas.com.my

Abstract. In a drilling process for oil and gas exploration, understanding of the interaction between the cutting tool and the rock is important for optimization of the drilling process using polycrystalline diamond compact (PDC) cutters. In this study the finite element method in ANSYS AUTODYN-2D is used to simulate the dynamics of cutter rock interaction, rock failure, and fragmentation. A two-dimensional single PDC cutter and rock model were used to simulate the orthogonal cutting process and to investigate the effect of different parameters such as depth of cut, and back rake angle on two types of rocks (sandstone and limestone). In the simulation, the cutting tool was dragged against stationary rock at predetermined linear velocity and the depth of cut (1,2, and 3 mm) and the back rake angles(-10°, 0°, and $+10^o$) were varied. The simulation result shows that the $+10^o$ back rake angle results in higher rate of penetration (ROP). Increasing depth of cut leads to higher ROP at the cost of higher cutting force.

1. Introduction

Due to depletion of hydrocarbon reserves at shallow depth, oil and gas industries attempt to drill for reserves at a greater depth. Exploration and production wells are drilled to produce these hydrocarbons. Unfortunately, it is difficult to exactly predict the interaction between the cutting tool and rock before the drilling operation. Understanding the interaction between the cutting tool and the rock is important to optimize the drilling process through higher rate of penetration (ROP) and better tool life [1, 2]. There are three types of drill bits used in oil and gas industry, namely: roller cone bits, fixed cutter bits and hybrid bits. The selection and use of these bits depend on several factors such as formation to be drilled, projected operating parameters, the capabilities of rig and the operator's past experience in offset wells. In this paper, PDC (Polycrystalline Diamond Compact) bit, which falls under fixed cutter category and one of most commonly used cutting tool in industry, is considered to assess the effect of different parameters on the drilling operation.

Experimental investigation of rock cutter interaction can help to understand the effects of different parameters. However, the high cost of conducting such experiments, the difficulty to exactly represent the downhole conditions in the experimental setup, time to conduct a number of experiments, and difficulty to directly observe the rock fragmentation and removal process leads to numerical simulation as a means to study the process at early stage. There are a number of factors which affect the performance of PDC bit with respect to ROP. Some of these parameters include weight on bit (WOB), depth of cut, rpm of the bit, rock hardness and bit design features. The bit design features include cutter material, number of cutters, cutting structure geometry (back rake angle and side rake angle), and other features.

The effect of back rake angle and side rake angle were studied by Gerbaud, Menand [2] (to estimate the cutting force. Rajabov, Miska [3] studied the effect of back rake angle and side rake angle on the mechanical specific energy (MSE) and aggressiveness of PDC cutters. Hareland, Nygaard [4] identified back rake angle, depth of cut and rock properties to be factors affecting the cutting efficiency using single PDC cutter. A variety of simulation codes and models have been used in the literature to study cutter-rock interaction. ANSYS AUTODYN has been used by [5, 6] to simulate the tunnel boring machine (TBM) and evaluate the cutter efficiency. LS-DYNA has been used by [1, 7] to study rock fragmentation. Some of the researchers [8-10] used discrete element simulation for rock-cutter interaction.

In this study, a two-dimensional code ANSYS AUTODYN-2D is used to simulate the cutter rock interaction, rock failure, fragmentation and study the effect of different parameters such as depth of cut, and back rake angle on two types of rocks namely sandstone and limestone. ANSYS AUTODYN [11] is an explicit analysis tool for modeling nonlinear dynamics of solids, fluids, gas, and their interaction. AUTODYN-2D is capable of simulating non-linear, dynamic fracture failure and fragmentation.

2. Numerical method

Numerical simulations were performed using an explicit non-linear finite element code AUTODYN-2D. Material models in AUTODYN consist of equation of state (EOS), strength model and failure model. Material models can either be selected from available models in the library or defined by the user. In this section, the numerical models for the cutter and rock block together with the boundary conditions imposed for different sets of simulation will be described.

2.1. Modeling the cutter

The PDC bits are designed and manufactured either as matrix body bits or steel body bits. Their selection depends on the application. Matrix body PDC bits, which are made from composite materials consisting of tungsten carbide grains bonded with metallic binder, are abrasion and erosion resistant compared to steel body bits. On the other hand, steel body bits are preferred for their strength and high resistance in impact loading. In this paper diamond impregnated bit with matrix body construction is considered. The cutter material is considered to be Tungsten, the properties of which are available in AUTODYN library. In selecting the cutter material the deformation of the cutting tool is negligible compared to the rock, and the wear or abrasion of the cutter is insignificant. Thus, no failure mode was defined for the cutter and it is assumed to be rigid

2.2. Modeling the rock

Two types of rock materials were considered in this study. The properties of the rocks are presented in table 1. As this rock models are not available in the material library of AUTODYN, new material models were defined. The rock models were assumed to be isotropic and homogenous. The rock models are defined with linear EOS. For strength model, the Drucker-Prager model is adopted. The Drucker-Prager model is used to represent the behavior of the rocks where the cohesion and compaction behavior result in an increasing resistance to shear up to a limiting value of yield strength as the loading increases. The model requires the compressive and tensile strengths of the rock. The yield point under uniaxial compression and tension were assumed to be 90% of the uniaxial compressive and Brazilian tensile strength [6]. For failure model, principal stress failure which is used to represent brittle failure in materials was used. The failure stress was defined by the Brazilian tensile strength.

Table 1. Properties of rock materials.

Rock Type	Density, ρ (Kg/m^3)	Uniaxial compressive strength, S_c (MPa)	Brazilian tensile strength, S_T (MPa)	Young's modulus, E (GPa)	Poison's ratio, ν	Bulk modulus, B(GPa) $B=\frac{E}{3(1-2\nu)}$	Shear modulus, s (GPa) $S=\frac{E}{2(1+\nu)}$
Sandstone	2000	170	15	15	0.14	6.94	6.58
Limestone	2700	210	25	30	0.28	22.73	11.72

2.3. Model set-up and boundary conditions

During drilling operation, the PDC cutter is dragged in a circular path due to the rotation of the bit. Figure 1(a) shows the sketch of a single cutter in a vertical plane together with the rock model and parameters. The cutter is tilted with an angle, ϕ (back rake angle) with respect to the rock and the side rake angle is assumed to be zero. An arbitrary relief angle (α) is used.

The cutter and rock models were virtually represented in AUTODYN as shown in figure 1(b). The rock was modeled as a rectangular block of 20 mm in length and 40 mm in height. The cutting tool is modeled as square element with side length of 10 mm. The top nodes of the rock were fully constrained in X- and Y-directions and the sides are constrained in the X-direction while the bottom side is free. The cutting tool was constrained in the Y-direction. The back rake angle of the cutting tool, ϕ, was varied between +10°, 0°, and -10°. The cutting tool was dragged against stationary rock at linear velocity of 100 mm/s. The sliding contact between the cutter tip and the rock was assumed to be frictionless. The depth of cut, *dc*, was varied between 1, 2, and 3 mm during simulation. A gauge was defined at the tip of the cutter to measure the different parameters such as force, pressure, stresses and strains during the cutting process. The interaction between the cutter and the rock is defined with small external gap before the simulation starts.

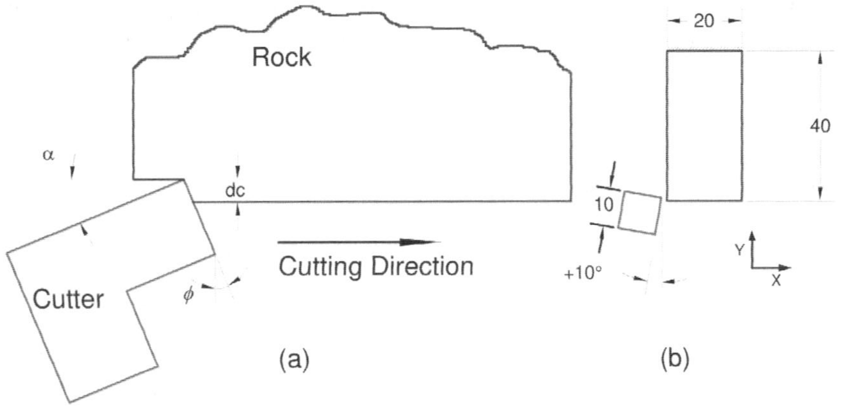

Figure 1. (a) Schematic representation of PDC cutter in a vertical plane with cutting parameters and (b) Cutter and rock model with +10^0 back rake angle.

3. Result and discussion

3.1. Fragmentation

The erosion option in AUTODYN is used to analyze the total mass of the debris after the cutting process. When the failure criterion is reached, the rock element will fail and remove from the rock block in the form of chip. Figure 2(a) shows the reduction in total mass due to failure for sandstone with 3 mm depth of cut and 0^0 rake angle for the entire cutting process along the length. Figure 2(b) shows a comparison of total mass removed for different rake angles and depth of cuts. The $+10^0$ rake angle has the highest rock mass reduction while 0^0 has the least. Based on this simulation $+10^0$ rake angle is recommended to get higher ROP. This is in good agreement with the result by [12] where the mean cutting force was reported to be the least at $+15^0$ rake angle compared with 0^0 and -15^0. Furthermore, the total mass reduction increases with increasing depth of cut which leads to higher ROP.

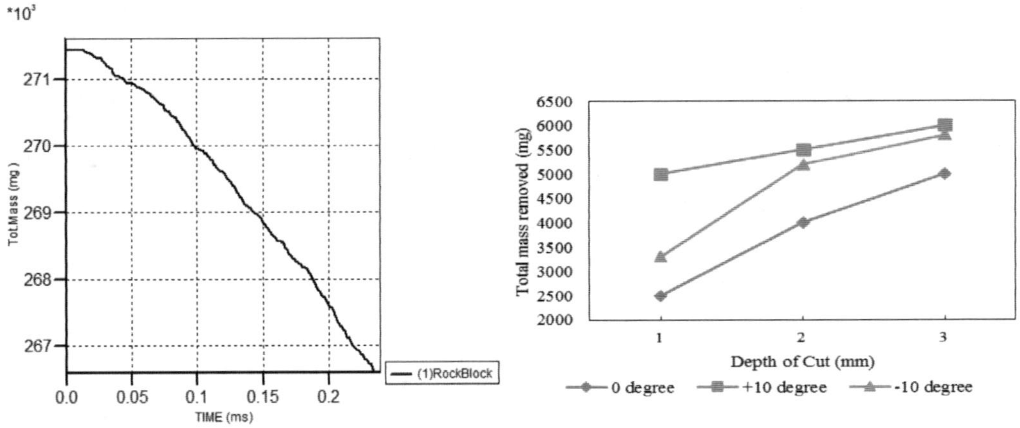

Figure 2. (a) Total mass of sandstone for 0^0 rake angle with 3 mm depth of cut and (b) Variation of total mass removed with depth of cut and rake angle.

Fragmentation option in AUTODYN was used to study rock fragments in terms of size and mass. The fragment plot only shows elements that contained material that has not failed. The fragment analysis at 1500 cycles for all simulation is as shown in figure 3. In all cases, sandstone has higher number of fragments and total mass compared to limestone. In addition, the number and mass of fragments increase with depth of cut.

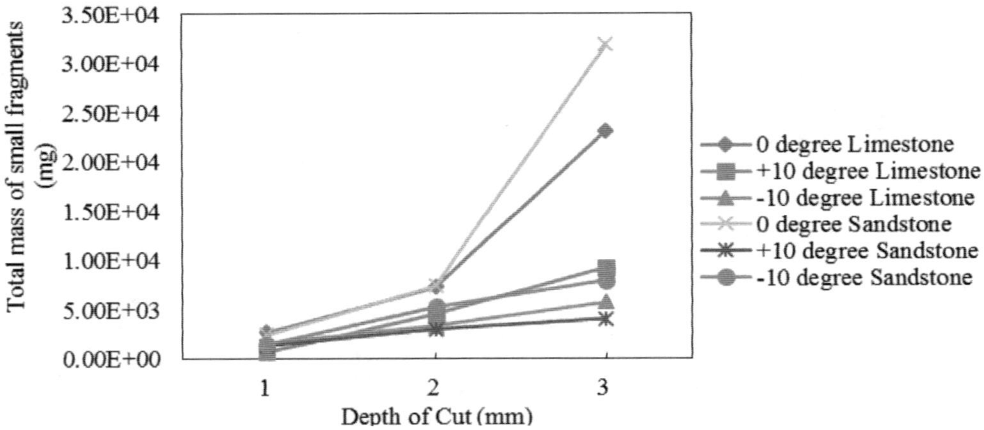

Figure 3. Variation of total mass of small fragments with depth of cut at 1500 cycles.

3.2. Cutting Force

The chip formation in the rock is due to shearing and crushing when the cutting tool moves against stationary rock. This phenomenon has been observed in the force vs cutting distance history as shown in figure 4 with a number of peaks and valleys. The peaks indicate the higher resistance from the rock followed by chip formation due to shearing and drop in force magnitude. The crushing mechanism is depicted with small variation in the force with relatively small magnitude in the valleys. Similar observations have been reported by Che et al. [13]using experimental data and simulation results from LS-DYNA. The cutting force increases with depth of cut for both sandstone and limestone. In all simulations, a higher magnitude of cutting force is observed when cutting limestone compared with sandstone which agree with the observation by [12].

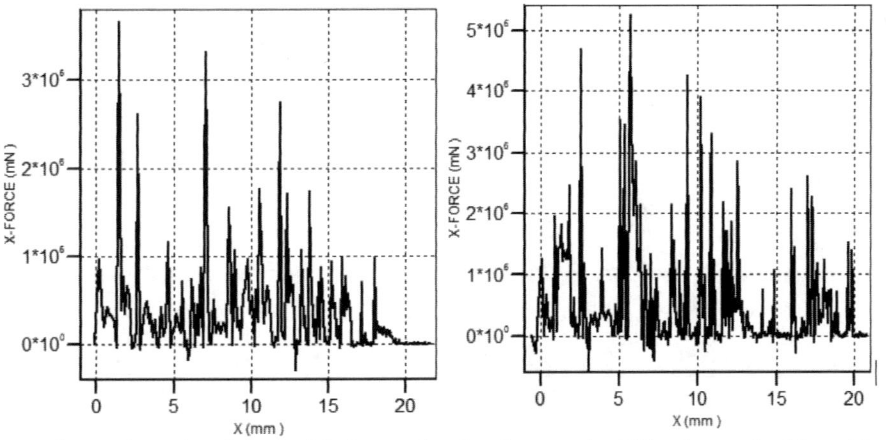

Figure 4. Horizontal cutting force with 3 mm depth of cut and +10° rake angle (a) sandstone and (b) limestone.

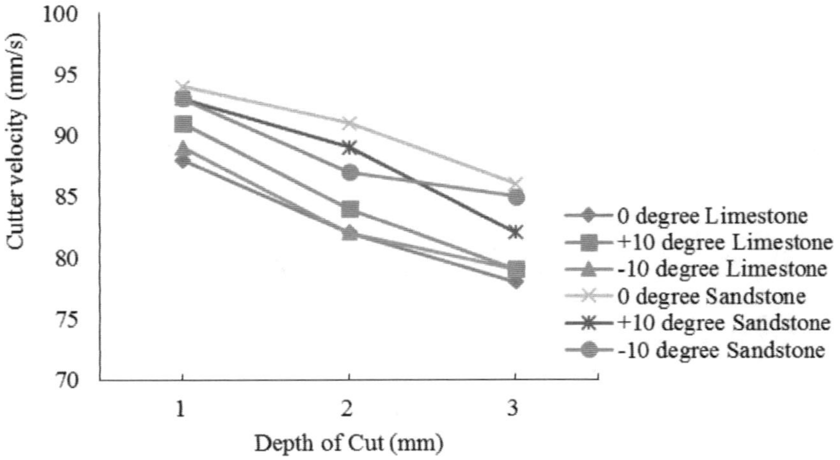

Figure 5. Variation of cutter velocity with depth of cut and rake angle.

The velocity profile of the cutter depicts similar trend with the cutting force. When there is higher resistance, the speed of the cutter decreases until the cutter force reaches peak value followed by chip formation due to shearing mechanism. This results in weaker rock with lower internal energy, hence the speed of the cutter increase again. This fluctuation in resistance and force leads to lower overall velocity of the cutter compared to the initial velocity. The variations in the final velocity of the cutter with depth of cut and rake angle is shown in figure 5. The velocity decreases with depth of cut which indicate higher cutting force as well. Similarly, limestone has higher velocity reduction which requires higher cutting force.

4. Conclusion
In this paper, it was attempted to simulate the cutting tool rock interaction for single cutter PDC bit. The effect of depth of cut and back rake angle on fragmentation, chip formation, and cutting force has been analyzed. These effects have been simulated for two types of rocks: sandstone and limestone. Based on this simulation $+10^{o}$ rake angle is recommended to get higher ROP with least mean cutting force. Furthermore, increasing depth of cut leads to higher ROP at the cost of higher cutting force. Future work includes three dimensional analysis and construction of experimental test rig to validate the simulation results.

Acknowledgement
This project was supported by Universiti Teknologi PETRONAS under Y-UTP project on “Bit Wear Vibration Study to Aid Drilling Optimization”.

References
[1] Menezes P L, Lovell M R, Avdeev I V, Higgs C F 2014 *International Journal of Rock Mechanics and Mining Sciences* **71** 131-42
[2] Gerbaud L, Menand S, Sellami H, PDC bits: all comes from the cutter rock interaction. IADC/SPE Drilling Conference; 2006.
[3] Rajabov V, Miska S Z, Mortimer L, Yu M, Ozbayoglu M E, The Effects of Back Rake and Side Rake Angles on Mechanical Specific Energy of Single PDC Cutters with Selected Rocks at Varying Depth of Cuts and Confining Pressures. IADC/SPE Drilling Conference and Exhibition; 2012: Society of Petroleum Engineers.
[4] Hareland G, Nygaard R, Yan W, Wise J 2009 *Journal of Canadian Petroleum Technology* **48** 60-5
[5] Cho J-W, Jeon S, Jeong H-Y, Chang S-H 2013 *Tunnelling and Underground Space Technology* **35** 37-54
[6] Cho J-W, Jeon S, Yu S-H, Chang S-H 2010 *Tunnelling and Underground Space Technology* **25** 230-44
[7] Che D, Han P, Peng B, Ehmann K F, Finite element study on chip formation and force response in two-dimensional orthogonal cutting of rock. ASME 2014 International Manufacturing Science and Engineering Conference collocated with the JSME 2014 International Conference on Materials and Processing and the 42nd North American Manufacturing Research Conference; 2014: American Society of Mechanical Engineers.
[8] Rojek J, Oñate E, Labra C, Kargl H 2011 *International Journal of Rock Mechanics and Mining Sciences* **48** 996-1010
[9] Rojek J 2014 *Computational Particle Mechanics* **1** 71-84
[10] Akbari B, Butt S, Munaswamy K, Arvani F, Dynamic Single PDC Cutter Rock Drilling Modeling and Simulations Focusing on Rate of Penetration Using Distinct Element Method. 45th US Rock Mechanics/Geomechanics Symposium; 2011: American Rock Mechanics Association.
[11] Inc A. ANSYS Autodyn User's Manual, Release 15.0. 2013.

Influence of resin system on the energy absorption capability and morphological properties of plain woven kenaf composites

S D Salman[1,3], Z Leman[1,2], M T H Sultan[2,3], M R Ishak[2,4] and F Cardona[2],
[1]Department of Mechanical and Manufacturing Engineering, Faculty of Engineering, Universiti Putra Malaysia, 43400 Serdang, Selangor, Malaysia
[2] Aerospace Manufacturing Research Centre (AMRC), Level 7, Tower Block, Faculty of Engineering, Universiti Putra Malaysia, 43400 Serdang, Selangor, Malaysia
[3] Materials Engineering Department, Faculty of Engineering, University of Mustansiriyah, Baghdad, Iraq
[4] Laboratory of Bio-Composites Technology, Institute of Tropical Forestry and Forest Products (INTROP), Universiti Putra Malaysia, 43400 UPM Serdang, Selangor, Malaysia

E-mail: suhaddawood2007@yahoo.com

Abstract. Due to both environmental and technical advantages, natural fibers are being used as reinforcement of polymeric composite in many industries. The flexibility of the most natural fibers is one of the important technical characteristic which allows them to resist impact forces. An investigation was carried out to compare the energy absorption capability of kenaf/PVB film and kenaf/epoxy composites. The hot and cold press techniques were used to fabricate the specimens with 35% kenaf fibre weight fraction. The charpy impact test was performed on forty notched specimens using a pendulum impact tester with different hammer energy. The results showed that the kenaf/PVB film composite has the highest energy absorption, strength and toughness compared with the epoxy composite. At high energy levels, the impact strength and toughness of the kenaf/PVB film was six times of kenaf/epoxy composite. In addition, the scanning electron microscopy was assessed to demonstrate the different failure in fracture surfaces. It was found that the kenaf/PVB film composite failed by fibre fracture while kenaf/epoxy composite failed by a combination of fibre pull-out and fibre fracture as well as crack propagations through the matrix.

1. Introduction

Capability of composite energy absorber is completely affected by a fibre and matrix type as well as fabrication and test conditions [1]. Generally, two categories of energy absorbers are being used, low and high speed impacts depending on the impact loading conditions [2]. In recent years, the use of polymeric composites reinforced with natural fibre is increasing due to their economical and environmental advantages as well as high impact resistance during their service life [3]. Among the natural lignocellulosic fibres, kenaf fibres possess moderately high specific strength and stiffness that could be utilized as reinforcing materials in polymeric resins to make useful structural composite material [4, 5].

Recent works on the impact resistance of curaua fibre reinforced either polyester [6] or epoxy [7] were investigated for continuous and aligned fibres. It has been reported that the toughness measured by the absorbed impact energy in Charpy test increases linearly up to volume fractions of 20-30% of fibres. In other studies, low impact tests were carried out on hemp reinforced unsaturated polyester composite specimens by Dhakal et al. [8, 9]. Likewise, Meredith et al. [10] investigated the unwoven hemp, flax and jute fibres reinforced composites and concluded that the unwoven hemp with its low embodied energy is a promising candidate for sustainable energy absorption structures. In an experimental study, Yan and Chouw [11] reported that flax fabric reinforced epoxy composite tube has the potential to be a useful energy absorber device. Similarly, Santulli and Caruso [12] carried out low velocity impact testing on the jute/epoxy and hemp/epoxy laminates. It has been concluded that the increase in fibre volume fraction results in an increased damage propagation phase thus increasing the impact resistance properties of the composite.

However, it is clear from the literature review that there are no reported works on the impact response of composites of plain woven kenaf reinforced PVB film. Hence, the objectives of the current study were to fabricate kenaf/PVB film and kenaf/epoxy composites and test using varying charpy hammer energy. Moreover, scanning electron microscopy (SEM) analysis of the fracture surface of the notched specimens was analyzed.

2. Materials and experimental procedure

To clarify the effect of the resin system with the energy absorption capability, kenaf/PVB film and kenaf/epoxy were fabricated and tested. Plain woven kenaf fibre is the main fibre that is used in this study with 2mm thickness, 1.2g/cm^3 density, and 100.64MPa average breaking strength [13]. PVB film (Polyvinyl butyral) is employed in a wide array of industrial and commercial applications as well as epoxy resin, which are commonly used in industry, were used to fabricate the composite samples. The epoxy used in this study is LY556 liquid with a density of 1.14 g/m^3, elongation at break 4.5%, and tensile strength 73.3MPa, with 1:2 HY951 hardener. While the tensile strength of the PVB film ≥ 20MPa and breaking elongation ≥ 200%, (manufacturer data sheet).

The composite samples were made with 35% kenaf fibre weight content by using a hot hydraulic press technique to reinforce PVB film, and cold hydraulic press-molded to reinforce epoxy resin. The plain woven kenaf fibre reinforced PVB film composite panels were prepared by compressing layers of kenaf fabric and PVB film stacking between the hot plates of a compression moulding press. The plates were electrically heated to a temperature of 170°C for 20 minutes, at a pressure of 8 MPa applied on the material. Same procedure was followed for the kenaf/epoxy composites but with cold hydraulic press. Plates of these composites with five layers of plain woven kenaf reinforced polymers were press-molded and allowed to complete inner cured for 4 hours. Twenty specimens were fabricated for each composite and energy, five replications were considered. Standard ASTM D 6110 Charpy notched specimens [14], 127 mm × 12.7 mm × 7 mm was cut from each composite plate and tested with different striking energy (0.5, 2.7, 5.4, 21.6) Joule. By using a wheel saw machine, the specimens were carefully cut and finished from the composite to the accurate size. They were notched with the radius of curvature 0.25R mm to 2.54 mm by motorised notch vis in the middle, according to the standard. The charpy impact tests were conducted by using a universal testing machine INSTRON, MECOMB in the Laboratory of Bio-Composites Technology, Institute of Tropical Forestry and Forest Products (INTROP), Universiti Putra Malaysia. A detail flow chart for processing method is explained in figure 1, which includes the fibre preparation, mold preparation (two stainless steel plates), fabricating process (different resins systems), composite sample preparation (notched) and charpy impact testing progress. While figure 2 shows the specimens after breaking.

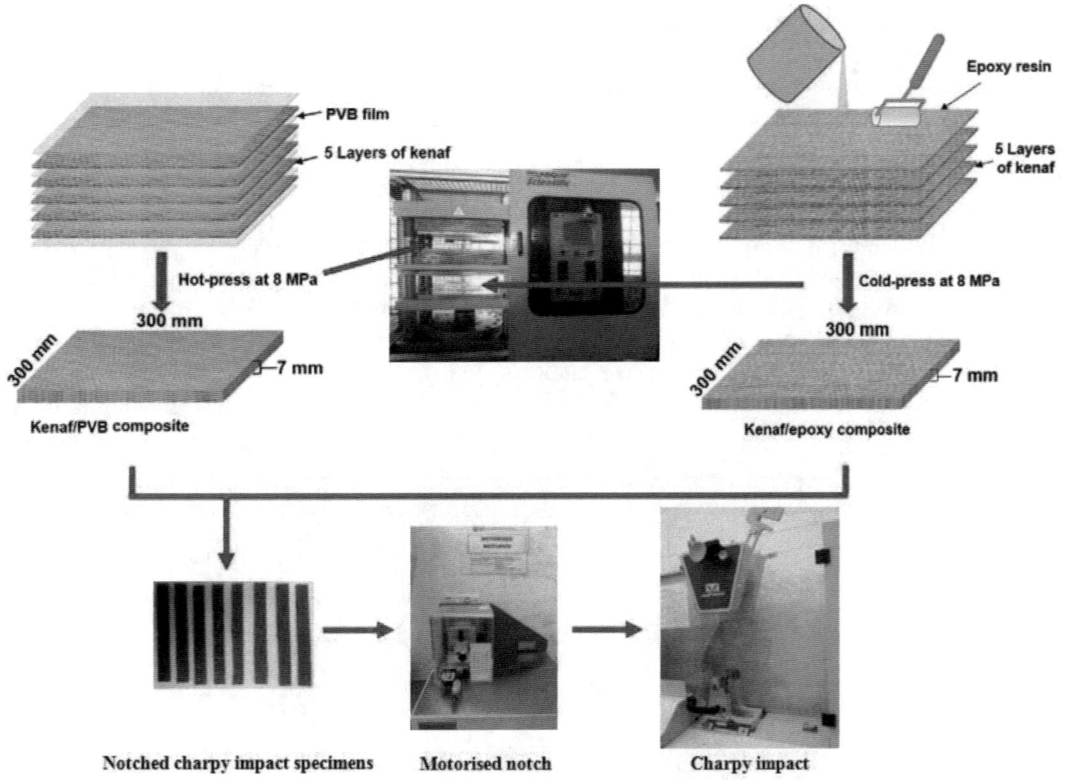

Figure 1. Drawing of the plain woven kenaf/PVB film and kenaf/epoxy composite laminates, by using hot press and hand lay-up techniques.

Figure 2. The composite specimens after Charpy impact tests.

3. Morphological Observation

The influence of the resin system on the fractured surface of the three types of composites after charpy impact tests were observed by using the scanning electron microscope (SEM) instrument model ZEISS SUPRA 35VP. All the fractured specimens were coated with a thin layer of gold to avoid electron charge accumulation, and subjected to a voltage of 10–15 kv.

4. Results and discussion

The variation of the charpy impact energy with the two types of kenaf/PVB film and kenaf/epoxy composites is shown in figure 3. In this Figure, a highly increment in the absorbed energy can be seen for kenaf/PVB film composite. While much less energy is absorbed by kenaf/epoxy composite at different energy levels. Furthermore, the charpy impact energy of kenaf/PVB film and kenaf/epoxy composites showed almost same behavior when tested with a low energy level, (0.5 J). The level of maximum impact energy for the kenaf/epoxy composite is significantly lower than the corresponding for kenaf/PVB composite, reached of the order of 0.895% and 3.995% at 21.6 J, respectively.

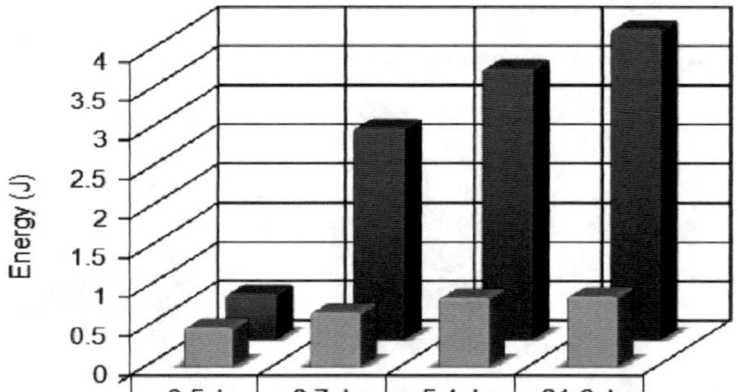

	0.5 J	2.7 J	5.4 J	21.6 J
Epoxy	0.497	0.691	0.877	0.895
PVB film	0.579	2.693	3.454	3.955

Figure 3. Charpy impact energy of the composites.

A similar finding was also observed in the average impact strength of the two types of composites, as shown in figure 4. It was clear that for the composite specimens fabricated using plain woven kenaf/PVB composite recorded the higher values of impact strength compared with the composite specimens fabricated using epoxy resin, at different energy levels. This may be explained by that PVB film has higher elongation properties compared to epoxy resin due to its low viscosity which help in penetrating fabrics easily. High strength and viscosity of the epoxy resin have increased the specimen stiffness and decreased from its ability to absorb impacts. Furthermore, using a hot press technique to fabricate the kenaf/PVB composite has affected the impact properties of the composites which lead to good resin/fabrics penetration. As research studies have reported, the type of polymer is a critical factor which has an important influence on the impact strength of the natural fibres composites [15-17]. A similar trend has been reported in the study done by Rassmann et al. [18] on kenaf fibres reinforced three different resins; namely polyester, vinyl ester and epoxy. The three different resins affect the impact strength differently depending on the fibre volume fraction. It was evident that the impact strength of the epoxy laminates is slightly lower than that of the polyester and vinyl ester laminates for each of the three fibre volume fractions. Generally, it can be seen that the impact energy strongly depends on the resin system. Another mechanism that may contribute to the increase in composite strength is the low interface shear stress between a hydrophilic lignocellulosic fiber and a hydrophobic polymeric matrix [19]. As reported for curaua fibers [7, 20], ramie fibers [21] and coir fibers [22] reinforced epoxy composites. The incorporation of lignocellulosic fibres offers an obstacle to the propagation of an initial crack at the specimen notch, then cause others cracks at the weak fibre/matrix interface propagate longitudinally through the interface following the specimen length direction. Consequently, this mechanism is associated with a higher absorbed impact energy resulting in corresponding higher notch strength of the composite. Ultimately, it is worth mentioning that among the literature review results, the present results on plain woven kenaf fibers reinforcing PVB composites are the ones with the highest impact properties.

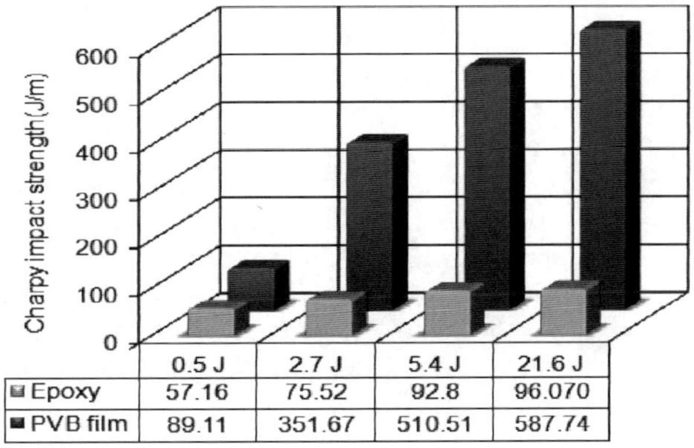

Figure 4. Charpy impact strength of the composites.

Similarly, the charpy impact toughness seems to have same impact energy and impact strength behaviors at different energy levels, as represented in the figure 5. It is clearly seen that the kenaf/PVB composite has greater impact toughness than the impact toughness of the kenaf/epoxy composite, especially at high energy levels. This means that the kenaf/PVB composite absorb the most energy before failure while the kenaf/epoxy composite absorb the least. The average impact toughness of two composites exhibited high difference when tested at high energy levels, approximately six times of the kenaf/epoxy composite value at both 5.4J and 21.6 J. While the average impact toughness of the two composites showed almost same behavior when tested at low energy levels. As stated by others [23, 24], the properties of the polymer were accountable for deciding the impact toughness of composites. This implies that both the effect of interlaminar delimitation and interfacial strength between fibre and matrix highly decide impact properties, especially in the case of weaving pattern as stated by researchers [25-27].

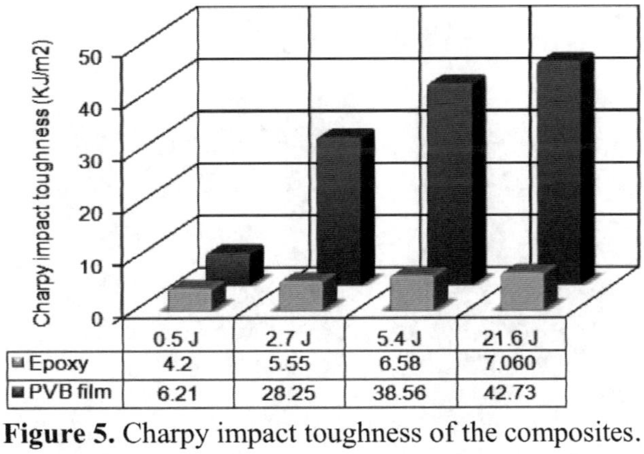

Figure 5. Charpy impact toughness of the composites.

5. Morphological Properties

SEM examinations of the charpy impact fractured surface of the kenaf/PVB film and kenaf/epoxy composites specimens are presented in figures 6 and 7. It was observed that for the two composites, there is no delamination between layers and the interfacial bonding between kenaf fiber and PVB film was the best. Similar failure modes have been reported for bamboo and coconut fibres reinforced epoxy composites [28]. Based on the morphology achieved, kenaf/PVB composite failed by fibre fracture while kenaf/epoxy composite failed by a combination of fibre pull-out and fibre fracture as well as crack propagations through the matrix. A tortuous path in crack propagation through the fibre/matrix interface and many hollow portions after the fracture can be seen in the micrograph of the kenaf/epoxy composite, indicating that the phenomenon of fibre pull-out occurred to a large extent. In figure 7 (b), the crack propagating through the epoxy matrix reaches a kenaf fiber causing decohesion and separation of fibrils and consequently reduced its strength. These failure mechanisms agree very well the impact damage observed by Aly et al. [25] for flax fibre reinforced epoxy composite. Generally, the strength of the composite is governed by the initiation and propagation of microscopic cracks through the matrix. This indicates that PVB film bonds better with the kenaf fibres (disorderly distributed in the matrix) than epoxy resin. The appearance of transverse cross sections of the kenaf fibre end reflects the absence of kenaf fibre pullout and ensures enhanced interfacial adhesion which leads to good impact resistance for the kenaf/PVB composite. Furthermore, both crosslinking of kenaf fibre and PVB film properties could create the complex load sharing behavior between the kenaf fibre and PVB film, as shown in figure 6 (a) and (b).

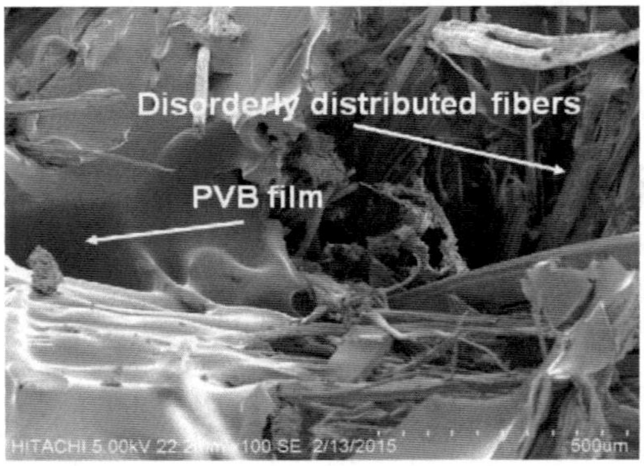

(a)

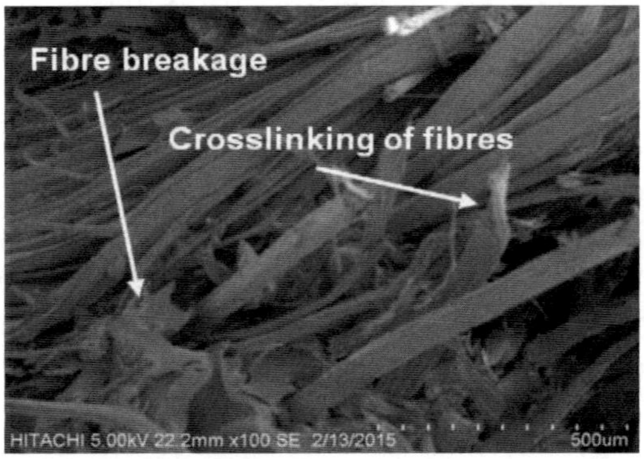

(b)

Figure 6. The SEM micrographs of the charpy impact failure surfaces of kenaf/PVB film composite.

(a)

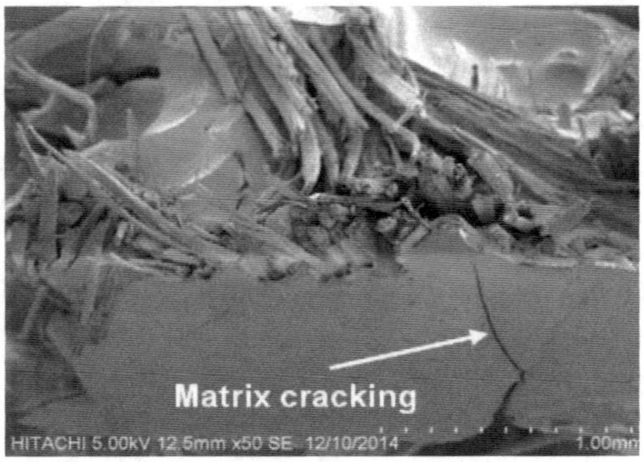

(b)

Figure 7. The SEM micrographs of the charpy impact failure surfaces of kenaf/epoxy composite.

6. Conclusion

Results of this study reveal that significant increases in energy absorption, strength and toughness of the kenaf/PVB composite were found at different energy levels. The impact strength and toughness of the kenaf/PVB film was six times of kenaf/epoxy composite, especially at high energy levels. SEM examinations of charpy impact test specimens show that there is no delamination between layers and the interfacial bonding between kenaf fiber and the two resins, but PVB film was the highest. In addition, kenaf/PVB composite failed by fibre fracture while kenaf/epoxy composite failed by a combination of fibre pull-out and fibre fracture as well as crack propagations through the matrix.

Acknowledgement

This work is supported by UPM under GP-IPM grant, 9422100 and GP-IPB grant, 9415402. The authors would like to express their gratitude and sincere appreciation to the Mechanical and Manufacturing Engineering Department and Aerospace Manufacturing Research Centre of the Universiti Putra Malaysia. Our appreciation and gratitude also extend to the Ministry of Higher Education & Scientific Research of Iraq and to the Material Engineering Department-College of Engineering - The University of Mustansiriyah for their scientific assistance and financial support.

References

[1] Jacob G C, Fellers J F, Simunovic S, Starbuck J M 2002 *Journal of composite materials* **36** 813-50
[2] Hamada H, Ramakrishna S 1996 *Journal of Thermoplastic Composite Materials* **9** 259-79
[3] Saravana Bavan D, Mohan Kumar G 2010 *Journal of Reinforced Plastics and Composites* **29** 3600-13
[4] Aji I, Sapuan S, Zainudin E, Abdan K 2009 *International Journal of Mechanical and Materials Engineering* **4** 239-48
[5] Samivel P, Babu A R 2013 *Int J Mech Eng Rob Res* **2** 348-60
[6] Ferreira A d S, Lopes F P D, Monteiro S N, Satyanarayana K G 2010 *Matéria (Rio de Janeiro)* **15** 131-7
[7] Monteiro S N, Lopes F P D 2007 *Proceedings of the 62nd International Congress of the Brazilian Association for Metallurgy and Materials* (Vitoria, ES, Brazil) 871-900
[8] Dhakal H N, Zhang Z Y, Richardson M O W, Errajhi O A Z 2007 *Composite Structures* **81** 559-67
[9] Dhakal H, Zhang Z, Bennett N, Reis P 2012 *Composite Structures* **94** 2756-63
[10] Meredith J, Ebsworth R, Coles S R, Wood B M, Kirwan K 2012 *Composites Science and Technology* **72** 211-7
[11] Yan L, Chouw N 2013 *Materials & Design* **51** 629-40
[12] Santulli C, Caruso A 2009 *Malaysian Polymer Journal* **4** 19-29
[13] Salman S D, Sharba M J, Leman Z, Sultan M, Ishak M, Cardona F 2015 *International Journal of Polymer Science* **2015**
[14] Venkatesh V, Izman S, Yap T, Brevern P, El-Tayeb N 2010 *International Journal of Precision Technology* **1** 287-301
[15] Aktaş M, Atas C, İçten B M, Karakuzu R 2009 *Composite Structures* **87** 307-13
[16] Sapuan S, Harun N, Abbas K 2007 *Journal of Tropical Agriculture* **45** 66-8
[17] Omar M F, Md Akil H, Ahmad Z A, Mazuki A A M, Yokoyama T 2010 *Materials & Design* **31** 4209-18
[18] Rassmann S, Paskaramoorthy R, Reid R G 2011 *Materials & Design* **32** 1399-406
[19] Nascimento D C O, Monteiro S N, Motta L C 2009 *Proceeding of the Characterization of Minerals, Metals & Materials-TMS Conference* 1-7

[20] Ferreira A S, Monteiro S N, Lopes F P D, Aquino R C M P 2008 *Proceedings of the 63rd International Congress of the Brazilian Association for Metallurgy and Materials* (Santos, SP, Brazil) 805-13

[21] Zhou N, Geng X, Ye M, Yao L, Shan Z, Qiu Y 2014 *Industrial Crops and Products* **56** 1-8

[22] Monteiro S N, Costa L L, Santafé H P G 2008 *Proceedings of the 18th Brazilian Congress on Materials Science and Engineering-CBECIMAT* 1-12

[23] Huber T, Bickerton S, Müssig J, Pang S, Staiger M P 2013 *Composites Science and Technology* **88** 92-8

[24] Joseph P V, Mathew G, Joseph K, Groeninckx G, Thomas S 2003 *Composites Part A: Applied Science and Manufacturing* **34** 275-90

[25] Aly N M, Saad M A, Sherazy E H, Kobesy O M, Almetwally A A 2013 *Journal of Industrial Textiles* **42** 204-18

[26] Sapuan S M, Leenie A, Harimi M, Beng Y K 2006 *Materials & Design* **27** 689-93

[27] Karimi S, Tahir P M, Karimi A, Dufresne A, Abdulkhani A 2014 *Carbohydrate Polymers* **101** 878-85

[28] Nuthong W, Uawongsuwan P, Pivsa-Art W, Hamada H 2013 *Energy Procedia* **34** 839-47

Numerical analysis of cocurrent conical and cylindrical axial cyclone separators

M A M Nor[1], H H Al-Kayiem[1] and T A Lemma[1]
[1]Department of Mechanical Engineering, Universiti Teknologi PETRONAS,32610 Bandar Seri Iskandar, Perak, Malaysia.

Email: azri.nor.my@gmail.com

Abstract. Axial concurrent liquid-liquid separator is seen as an alternative unit to the traditional tangential counter current cyclone due to lower droplet break ups, turbulence and pressure drop. This paper presents the numerical analysis of a new conical axial cocurrent design along with a comparison to the cylindrical axial cocurrent type. The simulation was carried out using CFD technique in ANSYS-FLUENT software. The simulation results were validated by comparison with experimental data from literature, and mesh independency and quality were performed. The analysis indicates that the conical version achieves better separation performance compared to the cylindrical type. Simulation results indicate tangential velocity with 8% higher and axial velocity with 80% lower recirculation compared to the cylindrical type. Also, the flow visualization counters shows smaller recirculation region relative to the cylindrical unit. The proposed conical design seems more efficient and suits the crude/water separation in O&G industry.

1. Introduction

Liquid-liquid cyclone separator works by inducing centrifugal forces (caused by the swirling fluid) to push denser liquid to the wall while lighter one migrates to the centre. The swirling motion is obtained by injecting the mixture tangentially into the cyclone body through small holes or deflecting the fluid using a swirl generator. The latter does not induce adequate swirling motion and the former is prone to droplet break ups due to high shear stress at the inlet but attains high swirl strength [1, 2]. Regardless of the inlet versions the unit is available as counter current or cocurrent, where the former and latter body shapes are cylindrical-conical and cylinder respectively. Axial cylindrical cocurrent cyclone is postulated to have a stable core, lower pressure drop and can operate at high capacity compared to the counter current version [3]. Though these sound promising, cylindrical cyclone suffers from recirculation region that promotes oil core-water mixing and break ups [4, 5]. It also suffers from swirl decay going downstream [3, 4, 6]. The use of conical body may reduce the recirculation region though this is proven in counter current version and is worth investigating for axial cocurrent [5]. Very little work however has been done on cocurrent cyclone such as by Schummer et al. [1], Dirkzwager [3], Slot [4], Rocha et al. [6] and Murphy et al. [7] in which cylindrical body is employed.

The literature review reveals that the application of axial inlet cocurrent conical cyclone separator using swirl generator is not yet available. The swirl generator could reduce droplet break ups by lowering shear stress and turbulence and produces symmetrical flow [3, 4]. The one suggested and tested by Slot has a cylindrical separation chamber which suffers from recirculation and lower downstream tangential velocity. This problem motivates us to propose modified technique that is the chamber has conical shape.

The objective of the present paper is to present a simulation procedure of the novel idea of axial cocurrent separator with conical chamber, and to discuss the simulation results. The conical shape separation section is employed to increase the tangential velocity to achieve high separation efficiency and smaller recirculation region. The paper presents and compares a preliminary numerical result of single phase flow of cylindrical and conical axial cocurrent cyclone separators using brine, and compare between the flow field parameters in the cyclone body. Several aspects studied are tangential and axial velocities, pressure distribution and recirculation region. This simulation is meant to visualize the flow field and improve the design of prototype for experimental investigations.

2. Numerical simulation

2.1. Description model

The cylindrical geometry employed is based on the unit used by Slot, figure 1 whom experimented and simulated water only liquid-liquid axial cyclone separator [4]. In the case of the conical geometry, the original cylindrical body tube is constricted 0.3° inwards axially to form a conical frustum but has the same separation length as the original. A small angle alteration is used to investigate the effect of having a conical body without altering too much of the original geometry. The swirl generator is excluded from the simulation and a swirling fluid is imposed at the entry at 62° deflection. The exclusion is found to have little effects on the overall flow properties of the cyclone [4].

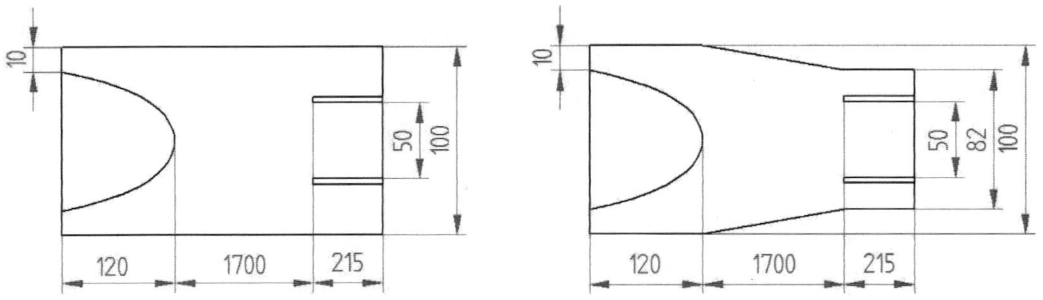

Figure 1. The left and right pictures are the cylindrical and conical cyclone respectively. (Drawn not to scale).

2.2. Mesh generation

Hexahedral mesh in ICEM is used as it is less diffusive than tetrahedral and reduces the cell count without compromising the accuracy. A mesh independent study is carried out for 800,000, 900,000 and 1.4 million elements. It is found that 900,000 produced acceptable results in relation to Slot experimental and simulation with reasonable computation cost as shown in figure 2.

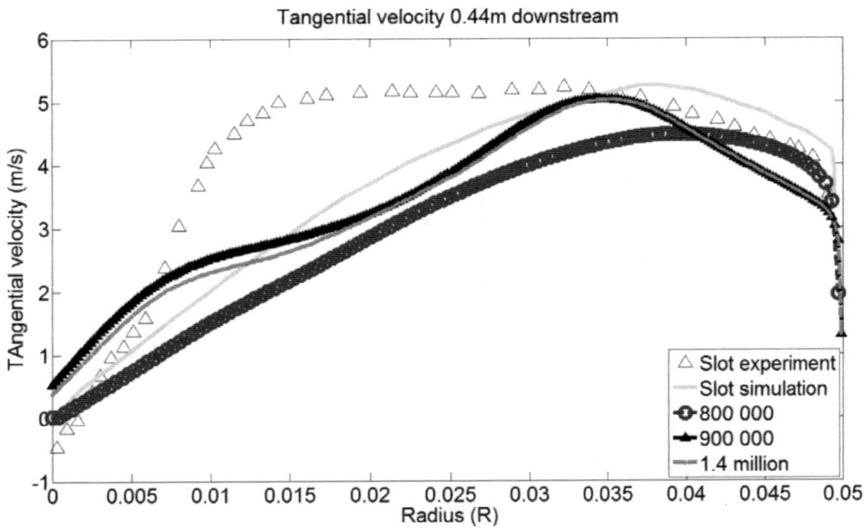

Figure 2. Mesh independence study denotes 900,000 is reasonably acceptable.

2.3. Boundary conditions

FLUENT with Reynolds stress model (RSM) is used as it is known to simulate swirling flow with acceptable accuracy in cyclonic flow. Linear pressure strain option is utilized and wall reflection terms option is disabled since it is found to acceptably predict the fluid motion [7]. SIMPLE is used for pressure-velocity coupling, PRESTO! for pressure and QUICK for the rest of the spatial discretizations [5]. The brine density and viscosity are 1067.80 kg/m3 and 1.18x10-3 kg/ms, respectively. Velocity inlet is used with tangential and axial velocity components at 5.71 m/s and 10.74 m/s, respectively. The central and annulus outlets are set as pressure outlets, with the central outlet flow split of 30%. The turbulent intensities at the inlet and outlets are set to 5% and 10% respectively. The hydraulic diameters at the inlet, central and annulus outlets are set to distances that define those boundaries. Standard wall function with no slip wall are implemented which could predict the near wall swirling flow properties satisfactorily [7, 8]. The simulation is carried out in transient of 0.5 ms and data sampling is enabled after the simulation has reached statistically steady for time averaged data for 1s.

2.4. Validation

The time averaged results are validated against Slot experimental and simulated data for the cylindrical version as in figure 3. A small gas core is observed (might lead to flow asymmetries) in Slot's experiment could cause the differences between the simulation and experimental data [4]. The average difference between simulated and experimental data is about 30%. This also might be caused by the fact that RANS models all turbulent length scales. The use of LES could lead to fewer discrepancies. Overall, the simulated axial and tangential velocity profiles have acceptable values in relation to experimental data and close to Slot simulation. The simulated and Slot simulation is close to one another which could be caused by the differences in meshing technique.

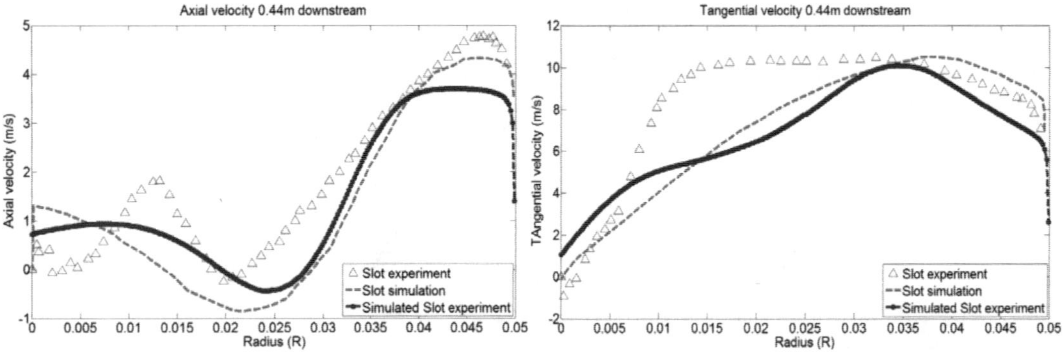

(a) (b)

Figure 3. Comparison of experimental and simulated data for the axial cyclone separator of (a) axial velocity and (b) tangential velocity. The sampling location is located 0.44m downstream from the swirl generator tail across the tube radius.

3. Results and discussion

3.1. Tangential velocity

Tangential velocities, V_{tan} at different axial locations are sampled at 0.6 and 1.4m from the swirl generator tail to fully capture the fluid behaviour downstream. The plots depict the formation of the Rankine Vortex in both cyclones which is favourable for efficient separation as in figure 4. The same profile exists regardless of counter current or cocurrent cyclone used [2, 3]. The conical also has 8% higher V_{tan} on average at all radial and axial locations relative to the cylindrical. The maximum V_{tan} at any axial location in a conical cyclone is higher and almost constant in magnitude compared to the cylindrical type. This similar-magnitude-behaviour is typically seen in cylindrical-conical counter current cyclone in the conical part [9] which prompts it's beneficial usage in a conical cocurrent cyclone.

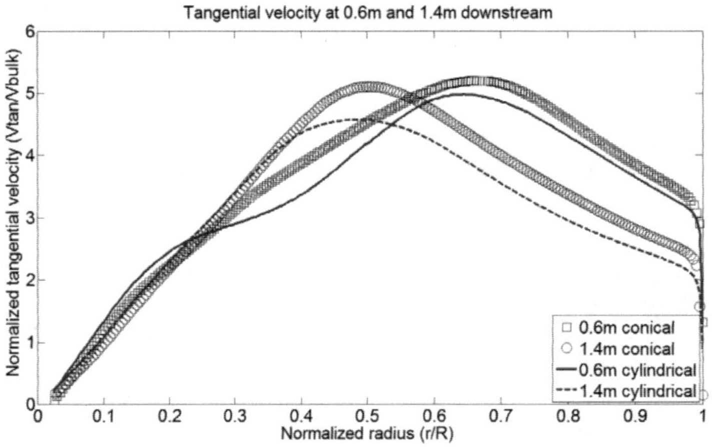

Figure 4. Tangential velocity plots for both cyclones at various downstream locations. The values are normalized against 2 m/s bulk velocity.

The maximas for both cyclones also appear to shift to the centre pipe due to swirl decay [7]. From the maximum Vtan to the wall, the conical has higher velocity which could be caused by the fact that the conical annulus outlet is smaller than the cylindrical's; also seen in a counter current cyclone as the underflow diameter is reduced [10]. Such event is favoured as higher centrifugal force can be imposed on the particle for better separation even at the downstream before exiting. In figure 5, both cyclones portray velocity decay downstream due to swirl decay caused by wall friction [7], but conical cyclone still attain higher Vtan (from the max Vtan to the wall) than cylindrical. It is evident that having a conical body can increase the Vtan along the axis for higher performance. The plot shows that conical cyclone can attain higher tangential velocity radially downstream than cylindrical version. The values are normalized against 2 m/s bulk velocity.

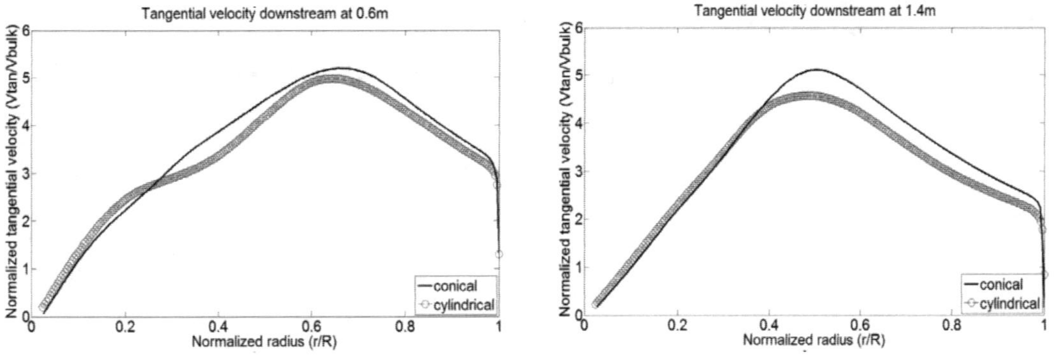

Figure 5. Normalized tangential velocity predicted by the simulation for the conical and the cylindrical types of the hydrocyclone, at 0.6 and 1.4m downstream of the swirl generator.

3.2. Axial velocity

Positive velocity in Figure 6 denotes favorable fluid flow towards the exits whereas negative dictates recirculation towards the inlet. The figure depicts the half W shaped profile (sampled across the radius) and clearly portrays the annulus recirculation denoted by the line plots below the blue line. This region has also been noticed by Kegge [8], Mattner et al. [11]and Ko [12]. Conical cyclone possesses better axial velocity, Vaxial profile than its cousin notably by having smaller recirculation flow magnitude at all axial lengths (80% average improvement). The reverse flow region is critical and must be minimized as it prevents droplets from migrating to the inner vortex for separation [5]. The lowered recirculation region also leads to higher Vtan and Vaxial which is evident from plots in figure 5 and figure 6. The iso-surface of -0.1m/s is generated for better visualisation of such region. Various iso-surface values are tested and the recirculation region is indeed thinner and shorter in conical cyclone than in cylindrical's which is continuous (figure 7).

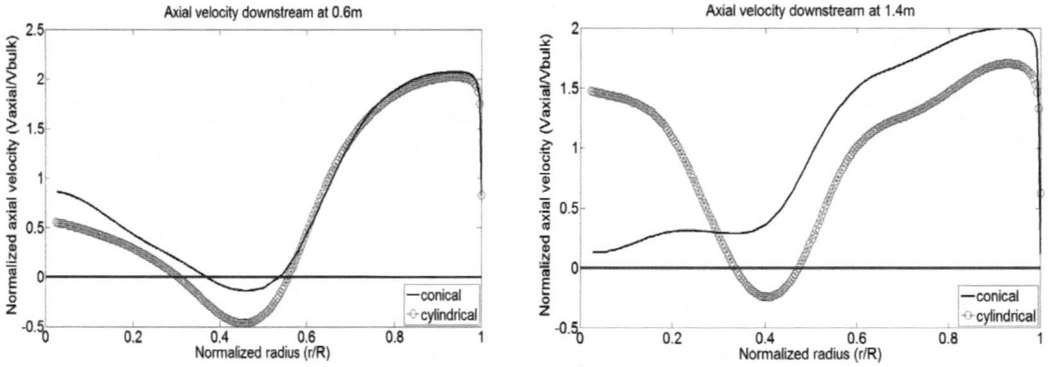

Figure 6. Axial velocity plots at various downstream locations between conical and cylindrical cyclones. The values are normalized against 2 m/s bulk velocity.

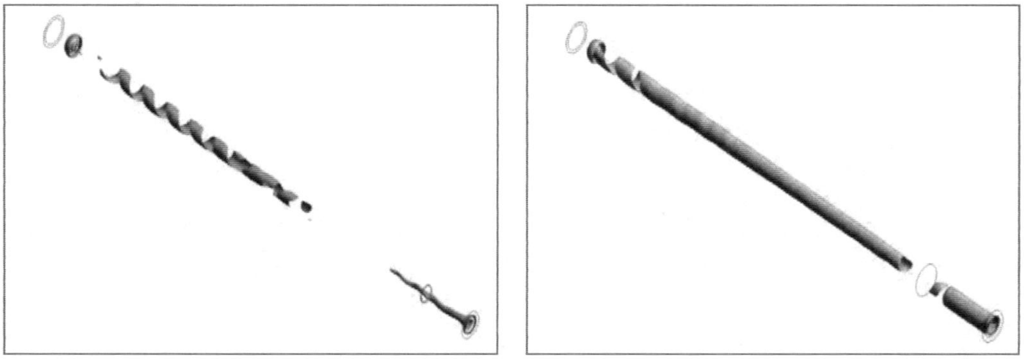

Figure 7. Iso-surface of Vaxial -0.1 m/s. The left and right are the conical and cylindrical units, respectively.

4. Conclusions

Numerical analyses of a single phase conical and cylindrical cocurrent cyclones have been carried out based on an experiment conducted by Slot. Based on the preliminary findings, it can be concluded that the conical cocurrent cyclone can achieve greater separation efficiency in comparison to the cylindrical type. This is achieved by having greater tangential velocity (8% maximum Vtan on average), lower reverse axial velocity (80% lower on average) and smaller recirculation region through the use of conical body. The next step is to carry out parametric study in search of the optimal design.

Acknowledgements

The authors would like to extend heartiest appreciation to the Universiti Teknologi PETRONAS for the financial and technical support for the production of this paper under YUTP - FRG fund 015-3AA-A72.

References

[1] Schummer P, Noe P, Baker M. LDV measurements in the vortex flow created by a rotating wall dewatering cyclone. In: Svarovsky L, Thew M T, editors. Hydrocyclones—Analysis and Applications, Kluwer. 12: Springer Netherlands; 1992. p. 359-76.

[2] Shi S-y, Xu J-y, Sun H-q, Zhang J, Li D-h, Wu Y-x 2012 *Chemical Engineering Research and Design* **90** 1652-9

[3] Dirkzwager M 1996 *A new acial cyclone design for fluid-fluid separation* TU Delft, Delft University of Technology)

[4] Slot J J. Development of a centrifugal in-line separator for oil-water flows [Ph.D. thesis]: University of Twente; 2013.

[5] Noroozi S, Hashemabadi S H 2011 *Chemical Engineering Research and Design* **89** 968-77

[6] Rocha A D, Bannwart A C, Ganzarolli M M 2015 *International Journal of Heat and Fluid Flow* **53** 81-90

[7] Murphy S, Delfos R, Pourquie M J B M, Olujić Ž, Jansens P J, Nieuwstadt F T M 2007 *Chemical engineering science* **62** 1619-35

[8] Kegge S J. Numerical simulation of an axial hydrocyclone [M. Sc. Thesis,]: Delft University of Technology; 2000.

[9] Saidi M, Maddahian R, Farhanieh B 2013 *Heat and Mass Transfer* **49** 247-60

[10] Elsayed K. Analysis and Optimazation of Cyclone Separators Geometry Using RANS and LENS Methodologies [PhD]: Vrije Universiteit Brussel; 2011.

[11] Mattner T W, Joubert P N, Chong M S 2002 *Journal of Fluid Mechanics* **463** 259-91

[12] Ko J, Zahrai S, Macchion O, Vomhoff H 2006 *Wiley Subscription Services, Inc, A Wiley Company* **52** 3334-44

Monotonic and fatigue properties of kenaf /glass hybrid composites under fully reversed cyclic loading

M J Sharba[1,3], Z Leman[1], M T H Sultan[2], M R Ishak[2,4] and M A A Hanim[1,4]
[1]Department of Mechanical and Manufacturing Engineering, Faculty of Engineering, Universiti Putra Malaysia, 43400 Serdang, Selangor
[2]Aerospace Manufacturing Research Centre (AMRC), Level 7, Tower Block, Faculty of Engineering, 43400 UPM Serdang, Selangor, Malaysia
[3]Mechanical Engineering Department, AL-Suwayrah Technical institute, Middle Technical University, Iraq.
[4]Laboratory of Bio-Composites Technology, Institute of Tropical Forestry and Forest Products (INTROP), Universiti Putra Malaysia, 43400 UPM Serdang, Selangor, Malaysia

E-mail: mohaimaneng@gmail.com

Abstract. The aim of this work is to investigate the effect of hybridization of kenaf-glass fibers reinforced unsaturated polyester on fatigue life. Three types of composites were fabricated using hands lay-up method, namely, kenaf, glass, and hybrid composites with 30% of weight fraction, the hybrid was mixed with a ratio of kenaf: glass 10:20. Monotonic tests were achieved (Tensile and compression) to determine the fatigue stress levels. Fully reversed fatigue loading was conducted with a stress ratio of -1 and stress levels 55-85 % of the ultimate static stresses, all tests were conducted at 10 Hz of frequency. The results proof a positive hybrid composite; also agree with the rule of mixture that can predict the final composite properties. Moreover, it's been observed an improvement in overall mechanical properties of hybrid compared to individual ones.

1. Introduction

Composites are an important category of materials for engineering applications. They form an essential part in the design process in many sectors, including the automotive, marine and aircraft industries. Over the past decade, there has been an increased demand for ''green'' or natural-fiber reinforced composites [1, 2].

Natural fiber reinforced plastics materials are low cost, light-weighted, have enhanced mechanical properties, free from health hazards, and thus have the potential for structural applications. Furthermore, studies were tried to extend the usage of natural fiber reinforced composites and improving their performance by using woven structure of natural fibers reinforced composites especially kenaf fiber [3, 4]. Despite the attractiveness of natural fiber reinforced polymer matrix composites, they suffer from lower modulus, lower strength, and relatively poor moisture resistance compared to synthetic fiber reinforced composites such as glass fiber reinforced plastics (GFRP) [5].

A combination of synthetic fiber with natural fibers in one resin to form a new hybrid laminated composites with an interesting balance to the material pros and cons of the individual ones is an alternative way to discover new materials to consumers. Hybrid composites are extensively used in many engineering applications [6].

Several studies reported the fatigue behavior of natural fiber reinforced composites and reported a considerable fatigue resistance. Abdullah et al. [7] investigated the fatigue life and fiber loading effect of unidirectional kenaf reinforced epoxy under tension-tension fatigue loading, used a stress ratio of 0.5 at 5 stress levels and 5 Hz of frequency. It was found that kenaf fiber with higher volume fraction showed better fatigue life. Kalam et al. [8] studied the effect of oil palm fruit punch fiber loading on fatigue properties of oil palm reinforced epoxy composite. The uniaxial tension-tension fatigue loading was conducted with 0.1 and 0.5 of stress ratios and 20 Hz of frequency. They concluded that increasing the fiber content reduced the fatigue life of composites. Liang et al. [9] reported the fatigue behaviors of flax reinforced epoxy under uniaxial tension-tension loading under 0.1 stress ratio and 5 Hz frequency. Recently, Abdullah et al. [10] investigated the fatigue endurance of unidirectional Arenga Pinnata fiber reinforced epoxy under tension-tension mode with stress ratio of 0.1 and 10 Hz frequency and found that the fatigue life improving by reducing the stress level.

Despite of the considerable fatigue endurance of natural fibers based composites, it can be found that only exists in low cyclic fatigue regime which limited the applications of natural fibers composites. As a result, a new interested trend of using natural fibers by hybridizes them with synthetic fibers to improve their strength and extend the range of their applications. Regarding fatigue life of natural-synthetic hybrid composites, it can be noticed that a few studies reported a fatigue results of natural-synthetic hybrid composites. Thew et al. [11] used Bamboo-glass reinforced Polypropylene hybrid composites in sandwich sequence, compared to pure bamboo it was found that the hybrid has shown better fatigue life and more resistant to environmental aging. Bagheri et al. [12] has selected flax-carbon reinforced epoxy hybrid composite for biomedical application. Even though, there was no complete comparison with individual fibers in this study, the results showed improvement in fatigue strength compared with previous work on pure flax.

This work presents the mechanical properties of non-woven kenaf; woven roving glass and kenaf-glass hybrid reinforced composites were studied and characterize the fatigue performance of composites through S-N diagram, specifically the effect of hybridization on uniaxial tension–compression cyclic loading. The results were analyzed and a comparison was made among the three campsites. The tension – compression mode was selected in fatigue test at a stress level of (-1) under four stress levels from 55-85 % of ultimate stress of composite.

2. Experimental details

The materials selected for this study were non-woven kenaf randomly oriented, woven roving, E- glass EWR600 type which used as a reinforcement, and unsaturated polyester with 2% of catalyst as a matrix as shown in figure 1. Both fibers were used as received without any chemical or thermal treatment.

Figure 1. Kenaf and Glass fibers used to fabricate composites.

Composite laminates were fabricated using hands lay-up technique, two steel plates covered with transparent paper to ensure a good surface finishing. The reinforcements were used without any pretreatment. Three types of composites were fabricated (pure glass, kenaf-glass and pure kenaf) with an orthopathalic unsaturated polyester mixed with 2 wt% Butanox M60 MEKP catalyst. A static weight of 15 kg was used for post curing for 24 hours before cutting the samples. The fiber weight fraction Wf of composites was calculated using the ratio of fiber weight to total composite weight, the weight fraction of glass/UP laminate was 45%, for kenaf/UP 20% and for hybrid composite 30% with a ratio of kenaf: glass 10:20 % of fiber weight, the final samples that have been prepared from composite laminates as shown in figure 2.

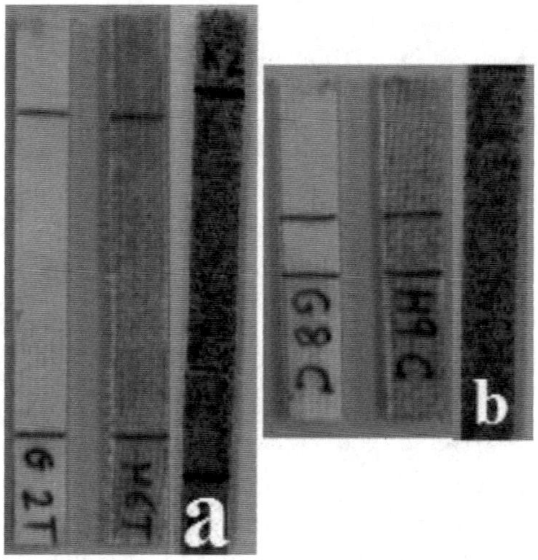

Figure 2. The final samples prepared for testing (a) tensile and (b) compression.

3. Mechanical tests

In order to achieve fatigue, stress levels, the ultimate tensile and compression stresses should be determined from monotonic tests. The ultimate tensile stress for composites was measured through tensile test and conducted according to ASTM D3039, on universal testing machine Instron 3366 equipped with 100kN load cell. Four specimens of (250 mm length and 25 mm wide) were tested with a crosshead speed of 2 mm/min and the average stress was taken.

The ultimate compressive stress for composites was calculated from the static compression test, which conducted according to ASTM D-3410, on universal instron 3366 of 100 kN the load cell. Four specimens (140mm length and 15 wide) were tested at a crosshead speed of 1 mm/min and the average compression stress was taken. A gauge length of 14 mm was used to minimize buckling effect, according to test standard recommendation.

Fatigue specimens were prepared and the fatigue test procedure was followed according to ISO 13003. Fully reversed loading condition (tension-compression) fatigue test was conducted by servo-hydraulic universal testing machine Instron 3374 equipped with 100 kN unit cell, the stress ratio of (-1) at 10 Hz of frequency. The test was performed at four stress levels started from 55-85 % of the static ultimate stress of composite. Eight samples (120 mm length and 15 mm wide with minimum gauge length to prevent buckling) were prepared for each group of material in terms of two samples per stress level.

4. Result and Discussion

The stress-strain results show in figure 3, the effect of hybridization of kenaf-glass reinforced unsaturated polyester. Glass fiber composite presents superior tensile properties with ultimate tensile stress 148 MPa, and poor ultimate tensile stress for kenaf mat fiber composite of 12 MPa. In between

10% of kenaf and 20% of glass fiber of fiber weight gives a significant improvement compared to pure kenaf to reach for 48 MPa of UTS for hybrid composite as shown in figure 4. This called positive effect of hybridization when the main parameter improved compared to individual material [13]. It's clear that the hybridization in this study follow the rule of mixture, especially for the tensile properties (UTS, elongation at break and the tensile modulus) and the range value of properties lie between the two individual ones, similar trend were reported in previous studies [14-17].

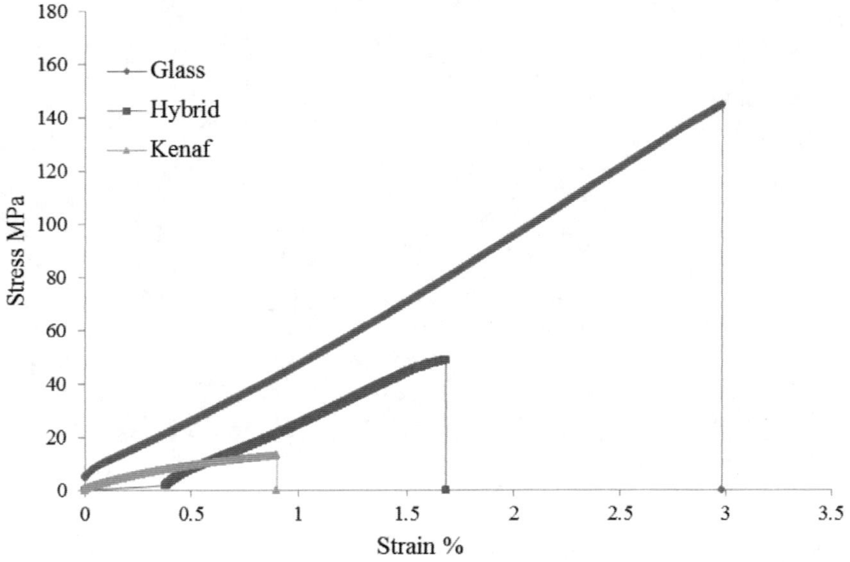

Figure 3.Stress-Strain relations of glass, hybrid and kenaf reinforced UP.

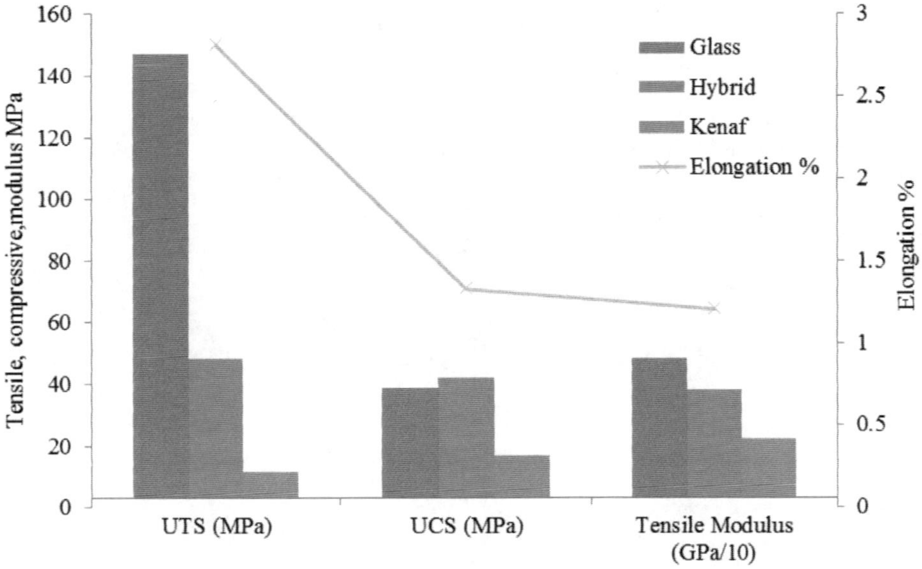

Figure 4. Comparison of mechanical properties of three types of composite.

On the other hand the compression results show the relatively lower value of the ultimate compression stress compared to tensile stress and it was just 38 MPa, the hybridization effect was more noticeable in compression test and shows higher values of average 42 MPa as shown in figure 5. This can be explained due to the high compression stress of kenaf fiber. The failure modes of composites studied, matrix cracking and fiber breakage was the general mode for glass and kenaf fiber composites, while the hybrid showed some fiber pull out and delamination near to failure zone, the main reason of this is the difference in strain of the two fibers that can delay or speed the fiber breakage failure [18].

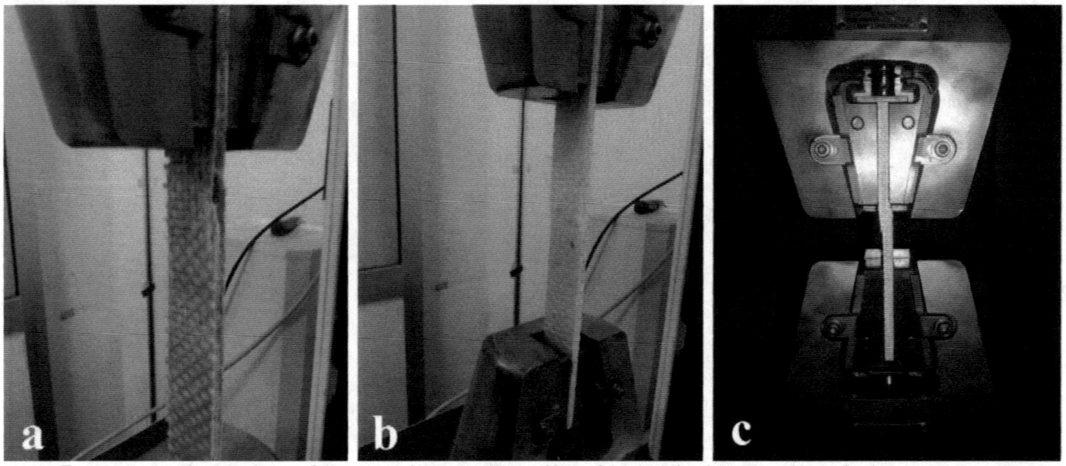

Figure 5. Monotonic testing of composites (a) Tensile of hybrid; (b) Tensile of glass; (c) compression of glass.

The fatigue S-N diagram as shown in figure 6 presents the results of three types of composites tested in this work under the most aggressive mode, fully reversed cyclic loading. The results display a normal scattering in their values, especially in hybrid and kenaf composites due to a variety of natural fiber behavior. The inconsistency in data scatter could be due to the presence of void in the specimen tested. Other data were scattered in a reasonable range [8]. It is been observed that the number of cycle to fail increases when maximum stress reduced.

The failure modes of the specimens under cyclic loading was monitored during its fatigue life and found there was a matrix cracking in the early stage of fatigue test. The final failure mode was observed in static tensile and tension-compression fatigue is similar, in brittle manner with a serrate fracture surface and showed small delamination area with fiber splitting for hybrid composite as shown in figure 7. Similar observations were also made by [19].

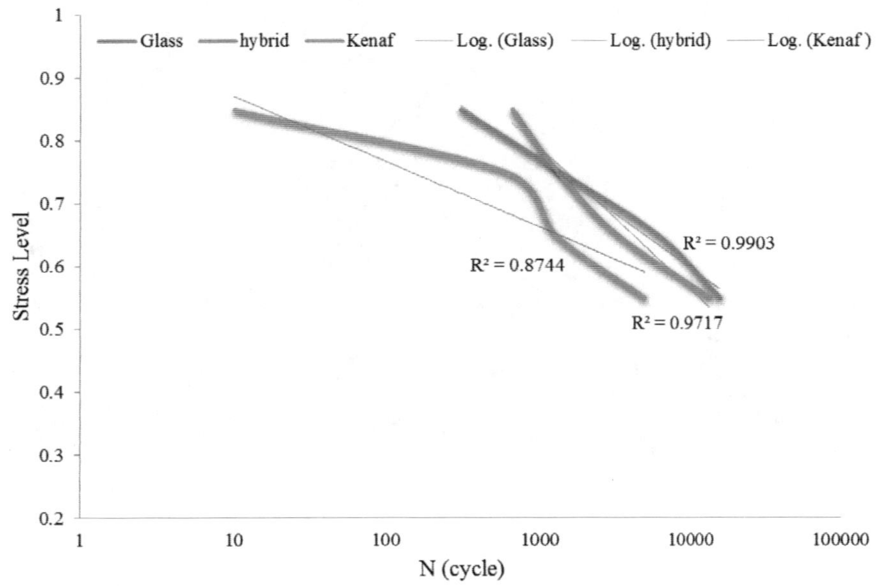

Figure 6. S-N curves of glass, kenaf, and hybrid composites

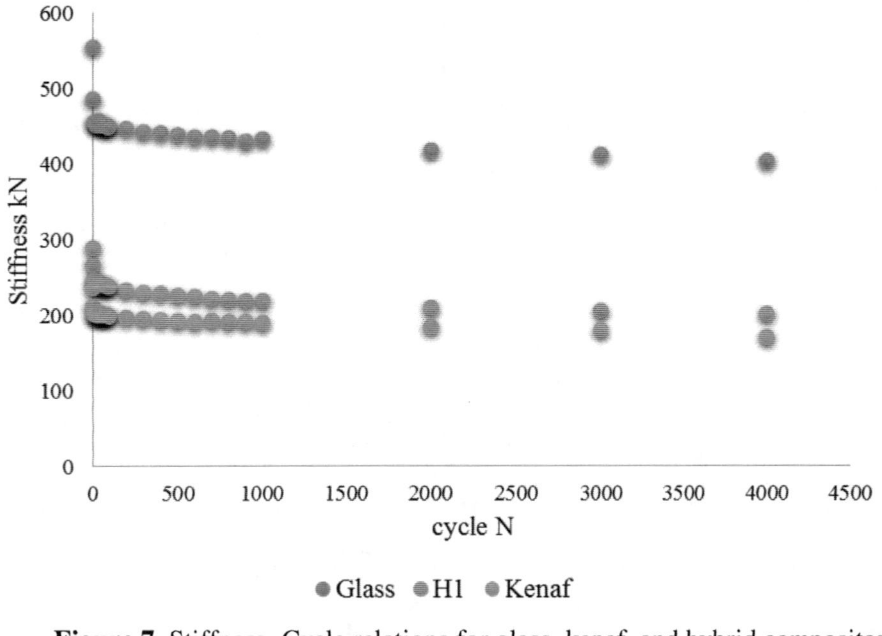

Figure 7. Stiffness- Cycle relations for glass, kenaf, and hybrid composites.

The figure 8 compared the stiffness degradation at the same duration of fatigue life for the three types of composites and it's surprising that stiffness degradation of kenaf composite slower than glass fiber composite, this can be explained due to the ability of natural fiber continuously transfer loads onto adjacent fibers, this is also supported by the observation reported by [20].

Figure 8. Fatigue test (a) kenaf at failure, (b) hybrid at 10^2 cycles and (c) hybrid at 10^4 cycles.

5. Conclusion

In conclusion, the main point of this study is to identify the hybridization effect on the fatigue life of natural- synthetic fibers reinforced unsaturated polyester. The results proof a positive hybrid composite; also agree with the rule of mixture that can predict the final composite properties. Kenaf-glass hybrid composite shows good tensile properties compared to pure kenaf and better compression stress compared to pure glass, which gives an overall improvement in mechanical properties. The kenaf fiber composite can be used for structural and aerospace applications due to cost-effective properties.

Acknowledgement

This work is supported by UPM under GP-IPS 9438714 grant. The authors would like to express their gratitude and sincere appreciation to the Ministry of Higher Education & Scientific Research of Iraq for its scientific assistance and financial support.

References

[1] Fotouh A, Wolodko J D, Lipsett M G 2014 *Composites Part B: Engineering* **62** 175-82
[2] Yahaya R, Sapuan S M, Jawaid M, Leman Z, Zainudin E S 2014 *Polymer Composites* n/a-n/a
[3] Salman S D, Sharba M J, Leman Z, Sultan M T H, Ishak M R, Cardona F *International Journal of Polymer Science*
[4] Dan-Mallam Y, Abdullah M Z, Yusoff P S M M 2014 *Polymer Composites* **35** 1900-10
[5] Jawaid M, Abdul Khalil H P S 2011 *Carbohydrate Polymers* **86** 1-18
[6] Yahaya R, Sapuan S M, Jawaid M, Leman Z, Zainudin E S 2014 *Materials & Design* **63** 775-82
[7] Abdullah A H, Alias S K, Jenal N, Abdan K, Ali A 2012 *Fatigue Behavior of Kenaf Fibre Reinforced Epoxy Composites* **16**
[8] Kalam A, Sahari B B, Khalid Y A, Wong S V 2005 *Composite Structures* **71** 34-44
[9] Liang S, Gning P-B, Guillaumat L 2014 *International Journal of Fatigue* **63** 36-45
[10] Abdullah A H, Mat M F, Fatigue Life of Arenga Pinnata/Epoxy Composites. Advanced Materials Research; 2015.
[11] Thwe M, Liao K 2003 *Journal of Materials Science* **38** 363-76
[12] Bagheri Z S, El Sawi I, Bougherara H, Zdero R 2014 *Journal of the Mechanical Behavior of Biomedical Materials* **35** 27-38
[13] Front matter. In: Harris B, editor. Fatigue in Composites: Woodhead Publishing; 2003. p. i-iii.
[14] Samal S K, Mohanty S, Nayak S K 2009 *Polymer-Plastics Technology and Engineering* **48** 397-414

[15] Kushwaha P K, Kumar R 2010 *Journal of Reinforced Plastics and Composites* **29** 1952-62
[16] Jarukumjorn K, Suppakarn N 2009 *Composites Part B: Engineering* **40** 623-7
[17] Naidu V N P, Kumar M A, Reddy G R, Khanam P N, Reddy M M, Chakradhar K V P 2011 *International Journal of Macromolecular Science* **1** 19-22
[18] Atiqah A, Maleque M A, Jawaid M, Iqbal M 2014 *Composites Part B: Engineering* **56** 68-73
[19] Shah D U, Schubel P J, Clifford M J, Licence P 2013 *Composites Science and Technology* **74** 139-49
[20] Liang S, Gning P B, Guillaumat L 2012 *Composites Science and Technology* **72** 535-43

IOP Conf. Series: Materials Science and Engineering **100** (2015) 012056 doi:10.1088/1757-899X/100/1/012056

Dynamic Response and Signal to Noise Ratio Investigation of NIR-FBG Dynamic Sensing System for Monitoring Thin-walled Composite Plate

Z M Hafizi[1, 2], J Epaarachchi[2]
[1]Advanced Structural Integrity and Vibration Research (ASIVR), Faculty of Mechanical Engineering, Universiti Malaysia Pahang (UMP)
26600 Pekan, Pahang, Malaysia.
[2]Faculty of Health, Engineering and Sciences, University of Southern Queensland
Toowoomba, 4350 QLD, Australia

Email: hafizi@ump.edu.my

Abstract. Optical fiber systems for dynamic response measurement become an attractive study nowadays especially in the field of Structural Health Monitoring (SHM). This work presents the investigation of an optical fiber sensor system utilizing Near Infra-Red Fiber Bragg Grating (NIR-FBG) sensor for SHM of a thin-walled composite structure. In this study, a comparison between the experimental result and simulation study using finite element analysis has been presented. By comparing both results, the FBG dynamic sensing system was shown to have an excellent capability in acquiring dynamic response due to flexural wave propagations. In the meantime, a signal to noise ratio (SNR) study was also performed to several FBG dynamic sensor systems; particularly to see the comparison between NIR-FBG sensor and 1550 nm FBG sensor. Furthermore, with a proper configuration, an NIR-FBG system was proved to have better performance than the 1550 nm based FBG sensor.

1. Introduction

Optical-based sensor system has a huge potential nowadays. One of the significant benefits of using these optical sensors is the ability of utilizing the advanced technology of optical transmission systems. In other words, with the technology, distance becomes less concerned. It is an undeniable fact that these light-wave communication systems are very successful, especially in recent telecommunications applications. Besides having low transmission loss, this system also offers impressive advantages; small size and less weight of optical fiber cable, compared with conventional wiring cable.

An FBG in a dynamic sensor system usually involves a special interrogation system where wavelength variations; due to strain changes or temperature variations, can be converted into an electrical signal. The FBG sensors are fabricated in the core region of specially fabricated single mode low-loss germanium doped silicate optical fibers. The grating is the laser-inscribed region which has a periodically varying refractive index. This region reflects only a narrow band of light corresponding to the Bragg wavelength λ_B, which is related to the grating period Λ_o [1],

$$\lambda_B = \frac{2n_o\Lambda_o}{k} \tag{1}$$

where k is the order of the grating and n_o is the initial refractive index of the core material prior to any applied strain. Due to the applied strain, ε, there is a change in the wavelength, $\Delta\lambda_B$, for the isothermal condition,

$$\frac{\Delta\lambda_B}{\lambda_B} = \varepsilon\, P_e \tag{2}$$

where P_e is the strain optic coefficient and is calculated as 0.793.

Based on available literatures since early 20th centuries, the basic design of an FBG dynamic sensing system can be can be summarized as shown in figure 1. There are four major components in a basic FBG dynamic sensing as labelled by numbers in figure 1. In the third component; an FBG can be placed on the structures under monitoring by surface glue or embedding between layers. Meanwhile the fourth element is the most complex part; since it is the part where the optoelectronic conversion took place. The design of this part depends on the type of light sources used, either broadband or narrowband light sources. The use of photo detectors in this crucial part provides a lot of optical noises, thus need a proper digital or analog signal filtering. However, some systems have a complex arrangement. Although it may achieve the required results, yet, the more complicated systems may result in extra optical losses and relatively higher cost. There are two popular interrogation methods that are available for high frequency vibration signals with FBGs; that are edge filter detection and power detection methods [2].

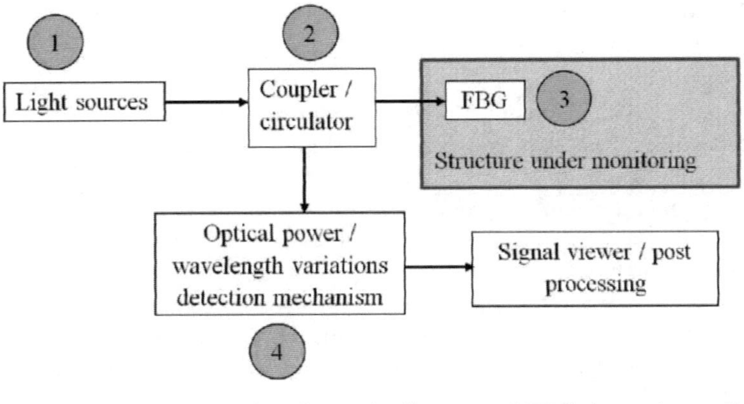

Figure 1. The basic schematic diagram of FBG dynamic sensing.

1.1. Noise in FBG dynamic system

The theory of the SNR has been explained in many texts [3, 4], as the ratio of signal power to the noise power. It should be noted that, the better the SNR of the system, the sensitivity to the higher frequency contents of wave propagations will also increase. There are many ways that have been investigated in order to improve the SNR of an FBG-based dynamic system; such as using smaller grating [5] and cantilever type FBG bonding [6]. However, one of the main criteria for having a sensitive system is the ability of providing higher optical power to the system. In terms of power, P, the SNR in decibel (dB) can be written as

$$SNR_{dB} = 10\log\frac{P_s}{P_n} \tag{3}$$

where P_s and P_n indicate the average power of the signal and noise, respectively.

Optical noise is necessary to be considered when designing the FBG dynamic sensing. There are a lot of possible noises that can occur in an optical based system. Based on Figure 1, there are two parts that might contribute to the possibilities of optical noises: the light sources (in element 1) and the photodiodes (in element 2). The noises from light sources can be neglected if LEDs types of light source are used. If using lasers, possible optical noises can be from the mode partition noise, laser intensity noise and laser phase noise [7]. These noises are relatively small and usually attenuate before reaching the receiver part. However, if they are able to reach the photodiodes, the noise signals can appear in the resultant output signals. Usually the use of appropriate laser driver can eliminate any significant noises associated with the lasers.

The biggest noises in an FBG dynamic sensing system are due to the photodiodes (PD) [1]. The use of PIN type of PD will reduce the amount of noise to only three kinds: thermal noise, quantum and dark current noise. However, the impact of the dark current noise is usually ignored; since its power is relatively so low [7]. Meanwhile, quantum noise is not to be a major concerned in PIN PD. However, the impact of the thermal noise cannot be neglected, since it is always visible in the signal viewer; even if optical signals are not present, as illustrates by figure 2.

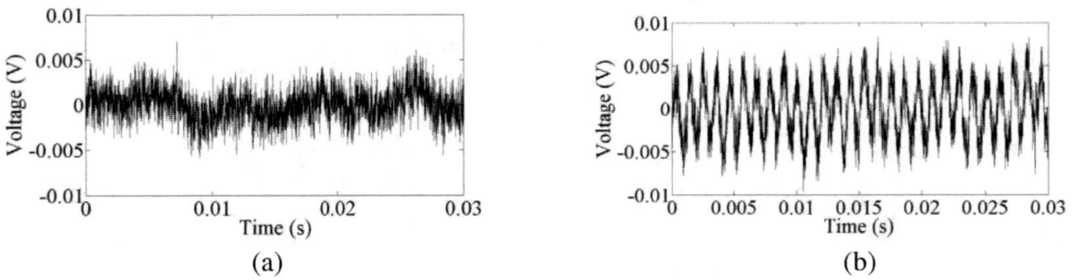

Figure 2. Thermal noise existence in an FBG dynamic system. (a) Output signal without any dynamic input and (b) Output signal due to a vibration.

The thermal noise is generated by the process of converting the photocurrent to voltage by the load resistor. The load resistor creates its own noise because of the random thermal motion of electrons [7]. Increasing the load resistor may reduce the noise and increase the receiver's sensitivity; however it will also affect the receiver bandwidth thus, causing limited bandwidth. In order to investigate the SNR of the FBG dynamic system; particularly in the thermal noise effect, an experiment with several types of FBG dynamic sensing systems was conducted. The case study will be explained in the next section.

1.2. NIR-FBG system

In this research, NIR-FBG sensor in FBG dynamic sensing system was also investigated. Through limited literatures available, an NIR-FBG sensor was shown to have a sufficient capability to perform as a static and dynamic sensor [8-11]. Meanwhile, [10] has successfully validated the capability of an NIR-FBG system to be applied in an SHM system. The NIR-FBG system was also revealed to have a relatively comparable performance with the conventional 1550nm-FBG sensor which has been widely used nowadays. Furthermore, in an NIR-FBG system, contain a relatively cheap silicon-based technology.

2. Experimentation

A thin glass / epoxy sample was prepared. The symmetric sample was manufactured with 8 plies of laminates; thickness 4 mm and has the orientation of [0°,90°,-45°,45°,45°,-45°,90°,0°]. Figure 3(a) shows the detailed geometry of the sample. Two FBGs with different wavelengths, 830 nm (NIR) and 1550 nm, were surface-glued by using the epoxy resin and were placed near to each other as illustrated by figure 3(b). A shaker model number V201-M4-CE, from the LDS Test and Measurement Ltd; was attached approximately at the middle of the composite specimen.

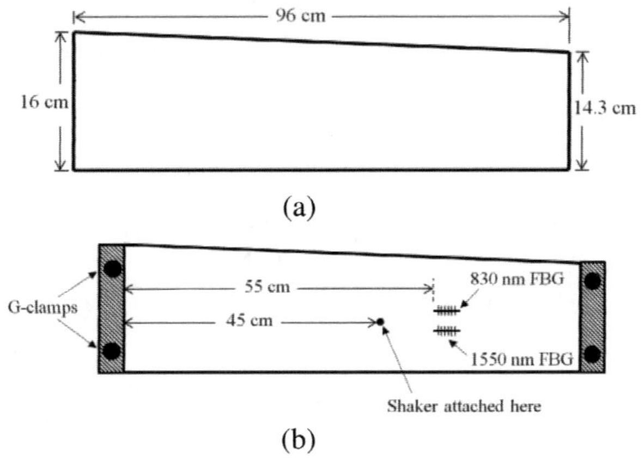

Figure 3. Thin plate specimen.

Each FBGs (NIR and 1550 nm), was connected with two different kinds of arrangements in order to measure and compare the SNR from each system. All four configurations of FBG dynamic sensing systems can be summarized as in figures 4; whereby the systems were labeled as System A, B, C and D. For System A, C and D a common configuration has been used; whereby this type of arrangement has been reported in several research works [12, 13]. Note that in System A, the NIR-FBG sensor was utilized, instead of the usual 1550 nm FBG sensor. Meanwhile System C and System D were distinguished by the use of two different optical components: the optical coupler and optical circulator, as illustrated by the figure 6 and 7. For System B, only single FBG sensor was used; which is the NIR-FBG. Its configuration and detail sensing mechanism has been explained thoroughly in [10], and it was sufficient for detecting the location of an impact on a thin composite plate. The System B had its own advantages, where it only required a 'stand-alone' NIR-FBG sensor.

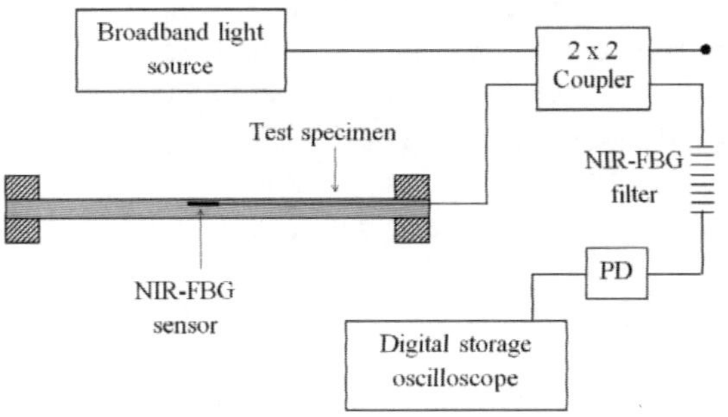

Figure 4. NIR-FBG dynamic sensing with a fixed FBG filter (System A).

IOP Conf. Series: Materials Science and Engineering **100** (2015) 012056 doi:10.1088/1757-899X/100/1/012056

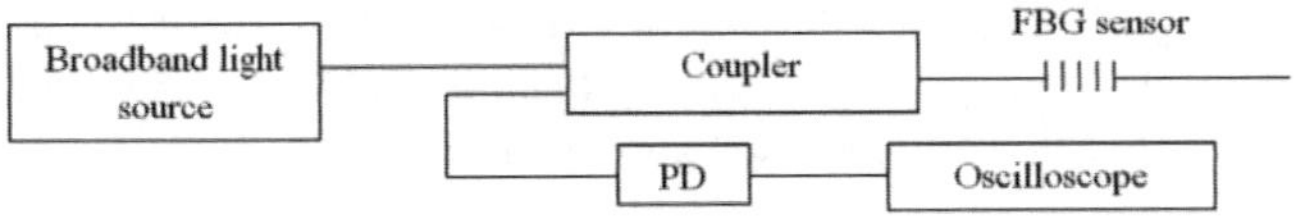

Figure 5. NIR-FBG dynamic sensing with a single FBG (System B).

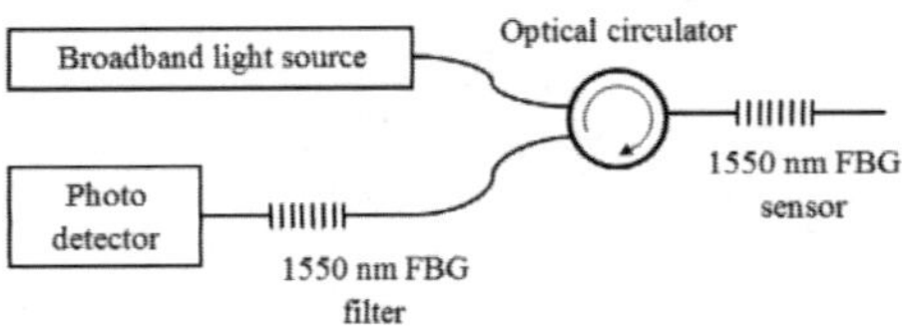

Figure 6. 1550-FBG dynamic sensing with a fixed FBG filters using optical coupler (System C).

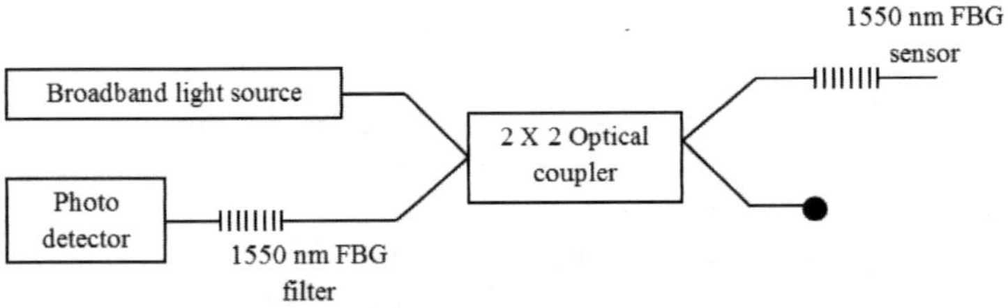

Figure 7. 1550-FBG dynamic sensing with a fixed FBG filters using optical circulator (System D).

A sinusoidal excitation with constant amplitude and frequency was applied to the composite sample. The response due to the excitation was measured by the Systems A, B, C and D. Prior to the shaker excitation, an impact with an impact hammer was done to the sample. The FFT result of the impact signal revealed some of the sample's natural frequencies, as shown in the figure 8. Note the impact signal was acquired by using system A (figure 4). The resulting frequency values as labeled by letter 'a' to 'g' are presented in the table 1. One of the natural frequencies, approximately 1100 Hz (Peak 'g'), was selected as the excitation frequency.

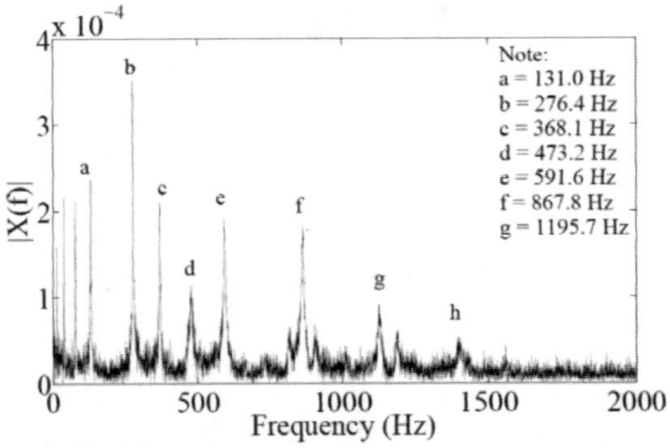

Figure 8. Frequency spectrums of the impact signal.

In the meantime, a free vibration analysis by Strand7 software was also done to the sample, for better understanding of its behaviour during the vibration. Strand7 is excellent finite element analysis (FEA) software that can provide a fully integrated visual environment, combined with powerful solvers. The software can be used for a variety of analysis such as linear and nonlinear static analysis; natural frequency, harmonic response and spectral response analysis; and steady-state and transient heat transfer analysis. For this study, Strand7 was used to determine the natural frequencies and mode shape of the composite specimen. Thus, the natural frequency solver was chosen, and set to calculate up to 40 mode shapes.

3. Results and discussion

All the 40 mode shapes were successfully calculated by the Strand7 software. However, only selected results were shown in this paper. Table 1 tabulated some of the natural frequencies by the theoretical results (Strand7) and its comparison to the experimental results (NIR-FBG dynamic sensor). Some of the selected mode shapes, specifically the ones which correspond to the peaks a – g, as shown in the figure 8, can be presented by the figures 9. The results reveal an important sensing characteristic of general FBG dynamic sensing, which has excellent response to the flexural type of wave propagations. However, this also as the result of the excitation of the shaker mainly excited the flexural waves all over the plate surface.

Table 1. List of selected peak frequencies as shown by the Figure 8.

Frequency peak	FEA (Hz)	NIR-FBG (Hz)	Error (%)
a	131.6	131.0	0.46
b	275.4	276.4	0.36
c	366.0	368.1	0.57
d	478.5	473.2	1.11
e	591.4	591.6	0.03
f	862.9	867.8	0.57
g	1126.0	1195.71	6.20

(a) Peak *a*: 131.6 Hz.

(b) Peak *b*: 275.4 Hz.

(c) Peak *c*: 366.0 Hz.

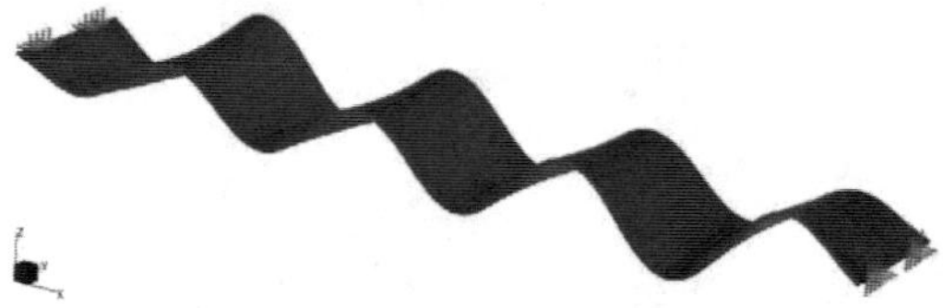

(d) Peak *d*: 478.5 Hz.

(e) Peak *e*: 591.4 Hz.

(f) Peak *f*: 862.9 Hz.

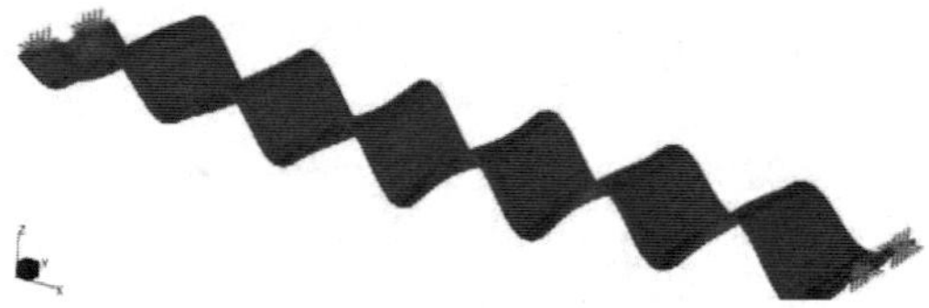

(g) Peak *g*: 1126.0 Hz.

Figure 9. Selected mode shapes which simulated by the Strand7 software.

Meanwhile, the figures 10 to 13 reveal the responses from all the FBG dynamic systems to the given 1100 Hz of excitation from shaker. In all figures, the plots can be divided to the signal with the vibration (a), signal without the vibration (b) and the FFT of the signal with the vibration (c), respectively. All the results are presented without signal filtering or denoising, in order to see the existence of noises. Although all systems were able to acquire the excitation frequency, 1100 Hz, however the noises from each system were noticeable. The system B and C have relatively weak responses among others. Although system B has a higher SNR than system C, yet, it was not able to show a stable response as obtained by using system C. Table 2 summarized the SNR values, for all systems. Note that, the SNR values of each system were determined by using equation (1).

IOP Conf. Series: Materials Science and Engineering **100** (2015) 012056 doi:10.1088/1757-899X/100/1/012056

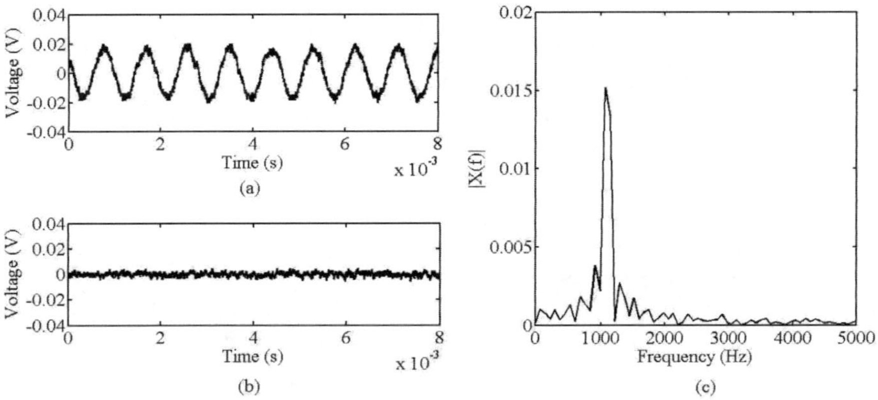

Figure 10. Response of System A.

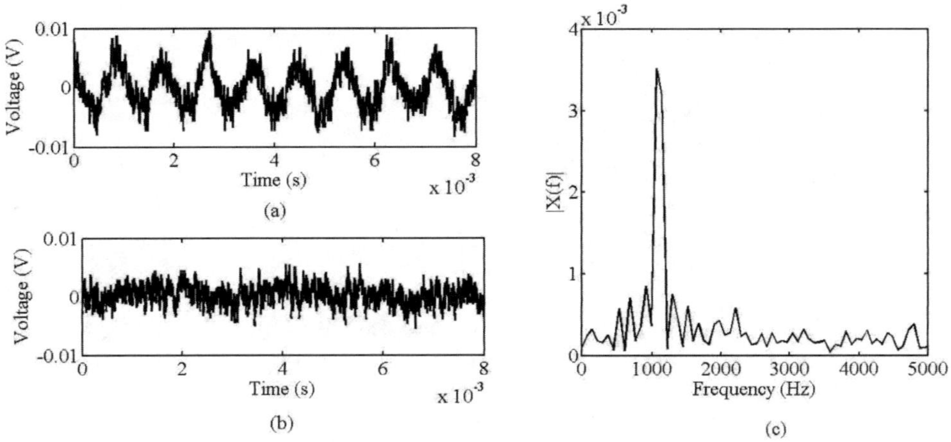

Figure 11. Response of System B.

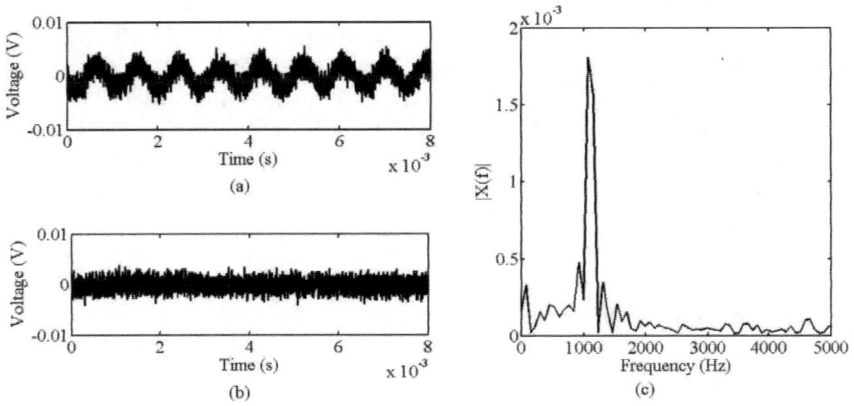

Figure 12. Response of System C.

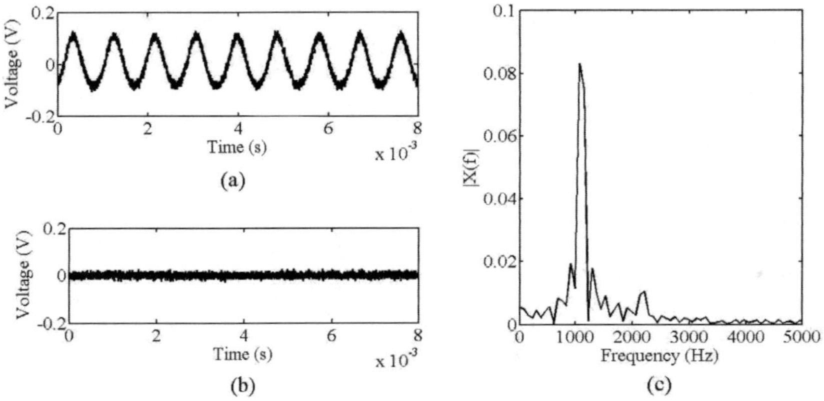

Figure 13. Response of System D.

Table 2. SNR of the FBG dynamic sensing systems.

	SNR (dB)
System A	17.97
System B	6.15
System C	4.74
System D	19.41

In the meantime, the results comparison between the 1550 nm based FBG dynamic systems, C and D, reveals a significant difference between the usage of two optical components; the optical circulator and optical coupler. Both components are known to be able to transmit the reflected light from the FBG sensor to the photo diodes. Yet, the optical powers received by the PDs, using an optical circulator are much greater than the one with an optical coupler.

For an FBG dynamic system with 50:50 ratio optical couplers, the optical power received by the PD is only about a quarter of the original power from broadband light source. Meanwhile, the light power in the circulator-based FBG dynamic system will remain at the same level until it reaches the FBG filter. As a result, a small wavelength variation due to dynamic strain at the FBG sensor give a big changes to the optical power level, before being received by the PD. Therefore it increases the SNR, thus raises the sensitivity of the sensor system. Note that, the System A or the NIR-FBG dynamic sensing with a fixed FBG filter actually utilized an optical coupler. However, its SNR value was as the same as the System D. Therefore, it is expected that if an optical circulator is used in the System B, instead of the coupler, the SNR and sensitivity of the system will increase.

However, the coupler-based system has its own benefit; whereby it allows many connections to be made through the system. Besides, the coupler also can perform as an optical divider. Consequently, the system can have several series of parallel systems, such as combining with FBG static measurement, or dynamic measuring system with several channels and etc. These extra capabilities would not be achievable with the circulator-based system.

4. Conclusions

This study has revealed that an FBG dynamic sensing system have an excellent capability to capture the acoustic waves that are consistent with the flexural wave propagations. It is supported by the small percentage of errors between results from FEA analysis and experimental results; which is below 1% for frequency not more than 1000 Hz. Meanwhile, based on the observation through SNR comparison for NIR-FBG and 1550 nm based systems; the NIR-FBG system was shown to have significantly higher SNR value, and therefore able to give more accurate results with less effort on the denoising. The outcomes of this study lead for a relatively cheaper option of an FBG system for dynamic response measurements.

Acknowledgements
The authors would like to thank the Faculty of Engineering and Surveying and the Centre of Excellence in Engineered Fiber Composites, University of Southern Queensland for providing laboratory facilities and financial support. They would also like to thank the interdisciplinary Photonics Laboratories (iPL), the University of Sydney for their support in fabricating the NIR FBGs. Finally, special thanks to Universiti Malaysia Pahang for providing financial assistance under project no. RDU1403155.

References

[1] Kashyap R 1999 *Fiber Bragg gratings* (San Diego: Academic Press)
[2] Wild G, Hinckley S. Optical fibre Bragg gratings for acoustic sensors. 2010.
[3] Semmlow J L 2005 *Circuits, Systems and Signals for Bioengineers: A Matlab-Based Introduction* Elsevier Inc.)
[4] Vasilescu G 2005 *Electronic Noise and Interfering Signals: Principles & Applications* Springer-Verlag Berlin Heidelberg New York)
[5] Lee J-R, Tsuda H 2005 *Scripta Materialia* **53** 1181-6
[6] Seo D-C, Yoon D-J, Kwon I-B, Lee S-S. Sensitivity enhancement of fiber optic FBG sensor for acoustic emission. 2009.
[7] Cvijetic M 2004 *Optical transmission systems engineering* (685 Canton Street, Norwood, MA 02062, USA: Artech House, Inc)
[8] Epaarachchi J A, Canning J, M. S 2010 *Journal of Composite Materials* **44** 809-19
[9] Epaarachchi J A, Clegg R E, Canning J. Investigation of the use of Near Infrared FBG (830nm) sensors for Monitoring the Structural Response of 0/90 Woven cloth/CSM Glass Fibre/Vynylester Composite under Static and Fatigue Loading. 2007.
[10] Hafizi Z M, Epaarachchi J, Lau K T 2015 *Measurement* **61** 51–7
[11] Hafizi Z M, Kahandawa G C, Epaarachchi J, Lau K T, Canning J, Cook K. NIR Fibre Bragg Grating as Dynamic Sensor: An Application of 1D Digital Wavelet Analysis for Signal Denoising. 2013.
[12] Hafizi Z M, Epaarachichi J, Lau K T, Leng J, Bar-Cohen Y, Lee I, et al. Non-velocity-based analysis of passive ultrasonic signal for source location detection in composite plates: a pilot study. 2012.
[13] Jian Z, Hong Z, Matched FBG Application Research on Dynamic Sensing and Demodulation. Photonics and Optoelectronics, 2009 SOPO 2009 Symposium on; 2009.

IOP Conf. Series: Materials Science and Engineering **100** (2015) 012057 doi:10.1088/1757-899X/100/1/012057

An investigation of cutting mechanics in 2 dimensional ultrasonic vibration assisted milling toward chip thickness and chip formation

I Rasidi I[1], N H Rafai[1], E A Rahim[1], S A Kamaruddin[1] H Ding[2] and K Cheng[3]

[1]Faculty of Mechanical & Manufacturing Engineering,
Universiti Tun Hussein Onn Malaysia, Parit Raja, Johor, Malaysia.
[2]School of Mechatronics Engineering, Harbin Institute of Technology,
Harbin 150001, China
[3]Advance Manufacturing Engineering Enterprise, School of Engineering and Design,
Brunel University, West London UB8 3PH, United Kingdom

Email: rasidi@uthm.edu.my

Abstract. The purpose of this paper is to investigate the effects of 2 dimensional Ultrasonic Vibration Assisted Milling (UVAM) cutting mechanics, considering tool path trajectory and the effect on the chip thickness. The theoretical modelling of cutting mechanics is focused by considering the trajectory of the tool locus into the workpiece during the machining. The studies found the major advantages of VAM are come from the intermittent tool tip interaction phenomena between cutting tool and workpiece. The reduction of thinning chip thickness formations can be identifying advantages from vibration assisted milling in 2 dimensional. The finding will be discussing the comparison between conventional machining the potential of the advantages toward the chip thickness and chip formation in conclusion.

1. Introduction

A modern industry nowadays is looking forward to miniaturize a precision component to support a high demand of micro and precise product such as electronic part, holographic optical element, micro-mechanical component etc. A high quality and precise machine and tool component is needed to produce the desire product. In order to manufacture in the shop floor machine factory, it significant to find the best solution with taking into account the tool condition, tool failure, thermal softening and surface finish etc. The solution for finding a best method of machining must be emphasize to overcome the demand of product without losing the profit margin, where the cost of tools or production is higher than the product selling.

A lot of efforts have put through over the last few decades and vibration assistance machining has been seen is the most relevant machining method to achieve the desire quality. Vibration assisted machining (VAM) is an addition of small amplitude and frequency into cutting tool. The purpose of the amplitude and frequency is for giving an extra displacement with higher reciprocating in circular or ellipse tool motion. However, there is a no significant comparative published between conventional cutting and 2D UVAM in term of mechanism and kinematics. Figure 1 shows a machine setup for the 2D UVAM preliminary cutting trial in the lab.

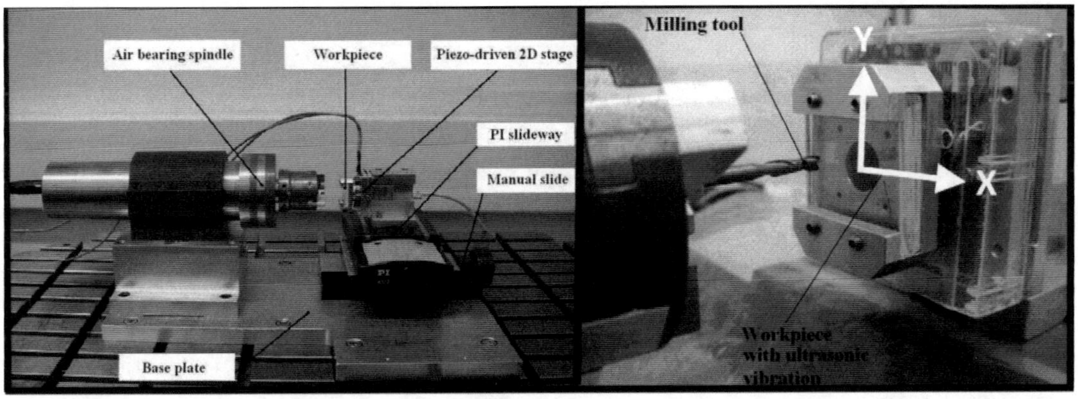

Figure 1. Ultrasonic Vibration Assisted Machining setup.

There are many researchers and company design and developed their own VAM for their application and product. Adachi and Arai designed an ultrasonic vibration cutting, low frequency vibration cutting has been found to prolong tool life and help reduce burr sizes in drilling [1]. They developed and electro-hydraulic servo system to generate vibration of 1,000 Hz in the spindle on conventional NC machine. Wang and Zhou proved by their experimental to cut hard and brittle material in high critical depth of cut. By applying at 2 μm amplitude ultrasonic in diamond tool tip a surface roughness of 100 nm has achieved [2]. The material has been cut was a fuse silica with cutting condition vibration amplitude 3 μm, vibration frequency 40 kHz, spindle speed 9,000 rpm, feedrate 5 μm/rev. Ding and Rasidi has modelled the 2 dimensional vibration assisted micro-milling based on the trajectory of tool path. It compose the vibration in X and Y directions relative to the workpiece when the X axis consist a feedrate as a feed direction and Y axis acting perpendicular to the normal motion of workpiece. Through their experimental studies about burr, the height of top burr decreases firstly in ploughing/rubbing--predominate stage and then increases in chip removal process with the increment of feed per tooth [3].

This paper intend explained the advantages of 2D UVAM in chip thickness and formation even there are many researchers proved they can get a good result but they did not emphasized the mechanisms to bring the benefit to this kind of phenomena.

2. Cutting mechanics in 2 dimensional VAM

2.1. Tool path trajectory

In conventional milling, the trajectory of the tool path depends on the spindle speed and feedrate relative to the workpiece. The spindle speed is the important part as it carries the tool tip edge to the round shape, but it is different and shape changing when the feedrate starts to move.

A preliminary concept of motion of 2D UVAM explained with the vibration amplitude in X and Y is the same displacement/amplitude, the vertical vibration (Y axis) increased faster than the circular one. The mathematical equation of these two models can be described,

$$\frac{x^2}{a^2}+\frac{y^2}{b^2}=1 \tag{1}$$

Defining the X and Y axis amplitude of the tool in equation (1). Combining for the 2 Dimensional cutting on both axes can be expressed as,

$$x = A\cos(2\pi ft)$$
$$y = b\sin(2\pi ft) \quad (2)$$

The centre of the endmill tool bar must be described in order to know the behaviour of the tool tip motion later. The reason because the boundary of the tool bar is taken as a reference to drive the tool tip motion [2]. It can be expressed as:

$$\begin{cases} x2 = ft + A\sin(2\pi f_x t + \phi_x) \\ y2 = B\sin(2\pi f_y t + \phi_y) \end{cases} \quad (3)$$

Bao and Tansel have modelled the tool tip trajectory and tried to investigate the chip thickness and difference from the Tlusty and Macneil model using equation (4) [4, 5].

$$\begin{cases} x = \dfrac{ft}{60} + r\sin\left(\omega t - \dfrac{2\pi z}{Z}\right) \end{cases}$$
$$\begin{cases} y = r\cos\left(\omega t - \dfrac{2\pi z}{Z}\right) \end{cases} \quad (4)$$

Liu and Cheng modelled a conventional of the tool path is simply traces by the spindle rotation angular velocity, tool cutter helix angle, regenerative cutter displacement and the cutting tool radius [5]. However, in 2D UVAM, the tool tip motion relative to the workpiece is complex and cannot be considered a helix shape as the vibration acts on the X and Y axes simultaneously.
So, the component of 2D UVAM can be obtained when the described components in equation (2) – (4) are combined together, which can be expressed by.

$$\begin{cases} x1 = ft + r\sin[\omega t - \dfrac{2\pi(z_i - 1)}{Z}] + A\sin(2\pi f_x t + \phi_x) \\ y1 = r\cos[\omega t - \dfrac{2\pi(z_i - 1)}{Z}] + B\sin(2\pi f_y t + \phi_y) \end{cases} \quad (5)$$

The trajectory of the tool locus in 2D UVAM is robust. Depending on the 4 important factors described in Figure 2, it contains:

- Vibration frequency X and Y axis
- Vibration amplitude X and Y axis
- Spindle speed
- Feed rate

The two dimensional vibrations assisted milling process is shown in Figure 2, where the X axis feed direction and amplitude is in the X axis, the Y axis is the normal direction and the Z axis is the depth of cut in order to determine the locus of the tool simulation relatively in to the work piece.

2.2. Intermittent cutting and chip thickness

Figure 2, generated by MatLab programming, revealed the cutting mechanics in 2D, with effect of vibration frequency and amplitude, promoted many cycles take places in a specified time interval. It is obviously explained by the fact that the cutting force is reduced when the tool point of the tool is in high force to remove the material in one single cut. In 2D UVAM, the vibration introduced non-contact within the tool and work piece, helping to cut the material to smaller pieces while opposing the frequency and amplitude.

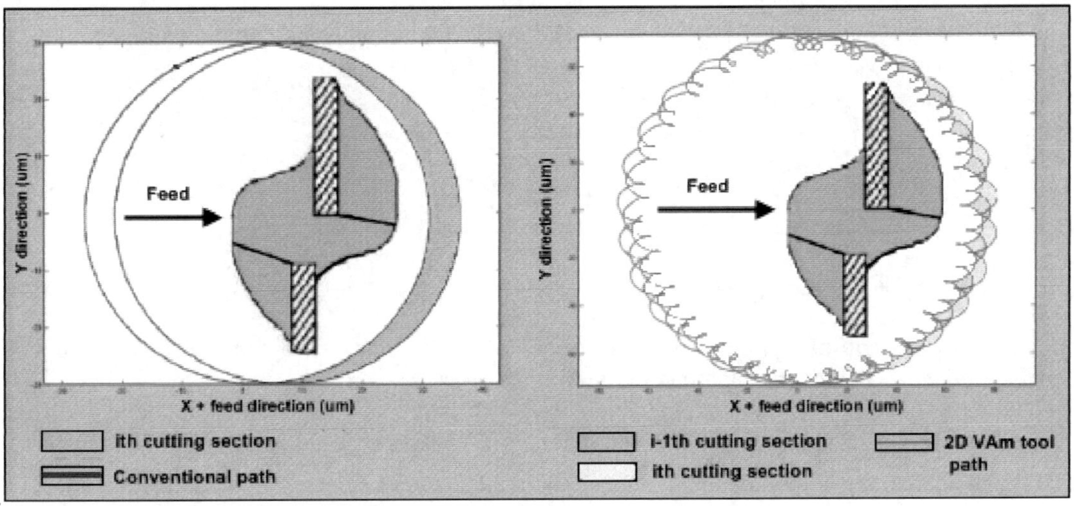

(a) Conventional machining (b) 2D VAM

Figure 2. Difference of chip thickness affecting cutting force reduction.

Therefore, in 2D UVAM much smaller and shorter chips are produced, instead of a large and continuous chip as in conventional cutting. In addition, the tool motion operates with lower average force for a much larger cumulative distance in repetitive passes, compared to conventional machining in the same amount of time.

Figure 3 shows the chips thickness performs by (a) without vibration and (b) 2D VAM by the test bed in UTHM laboratory. The depth of cut was 500 μm, vibration in X and Y was 3,000 Hz. Figure (a) obviously found the chip thickness is longer, thicker and bigger than (b). It is because the cutting force has been reduced by introducing the 2D VAM affect the discontinuous chip, smaller and thinner. The intermittent motion of tool play important role to get this type of chip. Thus, smaller and discontinuous chip has been preferred by the machinist similar in tool "chip breaker" end-mill type in common industries.

According to the theory of vibration assisted machining higher frequency applied during cutting, the thinner chip will produced but faster cutting process has been achieved. It will promote the higher material removal rate process.

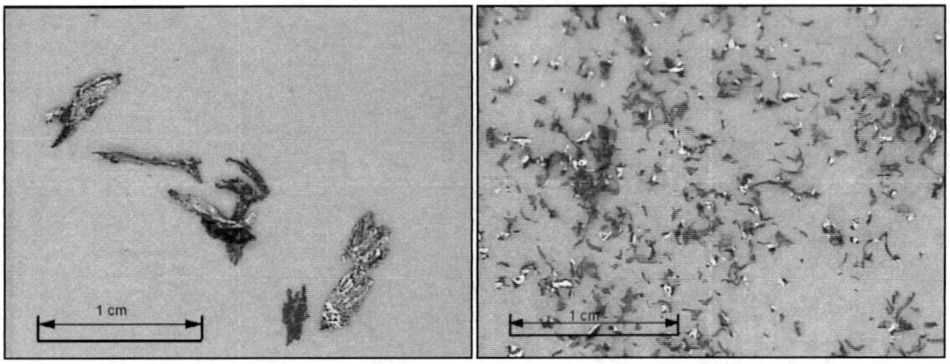

Figure 3. Chip thickness and formation (a) Conventional milling, (b) 2D VAM.

In the same volumetric material removal, the work performed by 2D UVAM is therefore much more consistent than with conventional machining. Summary of this paper is shown in table 1.

Table 1. Comparison investigation between Conventional cutting and UVAM.

	Conventional Miling	Ultrasonic Vibration Assisted Machining	Comments
Chip thickness	• Chip always thicker and depending on feedrate and spindle speed only.	• Chip thickness thinner and strongly depending on the amplitude, spindle speed, smaller amplitude, and thinner chip produced. • The consecutive overlapping toolpaths results in chips that are thinner than the depth of cut.	• Conv. machining cut a workpiece in each single rotation in one cycle, but 2D VAM cut the workpiece much smaller cause the frequency and amplitude effect
Chip formation	• Chip always continuous and depending on feedrate and spindle speed only.	• Discontinuous/shorter chip strongly depending on the frequency, amplitude and spindle speed. Higher frequency, smaller chip produced. • Instantaneous uncut chip thickness at each point in the tool path is less than the depth	• Conv. machining cut a workpiece in each single rotation in one cycle, but 2D VAM cut the workpiece much smaller cause the frequency and amplitude effect

3. Conclusion

From this paper we can conclude the 2D vibration assisted milling is obviously shown to promote a distinct advantages compared with conventional machining methods over a wide range of precision machining application whether in macro and micro. It is founding the major improvement in reduction cutting force, reduction in temperature, extended tool life, discontinuous, thinning and changes of chip formation and improving the surface roughness came from the effect of engaging and disengages of the tool related to the workpiece. The creating of gap between the rake face and workpiece also play important effect to this major improvement. These mechanics advantages include:

- Intermittent gap promote discontinuous, smaller and less chip thickness
- Frequency and amplitude helps tool cutting edge to cut in ductile region efficiently.
- Alternating cycles motion allows allow cutting force reduction whilst prolong tool wear.

Acknowledgements

The authors would like to thank Minister of Education for Research Acculturation Grant Scheme (RAGS), Advance Machining Research Group (AMRG) Universiti Tun Hussein Onn Malaysia, (UTHM) and to Cedrat Technology SA for providing the piezo-actuator system.

References

[1] Adachi K, Arai N, Harada S, Okita K, Wakisaka S 1987 *Bulletin of the Japan Soc. of Precision Engineering* **21** 258-64
[2] Wang X, Zhou M, Gan J-K, Ngoi B 2002 *The International Journal of Advanced Manufacturing Technology* **20** 99-102
[3] Ding H, Chen S, Ibrahim R, Cheng K 2011 *Proceedings of the Institution of Mechanical Engineers, Part B: Journal of Engineering Manufacture* **225** 2032-9
[4] Bao W, Tansel I 2000 *International Journal of Machine Tools and Manufacture* **40** 2155-73
[5] Liu X, Cheng K 2005 *International Journal of Machine Tools and Manufacture* **45** 1301-20

IOP Conf. Series: Materials Science and Engineering **100** (2015) 012058 doi:10.1088/1757-899X/100/1/012058

Surface Roughness and Tool Wear on Cryogenic Treated CBN Insert on Titanium and Inconel 718 Alloy Steel

S Thamizhmanii[1], R Mohideen[1], A M A Zaidi[2] and S Hasan[3]
[1]International College of Automotive,
Peramu Jaya, 26607, Pekan, Pahang, Malaysia.
[2]Universiti Pertahanan Nasional Malaysia (UPNM),
Kem Sungai Besi, 57000, Kuala Lumpur, Malaysia.
[3]University Tun Hussein Onn Malaysia,
Parit Raja, Batu Pahat, Malaysia.

Email: siva@icam.edu.my

Abstract: Machining of materials by super hard tools like cubic boron nitride (cbn) and poly cubic boron nitride (pcbn) is to reduce tool wear to obtain dimensional accuracy, smooth surface and more number of parts per cutting edge. wear of tools is inevitable due to rubbing action between work material and tool edge. however, the tool wear can be minimized by using super hard tools by enhancing the strength of the cutting inserts. one such process is cryogenic process. this process is used in all materials and cutting inserts which requires wear resistance. the cryogenic process is executed under subzero temperature -186º celsius for longer period of time in a closed chamber which contains liquid nitrogen. in this research, cbn inserts with cryogenically treated was used to turn difficult to cut metals like titanium, inconel 718 etc. the turning parameters used is different cutting speeds, feed rates and depth of cut. in this research, titanium and inconel 718 material were used. the results obtained are surface roughness, flank wear and crater wear. the surface roughness obtained on titanium was lower at high cutting speed compared with inconel 718. the flank wear was low while turning titanium than inconel 718. crater wear is less on inconel 718 than titanium alloy. all the two materials produced saw tooth chips.

1. Introduction

Machining of materials is recognized as removing unwanted materials by using different cutting inserts with variable cutting parameters. Cryogenic process is treating the work materials and cutting inserts at – 186° Celsius. CBN and PCBN cutting tools are used to machine difficult to cut materials like high strength alloy steels, stainless steel, Inconel 718, Titanium etc. Figure 1 shows generally accepted cutting speeds in high speed machining of various materials. High speed turning is recognized as a main manufacturing technology for higher productivity and throughput [1]. Barron [2] performed abrasive wear tests on a wide variety of steels and concluded that metals which can exhibit retained austenite at room temperature can have their wear resistance increased significantly, by subjecting them to cryogenic treatment. Seah, Rahman [3] studied and found that cryogenic treated tools improved wear resistance and overall tool lives of tungsten carbide tool inserts. Another studies compared the performance of cryogenically treated and untreated tungsten inserts during high speed milling of medium carbon steel. The cryogenically treated inserts exhibit better tool wear resistance than untreated.

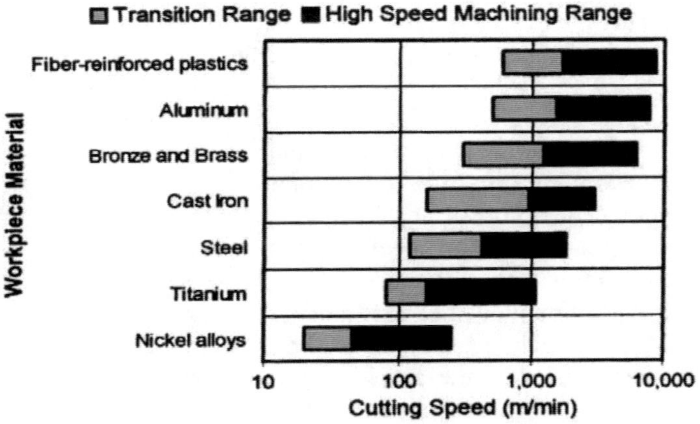

Figure 1. High-speed cutting ranges in machining of various materials [1].

2. Experimental Setup

2.1. Principle of deep cryogenic treatment

In this research, CBN inserts were cryogenically treated and the process is explained briefly. Cryogenic expresses study and use of materials at very low temperature, below – 186° Celsius. Liquid Nitrogen (LN) is the most commonly used element in cryogenics. Nitrogen melts at -201.01° Celsius and boils at – 198.79° Celsius, it is the most abundant gas, composing about four fifths (78.03 %) by volume of the atmosphere. It is colorless, odorless, tasteless and non-toxic gas. Deep cryogenic treatment comprises of cooling the material over a period for few hours to the sub-zero range, holding at this temperature for a long time and then returning to room temperature. The process is based on the predetermined thermal cycle that involves cooling of the engineering components/materials in a completely controlled cryogenic chamber. The material is slowly cooled to -196°Celsius and soaked at deep cryogenic temperature for 20 hours. The material is then allowed to return slowly to ambient temperature. The complete cryogenic cycle would take up to 25-30 hours. Thermal control is achieved by continuously monitoring inputs and regulating the flow of LN into the chamber and alternating the heat. Precise program control takes the cycle through its three phases of descend, soak and ascend. It is imperative that a slow descend is followed by soaking period for at least 24 hours at -186° Celsius and raised to room temperature with a slow ascend. In this process, no dimensional change and thermal crack occur. Strict computer control and precise processing profiles assure that optimum results are achieved with no dimensional change or thermal shock. The system is relatively simple and does not require large capital outlay.

2.2. Titanium (Ti-6Al-4V) alloy

Titanium alloy is an attractive material in many industries due to its unique and excellent combination of strength to weight ratio and their resistance to corrosion. However, because of its low thermal conductivity and high chemical reactivity, Ti-6Al-4V alloy is generally classified as a difficult to cut material that can be characterized by low productivity and rapid tool wear rate even at conventional cutting speeds. Table 1 and 2 show the mechanical and chemical properties of Titanium alloy and Inconel 718 alloy respectively. The chemical and mechanical properties details were given along with test certificates by the supplier.

2.3. Inconel 718

Inconel 718 is categorized as a nickel-based super alloy, which is extensively used in aerospace and energy industry, specifically in the high temperature area of engine, because of their excellent mechanical resistance to high temperature and corrosion. Many previous investigations have reported that nickel-based super alloys are difficult to machine [4, 5] due to the following reasons: (1) high hot hardness and strength causing deformation of the cutting tool; (2) the rapid work hardening causing severe notch wear; (3) presence of primary carbide particles encouraging abrasion tool wear; (4) low thermal diffusivity leading to extremely high local temperature on cutting edge; (5) welding of the work piece to the tool cutting edge forming unstable built-up edge (BUE) which deteriorate machined surface; and (6) possible chemical reactions accelerating tool wear. One of major issues in machining of nickel-based super alloys is short tool life and thereby low productivity. Welding and adhesion of work piece material onto the cutting tool frequently occur during machining causing severe notching as well as alterations of the tool rake face due to the consequent pull-out tool material [6, 7]. Surface quality is another important issue in the machining of nickel-based super alloys since critical components, such as discs for jet engine, are often made of this material. The main problems reported from previous investigations are surface tearing, cavities, cracking, metallurgical changes, plastic deformation and formation of residual stresses in the case of machining with carbide cutting tools [5, 8]. Sadat and Reddy [9] found that an increase in cutting speed reduced surface damages by reducing the cutting force due to an increase in cutting temperature and corresponding drop in work piece strength. Ezugwu, Bonney [5] reported that the change of microstructure in subsurface when cutting tool changed from a new tool to worn tool.

Table 1. Mechanical properties.

Grades	Tensile strength	Yield strength	% elongation
Inconel 718	1240 MPa	1036 MPa	12
Titanium	147.8 KSI	138.5 KSI	19.2

Table 2. Chemical properties.

Alloying elements	Titanium alloy	Inconel 718
Carbon	0.030-0.031	0.08 max.
Manganese	< 0.01	0.35 max.
Chromium,	--	17-21
Molybdenum	< 0.01	2.80-3.30
Titanium alloy	Balance	0.65-1.15
Nickel	50-55	--
Aluminum	6.51-6.57	0.20-0.80
Vanadium	4 - 4.07	--
Fe	0.17- 0.21	Balance
Silicon	< 0.01	0.35 max.
Copper	< 0.01	0.30 max.
Cobalt	--	1.00 max.
Phosphorus	--	0.015 max
Sulfur	--	0.015 max.

IOP Conf. Series: Materials Science and Engineering **100** (2015) 012058 doi:10.1088/1757-899X/100/1/012058

Table 3. Operating parameters.

Cutting speed m/min	Feed rate – mm/rev	Depth of cut –mm
30	0.05	0.50, 0.75 & 1.00
40	0.10	0.50, 0.75 & 1.00
50	0.15	0.50, 0.75 & 1.00

3. Results and discussion

3.1. Surface roughness

Formation of the machined surface is closely associated with the reaction of the edge of cutting tool and workpiece. Dhar N R [10] reported that the surface roughness was low with cryogenic treated inserts in comparison with non-treated inserts at all cutting speeds. This has been more or less true for all the work – tool combinations undertaken in this research. Chen [11] stated that the surface roughness would be low when the hardness of the material increased. Lateral plastic flow of the work material in front of the cutting edge increased roughness of the machined surfaces.Thakur, Ramamoorthy [12] turned Inconel 718 material by tungsten carbide tool and reported that the roughness value decreased at high cutting speed but increased as the feed rate increased. This was due to increased friction between work material and tool interface which eventually increases the temperature in the cutting zone. The surface roughness values are higher when lower feed rate was used, while the roughness values were smaller when higher feed rate was employed in the tests. This is presumably, is due to ploughing action caused by smaller uncut chip thickness [13, 14]. Rougher surface at small feed rate may associate with minimum chip thickness. Figure 2 (a to f) show graphical representation of surface roughness against operating parameters as given in the Table 3. The surface roughness obtained at cutting speed of 30, 40 and 50 m/min with at feed rate is low in turning Titanium than Inconel 718 alloy steel even after 500 mm length turning. Smooth roughness obtained at cutting speed of 50 m/min with feed rate 0.05 mm / rev. at 1 mm depth of cut. This was due to more tool plough effect and plastic deformation produced low surface roughness.

Thakur, Ramamoorthy [12] turned Inconel 718 material by tungsten carbide tool and reported that the roughness value decreased at high cutting speed but increased as the feed rate increased. This was due to increased friction between work material and tool interface which eventually increases the temperature in the cutting zone. The surface roughness values are higher when lower feed rate was used, while the roughness values were smaller when higher feed rate was employed in the tests. This is presumably, is due to ploughing action caused by smaller uncut chip thickness [13, 14]. Rougher surface at small feed rate may associate with minimum chip thickness. Figure 2 (a to f) show graphical representation of surface roughness against operating parameters as given in the table 3. The surface roughness obtained at cutting speed of 30, 40 and 50 m/min with at feed rate is low in turning Titanium than Inconel 718 alloy steel even after 500 mm length turning. Smooth roughness obtained at cutting speed of 50 m/min with feed rate 0.05 mm / rev. at 1 mm depth of cut. This was due to more tool plough effect and plastic deformation produced low surface roughness.

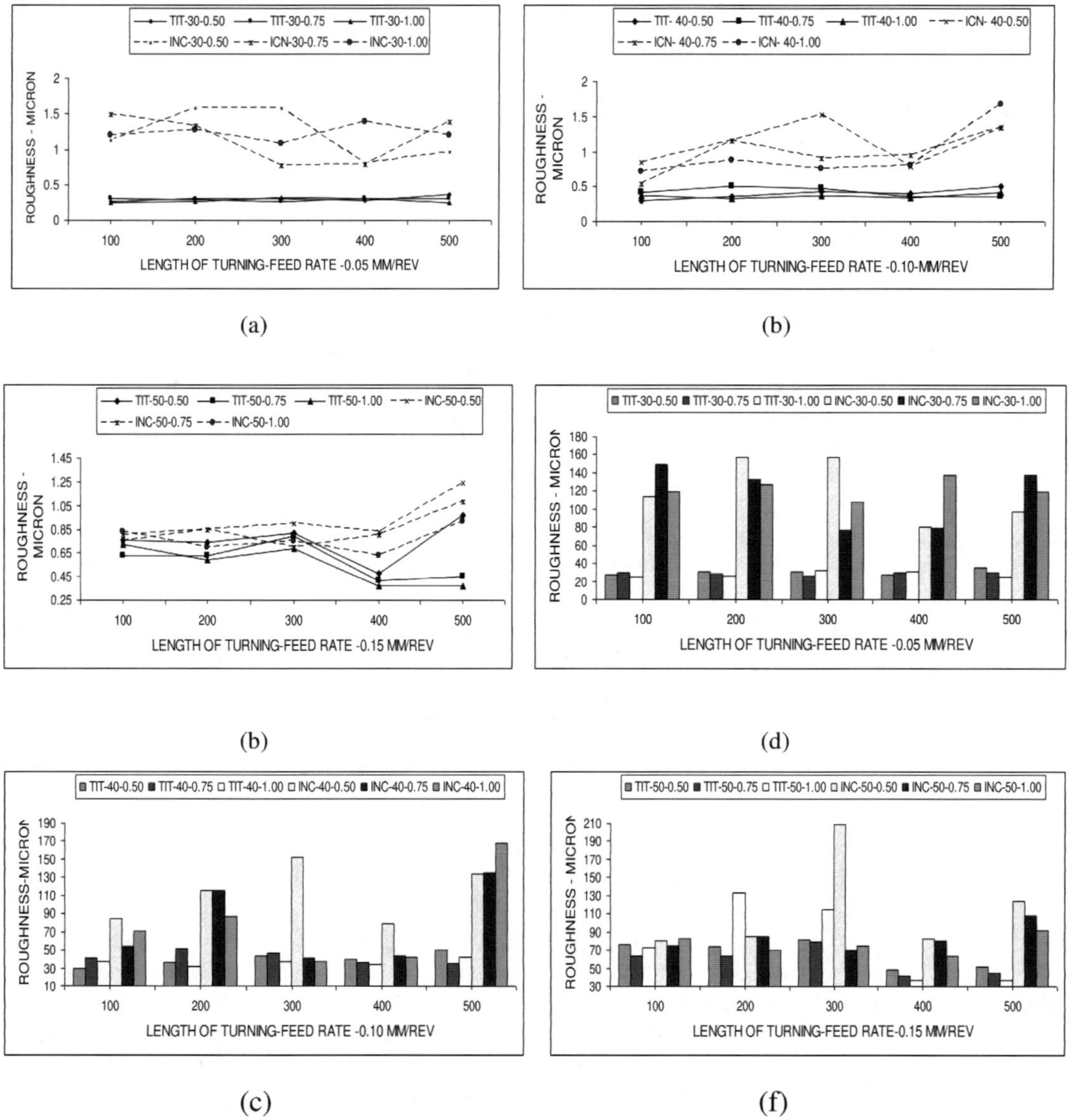

Figure 2. Graphs showing length of turning vs surface roughness: (a) 30 m/min cutting and feed rate of -0.50 (b) 40 m/min cutting speed with 0.10 feed rate (c) 50 m/min cutting speed with 0.15 feed rate, (d) DOC of 0.50 mm, (e) DOC of 0.75 mm and (f) DOC of 1.00.

3.2. Tool Flank Wear

Flank wear is formed on the flank side of the insert which is due to rubbing action of the insert and machining surface. Figure 3 shows the various tool wear that is likely to occur as per ISO 3685 of 1993 [15]. The tools often fail rapidly due to non-uniform flank wear, chipping, depth of cut notching, and plastic deformation of the cutting edge even at relatively low cutting. Low thermal conductivity (11.4 W/m K), leading to high cutting temperature in cutting and high strength under high temperature are responsible for its generally poor machinability [6]. In addition, the cutting forces generated are also very high around double that found when cutting medium carbon steels [7]. This, in combination with the relatively short chip contact length and high cutting temperature, induces very high stress concentration on the area of tool / work piece interaction and leads to highly plastics deformation in cutting edge and work piece material. Manu Dogra et al. [15] found that during the turning of hardened AISI 4340 steel using CBN-TiN coated carbide and PCBN tools, result revealed that tool life of CBN –TiN coated carbide inserts was approximately 18-20 min per cutting edge, whereas PCBN tools produced a tool life of 32 min. Flank wear is primarily attributed to rubbing of the tool along the machined surfaces, causing abrasive diffusive and adhesive wear mechanisms and also high temperature, which affect the tool material properties as well as work piece surface.

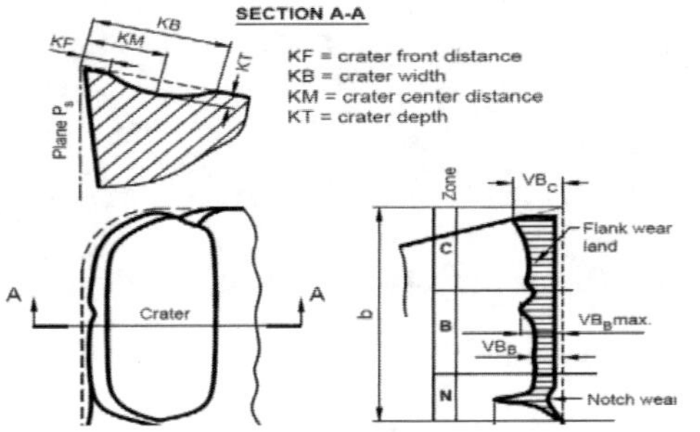

Figure 3. Various tool wear on a single point tool [15].

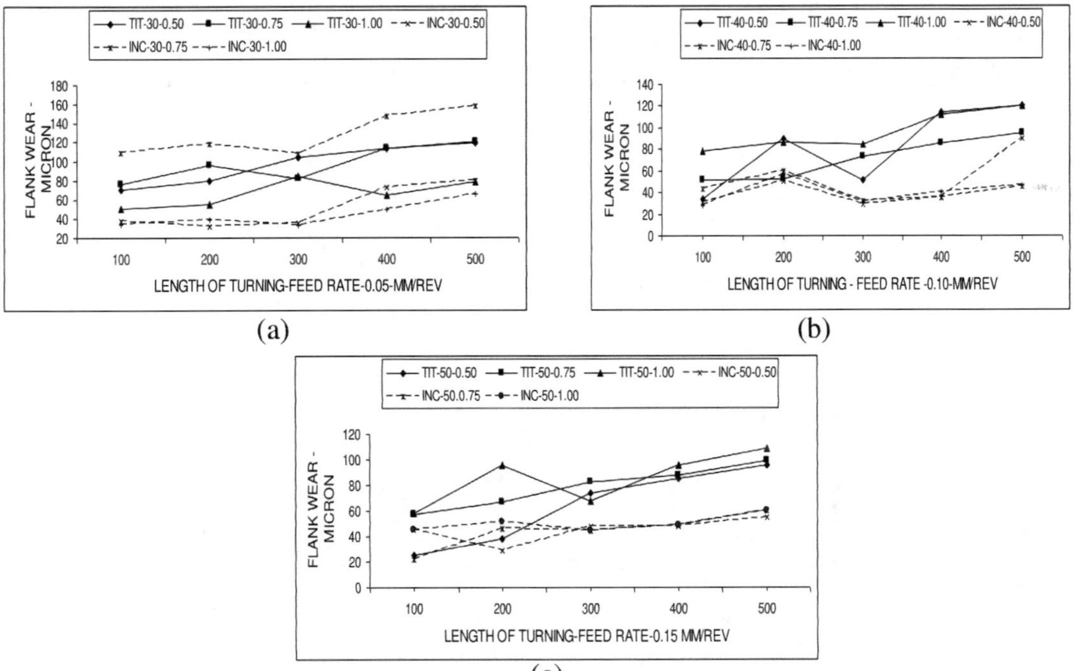

Figure 4. Graph showing length of turning Vs flank wear (a) 30 m/min cutting and feed rate of -0.50 (b) 40 m/min cutting speed with 0.10 feed rate and (c) 50 m/min cutting speed with 0.15 feed rate.

In all the cutting speeds, flank wear was formed more at low feed rate of 0.05 mm/rev. Titanium alloy steel. At 1 mm depth of cut in turning Inconel 718, the flank wear was low than 0.50 and 0.75 mm depth of cut. In this research, it was found that flank wear formation was high in Titanium than Inconel 718 and shows that Titanium alloy is difficult to cut metal. Figure 4 (a to c) show graphical representation of flank wear for all the three parameters. Figure 5 (a to i) shows SEM view on flank wear by titanium alloy and more abrasion occurred than Inconel 718. Figure 6 show SEM view on flank wear by Inconel 718 alloy steel which shows less flank wear.

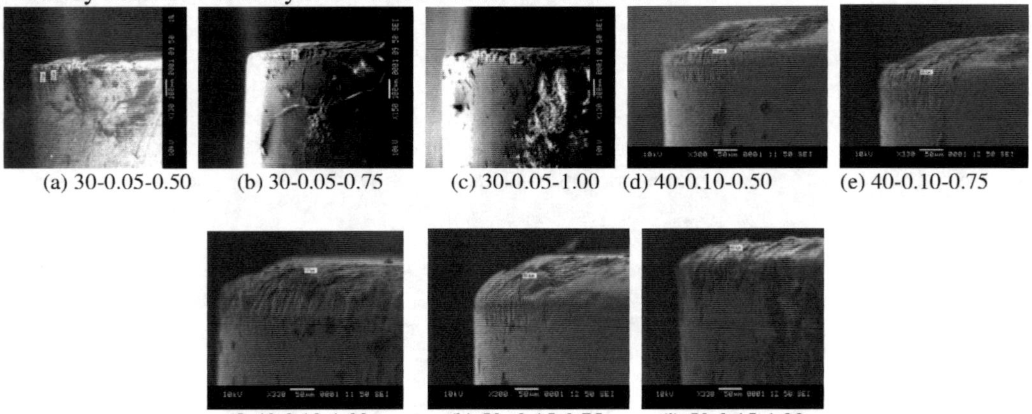

(a) 30-0.05-0.50 (b) 30-0.05-0.75 (c) 30-0.05-1.00 (d) 40-0.10-0.50 (e) 40-0.10-0.75

(f) 40-0.10-1.00 (h) 50- 0.15-0.75 (i) 50-0.15-1.00

Figure 5. SEM view on flank wear on Titanium alloy.

(a) 30-0.05-0.50 (b) 30-0.05-0.75 (c) 30-0.05-1.00 (d) 40-0.10-0.50 (e) 40-0.10-0.75 (f) 40-0.10-1.00

(g) 50-0.15-0.50 (h) 50- 0.15-0.75 (i) 50-0.15-1.00

Figure 6. SEM view on crater wear while turning Titanium.

(a) 30-0.05-0.50 (b) 30-0.05-0.75 (c) 30-0.05-1.00 (d) 40-0.10-0.50 (e) 40-0.10-0.75

(f) 40-0.10-1.00 (g) 50-0.15-0.50 (h) 50- 0.15-0.75 (i) 50-0.15-1.00

Figure 7. SEM view on flank wear on Inconel 718 alloy.

3.3. Crater wear

Crater wear is formed at rake face of the tool. Figure 3 shows the various crater wear formation which is represented by KT. The ISO 3685 of 1993 recommends the criterion of tool life due to crater wear and can be calculated by using the formula as given below:

$$K_T = (0.06) + 0.3f \tag{1}$$

where f is the feed rate and K_T is the depth of crater.

Crater wear is dished out section which develops on the rake face of the tool. The formation of crater wear occur little away from the cutting edges. In fact at low cutting speed, crater wear is usually insignificant compared with flank wear in normal operations. There is no standard available for maximum depth of the crater specification like flank wear. Deeper crater will lead to failure of the cutting edge. When machining using CBN, PCBN inserts and other high strength inserts the formation of wear take longer time. At high cutting speeds crater wear formation would be more severe and depth of crater KT will be deeper. While turning difficult to cut materials crater wear was more due to saw tooth chips. In turning Titanium alloy, the formation of crater wear was low than Inconel 718. The saw tooth chips rougher and deeper crater formed. Figure 6, 7 and 8 show the crater wear formed while turning Titanium and Inconel 718 respectively. More craters formed in turning Titanium than Inconel 718.

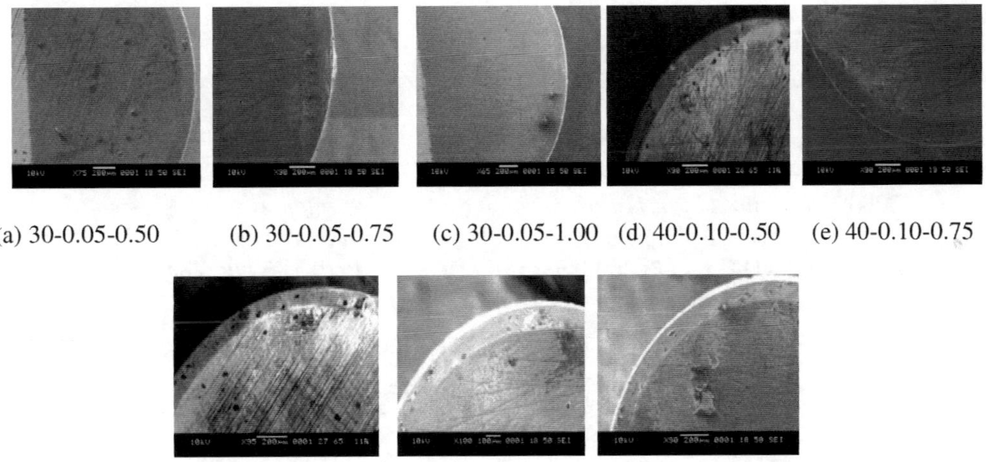

(a) 30-0.05-0.50 (b) 30-0.05-0.75 (c) 30-0.05-1.00 (d) 40-0.10-0.50 (e) 40-0.10-0.75

(f) 50-0.15-50 (g) 50-0.15-0.75 (h) 50-0.15-1.00

Figure 8. SEM view on crater wear on Inconel 718 alloy

4. Conclusions

The following conclusions were arrived in this research. They are:

- Surface roughness obtained in turning Titanium alloy was low at cutting speed 50 m/min than 30 and 40 m/min than Inconel 718.
- Surface roughness at depth of cut 1 mm is low than 0.50 and 0.75 mm in Titanium alloy than Inconel 718.
- Flank wear by Inconel 718 was low than Titanium. Flank wear at 1 mm depth of cut was low than 0.50 and 0.75 mm.
- Crater wear formed in turning Titanium alloy was un-uniform in wear. The crater wear by Inconel 718 was less deep and small in magnitude of KT.

- Cryogenic treated inserts performed very well by producing smooth surface roughness in Titanium alloy than Inconel 718. On the other hand, flank wear was low in turning Inconel 718 than Titanium alloy.

Acknowledgements
The authors would like to thank International College of Automotive (ICAM) and Ministry of Higher Education Malaysia for their financial support via Fundamental Research Grant Scheme (FRGS) Vot 0751.

References

[1] Ezugwu E, Wang Z, Machado A 1999 *Journal of Materials Processing Technology* **86** 1-16
[2] Barron R 1982 *Cryogenics* **22** 409-13
[3] Seah K, Rahman M, Yong K 2003 *Proceedings of the Institution of Mechanical Engineers, Part B: Journal of Engineering Manufacture* **217** 29-43
[4] Arunachalam R, Mannan M 2000 *Machining Science and Technology* **4** 127-68
[5] Ezugwu E, Bonney J, Yamane Y 2003 *Journal of materials processing technology* **134** 233-53
[6] Watmon T B, Ijeh A C, Coating Cutting Tools with Hard Substance Lowers Friction Co-efficient and Improves Tool Life-A Review. Proceedings of the international multiconference of engineers and computer scientists; 2010.
[7] Pawade R, Joshi S S, Brahmankar P, Rahman M 2007 *Journal of Materials Processing Tech.* **192** 139-46
[8] Sharman A, Dewes R C, Aspinwall D K 2001 *Journal of Materials Processing Technology* **118** 29-35
[9] Sadat A, Reddy M 1992 *Experimental mechanics* **32** 282-8
[10] Dhar N R P S, Chattopadhyay A B 2002 *Journal of Materials Processing Technology* **118** 9-35
[11] Chen W 2000 *International journal of machine tools and manufacture* **40** 455-66
[12] Thakur D, Ramamoorthy B, Vijayaraghavan L 2009 *Materials & Design* **30** 1718-25
[13] Zou B, Chen M, Huang C, An Q 2009 *Journal of Materials Processing Technology* **209** 5802-9
[14] Chou Y K, Song H 2004 *Journal of Materials Processing Technology* **148** 259-68

IOP Conf. Series: Materials Science and Engineering **100** (2015) 012059 doi:10.1088/1757-899X/100/1/012059

Numerical analyses for improved hydrodynamics of deep water torpedo anchor

B P Soh[1], W Pao[1] and H H Al-Kayiem[1]
[1]Department of Mechanical Engineering, Universiti Teknologi PETRONAS, 32610 Bandar Seri Iskandar, Perak, Malaysia

Email: William.Paokings@petronas.com.my

ABSTRACT. The literature shows a lack on well established guidelines on the design of the torpedo anchors. This research aims to investigate the effect of various geometric parameters in the torpedo anchors' design to optimize for higher terminal velocity. A CFD model was developed and simulations were carried out using ANSYS FLUENT commercial software. Consequently, correlations between various parameters and its resultant drag coefficient were developed. The parameters of interest included geometric changes of the original design, as well as sea water properties that reflect water depth in South China Sea. New design features are proposed and investigated in the overall parametric studies. As a result, it was found that the terminal velocity can be improved by sharper tip angle, greater aspect ratio, greater diameter ratio, and an optimum rear angle of 30°. Clear relationships between each factor and its resulted terminal velocity were developed. More importantly, the sensitivity of drag coefficient towards each of the parameters was predicted.

1. Introduction

Floating structures such as floating production storage and offloading unit (FPSO) and mobile drilling unit (MODU) must be anchored with robust mooring system. These anchorage solutions are such as the Suction Caisson Anchors, Vertical Loaded Anchors (VLA), Suction Embedded Plat Anchors (SEPLA) and Torpedo Anchors. Among them, the torpedo anchor, which was initially developed and patented by Petrobras [1], has several advantages over the others. For instance, torpedo anchors are highly economical because no external energy is required for its installation [2]. Besides, it was found that the deployment of torpedo anchors is much easier and faster as compared to similar solutions such as VLA and suction piles [3]. In essence, torpedo anchorage system has competitive edge in terms of cost reduction and simplified installation [4]. Its applications are also less affected by increasing water depth as compared to conventional anchoring concepts [5].

Typical sizes of torpedo anchors range from 10-20 meters in height and 0.325-1.2 meters in diameter. A single unit of torpedo anchor can have a dry weight of 40-100 tons. It is usually released from an installation vessel with towards the seabed till a drop height of approximately 50 meters is reached. Then, it will be released to fall vertically downward by gravitational pull. Through the free fall period, the anchor is able to achieve a very high speed, and subsequently penetrates into the seabed. However, there exists a threshold speed for torpedo anchor regardless of the drop height [6]. This point occurs when the downward acceleration is equal to zero. This particular speed limit is known as the terminal velocity. According to Raie and Tassoulas [7], higher terminal velocity will

IOP Conf. Series: Materials Science and Engineering **100** (2015) 012059 doi:10.1088/1757-899X/100/1/012059

consequently provide greater holding capacity for the platforms, by resulting in deeper penetration. This finding is in line with the results of tests conducted by Hasanloo et al. [8].

The achievable threshold speed during the free fall phase of torpedo anchor has to be pushed forward for many future applications. However, there is a lack of research conducted for improvement on the terminal velocity during the anchor drop down. Besides, there is no well-established guideline developed for the designs of torpedo anchors.

Fernandes et al. [9] conducted small scale laboratory tests by using scaled torpedo anchors according to their design ratio. It was determined that presence of rear lines could increase the drag acting against torpedo anchor while it is travelling vertically downward. Moreover, the absence of pulley can further reduce the drag, thus result in higher kinetic energy gained by the anchor. Besides, according to Hasanloo and Yu [2], there is a minimum weight required for the anchor to fall steadily at different water depth. At the same time, density of the anchor was found to have positive impact on its travelling velocity. On the other hand, aspect ratio is identified to have direct influence on the drag coefficient of cylindrical prototypes, utilized by the European Nuclear Energy Agency to study the feasibility of disposal of radioactive waste through free fall cylindrical projectiles into oceanic sediments. The relationship was categorized as followed [2].

$$0.030 + 0.0085\,(L/D) < C_d < 0.039 + 0.0109\,(L/D) \tag{1}$$

Furthermore, it was concluded that the embedment depth of torpedo anchor is directly proportional to its impact velocity [10]. The dependence of impact velocity on its geometry and mass are analyzed too. On the other hand, it was shown implicitly that embedment depth of torpedo anchor is dependent on its terminal velocity [11]; the variation of tip was illustrated too, but the resulting impact on its terminal velocity was not examined. CFD procedures were proposed for 3 major phases that the torpedo anchor will encounter, namely its installation, set-up by consolidation of soil, and pull out as reported by Raie [11]. Recently, Hasanloo et al. [8] used 7 prototypes of torpedo anchors with different densities, aspect ratio, scale ratio, and fin sizes to study their influence on falling velocity during acceleration. As a result, the relationship between drag coefficient, and Reynolds number, Re was plotted.

For sea water properties, the viscosity ranged from approximately 0.0010 N.s/m2 to 0.0015 N.s/m2 when water depth increases from 100 meters to 2000 meters [12]. Accordingly, sea water density changes from 1024 kg/m^3 to 1028 kg/m^3 within the same range of ocean depth. There were field tests being conducted to test the feasibility of using torpedo anchors for FPSO. It was found that torpedo anchor is well suitable for mooring of large FPSO in deep water; in this case it is the P-50 mooring system [3]. Specifically, a total of 10 units of T-98 torpedo anchors were used in this mooring system to provide necessary holding capacity for the floating structure. The T-98 torpedo design was done purposely for this FPSO operating in water depth of 1240 meters, in the Albacora Leste Field located in the Campos Basin, Brazil. According to Brandão et al. [3] this T-98 design has a total mass of 98 metric tons, diameter of 1.07 meters, and length of 17 meters with 4 wings to ensure its directional stability.

This research aims to propose designing methods for attaining higher terminal velocity. Besides, as the coefficient of drag is the determining factor for terminal velocity, correlations between geometric changes and its resulted drag coefficient will be developed.

2. Methodology

The present research makes extensive use of FLUENT for computational fluid dynamics simulations. The working fluid is a model of sea water, and the type of fluid flow is set to be turbulent due to the high velocities involved. Several assumptions are made such as the sea water is modeled as incompressible Newtonian fluid. The changes in temperature with increasing depth are neglected. The horizontal velocity of fluid flow is assumed to be zero in comparison to the vertical free fall velocity of the anchor. Consequently, the anchor is assumed to have perfect downward directional stability during its free fall period. The parameters of interest involve sea water density and viscosity variation, which represents the water depth in South China Sea from water surface to a depth of 2000 m as referred to Murray [12]. Besides, the effects of varied design features such as tip angle, aspect ratio, rear angle, and diameter ratio were studied comprehensively in the parametric studies. Due to the limitation of space, only the effects of tip and rear angles are discussed in this paper.

2.1. Governing Equations

The problem of interest is assumed incompressible, isothermal and furthermore, the viscosity is taken as constant throughout the fluid. These assumptions greatly simplify the Navier-Stokes equation to the form

$$\rho\left(\frac{\partial \mathbf{u}}{\partial t}+\mathbf{u}\cdot\nabla\mathbf{u}\right)=-\nabla p+\mu\nabla^2\mathbf{u}+\mathbf{f}\ , \tag{2}$$

where $\mathbf{u}$ is the velocity, ρ the density, p the hydrostatic pressure, μ is the viscosity and $\mathbf{f}$ the external body force. Equation (2) needs to be complemented by the mass conservation equation

$$m=\rho\mathbf{u}=\text{constant} \tag{3}$$

Taking the particle derivative of equation (3) yields

$$\frac{Dm}{Dt}=\frac{D\rho}{Dt}\mathbf{u}+\rho\frac{D\mathbf{u}}{Dt}=0\Rightarrow\nabla\cdot\mathbf{u}=0. \tag{4}$$

2.2. Model Development and Boundary Conditions

The main idea of simulation is that the anchor is set at a stationary position in the middle of the domain with fluid flowing upward through the inlet with pre-defined velocity. Effects of each factor were obtained by repeated simulations with varied values, at recurring different velocities for each set of parameters. The boundaries are designed to be far enough from the torpedo anchor, so that the analyses are not affected by its proximity. Meshing was done with pre-dominantly quadrilateral cells, with small portion of triangular cells for smooth transitions at regions of irregular geometry. The dimensions used as the datum of 2D model are shown in figure. 1.

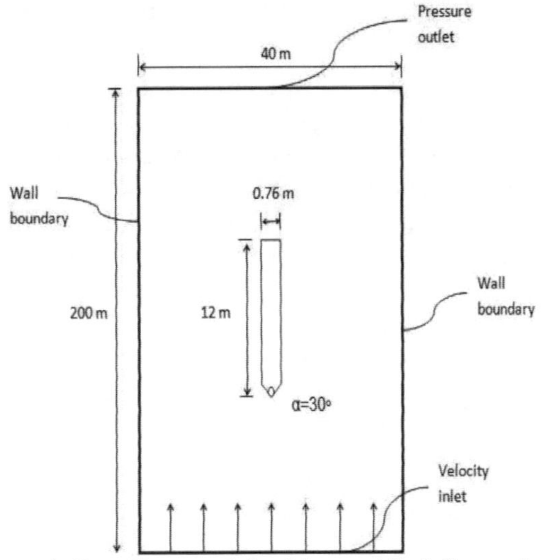

Figure 1. Baseline model with specified dimensions.

IOP Conf. Series: Materials Science and Engineering **100** (2015) 012059 doi:10.1088/1757-899X/100/1/012059

Besides examining the influences of varied aspect ratio and tip angle for the conventional torpedo anchor design, the effects of newly proposed design features are investigated too. Different geometries were created, while the same settings for meshing as well as its solution setup were integrated.

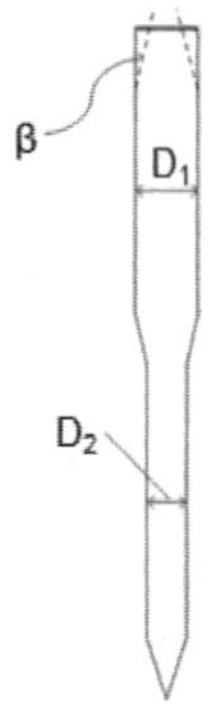

Figure 2. Illustration of rear angle, β and diameter ratio, D1/D2.

The examined values for the main parameters are presented in table 1.

Table 1. Parameters for torpedo anchor's simulation.

Parameters	Base model	Present study
Anchor weight (kN)	400	400
Diameter (m)	0.76	0.5,0.667,1.0
Length (m)	12	10
Water density (kg/m^3)	1024	**998.2-1027.3**
Viscosity (N.s m^{-2})	0.001005	**0.001-0.0015**
Tip angle, α (°)	30	**15, 30, 45, 60**
Aspect ratio (L/D)	15.79	**10, 15, 20**
Rear angle, β(°)	-	**0, 15, 30, 45**
Diameter ratio (D_1/D_2)	1	**1.5, 2.0, 2.5**
Fin	Finless	Finless

2.3. Mesh Dependency Check

To ensure that the solution is independent of the mesh configuration, the resultant drag coefficients versus different mesh density were recorded as shown in figure 3. It is evident that the drag coefficient tends towards a constant when mesh density is increased as depicted in figure 3. The final value of drag coefficient is independent of mesh density beyond a certain limit. The coefficient of drag converges towards stable value of about 0.24 when finer mesh is utilized. Thus, based on figure 3, all the simulations henceforth are conducted with mesh density of more than 4.25 triangles per unit area and beyond.

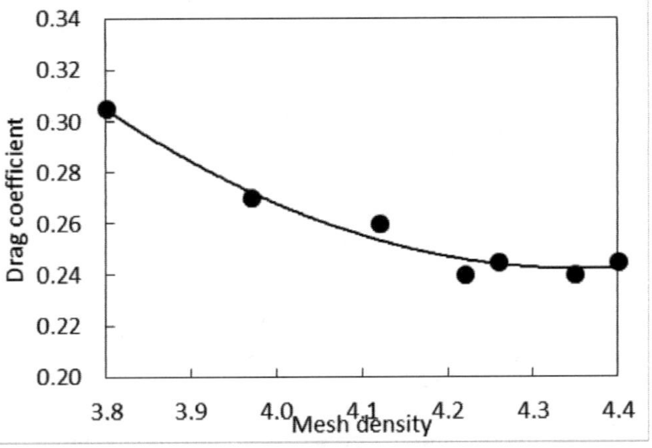

Figure 3. Drag coefficient versus mesh density (number of elements per unit area).

3. Validation

Firstly, the developed CFD model was compared with the results published by Raie, [11]. In line with the full scale field test performed by Petrobras [5], the simulation was done for a T-40 torpedo anchor. It was conducted by using a torpedo anchor made of steel with overall weight of 0.4 MN, length of 12 meters and diameter of 0.76 meters. Consequently, the percentage differences between obtained drag coefficient from CFD simulations and the reported values are 5.58% and 5.73%, for inlet velocity of 80 m/s and 90 m/s respectively. Furthermore, the calculated terminal velocity only deviates 3.76% from the reported value. Both reported values and results from CFD model are tabulated in Table 2. Another validation was done by comparison with laboratory test conducted by Hasanloo et al. [8] as shown in figure 4.

Table 2. Validation of model with full scale field test.

Reported result [11]	Present	% difference
C_d at 80 m/s : 0.2016	0.2134	5.58%
C_d at 90 m/s : 0.2007	0.2122	5.73%
Terminal velocity, V_T : 87.2 m/s	83.92 m/s (calculated)	3.76%

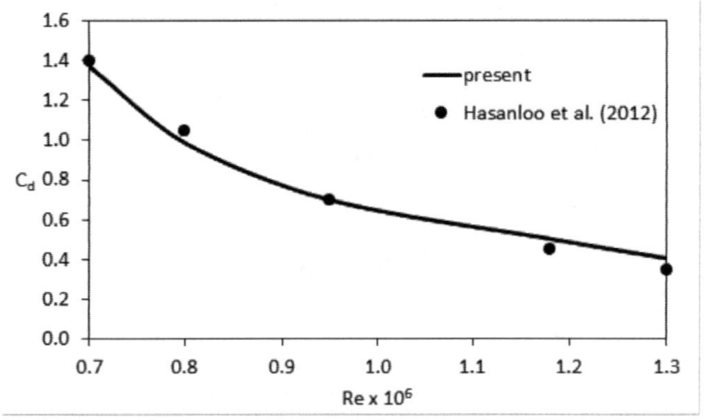

Figure 4. Validation of present model with laboratory test of torpedo anchor.

This validation was conducted according to the specified dimensions. However, it was scaled up 10 times as the prototypes used were 10 times smaller than actual units. The drag coefficients obtained by present model was plotted against Reynolds number. As shown in the Figure 4, the results acquired from present simulations were very close to the experimental results with an overall error below 5%.

4. Result and discussion

4.1. Effects of Tip Angle, α

Figure 5 shows the variation of torpedo's terminal velocity versus the tip angle. It can be observed that terminal velocity always increases as the tip angle of torpedo anchor decreases. In other words, the drag force acting upon the anchor increases as the anchor's tip become wider. As the graphs for different viscosity almost overlaps for the same density, it shows that the viscosity of sea water is less significant to influence its aerodynamics when tip angle is manipulated.

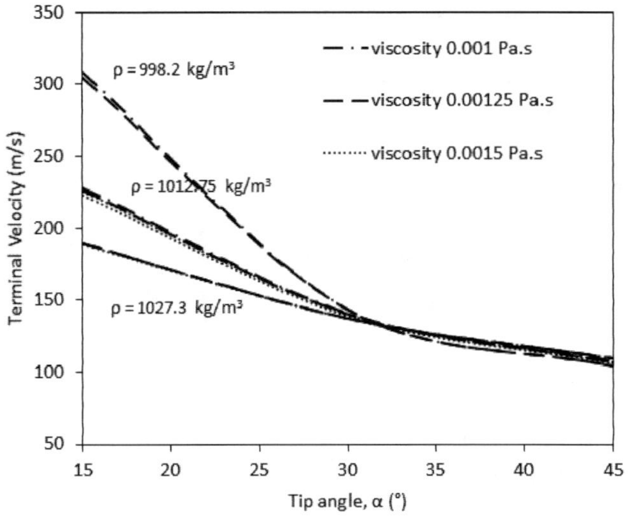

Figure 5. Effects of tip angle on terminal velocity at varied sea water density and viscosity.

To improve the anchor's terminal velocity significantly, a 15° tip angle can be implemented. It is noteworthy that there is an inflexion point for all conditions, which occurs at tip angle of 30°. Beyond this angle, the effect of tip angle on the terminal velocity is less significant, as the gradient become much smaller. Furthermore, as it can be observed from the gradient of each density, the influence of tip angle is more dominant in sea region of lower density. In other words, the significance of tip angle is greater in shallower sea region as compared to deep sea region of application, in achieving higher terminal velocity. In essence, in order to ensure higher terminal velocity, a tip angle of 30° or smaller should be utilized.

Figure 6 depicted the drag coefficient versus the Reynolds number for different tip angles varied from 15° to 60°. Tip angle of 60° has notably highest drag coefficient as compared to lower tip angles of 15°, 30°, and 45°. As tip angle varies from 15° to 45°, drag force acting on the anchor increases steadily towards constant value from 270.09 kN to 281.36 kN. Thereafter, more drastic changes in the resisting force can be observed. Furthermore, the anchor's terminal velocity decreases as tip angle increases from 15° to 60°. Thus, a design of torpedo anchor with tip angle beyond 45° is to be avoided. Sharper anchor tip would allow the torpedo to gain higher vertical downward speed with reduced drag coefficient.

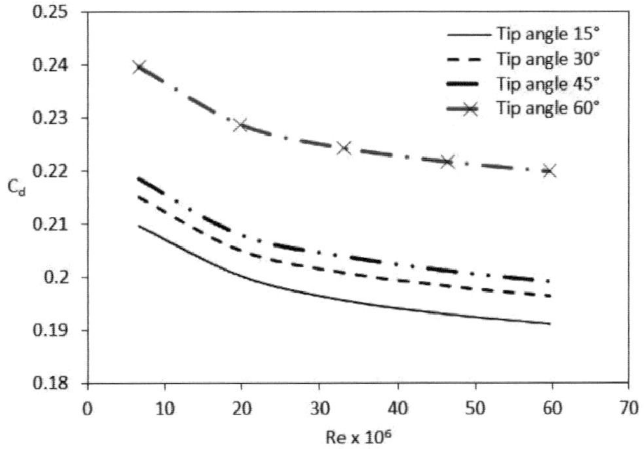

Figure 6. Drag coefficient versus Reynolds number for different tip angles.

4.2. Effects of Rear Angle, β

Figure 7 shows the effects of rear angle on terminal velocity for different densities and viscosities. It is obvious that in shallower water, terminal velocity reaches its optimum values at rear angle of 30°. The decrease in terminal velocity thereafter can be due to the vortices of fluid flow at the end of torpedo anchor, when the rear part becomes too sharp. This may also be resulted due to the presence of reversed flow, as rear angle is designed to be greater than 30°. However, for deeper sea water with higher density, the terminal velocity still does increase with rear angle greater than 30°, instead of decreased beyond that point. Thus, this finding should be taken into consideration for the anchorage systems of floating platforms at different depth. The implementation of new design feature: rear angle, turned out to be capable of improving the aerodynamics of torpedo anchor. As there is an angle at the end of torpedo anchor design, the drag force is reduced as compared to the original design. Subsequently, the greater downward acceleration is allowed to achieve higher terminal velocity. Nevertheless, there is an optimum rear angle which results in lowest drag coefficient and therefore smallest drag force, that is at 30° rear angle. The re-bounce of drag acting upon the anchor might be due to reversed flow or vortices of fluid at rear end of torpedo anchor. Anyhow, the inclusion of rear angle in torpedo anchor design is beneficial.

4.3. Sensitivity Analysis

Figure 8 shows the sensitivity of drag coefficients with respect to different parameters that were investigated. It shows that aspect ratio is the most influential factor that can be used to manipulate the drag coefficient effectively. This is followed by the rear angle, water viscosity, water density, anchor's diameter ratio, and lastly tip angle being the least dominant factor. Aside from the percentages of influence, some parameters have negative impact on the drag coefficient. The parameters that have negative correlations with drag coefficient is illustrated by darken background in figure 8 and vice versa.

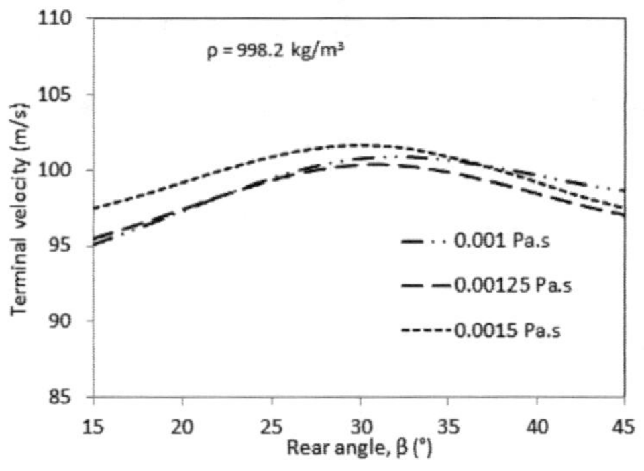

(a)

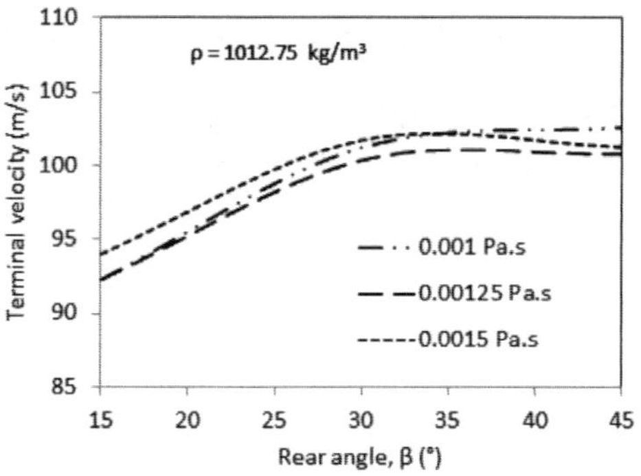

(b)

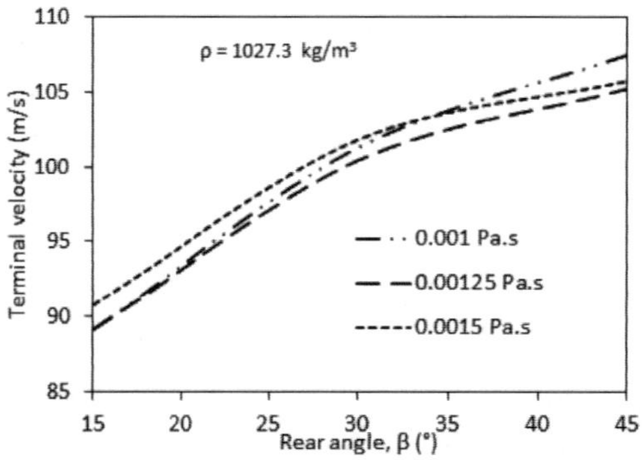

(c)

Figure 7. Effects of rear angle on terminal velocity with different viscosities at sea water density of (a) 998.2 kg/m^3, (b) 1012.75 kg/m^3 and (c) 1027.3 kg/m^3.

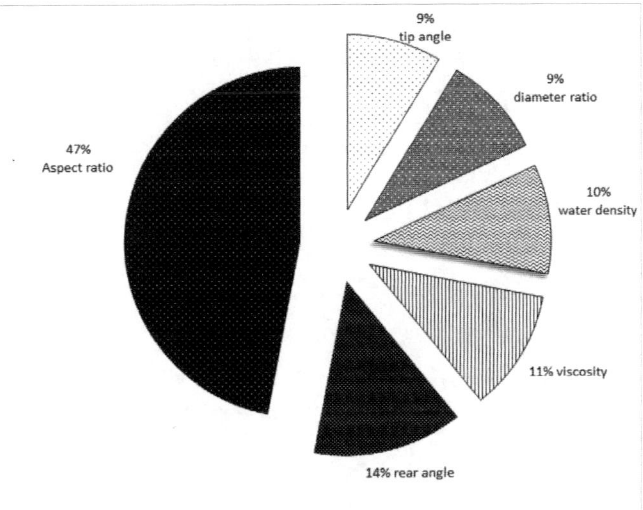

Figure 8. Affecting percentage of each studied parameters towards the drag coefficient.

5. Conclusion

Anchor will encounter higher drag force to achieve greater terminal velocity as the water depth increases. However, the effect of water depth is not significant for torpedo anchor, as its terminal velocity does not vary much with increased sea water density and viscosity. Higher terminal velocity can be achieved by implementing higher aspect ratio, lower tip angle, greater β, and greater diameter ratio into its design. It is noteworthy to recognize the optimum tip angle is 30°. In line with that, the effect of diameter ratio is not substantial, but it may be utilized as one of the cost reduction measure. Drag coefficient is the most sensitive towards changes in aspect ratio, and its influence can be as high as 47% quantitatively in comparison to any other parameters.

References

[1] Audibert J M, Movant M N, Jeong-Yun W, Gilbert R B, Torpedo piles: laboratory and field research. The Sixteenth International Offshore and Polar Engineering Conference; 2006: International Society of Offshore and Polar Engineers.

[2] Hasanloo D, Yu G, A Study on the Falling Velocity of Torpedo Anchors during Acceleration. The Twenty-first International Offshore and Polar Engineering Conference; 2011: International Society of Offshore and Polar Engineers.

[3] Brandão F, Henriques C, Araújo J, Ferreira O, dos Santos Amaral C, Albacora Leste field development-FPSO P-50 mooring system concept and installation. Offshore Technology Conference; 2006: Offshore Technology Conference.

[4] Ehlers C J, Young A G, Chen J-h, Technology assessment of deepwater anchors. Offshore technology conference; 2004: Offshore Technology Conference.

[5] Medeiros Jr C, Low cost anchor system for flexible risers in deep waters. Offshore Technology Conference; 2002: Offshore Technology Conference.

[6] JT L. Deep sea anchors. In Workshop on Deep Water Subsea Tie-back. 2011.

[7] Raie M S, Tassoulas J L 2009 *Journal of geotechnical and geoenvironmental engineering*

[8] Hasanloo D, Pang H, Yu G 2012 *Ocean Engineering* **42** 135-46

[9] Fernandes A C, de Araujo J B, de Almeida J C L, Machado R D, Matos V 2006 *Journal of offshore mechanics and Arctic engineering* **128** 286-93

[10] O'Loughlin C, Randolph M, Richardson M, Experimental and theoretical studies of deep penetrating anchors. Offshore Technology Conference; 2004: Offshore Technology Conference.
[11] Raie M S 2009
[12] JW M 2004 *Chapter 3: Properties of water and seawater* University of Washington)

Numerical simulation of damage evolution for ductile materials and mechanical properties study

A El Amri[1], I Hanafi[2], M E Y Haddou[1] and A Khamlichi[3]
[1]Condensed Matter, Radiation and Numerical Simulation Laboratory, Faculty of Sciences, M'hannech II 93002, Tetouan, Morocco
[2]National School of Applied Sciences (ENSAH), BP 03 Ajdir, Al Hoceima, Morocco
[3]Systems of Communications and Detection Laboratory, Faculty of Sciences, M'hannech II 93002, Tetouan, Morocco

E-mail: abdelouahid26@gmail.com

Abstract. This paper presents results of a numerical modelling of ductile fracture and failure of elements made of 5182H111 aluminium alloys subjected to dynamic traction. The analysis was performed using Johnson-Cook model based on ABAQUS software. The modelling difficulty related to prediction of ductile fracture mainly arises because there is a tremendous span of length scales from the structural problem to the micro-mechanics problem governing the material separation process. This study has been used the experimental results to calibrate a simple crack propagation criteria for shell elements of which one has often been used in practical analyses. The performance of the proposed model is in general good and it is believed that the presented results and experimental–numerical calibration procedure can be of use in practical finite-element simulations.

1. Introduction

There is a strong interest in the use of aluminium alloy sheet for vehicle applications, particularly the body, where resistance spot welding is the principal joining method. It is important that the particular discontinuities that are often found in aluminium alloy spot welds do not adversely affect the weld properties. The 5182 Aluminium alloy have a wide range of applications in the aerospace and automotive industries. This is mainly due to its advantageous characteristics that include high electrical and thermal conductivities, high ductility, high specific strength, ease of casting, low density, high strength to weight ratio and reasonable corrosion resistance [1]. The effect of high temperature on mechanical properties such as the tensile strength, the yield stress and young's modulus depend on temperature is linked to transformations of the material structure due to various processes imply that inelastic deformation can occur more easily at elevated temperatures, so more plastic deformation and creep occur in the plastic zone of a fatigue crack. In this work, the 5182 H111 Aluminium alloy was studied in order to find out the effect of tensile strength. Both the malleability and the ductility properties relate to the plasticity of the material. Malleability refers to the ability of plastic deformation under compressive loads, while ductility refers to plastic deformation under tensile loads.

Percentage elongation before the tensile test begins two punch marks are made on the stem of the tensile test piece, distance between these marks is noted and is known as gauge length 10. After the tensile test piece fractures in two pieces, the two pieces are retrieved and placed together as close to each other as possible. Now the distance between the two punch marks is measured and noted again. Let this distance be l1. The % elongation is calculated as

$$\frac{l_1 - l_0}{l_0} x100 \tag{1}$$

Elongations indicate that material is very ductile. Low values indicate that material is brittle and has low ductility. For mild steel, the percentage elongation usually is 20% or more. The aluminium can replace the steel in the construction of the hull structures in order to reduce the total weight of the ship [2], the aluminium and the steel behave as brittle and ductile material, respectively, and their impact strength should be compared to guarantee the safety of the design.

The progressive or sudden deterioration of their mechanical strength due to loading, thermal or mechanical effects; Young's modulus and also micro-hardness are among these mechanical properties. Hence, studying the variation of young's modulus and micro-hardness are two of the proposed indirect methods that can be used to measure the damage parameters in ductile metals [3]. The numerical simulation is a widely used tool to optimize the sheet metal forming process and to reduce the development time and therefore the finale cost of products in manufacturing such as, automotive and aerospace. The identification step precedes the design of the manufacturing process.

Ductile fracture process is controlled by nucleation, growth and coalescence of micro voids, so it is natural to link material fracture behaviour to the parameters that describe the evolution of micro voids rather than conventional global fracture parameters [4]. Damage parameter is not incorporated into the constitutive equation and it is assumed that presence of voids does not significantly alter the behavior of the material. The von Mises criterion is most frequently used as yield criterion in uncoupled models, Damage parameter is incorporated into constitutive equation and crack growth simulation is automatically performed using a complete deterioration of elements in front of the crack tip [5].

Finite element modelling (FEM) has been widely introduced into the design of manufacturing products because of its high efficiency in predicting several problems and major defects occurring in sheet metal forming manufacturing process like necking. In order to predict those defects within a virtual manufacturing system [6, 7], an accurate description of the sheet metal mechanical behavior is an essential requirement. One of the most widely used models is Johnson-Cook model for ductile materials. Numerical simulation has been used to study the mechanical properties such as the tensile strength, the yield strength and Young's modulus depend the temperature because Young's modulus of some tempered steels increases slightly at mid temperatures before decreasing at high temperature on mechanical properties is linked to transformations of the material structure due to various processes [8, 9]. In order to describe the cyclic behaviour of the material for analysis with finite element method (FEM) based analysis code ABAQUS, the test data, i.e. stress-strain curves, have to be processed. This model describing fatigue behaviour of 5182 H111 Aluminium alloy under cyclic loading and their applicability for modelling of low-cycle-fatigue are discussed in this report.

2. Experimental details

The material considered in this study is the 5182 H111 Aluminium alloy. The main mechanical and thermal properties of tested specimen are summarized in table 1. The specimen geometry is shown in figure 1. Specimen is specially prepared for thermographic measurements. It is polished and cleaned to remove any oxide and grease.

The experimental setup for testing the tensile feature of specimen is illustrated in down image tensile testing and thermographic measurements were performed simultaneously. The testing of specimens was carried out on the electromechanical testing machine HOYTON TN-D, with the maximum displacement 210 mm and the strain control at room temperature.

The extension was registered using a double extensometer. The precision of the extensometer measurement is ± 0.001 mm. The therma CAM FLIR T335 infrared camera has been used for recording thermograms. The camera resolution is 320x240 pixels. The thermal measurement is performed at a frequency of 150 Hz. It was positioned at a distance of 0.5m from the sample surface. The camera sensitivity is 50 mK at 30 °C. The camera is provided with the automatic correction of emissivity and atmospheric transmission based on the distance, atmospheric temperature and relative humidity. It simultaneously makes video and thermographic recording or tracks.

The FLIR Quick Report 1.2 SP2 is capable of measuring temperature at sports, on lines and in select areas of various shapes and dimensions, as well as of showing isotherms using the gradation of grey or the palette of various colours and shades.

3. Numerical method

In the context of thermo mechanical model with internal variables, and to get a more complete understanding of the Mechanical properties of materials during testing, numerical simulations of the aluminium 5182 H111 alloy were performed [8], to evaluate thermal loads at different locations on the surfaces and at different set as well as within the specimen. The numerical analysis using explicit dynamic finite element based on the ABAQUS Software, in order to reproduce the surface thermal loads and to obtain the temperature evolution in the specimen. The material considered in this study are the 5182 H111 Aluminium alloy, with length L= 250 mm and the depth D=5mm and the width 57 mm. Finite element modelling is performed by assuming 3D deformation. The main mechanical and thermal Properties of tested specimens are summarized in table 1 but the specimen's geometries is shown in figure 1 and its meshing is shown in figure 2.

Table 1. Mechanical and thermal properties of the 5182 Aluminium alloy.

Tensile modulus MPa	2700
Tensile strength (Nmm^{-2})	284
Yield stress (MPa)	154
Mass thermal capacity ($Jkg^{-1}K^{-1}$)	909
Thermal conductivity ($Wm^{-1}K^{-1}$)	209-232

To obtain a thermal stress loading under homogeneous uniform temperature distribution, the 5182 H111 alloy was restrained against axial expansion by creating an interaction boundary condition at its outside edge. In order to determine thermal strain in the analysis we have need to the thermal expansion coefficient, α [10]. Poisson's coefficient does not depend on temperature and takes the constant value, $\upsilon = 0.3$.

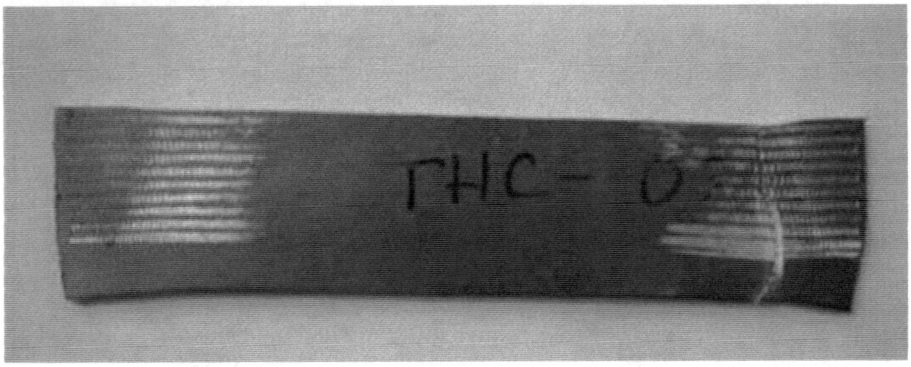

Figure 1. Geometry of the tested specimens 5182 H111 Aluminium alloy.

Fine meshes with 0.0012 mm elements and coarse meshes with 0.012 mm elements were used. The two meshes used are shown in (figure 2).

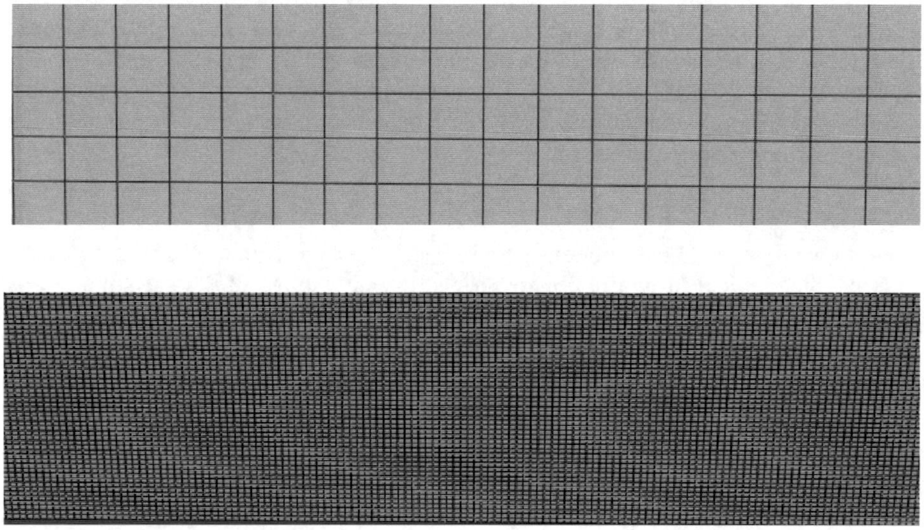

Figure 2. Fine meshes 0.012 (upper) and coarse meshes 0.0012 (lower).

4. Load and mechanical analysis of numerical simulation

A displacement boundary condition was applied at the outside extremity of the tensile specimen. The displacement was smoothly ramped up in the first portion of the test and then held constant as shown in figure 3.

Figure 3. Load and boundary condition.

5. Results and discussion

The tensile behavior of the specimens was determined as the load–time relation during tensile loading. Furthermore, a change in the loading curve during time was obtained. Load start, i.e. elastic strain start, then plastic strain start point where the maximum force is reached, i.e. the end of homogenous plastic strain. The results presented in figure 6 prove that the maximum elongation before final failure line started. The results presented in figure 5 prove that in the first 600 seconds the force reaches 70kN.

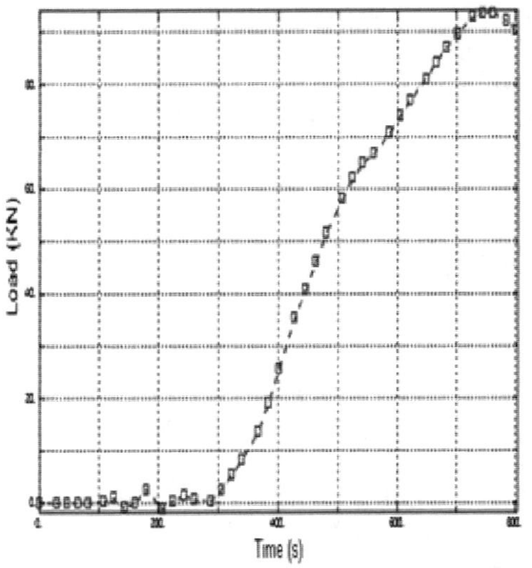

Figure 5. load vs. time.

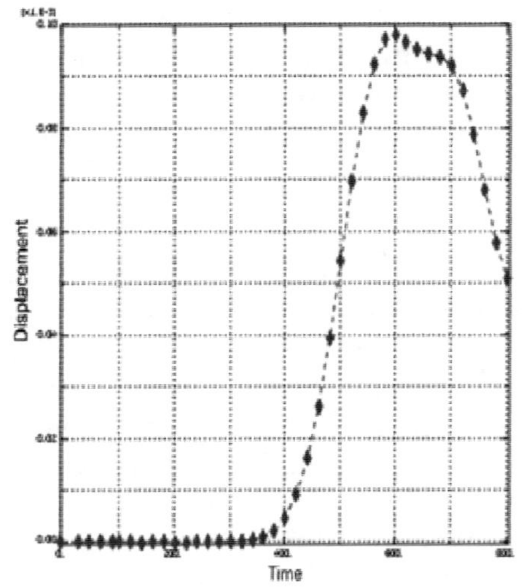

Figure 6. displacement vs. time.

The beginning of plastic deformation, reaching maximum, force 95 KN, the homogeneous plastic deformation to the final fracture of the specimen. The analysis of the results presented in Figure 5 show that in the first 350 seconds the force reaches 15 kN which the elastic strain start, The maximum tensile force is about 95 kN , point that the plastic strain and final failure start. Finally, the figure 7 (a) shows that the increase damage criterion at different temperature with the time additionally the tensile temperature increase gradually with increase in the time figure 7(b).

The performance of the investigated crack propagation criteria is in general very good compared with the experimental results [8].

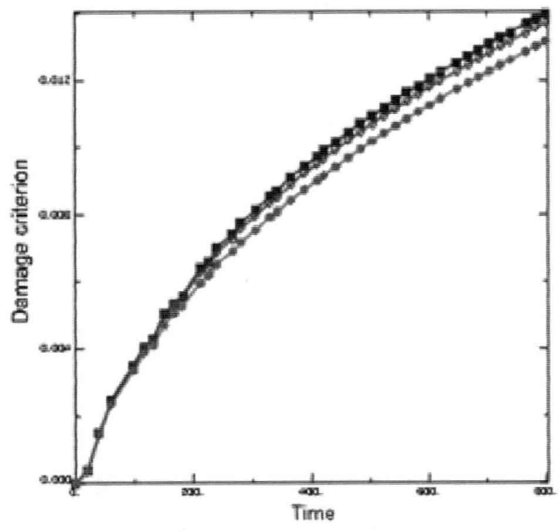

Figure 7(a). Damage criterion vs. time.

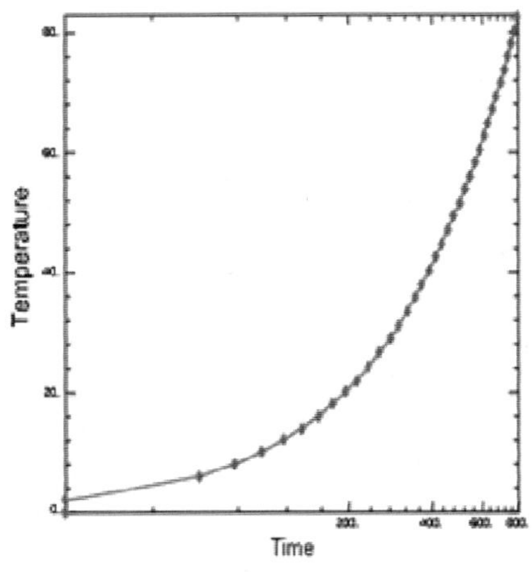

Figure 7 (b). Temperature vs time.

6. Conclusion

This paper has presented a comprehensive experimental and numerical study aimed at developing and validating a predictive model of fracture damage in plate structures under plastic conditions. The aim of the work has been to develop and use a coupled experimental–numerical procedure where the effect and modelling of fracture is not disturbed by other complex effects like buckling, contact and friction. The obtained results confirm that it is very useful to use Johnson-Cook model based on ABAQUS

IOP Conf. Series: Materials Science and Engineering **100** (2015) 012060 doi:10.1088/1757-899X/100/1/012060

software for early diagnostics of complex metal structures in the exploitation or service conditions. The obtained results prove that the Johnson-Cook model offers the possibility of non-destructive and real time testing to observe the physical process of metal degradation and to detect the occurrence of energy dissipation. A large-scale grounding experiment is also simulated showing very good agreement with measurements.

References

[1] Ilangovan S, Srikanthan R, Vyass G V *International Journal of Research in Engineering and Technology* **5** 57-61
[2] Liu B, Villavicencio R, Soares C G 2013 *Analysis and Design of Marine Structures* 167
[3] Lemaitre J 2012 *A course on damage mechanics* Springer Science & Business Media)
[4] Abendroth M, Kuna M 2003 *Computational Materials Science* **28** 633-44
[5] Velay V, Persson A, Bernhart G, Bergström J, Penazzi L 2002 *Proc. of the 6th International Tooling Conference Karlstad* 667
[6] Bauvineau L, Burlet H, Eripret C, Pineau A 1996 *Le Journal de Physique IV* **6** C6-33-C6-42
[7] Junis B, Rakin M, Međo B, Sedmak A 2010 *FME Transactions* **38** 197-202
[8] Cabrera F, Manzanares J, Hanafi I, Arenas A, Garrido I, Toledano V, et al. 2014 *International Review of Applied Sciences and Engineering* **5** 1-8
[9] Smerd R. Constitutive behavior of aluminum alloy sheet at high strain rates [Master Thesis]: University of Waterloo; 2005.
[10] Zineb A, Abdellatif K, Cabrera F M 2010 *American Journal of Engineering and Applied Sciences* **3** 740-8

Engine performance and emission of compression ignition engine fuelled with emulsified biodiesel-water

W N Maawa[1], R Mamat[1], G Najafi[2], O Majeed Ali[1], and A Aziz[1]
[1]Faculty of Mechanical Engineering, Universiti Malaysia Pahang, 26600 Pekan, Pahang, Malaysia
[2]Dept. of Mechanical Engineering, Tarbiat Modares University, P.O. Box 14115-111, Tehran, Iran

Email: maawa.ghazali@gmail.com

Abstract. The depletion of fossil fuel and environmental pollution has become world crucial issues in current era. Biodiesel-water emulsion is one of many possible approaches to reduce emissions. In this study, emulsified biodiesel with 4%, 6% and 8% of water contents were prepared to be used as fuel in a direct injection compression ignition engine. The performance indicator such as brake power, brake specific fuel consumption (BSFC) and brake thermal efficiency (BTE) and emissions such as NO_x and particulate matter (PM) were investigated. The engine was set at constant speed of 2500 rpm and load from 20% to 60%. All the results were compared to B5 (blend of 95% petroleum diesel and 5% palm oil biodiesel) biodiesel. At low load, the BSFC decrease by 12.75% at 4% water ratio and decreased by 1.5% at 6% water ratio. However, the BSFC increases by 17.19% with increasing water ratio to 8% compared to B5. Furthermore, there was no significant decrease in brake power and BTE at 60% load. For 20% and 40% load there was some variance regarding to brake power and BTE. Significant reduction in NO_x and PM emissions by 73.87% and 20.00% respectively were achieved with increasing water ratio to 8%. Overall, it is observed that the emulsified of biodiesel-water is an appropriate alternative fuel method to reduce emissions.

1. Introduction

The global energy consumption has been increasing multi-fold due to rapid population growth and economic development. This scenario occurred after the commencement of industrial revolution in the late 18th and early 19th century. Currently, fossil fuels are the primary source that powers our modern world and the reserves are likely to be depleted in 2052 [1]. This issue has led to increased exploration for alternative renewable energy. Recently, biodiesel has been considered as an important renewable energy source because of its potential to fulfill the energy demand, and reduce greenhouse gas emissions [2, 3]. In different parts of the world, depending upon climatic and agricultural conditions, different feedstocks are chosen for biodiesel production. As the world's second largest producer of palm oil in the world, biodiesel from palm oil based appears to be a practicable long-term solution for Malaysia [4]. Several studies reported that pure diesel and palm oil methyl ester (POME) blends produce comparative results in terms of engine performance. A series of tests done by Ndayishimiye and Tazerout [5] to study the performance and emissions of preheated POME. They concluded that 27% of (BTE) was improved when using palm oil and diesel blends compared to that of the engine operated with conventional diesel.

IOP Conf. Series: Materials Science and Engineering **100** (2015) 012061 doi:10.1088/1757-899X/100/1/012061

Diesel emulsion is one of many possible approaches to reduce emissions of both oxidises of nitrogen (NO_x) and PM from diesel engines [6]. It was found that the presence of water vapor in reactants influences the physics and chemical kinetics of combustion [7]. A recent study by Raheman and Kumari [8] showed that the addition of water in the form of emulsion improves combustion efficiency and emissions. The engine torque, power and BTE increase as the water percentage in the emulsion increases. Sudrajad, Hirotsugu [9] conducted a series of experiment on a single cylinder water cooled diesel engine by using 10% of water blended with pure diesel. They found that the brake specific fuel consumption (BSFC) was decreased and the there was reduction in carbon monoxide (CO), nitrogen monoxide (NO) and sulphur dioxide (SO2). Subramanian [10] compared the effects of water–diesel emulsion and water injection into the intake manifold on performance, combustion and emission characteristics of a direct injection diesel engine under similar operating conditions and found that the BTE is reduced at all outputs below diesel values with water injection due to poor combustion. Hydrocarbon (HC) and CO levels are lower at low loads with water injection as compared to the emulsion. Nguyen, Dan [11] investigated the effects of double injection on combustion, performance and emissions of Jatropha water emulsion in a diesel engine. They found that a drastic increase in the ignition delay, combustion duration, and a shift in the combustion center toward the later combustion stage, as well as increased exhaust gas temperatures, and reduced brake thermal efficiency compared with the Light oil. Ithnin, Ahmad [12] tested the water-in-diesel (W/D) fuel with different water percentages (5%, 10%, 15% and 20%) on a direct injection air-cooled diesel engine. NOx and PM are found to be reduced for all types of W/D.

The aim of the present work is to study the effect of presence of water in the form of emulsion with blended biodiesel-diesel fuel B5. The engine performance parameters and emissions were measured experimentally and the results evaluated compared to blended fuel B5.

2. Material and method

2.1. Engine Experimental Setup

The experimental setup consists of four cylinder diesel engine, an engine test bed and a gas analyzer. The schematic diagram and real picture of the experimental setup is shown in figure 1 (a) and (b). Two separate fuel tanks equipped with stirrer and fuel valve systems were arranged to carry out the experiments, one is for diesel-biodiesel fuel and the other one is for emulsified biodiesel-water.

The engine used for testing was naturally aspirated four-stroke direct injection water-cooled Mitsubishi 4D68 inline four-cylinder engine that utilizes a high pressure common rail fuel system, variable geometry turbocharger and dual overhead camshafts. The basic specifications of the engine are shown in table 1. A Kistler 6041A piezoelectric pressure transducer was installed to the first cylinder of the engine through glow plug adapter to measure the pressure inside the cylinder. The speed of the engine is measure using a non-contact proximity sensor Kistler type 2613B while torque is measured using strain gauge bases load cell.

Table 1. Technical features of the test engine.

Engine model	Mitsubishi 4D68
Number of cylinders	4
Rated output, (kW)	64.9 at 4500 rpm
Maximum Torque, (Nm)	177 at 2500 rpm
Bore/stroke, (mm)	42.7/93
Engine displacement, (cc)	1998
Compression ratio	22.4:1
Combustion chamber	Swirl chamber

Table 2. Properties of B5 fuel [13].

Properties	B5
Density at 15°C (kg/l)	0.842 - 0.846
Viscosity at 40°C (mm^2/s)	4.136 - 4.549
Flash point (°C)	75.0 - 81.0
Cloud point (°C)	14 - 16
Pour point (°C)	9 - 12
Sulphur content (mg/kg)	0.17 - 0.18
Cetane number	54.8 - 61.5
Higher Heating Value, HHV (MJ/kg)	44 - 47

The test engine was coupled to a 150 kW eddy current type water-cooled, dynamometer controlled by Dynalec controller. The thermocouples used in measuring the necessary temperature at different locations during the test were the K type thermocouples. Fuel flow rate and airflow were measured by AIC fuel flow meter model 1204 HR 2000 and CENTERTEK blade anemometer respectively. Kane 900 gas analyzer was used to measure NO_x and PM emissions. The high speed data acquisition system consists of DEWE5000 is connected to the DEWESoft and DEWECa software for data logging.

2.2. Preparation of Emulsified Biodiesel

Palm oil based B5 biodiesel, 5% palm (*Elaeis guineensis*) and 95% diesel is readily available at most petrol stations in Malaysia since 2011 [14]. The properties of biodiesel B5 have been shown in Table 2 [13]. There are two distinctive emulsification techniques for water and oil; water-in-oil (W/O) and oil-in-water (O/W). However it is mostly restricted to use two-phase water-in-oil (W/O) technique for internal combustion engines compared to oil-in-water (O/W) technique [14].

The emulsified fuel was prepared using a mechanical homogenizing stirrer machine. The fuels were blended by using a homogenizer device at a speed of 800 rpm for 15 min. Lin and Wang [15] showed that when the rotational speed increases (in the range from 300 rpm to 1400 rpm), the stabilization increases as well. However, when the rotational speed was further increased (>1000 rpm), only a small effect on stabilization was observed. Emulsion stability increased with stirring time up to 15 min, beyond which the stability decreased with further increase in stirring time due to the drop-out of emulsifier from oil-water interface [16].

By using the method that utilized by Aziz, Jusoh [17], both surfactant Span 80 and Tween 80 were added into the mixture of biodiesel B5 and water to increase the affinity and reduce the interfacial between the water and oil phases. The test fuels were biodiesel (B5), B5 emulsified with 4% water (BW4), B5 emulsified with 6% water (BW6), and B5 emulsified with 8% water (BW8). The quantity of the each surfactant was set at 1 % of the total volume.

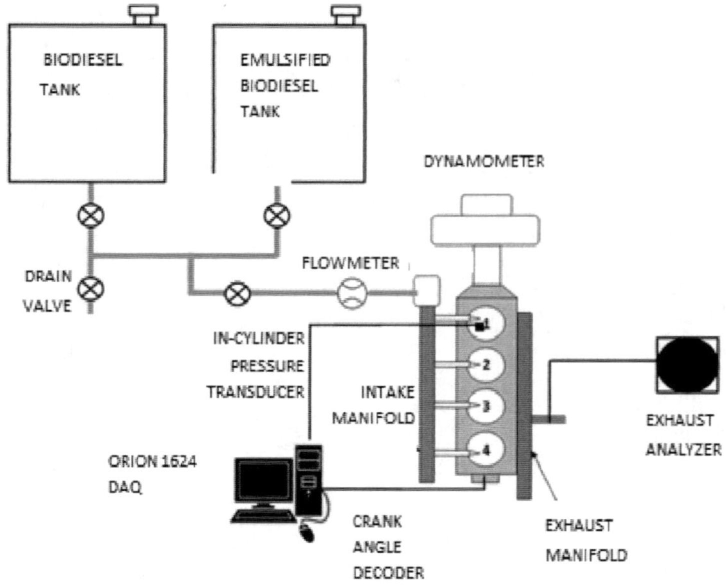

(a)

(b)

Figure 1. (a) Schematic diagram of experimental set up and (b) real picture of experimental setup.

3. Result and discussion

3.1. Engine Performance

The brake power generated by the engine under different load conditions starting from 20% to 60% at constant speed 2500 rpm is illustrated in figure 2. As the load increases, the brake power produced by the engine also increases for all blends of fuel. At 60% load, the power generated for fuel B5 is slightly higher than other water-emulsified fuels. This is due to the higher energy value of B5 compared to emulsified biodiesel. The power generated for fuel B5, BW4, BW6 and BW8 at 60% load were 28.067 kW, 27.63 kW, 27.7 kW and 27.95 kW. The trend has resulted from the energy value of the fuels. From the results it is concluded that the power produced by emulsified fuels were comparable to that of B5 at higher load.

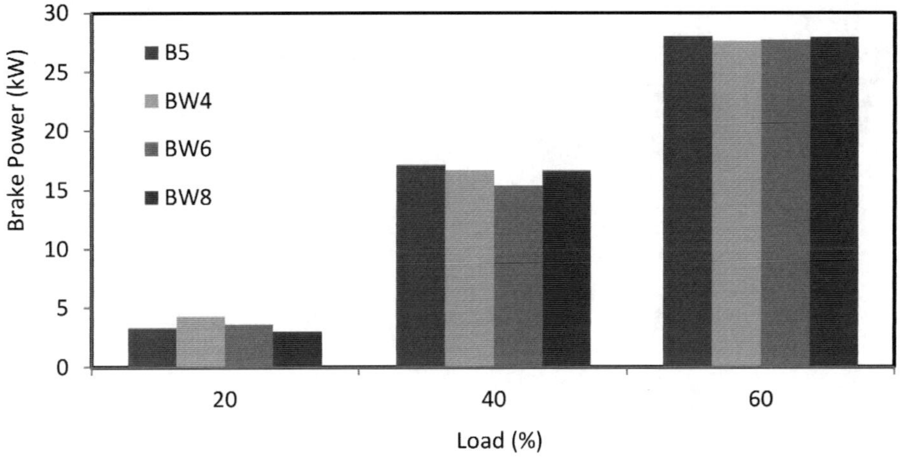

Figure 2. Comparison of brake power vs.Load.

The variation of brake specific fuel consumption (BSFC) with respect to load is presented in figure 3(a). The BSFC was found to decrease with increase in load for all tested fuels. This is due to the higher percentage increase in brake power with load as compared to the increase in fuel consumption. From the graph, it can be noticed that the BSFC were the lowest at 40% of load. That is because the engine were operating at its optimal condition.

Figure 3(b) shows the variation of BTE with respect to load for different fuels. HHV is used to evaluate the BTE. There was no significant variation of BTE with respect to emulsified biodiesel. In general, thermal efficiency increased within the engine speed until it reaches a maximum value, beyond which the thermal efficiency decreases. As known at low speeds, the time available for heat to be transferred to the cylinder walls is relatively long, and hence significant amount of heat loss occurs. As speed increases the brake power increases, leading to a higher BTE. However, increasing friction and inertia of moving parts due to high speed lead to decrease in BTE [18, 19].

3.2. Emissions

Figure 4(a) shows the comparison of NO_x vs. Load. The NO_x emissions were increased when the load increase. By utilizing water in emulsified biodiesel, the NO_x emissions were decrease significantly. The main reason is that the heat absorption by water vaporization causes a reduction in the peak ignition temperature which then reduces the NO_x formation [20]. The maximum reduction is 73.87%. The variation of PM emissions with respect to load is illustrated in figure 4(b). It can be identified that the PM emissions reduced when the water percentage increased. This was attributed to the enhancement of combustion due to the presences of micro emulsion [21]. The maximum reduction for PM was 20.00%.

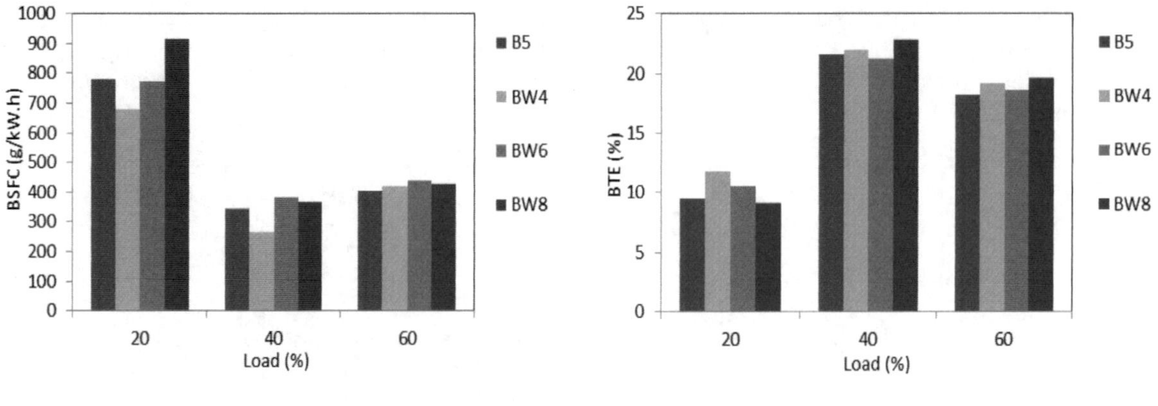

(a) (b)

Figure 3. Comparison of BSFC (a) and BTE (b) vs. load.

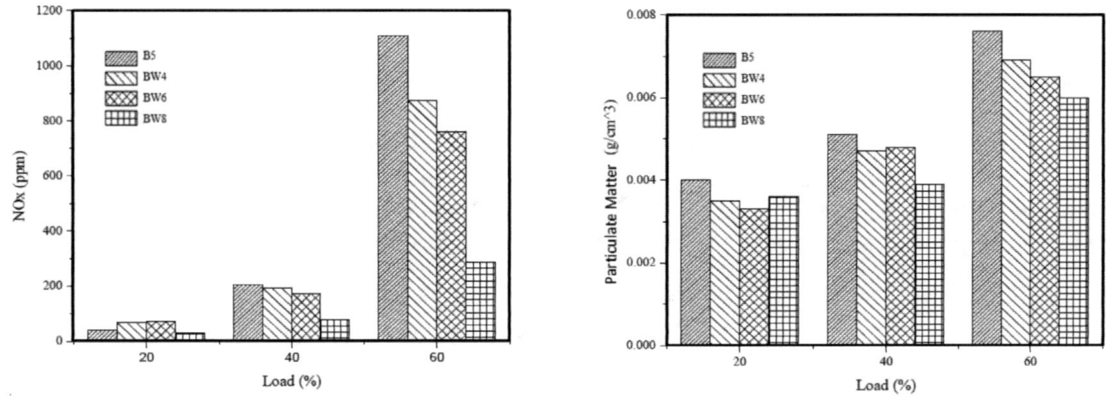

(a) (b)

Figure 4. Comparison of NO_x (a) and PM (b) vs. Load.

4. Conclusion

Emulsified biodiesel containing 4%, 6% and 8 % water were prepared and tested for engine performance and exhaust emissions in a 4-cylinder diesel engine. The results were compared with biodiesel blends (B5) as base fuel. Based from the results, several conclusions can be made which are follows;

1. The result shows no significant differences between the engine brake power and BTE measured for all emulsified biodiesel compared to B5 at 60% load.
2. At low load, the BSFC decrease by 12.75% at four percentage water ratio. However, the BSFC increases to 17.19% with increasing water ratio to 8% compared to B5.
3. The NO_x and PM emissions were reduced up to 73.87% and 20.00% respectively at 8% of water.

Acknowledgement

The author would like to thank to member of technical expert in Mechanical Laboratory, Universiti Malaysia Pahang for support and encouragement.

References

[1] Randers J 2012 *2052: A global forecast for the next forty years* (Norwegian: Chelsea Green Publishing)
[2] Hassan M H, Kalam M A 2013 *Procedia Engineering* **56** 39-53
[3] Thomas V, Choi D, Luo D, Okwo A, Wang J 2009 *Chemical Engineering Research and Design* **87** 1140-6
[4] Darshini D, Dwivedi P, Glenk K 2013 *Energy Policy* **62** 1128-37
[5] Ndayishimiye P, Tazerout M 2011 *Energy* **36** 1790-6
[6] Alahmer A, Yamin J, Sakhrieh A, Hamdan M 2010 *Energy Conversion and Management* **51** 1708-13
[7] Samec N, Kegl B, Dibble R W 2002 *Fuel* **81** 2035-44
[8] Raheman H, Kumari S 2014 *Biosystems engineering* **123** 29-39
[9] Sudrajad A, Hirotsugu F, Ali I 2011 *Modern applied science* **5** p73
[10] Subramanian K 2011 *Energy Conversion and Management* **52** 849-57
[11] Nguyen K-B, Dan T, Asano I 2015 *Energy* **80** 746-55
[12] Ithnin A M, Ahmad M A, Bakar M A A, Rajoo S, Yahya W J 2015 *Energy Conversion and Management* **90** 375-82
[13] Foon C S, May C Y, Liang Y C, Ngan M A, Basiron Y 2005 *Malaysian Palm Oil Board, Palm Oil Developments* 28-34
[14] Yusoff M H M, Abdullah A Z, Sultana S, Ahmad M 2013 *Energy Policy* **62** 456-62
[15] Lin C-Y, Wang K-H 2004 *Fuel* **83** 507-15
[16] Liu Y, Jiao W, Qi G 2011 *Renewable Energy* **36** 1463-8
[17] Aziz A, Jusoh A, Mamat R, Abdullah A A, Effect of Water Content and Tween 80 to the Stability of Emulsified Biodiesel. Applied Mechanics and Materials; 2014: Trans Tech Publ.
[18] Chen G, Tao D 2005 *Fuel processing technology* **86** 499-508
[19] Alahmer A 2013 *Energy Conversion and Management* **73** 361-9
[20] Lin C-Y, Chen L-W 2008 *Fuel* **87** 2154-61
[21] Lin C-Y, Chen L-W 2008 *Fuel* **87** 2154-61

IOP Conf. Series: Materials Science and Engineering **100** (2015) 012062 doi:10.1088/1757-899X/100/1/012062

Simulation study of electromagnetic circuit design in laminated magnetorheological elastomer isolator

N A A Wahab[1], S A Mazlan[1], K Hairuddin[1] and H Zamzuri[1]
[1]Malaysia-Japan International Institute of Technology,
Universiti Teknologi Malaysia, 54100 Jalan Semarak, Kuala Lumpur, Malaysia

E-mail: amri.kl@utm.my

Abstract. This paper presents a simulation study of magnetic flux density distribution in a laminated magnetorheological elastomer (MRE) isolator. The performance of laminated MRE isolator depends on the particular magnetic properties of materials. Therefore, the electromagnetic simulation using Finite Element Method Magnetic (FEMM) had been carried out in designing and optimizing the MRE isolator. It is apparent that the capability of magnetic flux penetrated through the MRE depends on the design parameters approach including the selection of the materials, shape of MRE, thickness of MRE and steel sheet.

1. Introduction

Magnetorheological elastomer (MRE) is a smart composite material and considered as a new generation of magnetorheological (MR) materials which composed of rubber and magnetic particles with property in such a way that the stiffness and damping are controllable with the application of magnetic field [1-4]. The innovative designs of devices that incorporate MRE for various applications such as vehicle seat vibration suspension [5, 6], and base isolation have been studied by many researchers [7-9]. Recently, the study about MRE isolator in seismic isolation system which categorized as a semi active isolation devices have been explored to reduce an earthquake vibration of civil structures [1]. Therefore, the isolation system using MRE could be considered as a potential isolator for improving seismic protection capability of structure [9, 10].

Conventionally, rubber matrix has a constant stiffness characteristic, once designed and installed to the structures, behaves as a passive type of isolation system. The development and characteristic of MRE were investigated to increase the adaptability of such passive base isolation system [8, 11, 12]. For example, Zhou et al. [8] investigated the performance comparison between an MR fluid damper and an MRE isolator in building structure. Behrooz et al. [13, 14] proposed a variable stiffness and damping isolator (VSDI) to be used in vibration mitigation of scaled building structure and the study started from the development of the MRE until the characterization of the VSDI as an isolator. Li et al. [15-17] proposed adaptive seismic isolator and tested to evaluate and characterize the behaviour of the MRE seismic isolator. However, most of the researchers focused on the development of material properties for specific applications and mechanical design of MRE isolator.

Other factors such as a design concept and an advancement of electromagnetic properties are not properly studied. Therefore, in order to predict the magnetic field strength across the MRE, an axial-symmetrical model of laminated MRE isolator was simulated using finite element method magnetics (FEMM). Furthermore, the geometrical parameters in designing the laminated MRE isolator was determined and considered to optimize the performance of the isolator.

2. Design concept of laminated MRE isolator

The conceptual design of laminated MRE isolator is based on vertical direction to absorb the vibration produced by movement of traffics. Moreover, this concept could work in horizontal direction to reduce the vibration under the seismic impact loading or shock impact from a vehicle accident. There are five main parts consisted in this design which are MRE element, steel plate, electromagnetic coil and mounting plate on the top and bottom of isolator as shown in figure 1. In order to guide the magnetic flux density across the laminated MRE isolator, some parameters have been considered during designing the device such as the shape of MRE, the materials selection for other parts of laminated MRE isolator and the type of coil. The design concept has been modified from the conventional laminated rubber bearing by adding more steel plates and an electromagnetic coil to generate magnetic field to go through the laminated MRE isolator. For increasing the magnetic flux distribution, the laminated MRE isolator is designed with 25 layers of thin MRE and 24 layers of steel sheets thickness.

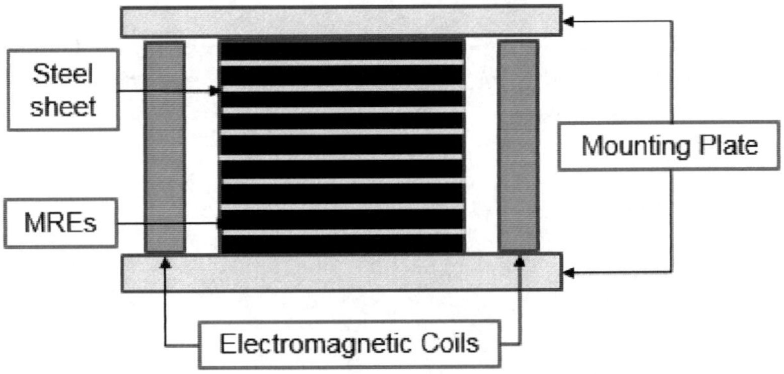

Figure 1. Conceptual design of the laminated MRE isolator.

The new concept of laminated MRE isolator consists of six different geometrical dimension parameters. They are the size of laminated MRE isolator, thickness of the gap between top plate and coil bobbin, thickness of the MRE and steel plate, number of layers for MRE and steel plates in the isolator, size of upper and bottom plate and size of coil bobbin. The magnetic field generated by the coils could be controlled by the current from the external direct current power source, while the shear modulus is determined by the strength of magnetic field. The shear modulus is used to determine the stiffness of the laminated MRE isolator.

3. Simulation

The electromagnetic circuit design is one of the important criteria to be considered in the proposed design of laminated MRE isolator. In this study, FEMM is used to analyse the magnetic field strength in laminated MRE isolator since it is useful for analysing the magnetic behaviour [11, 18]. In order to carry out the FEMM simulations, a two-dimensional cross section of the isolator is drawn within the FEMM, which incorporates steel plates, coils, coil cover, housing and MRE. During the initial stage, laminated MRE isolator consisted of four parts in which part 1 was the housing, part 2 was the steel plate, part 3 was the coils and part 4 was the coil cover. Part 1 and part 2 were made of magnetic material while part 4 was made of non-magnetic materials and part 3 was made of copper wire. After a few modifications, 1020 steel is used for magnetic materials, aluminium 6061-T6 for non-magnetic materials and 18 SWG coated copper wire for coil is selected to develop the laminated MRE isolator. The selection of coil in this study is selected based on the highest vanes of magnetic field intensity, *H*. Details of each part are shown in table 1. Therefore, 18 SWG was selected with 3000 number of turns and 5 A of applied current.

Table 1. List of parts and material for laminated MRE isolator.

Part no	Part name	Type of material	Material	Type
1	Housing	Magnetic	Low carbon steel	1020
2	Steel plate	Magnetic	Low carbon steel	1020
3	Coils	Non-magnetic	Copper wire	18SWG
4	Coil cover	Non-magnetic	Aluminum	6061-T6

For an asymptotic boundary condition, the coefficients in a boundary condition can be represented by

$$\frac{1}{\mu_r \mu_o}\frac{\partial A}{\partial n}+c_o A+c_1=0 \quad (1)$$

Where A is magnetic vector potential, μ_r is the relative magnetic permeability of the region adjacent to the boundary, μ_o is the permeability of free space, and n represents the direction normal to the boundary

$$c_o=\frac{1}{\mu_r \mu_o R} \quad (2)$$

$$c_1=0 \quad (3)$$

Where R is the outer radius of a sphere problem domain.

4. Results and discussion

The purpose of electromagnetic circuit design in the laminated MRE isolator is to generate the highest possible magnetic flux density across the MRE. The laminated MRE isolator is modeled with asymmetric model. After the asymmetric drawing was sketched as shown in figure 2 (a), meshing area was created as shows in the figure 2 (b), the analysis results displayed the magnetic flux line as in figure 2 (c) and the flux density values in figure 2 (d).

The design parameters that considered in this model are selection of materials, diameter of coils, number of coil turns and current value that affected the magnetic field intensity H, thus related to magnetic flux density B [19]. The constitutive relationship between B and H can be represented by

$$B = \mu H \quad (4)$$

Where B is the magnetic flux density, H is the field intensity, and μ represents the material dependent parameter called permeability.

Figure 3 shows the area of MRE, which has been analyzed as the magnitude of the flux density. The average value of magnetic flux density is indicated by the effective area as shown in figure 3. The pattern of magnetic flux lines seemed to penetrate the MRE and formed a closed loop surrounding the electromagnetic coil. This is due to the presence of magnetic particles inside the MRE. The magnetic field strength is measured through various applied current from 1 A up to 5 A. Figure 4 shows magnetic flux density in the effective area based on the currents applied. According to the graph, it can be observed that the flux lines have passed across the effective region of the MRE isolator. Therefore, the MRE in the effective area can be rheologically influenced by changing of the current input. As expected, when the applied current has been increased, the magnetic flux density also increases. The average magnetic flux at the maximum current of 5 A is about 0.53T.

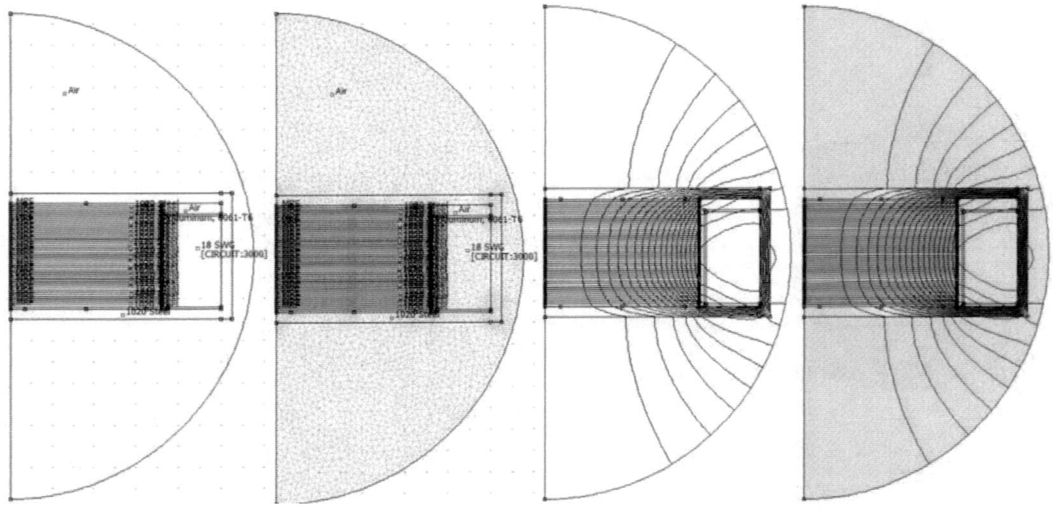

Figure 2. Model of laminated MRE isolator in simulation software at (a) designing stage (b) Meshing area of the model (c) direction of magnetic flux lines and (d) density plot.

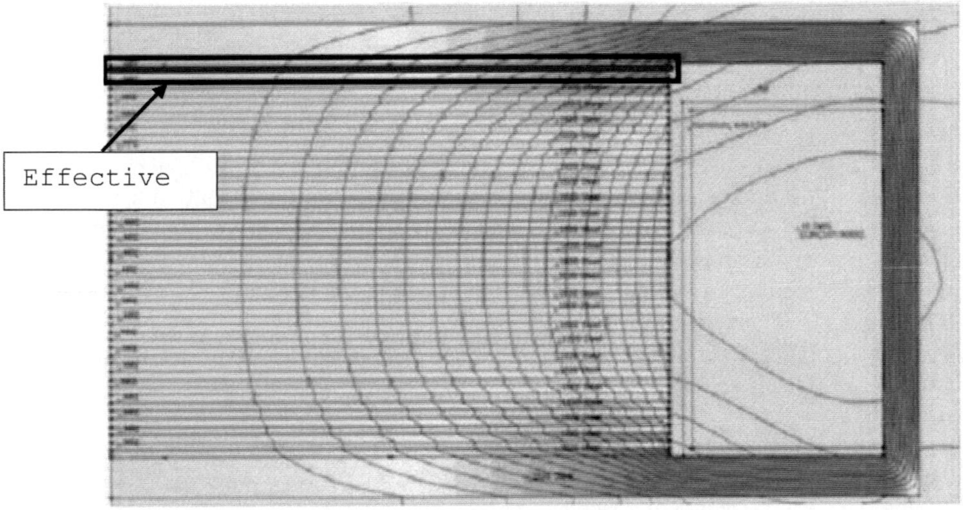

Figure 3. The area of magnetic flux density of laminated MRE isolator.

The result from figure 3 and 4 showed that the magnetic flux density is increased until the end of measurement points due to the magnetic concentration and permeability is higher at the end of measurement area which is the nearest to magnetic material. Therefore, the proposed laminated MRE isolator is developed with the highest possible magnetic flux density distribution through MRE in order to achieve maximum stiffness from the current value. As discussed by Li et al [3], the lateral stiffness of MRE increased when the magnetic field increased. The final geometrical dimension of the laminated MRE isolator achieved after a few modifications have been made to attain uniform magnetic field strength as shown in table 2. The size of laminated MRE isolator designed is 200 mm for width, 210 mm for length and 60 mm for thickness.

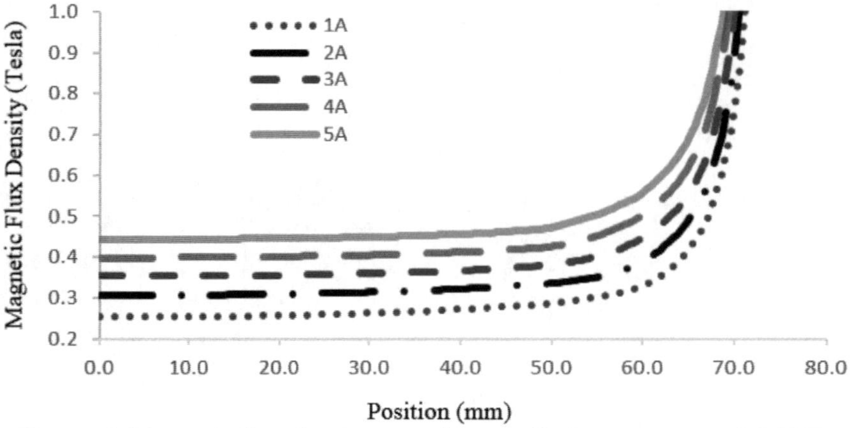

Figure 4. Magnetic flux density at various applied current through MRE.

Table 2. Structural parameters of laminated MRE isolator.

Parameters	Number of sheet	Thickness of sheet (mm)
MRE	25	1
Steel	24	1
Mounting plate	2	5
Size of coil bobbin	25 mm	

5. Conclusion

In this paper, a laminated MRE isolator was designed to be a smart device in isolation system for civil engineering application. The laminated MRE isolator was designed based on the optimization of magnetic flux density across the MRE after a several modification of magnetic material, non-magnetic material and electromagnetic circuit by using FEMM. The proposed laminated MRE isolator could generate higher magnetic field strength at MRE in the shear position of MRE. It was proven that the MRE could efficiently operate for the designed laminated MRE isolator as a seismic isolator.

Acknowledgements

The research described in this paper was supported by FRGS (vote no 4L094) from Ministry of Higher Education, Malaysia and RUG from UTM (vote no 06H06). The authors are grateful for the support.

References

[1] Li Y, Li J, Li W, Du H 2014 *Smart Materials and Structures* **23** 123001
[2] Usman M, Sung S, Jang D, Jung H, Koo J 2009 *Journal of Physics: Conference Series* vol 149 IOP Publishing) p. 012099
[3] Li Y, Li J, Li W, Samali B 2013 *Smart Materials and Structures* **22** 035005
[4] Behrooz M, Wang X, Gordaninejad F 2014 *Smart Materials and Structures* **23** 045014
[5] Li W, Zhang X, Du H 2012 *Journal of Intelligent Material Systems and Structures* 431-5
[6] Yang J, Gong X, Deng H, Qin L, Xuan S 2012 *Smart Materials and Structures* **21** 125015
[7] Behrooz M, Wang X, Gordaninejad F 2014 *Smart Materials and Structures* **23** 045013
[8] Zhou Y, Li W H, Hadi M N 2010 *Applied Mechanics and Materials* vol 37 Trans Tech Publ) p. 862-5

[9] Li J, Li Y, Li W, Samali B 2013 *SPIE Smart Structures and Materials+ Nondestructive Evaluation and Health Monitoring* International Society for Optics and Photonics) p. 86920H-H-12
[10] Eem S-H, Jung H-J, Koo J-H 2011 *Magnetics, IEEE Transactions on* **47** 2901-4
[11] Yu M, Wang S 2010 *Smart Materials and structures* **19** 065023
[12] Deng H-x, Gong X-l 2008 *Communications in nonlinear science and numerical simulation* **13** 1938-47
[13] Behrooz M, Wang X, Gordaninejad F 2012 *SPIE Smart Structures and Materials+ Nondestructive Evaluation and Health Monitoring* International Society for Optics and Photonics) p. 83411I-I-9
[14] Behrooz M, Wang X, Gordaninejad F 2013 *SPIE Smart Structures and Materials+ Nondestructive Evaluation and Health Monitoring* International Society for Optics and Photonics) p. 86880Z-Z-7
[15] Yang J, Du H, Li W, Li Y, Li J, Sun S, et al. 2013 *Smart Materials and Structures* **22** 117001
[16] Li Y, Li J, Tian T, Li W 2013 *Smart Materials and Structures* **22** 095020
[17] Li Y, Li J, Li W 2013 *Advanced Intelligent Mechatronics (AIM), 2013 IEEE/ASME Int. Conf. on* IEEE) p. 381-6
[18] Yang C, Fu J, Yu M, Zheng X, Ju B 2014 *Journal of Intelligent Material Systems and Structures* 1045389X14541492
[19] Hairuddin K, Mazlan S A, Zamzuri H, Nor N M 2014 *Applied Mechanics and Materials* vol 660 Trans Tech Publ) p. 763-7

Numerical simulation of a novel expanded metal tubular structure for crashworthiness application

A H A Abdelaal[1] and F Tarlochan[1]
[1]Mechanical and Industrial Engineering Department, College of Engineering, Qatar University, PO Box 2713 Doha, Qatar

E-mail: faris.tarlochan@qu.edu.qa

Abstract. Search for new geometries and materials that would serve in crashworthiness applications is a cumulative process. Recent studies investigated the performance of expanded metal tubes and the possible ways to enhance its energy absorption capability. The aim of this work is to investigate the crashworthiness characteristics of new concept is proposed where expanded metal tube is suited into a double-walled tube made of the same material to form one structure. The tube was then numerically tested through a verified model using ABAQUS software. Moreover, the influence of the size of the expanded metal cell was also investigated in the present study. The new concept showed an enhanced energy absorption characteristics related to the change in the mass of the tubular structure. The enhancement was related to both the change in deformation pattern, and the increase in crushed mass.

1. Introduction

Number of factors makes thin-walled structures one of the most widely used structures for crashworthiness applications. These structures are relatively cheap, light and capable of absorbing high amount of energy through a stable progressive collapse. During collision thin-walled structures are plastically deformed and fractured what leads to the dissipation of collision's kinetic energy [1]. The term "Crashworthiness" was adopted by the automotive safety community to describe the occupant's safety performance afforded by all types of motor vehicles in different accident types. Meeting automotive industry standards made researchers and engineers look for methods to enhance the energy absorption capacity without increasing structure weight in their design of the energy absorbers. For this reason, maximizing specific energy absorption (SEA, amount of energy absorbed per unit mass) and attaining a unity value for crush force efficiency (CFE, the ratio of the average crushing force to the maximum force) are namely the main objectives in the design of thin-walled structures for crashworthiness applications [2]. A set of factors influence these performance indicators that can be summarized in first: the material properties represented in Young's modulus, yield stress, and strain hardening. Second factor is the geometry of the energy absorber. The geometry parameters could be the walls' thickness, cross-section width or diameter. Other factor is the initial imperfections like having patterns, corners, grooves or indentations in the structure [3].

Numerous studies have been published on the crushing response of empty and foam-filled tubular structures. The effect of internal geometries on the energy absorption performance of simple extruded profiles was explained thoroughly in studies [4-13]. Introducing multi-cellular structures to the column generally enhance the energy absorption performance of tubular structures [8]. Multi-cellular structures are mostly simple extruded profiles that can fill the tube partially or fully.

In studies [12-21] the influence of outer geometry of simple thin-walled structures on energy absorption characteristics was studied. Such geometries are mainly classified into prismatic and non-prismatic members. The studies have shown that the energy absorption of thin-walled structures depends greatly on the cross-section of structure. In addition, increasing the perimeter to a certain limit shows higher energy absorption capability in most of the tested profiles [1].

Studying the energy absorption characteristics of thin-walled structures under impact was conducted both numerically and experimentally. Studies [22-24] showed that findings of numerical and experimental studies had almost matching results with inconsiderable variation.

The aforementioned solutions are limited to the available manufacturing techniques commonly used in producing thin-walled structures. Even the most recent studies in the area of crashworthiness applications are still focused on simple geometries. With the advancement in manufacturing techniques like Direct Metal Laser Sintering (DMLS) it is now possible to produce complex geometries that can be integrated to the thin-walled structures. DMLS functions by melting metal powder layer by layer in a high vacuum with electron beam. This technology produces dense parts from metal powder with characteristics of target material what makes it a powerful tool in producing complex 3D shapes and parts.

A continuous effort is made in investigating new materials and geometries that could improve the performance of tubular structures in crashworthiness applications. Very few information about the performance of expanded metal tubes is found in the literature. The process of manufacturing expanded metal sheets is achieved through an in-line expansion of partially slit metal sheets. The outcome of this process is a diamond-like pattern shown in Figure 1. Paper [25] summarizes the international patents on expanded metal. Major advantage of expanded metal sheets is its ability to absorb energy under different loading conditions like bending, shear, or axial compression. The use of concentric tubes in [3] showed a significant enhancement to both energy absorption capability and mean load due to the interaction between both tubes during collision. The aim of the present work is to investigate other possible mechanisms to further enhance the energy absorption of expanded metal tubes.

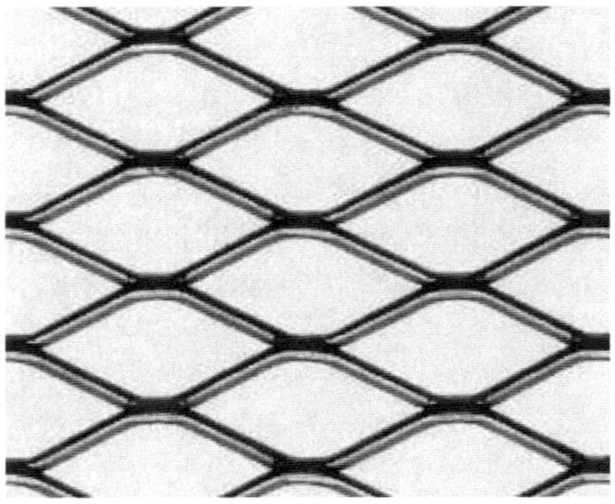

Figure 1. Typical expanded metal sheet.

1.1. Crashworthiness Indicators

To indicate the crashworthiness performance the overall crash response was considered. Different variables were obtained to characterize the overall crash response:

1.1.1. Peak Load. Peak load is the maximum force observed in the effective stroke of the tubular structure during impact. It is important to determine this value as it indicates possible consequences or damages that could be seen by the body being protected.

1.1.2. Energy Absorption. The total energy absorption E in a crash test equals the area under the load-displacement curve. It is defined as equation (1).

$$E = \int_0^{d_{max}} F ds \tag{1}$$

For simplicity, d_{max} or stroke efficiency is taken to be 0.75L where L is the tube length. To relate how much energy a structure can absorb to it mass, specific energy absorption is obtained. It is defined as equation (2).

$$SEA = \frac{E}{m} \tag{2}$$

Where m is the crushed mass of the component.

1.1.3. Crushing Force Efficiency. The crushing force efficiency is a ratio between the mean and peak crushing force, the two parameters that are related to the declaration occurring during a crash. It is defined as equation (3).

$$CFE = \frac{P_{mean}}{P_{max}} \tag{3}$$

An ideal energy absorber would have a CFE value close to one. That is because an ideal absorber would preserve a peak load for its entire crushed length.

2. Design Methodology

Findings of [3] showed that the deformation of expanded metal tubes lead to unstable structural response. In an attempt to solve this issue, expanded metal tube was fit in-between two metal sheets as shown in figure 2. A schematic view of an expanded metal cell is shown in figure 3. The major dimensions are the two orthogonal axis, l_1 and l_2, width (w), and thickness (t) of the expanded metal cell. The values are tabulated in table 1 for the different proposed designs. The thin-walled tube has a cross-sectional area of 80x80 mm and an overall thickness of 2.0 mm. The tube's length is chosen as 350 mm to assure the full potential deformation of the tube when it is crashed. Material assigned to all designs is A36 mild-steel. The test is conducted through an axial impact on the tubular structure using a striker of 275 kg mass with an initial velocity of 15.6 m/s as shown in figure 4.

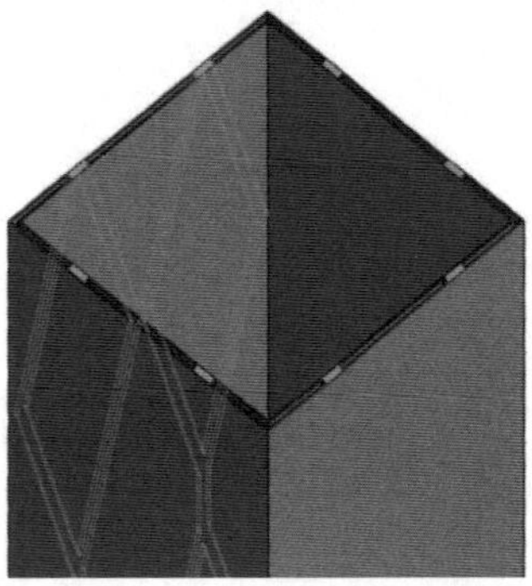

Figure 2. Design concept.

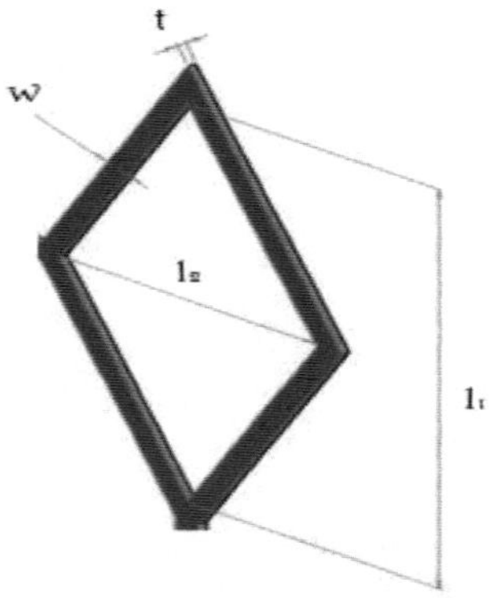

Figure 3. Schematic view of an expanded metal cell.

Table 1. Geometry and dimensions of structures under study.

	Expanded Metal Sheets Dimensions				Tube Dimensions			
Specimen ID	**l_1 (mm)**	**l_2 (mm)**	**w (mm)**	**t (mm)**	**Length (mm)**	**Perimeter (mm)**	**Mass (g)**	**Profile**
S1	35.68	31.22	2.74	1	350	320	1000	
S2	72.60	32.50	2.74	1	350	320	919	
S3	108.97	32.78	2.74	1	350	320	976	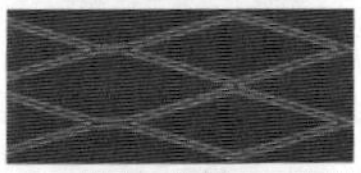
S4	146.91	32.88	2.74	1	350	320	973	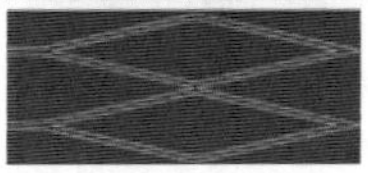

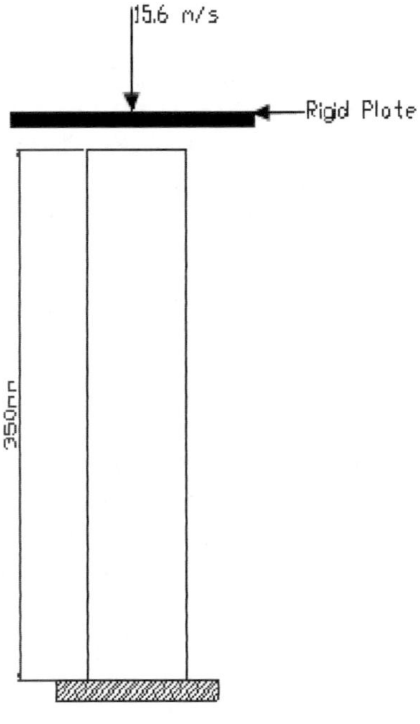

Figure 4. Simulation test setup.

2.1. Finite Element Model

To find the continuum point's time dependent deformation equation (4) is used.

$$\sigma_{ij} + \rho f_i = \rho \ddot{x}_i \quad (4)$$

Where, σ_{ij} represents the Cauchy stress, ρ is the density, f_i is the body force and $\ddot{x}_i$ is the acceleration. Using the divergent theorem equation (4) can be transformed into the virtual work principle given in equation (5).

$$\int_V \rho x_i \delta \ddot{x}_i dV + \int_V \sigma_{ij} \delta x_{ij} dV - \int_V \rho f_i \delta x_i dV - \int_{S^2} t_i \delta x_i dS = 0 \quad (5)$$

Equation (5) can be written in matrix form as shown in equation (6).

$$\sum_{i=1}^{n} \left\{ \int_V \rho N^t N a dv + \int_V B^t \sigma dv - \int_V \rho N^t b dv - \int_A N^t F dA + \int_s N^t F_c ds \right\}^i = 0 \quad (6)$$

Where the number of elements is expressed as n, the stress column vector as σ, the interpolation matrix as N, the nodal acceleration vector as a, the body load column vector as b and the applied traction force as F. A general way to explain the matrix form is stated in equation (7).

$$[M]\left[\frac{d^2u}{dt^2}\right] + [C]\left[\frac{du}{dt}\right] + [K]\{U\} = [F(t)] \quad (7)$$

Where M, C and K are the mass, damping and stiffness matrices respectively.

First the initial and boundary condition are considered. Then the displacements are solved according to these conditions. After the displacements solution is obtained, the computation of other variables such as plastic strains, force and energies starts. The mentioned equation is usually solved by finite element software using implicit methods. However, explicit methods are desired for nonlinear dynamic problems like the crash test in our case. The main idea of such explicit methods is the division of total time into small time intervals or steps. Desired dynamic equations are solved and their variables are evaluated at increment t+1 based on knowing their previous values at t. The major difference between the explicit and implicit methods is that current step dependency. In implicit methods the evaluation of a step one relies on the information of the previous step and the current one where in explicit methods it only depends on the previous step value. This dependency on the current step value makes solving such equations harder, thus explicit methods are used in such cases.

In this work the finite element model for thin-walled tube with the employed complex geometries were generated using ABAQUS. The non-linear finite element code ABAQUS explicit was used to predict the response of the tube and its complex structures after it is impacted by a plate of certain mass with certain velocity. To model the first section of the tubular structure, a square tube of 80.0 x 80.0 mm, 350.0 mm length and 2.0 mm thickness is first modeled with a groove that goes through the total length on the tube with a thickness of 1.0 mm, inside this groove an expanded metal tube is modeled. A better picturing of the tube would be double walled tube with expanded metal cells in-between. The thin walled tube was modelled using 4 node shell continuum (S4R) element with reduced integration points. The shell elements size is selected to be 2.50 mm. The top plate (striker) was modeled as a rigid body with a reference point mass of 275 kg and with a predefined velocity field of 15.6 m/s on one of its reference point. The motion of the top plate was restricted to only translational motion along the tube's axes. The thin walled tube is fixed on the lower rigid plate using a constraint between their two surfaces. To account for the contact between the two plates and the tube, a general contact algorithm is chosen with a penalty coulomb friction coefficient of 0.2 [26]. This is to avoid the interpenetration between the tube walls.

To account for the structure materials, the A-36 mild-steel models were adopted and verified from [1]. The steel verified mode showed 2% deviation from the one in [1], and thus it was used in the simulation for this study. The mild-steel has a young modulus E=200 GPa and Poisson's ratio of 0.3.

The plastic behavior of the steel was modeled using Johnson-Cook model with the parameters shown in table 2.

Table 2. Johnson-Cook parameters [1].

Parameters	Values
A	146.7 MPa
B	896.9 MPa
Strain Power Coefficient, N	0.32
C	0.033
Temperature Power Coefficient, M	0.323
Reference Strain Rate	1.0 s^{-1}
Density	7850 kg/m^3
Melting temperature, Tm	1773 K
Specific Heat, Cp	486 J/kg-K

3. Results and Discussion

3.1. Force-displacement characteristics and deformation patterns.

Figures 5 and 6 show the force-displacement curves for the four numerically tested samples of the present study. The displacement represents the change in position of the striker which was brought to a full contact with tube at the test beginning. Samples with the same force-displacement curve features were grouped together to simplify the analysis.

The deformation pattern of S2 is very similar to S4 with a slight phase-out of the location at which folds starts to form. Force of S2 is generally larger than S4 as more interaction is happening between the expanded metal sheets and the double-walled tube.

The behavior of S1 and S3 is different as the number of expanded metal cells is higher in the case of S1 sample leading to formation of more folds. Observing the deformed tubes shows that S1 had the most stable collapse among tested samples. This is also evident in the value of CFE. Compared to models where expanded metal sheets is solely used in forming the tubular structure, the outward buckling has not occurred for any of the tested tubes as the double-walled tube controls the deformation of the expanded metal sheets. In all tested sample, progressive collapse starts with the formation of regular lobes then some irregularity starts to appear at the end of the effective stroke length.

Table 3. Summary of crashworthiness parameters for studied concepts.

	E (kJ)	SEA (kJ/kg)	P_{Mean} (kN)	P_{Max} (kN)	CFE (%)
S1	12.1	16.1	46.4	121.1	38.3
S2	5.2	7.6	20.1	118.4	17.0
S3	10.7	14.6	41.2	129.8	31.7
S4	9.9	13.6	38.0	130.4	29.1

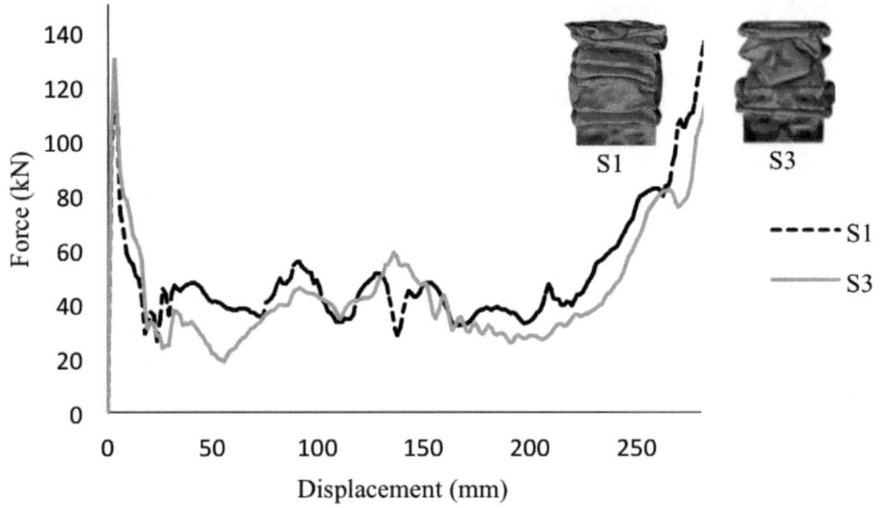

Figure 5. Crushing force vs. displacement curves for S1 and S3 models.

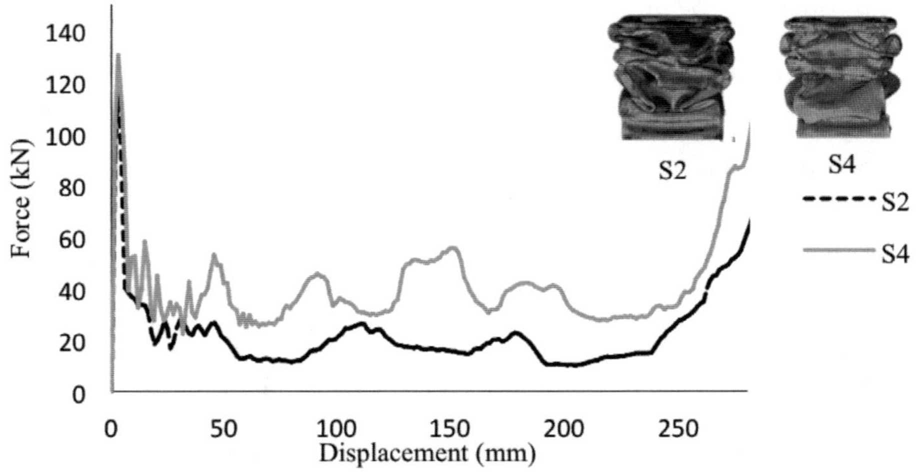

Figure 6. Crushing force vs. displacement curves for S2 and S4 models.

3.2. Energy Absorption

In figure 7, the energy absorption is plotted as a function of the deformation length of the different structural design concepts. From these figures, it can be seen that S1 with the smallest cell size attained the highest energy absorption value among tested samples.

The relation between cell size and energy absorption isn't clear in figure 7. However, a closer look at other factors affecting the energy absorption shows that as the total mass of the tube increases the energy absorption increases as well. This is demonstrated in figure 8. In this particular case, increasing the mass of the tube increases the deformed mass and enhances the energy absorption performance of the tube with a minimum influence of cell size on the energy absorption characteristics. Finally, comparing the present models with the ones presents in [21] and of the same mass and overall thickness shows a significant enhances in energy absorption capabilities of the tubular structure.

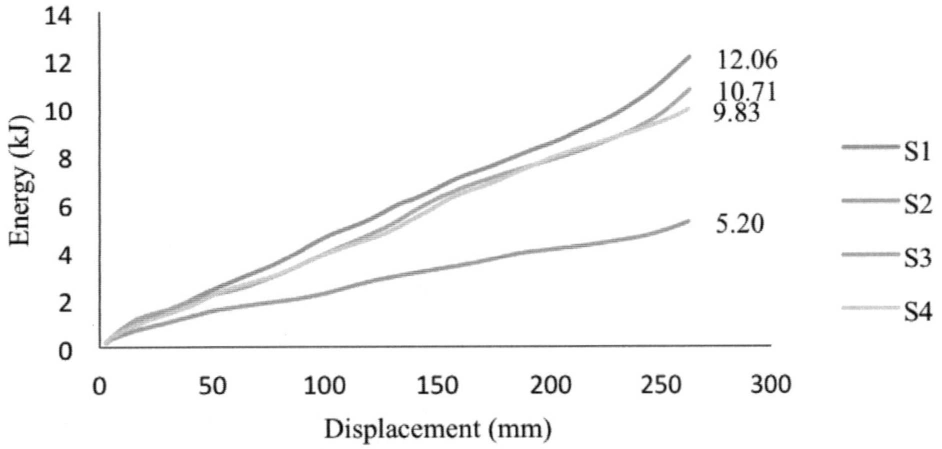

Figure 7. Energy vs. displacement for studied concepts.

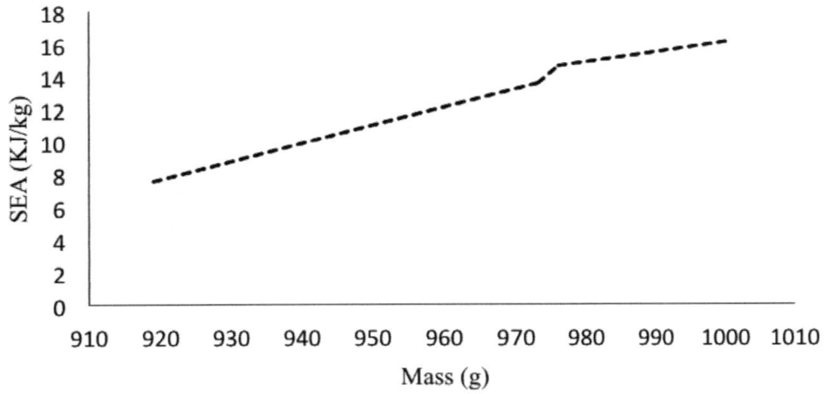

Figure 8. SEA vs. tube mass of studied concepts.

4. Conclusion

In this study, expanded metal sheet was placed between the two sides of a double-walled tube made of the same material. Crushing of the tubes was conducted numerically through a verified model on ABAQUS software. It was observed that the proposed design enhanced the energy absorption over ordinary expanded metal tubes. The major reason for that is the change in the deformation modes that leads to a progress folding during the collision instead of outwards buckling. Such behavior results from the restriction caused by the double-walled tube. Increasing the cell size by varying the two orthogonal axis showed no correlation to the energy absorption; however, the tubes with more mass regardless of the size of their cells showed better energy absorption compared to the ones with less mass. Further work needs to be done to understand the influence of expanded metal cells thickness on the performance of the tubes for the same cell size and double-walled tube overall wall thickness.

Acknowledgement

The authors would like to appreciate the efforts of Qatar University in providing all necessary support.

References

[1] Tarlochan F, Samer F, Hamouda A, Ramesh S, Khalid K 2013 *Thin-Walled Structures* **71** 7-17
[2] Tang Z, Liu S, Zhang Z 2013 *Thin-Walled Structures* **62** 75-84
[3] Smith D, Graciano C, Martínez G 2014 *Thin-Walled Structures* **84** 170-6

[4] Ajdari A, Nayeb-Hashemi H, Vaziri A 2011 *International Journal of Solids and Structures* **48** 506-16
[5] Hanssen A, Langseth M, Hopperstad O 2001 *International Journal of Mechanical Sciences* **43** 153-76
[6] Hanssen A G, Langseth M, Hopperstad O S 2000 *International Journal of Impact Engineering* **24** 347-83
[7] Sun G, Li G, Hou S, Zhou S, Li W, Li Q 2010 *Materials Science and Engineering: A* **527** 1911-9
[8] Zhang X, Cheng G 2007 *International Journal of Impact Engineering* **34** 1739-52
[9] Mahdi E, Sebaey T 2014 *Thin-Walled Structures* **76** 8-13
[10] Tasdemirci A, Kara A, Turan K, Sahin S 2015 *Thin-Walled Structures* **91** 116-28
[11] Sharifi S, Shakeri M, Fakhari H E, Bodaghi M 2015 *Thin-Walled Structures* **89** 42-53
[12] Nia A A, Parsapour M 2014 *Thin-Walled Structures* **74** 155-65
[13] Liu S, Tong Z, Tang Z, Liu Y, Zhang Z 2015 *Thin-Walled Structures* **88** 70-81
[14] Shahi V J, Marzbanrad J 2012 *Thin-Walled Structures* **60** 24-37
[15] Nagel G, Thambiratnam D 2004 *International journal of mechanical sciences* **46** 201-16
[16] Reddy S, Abbasi M, Fard M 2015 *Thin-Walled Structures* **94** 56-66
[17] Abbasi M, Reddy S, Ghafari-Nazari A, Fard M 2015 *Thin-Walled Structures* **89** 31-41
[18] Malekzadeh P, Daraie M 2014 *Thin-Walled Structures* **84** 1-13
[19] Zhang X, Wen Z, Zhang H 2014 *Thin-Walled Structures* **84** 263-74
[20] Tanlak N, Sonmez F O 2014 *Thin-Walled Structures* **84** 302-12
[21] Smith D, Graciano C, Martínez G, Teixeira P 2014 *Thin-Walled Structures* **85** 42-9
[22] Yin H, Wen G, Liu Z, Qing Q 2014 *Thin-Walled Structures* **75** 8-17
[23] Abedrabbo N, Mayer R, Thompson A, Salisbury C, Worswick M, Van Riemsdijk I 2009 *International Journal of Impact Engineering* **36** 1044-57
[24] Ahmad Z, Thambiratnam D, Tan A 2010 *International Journal of Impact Engineering* **37** 475-88
[25] Smith D, Graciano C, Martínez G 2009 *Recent Patents on Materials Science* **2** 209-25
[26] Dehghan-Manshadi B, Mahmudi H, Abedian A, Mahmudi R 2007 *Materials & design* **28** 8-15

Influence of damping coefficient from permanent magnets on chatter formation during end milling of titanium alloy (Ti-6Al-4V)

S A Sulaiman[1] and A N Amin[2]
[1] Faculty of Mechanical Engineering, Universiti Teknologi MARA Johor, 81750 Masai, Johor, Malaysia
[2] Departments of Material and Manufacturing, Kulliyyah of Engineering, Jalan Gombak, 53100 Kuala Lumpur, Malaysia

Email: syidatul_akma7091@johor.uitm.edu.my

Abstract. In machining operations, chatter is undesirable due to its adverse effects on the product quality, operation cost, machining accuracy and machine tool life. It is also responsible for reducing output. Chatter is a self-excitation phenomenon occurring in machine tools, in which the cutting process tends to lower the damping capacity of the machine structural components ending in an unstable behavior of the system. Chatter arises due to resonance when the vibrations of the instability of chip formation and the natural vibration modes of the machine-system components coincide. This paper focuses on the influence of damping coefficient from permanent magnets on chip serration frequency as an approach of minimizing chatter in end milling of Titanium alloy (Ti6Al4V). The method consists of two ferrite permanent magnet bars (dimensions: 1″ x 6″ x 3″), mounted 5mm from the cutting tool using a specially designed fixture which provided a uniform magnetic field of 2500-2700 Gausses. A titanium alloy Ti6Al4V block was then end milled using uncoated WC-Co inserts. A sequence of 15 experimental runs was conducted based on a small Central Composite Design (CCD) model in Response Surface Methodology (RSM). The primary (independent) parameters were: cutting speed, feed, and depth of cut. The data acquisition system comprised a vibration sensor (accelerometer) and a signal conditioning unit was used to measure the vibration data. The resultant vibrations were then analyzed using the DASYLab 5.6 software. Machining tests were conducted for two different conditions – with and without the application of magnets. Scanning Electron Microscope (SEM) was used to measure the chip segmentations. The SEM analysis of chip serrations demonstrated that the chip serration frequency were more stable while cutting under the presence of permanent magnets due to lower intensity of chatter.

1. Introduction

Milling process could be considered the most efficient method of metal cutting which provides high material removal rate while maintaining a high quality level. The most critical limitation in machining productivity and part quality is the occurrence of the instability phenomenon called regenerative chatter [1]. Several hypotheses have been developed, over the years, to address chatter formation phenomenon and its suppression methods. Piezoelectric actuators [2], magnetostrictive actuators [3] and multiple time-varying parameters (MTVP) [4] have been employed to reduce chatter in turning. Stone [5] extended the non-uniform pitch concept by proposing end mills with a different helix on adjacent flutes. Alternating the helix increased the pitch variations along the axial depth of cut hence both the chatter free performance and the speed range were improved. Although, chatter has been extensively investigated over the past 70 years and several hypotheses and theories have been developed, the root cause of chatter still remains controversial [6]. Thus the use of permanent magnets has become a new approach to control chatter.

Identification of chip frequency is important since chatter is thought to occur when the chip serration frequency (primary or secondary) coincides with or get close to, or becomes an integer multiple of any of the natural frequencies of a system [7] . Chatter occurred at nearly constant frequencies, close to the natural frequencies of prominent system components over a range of cutting speed for a given set of cutting conditions [8] Chatter is an abnormal tool behavior which it is one of the most critical problems in machining process and must be avoided to improve the dimensional accuracy and surface quality of the product [9]. Chatter arising during end milling and turning is a result of resonance, caused by mutual interaction of the vibrations due to serrated elements of the chip and the natural vibrations of the system components, e.g. the spindle and the tool holder. Moreover, serrated behavior is the outcome of the discrete nature of the chip formation process that exists in the entire cross section of the chip. Serrated edges appear either as primary and/or secondary saw/serrated teeth depending on the material properties and cutting conditions [10, 11]. The main objective of this research is to study the influence of magnetic field from permanent magnets in end milling of titanium alloy on chip serration frequency. Investigation on method of chatter suppression is much desired to improve the efficiency of the milling operation. The use of permanent magnet has potential application in controlling chatter. Since magnets have strong attractive forces, they were applied to increase the damping capacity of the vibration component/system.

2. Methodology

Experimental test was conducted on Vertical Machining Center (VMC ZPS, Model: 1060) with a maximum spindle speed of 8000 rpm under the influence of magnetic field from permanent magnets. Two ferrite magnets bar with dimension of 1 inch x 6 inch x 3 inches were mounted 5mm from the cutting tool by specially designed fixture. The strength of these magnets was 2500-2700 Gauss. Titanium alloy (Ti6Al4V) block was used as a workpiece material. Machining was performed with a 20 mm diameter end mill tool holder fitted with one uncoated WC-Co insert. Machining process was done under two different conditions. First condition is under room temperature and the other one is under the influence of magnetic field from permanent magnets bar. Scanning Electron Microscope (SEM) was employed in order to investigate the chip segmentations. Figure 1 includes the photograph of the experimental setup

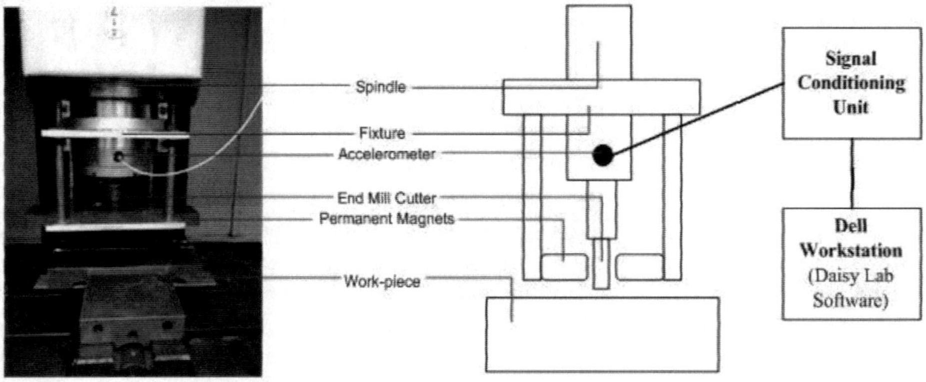

Figure 1. Fixture mounted on the spindle.

The experiments were designed based on the Response Surface Methodology (RSM) approach using DESIGN EXPERT (DOE) software. Three independent variable are tested which are cutting speed, feed and depth of cut. The ranges for the independent variable are 19.55≤cutting speed≥120.45m/min, 0.05mm/tooth≤feed≥0.18mm/tooth and 0.24mm≤depth of cut≥1.51mm. The effect of different cutting parameters on chip serration frequencies and its interaction with the mode frequencies of the end milled components has been investigated considering the vibration data that are recorded during machining using an online data acquisition system. Initially, knocking test has been done to measure the natural frequency of the system component. The accelerometers are connected to the component and the natural mode frequency data from the FFT graph were recorded by knocking the spindle using impact hammer as shown in figure 2.

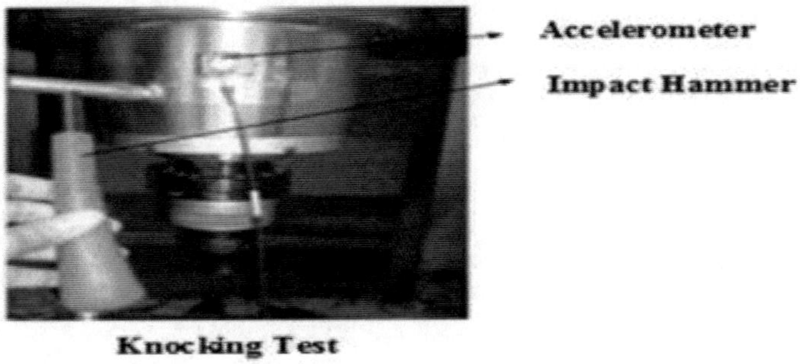

Figure 2. Knocking test to determine the natural frequencies of the spindle.

3. Results and Discussion

Experimental studies are conducted on the instabilities of chip formation in end milling operations to investigate the mechanisms of formation of chip serration under different cutting conditions. The details calculation procedure for chip serration frequencies in end milling of titanium alloy for two different cutting conditions (magnets and without magnets) are shown in table 1.

Sample SEM photographs of the top view of the chips formed during these two operations are presented to have a clear understanding of the mechanism of serrated chip formation. During end milling operation there is instability of chip formation in the form of serrated teeth and the serrated teeth are observed on both the lengthwise edges of the chip. The calculated chip frequency and the recorded vibration amplitude for all runs were presented in table 2.

Result shows that the calculated chip serration frequencies are higher when magnets were added to the system compared to those under normal cutting. Cutting under the presence of magnets made the chips more stable with comparatively higher frequency of primary serration teeth as compared to normal cutting. From this observation, chatter appears in the system when these serrated teeth get amplified and their frequencies are lowered and coincides with the natural frequency of any one of the system components, leading to resonance [9]. The increment in chip serration frequency is due to the damping effect of the magnets on the oscillatory motion of the tool. This results due to the retarding forces of attraction from the magnets, located in diagonally opposite direction, on the vibrating tool.

3.1. Resonance theory without application of magnets

This section is mainly designed to investigate the chatter formation as an outcome of resonance effect due to interaction between the chip serration frequency and prominent mode frequencies of the spindle. Initially, the knocking test was done to find the natural frequency of the spindle with and without fixture attachment. From knocking test (without fixture) the natural frequencies (f_n) for spindle are appeared at 8,758 Hz and 11,291 Hz as shown in figure 3. But since the highest peak amplitude was at 11,291Hz, thus it is selected as the prominent natural frequency of the spindle. According to the resonance theory of chatter; chatter occurs when the chip serration frequency approaches the prominent natural frequency of a component or one of its higher harmonic frequencies.

Table 1. Chip serration frequency calculation for end milling of titanium alloy under the influence of magnetic field and normal cutting.

Cutting without magnets	Cutting with magnets
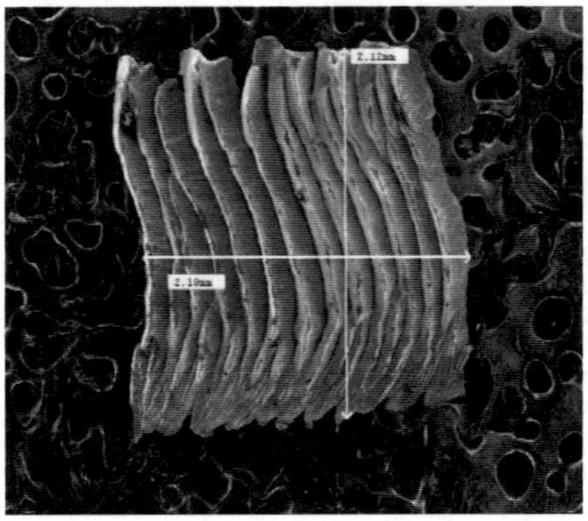	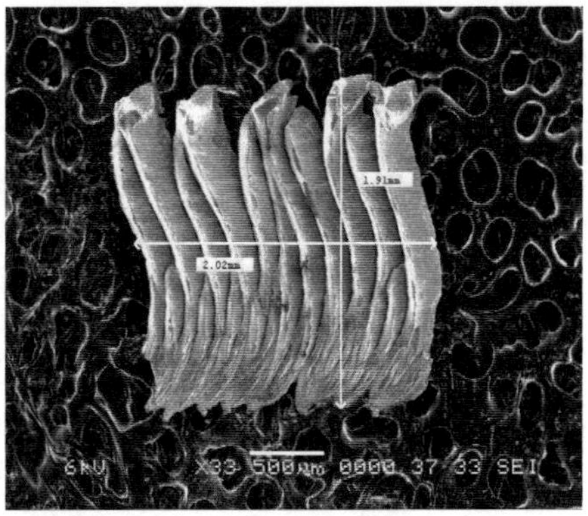
Cutting Speed(V)=70m/min No of serrated teeth(n) = 13 Length in SEM (L) = 2.18mm Chip shrinkage coefficient (K) = 1.205	Cutting Speed(V)=70m/min No of serrated teeth(n) =11 Length in SEM(L) = 2.02mm Chip shrinkage coefficient(K)=1.048

Detemination of chip serration frequency considering the SEM view

$fc = \frac{70x1000x13}{60(2.18x1.205)} = 5{,}773\text{Hz}$	$fc = \frac{70x1000x11}{60(2.02x1.048)} = 6{,}062\text{Hz}$

Where

f_c = Chip Frequency

To investigate the resonance effect, the calculated primary chip serration frequencies for five cutting conditions are compared with the dominating mode frequencies of the spindle. The summary of the relationship between calculated chip serration frequency and vibration amplitude are presented in table 3. From the experimental results (without magnet) the highest vibration amplitude (run 4) was observed when f_c was nearly equal to f_n or $f_c/f_n \approx 1$ (see table 3).

For clear understanding, the relationship of vibration amplitude and chip serration frequency over natural frequency (*fc/fn*) is plotted in figure 4. It has been observed that when the ratio of the primary chip serration frequency and the dominating mode frequency of the spindle is close to one, there is a peak in acceleration amplitude (0.03703volt) which proves that chatter is formed as a result of resonance.

Table 2. Chip serration frequency for normal cutting and cutting under the influence of magnetic field from permanent magnets.

Run Number	Cutting Speed (m/min)	Feed (mm/tooth)	Depth of Cut (mm)	Chip frequency (Hz)		Vibration Amplitude (V)	
				Without magnets	With magnets	Without magnets	With magnets
1	70	0.1	1.51	14,625	15,420	0.02689	0.02004
2	70	0.1	0.24	9,845	10,465	0.00975	0.00731
3	100	0.05	1.25	11,241	17,074	0.01483	0.01109
4	**100**	**0.15**	**0.5**	**11,182**	**14,405**	**0.03703**	**0.01482**
5	70	0.1	0.88	8,199	9,512	0.01241	0.00943
6	70	0.1	0.88	9,841	10,762	0.01241	0.00939
7	70	0.18	0.88	5,773	6,062	0.02249	0.01689
8	70	0.1	0.88	9,427	10,179	0.01102	0.00851
9	40	0.15	1.25	5,116	5,146	0.01336	0.00913
10	70	0.02	0.88	14,319	18,884	0.00767	0.00529
11	120.45	0.1	0.88	15,840	16,672	0.01577	0.01168
12	40	0.05	0.5	12,622	14,759	0.00439	0.00346
13	19.55	0.1	0.88	6,101	6,758	0.00301	0.00213
14	70	0.1	0.88	9,410	10,076	0.01168	0.00943
15	70	0.1	0.88	9,986	10,259	0.01474	0.01003

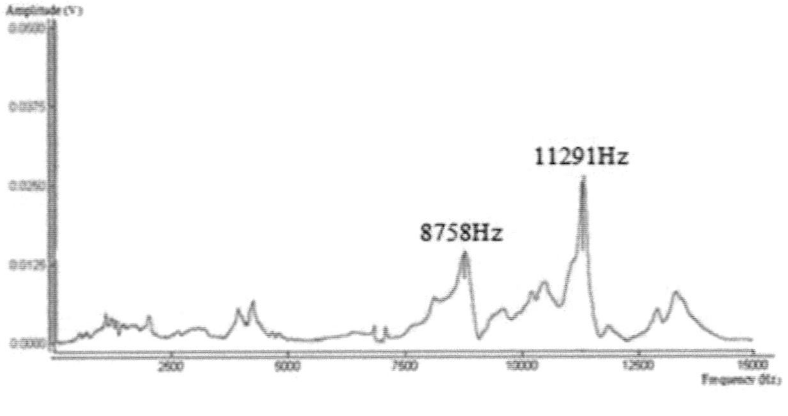

Figure 3. FFT for natural frequency of spindle (without fixture attachment).

IOP Conf. Series: Materials Science and Engineering **100** (2015) 012064 doi:10.1088/1757-899X/100/1/012064

Table 3. Relation of *fc/fn* with vibration amplitude (without magnets)

Natural frequency (f_n)	Chip serration frequency (f_c)	f_c/f_n	Vibration amplitude	Run number
11291	5116	0.453104	0.01241	9
11291	9845	0.871933	0.00723	2
11291	11241	0.995572	0.03703	4
11291	12622	1.117881	0.00541	12
11291	15840	1.402887	0.00922	11

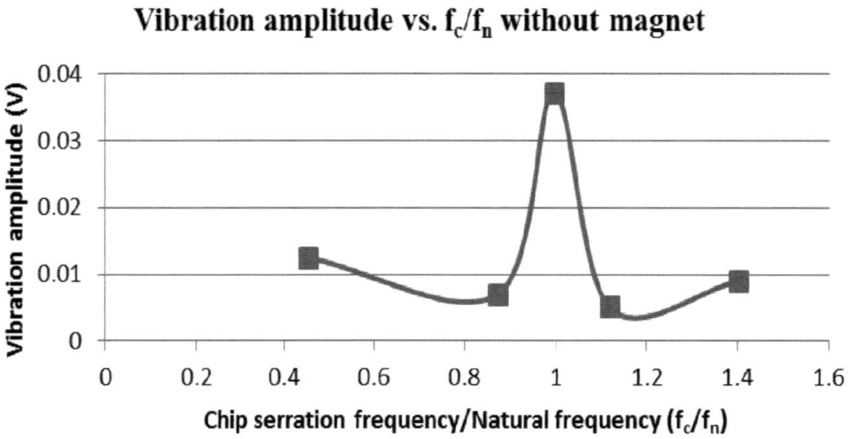

Figure 4. Effect of chip serration frequency/natural frequency ratio (without magnets) on vibration amplitude.

3.2. Resonance theory with application of magnets

The fixture with the magnet attachment has its own mass which reduces the natural frequency of the spindle. From FFT plot in figure 5, the prominent natural frequency was shifted to the left as compare with the FFT for natural frequency in figure 3. This happened due to the addition of mass from fixture when it was clamped on the spindle. From knocking test it was found that the spindle with fixture attachment had natural frequencies as shown in table 4.

From table 4, it can be inferred that the second peak of natural frequency had more probability of interacting with the calculated f_c as the first one was too low and the vibration amplitude for the third peak was too low as well. From the calculated primary chip frequency (with magnet), chip serration frequency (f_c) starts from more than 5000 Hz. So, resonance in this case was seen whenever f_c approached any higher harmonic of the second f_n peak of 4,500 Hz. The higher harmonics after 4,500Hz are at 9,000 Hz (double) and 13,500 Hz (three times higher).

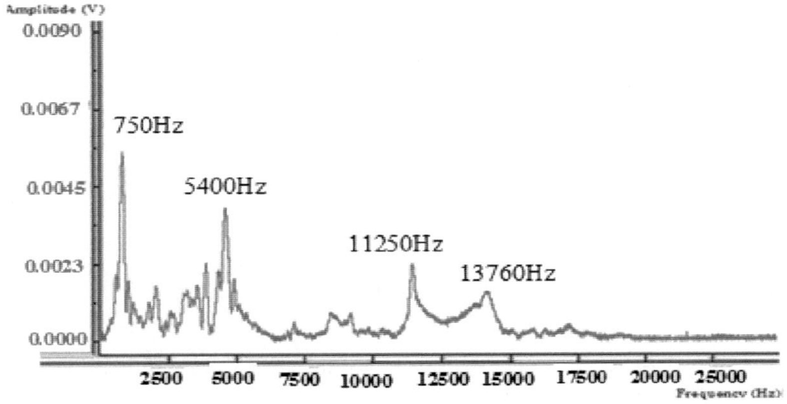

Figure 5. FFT for natural frequency of the spindle with fixture attachment.

Table 4. Natural frequency of the spindle (with fixture attachment).

Natural frequency of the spindle (Hz)	
First peak	750
Second peak	4500
Third peak	11250
Fourth peak	13670

From table 5 and figure 6, it has been observed that chatter occurred when chip serration frequency was equal to 9,512 and 14,759 Hz, which are near to the higher harmonic (9,000 and 13,500Hz) of natural frequency 4,500. However, the peak amplitudes are much smaller with the magnets, with the maximum amplitude of 0.009 volts with magnets against 0.037 volts without magnet, showing the damping action of the magnets.

Table 5. Relation of f_c/f_n with vibration amplitude (with magnets).

Natural frequency (f_n)	Chip serration frequency (f_c)	f_c/f_n	Vibration amplitude	Run number
4500	5146	1.143556	0.00471	9
4500	6062	1.347111	0.00560	7
4500	9512	2.113778	0.00817	5
4500	10762	2.391556	0.00657	6
4500	14759	3.279778	0.00913	12
4500	17074	3.794222	0.00531	3

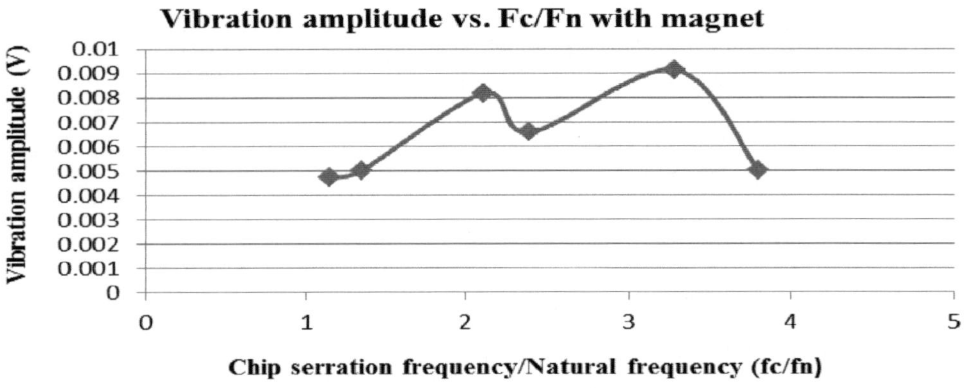

Figure 6. Effect of chip serration frequency/natural frequency ratio (with magnets) on vibration amplitude.

Chatter occurred due to the resonance between the chip serration frequencies with any prominent mode frequency of the different machine components [9]. Since these high peaks are caused by resonance and the amplitude of vibration in resonance is a function of the damping coefficient of the system, so any factor that would lead to the increase in the damping factor is expected to lower the amplitude of chatter. These findings have once again proven the resonance theory of chatter formation. figure 7 illustrates the frequency response of a system during resonance. The maximum response amplitude during resonance depends on how much damping the system has. The less the damping there is (i.e. the smaller is ζ) the larger it gets, and if there is more damping in the system, then the amplitude will be at minimum level.

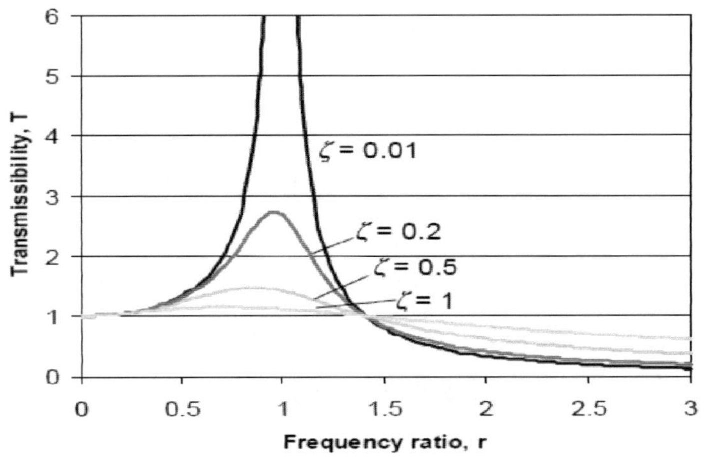

Figure 7. Response amplitude of a Damped system with an applied force as a function of the frequency ratio.

Hence the application of magnets definitely increased the damping coefficient, i.e. increased the ζ value of the system which resulted in approximately 4 times lower value of the maximum acceleration amplitude peak during machining.

IOP Conf. Series: Materials Science and Engineering **100** (2015) 012064 doi:10.1088/1757-899X/100/1/012064

4. Conclusion

The main objective of this research is to find the effect of application of permanent magnets on chip serration frequencies. Based on the findings on end milling of Titanium alloy (Ti6Al4V) it has been concluded that machining under the influence of magnetic field resulted in suppression of chatter. This is because, cutting under the presence of magnets made the chips more stable with comparatively higher frequency compared to those under normal cutting, due to the absence or reduction of chatter amplitude. This phenomenon occurred because of the damping effect of the magnets on the oscillatory motion of the tool and the formation of unbiased serrated teeth, characteristic of the materials and the applied cutting parameters. It was also concluded that when the ratio of the primary chip serration frequency and the dominating mode frequency of the spindle are close to 1 or higher integer number (2,3), there is a peak in acceleration amplitude which proves that chatter is formed as a result of resonance. Since chatter occurs due to the resonance any factor that would lead to the increase in the damping factor of the system is expected to lower the amplitude of chatter. This is why the application of magnet contributed to a drastic reduction (4 times) of chatter acceleration amplitude.

References

[1] Altintas Y, Chan P K 1991 *International Journal of Machine Tools and Manufacture* **32** 329-47

[2] Tewani S G, Rouch K E 1995 *International Journal of Machine Tools and Manufacture* **35** 91-108

[3] Pratt J R, Nayfeh A H 2001 *Philosophical Transactions: Mathematical, Physical and Engineering Science, Nonlinear Dynamics in Metal Cutting* **359** 759-92

[4] Yang F, Yu J 2003 *Journal of Materials Processing Technology* **3** 431-8

[5] Stone B J 1970*, Proceeding of the 11th International MTDR Conference* 169-80

[6] Anayet M U P, Amin A K M N, Faris W, Alam S 2008 *Proceedings to third international conference on Mechatronics, International Islamic University of Malaysia* **416-421**

[7] Amin A K M N, Anayet M U P, Sharulhazrin M S, Hafizuddin I 2010 *International Conference on Industrial Engineering and Operations Management Dhaka, Bangladesh*

[8] Soo S L, Aspinwall D K 2007 *Proceedings of the Institution of Mechanical Engineers, Part L: Journal of Material: Design and Applications* **221** 197-211

[9] Sulaiman S A, Amin A K M N 2014 *Applied Mechanics and Materials* **554** 170-4

[10] Nurul Amin A K M, Arif M D, Ahmad Mohrodi M A H B, Jaafar I H 2013 *Applied Mechanics and Material* 211-6

[11] Mohammad U A K B, Amin A K M, Ahamad Mahrodi M A H, Arif M D 2013 *Applied Mechanics and Material* **393** 211-6

Immersed boundary-lattice Boltzmann method for simulation of muco-ciliary transport: effect of mucus depth at various amounts of cilia beat frequency

M M Shahmardan[1], M H Sedaghat[1], M Norouzi[1] and M Nazari[1]
[1]Department of Mechanical Engineering, University of Shahrood, Shahrood, Iran

Email: mmshahmardan@shahroodut.ac.ir

Abstract. Numerical simulation based on immersed boundary-lattice Boltzmann method has been employed to study 2D muco-ciliary transport problem. The periciliary liquid (PCL) and mucus layers in this study are considered as the Newtonian and viscoelastic fluid respectively. An Oldroyd-B model is used as the constitutive equations of mucus layer. To simulate accurate effects of the cilia and PCL-mucus interface on the fluid, immersed boundary method is used. Numerical simulations have been performed to investigate the effects of mucus depth on the muco-ciliary clearance at various values of cilia beat frequencies. Our results show that, by increasing mucus depth, which results from air pollution and smoking, mean mucus velocity decreases. But it can be completely modified by increasing cilia beat frequency and the cilia beat frequency has great effect on the muco-ciliary clearance.

1. Introduction

The mechanics of muco-ciliary transport is one of the fundamental problems in respiratory mechanics and the study of this problem have attracted a great deal of research efforts. The airways surface liquid layer (ASL) of the human respiratory system is normally from 5 to 10 μm deep [1]. This liquid consists of two sub layers: the innermost watery periciliary liquid layer (PCL) and the outermost viscoelasitc mucus layer [2]. The main function of the mucus layer is in cleaning the inspired air of unwanted particles, cellular debris and secretions from the airways of the lung [3]. The airways of the lung are lined with a dense mat of cilia which beat back and forth in a co-ordinated wave and hence the mucus layer together with entrapped particles is propelled out of the lungs [4]. These Cilia have two quite distinct stages of their beating cycle in which the effective stroke when the cilium is extended out of the cell surface and the recovery stroke when it moves back slowly to the start of its effective stroke. The effective stroke occupies a shorter period of the beat than the recovery stroke so the work done during the effective stroke is several times the amount of work done performed during the recovery stroke [5].

Structural and fuctional disorders of cilia may be caused by genetic disorders such as Primary Ciliary Dyskinesia (PCD) and some acquired insults such as exposing to toxins and pollutants leading to inefficiency in mucu-ciliary transport [6]. The PCD is a cilia-related disease in which cilia beat in abnormal patterns or beat slowly or become completely stationary, results from some defects in the components of the axonemes [7]. Due to disorder in cilia motion, the velocity of mucus layer reduces and mucus would accumulate with airways which leads to airway blockage, lung damage and infections [8]. PCD lead to the abnormalities in the cilia beat frequency (CBF), cilia beat pattern

(CBP), and Nitric Oxide (NO) diseases which have been reported in the literature. Rossman et al. [9] reported two patterns of abnormal ciliary beat (CBP); an oscillating and a rotating type of motion. Their results show that this abnormal motion was present in up to 40 percent of cells and the remainder were totally immotile. Another abnormal CBP is reported in Rutland et al. [10] which is called "windscreen wiper" motion. Chilvers et al. [11] stated that abnormalities in the cilia beat frequency (CBF) is the result of nasal rhinorrhea or blockage and moist-sounding cough which is results from PCD. Their study [11] show that determining ciliary beat pattern related to specific ultrastructural ciliary defects might help in the diagnosis of PCD. In addition, some investigation [12-14] stated that nasal NO is found in very small quantities in PCD patients. Variation in CBF play vital role on muco-ciliary transport process [15, 16]. Although smoking has no effect [17] or has a less effect [10] on CBF, CBF is increased by ethanol [17]. The characteristics of the PCL layer such as temperature, pH, tonicity, viscocity, relative humidity, and pressure influence CBF [6].

1.1. Problem Statement
In this study, two-layer model consisting of an upper viscoelastic mucus layer and a lower Newtonian periciliary layer (PCL) is developed to simulate the effects of mucus depth on the muco-ciliary transport process at three different cilia beat frequency. A schematic of the problem is shown in figure 1.

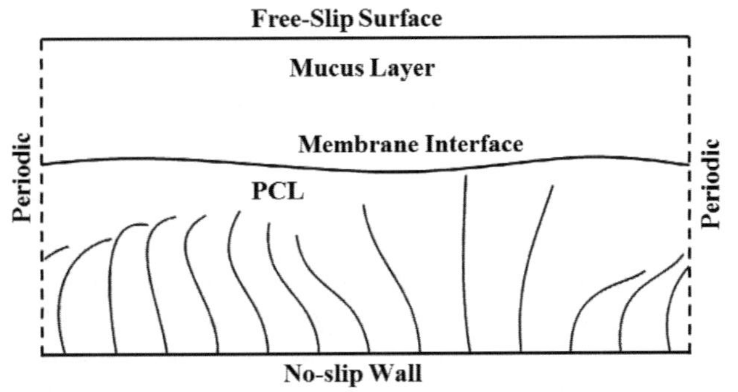

Figure 1. A schematic geometry for the muco-ciliary transport.

1.2. Significance of the Study
As metioned before CBF has a great effect on the muco-ciliary clearance. On the other hand many air pollutants, especially tobacco smoke, have a detrimental effect on the mucus transport [18]. One reason is increasing the depth of mucus layer. In this study the effect of mucus depth at various amounts of CBF has been studied. The main contribution of the present study is consideriing mucus as an Oldroyd-B fluid for the first time. This model is discussed in term of the convected components of the stress tensor and the metric coefficients of the convected coordinate system. Material constants appears in this constitutive equation and also temperature history may be included if it is desired to account for nonisothermal effects [19].

2. Mathematical formulation
In this part the governing equations and the numerical method for solving the muco-ciliary problem process is defined. As Figure 1 shows, in this problem, ASL is considered as two separate layers, the lower periciliary liquid layer consists of watery and nearly Newtonian liquid and the upper mucus layer is a viscoelastic fluid. Immersed boundary method [20] is used to simulate precise effects of the cilia and PCL-mucus interface on the fluid. At the beginning of this section, the standard parameter set which we use for reference is defined as follows [15, 16]:

Length of cilia is L=6μm, the spacing between any two neighboring cilia along the epithelium wall is chosen as 3 μm, the standard depth of PCL as LPCL=6μm, and the depth of the mucus layer as LM=4μm, the density of the fluid is. The stiffness constant of the virtual membrane is T0=32×10-3

N/m, the cilia beat frequency is set at σ=60 rad/s, the viscosity of PCL is ηPCL=0.001 Pa.s, and the viscosity of mucus layer is ηM=0.0482 Pa.s. As mentioned before, is decomposed into viscous and elastic parts. Since a full viscoelastic characterization for mucus is not yet available, we assume that the standard viscosity of Newtonian part of mucus is as the same as the viscosity of PCL i.e. ηM,N=0.001 Pa.s. For the reference parameters the viscosity ratio is β=0.98. The standard value of relaxation time of mucus is about λ=0.034 s.

2.1. Governing Equations

The governing equations of incompressible fluid flow in a model which is shown in figure 1 can be written as follows [15]:

$$\vec{\nabla}\cdot\vec{u}=0 \tag{1}$$

$$\frac{\partial\vec{u}}{\partial t}+\vec{u}.\vec{\nabla}\vec{u}+\frac{1}{\rho}\vec{\nabla}p=\frac{1}{\rho}\vec{\nabla}.\sigma+f \tag{2}$$

$$\vec{f}(\vec{x},\mathrm{t})=\int_{\Gamma}\vec{\mathrm{F}}(\mathrm{s},\mathrm{t})\delta(\vec{\mathrm{x}}-\vec{\mathrm{X}}(\mathrm{s},\mathrm{t}))\mathrm{ds} \tag{3}$$

$$\frac{\partial\vec{\mathrm{X}}(s,t)}{\partial t}=\vec{U}(\vec{X}(s,t),t)=\int_{\Omega}\vec{u}(\vec{x},t)\delta(\vec{x}-\vec{X}(s,t))d\vec{x}. \tag{4}$$

Where $\vec{x}$ is the Eulerian coordinates, $\vec{u}$ is the velocity vector, ρ is density, p is static pressure, t is time, σ is the stress tensor and f is the boundary force acting on the fluid field. Also $\vec{X}$ is the Lagrangian coordinates, $\vec{U}$ is the velocity vector of Lagrangian nodes, s is the arc length of Lagrangian nodes and $\vec{\mathrm{F}}$ is the boundary force density which contains cilia and membrane forces. In addition $\delta(\vec{x}-\vec{X}(s,t))$ is a Dirac delta function. Equation (1) and (2) describe the interaction between the immersed boundary and the fluid by distributing the boundary force at the Lagrangian points to Eulerian points and interpolating the velocity at the Eulerian points to Lagrangian points.

The cilia have a cyclic motion in PCL. There is no interface material between the PCL and mucus layers so the cilia can penetrate the mucus-PCL interface. In addition effect of the surface tension is added by the force generated due to an imaginary elastic membrane. In this study the flow is assumed to be periodic in the horizontal direction. The bottom boundary is assumed to be no-slip wall and the top boundary is assumed to free-slip boundary [15].

2.1. Constitutive Equations

Previous experimental studies have concluded that viscoelasticity play a vital role on effective transport of mucus [1,16]. In this study an Oldroyd-B model is used as a constitutive equation for mucus. In this model, the stress tensor σ can be decomposed into two parts [19]:

$$\sigma_M=\sigma_{M,N}+\sigma_{M,E} \tag{5}$$

$\sigma_{M,N}$ is related to Newtonian part of stress tensor and the second term ($\sigma_{M,E}$) is related to the elastic contribution of it which can be calculated as:

$$\sigma_{M,N}=2\eta_{M,N}D \tag{6}$$

$$\sigma_{M,E}+\lambda\overset{\nabla}{\sigma}_{M,E}=2\eta_{M,E}D \tag{7}$$

In equation (7), λ is the relaxation time, which gives an indication of the magnitude of the elastic nature of the fluid. The viscosity of mucus also decomposed into Newtonian ($\eta_{M,N}$) and elastic ($\eta_{M,E}$) viscosities as follows:

$$\eta_M = \eta_{M,N} + \eta_{M,E}, \ \beta = \frac{\eta_{M,E}}{\eta_M} \tag{8}$$

and the upper convected derivative of $\sigma_{M,E}$ is defined as:

$$\overset{\nabla}{\sigma}_{M,E} = \frac{\partial \sigma_{M,E}}{\partial \tilde{t}} + \vec{u}.\nabla\sigma_{M,E} - \sigma_{M,E}.\nabla\vec{u} - \nabla\vec{u}^T.\sigma_{M,E} \tag{9}$$

In equation (3) and (4) D is the rate of deformation which is defined as:

$$D = \frac{1}{2}\left(\nabla V + \nabla V^{\mathrm{T}}\right) \tag{10}$$

The Oldroyd-B constitutive equation can be derived from a molecular model in which the polymer molecule is idealized as an infinitely extensible Hookean spring connecting two Brownian beads [19].

2.2. *Numerical Method*

In this part 2D numerical model based on immersed boundary-lattice Boltzmann method has been used to simulation of muco-ciliary clearance problem (figure 1). This method uses a fixed Cartesian mesh to represent fluid phase, which is composed of Eulerian points. For the boundary immersed in the fluid, a set of Lagrangian points are used to represent it [20].

At the beginning of this part to simulate real rheology properties of the mucus layer and also the real dimension of the geometry, transformation coefficients form the physical domain to the lattice Boltzmann domain are introduced as:

$$C_{h_x} = \frac{\Delta x^{\mathrm{phy}}}{\Delta x^{\mathrm{lb}}}, \ C_{h_y} = \frac{\Delta y^{\mathrm{phy}}}{\Delta y^{\mathrm{lb}}}, \ C_t = \frac{\Delta t^{\mathrm{phy}}}{\Delta t^{\mathrm{lb}}}, \ C_\rho = \frac{\rho^{\mathrm{phy}}}{\rho_0^{lb}} \tag{11}$$

In equation (11) Δx^{phy} and Δy^{phy} are gird spacing in x and y direction, Δt^{phy} is time spacing and ρ^{phy} is density for physical domain. In addition Δx^{lb} and Δy^{lb} are grid spacing, Δt^{lb} is time spacing and ρ_0^{lb} is referenced density in LBM domain. In addition C_{h_x}, C_{h_y}, C_t and C_ρ are related to space in x and y direction, time and density transformation coefficients respectively. In this study "lb" superscript is related to LBM variables and for simplicity, physical variables are shown without any superscript.

In equation (11) by considering $\Delta x = \Delta y$ and $\Delta x^{\mathrm{lb}} = \Delta y^{\mathrm{lb}}$ we can write $C_{h_x} = C_{h_y} = C_h$. The other transformation coefficients can be obtained as follows:

$$C_u = \frac{\vec{u}}{\vec{u}^{lb}} = \frac{C_h}{C_t}, \ C_v = \frac{v_N}{v^{lb}} = \frac{C_h^{\ 2}}{C_t}, \ C_P = \frac{P}{P^{lb}} = C_\rho\left(\frac{C_h}{C_t}\right)^2 \tag{12}$$

where v_N is the kinematic viscosity of Newtonian part of the fluid. C_u, C_v and C_P are velocity, Newtonian kinematic viscosity and pressure transformation coefficients respectively.

The lattice Boltzmann equations in fluid points can be shown as follows [26]:

$$f_\alpha\left(\vec{x}^{lb} + \vec{e}_\alpha \Delta t^{\mathrm{lb}}, t^{lb} + \Delta t^{\mathrm{lb}}\right) - f_\alpha\left(\vec{x}^{lb}, t^{lb}\right) = -\frac{1}{\tau}\left(f_\alpha\left(\vec{x}^{lb}, t^{lb}\right) - f_\alpha^{eq}\left(\vec{x}^{lb}, t^{lb}\right)\right) + \vec{F}_\alpha \Delta t^{\mathrm{lb}} \tag{13}$$

$$F_\alpha = \omega_\alpha\left(1 - \frac{1}{2\tau}\right)\left(\frac{\vec{e}_\alpha - \vec{u}^{lb}}{c_s^2} + \frac{\vec{e}_\alpha . \vec{u}^{lb}}{c_s^4}\vec{e}_\alpha\right) \cdot \vec{f}^{lb} \tag{14}$$

$$\rho^{lb} = \sum_\alpha \vec{e}_\alpha f_\alpha \tag{15}$$

$$\rho^{lb}\vec{u}^{lb} = \sum_{\alpha} \vec{e}_{\alpha} f_{\alpha} + \frac{\rho^{lb}}{2} \vec{f}^{lb} \Delta t^{lb} \tag{16}$$

where f_{α} is the distribution function, f_{α}^{eq} is its corresponding equilibrium distribution function for the discrete velocity $\vec{e}_{\alpha}$, F_{α} is the discrete force term, $\vec{f}^{lb}$ is the force density action on the fluid field, w_{α} is weighting coefficient which depends on the selected lattice velocity model, τ is the single relaxation time, ρ^{lb} is the density of fluid in LB domain,$c_s=c/\sqrt{3}$ is the sound speed and $c=\Delta x^{lb}/\Delta t^{lb}$ is the lattice speed where Δx^{lb} and Δt^{lb} are the lattice constant and the time step size, respectively. f_{α}^{eq} is the equilibrium distribution function (EDF) defined as:

$$f_{\alpha}^{\ eq} = \rho^{lb}\omega_{\alpha}\left[1+\frac{3(\vec{e}_{\alpha}\cdot\vec{u}^{lb})}{c^2}+\frac{9(\vec{e}_{\alpha}\cdot\vec{u}^{lb})^2}{2c^4}-\frac{3\left|\vec{u}^{lb}\right|^2}{2c^2}\right] \tag{17}$$

Here, the two-dimensional nine velocities (D2Q9) model is used. The particle velocity $\vec{e}_{\alpha}$ and weighting coefficients w_i may be written as:

$$\vec{e}_{\alpha} = \begin{cases} [0,0] \\ c[\cos(\frac{(\alpha-1)\pi}{2}),\sin(\frac{(\alpha-1)\pi}{2})] \\ \sqrt{2}c[\cos(\frac{(\alpha-5)\pi}{2}+\frac{\pi}{4}),\sin(\frac{(\alpha-1)\pi}{2}+\frac{\pi}{4})] \end{cases}, \quad \omega_{\alpha} = \begin{cases} \frac{4}{9} & \alpha=0 \\ \frac{1}{9} & \alpha=1\text{-}4 \\ \frac{1}{36} & \alpha=5\text{-}8 \end{cases} \tag{18}$$

The single relaxation time in LB equations (τ) is calculated as:

$$\tau = \frac{\nu^{lb}}{c_s^2 \Delta t^{lb}} + 0.5 \tag{19}$$

where ν_{lb} is the kinematic viscosity in LBM domain.

In equation (14), $\vec{f}^{lb}$ consists of three different forces as follows:

$$\vec{f}^{lb} = \vec{f}_E^{lb} + \vec{f}_{Cilia}^{lb} + \vec{f}_{Mem}^{lb} \tag{20}$$

$\vec{f}_E^{lb}$ is related to the elastic part of the stress tensor in mucus layer, $\vec{f}_{Cilia}^{lb}$ is related to the force that each cilium impose to the fluid and $\vec{f}_{Mem}^{lb}$ is related to the elastic force on the interface between the mucus and PCL. The first force which should be added to LB equations due to elastic part of the stress tensor of mucus. For calculating this force using the physical velocity field at the time level $t=t_n$ and using finite difference method, the elastic part of the stress tensor (equation (7)) is solved and the value of σ_E in each time step is calculated. So the elastic force of stress tensor can be calculated as:

$$\vec{f}_E = \frac{1}{\rho}\vec{\nabla}.\sigma_{M,E} \tag{21}$$

Since $\vec{f}_E$ is defined as the physical variables, so it should be multiplied by force transformation coefficient to transform into LB force as follows:

$$\vec{f}_E^{lb} = \vec{f}_E \frac{C_t^2}{C_h} \tag{22}$$

Immersed boundary method has been employed to simulate cilia forces. In this method, the interpolated velocity of fluid at the cilia points is enforced to be equal to the velocity of the cilia at the same position at every evolution time step by a set of velocity correction $\delta\vec{U}^{lb}$. For calculating this velocity correction the following system of equations should be solved [20]:

$$A\vec{X}=\vec{B} \tag{23}$$

$$X=\left\{\delta\vec{U}_1^{lb},\delta\vec{U}_2^{lb},\cdots,\delta\vec{U}_m^{lb}\right\}^{\mathrm{T}} \tag{24}$$

$$A=\begin{bmatrix}\delta_{11} & \delta_{12} & \cdots & \delta_{1n}\\ \delta_{21} & \delta_{22} & \cdots & \delta_{21}\\ \vdots & \vdots & \ddots & \vdots\\ \delta_{m1} & \delta_{m2} & \cdots & \delta_{mn}\end{bmatrix}\begin{bmatrix}\delta_{11}^B & \delta_{12}^B & \cdots & \delta_{1m}^B\\ \delta_{21}^B & \delta_{22}^B & \cdots & \delta_{2m}^B\\ \vdots & \vdots & \ddots & \vdots\\ \delta_{n1}^B & \delta_{n2}^B & \cdots & \delta_{nm}^B\end{bmatrix}$$

$$B=\begin{Bmatrix}\vec{U}_1^{lb}\\ \vec{U}_2^{lb}\\ \vdots\\ \vec{U}_m^{lb}\end{Bmatrix}-\begin{bmatrix}\delta_{11} & \delta_{12} & \cdots & \delta_{1n}\\ \delta_{21} & \delta_{22} & \cdots & \delta_{21}\\ \vdots & \vdots & \ddots & \vdots\\ \delta_{m1} & \delta_{m2} & \cdots & \delta_{mn}\end{bmatrix}\begin{Bmatrix}\vec{u}_1^{*lb}\\ \vec{u}_2^{*lb}\\ \vdots\\ \vec{u}_n^{*lb}\end{Bmatrix} \tag{25}$$

Here, m is the number of Lagrangian points on the cilia, and n is the number of Eulerian points. $\vec{U}_l^{lb}$ $(l=1,2,...,m)$ is the velocity vector of Lagrangian (cilia) points and $\delta\vec{U}_l^{lb}$ $(l=1,2,...,m)$ is the unknown velocity correction vector at the Lagrangian points. In equation (25), $\delta_{ij}=D_{ij}(x_{ij}^{lb}-X_l^{lb})\Delta x^{lb}\Delta y^{lb}$ and $\delta_{ij}^B=D_{ij}(x_{ij}^{lb}-X_l^{lb})\Delta s_l^{lb}$ where Δs_l^{lb} is the arc length of cilia elements and $D_{ij}\left(x_{ij}^{lb}-X_l^{lb}\right)$ is Delta function which defined as [7]:

$$D_{ij}\left(x_{ij}^{lb}-X_l^{lb}\right)=\delta(x_{ij}^{lb}-X_l^{lb})\delta(y_{ij}^{lb}-Y_l^{lb}) \tag{26}$$

$$\delta(r)=\begin{cases}1-|r|, & |r|\le 2\\ 0, & |r|>2\end{cases} \tag{27}$$

In this method cilia forces at Lagrangian points ($\vec{F}_{Cilia}$) can be calculated as [25]:

$$\vec{F}_{Cilia}^{lb}=\frac{2\delta\vec{U}^{lb}}{\delta t} \tag{28}$$

By interpolating the value of cilia forces from Lagrangian (cilia) nodes to Eulerian (fluid) points the cilia force on each node of fluid can be calculated as:

$$\vec{f}_{Cilia}^{lb}=\sum_l \vec{F}_{Cilia}^{lb}D_{ij}(x_{ij}^{lb}-X_l^{lb}\Delta s_l^{lb}) \tag{29}$$

F_{Mem} is the elastic force exerted by the membrane interface of PCL and mucus to the fluid is calculated as [15]:

$$F_{Mem}=\frac{1}{\rho}\frac{\partial}{\partial s}\left[T(s)\tau(s)\right] \tag{30}$$

Where T(s) and τ(s) are defined as:

$$T(s)=T_0\left(\frac{\partial X_{Mem}}{\partial s_0}-1\right) \tag{31}$$

$$\tau(s)=\frac{\partial X_{Mem}}{\partial s}\Bigg/\left|\frac{\partial X_{Mem}}{\partial s}\right| \tag{32}$$

In equation (32) $\tau(s)$ is the unit tangential vector to the PCL–mucus membrane interface. The arc-lengths s and s_0 are measured along the current and initial configuration of the membrane interface, respectively. The scalar T_0 is the stiffness constant which describes the elastic property of the flexible boundary.

Using force transformation coefficients, F_{Mem} can be transformed into LB force as follows:

$$F^{lb}_{Mem} = F_{Mem}\frac{C_t^2}{C_h} \tag{33}$$

By interpolating the value of membrane force (F^{lb}_{Mem}) from Lagrangian (membrane) nodes to Eulerian (fluid) points the membrane force on each node of fluid can be calculated as:

$$\vec{f}^{lb}_{Mem} = \sum_{l}\vec{F}^{lb}_{Mem}D_{ij}(x^{lb}_{ij} - X^{lb}_{Mem,l}\Delta s^{lb}_{Mem,l}) \tag{34}$$

Lastly the membrane velocity $U_{Mem}\left(X_{Mem}\right)$ interpolated from the neighboring grid points. Subsequently, the location of the membrane interface is updated as:

$$\frac{\partial X_{Mem}}{\partial t} = U_{Mem} \tag{35}$$

3. Results and discussion

In this section using the numerical scheme introduced in mathematical formulation, 2D numerical simulation has been employed to study the effects mucus depth at various values of cilia beat frequency on the muco-ciliary clearance problem. The main feather of the present study is using an Oldroyd-B model for simulating mucus, in which the effect of mucus theology can be considered more accurate. In addition, for simulating cilia forces and also mucus-PCL interface an immerse boundary method is used.

The cilium beat pattern in our study is the cilium beat cycle reported in Folfurd and Blake [4] which is shown in figure 2(a). In our numerical model, there are 13 cilium and each cilium has a cyclic motion. The cilia arrangements for t= 8T/13 is plotted in figure 2(b) Where T is the consecutive beat cycle.

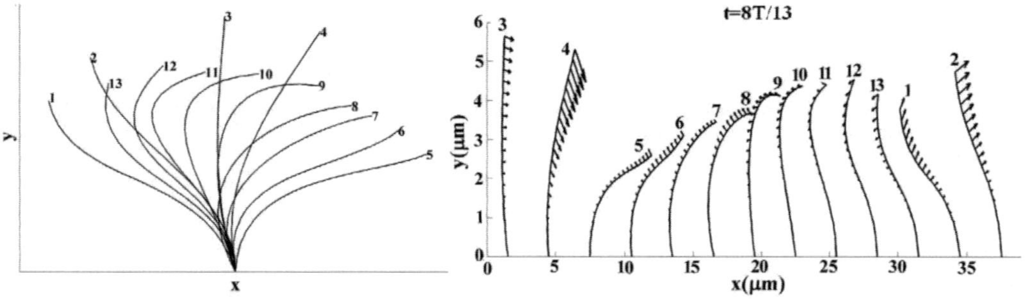

(a) (b)

Figure 2. (a)The beat cycle derived by Fulford and Blake [4] and (b) Arrangement of the cilia for t=8T/13

The time average mucus velocity at the outlet boundary of the domain calculated as:

$$u_0 = \frac{1}{T(H_2 - H_1)} \int_0^T \int_{H_1}^{H_2} u_M \, dy dt \qquad (36)$$

In equation (36) $H_2 - H_1$ is the depth of mucus layer and T is the consecutive beat cycle. The experimental investigation of Matsui *et al.* [21] showed a mean mucus transport of $39.2\ \mu m/s$. ICRP [22] reported a wide range of values depending upon disease, ambient conditions and other factors. For healthy subjects, values of 70 and $92\ \mu m/s$ for tracheal transport, and $40\ \mu m/s$ for bronchial transport were reported. Smith *et al.* [16] showed a mean mucus velocity of $38.3 \mu m/s$. Also numerical simulation of Lee *et al.* [15] and Jayathilake *et al.* [2] predicts a mean mucus velocity of $44.38\ \mu m/s$ using the standard parameter set. Our simulation using the standard parameter set predicts a time average mucus velocity of $u_0 = 44.07\ \mu m/s$ which is in a reasonable agreement with the previous reported results. The velocity field and the beating cilia at $t=4T/13$ for the standard parameter set have been plotted in figure 3. This Figure shows a high forward velocity around the cilium at its effective stroke and the largest mucus velocity occurs close to the mucus-PCL interface.

Table 1 shows the variation of mean mucus velocity as a function of mucus depth at three different values of cilia beat frequency. In this table the other numerical settings are equal to the standard setting. This table indicates that CBF has great influence on mucus flow and increasing in CBF as indicated in the literature leads to increase in mean mucus velocity. This table also shows that increasing mucus depth does not have a great effect on the mucus flow.

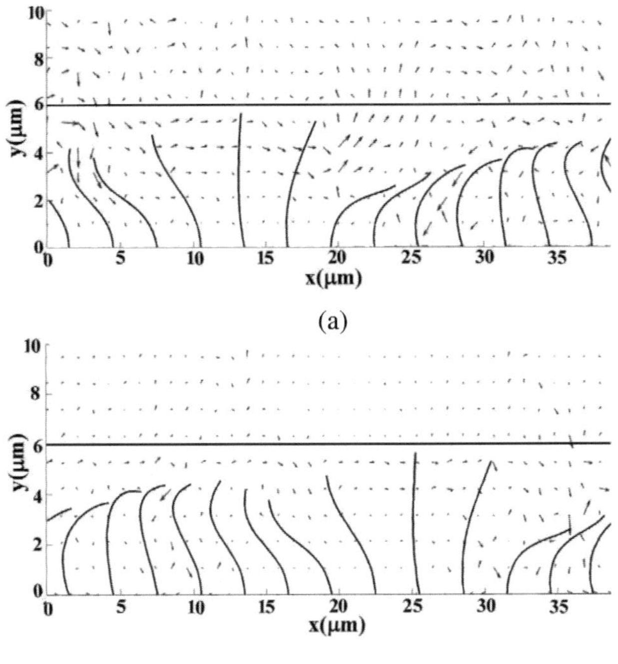

(a)

(b)

Figure 3.The velocity field and the beating cilia at (a) $t=4T/13$ and (b) t=T for the standard parameter set.

Table 1. Variation of mean mucus velocity (μm/s) with respect to the mucus layer depth for various amounts of CBF.

L_M(μm) / σ(rad/s)	3	3.5	4	4.5	5	5.5	6
40	38.41	33.46	33.28	33.48	33.78	32.55	31.65
60	50.6	44.64	44. 07	44.49	44.27	43.06	41.91
80	63.7	56.67	55.52	56.36	55.66	54.6	52.64

For better understanding these variations, non-dimensional mean mucus velocity as a function of mucus depth and CBF has been plotted in figure 4. As this figure indicates that, at lower values of mucus depth because of lower volume of mucus above the PCL, mucus velocity increases significantly. Changing the depth of mucus from 3.5μm to 5μm does not have a great effect on mucus velocity, but by increasing the depth of mucus from 5μm since the cilia are propelling a considerably larger volume of mucus, leads to slightly slower mucus transport and this variation is same for various amounts of CBF. This figure also illustrates that, the effect of CBF is more than mucus depth and decreasing mean mucus velocity because of increasing in mucus depth (which is results from air pollution or smoking) can be simply modified by increasing in CBF. For example, for σ=80rad/s at the worst condition for mucus depth when L_M=6μm, mean mucus velocity is almost 1.2 times the standard value. On the other hand by reducing the value of CBF, e.g. when σ=40rad/s for the lowest value of mucus depth, i.e. L_M=3μm, u_M/u_0 is about 0.9 and it reduces to 0.7 by increasing mucus depth.

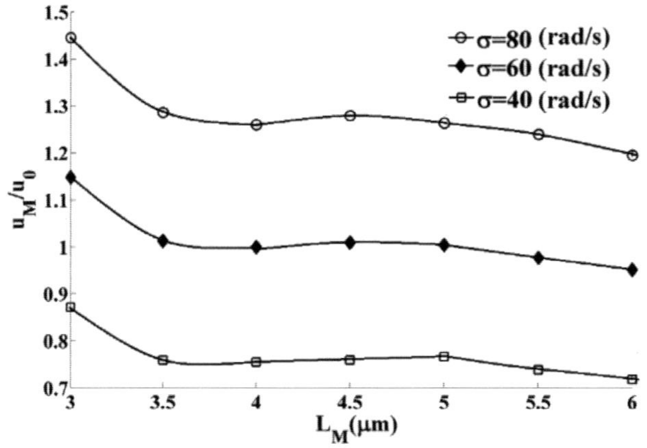

Figure 4. Mean mucus velocity for various amounts of depth of mucus layer at different values of CBF.

4. Conclusions

In this study a 2D numerical code based on the immersed boundary-lattice Boltzmann method has been developed to study the effect of mucus depth on muco-ciliary transport for various amounts of CBF. The main feature of the present investigation is considering mucus layer as a viscoelastic fluid and an Oldroyd-B model is used as the constitutive equation of it. So mucus could be modeled more realistically compared with the previous investigations reported in the literature. Interacting forces are modelled using the Immersed Boundary Method. Our results show that at lower values of mucus layer depth, mean mucus velocity increases and by increasing mucus depth of 3.5μm to 5μm, mean mucus velocity remains almost constant and by increasing this value from 5μm mucus flow reduces. The results also indicate that although increasing in mucus depth leads to decrease in mucus flow, increasing CBF can completely modify it and CBF has more effect on mucus flow. In future studies, we will include the nonlinear differential models as the constitutive relation of mucus to predict more important characteristics of real fluid material functions and flow behaviour.

Acknowledgements
The authors would like to acknowledge the Shahrood University for providing high performance computer. The authors also thank Dr. Mojtaba Heydari for his valuable comments regarding this work.

References

[1] Smith D, Gaffney E, Blake J 2008 *Respiratory physiology & neurobiology* **163** 178-88
[2] Jayathilake P, Le D, Tan Z, Lee H, Khoo B 2014 *Computer methods in biomechanics and biomedical engineering* 1-8
[3] Blake J 1984 *IMA journal of applied mathematics* **32** 69-87
[4] Fulford G, Blake J 1986 *Journal of theoretical Biology* **121** 381-402
[5] Blake J 1974 *Journal of theoretical biology* **45** 183-203
[6] Thomas B, Rutman A, Hirst R A, Haldar P, Wardlaw A J, Bankart J, et al. 2010 *Journal of Allergy and Clinical Immunology* **126** 722-9. e2
[7] Leigh M W, Pittman J E, Carson J L, Ferkol T W, Dell S D, Davis S D, et al. 2009 *Genetics in Medicine* **11** 473-87
[8] Bush A, Cole P, Hariri M, Mackay I, Phillips G, O'callaghan C, et al. 1998 *European Respiratory Journal* **12** 982-8
[9] Rossman C, Forrest J B, Lee R, Newhouse M 1980 *CHEST Journal* **78** 580-2
[10] Rutland J, Cole P 1980 *The Lancet* **316** 564-5
[11] Chilvers M A, Rutman A, O'Callaghan C 2003 *Journal of Allergy and Clinical Immunology* **112** 518-24
[12] Narang I, Ersu R, Wilson N, Bush A 2002 *Thorax* **57** 586-9
[13] Wodehouse T, Kharitonov S, Mackay I, Barnes P, Wilson R, Cole P 2003 *European Respiratory Journal* **21** 43-7
[14] Corbelli R, Bringolf-Isler B, Amacher A, Sasse B, Spycher M, Hammer J 2004 *CHEST Journal* **126** 1054-9
[15] Lee W, Jayathilake P, Tan Z, Le D, Lee H, Khoo B 2011 *Computers & Fluids* **49** 214-21
[16] Lubkin D, Gaffney E, Blake J 2007 *Bulletin of mathematical biology* **69** 289-327
[17] Stanley P, Wilson R, Greenstone M, MacWilliam L, Cole P 1986 *Thorax* **41** 519-23
[18] Blake J 1973 *Mathematical Biosciences* **17** 301-13
[19] Bird R B, Armstrong R C, Hassager O, Curtiss C F 1977 *Dynamics of polymeric liquids* **1** Wiley New York)
[20] Wu J, Shu C 2010 *Communications in Computational Physics* **7** 793
[21] Matsui H, Randell S H, Peretti S W, Davis C W, Boucher R C 1998 *Journal of Clinical Investigation* **102** 1125
[22] Protection I C o R, ICRP 1995 *ICRP Publication 66: Human Respiratory Tract Model for Radiological Protection* Elsevier Health Sciences)

Numerical prediction of flow induced fibers orientation in injection molded polymer composites

A N Oumer[1], N M Hamidi[1] and I Mat Sahat[1]
[1]Faculty of Mechanical Engineering, Universiti Malaysia Pahang,
26600 Pekan, Pahang, Malaysia

E-mail: nurye@ump.edu.my

Abstract. Since the filling stage of injection molding process has important effect on the determination of the orientation state of the fibers, accurate analysis of the flow field for the mold filling stage becomes a necessity. The aim of the paper is to characterize the flow induced orientation state of short fibers in injection molding cavities. A dog-bone shaped model is considered for the simulation and experiment. The numerical model for determination of the fibers orientation during mold-filling stage of injection molding process was solved using Computational Fluid Dynamics (CFD) software called MoldFlow. Both the simulation and experimental results showed that two different regions (or three layers of orientation structures) across the thickness of the specimen could be found: a shell region which is near to the mold cavity wall, and a core region at the middle of the cross section. The simulation results support the experimental observations that for thin plates the probability of fiber alignment to the flow direction near the mold cavity walls is high but low at the core region. It is apparent that the results of this study could assist in decisions regarding short fiber reinforced polymer composites.

1. Introduction

Nowadays, fiber reinforced plastic composites are replacing metals which are being used for many years. This is due to the fact that fiber reinforced plastics have high strength to weigh ratio, low cost compared to metals, and high resistance to corrosion. However, the production of the composites nowadays is very challenging to meet the market requirement. Moreover, the study of the flow of fiber-filled polymers in injection molding process is quite complex due to the fact that the flow of fiber filled thermoplastics in the molten state is modified by the presence of fibers and vice versa, i.e., the fiber motion and rotation are also affected by the flow. These add considerable complexities to the flow study of the polymer matrix. Above all, the study of the flow properties of concentrated suspensions in polymer melts remains most challenging. Therefore, it is very important to fully understand the flow behavior of the polymer-fiber mixture inside the injection molding cavity in order to be able to accurately predict the orientation state of the fibers.

To date numerous researchers have reported on the patters of fibers orientation in injection molded fiber reinforced polymer composites [1-4]. The first author who studied the flow of fiber suspensions and found distribution function of the orientation angle of the particle is Jeffery [5]. In his model, Jeffery assumed that the particle is rigid, neutrally buoyant, axisymmetric, and large enough so that Brownian motion (motion arising from collision of molecules in the fluid medium with the suspended

IOP Conf. Series: Materials Science and Engineering **100** (2015) 012066 doi:10.1088/1757-899X/100/1/012066

particles) can be negligible. However, most commercially available reinforced thermoplastics are concentrated suspensions as they contain around 30% by weight, or 15% by volume of fibers [6].

Hence, some modifications to Jeffery's model are necessary to account for the fiber-fiber interactions for concentrated fiber suspensions. Folgar and Tucker [7] have modified the Jeffery's model by introducing a fiber-fiber interaction coefficient. Moreover, the flow and fiber motion calculations are coupled through the stress terms. One of the well-known numerical methods for the stress contribution from the fiber suspension is solving of the stress tensor in the flow field. In this case, the fiber stress is modeled by the constitutive equation of the fiber orientation tensor. Instead of calculating the orientation distribution function directly, the orientation tensor method is used. However, tensor approximation inevitably encounters a closure problem of a higher-order orientation state. To tackle this problem, many researchers have developed different closure models [8, 9].

The aim of the paper is to numerically characterize the flow induced orientation state of short fibers in three dimensional injection molding cavities. The data obtained from the simulation are compared to qualitative experimental data. A bone shaped model was considered for the simulation and experiment. The model is intended to predict the fiber orientation across the through thickness of the specimen at various locations along its length.

2. Mathematical model

The mould cavity considered for this study is a bon-shaped tensile testing specimen cavity as shown in figure 1 with dimensions 164 mm x 18 mm x 3 mm. Highly viscous nylon/glass fiber mixture enters the cavity through the origin at the right side of the cavity.

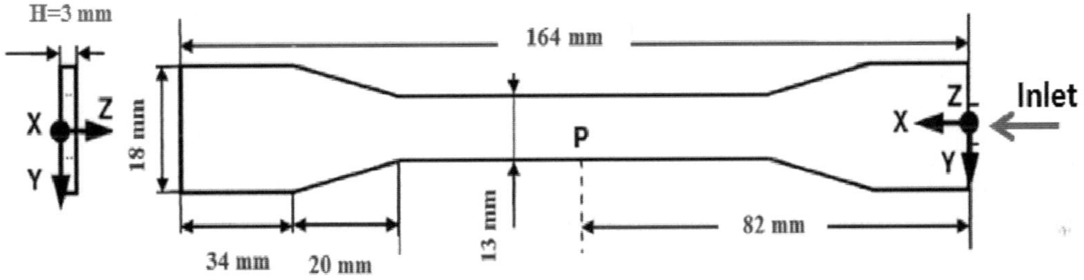

Figure 1. Schematic of the mold cavity

During the filling stage, the polymer/fiber mixture flow field is considered to be incompressible, non-isothermal and behaves as non-Newtonian fluid. Since the mixture under consideration is highly viscous, the inertia and body force terms in the momentum equation are neglected. Therefore, the general form of the mass, momentum and energy equations can be written as equation (1), equation (2) and equation (3), respectively [10]:

$$\nabla \cdot (\boldsymbol{u}) = \boldsymbol{0} \tag{1}$$

$$0 = -\nabla p + \nabla \cdot \tau \tag{2}$$

$$\rho C_p \left(\frac{\partial T}{\partial t} + \boldsymbol{u} \cdot \nabla T \right) = k \nabla^2 T + \dot{\gamma} \cdot \tau \tag{3}$$

where $\boldsymbol{u}$ is the velocity vector, ρ is the density with T being the temperature and p the hydrostatic pressure. The terms C_p, k, $\boldsymbol{\tau}$ and $\dot{\gamma}$ represent the specific heat, the thermal conductivity, the total extra stress tensor, and the shear rate of the suspension, respectively.

For this study, the density, thermal conductivity and specific heat capacity values are assumed to be constant. Moreover, the melt is assumed to be a non-Newtonian fluid exhibiting a shear thinning behavior with increasing shear rate and also its viscosity depends on the shear rate and temperature. To account for the dependence of the viscosity on the shear rate, the cross model is used.

$$\eta = \frac{\eta_0}{1+\left(\eta_0 \dot{\gamma} / \tau^* \right)^{1-n}} \tag{4}$$

where η_o, τ^*, *and* n are empirical constants.

The total stress for the suspension can be obtained by combining the contribution of the suspending fluid $\boldsymbol{\tau}^s$ and the contribution of the fiber $\boldsymbol{\tau}^f$ to the total extra stress. The fiber contribution to the total stress is [11],

$$\boldsymbol{\tau}^f = \eta \xi \boldsymbol{N} \frac{\partial u_k}{\partial x_l} A_{ijkl} \tag{5}$$

where η_s, *and* A_{ijkl} represent the suspension viscosity and the fourth order orientation tensor respectively. The coefficient $\boldsymbol{N}$ is a dimensionless number, called the particle number, which can be described in terms of the volume fraction and fiber aspect ratio. Physically, it represents the magnitude by which the suspension resists elongation parallel to the fiber axial direction. Similarly, the suspending fluid contribution to the total stress is,

$$\boldsymbol{\tau}^s = 2\eta\dot{\gamma} = \frac{1}{2}\eta\left(\mathbf{u} - \mathbf{u}^{\mathrm{T}}\right) \tag{6}$$

Therefore the total stress becomes

$$\boldsymbol{\tau} = \boldsymbol{\tau}^s + \boldsymbol{\tau}^f \tag{7}$$

In addition to the flow governing equation, one additional equation for the fiber orientation to be coupled with the flow equations is required. Assuming that the fibers are rigid cylinders, uniform in length and diameter, and that the number of fibers per unit volume is uniform, the equation change for the second order orientation tensor proposed by Advani and Tucker [12] is used.

$$\begin{aligned} \frac{Da_{ij}}{Dt} = \frac{\partial a_{ij}}{\partial t} + u_k \frac{\partial a_{ij}}{\partial x_k} = -\left(\omega_{ik} a_{kj} - a_{ik}\omega_{kj}\right) + \\ \lambda\left(\dot{\gamma}_{ik} a_{kj} + a_{ik}\dot{\gamma}_{kj} - 2\dot{\gamma}_{kl} A_{ijkl}\right) + 2C_I \dot{\gamma}\left[\delta_{ij} - 3a_{ij}\right] \end{aligned} \tag{8}$$

where u_k is the velocity component, δ_{ij} is a unit tensor, and ω_{ij} and $\dot{\gamma}_{ij}$ are the rotation rate and rate of deformation tensors respectively. λ is a constant that depends on the geometry of the fibers, defined as $\lambda = (r_e^2 - 1)/(r_e^2 + 1)$ with r_e being fiber aspect ratio. C_I is a dimensionless interaction coefficient which represents the degree of interaction between short fibers. The fourth order orientation tensor evolved in equation (5) and equation (8) needs to be approximated in terms of the second order tensor. In this paper, the hybrid closure approximation method proposed by Advani and Tucker is used [10]. For the numerical simulation, MoldFlow software is used. All the data required for the simulation, such as the type of polymer suspension to be used, glass fiber content and others parameters are provided as an input.

3. Model validation

The numerical model is validated by comparing the simulation results with qualitative experimental data. The materials used for the experiment is nylon polymer reinforced with 30% short glass fibers. Once the test samples are produced, small sections are cut out from the final molded specimens at three different positions along its length to examine the orientation states of the fibers. The three positions considered are near the mold inlet (10 mm from inlet), at the middle section of the specimen (82 mm from inlet) and far from inlet (150 mm from inlet) as shown in figure 2. Then the fibers distribution on the cut surfaces are examined by taking pictures using a Scanning Electron Microscope (SEM).

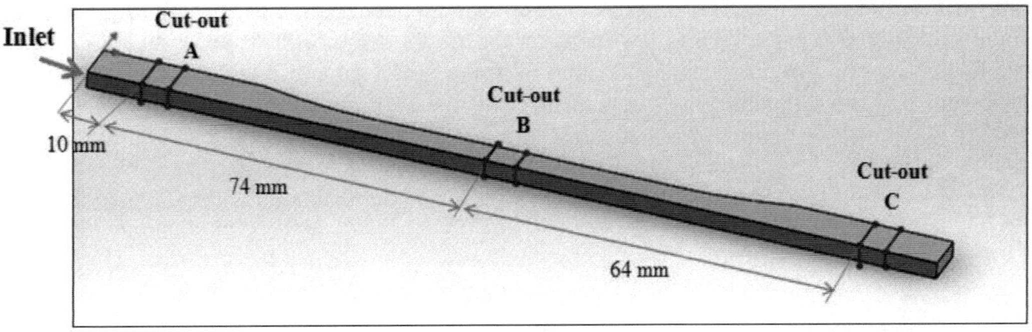

Figure 2. Cut out locations on the final molded sample for SEM observation.

4. Results and discussion

Figure 3 shows the fiber orientation across the sample thickness for the cut out position B which is at the middle of the sample. As can be clearly seen from the SEM image, two different regions (or three layers of orientation structures) across the thickness of the sample part can be found: a shell region which is near to the mold cavity wall, and a core region at the middle of the cross section. The figure shows that there are grooves and holes (circular sections) visible in the region close to the wall, which indicates that the fibers are oriented in random fashion. On the other hand there are more grooves than holes in the shell region. This indicates that most of the fibers in this region are oriented in the flow direction (x-direction). In the core region, circular sections are dominant indicating that the fibers are predominantly oriented perpendicular to the flow direction.

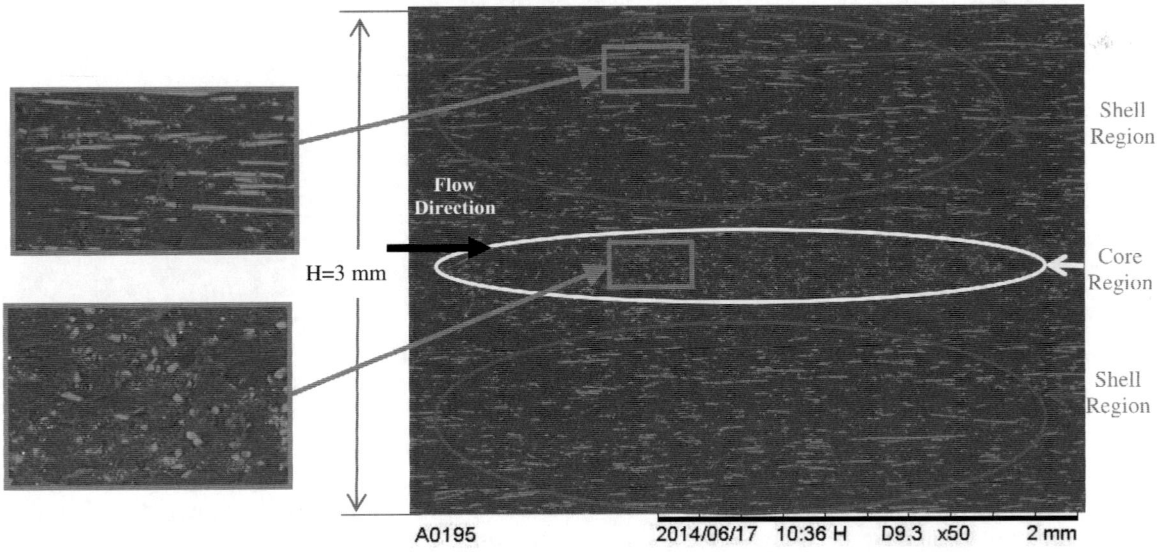

Figure 3. SEM micrograph of cut out B in the molded dog-bone shaped sample.

Figure 4 shows comparison of experimental and simulation results of fibers orientation across the thickness of the specimen near the mold inlet. From equation (8), the first principal values of the fiber orientation tensor, a_{xx}, are used to show the probability of fibers alignment in the flow direction. In this case a value of a_{xx} close to 1 indicates that there is high probability of fibers alignment in the flow direction whereas a value close to 0 indicates a high probability of fibers alignment perpendicular to the flow direction or offset orientation. Therefore, as can be seen from the simulation results in figure 4(b), the flow direction orientation component (a_{xx}) is maximum near the mold cavity walls whereas minimum at the mid-plane of the thickness. The minimum values of the flow direction orientation component are around 0.1 indicating that most of the fibers are oriented perpendicular to the flow direction in this region. On the other hand, the maximum value of a_{xx} near the mold cavity walls is 0.84 indicating that the probability of fibers alignment the flow direction is high in the shell region. The SEM image shown in figure 4(a) shows similar behavior indicating that the simulations results are in good agreement with the experimental data. Moreover, these arguments are in line with other similar research works [10, 13].

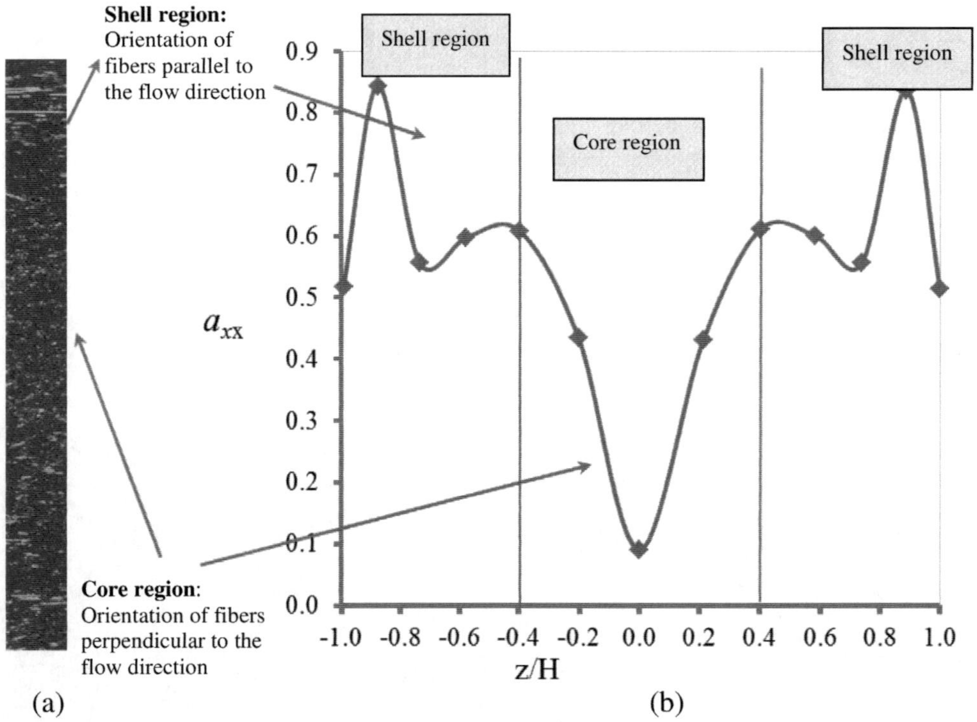

Figure 4. (a) SEM image for fibers orientation and (b) simulation results of flow directional orientation tensor component (a_{xx}) across the thickness of the specimen near the mold inlet.

Similarly, the simulation results of the flow orientation component a_{xx} for the cut-out positions A, B, and C across the sample thickness are shown in figure 5. The results show that the value of a_{xx} at the middle section (84 mm from inlet) of the specimen is 0.9 in the shell region and 0.84 in the core region which indicates that the fibers orient entirely to the flow direction at the middle position of the sample. This might be due to the reason that the cross section at the middle section is smaller than that of near inlet. During the mold filling stage, as the molten nylon/fiber mixture approaches the other end of the mold cavity (far from inlet), the probability of fibers alignment to the flow direction is higher than near inlet position. This is due to the reason that the initial orientation value given for the simulation is random orientation. Therefore, the fibers will need to travel longer distance in order to be aligned to the flow direction.

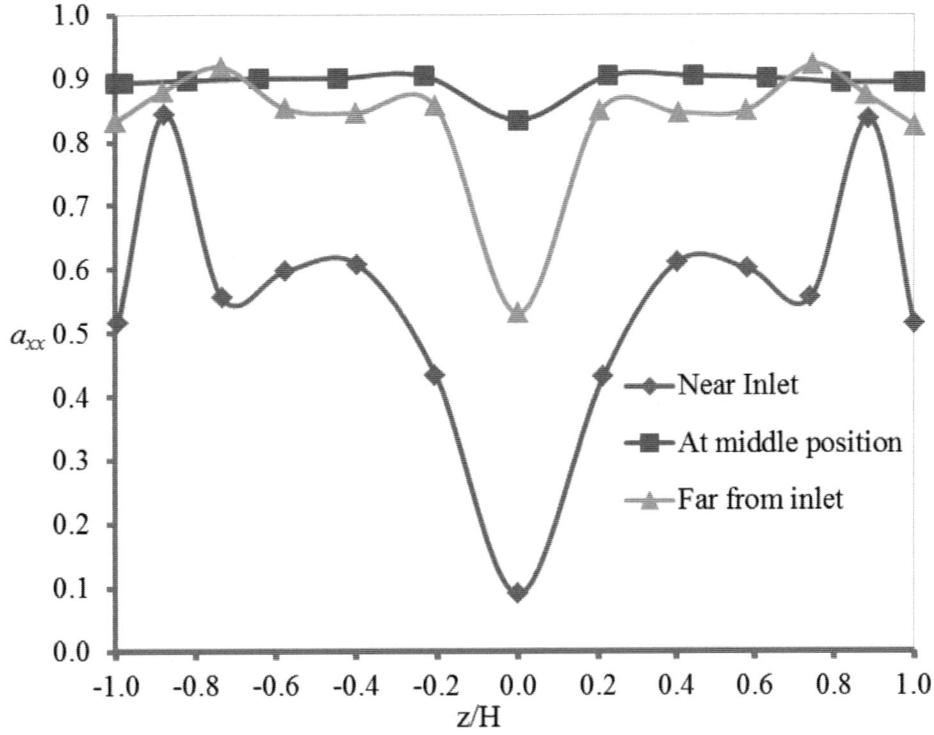

Figure 5. Simulation results for Flow directional orientation tensor component (a_{xx}) at various cut-out positions across the thickness of the specimen.

5. Conclusion

In this paper the numerical formulation used to simulate the degree of flow induced fiber orientation in dog-bone shaped specimen during injection molding has been presented. The simulation model has been validated by performing experiments and compared the experiments qualitative data with quantitative simulation results. Both the experiment and simulation results have showed that three layer fiber orientation structures (two shell regions and one core region) formed during mold filling. The simulation results support the experiment that for thin plates the probability of fiber alignment to the flow direction near the mold cavity walls is high. On the other hand the probability of fibers alignment to the flow direction is low at the core region. From the simulation and experimental results it can be concluded that (since the flow is parallel to the length of the specimen) the final dog-bone shaped product can withstand both tensile and compression force. The shell layer which has high probability of fibers alignment in the flow direction is strong along the length whereas the core region is strong across the length. In general, the results of this study could assist in decisions regarding the short fiber reinforced polymer composites.

Acknowledgements

The authors would like to thank University Malaysia Pahang for providing the necessary facilities and financial assistance for this study under project number RDU130353.

References

[1] Mazahir S M, Vélez-García G M, Wapperom P, Baird D 2015 *Journal of Non-Newtonian Fluid Mechanics* **216** 31-44

[2] Hartl A M, Jerabek M, Lang R W 2015 *Polymer Testing* **43** 1-9
[3] Nagy J, Reith L, Fischlschweiger M, Steinbichler G 2015 *Chemical Engineering Science* **128** 1-10
[4] Oumer A N, Mamat O 2014 *Int J Computational Science and Engineering* **9** 247-56
[5] Jeffery G B 1922 *Proceedings of the Royal Society of London Series A, Containing Papers of a Mathematical and Physical Character* **102** 161-79
[6] Vincent M, Devilers E, Agassant J-F 1997 *Journal of Non-Newtonian Fluid Mechanics* **Vol. 73** 317-26
[7] Folgar F, Tucker C L 1984 *Journal of Reinforced Plastics and Composites* **3** 98-119
[8] Qadir N, Jack D A 2009 *Composites Part A: Applied Science and Manufacturing* **40** 1524-33
[9] Dray D, Gilormini P, Re´gnier G 2007 *Composites Science and Technology* **67** 1601-10
[10] Park J M, Park S J 2011 *Mathematical Problems in Engineering* **2011** 14
[11] Kietzmann C v L, Walt J P v d, Morsi Y S 1998 *International Journal for Numerical Methods in Engineering* **41** 253-69
[12] Advani S G, Tucker C L 1987 *Journal of Rheology* **31** 751-84
[13] Oumer A N, Mamat O 2012 *Composites Part B: Engineering* **43** 1087-94

Mediator effect of statistical process control between Total Quality Management (TQM) and business performance in Malaysian Automotive Industry

M F Ahmad [1], R Z Rasi [1], N Zakuan[2], and M N N Hisyamudin[3]
[1]Department of Production and Operation, Faculty of Technology Management and Business, Universiti Tun Hussein Onn Malaysia (UTHM), Parit Raja, Batu Pahat, Johor, Malaysia
[2]Department of Business, Faculty of Management, Universiti Teknologi Malaysia (UTM), Johor Bahru, Johor, Malaysia
[3]Department of Materials and Design Engineering, Faculty of Mechanical Engineering, Universiti Tun Hussein Onn Malaysia (UTHM), Parit Raja, Batu Pahat, Johor, Malaysia

E-mail: mohdfauzi@uthm.edu.my

Abstract. In today's highly competitive market, Total Quality Management (TQM) is vital management tool in ensuring a company can success in their business. In order to survive in the global market with intense competition amongst regions and enterprises, the adoption of tools and techniques are essential in improving business performance. There are consistent results between TQM and business performance. However, only few previous studies have examined the mediator effect namely statistical process control (SPC) between TQM and business performance. A mediator is a third variable that changes the association between an independent variable and an outcome variable. This study present research proposed a TQM performance model with mediator effect of SPC with structural equation modelling, which is a more comprehensive model for developing countries, specifically for Malaysia. A questionnaire was prepared and sent to 1500 companies from automotive industry and the related vendors in Malaysia, giving a 21.8 per cent rate. Attempts were made at findings significant impact of mediator between TQM practices and business performance showed that SPC is important tools and techniques in TQM implementation. The result concludes that SPC is partial correlation between and TQM and BP with indirect effect (IE) is 0.25 which can be categorised as high moderator effect.

1. Introduction

The concept of total quality management (TQM) has been developed as a result of intense global competition [1, 2]. Firms that manage the international trade in global competition have put emphasis on TQM philosophy, procedures, tools and techniques. Juran [3] defines TQM as philosophy aimed at achieving business excellence through the application of tools and technique, as well as the management of soft aspects, such as human motivation in work. Furthermore, Demirbag et al. [4] defines TQM as a management philosophy aims to contribute continuous improvement in the organization with the participation of all employees to achieve customer satisfaction by producing better, cheaper, faster and safer than competitors. The role of TQM in improving business

IOP Conf. Series: Materials Science and Engineering **100** (2015) 012067 doi:10.1088/1757-899X/100/1/012067

performance is broadly agreed in the literature and empirical study[5-8]. TQM helps to manage the firm to improve the effectiveness and business performance to achieve world class status for the past two decades [9-11]. However, the study of mediators is neglected and is referred to less frequently in literature review. The fundamental systems-interactive paradigm of organisational analysis features the continual stages of input, throughput (processing), and output, which demonstrate the concept of openness and closeness. Processing is the process of changing from one "look" to another, or one culture to another [12-14]. In this study, the author defines input as TQM; processing as application tools and techniques namely SPC; and output as business performance. Thus, one of the objectives of this study is to empirically analyse the impact of SPC between TQM and business performance.

2. Literature review

Statistical process control (SPC) was pioneered by Walter A. Shewhart in the early 1920's. Then, W. Edwards Deming applied SPC methods in the United States during World War II to improve quality in the military products [15]. Deming also introduced SPC to Japan after the war had ended and he developed the control chart and the concept of statistical control in the production process [15]. SPC is used in order to observe a process to ensure the process conforming the specification with a minimum of recovery [16-18]. In particular, SPC allows sources of variation to be detected and measured, then amenable for correction. SPC emphasises on early detection and prevention of problems rather than the correction of problems after they have occurred. SPC eliminates reject product and identifies bottlenecks, waiting times and other sources of delays within the process [3]. Thus, SPC removes the need for post-manufacture inspection. However, the application of SPC relies not only on the tools with which it is applied, but also on how to do correction [19, 20]. Shewhart notes that every process has variation. Some variation that is controlled and natural to the process is called common sources of variation. In contrast, variation that is not controlled and not present at all times is called assignable sources of variation [17]. The assignable sources of variation can be detected, identified, and removed to ensure the process is stable within a value of limits [18]. In this study, author defines SPC as a method of statistical techniques for monitoring and controlling the variation in a process to produce product which meets specification. Rohani et al. [21] conduct study of relationship between SPC and business performance. SPC is a tool for TQM implementation. However, only few previous studies have examined the mediators and moderators between TQM and business performance [22-24]. The inconsistent results between TQM and BP are because the mediators and moderators have been overlooked in research designs [25-27]. This present research proposed a TQM performance model with mediator effect of TQM with structural equation modelling, which is a more comprehensive model for developing countries, specifically for Malaysia.

H1: Relationship between TQM and Business Performance. Most previous studies indicate a significant relationship between TQM practices and business performance [28-31]. In contrast, other studies show that TQM does not improve business performance [4, 32]. Other findings show partial correlation between TQM practices and business performance [4, 9, 33]. Accordingly, the author proposes that: H1: TQM practices are positively significant and have direct effects on business performance. H2a: Relationship among TQM and SPC. The relationship between TQM and SPC will be first reviewed. SPC can assume a stable process, but predictions cannot be made until the process is fully defined and controlled [34]. TQM provides infrastructure such as top management support, employee participation, and improvement to implement tools and techniques; thus, TQM should be implemented before the execution of SPC [3]. Ahmad et al., [16] indicate that TQM practices have impact on SPC. Thus, TQM practices are positively correlated with SPC. H2a: TQM practices are positively significance and direct effect on SPC. SPC aims to reduce process variance and significantly impacts performance improvement, cost savings, and customer satisfaction [35]. Morgan & Dewhurst [36] indicate that control charts could be adapted to effectively monitor supplier performance. Rohani et al. [21] conduct a survey on 326 respondents from the Malaysian automotive industry and find a positive relationship between SPC and business performance. Information provided by the SPC system enhances the ability of top management to make decisions, in turn increasing business performance based on SPC data [17, 37, 38]. Thus, SPC increases productivity by reducing waste and improving quality for short- and long-term benefits; this means that SPC practices are positively correlated with business performance. Accordingly, the author proposes that: H2b: SPC practices are positively significance and direct effect on business performance. There are lacks of empirical evidence of SPC as mediator between TQM and business performance in previous work [22]. In this study, SPC acts as mediator between TQM and business performance. Accordingly, the author proposes that: H13: SPC is a mediator between TQM and business performance.

3. Methodology

A seven-point Likert scale was used for the purpose of scoring TQM, SPC and business performance. The questionnaire had been validated by academician and quality experts. It was pilot tested and finalised. A sample of 1500 related to automotive industry companies, was selected from the directories of the Federation of Malaysian Manufacturers (FMM) and the foreign companies directory list in Malaysia. The question has been sent to product quality assurance department (PQA) or quality control manager as the target respondent.

4. Results

Response Rate: Of the 1500 surveys, 327 surveys, which were equivalent to 21.8% response rate, were returned. Of the 327 surveys, 6 surveys were found to have more than 10 percent of unanswered items and 2 surveys were excluded because respondents provided the same responses to all questions in the survey, resulting in an effective sample of 319 usable completed surveys (21.3 percent usable response rate). Convergent validity: All average variance extracted (AVE) values in TQM, BP and SPC constructs were above 0.50 and this evidence supported the convergent validity of the measurement model of TQM [39], as shown in table 1. The results showed that both Cronbach's alpha and composite reliability for the constructs achieved 0.7, as suggested by Nunally [40] and Hair [39].

Table 1. Convergent validity and composite reliability for second order measurement model.

Construct	Item	Loading (L), L>0.6	Cronbach's alpha α>0.7	Composite reliability (CR) CR>0.7	Convergent validity (AVE) AVE>0.5
TQM	TQM1	0.77	0.941	0.938	0.708
	TQM2	0.82			
	TQM3	0.79			
	TQM4	0.82			
	TQM5	0.72			
	TQM6	0.83			
	TQM7	0.83			
	TQM8	0.82			
	TQM9	0.81			
	TQM10	0.69			
SPC	SPC1	0.79	0.872	0.872	0.696
	SPC2	0.89			
	SPC4	0.84			
BP	BP1	0.85	0.934	0.938	0.715
	BP2	0.79			
	BP3	0.88			
	BP4	0.87			
	BP5	0.83			
	BP6	0.81			

Discriminant Validity: The results showed that AVE values were greater than the square of correlation among the constructs, as shown in table 2. Thus, it can be concluded that discriminant validity was supported.

Table 2. Discriminant validity for second order measurement model.

	TQM	SPC	BP
TQM	**0.708**		
SPC	0.569	**0.696**	
BP	0.599	0.598	**0.715**

Structural equation modeling (SEM): Same procedure of testing for mediator was conducted as Hypothesis H12. SPC as the mediator variable was included into the model, as shown in figure 1. The result showed that the relationship between TQM and BP was reduced from rc=0.81 (CR=13.177, $p<0.01$) to 0.55 (CR=6.531, $p<0.01$), but still significant, as shown in Table 3. The result also showed that TQM had a significant and direct effect on SPC with rc=0.81 (CR=13.967, $p<0.01$), and SPC also had a significant and direct effect on BP with rc=0.31 (CR=3.748, $p<0.01$). The goodness-of-fit indices for the structural model ($\chi 2/df=2.357$, GFI=0.930, AGFI=0.905, TLI=0.951, CFI=964, and RMSEA=0.065) were well within the generally accepted limits, indicating a good fit to the data. Thus, it can be concluded that SPC partially mediated the relationship between TQM and BP.

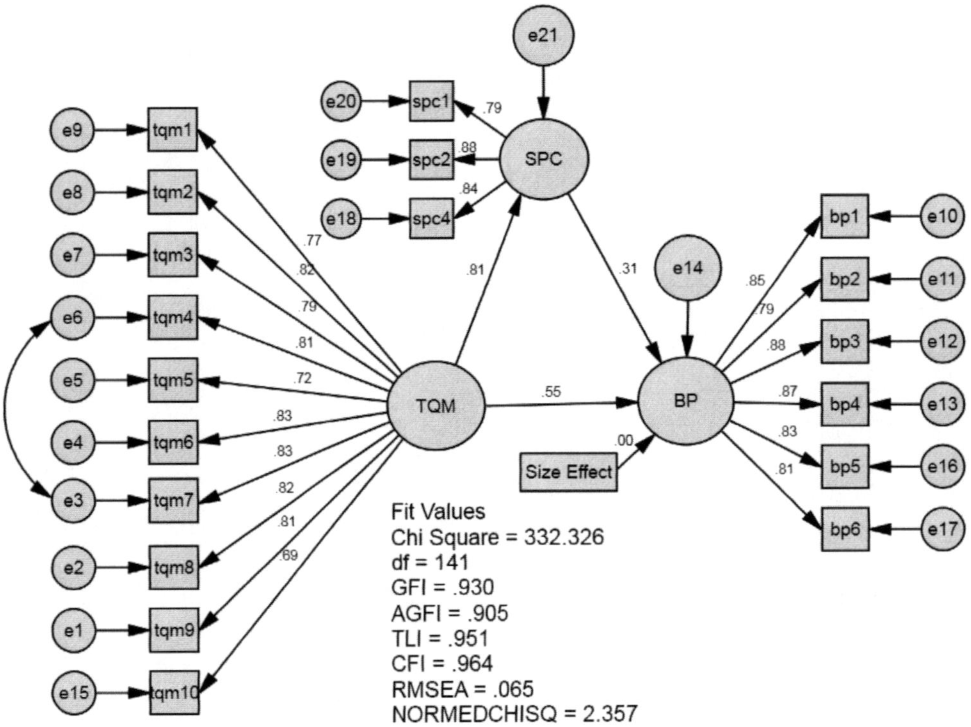

Figure 1. Mediator testing for SPC between TQM and BP.

Table 3. Mediators testing result for SPC between TQM and BP.

No.	Hypotheses	Links in the model	Standardised Estimate (rc)	CR	p-value	Result	Remark
1.	H11	TQM → BP	0.81	13.177**	0.000	Supported	Without SPC
2.	H11	TQM → BP	0.55	6.531**	0.000	Supported	With SPC
	H13a	TQM → SPC	0.81	13.967**	0.000	Supported	With SPC
	H13b	SPC → BP	0.31	3.748**	0.000	Supported	With SPC
		H13: SPC partially mediates the relationship between TQM and business performance.					

Note: *p<0.05; **p<0.01 (one-tailed test)

Table 4. Direct effect and indirect effect for SPC as mediator.

No.	Effect	Links in the model	Standardised Estimate (rc)	Result	Remark
1.	Direct Effect	TQM → BP	0.55	Supported	
2.	Indirect Effect	TQM → SPC → BP	0.25***	Supported	**High Effect**
	Total effect	TQM → SPC → BP	0.80		

Note: *IE>0.01 (Low); **IE>0.09 (Moderate); ***IE>0.250 (High)

Table 3 shows that the standard indirect effect (IE) of TQM to BP was 0.254, which can be categorised as a high effect of mediation [41]. However, table 4 shows the direct effect and indirect effect for SPC as mediator.

5. Discussion

It was expected that SPC was a mediator between TQM and business performance. According to Baron & Kenny (1986), mediator approach research focuses to identify mechanism as a process between TQM and business performance, besides direct impact of TQM towards performance. Mediator functions to explain "how" or "why" the relationship happened between TQM and business performance [26, 27]. The structural relationships in the structural models were used to test the mediating effect. The results showed that SPC mediated the relationship between TQM and business performance in automotive industries in Malaysia. Thus, SPC can explain the relationship between TQM and business performance, i.e., how to achieve business performance through implementing TQM. However, the mediator test showed partial mediation, explaining that the tool or technique alone did not completely explain the relationship between TQM and business performance [42]. The reason was that there were various tools and techniques that contributed to the business performance.

The results proved that SPC was a mediator between TQM and business performance. This finding is supported by Morgan & Dewhurst [36], Rohani et al., [21], Ahmad et al.,[16], Rahman et al., [43], and Taj & Morosan [35]. According to Rahman et al., [43], SPC is a useful tool to improve quality process, to timely detect abnormality, to check critical parameters, to reduce variations, and to maintain the stability of process. Rohani et al., [21] explored the relationships between SPC and performance from 326 companies in Malaysian automotive industry by using SEM and the result was significant. Many researchers view SPC as a monitoring tool to ensure that the output of a process conforms to the intended design [21, 36]. However, SPC also can be applied in outside production process to effectively improve supplier's delivery performance to the buyers [36]. The study indicates that SPC approach is effective in monitoring supplier performance through establishing achievable performance targets and meaningful data.

6. Conclusion and future research

The main objectives of this study are to examine the impact of SPC between TQM and business performance. The result concludes that SPC is partial correlation between and TQM and BP with indirect effect (IE) is 0.25 which can be categorised as high moderator effect. It has proved that the impact of SPC based on system theory.

The structural model gave an rc2=71.0%, which means that 71.0% of business performance variance was accounted by TQM and SPC, leaving 29.0% was accounted by unexplained factors that should be explored in future studies. It may include other tools and techniques such as Total Productive Maintenance (TPM), lean production and ISO 9001 that should thus be examined. There may be other critical factors that influence the automotive industries to be included and considered in TQM and business performance.

Acknowledgements

The authors would like to be obliged to UTHM for supporting this project.

References

[1] Laosirihongthong T, Dangayach G S 2005 *Journal of Manufacturing Systems* **24** 131-43
[2] Ahmad M F, Zakuan N, Ahmad J, Takala J 2015 *International Journal of Industrial and Systems Engineering* **20** 155-64
[3] Juran J M 1998 *Juran's Quality Handbook* (Singapore: Mc Graw Hill)
[4] Demirbag M, Tatoglu E, Tekinkus M, Zaim S 2006 *Journal of Manufacturing Technology Management* **17** 829-47
[5] Fynes B, de Búrca S, Mangan J 2008 *International Journal of Production Economics* **111** 56-69
[6] Prajogo D I, Hong S W 2008 *Technovation* **28** 855-63
[7] Sila I 2007 *Journal of Operations Management* **25** 83-109
[8] Ahmad M F B, Yusof S r M 2010 *Total Quality Management* **21** 11-20
[9] Arumugam V, Ooi K B, Fong T C 2008 *The TQM Journal* **20** 636-50
[10] Ahmad M F, Aziati A H N, Bon A, Ngadiman Y, Chan S. Review Relationship TPM as Mediator between TQM and Business Performance. In: Gen M, Kim K J, Huang X, Hiroshi Y, editors. Industrial Engineering, Management Science and Applications 2015. Lecture Notes in Electrical Engineering. 349: Springer Berlin Heidelberg; 2015. p. 991-6.
[11] Ahmad M F, Zakuan N, Jusoh A, Yusof S M, Takala J 2014 *Procedia - Social and Behavioral Sciences* **129** 244-9
[12] Rouse W B 2005 *Systems Engineering* **8** 279-95
[13] Ahmad M F, Zakuan N, Jusoh A, Tasir Z, Takala J 2013 *IOP Conference Series: Materials Science and Engineering* **46** 012020
[14] Ahmad M F B, Yusof S r M, Yusof N M 2007 *Jurnal Teknologi* **47** 75-89
[15] Deming E 1986 *Out of the Crisis* (Massachusetts: MIT: Cambridge)
[16] Ahmad M F, Zakuan N, Jusoh A, Takala J 2012 *Procedia - Social and Behavioral Sciences* **65** 186-91
[17] Besterfield D H 2009 *Quality Control* (New Jersey: Pearson Prentice Hall)
[18] Rohani J M, Yusof S M, Mohamad I 2010 *Jurnal Mekanikal* **30** 1-16
[19] Imai M 1986 *Kaizen: The Key to Japan's Competitive Success* (New York: McGraw-Hill)
[20] Nakajima S 1998 *Introduction to Total Productive Maintenance (TPM)* (Cambridge: Productivity Press)
[21] Rohani J M, Yusof S M, Mohamad I, The relationship between statistical process control critical success factors and performance: A structural equation modeling approach. IEEE Int Conf Industrial Engineering Management; 2009.
[22] Sadikoglu E, Zehir C 2010 *International Journal of Production Economics* **127** 13-26

[23] Ahmad M F, Arif M S M, Zakuan N, Rahman S, Latif M, Khalid M, The Mediator Effect of Customer Satisfaction between Quality Management Practices and Communication Behavior amongst Malaysia Hajj Pilgrims: Survey Result. Applied Mechanics and Materials; 2014.
[24] Arif M, Zakuan N, Rahman S, Abdullah T, Fadzil N 2014 *Applied Mechanics and Materials* **660** 1000-4
[25] Fotopoulos C V, Psomas E L 2010 *The TQM Journal* **22** 539-52
[26] Bennett J A 2000 *Res Nurs Health* **23** 415–20
[27] Baron R M, Kenny D A 1986 *Journal of personality and social psychology* **51** 1173-82
[28] Jun M, Cai S, Shin H 2006 *Journal of Operations Management* **24** 791-812
[29] Bou J C, Beltrán I 2005 *Total Quality Management & Business Excellence* **16** 71-86
[30] Gunday G, Ulusoy G, Kilic K, Alpkan L 2011 *International Journal of Production Economics* **133** 662-76
[31] Miyagawa M, Yoshida K 2010 *International Journal of Quality & Reliability Management* **27** 736-55
[32] Corredor P, Goñi S 2011 *Journal of Business Research* **64** 830-8
[33] Feng J, Prajogo D I, Tan K C, Sohal A S 2006 *European Journal of Innovation Management* **9** 269-78
[34] Flott L W 2002 *Met Finish* **100** 112-4
[35] Taj S, Morosan C 2011 *Journal of Manufacturing Technology Management* **22** 223-40
[36] Morgan C, Dewhurst A 2008 *International Journal of Production Economics* **111** 13-26
[37] Sanjoy K, Survey of various statistical process control methods. Symposium A Quarterly Journal In Modern Foreign Literatures; 1991.
[38] Oakland J S, Porter J L 1995 *Total Quality Management: Text with Cases* (Oxford: Butterworth Heinemann)
[39] Hair J F 2010 *Multivariate Data Analysis* (New York: Pearson Prentice Hall)
[40] Nunally J C 1978 *Psychometric Theory* (New York: McGraw-Hill)
[41] Cohen J 1988 *Statistical Power Analysis for the Behavioral Sciences* (Hillside: NJ:Lawrence Erlbaum)
[42] Awang Z 2012 *A Handbook on SEM* (Kota Bharu: UITM)
[43] Rahman M N A, Zain R M, Nopiah Z M, Ghani J A, Deros B M, Mohamad N, et al. 2009 *European J Sci Res* **26** 453-64

Behaviour of hybrid jute-glass/epoxy composite tubes subjected to lateral loading

A A Khalid[1]
Department of Mechanical Engineering, Faculty of Engineering, Institut Teknologi Brunei, Gadong, BE 1410, Brunei Darussalam

E-mail: asad.khalid@itb.edu.bn

Abstract. Experimental work on hybrid and non-hybrid composite tubes subjected to lateral loading has been carried out using jute, glass and hybrid jute-glass/epoxy materials. Tubes of 200 mm length with 110 mm inner diameter were fabricated by hand lay-up method to investigate the effect of material used and the number of layers on lateral-load-displacement relations and on the failure mode. Crush force efficiency and the specific energy absorption of the composite tubes were calculated. Results show that the six layers glass/epoxy tubes supported load higher 10.6% than that of hybrid jute-glass/ epoxy made of two layers of jute/epoxy four layers of glass/epoxy. It has been found that the specific energy absorption of the glass/epoxy tubes is found higher respectively 11.6% and 46% than hybrid jute-glass/epoxy and jute/epoxy tubes. The increase in the number of layers from two to six increases the maximum lateral load from 0.53KN to 1.22 KN for jute/epoxy and from 1.35 KN to 3.87 KN for the glass/epoxy tubes. The stacking sequence of the hybrid tubes influenced on the maximum lateral load and the absorbed energy. The maximum load obtained for the six layers jute-glass/epoxy tubes of different staking sequence varies between 1.88 KN to 3.46 KN. Failure mechanisms of the laterally loaded composite tubes were also observed and discussed.

1. Introduction

Composite materials have a wide range of applications because of their high stiffness and strength with respect to weight. In addition, composite materials have high corrosion resistance, thermal expansion, thermally resistive and considered as non-conductive materials. Collapse behaviour of thin-walled tubular structures made of metals and composites has received considerable attention for their application to the design for crash-worthiness of energy absorbing devices [1]. The crush behaviour of composite structures offers distinct advantages for automotive applications and has been the subject of numerous investigations [2]. Several studies were carried out to investigate the energy absorption capability and failure mechanism of composite structures like tubes [3-14] and cones [15-22] at different testing conditions.

Composite tubes under axial loading are widely used by researchers as energy absorption structures due to the high energy absorption capability of these tubes. However the axially loaded tubes can exhibit unstable deformation with large fluctuations compared to the tubes under lateral loading. The energy absorption capability of laterally loaded tubes is better than the tubes under lateral indentation [3]. Eyvazian et al. [5] performed an experimental study on corrugated aluminium tubes under lateral loading. The authors concluded that the crushing load and the energy absorption are improved with the corrugations. The highest energy attained is by using tubes with inner and outer corrugations [5]. Baroutaji et al. [6] studied the behaviour of oblong tube subjected to quasi static lateral loading. The

IOP Conf. Series: Materials Science and Engineering **100** (2015) 012068 doi:10.1088/1757-899X/100/1/012068

authors used three indenter types; flat plate, cylindrical and a point load with two constraints; inclined and vertical. They concluded that the unconstrained oblong tube laterally loaded by flat plate indenter show nearly perfect response with maximum efficiency. In fact, unstable deformation is produced with low energy absorption when the cylindrical and point indenters are used [6].

It is found that the energy absorption of composite and non-composite tubes under lateral will be enhanced when polyurethane foam is used to fill these tubes. In addition, more stable deformation can be achieved with using the foam filled tubes compared to empty tubes under lateral loading [7-9]. Zuraida et al. [10] investigated the behaviour of hybrid filament wound composite tubes under quasi static lateral indentation. The authors found that the stacking sequence of hybrid composite tubes under lateral indentation can be used to improve the structure strength and performance of these tubes. It is also proved that the circumferential and axial strains for the inner and outer sides of hybrid tube surface are dependent on the type of fibre used and on the stacking sequence [10].

The main focus of this study is to determine the load-displacement response and the energy absorption of jute, glass and hybrid jute-glass/ epoxy composite tubes under lateral loading. Furthermore, the study aims to investigate the effect of hybridization and stacking sequence of composite layers on the performance of the composite tubes. The failure mechanisms of the laterally crushed tubes is also observed and discussed.

2. Experimental work

2.1. Materials and fabrication

Composite tubes were fabricated by hand lay-up method using woven roving [±45°] jute/epoxy and glass/epoxy materials. A PVC tube of circular cross-section of 110 mm diameter was used as a mould to fabricate the composite tubes. Two circular plywood plates of 10 mm thickness were fitted inside the hollow tubes at each side. The plywood was used in order to extract the fabricated composite tubes from these mandrels after curing. Table 1 shows the dimensions of the fabricated composite tube and cone specimens. Samples of the fabricated specimens are shown in figure 1. The terms J and G are respectively jute and glass fibres. The numbers 2, 4 and 6 referred to the number of layers of the composite laminate structures.

Table 1. Dimensions of the composite tubes.

Specimen type	Composite tube			
	Inner diameter D (mm)	No of layers	Wall Thickness t (mm)	Length
J2	110	2	2.15	200
J4	110	4	3.94	200
J6	110	6	5.82	200
G2	110	2	2.46	200
G4	110	4	4.13	200
G6	110	6	6.26	200
J2G2J2	110	6	5.93	200
G3J3	110	6	6.25	200
J3G3	110	6	6.42	200
JGJGJG	110	6	5.96	200
G2J2G2	110	6	6.41	200

IOP Conf. Series: Materials Science and Engineering **100** (2015) 012068 doi:10.1088/1757-899X/100/1/012068

Figure 1. Sample of the fabricated composite tubes.

The mechanical properties of the jute and glass/ epoxy materials were obtained from the literatures [23] and listed in table 2.

Table 2. Mechanical properties of jute and glass/epoxy lamina.

Property	Description	Woven roving [±45]	
		jute/ epoxy	Glass/ epoxy
ρ	Density (kg/mm^3)	1.6e-6	2.1e-6
E_{11}	Young's modulus -longitudinal direction (MPa)	17680	52250
E_{22}	Young's modulus -transverse direction (MPa)	17680	52250
G_{12}	In-plane shear modulus (MPa)	3002	3080
G_{23}	Out of plane shear modulus (MPa)	2860	4200
v_{21}	Minor Poisson's ratio	0.29	0.35
X_t	Longitudinal tensile strength (MPa)	560	750
X_c	Longitudinal compressive strength (MPa)	380	430
Y_t	Transverse tensile strength (MPa)	29	38
Y_c	Transverse compressive strength (MPa)	98	136
S_c	In-plane shear strength (MPa)	45	67

2.2. Testing procedure

Constant slow speed compression testing was performed using a computer- controlled servo-hydraulic Instron machine type 5584. The cross-head speed was adjusted at 2.5 mm/min. Composite tubes were laterally crushed between two parallel steel flat platens, one static and one moving. The fixed platen was fitted with a load cell from which the load signal was taken directly to the computer. For each test, the lateral load was plotted on the Y-axis and the crosshead displacement on the X-axis.

3. Results and discussion

3.1. Lateral loading characteristics

In this section results of the laterally loaded composite tubes are presented. Figures 2, 3 and 4 show the lateral load-displacement graphs of the composite tubes using jute (J), glass (G) and hybrid jute-glass (JG)/epoxy respectively. In figures 2,3,4 and 6 are referred to the number of layers. Study shown that the load increased linearly till the initial failure then the load fluctuates with a sharp drop till failure. The glass/ epoxy tubes supported higher lateral load than the hybrid jute-glass/epoxy followed by the jute/epoxy tubes. In general, increase in number of layers from two to six show a significant increase in the maximum load obtained for the jute and glass/epoxy tubes as shown in figures 2 and 3.

It can be seen from figure 4 that the stacking sequence of hybrid composite tubes can be used to improve their strength and performance. For hybrid tubes with the external layers of glass/epoxy show higher strength and supported higher load than hybrid tubes with internal layers of glass/epoxy.

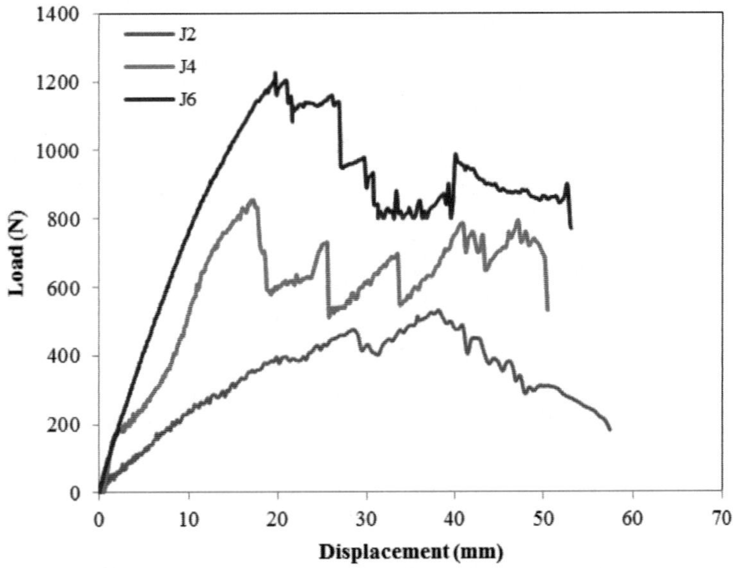

Figure 2. Load-displacement relation for jute/epoxy tubes.

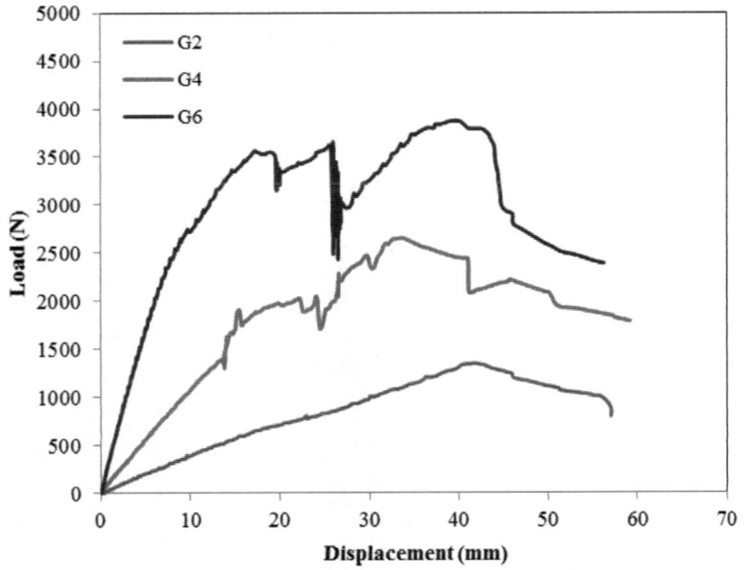

Figure 3. Load-displacement relation for glass/epoxy tubes.

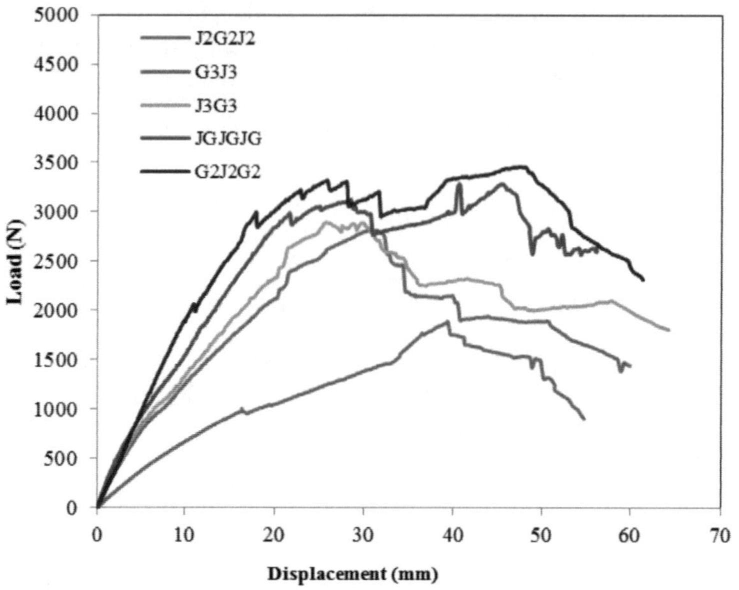

Figure 4. Load-displacement relation for jute-glass/epoxy tubes.

Figure 5 shows the effect of material used on maximum load obtained from the load-displacement graphs. This figure explains that the glass/epoxy tubes made of six layers supported the highest lateral load compared to the other tested tubes. The maximum lateral load obtained for the hybrid jute-glass/epoxy tubes made of six layers is significantly affacted by the stacking sequence of the jute and glass fibers. In general, the hybrid jute-glass/epoxy tubes supported moderate lateral load compared to the load obtained from glass/ epoxy tubes.

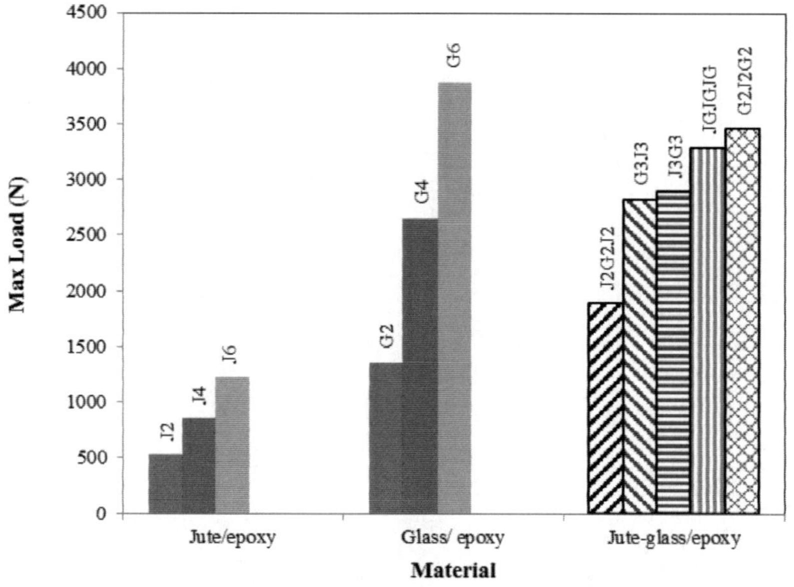

Figure 5. Maximum load of the laterally crushed composite tubes.

3.2. Lateral loading characteristics

The crush force efficiency (CFE) is an important factor to measure the crush performance and to evaluate the crashworthiness of energy absorber components. The CFE for the composite tubes under lateral loading is determined using equation 1. Results show that the crush force efficiency for hybrid jute-glass/epoxy tubes and glass epoxy are found higher than jute/epoxy.

$$CFE = \frac{F_{mean}}{F_{max}} \qquad (1)$$

where, F_{mean} and F_{max} are the mean and maximum crush failure loads respectively

3.3. Specific Energy Absorption

The specific energy absorption (SEA) for the composite tubes under lateral loading can be calculated as follows:

$$SEA = \frac{E_t}{m} \qquad (2)$$

where, E_t is the total absorbed energy represented by the area under the load-displacement graph and m is the mass of the tube. Table 3 shows the results of *SEA*, mean and maximum crush loads of the composite tubes subjected to lateral loading. By referring to table, *SEA* is improved by using the hybrid composite tubes.

Table 3. Test results of composite tubes under lateral loading.

Specimen type	Mass (g)	Mean load P_{mean} (KN)	Maximum load P_{max} (KN)	Crush force efficiency (CFE)	Specific energy absorption SEA (J/Kg)
J2	110	0.38	0.53	0.72	143
J4	230	0.63	0.85	0.74	210
J6	337	0.92	1.22	0.75	360
G2	125	1.16	1.35	0.86	250
G4	256	2.08	2.65	0.78	380
G6	366	3.44	3.87	0.89	795
J2G2J2	362	1.54	1.88	0.82	460
G3J3	376	2.61	2.82	0.93	580
J3G3	384	2.70	2.89	0.93	596
JGJGJG	368	3.08	3.28	0.94	670
G2J2G2	375	3.31	3.46	0.96	705

3.4. Failure mode

One of the important factors to evaluate the crashworthiness of the composite structures is the failure mechanisms. Composite tubes were fixed between two flat steel plates as shown in figure 6 where the load applied laterally on the tubes. Photographs of compression process leads to failure were taken to observe the progressive crushing mechanism on the composite tubes. In general, the tested tubes fractured at the centre position where cracks propagated at the top and bottom sides along the longitudinal direction of the tube. It can be seen that four longitudinal fracture lines are developed during the crushing process. Two fracture lines are formed close to the lines of contact between the tube and the flat platens. With further increase in the crushing distance and upon the complete of the compression process, another two fracture lines are formed diametrically opposite to each other.

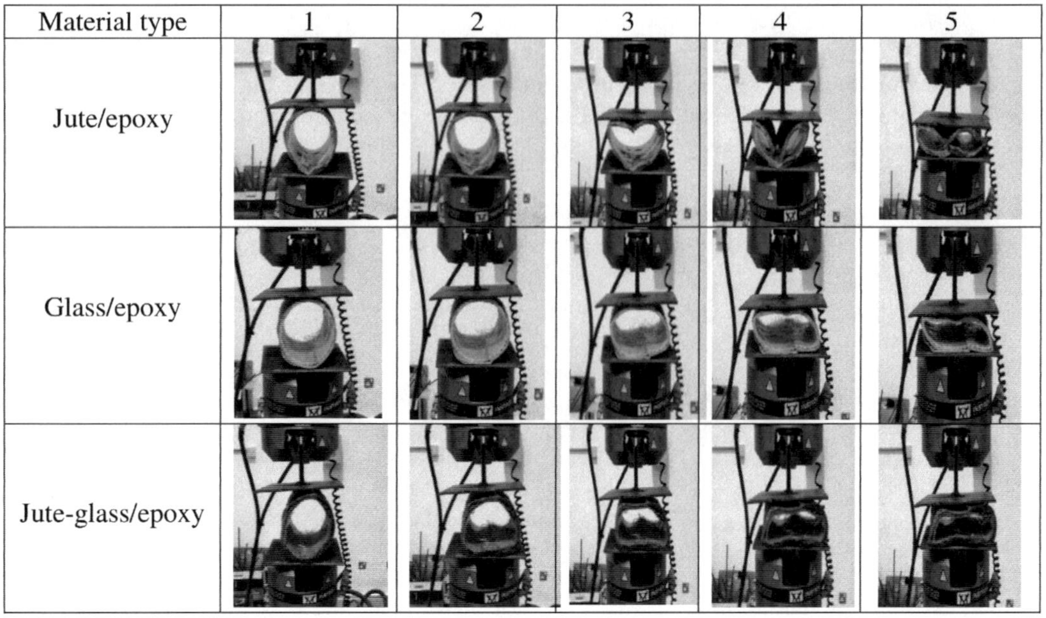

Figure 6. Failure mode for composite tubes under lateral loading.

4. Conclusion

In this paper, experimental tests were carried out to investigate the effect of lateral loading on the load-displacement relationship, specific energy absorption and failure mode of hybrid and non-hybrid composite tubes. Hand lay-up method is used to fabricate carefully the composite tubes of 2, 4 and 6 layers carefully. Results of the load-displacement graphs show that the maximum lateral load obtained for six layered glass/epoxy is found higher 10.6% than hybrid jute-glass/epoxy. The specific energy absorption of the glass/epoxy tubes is found higher respectively 11.6% and 46% than hybrid jute-glass and jute/epoxy tubes. The increase in the number of layers from two to six increased the maximum lateral load from 0.53KN to 1.22 KN jute/epoxy tubes. The increase was from 1.35 KN to 3.87 KN for the glass/epoxy tubes. The stacking sequence of jute and glass fibers has an effect on the performance of the composite tubes under lateral loading. The maximum lateral load varied from 1.88 KN to 3.46 KN with using different stacking sequence of the jute and glass fibers. In general, similar failure mechanism has been obtained from the fractured composite tubes where cracks propagate at the center position of the tube at the top and bottom sides of the tube. The post peak failures showed a significant deformation at the crack line position and then lead to the final fracture.

References

[1] Gupta N, Abbas H 2000 *International Journal of Impact Engineering* **24** 329-46
[2] Hull D 1991 *Composites science and technology* **40** 377-421
[3] Baroutaji A, Gilchrist M, Smyth D, Olabi A 2015 *Thin-Walled Structures* **86** 121-31
[4] Oshkovr S, Eshkoor R, Taher S, Ariffin A, Azhari C 2012 *Composite Structures* **94** 2337-42
[5] Eyvazian A, Akbarzadeh I, Shakeri M 2012 *Proceedings of the Institution of Mechanical Engineers, Part L: Journal of Materials Design and Applications* 1464420712437307
[6] Baroutaji A, Morris E, Olabi A-G 2014 *Thin-Walled Structures* **82** 262-77
[7] Niknejad A, Elahi S A, Liaghat G H 2012 *Materials & Design* **36** 24-34
[8] Niknejad A, Assaee H, Elahi S A, Golriz A 2013 *Composite Structures* **100** 479-92
[9] Yan L, Chouw N, Jayaraman K 2014 *Composites Part B: Engineering* **63** 15-26
[10] Zuraida A, Khalid A A, Ismail A 2007 *Materials & design* **28** 71-7
[11] Yan L, Chouw N, Jayaraman K 2014 *Materials & Design* **56** 528-41

[12] Rouzegar J, Assaee H, Niknejad A, Elahi S A 2015 *Materials & Design* **65** 343-59
[13] Khalid A A, Sahari B, Khalid Y 2000 *Journal, Institution of Engineers, Malaysia* **61** 13-22
[14] Eshkoor R, Oshkovr S A, Sulong A, Zulkifli R, Ariffin A K, Azhari C H 2013 *Materials & Design* **47** 248-57
[15] Khalid A A, Sahari B, Khalid Y A 2002 *Composites science and technology* **62** 17-27
[16] Kathiresan M, Manisekar K, Manikandan V 2012 *Composite Structures* **94** 3510-9
[17] Rezadoust A, Esfandeh M, Sabet S 2008 *Polymer-Plastics Technology and Engineering* **47** 147-51
[18] Ahmad Z, Thambiratnam D P 2009 *Computers & Structures* **87** 186-97
[19] Gupta N 2007 *International journal of impact engineering* **34** 708-20
[20] Kathiresan M, Manisekar K, Manikandan V 2014 *Composite Structures* **108** 584-99
[21] Price J, Hull D 1987 *Composites science and technology* **28** 211-30
[22] Gupta P, Gupta N 2013 *Thin-Walled Structures* **62** 85-95
[23] Khalid A A 2006 *Materials & design* **27** 499-506

IOP Conf. Series: Materials Science and Engineering **100** (2015) 012069 doi:10.1088/1757-899X/100/1/012069

Support vector machine to predict diesel engine performance and emission parameters fueled with nano-particles additive to diesel fuel

M Ghanbari[1], G Najafi[1*], B Ghobadian[1], R Mamat[2], M M Noor[2,3*] and A Moosavian[1]

[1]Mechanics of Biosystem Engineering Department, Tarbiat Modares University, Tehran, Iran

[2]Faculty of Mechanical Engineering, University Malaysia Pahang, Malaysia

[3]Department of Mechanical Engineering, University of Southern Queensland, Australia

E-mail: najafi14@gmail.com

Abstract. This paper studies the use of adaptive Support Vector Machine (SVM) to predict the performance parameters and exhaust emissions of a diesel engine operating on nanodiesel blended fuels. In order to predict the engine parameters, the whole experimental data were randomly divided into training and testing data. For SVM modelling, different values for radial basis function (RBF) kernel width and penalty parameters (C) were considered and the optimum values were then found. The results demonstrate that SVM is capable of predicting the diesel engine performance and emissions. In the experimental step, Carbon nano tubes (CNT) (40, 80 and 120 ppm) and nano silver particles (40, 80 and 120 ppm) with nano-structure were prepared and added as additive to the diesel fuel. Six cylinders, four-stroke diesel engine was fuelled with these new blended fuels and operated at different engine speeds. Experimental test results indicated the fact that adding nano particles to diesel fuel, increased diesel engine power and torque output. For nano-diesel it was found that the brake specific fuel consumption (bsfc) was decreased compared to the net diesel fuel. The results proved that with increase of nano particles concentrations (from 40 ppm to 120 ppm) in diesel fuel, CO_2 emission increased. CO emission in diesel fuel with nano-particles was lower significantly compared to pure diesel fuel. UHC emission with silver nano-diesel blended fuel decreased while with fuels that contains CNT nano particles increased. The trend of NOx emission was inverse compared to the UHC emission. With adding nano particles to the blended fuels, NOx increased compared to the net diesel fuel. The tests revealed that silver & CNT nano particles can be used as additive in diesel fuel to improve complete combustion of the fuel and reduce the exhaust emissions significantly.

1. Introduction

A lot of studies have been conducted for predicting various engine parameters. Some of the works have concentrated on performance parameters, some of them on exhaust emissions and remaining on both of them. In these researches, spark ignition (SI) and compression ignition (CI) engines with traditional fuels and biofuels have been investigated. However, due to the development of biofuels and our requirements to these fuel types for reducing pollutions and saving the fossil sources, studies about them can be more useful. Most studies are based on the use of traditional artificial neural network (ANN) for prediction. For example, Sharon et al. (2012) [1] employed ANN to predict performance and exhaust emissions of a diesel engine with 8 different blends of diesel and biodiesel fuels. Cay et al. (2013) [2] experimentally investigated the use of ANN for predicting performance and emission characteristics of an engine fuelled with methanol and gasoline. The inputs of ANN were fuel type, engine speed, torque and fuel flow. In another research, Cay (2013) [3] modeled engine performance by using ANN. The used fuel was gasoline, and three parameters of specific fuel consumption, engine power and exhaust temperature were predicted. Because ANNs follow empirical risk minimization (ERM) principle, they have some drawbacks like poor generalization, greater computational burden and proneness to over fitting [4, 5].

The disadvantages cause a performance reduction in modelling. In the literature, there are a few works that have used other machine learning methods which have some advantages over ANNs. In addition, the investigation on the performance comparison of learning methods in prediction of engine parameters is limited. Hosoz et al. (2013) [6] investigated the applicability of adaptive neuro-fuzzy inference system (ANFIS) for predicting several parameters of a diesel engine fuelled with diesel and biodiesel blends. Wong et al. (2014) [4] proposed a model based on sparse Bayesian extreme learning machine (SBELM) for engine performance prediction. A comparison was made between SBELM, extreme learning machine (ELM), Bayesian ELM and back-propagated neural network (BPNN) methods. The predicted engine parameters were fuel consumption, Lambda value, HC, CO and CO_2 emissions. In another research, Wong et al. (2015) presented a biodiesel engine modelling by kernel-based extreme learning machine (K-ELM). The K-ELM model was compared with least-square support vector machine (LS-SVM) model [7-16]. Previous researches using nanodiesel have been mentioned that blending the nano particles with diesel fuel improves the combustion parameters of CI engine [17-35].

In this study, support vector machine (SVM) has been employed to determine the engine power, torque, brake specific fuel consumption (bsfc), and emission components based on various diesel–biodiesel and nano particles blends and speeds. The main objective of this research is to investigate the ability of SVM for predicting engine performance parameters and exhaust emissions, and also compare their performances in modelling. In this research, the performances of SVM method were thoroughly evaluated.

2. Experimental setup

2.1. Methodology

In this study, the experiments were performed on a CI engine, 6 Cylinder. A 190 kW SCHENCK-WT190 eddy–current dynamometer was used in the experiments, specifications of diesel engine have been described in Table 1. Engine speeds varied in the range of 700-1000 rpm. Fuel consumption rate was measured in the range of 0.4–45 kg/h by using laminar type flow meter, Pierburg model. The emission parameters (CO, CO_2, HC and NOx) from an online and accurately calibrated exhaust gas analyser AVL DIGAS 4000 were recorded (figure 1). Considering the accomplished researches about nano fuels and diesel fuel nano additives, two silver nano-particles (Ag) and carbon nano tubes (CNT) were applied as nano additives to these fuels. Furthermore based on researches conducted about the effect of concentrations of used nano-particles in reduction of exhaust emissions, in this study three concentrations (40, 80 and 120 ppm) were applied. In order to ensure the validity of nano-particles utilized in this research, SEM and TEM pictures were taken (figure 2). Stability of nano fuel was tested at standard conditions. Also the use of carbon nano tubes (CNT) and silver nano particles in neat diesel blend has the tendency to settle down at the fuel tank. An ultrasonic processor (UP400S, Hielscher, USA) was used to perform the reaction and even mixing nano-particles with diesel fuel before the engine tests. The processor operated at 400 W and 24 kHz frequency (figure 2).

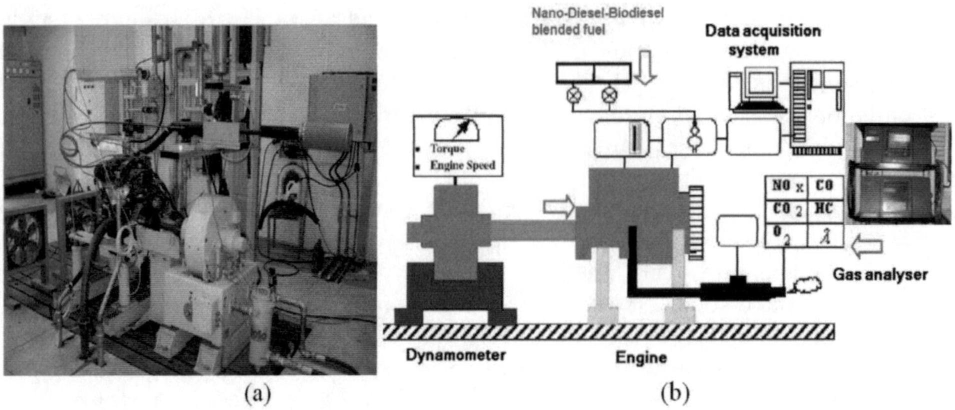

Figure 1. Engine test set-up and test instruments (a) real and (b) schematic.

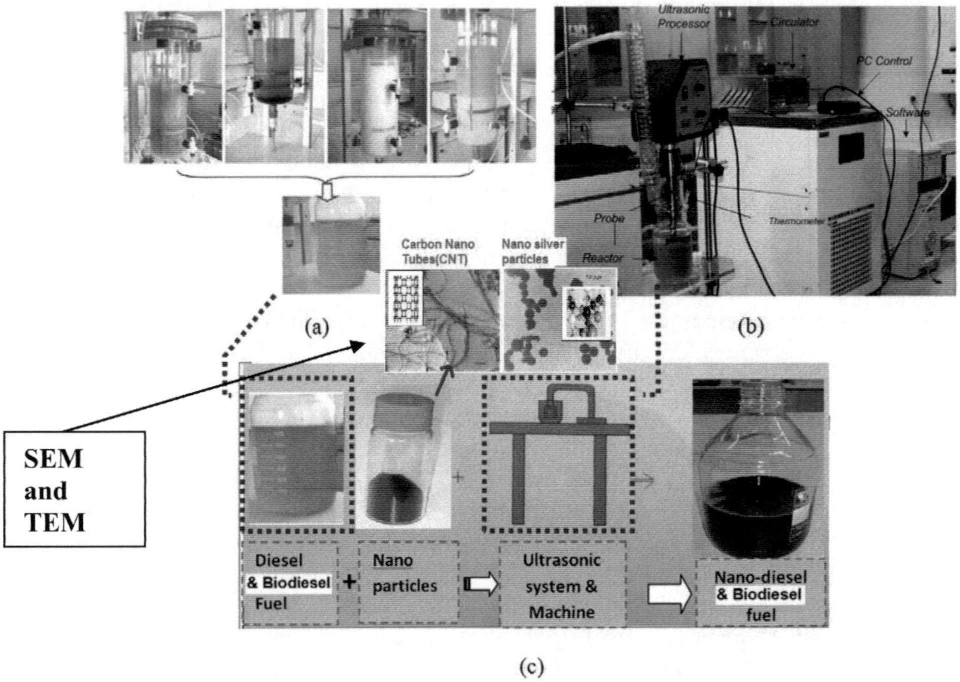

Figure 2. The Set up for ultrasonic-assisted nano-diesel production process (a)reaction, (b) ultrasonic set-up and (c) nano-diesel blend.

Table 1. Main characteristics of the test engine

Engine Type	CI engine, 6 Cylinder
Combustion Order	1-5-3-6-2-4
Bore ×Stroke(mm)	98.6 * 127
Displacement Volume (Lit)	5.8
Max. Torque (N.m/rpm)	376 / 1300
Max. Power(kW/rpm)	82 / 2300

In this paper, the quantity BX represents a blend consisting of X% biodiesel by volume, e.g., B20 indicates a blend consisting of 20% biodiesel in 80% diesel. Laboratory tests were then carried out using ASTM test standards to determine the fuel properties.

SVM. This is an engine simulation tool. SVM is a supervised learning method developed by Vapnik based on statistical learning theory. Nowadays, SVM is successfully used for many applications like prediction, pattern detection and classification, due to its high ability and capability in solving learning problems. SVM has good generalization ability because it follows the structural risk minimization (SRM) principle which minimizes an upper bound on the generalization error. SVM can be considered as a regressive model. Suppose a set of training data $\{x_i, t_i\}$, $i = 1, 2, ..., n$, $x_i \in R^d$, $t_i \in R$, where xi is input data and ti is the corresponding target value. SVM tries to estimate target values by the linear equation as follows:

$$y_i = w^T x_i + b \tag{1}$$

where y_i is the output of SVM, w is d-dimensional vector and b is a scalar. By using a ε–insensitive loss function, the optimal linear regression function can be obtained by solving the following optimization problem.

$$\frac{1}{2} w^T w + C \sum_{i=1}^{n} (\xi_i^+ + \xi_i^-) \tag{2}$$

subject to

$$\begin{cases} -\varepsilon - \xi_i^- \le t_i - y_i \le \varepsilon + \xi_i^+ \\ \xi_i^+ \ge 0 \\ \xi_i^- \ge 0 \end{cases} \quad \forall i \tag{3}$$

where C is penalty parameter, ε is a pre-specified value and ξ_i^+, ξ_i^- are slack variables representing upper and lower constraints on the output of the system. For nonlinear regression, SVM maps the input data into a high-dimensional feature space in which linear regression is performed. This is done by kernel function which is stated as $K(x_i, x_j) = \varphi(x_i).\varphi(x_j)$, where φ is a nonlinear function. Hence, the estimation of SVM is obtained by the following equation.

$$y_i = \sum_{i=1}^{n} (\alpha_i^+ - \alpha_i^-) K(x_i, x) + b \tag{4}$$

where α_i^+, α_i^- are Lagrange multipliers. The criterions used in this research are correlation coefficient (R), root mean square error (RMSE), mean relative error (MRE) and accuracy. The correlation coefficient (R) which evaluates the strength of the relationship between the experimental and predicted results is defined as:

$$R(a, p) = \frac{\text{cov}(a, p)}{\sqrt{\text{cov}(a, a)\,\text{cov}(p, p)}} \tag{5}$$

where cov(a,p) is covariance between a and p sets. a and p denotes the actual output and predicted output sets. RMSE is determined by:

$$RMSE = \sqrt{\frac{1}{n}\sum_{i=1}^{n}(a_i - p_i)^2} \tag{6}$$

where n is the number of the points in the data set. MRE which indicates the mean ratio between the error and the experimental values, is obtained by:

$$MRE(\%) = \frac{1}{n}\sum_{i=1}^{n}\left|100 \times \frac{(a_i - p_i)}{a_i}\right| \tag{7}$$

Accuracy which is a simple representation of prediction performance is calculated as:

$$Accuracy = \left(1 - \frac{|a_i - p_i|}{a_i}\right) \times 100 \tag{8}$$

3. Results and discussion

Figure 3 illustrates the performance of SVM in the brake power prediction under various numbers of the two parameters. The performance of SVM improved by increasing the training epoch. After determining the optimum numbers, the ability of SVM in prediction of the engine parameters was evaluated by calculating the criterions of correlation coefficient (R), RMSE, MRE and accuracy.

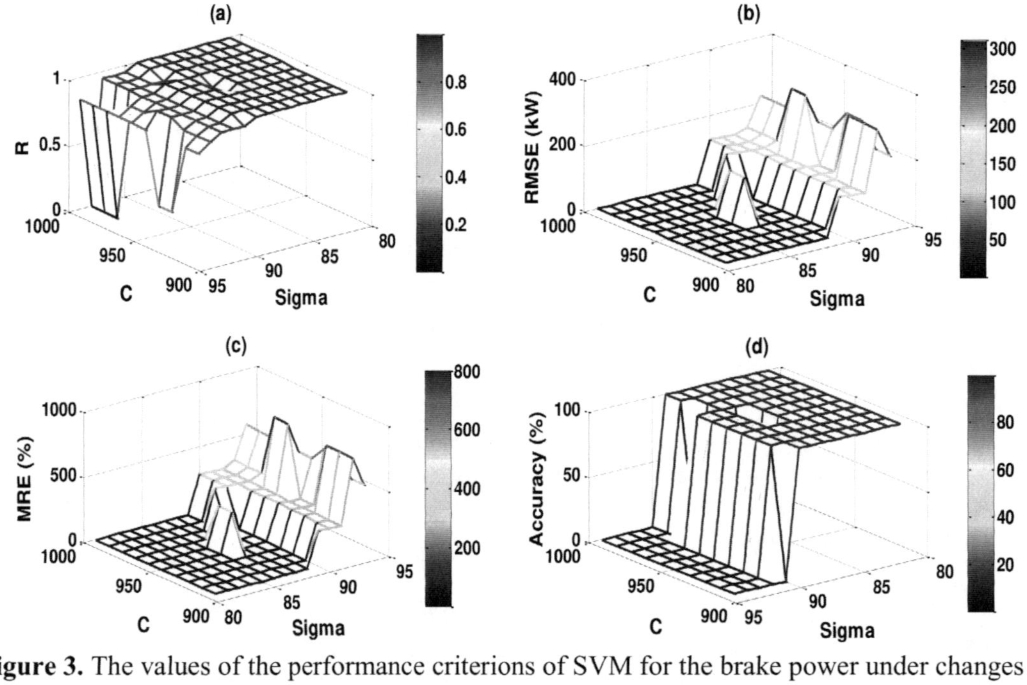

Figure 3. The values of the performance criterions of SVM for the brake power under changes of sigma and C: (a) R, (b) RMSE, (c) MRE and (d) Accuracy

Figure 4(b) shows the predicted versus experimental values for the brake power. SVM predictions for the brake power yielded a correlation coefficient (R) of 0.999, RMSE of 0.198 kW, MRE of 0.43% and accuracy of 99.57%. SVM predictions for the engine torque yielded a correlation coefficient (R) of 0.985, RMSE of 2.373 Nm, MRE of 0.511% and accuracy of 99.489% (figure 4(a)). In predicting the bsfc, SVM resulted in a correlation coefficient of 0.824, RMSE of 3.079 g/kWh, MRE of 1.103% and accuracy of 98.897% (figure 4(c)). Figure 5 shows the predicted versus experimental values for the exhaust emission parameters. SVM predictions for the CO, CO_2, HC and NO_X yielded a correlation coefficient (R) of 0.965, 0.951, 0.803 and 0.631, respectively. It was found that the RMSE values were 0.063 %V, 0.309 %V, 2.107 ppm and 185.522 ppm for the CO, CO_2, HC and NO_X, respectively. SVM predictions for the CO yielded a MRE of 3.421% (figure 5(a)). MRE value for the CO_2 was 2.448% (figure 5(b)). Figure 5(c) shows the predicted versus experimental values for the HC with MRE of 2.107%. According to figure 5(d), MRE of the NO_X was 9.346%. SVM accuracy for prediction of the CO, CO_2, HC and NO_X was 96.579%, 97.552%, 90.211% and 90.654%, respectively. The above results demonstrate that SVM can predict the exhaust emission parameters well.

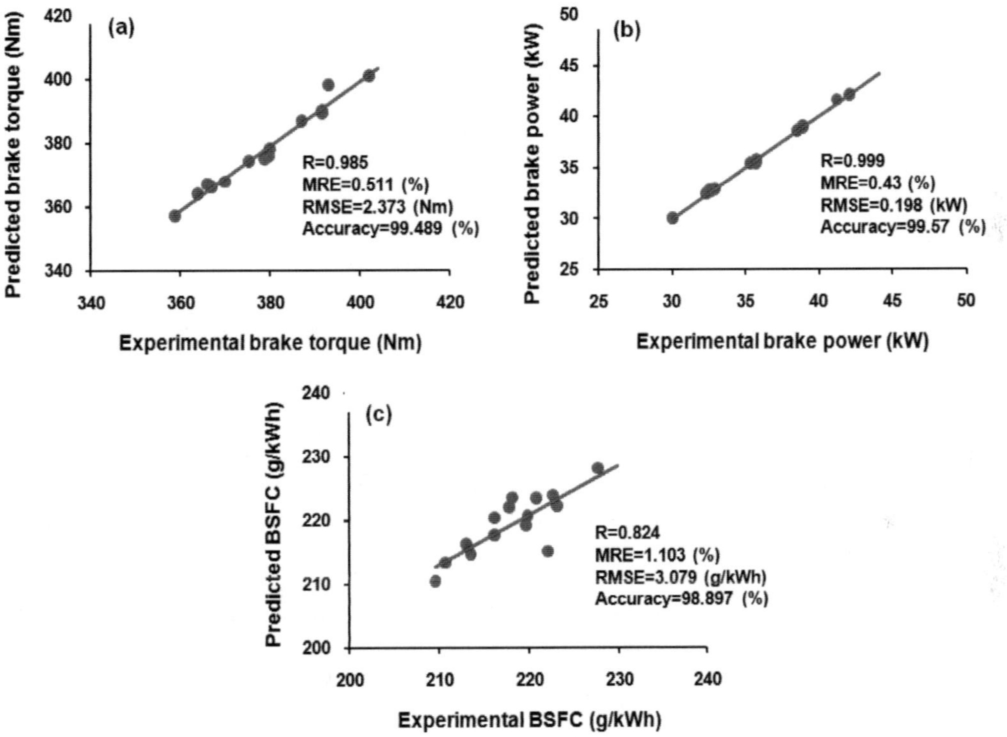

Figure 4. SVM predictions for the (a) brake torque, (b) brake power and (c) BSFC versus experimental values

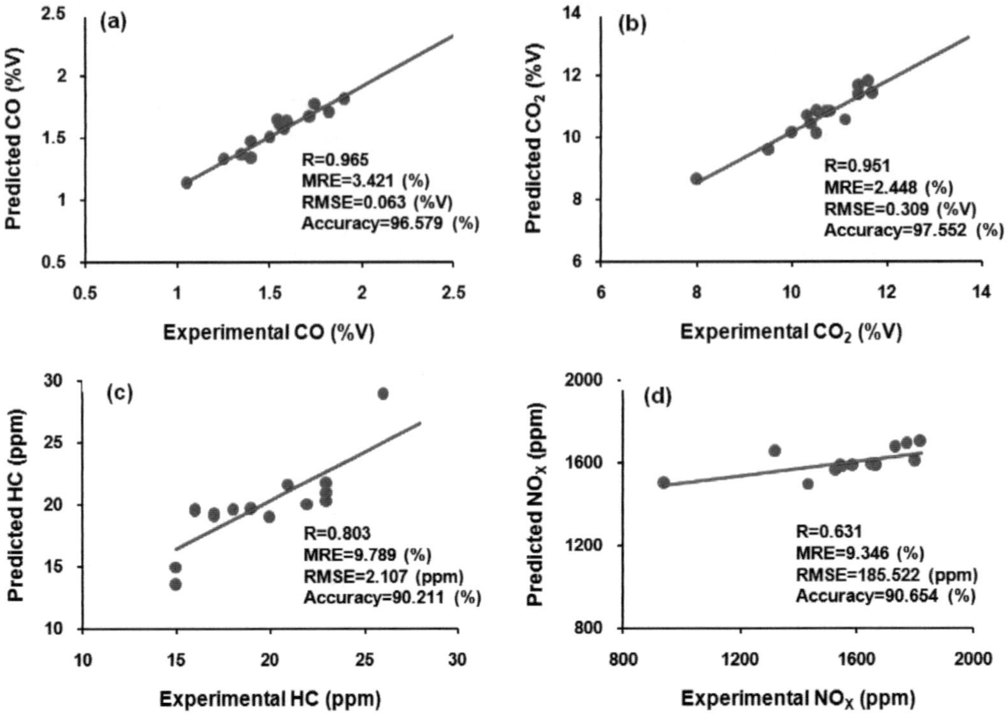

Figure 5. SVM predictions for the (a) CO, (b) CO2, (c) HC and (d) NO_X versus experimental values.

4. Conclusions

The results of this research study demonstrated that nanodiesel could be efficiently utilized as blended fuel in diesel engine. The produced nanodiesel was blended with biodiesel and diesel fuels in different concentrations and the CI engine performance and emission parameters were evaluated with these new blended fuels. Adding nano particles in diesel fuel can improve engine performance and reduce CO and HC emissions. But it can cause an increase in CO_2 and NO_X emissions. SVM can be employed to predict the engine performance and exhaust emissions.

SVM can be employed to predict the engine performance and exhaust emissions. But there is a need for finding the optimum values of its parameters. The SVM results were very good, R values in this model are very close to one, while root mean square error (RMSE) was found to be very low. The results showed that SVM predicted the engine performance and exhaust emissions with correlation coefficient (R) and accuracy in the range of 0.66–1 and 65–99.5%. The results demonstrate that SVM is capable of predicting the CI engine performance and emissions. Analysis of the experimental data by the SVM method revealed that there is a good correlation between the SVM predicted results and the experimental data. Therefore SVM proved to be a useful tool for correlation and simulation of engine parameters.

Acknowledgements

The authors wish to thank the Iranian Fuel Conservation Organization (IFCO) of NIOC for the research grant provided (Grant no.:1393-1) and UMP (Grant No.:RDU130313) and Mega Motor Company for providing of laboratory facilities.

References

[1] Sharon H, Jayaprakash R, Sundaresan A, Karuppasamy K 2012 *Fuel* **99** 197-203
[2] Çay Y, Korkmaz I, Çiçek A, Kara F 2013 *Energy* **50** 177-86
[3] Cay Y 2013 *Fuel* **111** 324-31

[4] Wong K I, Vong C M, Wong P K, Luo J 2014 *Neurocomputing*
[5] Tu J V 1996 *Journal of clinical epidemiology* **49** 1225-31
[6] Hosoz M, Ertunc H M, Karabektas M, Ergen G 2013 *Applied Thermal Engineering* **60** 24-32
[7] Wong P K, Wong K I, Vong C M, Cheung C S 2015 *Renewable Energy* **74** 640-7
[8] Vapnik V 2013 *The nature of statistical learning theory* Springer Science & Business Media)
[9] Widodo A, Yang B-S 2007 *Mechanical Systems and Signal Processing* **21** 2560-74
[10] Brereton R G, Lloyd G R 2010 *Analyst* **135** 230-67
[11] Jang J-S R 1993 *Systems, Man and Cybernetics, IEEE Transactions on* **23** 665-85
[12] Reddy M J, Mohanta D K 2007 *Electric Power Components and Systems* **35** 1385-407
[13] Sugeno M, Kang G 1988 *Fuzzy sets and systems* **28** 15-33
[14] Şişman-Yılmaz N A, Alpaslan F N, Jain L 2004 *Neurocomputing* **61** 139-68
[15] Sayin C, Ertunc H M, Hosoz M, Kilicaslan I, Canakci M 2007 *Applied thermal engineering* **27** 46-54
[16] Widodo A, Yang B-S 2011 *Expert Systems with Applications* **38** 8430-7
[17] Ma F, Hanna M A 1999 *Bioresource technology* **70** 1-15
[18] Schumacher L, Peterson C, Grepen J, Fuelling direct diesel engines with 2% biodiesel blend. Written for presentation at the 2001 annual international meeting sponsored by ASAE; 2001.
[19] Ghobadian B, Rahimi H, Biofuels-past, present and future perspective. International Iran and Russian Congress of Agricultural and Natural Science Shahre-Kord University, Shahre Kord, Iran; 2004.
[20] Fukuda H, Kondo A, Noda H 2001 *Journal of bioscience and bioengineering* **92** 405-16
[21] Lee S W, Herage T, Young B 2004 *Fuel* **83** 1607-13
[22] Najafi G, Ghobadian B, Tavakoli T, Buttsworth D, Yusaf T, Faizollahnejad M 2009 *Applied Energy* **86** 630-9
[23] De Caro P S, Mouloungui Z, Vaitilingom G, Berge J C 2001 *Fuel* **80** 565-74
[24] Xing-cai L, Jian-Guang Y, Wu-Gao Z, Zhen H 2004 *Fuel* **83** 2013-20
[25] Ghobadian B, Rahimi H, Khatamifar M, Evaluation of engine performance using net diesel Fuel and biofuel blends. The First Combustion Conference of Iran (CCT1) Tarbiat Modares University February; 2006.
[26] Elawad M, Yusaf T 2004 *Journal of Energy & Environment* **3** 61-8
[27] Li D-g, Zhen H, Xingcai L, Wu-gao Z, Jian-Guang Y 2005 *Renewable energy* **30** 967-76
[28] Ajav E, Singh B, Bhattacharya T 1998 *Biomass and Bioenergy* **15** 493-502
[29] Abu-Qudais M, Haddad O, Qudaisat M 2000 *Energy Conversion and Management* **41** 389-99
[30] Noguchi N, Terao H, Sakata C 1996 *Bioresource Technology* **56** 35-9
[31] Gerdes K, Suppes G 2001 *Industrial & engineering chemistry research* **40** 949-56
[32] He B-Q, Shuai S-J, Wang J-X, He H 2003 *Atmospheric Environment* **37** 4965-71
[33] Hansen A C, Zhang Q, Lyne P W 2005 *Bioresource technology* **96** 277-85
[34] Can Ö, Celikten I, Usta N 2004 *Energy Conversion and Management* **45** 2429-40
[35] Ajav E, Singh B, Bhattacharya T 2000 *Energy conversion and management* **41** 1533-41

IOP Conf. Series: Materials Science and Engineering **100** (2015) 012070 doi:10.1088/1757-899X/100/1/012070

Adaptive neuro-fuzzy inference system (ANFIS) to predict CI engine parameters fueled with nano-particles additive to diesel fuel

M Ghanbari[1], G Najafi[1], B Ghobadian[1], R Mamat[2], M M Noor[2,3] and A Moosavian[1]

[1]Mechanics of Biosystem Eng. Department, Tarbiat Modares University, Tehran, Iran
[2]Faculty of Mechanical Engineering, University Malaysia Pahang, Malaysia
[3]Department of Mechanical Engineering, University of Southern Queensland, Australia\

E-mail:najafi14@gmail.com

Abstract. This paper studies the use of adaptive neuro-fuzzy inference system (ANFIS) to predict the performance parameters and exhaust emissions of a diesel engine operating on nanodiesel blended fuels. In order to predict the engine parameters, the whole experimental data were randomly divided into training and testing data. For ANFIS modelling, Gaussian curve membership function (gaussmf) and 200 training epochs (iteration) were found to be optimum choices for training process. The results demonstrate that ANFIS is capable of predicting the diesel engine performance and emissions. In the experimental step, Carbon nano tubes (CNT) (40, 80 and 120 ppm) and nano silver particles (40, 80 and 120 ppm) with nano-structure were prepared and added as additive to the diesel fuel. Six cylinders, four-stroke diesel engine was fuelled with these new blended fuels and operated at different engine speeds. Experimental test results indicated the fact that adding nano particles to diesel fuel, increased diesel engine power and torque output. For nano-diesel it was found that the brake specific fuel consumption (bsfc) was decreased compared to the net diesel fuel. The results proved that with increase of nano particles concentrations (from 40 ppm to 120 ppm) in diesel fuel, CO_2 emission increased. CO emission in diesel fuel with nano-particles was lower significantly compared to pure diesel fuel. UHC emission with silver nano-diesel blended fuel decreased while with fuels that contains CNT nano particles increased. The trend of NO_x emission was inverse compared to the UHC emission. With adding nano particles to the blended fuels, NO_x increased compared to the net diesel fuel. The tests revealed that silver & CNT nano particles can be used as additive in diesel fuel to improve combustion of the fuel and reduce the exhaust emissions significantly.

1. Introduction

Effects of cerium oxide nano-particles addition in diesel and diesel-biodiesel-ethanol blends on performance and emission characteristics of a CI engine has been studied and results showed that the cerium oxide acts as an oxygen donating catalyst and provides oxygen for the oxidation of CO or absorbs oxygen for the reduction of NO_x. The tests revealed that cerium oxide nano-particles can be used as additive in diesel and diesel-biodiesel-ethanol blend to improve complete combustion of the fuel and reduce the exhaust emissions significantly [1]. The investigation on the performance comparison of learning methods in prediction of engine parameters is limited. Hosoz, Ertunc [2] investigated the applicability of adaptive neuro-fuzzy inference system (ANFIS) for predicting several parameters of a diesel engine fueled with diesel and biodiesel blends. Yücesu, Topgül [3] proposed a model based on sparse Bayesian extreme learning machine (SBELM) for engine performance prediction. A comparison was made between SBELM, extreme learning machine (ELM), Bayesian ELM and back-propagated neural network (BPNN) methods. The predicted engine parameters were fuel consumption, Lambda value, HC, CO and CO_2 emissions. In another research, Wong, Wong [4] presented a biodiesel engine modelling by kernel-based extreme learning machine (K-ELM). The used fuel was gasoline, and three parameters of specific fuel consumption, engine power and exhaust temperature were predicted. Because ANNs follow empirical risk minimization (ERM) principle, they have some drawbacks like poor generalization, greater computational burden and proneness to over fitting [5]. The disadvantages cause a performance reduction in modelling. In the literature, there are a few works that have used other machine learning methods which have some advantages over ANNs. In addition, the investigation on the performance comparison of learning methods in prediction of engine parameters is limited. The K-ELM model was compared with least-square support vector machine (LS-SVM) model [4, 6-15]. Previous researches using nano diesel have been mentioned that blending the nano particles with diesel fuel improves the combustion parameters of CI engine [16-33].

In this study, Adaptive Neuro-Fuzzy Inference System (ANFIS) has been employed to determine the engine power, torque, brake specific fuel consumption (bsfc), and emission components based on various diesel–biodiesel and nano particles blends and speeds. The main objective of this research is to investigate the ability of ANFIS for predicting engine performance parameters and exhaust emissions, and also compare their performances in modelling. In this research, the optimum values for the parameters of ANFIS are found, and then the performances of this method are thoroughly evaluated.

2. Experimental Setup

2.1. Methodology

There In this study, the experiments were performed on a CI engine, 6 Cylinder. A 190 kW SCHENCK-WT190 eddy- current dynamometer was used in the experiments. Fuel consumption rate was measured in the range of 0.4- 45 kg/h by using laminar type flow meter, Pierburg model. The emission parameters (CO, CO_2, HC and NO_x) from an online and accurately calibrated exhaust gas analyser AVL DIGAS 4000 were recorded (figure 1). Considering the accomplished researches about nano fuels and diesel fuel nano additives, two silver nano-particles (Ag) and carbon nano tubes (CNT) were applied as nano additives to these fuels. Furthermore based on researches conducted about the effect of concentrations of used nano-particles in reduction of exhaust emissions, in this study three concentrations (40, 80 and 120 ppm) were applied. Stability of nano fuel was tested at standard conditions. In order to ensure the validity of nano-particles utilized in this research, SEM and TEM pictures were taken (Figure 2). Also the use of carbon nano tubes (CNT) and silver nano particles in neat diesel blend has the tendency to settle down at the fuel tank. An ultrasonic processor (UP400S, Hielscher, USA) was used to perform the reaction and even mixing nano-particles with diesel fuel before the engine tests. The processor operated at 400W and 24 kHz frequency (figure 2). Main characteristics of the test engine has been described in table 1. Engine speeds varied in the range of 700-1000 rpm.

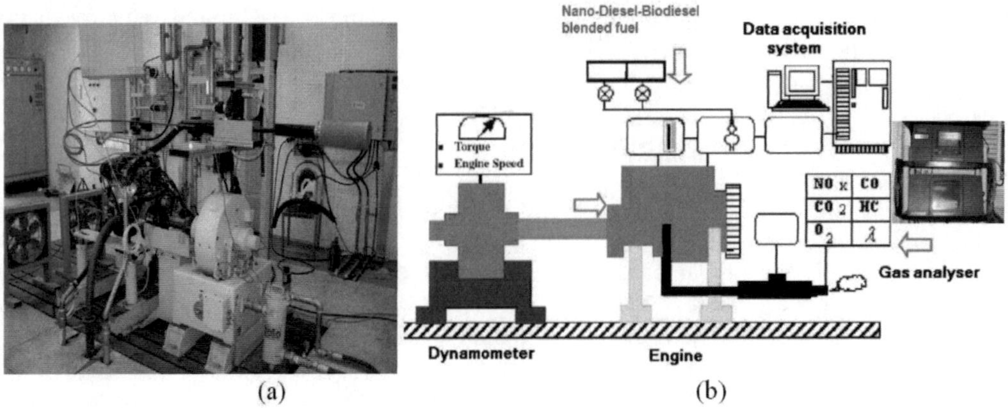

Figure 1. Engine test set-up and test instruments (a) real and (b) schematic.

Table 1. Main characteristics of the test engine.

Engine Type	CI engine, 6 Cylinder
Combustion Order	1-5-3-6-2-4
Bore ×Stroke(mm)	98.6 * 127
Displacement Volume (Lit)	5.8
Max. Torque (N.m/rpm)	376 / 1300
Max. Power(kW/rpm)	82 / 2300

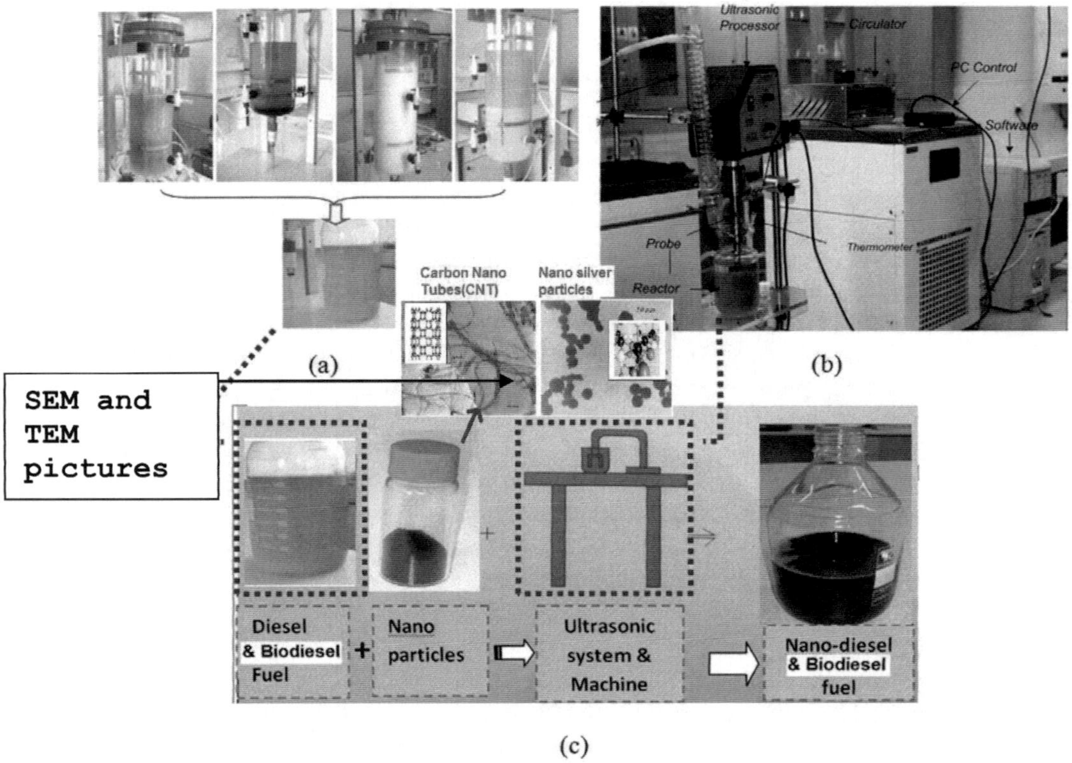

Figure 2. The Set up for ultrasonic-assisted nano-diesel production process (a)reaction, (b) ultrasonic set-up and (c) nano-diesel blend.

In this paper, the quantity B_X represents a blend consisting of X% biodiesel by volume, e.g., B20 indicates a blend consisting of 20% biodiesel in 80% diesel. Laboratory tests were then carried out using ASTM test standards to determine the fuel properties.

2.2. ANFIS.

ANFIS model introduced by Jangis a model consisting of fuzzy logic and ANNs. ANFIS which is based on the Sugeno fuzzy inference model (SFIM),constructs an input–output mapping according to both the fuzzy if–then rules and the stipulated input–output data pairs. The if-then rules of fuzzy system are usually employed to obtain the inference of imprecise model, which can process information in system as well as the human experience. Using stipulated input–output training data pairs by good membership functions, the if-then rules are built. ANFIS can adjust the membership function using back propagation gradient descent. ANFIS simulates Takagi-Sugeno-Kang (TSK) fuzzy rule of type-3 where the consequent part of the rule is a linear combination of the inputs and a constant. The final output of the system is weighted by average of each rule's output. The form of the type-3 rule simulated in the system is as follows:

If x_1 is A_1 and x_2 is A_2 and … and xp is Ap then $y=c_0+c_1x_1+c_2x_2+...+c_px_p$

where $[x_1,x_2,\ldots,x_p]$ are the input variables, $[A_1,A_2,\ldots,A_p]$ are fuzzy sets which are determined during the training process, $[c_0,c_1,\ldots,c_p]$are the consequent parameters and y is the output variable. In order to evaluate the performance of a method in prediction, several criterions are calculated. The criterions used in this research are correlation coefficient (R), root mean square error (RMSE), mean relative error (MRE) and accuracy. The correlation coefficient (R) which evaluates the strength of the relationship between the experimental and predicted results is defined as:

$$R(a,p)=\frac{\operatorname{cov}(a,p)}{\sqrt{\operatorname{cov}(a,a)\operatorname{cov}(p,p)}} \tag{1}$$

where cov(a,p) is covariance between a and p sets. a and p denotes the actual output and predicted output sets. RMSE is determined by:

$$RMSE=\sqrt{\frac{1}{n}\sum_{i=1}^{n}(a_i-p_i)^2} \tag{2}$$

where n is the number of the points in the data set. MRE which indicates the mean ratio between the error and the experimental values, is obtained by:

$$MRE(\%)=\frac{1}{n}\sum_{i=1}^{n}\left|100\times\frac{(a_i-p_i)}{a_i}\right| \tag{3}$$

Accuracy which is a simple representation of prediction performance is calculated as:

$$Accuracy=\left(1-\frac{|a_i-p_i|}{a_i}\right)\times 100 \tag{4}$$

3. Flow dynamic of LLHC

The performance of ANFIS is highly depend on the type and number of membership function. In this research after testing several membership functions, it was found that ANFIS based on Gaussian curve membership function (gaussmf) had better performance. Figure 3 illustrates the performance of ANFIS in the brake power prediction under various numbers of the two parameters. It is seen that ANFIS performance with two numbers of membership function is higher. Moreover, the performance of ANFIS improved by increasing the training epoch number up to 200, but the performance did not change with further increase. Hence, the number of 200 is selected as an optimum number for the training epoch. After determining the optimum numbers, the ability of ANFIS in prediction of the engine parameters was evaluated by calculating the criterions of correlation coefficient (R), RMSE, MRE and accuracy.

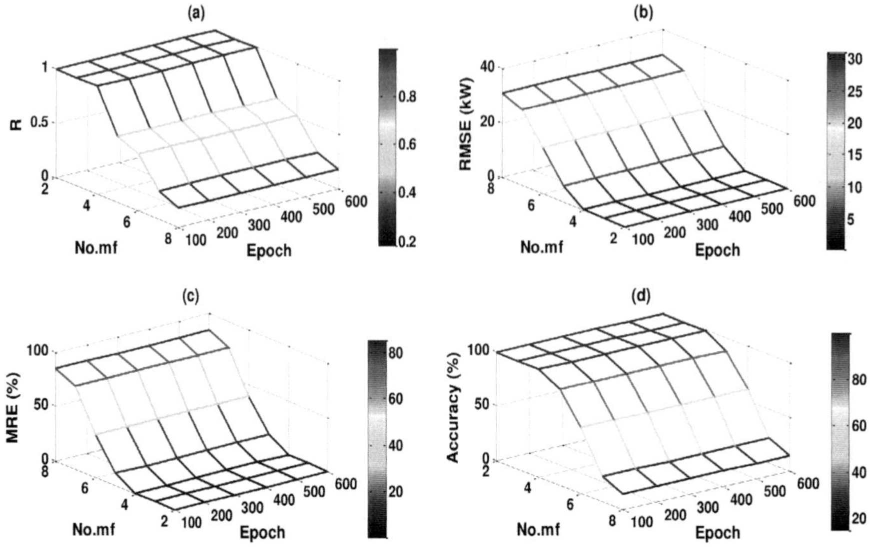

Figure 3. The values of the performance criterions of ANFIS for the brake power under different numbers of membership function and training epoch: (a) R, (b) RMSE, (c) MRE and (d) Accuracy.

Figure 4(b) shows the predicted versus experimental values for the brake power. ANFIS predictions for the brake power yielded a correlation coefficient (R) of 1, RMSE of 0.119 kW, MRE of 0.261% and accuracy of 99.739%. ANFIS predictions for the engine torque yielded a correlation coefficient (R) of 0.996, RMSE of 1.323 Nm, MRE of 0.274% and accuracy of 99.726% (Figure 4a). In predicting the bsfc, ANFIS resulted in a correlation coefficient of 0.949, RMSE of 1.973 g/kWh, MRE of 0.805% and accuracy of 99.195% (Figure 4(c)).

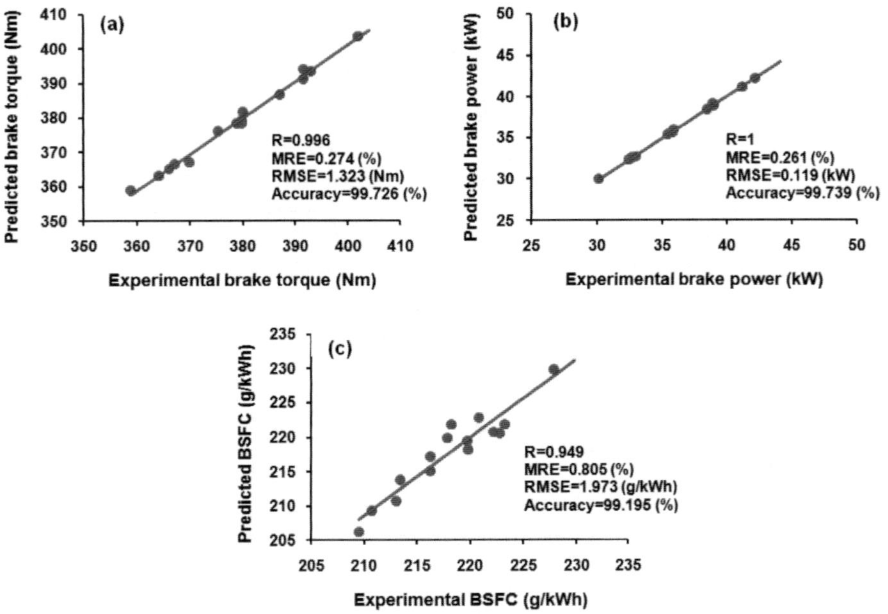

Figure 4. ANFIS predictions for the (a) brake torque, (b) brake power and (c) BSFC versus experimental values.

Figure 5 shows the predicted versus experimental values for the exhaust emission parameters. ANFIS predictions for the CO, CO_2, HC and NO_X yielded a correlation coefficient (R) of 0.945, 0.934, 0.966 and 0.967, respectively. It was found that the RMSE values were 0.072 %V, 0.354 %V, 0.916 ppm and 62.315 ppm for the CO, CO_2, HC and NO_X, respectively. ANFIS predictions for the CO yielded a MRE of 3.547% (Fig. 5(a)). MRE value for the CO_2 was 2.844% (Fig. 5(b)). Fig. 5c shows the predicted versus experimental values for the HC with MRE of 4.186%. According to Fig. 5(d), MRE of the NO_X was 3.229%. ANFIS accuracy for prediction of the CO, CO_2, HC and NO_X was 96.453%, 97.156%, 95.814% and 96.771%, respectively. The above results demonstrate that ANFIS can predict the exhaust emission parameters well.

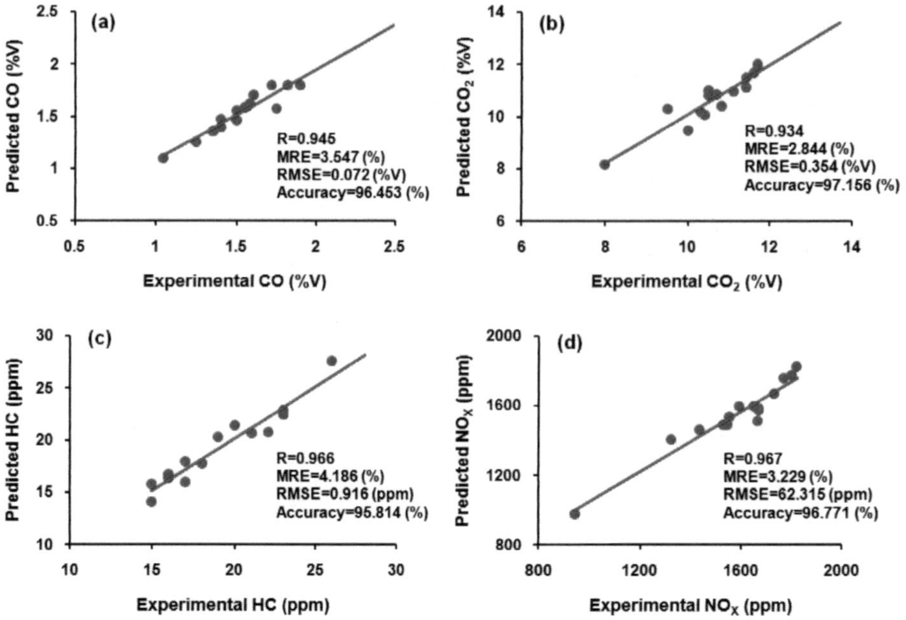

Figure 5. ANFIS predictions for the (a) CO, (b) CO2, (c) HC, (d) NO_X versus experimental values.

4. Conclusions

According to experiments and results obtained by the proposed methods, findings can be summarized as follows: The results of this research study demonstrated that nanodiesel could be efficiently utilized as blended fuel in diesel engine. The produced nanodiesel was blended with biodiesel and diesel fuels in different concentrations and the CI engine emission parameters were evaluated with these new blended fuels. Adding nano particles in diesel fuel can improve engine performance and reduce CO and HC emissions. But it can cause an increase in CO_2 and NO_X emissions. SVM can be employed to predict the engine performance and exhaust emissions.

ANFIS can be employed to predict the engine performance and exhaust emissions. But there is a need for finding the optimum values of its parameters. The ANFIS results were very good, R values in this model are very close to one, while root mean square error (RMSE) was found to be very low. The results showed that ANFIS predicted the engine performance and exhaust emissions with correlation coefficient (R) and accuracy in the range of 0.6–1 and 60–99.9%. The results demonstrate that ANFIS is capable of predicting the CI engine performance and emissions. Analysis of the experimental data by the ANFIS method revealed that there is a good correlation between the ANFIS predicted results and the experimental data. Therefore ANFIS proved to be a useful tool for correlation and simulation of engine parameters.

Acknowledgments

The authors wish to thank the Iranian Fuel Conservation Organization (IFCO) of NIOC for the research grant provided (Grant no.:1393-2) and UMP (Grant No.:RDU130313) to complete this project and Mega Motor Company for providing of laboratory facilities..

References

[1] Selvan V A M, Anand R, Udayakumar M 2009 *J Eng Appl Sci* **4** 1819-6608
[2] Hosoz M, Ertunc H M, Karabektas M, Ergen G 2013 *Applied Thermal Engineering* **60** 24-32
[3] Yücesu H S, Topgül T, Cinar C, Okur M 2006 *Applied Thermal Engineering* **26** 2272-8
[4] Wong P K, Wong K I, Vong C M, Cheung C S 2015 *Renewable Energy* **74** 640-7

[5] Tu J V 1996 *Journal of clinical epidemiology* **49** 1225-31
[6] Vapnik V N 2000 *Springer-Verlag, New York*
[7] Widodo A, Yang B-S 2007 *Mechanical Systems and Signal Processing* **21** 2560-74
[8] Gunn S R 1998 *ISIS technical report* **14**
[9] Jang J-S R 1993 *Systems, Man and Cybernetics, IEEE Transactions on* **23** 665-85
[10] Reddy M J, Mohanta D K 2007 *Electric Power Components and Systems* **35** 1385-407
[11] Sugeno M, Kang G 1988 *Fuzzy sets and systems* **28** 15-33
[12] Şişman-Yılmaz N A, Alpaslan F N, Jain L 2004 *Neurocomputing* **61** 139-68
[13] Sayin C, Ertunc H M, Hosoz M, Kilicaslan I, Canakci M 2007 *Applied thermal engineering* **27** 46-54
[14] Widodo A, Yang B-S 2011 *Expert Systems with Applications* **38** 8430-7
[15] Ma F, Hanna M A 1999 *Bioresource technology* **70** 1-15
[16] Schumacher L, Peterson C, Grepen J, Fuelling direct diesel engines with 2% biodiesel blend. Written for presentation at the 2001 annual international meeting sponsored by ASAE; 2001.
[17] Ghobadian B, Rahimi H, Biofuels-past, present and future perspective. International Iran and Russian Congress of Agricultural and Natural Science Shahre-Kord University, Shahre Kord, Iran; 2004.
[18] Fukuda H, Kondo A, Noda H 2001 *Journal of bioscience and bioengineering* **92** 405-16
[19] Lee S W, Herage T, Young B 2004 *Fuel* **83** 1607-13
[20] Najafi G, Ghobadian B, Tavakoli T, Buttsworth D, Yusaf T, Faizollahnejad M 2009 *Applied Energy* **86** 630-9
[21] De Caro P S, Mouloungui Z, Vaitilingom G, Berge J C 2001 *Fuel* **80** 565-74
[22] Xing-cai L, Jian-Guang Y, Wu-Gao Z, Zhen H 2004 *Fuel* **83** 2013-20
[23] Ghobadian B, Rahimi H, Khatamifar M, Evaluation of engine performance using net diesel Fuel and biofuel blends. The First Combustion Conference of Iran (CCT1) Tarbiat Modares University February; 2006.
[24] Elawad M, Yusaf T 2004 *Journal of Energy & Environment* **3** 61-8
[25] Li D-g, Zhen H, Xingcai L, Wu-gao Z, Jian-Guang Y 2005 *Renewable energy* **30** 967-76
[26] Ajav E, Singh B, Bhattacharya T 1998 *Biomass and Bioenergy* **15** 493-502
[27] Abu-Qudais M, Haddad O, Qudaisat M 2000 *Energy Conversion and Management* **41** 389-99
[28] Noguchi N, Terao H, Sakata C 1996 *Bioresource Technology* **56** 35-9
[29] Gerdes K, Suppes G 2001 *Industrial & engineering chemistry research* **40** 949-56
[30] He B-Q, Shuai S-J, Wang J-X, He H 2003 *Atmospheric Environment* **37** 4965-71
[31] Hansen A C, Zhang Q, Lyne P W 2005 *Bioresource technology* **96** 277-85
[32] Can Ö, Celikten I, Usta N 2004 *Energy Conversion and Management* **45** 2429-40
[33] Ajav E, Singh B, Bhattacharya T 2000 *Energy conversion and management* **41** 1533-41

Institute of Physics
Dirac House, Temple Back
Bristol BS1 6BE UK

ISSN: 1757-8981
ISBN 978-1-5108-1809-5

CURRAN ASSOCIATES INC.
proceedings.com